CONJUGATED POLYMERS

Conjugated Polymers

The Novel Science and Technology of Highly Conducting and Nonlinear Optically Active Materials

Edited by

J.L. Brédas

Laboratory for Chemistry of Novel Materials, University of Mons-Hainaut, Mons, Belgium

and

R. Silbey

Department of Chemistry, Massachusetts Institute of Technology, Cambridge, U.S.A.

SPRINGER SCIENCE+BUSINESS MEDIA, B.V.

Library of Congress Cataloging-in-Publication Data

Conjugated polymers : the novel science and technology of highly conducting and nonlinear optically active materials / edited by J.L. Brédas and R. Silbey.
p. cm.
Includes bibliographical references and index.
ISBN 978-0-7923-1403-5 ISBN 978-94-011-3476-7 (eBook)
DOI 10.1007/978-94-011-3476-7
1. Polymers--Electric properties--Congresses. I. Brédas, J. L. (Jean Luc), 1954- . II. Silbey, Robert J.
QD381.9.E38C665 1991
620.1'9297--dc20 91-27648

ISBN 978-0-7923-1403-5

Printed on acid-free paper

TABLE OF CONTENTS

PREFACE

The field of conjugated and conducting polymers has emerged in the last decade as one of the most exciting and interdisciplinary areas of physical science. Synthetic chemists and condensed-matter physicists, electrochemists and polymer scientists, physical chemists and materials scientists are at home here, each bringing a specific approach and particular concepts, thereby enriching the whole. A common language developed to incorporate diverse areas: for instance, the ideas of semiconductor doping had to be reworked using the concepts of redox or acid-base chemistry; solitons, polarons, or bipolarons from solid-state physics were combined with π-radicals, cations, or dications from the organic chemistry lexicon. All of these brought a fundamentally new perception of the organic solid state.

The field keeps progressing rapidly, as evidenced by the many International Conferences held in the last few years (Kyoto 1986, Santa Fe 1988, Mons 1989, Tübingen 1990, Luleå 1991). The advances up to the mid-eighties were reviewed in the "Handbook of Conducting Polymers" edited by T. Skotheim. At that time, workers were still plagued by poorly characterized systems that were, for the most part, insoluble and infusible. Since then, there has been major progress on all fronts:

(i) the development of novel synthetic routes that lead to better characterized polymers with much improved mechanical properties and to conducting ultra thin films;

(ii) the discovery of soluble polymers, making easier characterization and processability, both essential for technological applications;

(iii) the renewed interest in the chemistry and physics of polyanilines;

(iv) the production and characterization of very highly conducting polyacetylene;

(v) the fine understanding of the electronic, vibronic, and vibrational structure;

(vi) the realization that these conjugated systems can be incorporated as active elements in solid-state devices and possess very large nonlinear optical susceptibilities.

All of these have led to a new vitality that, we believe, is well represented by the current volume; it would have taken a much larger book to cover the entire field and so difficult choices had to be made: these reflect the editors' desire to include something from each of the major new developing areas mentioned above. We hope that the present volume will give the novice an in-depth view of the beauty and the excitement of this field and, at the same time, provide the expert with insight into the latest advances.

J.L. Brédas and R. Silbey
Mons, March 1991.

CONJUGATED POLYMERS: THE INTERPLAY BETWEEN SYNTHESIS, STRUCTURE, AND PROPERTIES

C. B. GORMAN and R. H. GRUBBS
Division of Chemistry and Chemical Engineering
California Institute of Technology
Pasadena, California 91125

J.L. Brédas and R. Silbey (eds.), Conjugated Polymers, 1–48.

1. Introduction

Polymers containing highly loosely held electrons in their backbones, often referred to as conjugated polymers or conducting polymers, have attracted much research effort in the last ten years or so. Many investigations were first motivated by the observation that oxidizing or reducing the backbone (doping) of polyacetylene at a level equivalent to removing/adding an electron in one out of every 5-15 repeat units, increased the conductivity of the matrix by many orders of magnitude.[1-3] Because the electrons in these delocalized systems are also easily polarized by an external electric field such as that found in light, these types of polymers have also attracted interest for their nonlinear optical properties.[4-6] These polymers have also been proposed as "molecular wires" for nanotechnology.[7]

It is ironic that one of the lures towards the synthesis of polymers: the ability to cheaply produce a material that is easily cast or molded into the shape of choice either from the melt or from solution, is not commonly achievable with most conjugated polymers. In contrast, a polymer such as polyethylene is synthesized commercially in the amount of some 10 billion pounds a year[8] commonly using a titanium-based catalyst system that can produce 10^6 grams of polyethylene per gram of catalyst.[9] This polymer is practically free of catalyst and can be dissolved or, more commonly, melted and cast into most any shape. The most common synthesis of polyacetylene employs a very similar titanium-based system, but requires a substantial amount of catalyst mixture (> 5% weight catalyst/weight polymer) and is an insoluble polymer that cannot be melted. Moreover it is difficult, even with repeated washings, to remove the last 1-2% of catalyst residues from the polymer.[10] Although the catalyst systems vary for other conjugated polymer syntheses, all of the polymers in figure 1 are insoluble in organic solvents. This insolubility is not surprising since these polymers tend to be conformationally rigid, at least over the length of several monomer units, in order to maintain the conjugation of the electrons in the backbone.

In this chapter, approaches towards the synthesis of conjugated polymers will be discussed with particular emphasis on what control, if any, the method offers over the molecular weight and its distribution for the polymer. Since the form of most conjugated polymers cannot be changed once they are synthesized, several clever procedures have been developed to produce the polymer in a desirable macroscopic form such as a fiber or sheet. Many of the polymers have been derivatized with the goal of controlling their materials properties, most notably their solubility, and strategies and results of synthesizing soluble derivatives of the polymers will be discussed. Many conjugated polymers are synthesized via soluble precursor polymers, and these will be discussed in their own section. These precursors can be cast, coated, etc. and subsequently converted to the conjugated form.

2. Structural features of conjugated polymers

The structure of all conjugated polymers have the same signature. Each atom along the backbone is involved in a π bond which is much weaker than the σ bonds that hold the atoms in the polymer chain together.† Placed side by side (conjugated), these π bonds can delocalize over all the atoms. The extent of delocalization of an electron in an extended π system is a matter of some interpretation and debate: although every electronic wavefunction is defined for any point in space, the majority of electron density is smeared over a relatively small volume.[11] Polymers are not completely straight and flat over an infinite distance. Single (i.e. C-C) bonds, even when they are part of a conjugated system, can rotate given the thermal energy available at room temperature. The rotational potential of a single bond has been calculated to be about 6 kcal/mol in polyacetylene and 3 kcal/mol in polythiophene (gas phase calculations).[12] These rotations reduce the conjugation (delocalization) of the electrons in the π-system.

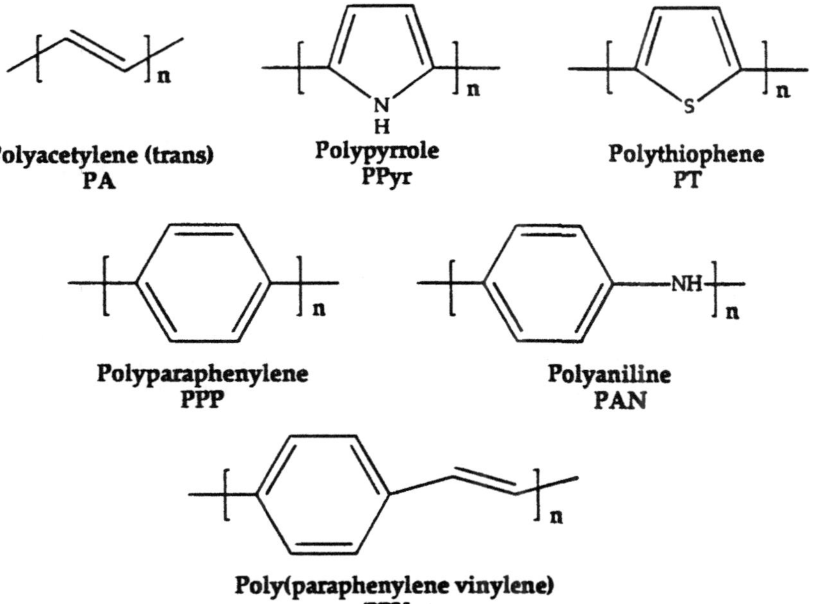

Figure 1. Structure of the repeat unit of several conjugated polymers including their common abbreviation

This effect can be observed, for example, in the thermochromism of polythiophene derivatives. The lowest energy optical absorbance in a conjugated chain typically arises from a $\pi \rightarrow \pi^*$ transition. The more conjugated the double bonds are in the polymer, the lower the energy of this absorption will be. Thus, twisting about the chain will reduce the amount of conjugation and increase the energy (i.e. blue shift) of this absorption.[13-15] This discussion does not address the problem of chemical (sp^3, conjugation disrupting)

† A notable exception is the polysilanes which are conjugated through their σ (i.e. single-bonded) system.

defects that can be formed in the polymer backbone during synthesis or subsequent manipulation.[16-19]

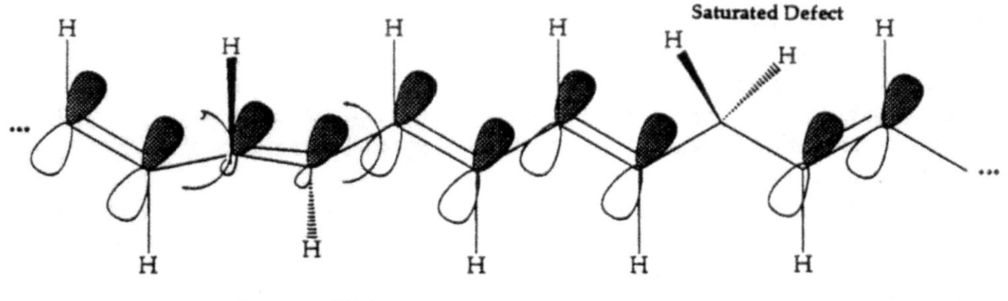

Figure 2. Schematic of a π-system including a conformational (rotational) defect and a saturated (sp^3) defect.

3. Polymer synthesis: Basic methods

This brief treatment is designed to contrast the two basic mechanisms involved in polymer synthesis: step-growth (condensation) or chain-growth (addition). The terms in parenthesis are often used in place of the more accurate terms. These terms actually refer to earlier observations about polymerizations and, for example, "condensation" is not always appropriate for every step-growth polymerization. The mechanism that a polymerization reaction follows will determine the final molecular weight (M_n)[††] and the molecular weight distribution of polymer chains in the final product. Unless an experiment is designed to examine the effect of chain length on some physical property of the polymer, it is rarely desirable to have a mixture of short chains in the sample. Optical properties such as absorption and Raman scattering will be dependant upon the number of units in the polymer until a relatively high molecular weight has been reached (≈ 50 monomer units is a round number) at which point these properties become much less dependent upon chain length. Particularly the oxidation/reduction potentials for generation of mobile charge carriers (solitons, polarons, bipolarons, etc.) and their spectral properties will vary for short chains and become relatively invariant for longer chains. Likewise, samples possessing a widely varying collection of chain lengths (i.e. a large polydispersity, defined as M_w/M_n) often have a dispersion of different conjugation lengths, especially when the chain lengths are short, and these complicate characterization of the polymer. In many cases, conjugated polymers are found to possess a bimodal molecular weight distribution. In other words, instead of having a distribution of molecular weights around some average, there is a distribution of high molecular weight chains, and a distribution of lower molecular weight chains, often oligomers. Again, especially when oligomers are present, it is difficult or

---

[††] This is a number average molecular weight. The distribution of molecular weights in the polymer sample can be weighted by their number fraction (M_n), their weight fraction (M_w) their viscosity fraction (M_v), etc.

impossible to determine the physical properties of the high molecular weight polymer, and many experimental efforts have been challenged on this point.

Both polymerization mechanisms can result in high molecular weight polymers with a relatively narrow molecular weight distribution. However, mechanism can be very important. For example, in the synthesis of an insoluble conjugated polymer, one type of polymerization might result in very low molecular weight oligomers precipitating from solution and the other could result in synthesis of high molecular weight polymer before precipitation occurs.

3.1 STEP-GROWTH POLYMERIZATION

The most general reaction for two molecules can be represented as:

$$A + B \rightarrow A\text{-}B$$

and, in principle, any reaction that links two units together can be used to form a polymer. In general, we can write:

$$A\text{-}A + B\text{-}B \rightarrow \text{-[-}A\text{-}A\text{-}B\text{-}B\text{-]-}$$

or

$$nA\text{-}B \rightarrow \text{-[-}A\text{-}B\text{-]-}$$

A common example of this type of polymerization is the synthesis of polyesters which takes advantage of the fact that an alcohol and an acid can be condensed together (with the elimination of water, thus the origin of the term "condensation") to form an ester:

$$n\text{HO-R-OH} + n\text{HOOC-R'-COOH} \rightarrow$$
$$\text{H-(-O-R-OOC-R'-CO-)}n\text{-OH} + (2n\text{-}1)\text{H}_2\text{O}$$

Polyamides, polyurethanes, polycarbonates, and many other polymers are routinely synthesized in this manner. This polymerization, in principle, is general since many synthetic organic reactions can be employed using difunctional units. This reaction takes place in a number of statistical steps, and, although the rate of bond formation can be very fast, the rate of molecular weight increase is initially very slow. First, most of the monomers will react to form dimers, and statistically, only after most of the dimers have formed will dimers start to react with monomers or dimers to form trimers or tetramers. Already 50% of the bonds have formed, but the degree of polymerization is only n=2. This slow buildup of molecular weight is the primary disadvantage of step-growth polymerization. If an organic reaction has a 50% yield, and it is used in a step-growth polymerization, dimers and trimers will dominate the products. One can show that at 90% conversion, the average degree of polymerization is only 10, at 95% conversion it is 20, and only above 98% conversion (the last 2% of the reaction) do long chains start to reach to form high molecular weight polymer (high polymer). Most organic reactions do not have a 98% yield, so in practice, this method is limited.

Many conjugated polymers are synthesized by step-growth. If the polymer is soluble and the reaction employed is of high yield, high polymer can be expected. However, if oligomers precipitate before they react with one another, or if the yield of the reaction is low, the degree of polymerization can be very low. This problem is often encountered in conjugated polymer synthesis.

3.2 CHAIN-GROWTH POLYMERIZATION

The second mechanism of polymerization, chain-growth, involves three distinct steps: initiation, propagation, and termination. An initiator reacts with a monomer to produce a new species which can then react with another monomer, and another, until the monomer is depleted or the growing polymer chain undergoes a termination reaction. This procedure is shown schematically below where * refers to a species that can add another monomer. This species is often a radical, cation, anion, or attached transition metal complex that can insert a monomer between itself and the growing polymer chain.

$$I \rightarrow I^*$$
$$I^* + M \rightarrow I\text{-}M^* \qquad \text{(initiation)}$$

$$I\text{-}(\text{-}M\text{-})^*_{n-1} + M \rightarrow I\text{-}(\text{-}M\text{-})^*_n \qquad \text{(propagation)}$$

$$I\text{-}(\text{-}M\text{-})^*_n + T \rightarrow I\text{-}(\text{-}M\text{-})^*_n\text{-}T \qquad \text{(termination)}$$

This reaction is most commonly observed with vinyl monomers, yielding polyethylene, polystyrene, polyacrylates, etc.

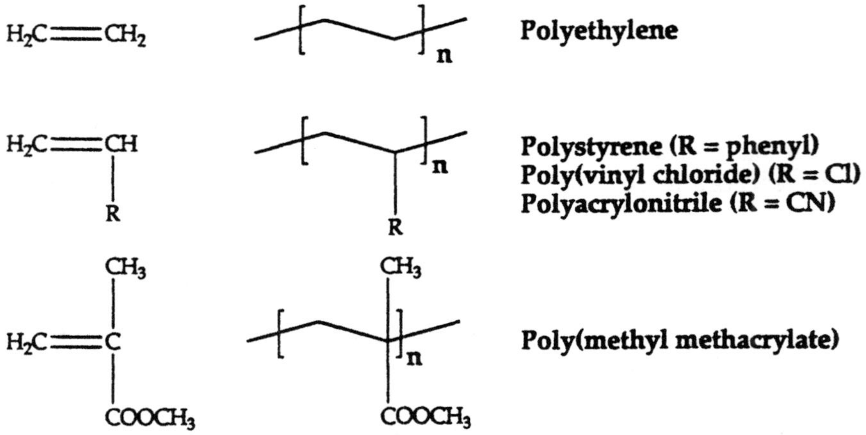

Figure 3. Common vinyl polymers synthesized by chain-growth polymerization.

The rate at which the molecular weight of a chain-growth polymerization increases depends on three rates: the rate of initiation, propagation, and termination (R_i, R_p, R_t). In many cases, once initiated, the polymerization proceeds very quickly, and the propagating chain end "lives" for only a short time before terminating. Usually, the propagating species forms a high polymer during this time. The most noticeable feature in this polymerization ($R_p >> R_i$) is that high polymer forms very quickly, an advantage if the polymer is insoluble and precipitates from the reaction solution quickly as well.

MODES OF POLYMERIZATION

MECHANISM

MOLECULAR WT. CONTROL

$$\overline{M}_n = \overline{X}_n \times \text{Mol. Wt. Monomer}$$

STEP GROWTH

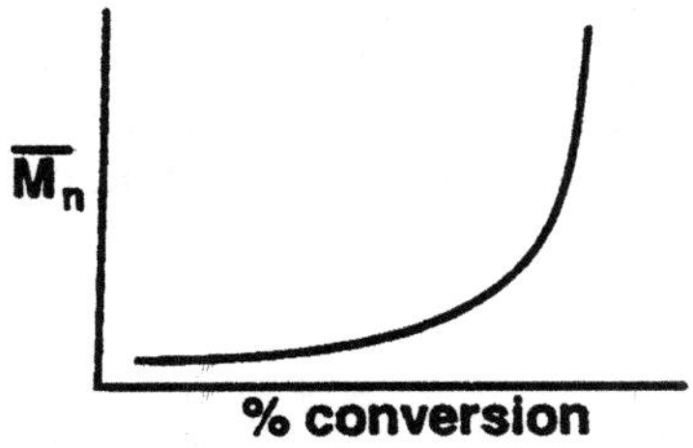

$$\overline{X}_n = \frac{1}{1-\rho}$$

ρ = % functional groups reacted/100

CHAIN GROWTH

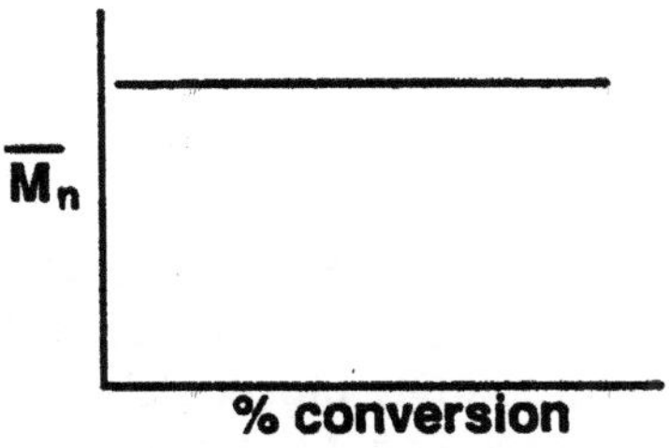

$$\overline{X}_n = \frac{1}{1-\rho}$$

$$\rho = \frac{R_p}{R_p + R_t + R_{ct}}$$

LIVING

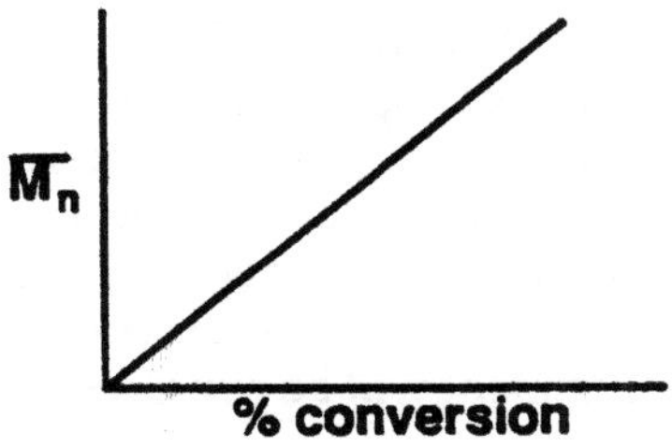

$$X_n = \frac{\text{moles monomer consumed}}{\text{moles of initiator}}$$

$R_i >> R_p$; $R_t \approx 0$

Figure 4. Modes of polymerization

Not much control over molecular weight is available in this scenario. However, if $R_i >> R_p$, all the chains will initiate at about the same time and, if there is no chain termination, the molecules will continue to grow until all the monomer is depleted. At this time, since the propagating species is still active at the end of the polymer, a second monomer may be added which will then polymerize to give a polymer containing two "blocks" (a block copolymer). This special case, referred to as a "living polymerization", allows the ultimate control over molecular weight and provides the most narrow molecular weight distribution available. A common example is the living anionic polymerization of styrene using alkyl lithium reagents (RLi).

3.3 RING-OPENING POLYMERIZATION

Ring-opening polymerization can follow either a step-growth or chain-growth mechanism and involves the breaking of a bond in a ring to form an open-chain intermediate. In step-growth, these intermediates can react with each other to form polymer. In chain-growth, typically a few monomers ring-open initially, either with or without an initiator, to form a propagating species which then reacts with another monomer, opening it to form a new propagating species. Several living ring-opening polymerizations are known.

Figure 5. General scheme for a ring-opening polymerization.

4. Direct Synthetic Methods

Since most conjugated polymers cannot be dissolved or melted, they must be synthesized directly in the desired shape and location. This necessity has frustrated many researchers since the form, most importantly the morphology of the conjugated polymer has one of the greatest influences on many of the properties, most notably the electrical conductivity of the doped polymer. Theory and experiment have placed much emphasis on justifying the formation of mobile, delocalized carriers within the polymer chains (polarons, bipolarons, solitons).[20, 21] However, since no one polymer strand is long enough to persist over a macroscopic length, the measured conductivity of a polymer sample requires that carriers hop between polymer strands. This hopping is widely believed to limit the bulk conductivity.[22] Synthesis of highly conductive polymer samples has become an art in some circles since small changes in synthesis, catalyst removal, or doping can dramatically affect sample morphology and result in wide variation in conductivity. Thus, for example, it is important to understand the effects of polymerization conditions on polyacetylene morphology.[23] Deliberate changes in polymer morphology resulting from chain alignment will be reviewed in section 7.

4.1. ELECTROCHEMICAL SYNTHESIS[24]

Most conjugated polymers are synthesized by electrochemical means. Pyrrole, thiophene and their derivatives are often polymerized in this manner as are carbazoles, azulenes and pyrenes.[25] Typically, this oxidation (2 C-H bonds are converted to a C-C bond plus two H^+) is accomplished by placing the monomer and electrolyte (e.g. $AgClO_4$) in a suitable solvent such as acetonitrile and oxidizing at a mild potential (e.g. 0.5 V). A polymer film grows at the anode in a reaction that is thought to occur as illustrated in figure 6. An example termination step is shown. In principle, any trace species that can react with the propagating radical cation can terminate one end of the polymer. Any polymer chain that does not re-oxidize is effectively terminated as well. The kinetics of this polymerization are expected to be complex. Among other factors, the reaction occurs in a heterogeneous environment (the electrode surface), and the polymerization is based upon coupling of the radical cations which can occur between the oxidized forms of any combination of monomers and polymer chains.

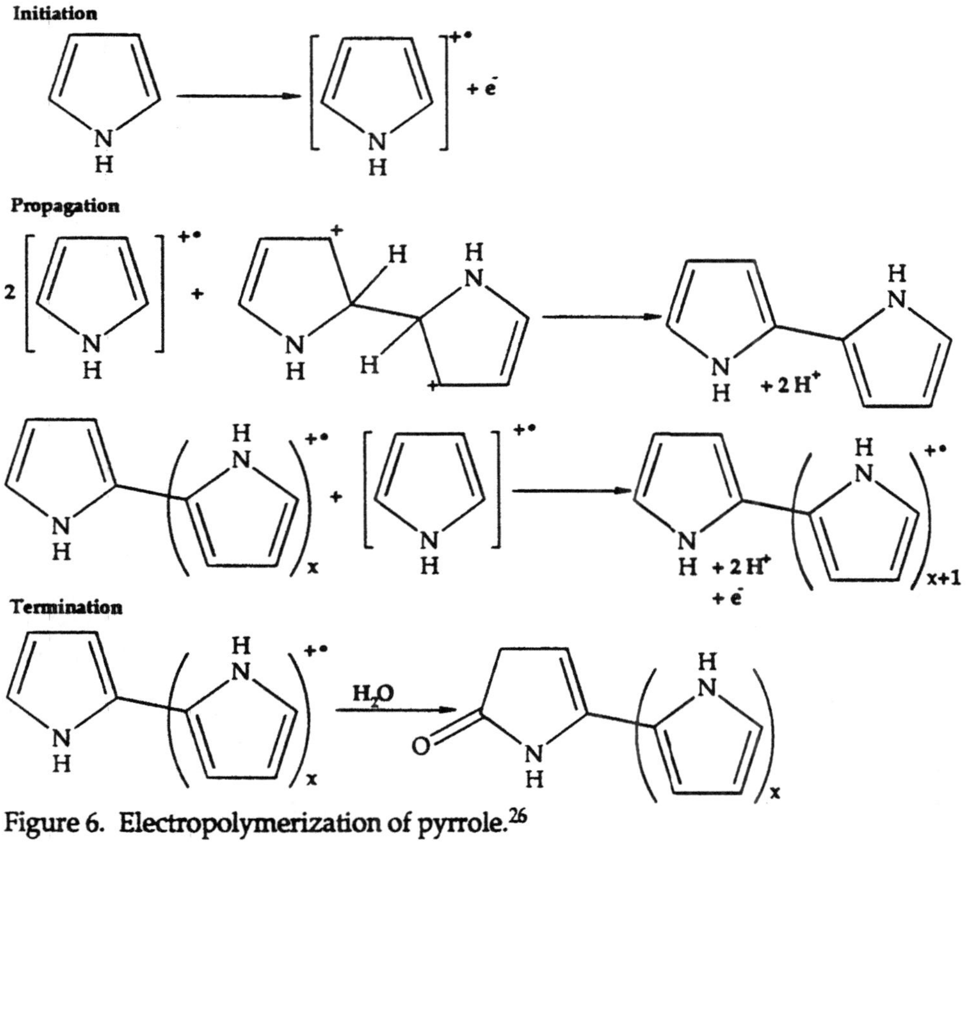

Figure 6. Electropolymerization of pyrrole.[26]

Although it is thermodynamically favorable for the linkages to be formed at the 2,5 positions of the monomer units as shown above, there is concern that the electrochemical polymerization results in a significant number (perhaps 30%) of linkages that are not of this 2,5 type. 2,3 Couplings do not extend conjugation, in fact, branching of this type actually reduces the conjugation length of the main chain of the polymer.[27]

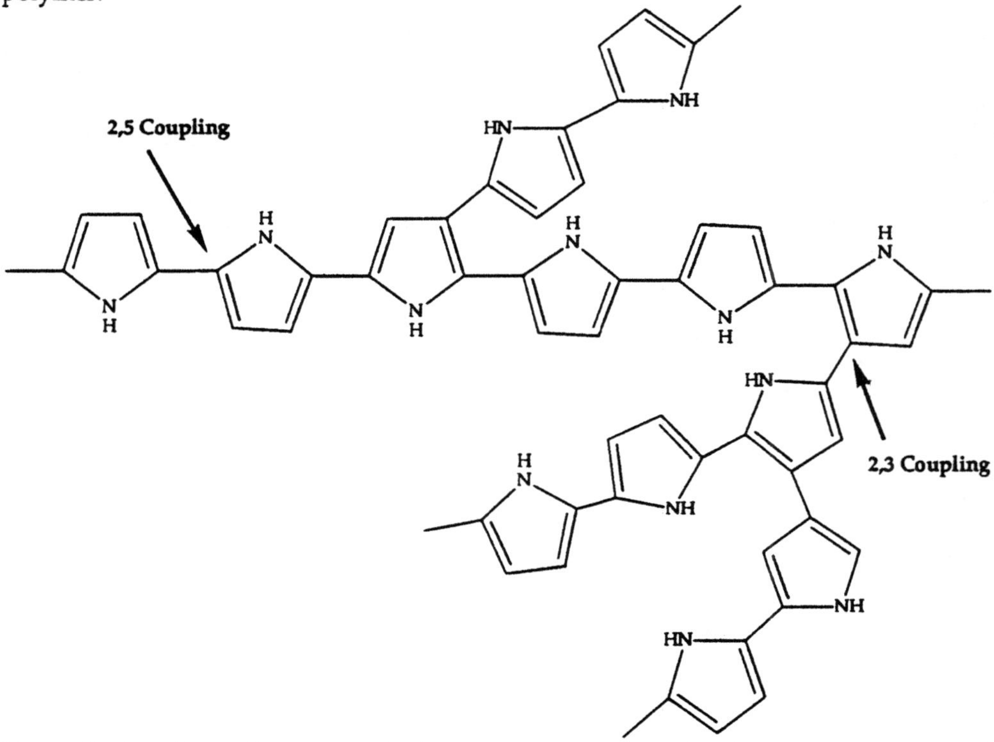

Figure 7. Possible differences in the linkages in the electrochemical polymerization of pyrrole.[28]

The anodically prepared polymer film is already doped by the excess of charge that is passed in the polymerization reaction and contains the counterion (in this example, ClO_4^-) from the electrolyte solution. The film can usually be compensated (undoped) by passing current in the opposite direction. This method allows for flexibility in the choice of the counterion and has the advantage that the counterion is homogeneously dispersed in the polymer film.[29-35] Moreover, since the reaction place electrochemically, there are methods for exploring the kinetic parameters for doping.[36] Nevertheless, the degree of polymerization is generally not known. Some oligomeric material is present.[37] By observation of tritium loss from 1,2-tritiated monomers, a degree of polymerization of about 750 was determined for poly(3,4-dimethyl pyrrole perchlorate).[38]

New polymers synthesized by this procedure continue to be reported as new oxidizable conjugated units become available. For example, the fused monomers shown in figure 8 have a higher degree of conjugation and are easier to oxidize than their smaller counterparts.

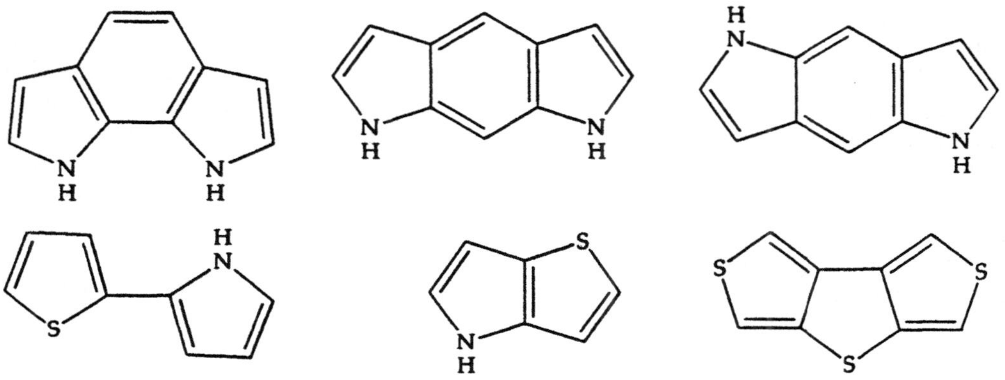

Figure 8. Sample monomers that have been oxidatively polymerized either chemically or electrochemically.[39-42]

Previously, this synthesis had the disadvantage that the size of the film was limited by the size of the anode. However, Naarman et al. have introduced a rotating drum electrode that allows the continuous synthesis of films.[43, 44] Other electrode modifications include the use of a needle-to-circle electrode configuration to grow visually striking 2-D and 3-D fractal patterns of polypyrrole,[45] and precoating of the electrode followed by polymerization to synthesize conducting polymer/plastic polymer composites[46] or blends.[47] A polymer-covered electrode with incorporated redox or other catalytic sites can be used as a surface-modified electrode.[48-52]

4.2. SYNTHESIS BY STEP-GROWTH POLYMERIZATION

Probably the most common method of *chemical* synthesis of conjugated polymers is via a step-growth (condensation) polymerization. Before the advent of the continuous electrochemical synthesis (section 4.1), this method was the only way to make large amounts of polymers such as polythiophene. As mentioned previously, these polymerizations require a high-yield reaction to proceed to a high degree of polymerization. However, in principle, any monomer that can be oxidatively polymerized electrochemically can be polymerized using a chemical oxidant (e.g. $FeCl_3$) as well.[53, 54] If the resulting polymer is insoluble, the end of the polymerization may occur in the solid state. This serious limitation is avoided in the synthesis of soluble conjugated polymers.

4.2.1. *Polyaniline (PAN)* is a unique conjugated polymer because it can exist in a variety of structures depending on the value of y in the general formula of the polymer shown in figure 9. When y is not equal to 0 or 1, these intermediate forms of polyaniline can be rendered conductive by protonating (proton doping) the imine nitrogens, formally creating radical cations on these sites.[55] This doping introduces a counterion (e.g. Cl^- if HCl was used as the dopant), and recently, the counterion was affixed to the parent polymer by partially sulfonating the benzene rings in the polymer, resulting in a so-called "self-doped" polymer.[56] Although aniline[57] and its derivatives[58, 59] can be polymerized electrochemically, polymers are often synthesized by step-growth polymerization in an aqueous solution of the monomer, an oxidant such as $(NH_4)_2S_2O_8$, and HCl.[60] This polymer is soluble in *N*-methylpyrrolidinone and 97% sulfuric acid[61, 62] and has a high molecular weight (≈ 325,000)[63] but shows a bimodal molecular weight distribution by gel permeation chromatography.[55] The proposed para linkage of the aniline units has been somewhat justified by the favorable comparison of the infrared spectra of polyaniline synthesized as above and polyaniline synthesized via a well-defined precursor polymer.[64] There is interest in the use of polyaniline in rechargeable batteries,[65] and repeated doping/undoping of the polymer has been undertaken to change the film morphology, resulting in high selectivities in the separation of gases (O_2/N_2 separation factor = 53).[66]

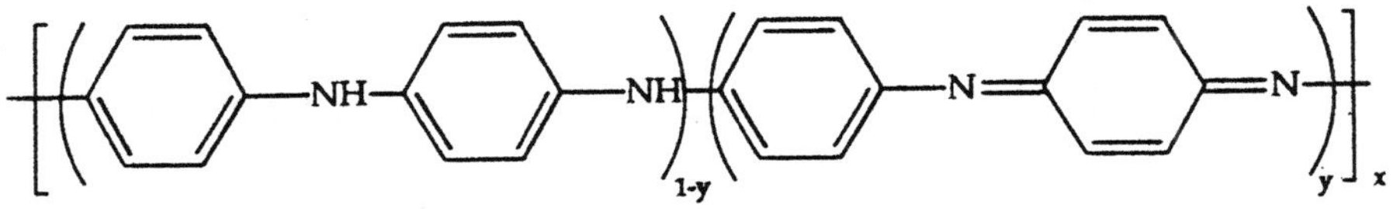

Figure 9. General formula of polyaniline.

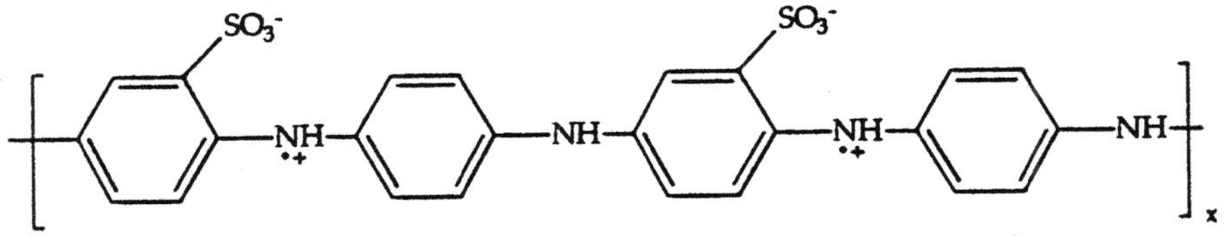

Figure 10. Representative formula of counterion affixed polyaniline.

4.2.2. *Poly(phenylene sulfide)* is a polymer structurally related to the amine form of polyaniline and is typically made by condensation of dichlorobenzene with sodium sulfide in *N*-methylpyrrolidinone. The polymer precipitates as a white powder which is soluble in high boiling solvents. Although the degree of polymerization is moderate (M_n = 11,000), the polymer can be p-doped with SO_3 to a conductivity of 80 S/cm [67] and has the advantage of being environmentally stable and easily processable.[68] It is currently sold by the Phillips Petroleum Company under the trade name of "Ryton".[69, 70] The related polymer, poly(phenylene oxide) dopes to a much lower conductivity ($\sigma \approx 10^{-3}$ S/cm) and has been studied much less.[71] Phenylated poly(phenylene sulfide) has been photolyzed to produce a "photo-doped" polymer of modest conductivity ($\sigma \approx 10^{-2}$ S/cm).[72]

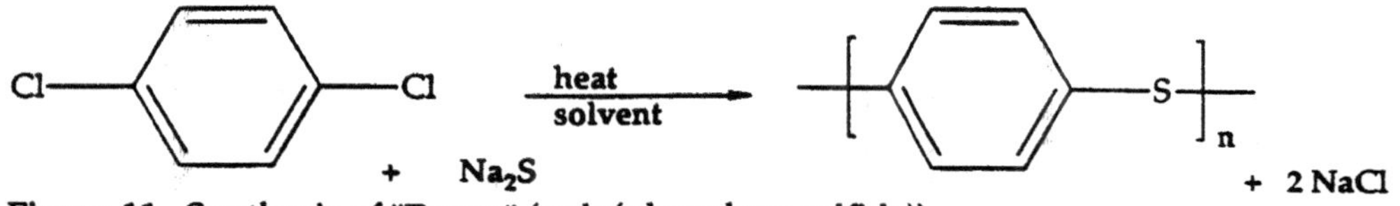

Figure 11. Synthesis of "Ryton" (poly(phenylene sulfide)).

4.2.3. *Polythiophene and its derivatives.* A large amount of work has centered around polythiophene, particularly since many soluble derivatives of the polymer are known (see below). Most workers synthesize polythiophene electrochemically, despite the uncertainty in the structure of electrochemically synthesized polymers (section 4.1) and concern that the electrochemical synthesis of polythiophene results in overoxidation of the polymer.[73] The chemical synthesis of polythiophene is most commonly performed using a nickel-catalysed coupling of the di-Grignard of thiophene (Figure 12)[74, 75] although organocuprate couplings,[76] a polymerization based on Friedel-Crafts alkylation,[77] coupling of the di-halide using a Ni(0) catalyst,[78] and direct oxidation with $FeCl_3$ are employed as well.[79] The first two (di-halo) coupling schemes are preferred because they ensure that the linkages are of the 2,5 type: direct chemical coupling does not necessarily guarantee this linkage. All of these syntheses suffer to some extent from the fact that it is not easy to remove all of the metal (Ni, Mg, Cu) employed, especially in the synthesis of the parent (insoluble) polymer.

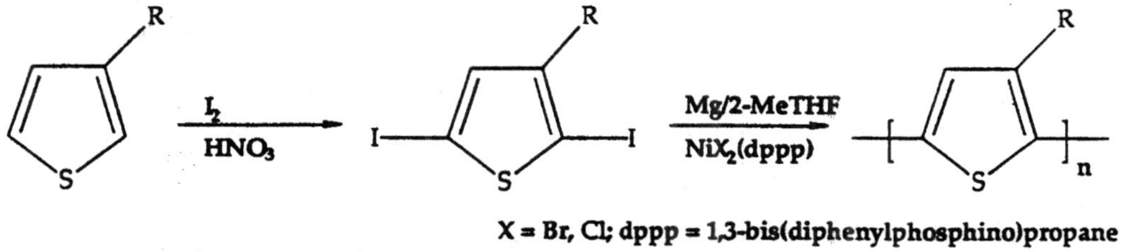

Figure 12. Nickel-catalysed polymerization of thiophene.

In the area of molecular technology, a switch based on two orthogonally fused polymer chains, one doped (conductive) and the other undoped (insulating) has been proposed,[80, 81] and an oligomer has been synthesized.[82] Many tasks remain here, such as incorporation of this unit into a device of molecular dimensions and discovering a way to selectively switch the unit.

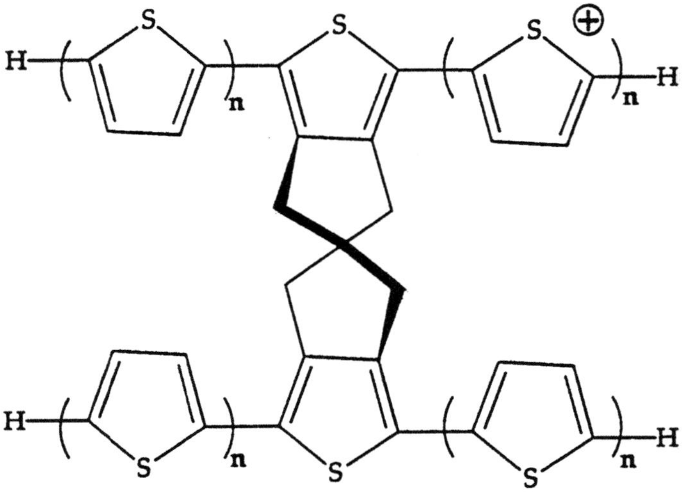

Figure 13. Model for two orthogonally fused conductive polymer chains. An oligomeric form of this model has been synthesized. Oxidation of only one of the two chains has not yet been realized.

As mentioned above, synthesis of an insoluble polymer by this route is not expected to produce a very high degree of polymerization. However, polymerization of a thiophene monomer with a floppy *n*-alkyl tail [$(\text{-}CH_2\text{-})_nCH_3$] at the 3 position produced a soluble polymer.[83-87] These polymers, particularly with *n*-alkyl tails of 6 or 8 carbons (poly(*n*-hexyl- or *n*-octyl thiophene)) have been well studied and have molecular weights of $M_w = 48{,}000$ and $M_w/M_n \approx 2$ for poly(*n*-hexyl thiophene) synthesized electrochemically,[88] and $M_w/M_n = 15{,}000\text{-}30{,}000/10{,}000 = 2\text{-}3$ for nickel-coupled poly(*n*-octyl thiophene) and $M_w/M_n = 150{,}000/30{,}000 = 5$ for $FeCl_3$ coupled 3-octylthiophene.[89] In this study, a high molecular weight shoulder in the $FeCl_3$ coupled polymer convinced the authors that chain-branching had taken place, and although this observation cast doubt on the exact nature of the material, it did improve its mechanical properties. This strategy has been extended to the synthesis of soluble poly(3-octyl pyrrole) although little subsequent work has been performed on this polymer.[90]

The side groups that solubilize the polymer tend to render the matrix amorphous, and this limitation appears to preclude them as highly conductive materials ($\sigma >$ approx. 10^2 S/cm) even when the polymer is subjected to stretch alignment (section 7).[91] However, there have been a number of applications such as photocells and antistatic coatings that do not require polymers with extremely high conductivities. Moreover, the presence of the side groups does not appear to reduce the third-order nonlinear susceptibility (χ^3) of the polymer.[92-95] This solubility has also allowed for study of the polymer conformation in solution using such techniques as light-scattering and small-angle neutron scattering, providing an opportunity to link the thermochromism and solvatochromism of these polymers (section 2) with changes in the polymer backbone conformation.[96, 97] These polymers have also been doped in solution and the spectroscopic properties of the doped species were studied.[98] Variation of the group attached at the 3-position of the thiophene monomer has resulted in the synthesis of a variety of derivatives (figure 14). These include poly(3-alkoxy thiophenes)[99-103] which, for reasons that are poorly understood, have slightly reduced bandgaps compared to the poly(3-alkyl thiophenes,[104] ethylmercapto-substituted polythiophenes,[105, 106] chiral polythiophenes which can stereoselectively recognize doping agents,[107] a water-soluble derivative which also contains an attached counterion useful in doping,[108, 109] polythiophene containing a pendant viologen redox group which has been used in a microelectrochemical transistor,[110] as well as poly(3-(ω-haloalkyl) thiophenes) which can act as a very convenient starting point for further synthetic effort.[111] Longer side groups (n = 12) have also reduced the melting point of the polymer and allowed for melt-processing.[89, 112, 113] Several of these soluble derivatives have been used in polymer blends[114] with polystyrene[115, 116] and polyethylene/vinyl acetate.[117] Most of these derivatives have been polymerized electrochemically. Their chemical polymerization often produces lower molecular weight materials.

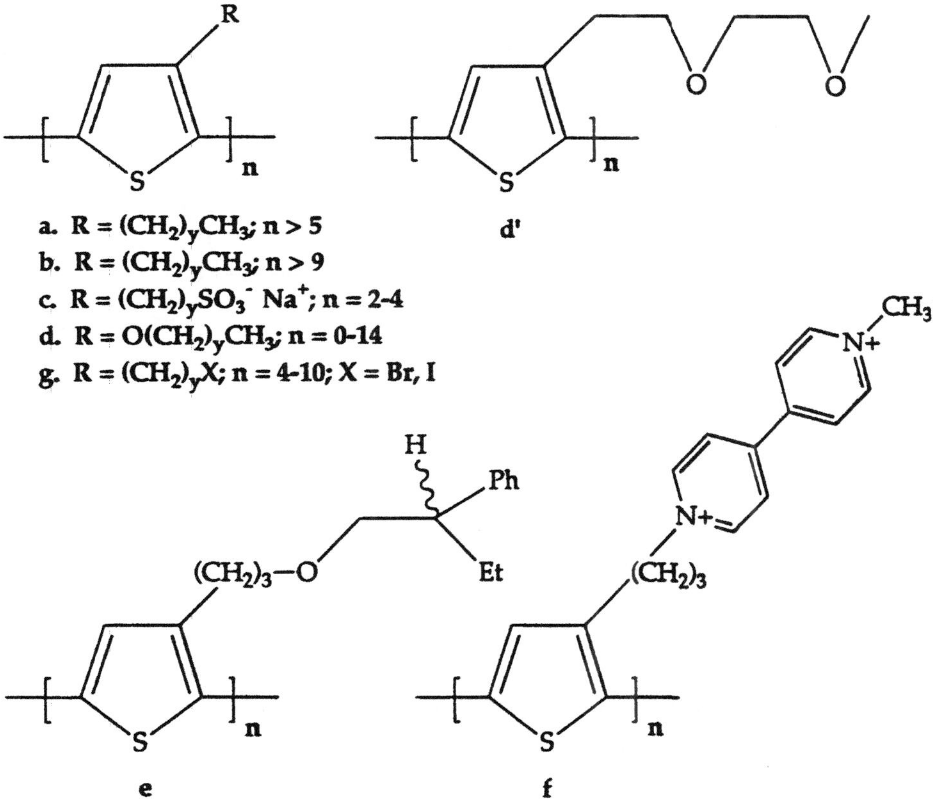

Figure 14. Structural variation of poly(alkyl thiophenes) allows for (a) organic-soluble (b) melt processible (c) water-soluble (d) reduced band-gap (e) chiral (f) pendant redox-active and (g) synthetically useful derivatives.

4.2.4. *Other 5-membered Heterocyclic Derivatives.* Several other classes of 5-membered heterocyclic derivatives exist. Most are polymerized via a step-growth mechanism. A dibenzo fused derivative of polythiophene (figure 15), which is also partially formed upon doping of poly(phenylene sulfide), can also be synthesized by reaction of dibromodibenzothiophene and sodium sulfide. Carbazoles have been polymerized by *in situ* coupling of their Grignards as well by as coupling of diiodocarbazole in molten iodine.[68, 118, 119] Thin films of carbazole polymers have been prepared by electrochemical oxidation of vacuum-deposited carbazole.[120] The most common use of carbazoles, however, is as pendant groups on other polymers. These polymers photoconduct by forming an excited-state that can then hop to another, unexcited carbazole moiety in the polymer. This form of energy transfer is not unrelated to electron hopping in conductive conjugated polymers, and the behavior of these polymers concerns workers in this area, especially when evaluating carrier transport.[121-124] Recently the resonant third-order nonlinear optical susceptibility (χ^3) of a composite containing poly(*N*-vinyl carbazole) and an electron acceptor (2,4,7-trinitrofluorenone) has been measured.[125] Use of this charge-transfer composite is not a recent development, however. In the form of a coated film, it has been used as the photoconductor in reprographic machines marketed by IBM.[126]

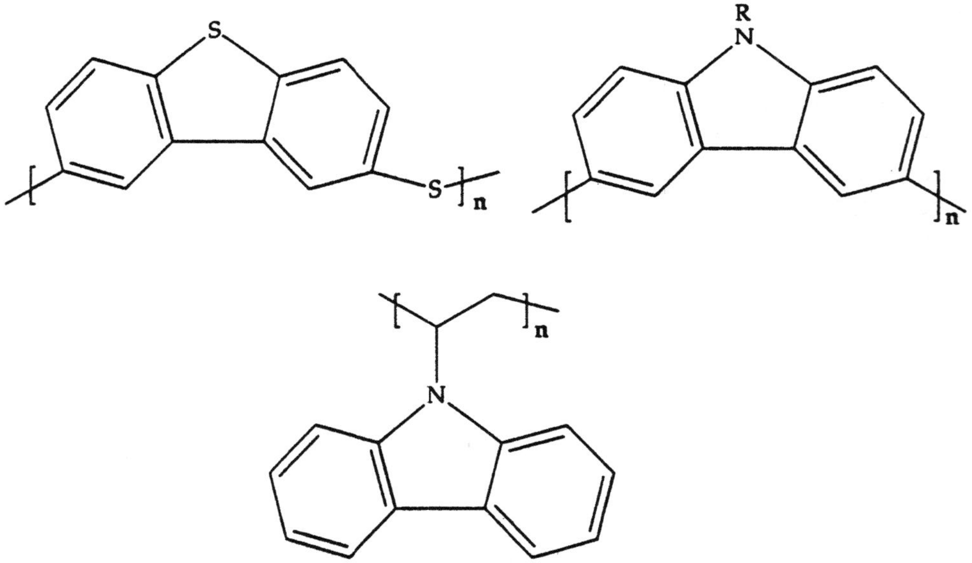

Figure 15. Poly(dibenzothiophene sulfide), polycarbazole, and poly(*N*-vinylcarbazole).

These examples involve pendant groups that interact with each other. There are, of course, several examples of polymers containing pendant chromophores, particularly as liquid crystal mesogens, that have been investigated as nonlinear optical materials. In these materials, the electronic interaction between chromophores does not generally contribute to the material's nonlinear optical susceptibility.[127-129]

There are considerable differences between *cis*-polyacetylene and polythiophene due to the presence/absence of a heteroatom. This heteroatom, in addition to fixing the geometry of some of the bonds in the polymer, influences the electronic structure of the polymer. The changes in properties corresponding upon structural variation have been explained in a very intuitive way using orbital-mixing arguments.[130, 131] There continues to be a synthetic challenge in the further variation of the heteroatom. For example, substitution by phosphorous is of interest, yet, only oligomers of the phosphorous analog of polythiophene have been synthesized.[132]

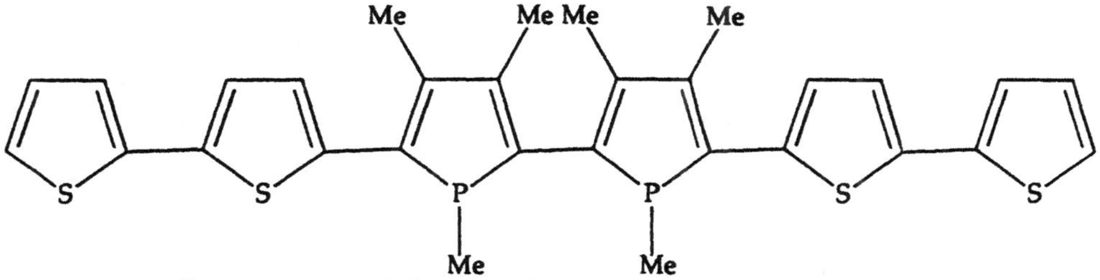

Figure 16. A first step toward the phosphorous analogues of polythiophenes.

4.2.5. *Polyparaphenylene (PPP).* Despite the high oxidation potential of benzene, polyparaphenylene has been prepared electrochemically using a variety of solvent systems, most of them highly acidic.[25] Although it is often impossible to comment on the structure of the resulting material, electrochemical syntheses of PPP continue to be reported in such exotic solvents as SbF_5-SO_2,[133] butylpyridinium chloride/$AlCl_3$ melt,[134] and trifluoromethane sulfonic acid.[135] It is believed that these cations are required to complex or protonate the benzene before it can be oxidized.[136] PPP also can be prepared in nitrobenzene using potentials as high as 30V. Also this potential is well above the stability window for the solvent, the films are synthesized before much degradation of the solvent occurs.[137]

Early chemical syntheses of PPP involved the oxidative coupling of benzene using an $AlCl_3/CuCl_2$ catalyst.[138] Like polythiophene, later chemical syntheses involve coupling of benzene para-dihalide or the di-grignard, or di-lithio derivative formed from it using a variety of metals or metal salts, mostly based on nickel or copper.[139] Not unexpectedly, different syntheses produce samples with very different morphologies.[140] This type of synthesis has been extended to the coupling of 2,5-di-*n*-hexylbenzene derivatives to make soluble polyparaphenylenes (Figure 17).[141-143] Although these new polymers are promising, they all still have relatively low degrees of polymerization ($n \leq 30$). Thus, the synthesis of high molecular weight polyparaphenylene via a precursor method has received much attention (section 5.1).

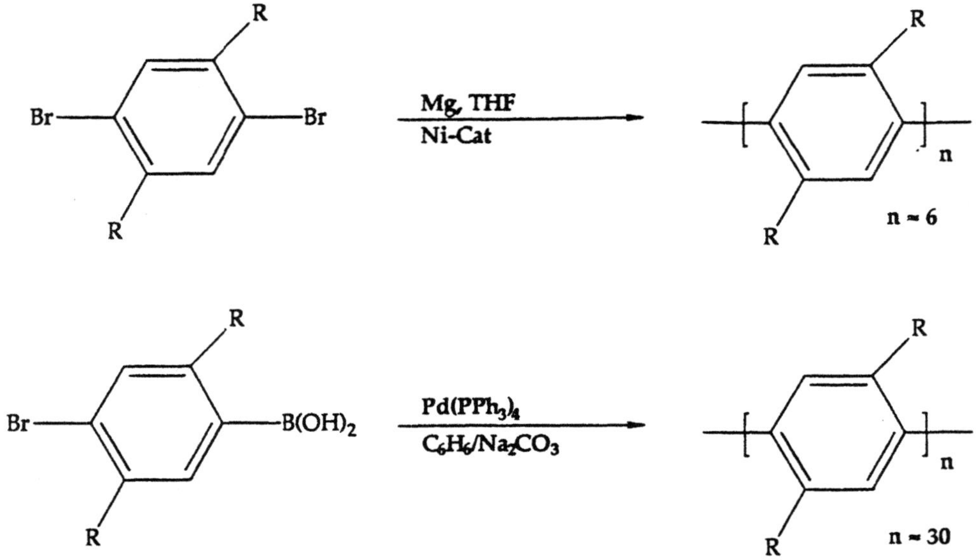

Figure 17. Syntheses of substituted PPP's.

Some heteroatom-substituted derivatives are known. Pyridines[144, 145] and quinolines[139] have been polymerized. An example, synthesized by step-growth polymerization, is poly(*p*-phenylene-*co*-2,5-pyrazine) shown in figure 18.[146, 147]

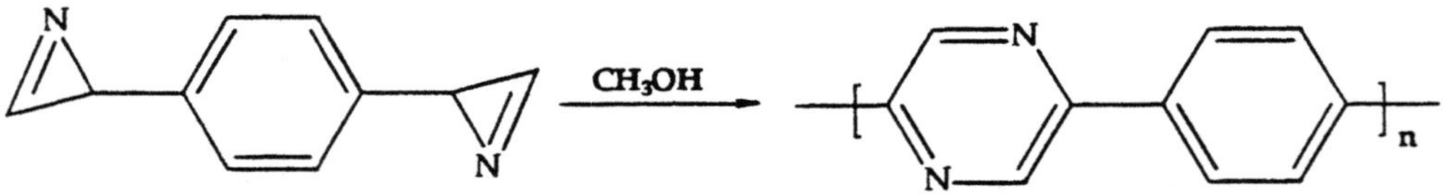

Figure 18. Poly(*p*-phenylene-*co*-2,5-pyrazine).

4.2.6. *Polysilanes.* Polysilanes, in which the conjugation is through the single bonds in the polymer backbone, have attracted much attention because of their interesting UV-spectroscopy and photophysics as well as their photochemistry, photoconduction, nonlinear optical properties and use in microlithography.[148] The backbone of the polymer is "conjugated" (using single or σ bonds in the polymer) but is still relatively floppy, so these polymers are often soluble and processible. The most common synthesis of polysilanes is a rather brutal Wurtz-coupling of R_2SiCl_2 units with sodium: a step-growth polymerization although it is complicated by the fact that it takes place in a heterogeneous environment (the sodium surface). This reaction does not tolerate functionality on the polysilane, but several routes for functionalizing the polymer after synthesis have been devised. This synthesis generally produces high molecular weight polymers, although the dispersity of molecular weights is sometimes polymodal, probably reflecting the heterogeneous nature of the reaction.

Figure 19. Polysilane polymerization via a Wurtz-coupling mechanism

Polysilanes have also been synthesized by dehydrogenative polymerization of silanes (i.e. $SiH_2R_2 \rightarrow$ -[-Si(R_2)-]-) using zirconocene and hafnocene catalysts. So far, these methods have only resulted in oligomers but are still an interesting application of the organometallic "σ bond metathesis" reaction to the synthesis of these main-group polymers.[149-151] Further development of this methodology is expected. Anionic ring-opening polymerization, although it does not proceed for several cyclic silanes,[152] has resulted in high polymer for 1,2,3,4-tetramethyl-1,2,3,4-tetraphenylcyclotetrasilane ($M_n \approx$ 30,000; $M_w/M_n \approx 2$).[153] This reaction follows chain-growth kinetics and has the potential to be modified to become a living polymerization (section 3.2). Silanes copolymers with π conjugated units (phenyl[154] and thiophene[155]) have also been prepared, allowing for the study of mixed π/σ conjugated systems.

4.2.7. *Polymers of Phthalocyanines.* A number of investigations of polymers linked through metals have been undertaken. Polymers of phthalocyanines (figure 20) have been synthesized using a variety of metals and non-metals in the center of the macrocyclic ring and a variety of linkages between these rings.[156, 157] In some cases, the linkage is conjugated with the metal and there is a delocalized path in the center of the polymer.[158] In many cases, however, the -[-M-X-]- linkage in the center of the polymer is non-conjugated, and delocalization arises from the overlap of the closely-spaced macrocyclic rings.[159-161] Partial oxidation of the polymer (doping) increases the conductivity of the material. However, conductivities are generally modest (≤ 1 S/cm). The monomeric macrocycle is typically very insoluble but nevertheless is used commercially for pigments and semiconductor applications. The polymers are also generally insoluble, but molecular weight determinations of some oxygen-linked phthalocyanines (M = Si, Ge, Sn) by end-group analysis and radiotracer techniques find degrees of polymerization of 70-120.[162] These materials are similar to charge-transfer complexes in that they both contain mixed valence stacks of conjugated planar molecules.[163, 164] However, these polymers are covalently linked. In contrast, charge-transfer complexes can easily dissociate upon dissolution or evaporation, thus losing extended delocalization. Phthalocyanines are only one of a number of macrocycles that can be stacked or polymerized in this manner. Many structurally similar systems can be found in the literature.[165]

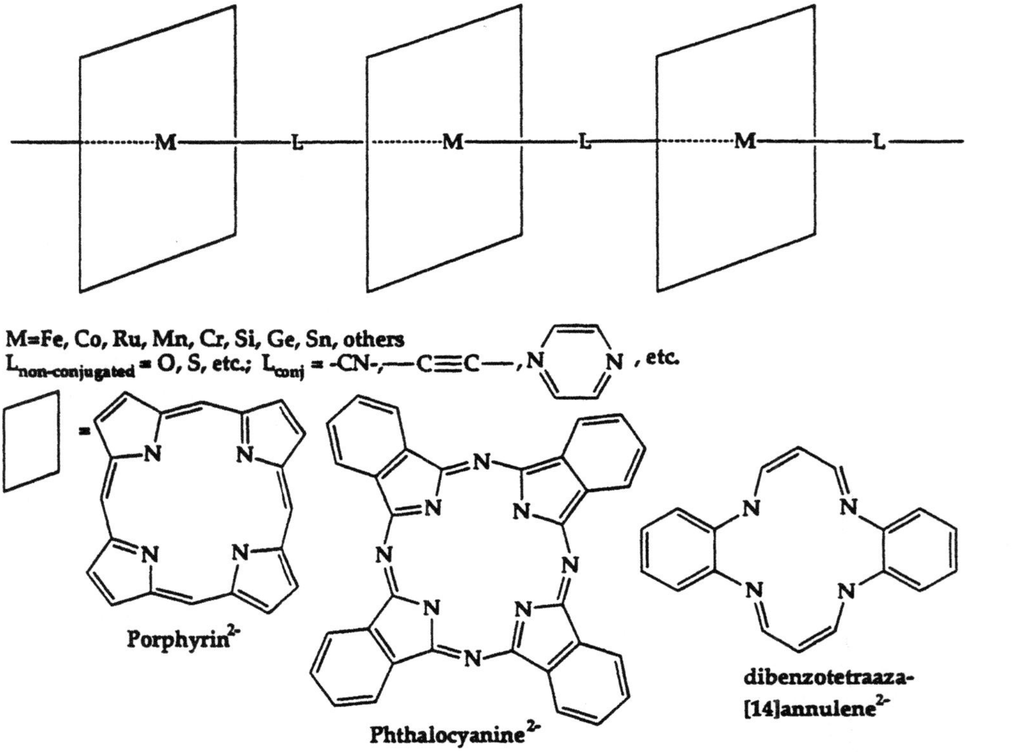

Figure 20. Metal phthalocyanine polymers can be linked via both conjugated and non-conjugated linkages. Only a few of the metals, linkages, and macrocycles are shown here.

4.2.8. *Other Conjugated Metal Coordination Polymers.* Metals are also incorporated into polymers via conjugated coordinative linkages such as di-thiolenes. These complexes, originally investigated as monomeric units that stacked to form one-dimensional systems along the stacking direction,[166] have been investigated in polymeric systems using a variety of counterion ligands.[167-170] Typically these are synthesized by reacting the conjugated bridge (typically a dianion or tetraanion) with a metal salt (e.g. NiX_2, X = Cl, Br, etc.). Most of these systems are insoluble and oligomeric in nature and have poor conductivities, even upon doping (values up to 20 S/cm reported). However, these polymers have recently been synthesized with more flexible linkages, allowing for the study of the effect of this linkage on solubility and electronic properties.[171, 172] None of the properties of these polymers have rivaled the all-organic conjugated polymers. They do, however, stand as examples of the role that a transition metal can play as part of a delocalized, conjugated system.

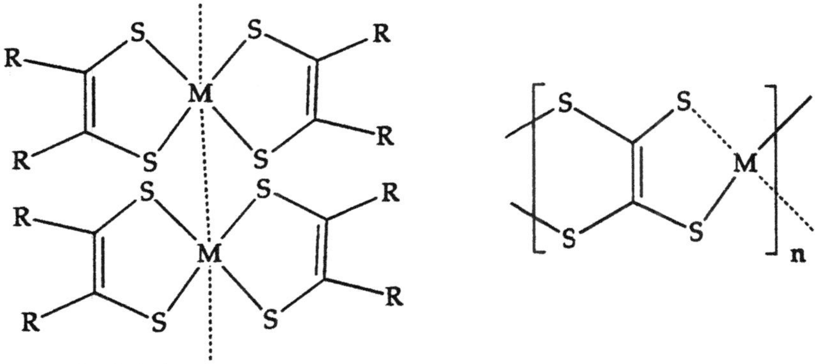

Figure 21. Stacked dithiolene complexes and linear dithiolene polymer (M = Ni). This example illustrates only one of the many conjugated counterions that have been employed.

4.2.9. *Ladder Polymers.* Another important structure in the realm of conjugated polymers is that of the ladder polymer where monomer units are joined by more than one bond (figure 22). The parent of this series, polyacene, would be expected to be non-bond alternate (section 6) and an intrinsic conductor much like graphite. Although there have been claimed syntheses of this polymer, from the photopolymerization of diacetylenes,[173] from oxidation of cyclized 1,2-polybutadiene,[174] and from pyrolysis of a phenol-formaldehyde polymer,[175] this system becomes insoluble and very air-sensitive as the number of fused benzene rings increases,[131] and adequate structural characterization is not yet available. Elemental analyses, where given, also suggest that these structures are not completely fused together as drawn. Since these polymers are linked as ribbons rather than chains, they all tend to suffer from insolubility at low degrees of polymerization.

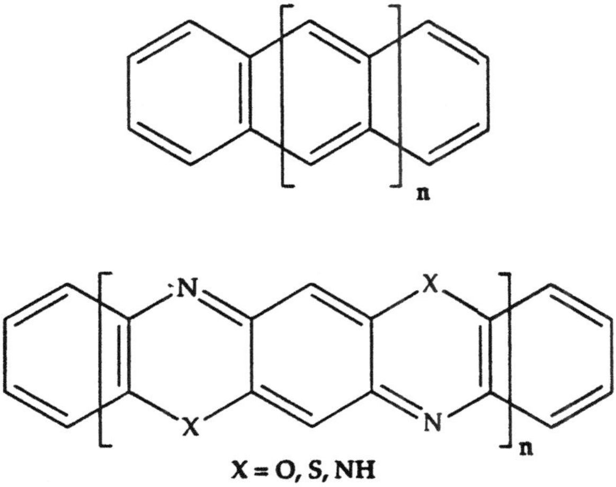

Figure 22. Polyacene and a representative ladder polymer.

Nevertheless, the promise of a thermally stable material prompts continued synthetic endeavor, including pyrolysis of polyacrylonitrile and polycyanoacetylene[176-178] (figure 23), explorations of electrochemical doping reactions on ladder polymers,[179] and work on the synthesis of ladder polymers with alkyl substituents to enhance solubility.[180] Some of these polymers have been investigated for their nonlinear optical properties.[181, 182]

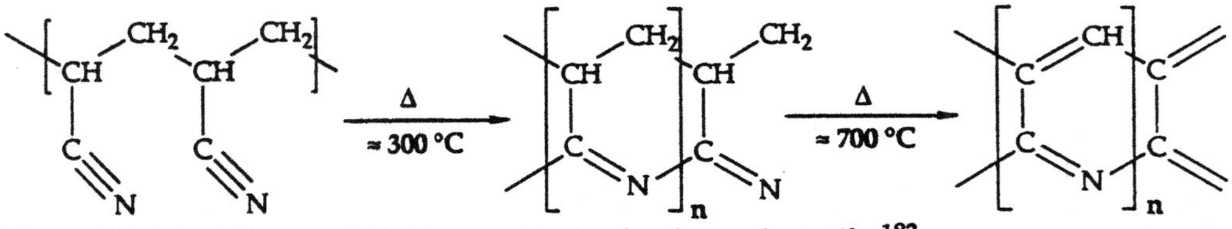

Figure 23. Possible events in the pyrolysis of polyacrylonitrile.[183]

Model compounds of ladder polymers are generally much better characterized and have some unique properties of their own. Oligomers of polyacenequinones have been reported as "molecular lines" since they are so rigid and can be synthesized with well-defined lengths.[184, 185] Related oligomers have been shown to form oriented Langmuir-Blodgett films,[186] and synthesis of thiol-terminated molecular lines allows them to self-assemble on a gold surface and become electrochemically doped.[187]

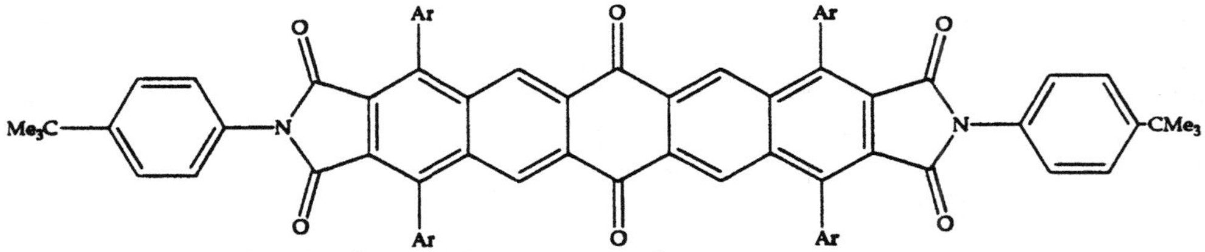

Figure 24. A "molecular line" 3.06 nanometers long.

Loss of some of the ladder-type linkages is found in polymers such as poly(*p*-phenylene-2,6-benzobisthizolediyl) (PBT, figure 25). These polymers have been electrochemically doped to a conductivity of 20 ± 10 S/cm,[188] and they are also of interest for their nonlinear optical properties.[189-192] Complexation of these polymers with Lewis acids (e.g. $AlCl_3$) renders them soluble.[193]

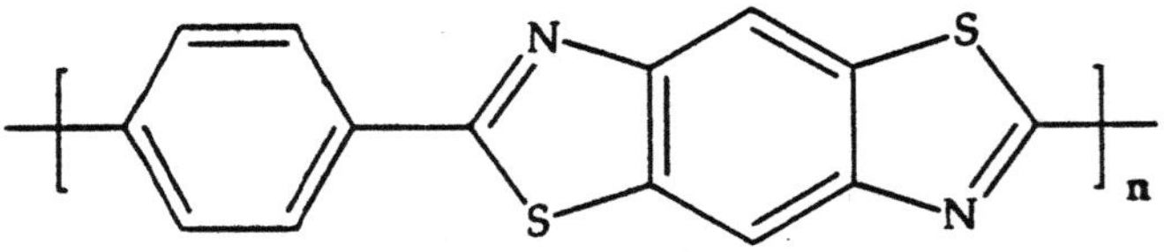

Figure 25. Poly(*p*-phenylene-2,6-benzobisthizolediyl) (PBT).

This section touches upon the more extensively studied conjugated polymers formed by step-growth polymerization. There exists substantially more structural variation than is outlined here. The reader is referred to an excellent reference for more examples, including general strategies for synthesis of conjugated polymers.[183]

4.3. THE UNUSUAL TOPOCHEMICAL POLYMERIZATION TO FORM POLYDIACETYLENES[194]

Single crystals of diacetylenes can be polymerized with light[195] or gamma irradiation[196] to produce crystals of polydiacetylene (figure 26). Naturally, this reaction requires the reacting units to be oriented in a precise way since molecular motion is limited in the crystalline state. Thus, only certain diacetylene monomers can be polymerized in this manner.[197] However, it has been found that several types of diacetylenes can be polymerized when ordered in multilayers using Langmuir-Blodgett techniques,[198, 199] and additional work has shown that these polymerizations can take place in the Langmuir-Blodgett trough directly and can be monitored by visible absorption spectroscopy.[200, 201] They have also been polymerized in lipid vesicles, and the structure of the bilayer is retained after the polymerization.[202-206]

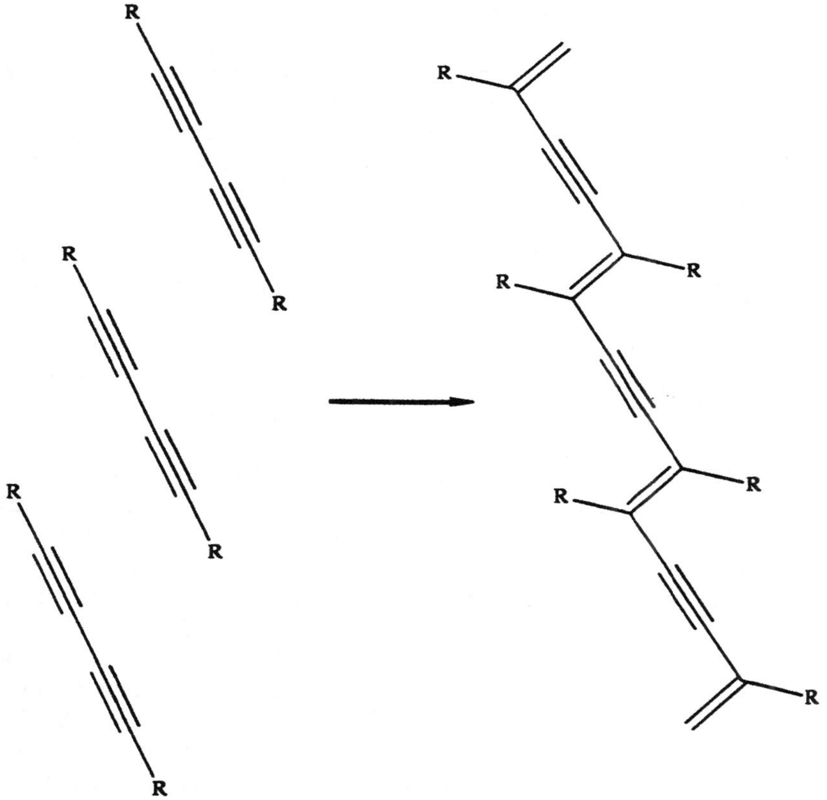

Figure 26. Generalized Scheme for the Topochemical Polymerization of Diacetylenes

Although most of these polymers are insoluble, several soluble polydiacetylenes have been synthesized. The most studied are those containing butoxycarbonylmethyleneurethane (BCMU) groups.[207, 208] Although not especially good electrical conductors, these polymers are sensitive to changes in environment such as solvent and temperature,[209-212] and mechanical stress ("Mechanochromic Behavior"),[213] undergoing conformational changes[214, 215] and/or aggregation which lead to striking changes in color. Some of these polymers have large third-order nonlinear optical susceptibilities (χ^3),[216] leading to interest in their use in devices such as active waveguides.[217, 218]

4.4. CHAIN-GROWTH POLYMERIZATIONS

Chain-growth polymerizations are useful in the synthesis of conjugated polymers because polymer properties can often be tailored by the selection of catalyst system and because higher molecular weight polymers can be synthesized at a lower degree of conversion of monomer, a point which is particularly attractive when it is necessary to form an insoluble polymer. It has already been pointed out that several conjugated "polymers" are suspected to be oligomers, particularly oxidatively coupled polyparaphenylene, and it is frustrating when the validity of a study must be questioned because it is unclear whether the material under study is really polymeric in nature. Although many would argue that molecular weight has little to do with polymer properties, this premise has little or no experimental verification. Moreover, some of these conjugated polymers are attracting interest as high strength materials,[219] and polymer strength is certainly dependent upon molecular weight.

4.4.1. *Polyacetylene via Ziegler-Natta polymerization.* Acetylene was first polymerized by Natta et al. by bubbling the gas through an organotitanium/trialkylaluminum catalyst mixture, resulting in a black, powdery precipitate.[220] It was not, however, until Shirakawa and co-workers discovered how to produce free-standing films[221] and optimized this procedure[222] that it was possible to handle this material easily. These continuous, highly crystalline films made possible the first observations of high conductivity upon exposure to strong oxidants[1, 2, 3] and later reductants. This development sparked a large amount of subsequent research on polyacetylene and other conjugated polymers. Shirakawa's method continues to be the most common way to synthesize polyacetylene, and the material produced from this synthesis is often referred to as S-PA (Shirakawa polyacetylene).[223] A detailed procedure for this synthesis is available,[10] but briefly, the polymer is synthesized by coating the walls of a Schlenk (inert atmosphere) tube with a slurry of $AlEt_3/Ti(O\text{-}n\text{-}Bu)_4$ (4:1) in toluene followed by admission of acetylene gas which polymerizes on the surface of the catalyst to produce a shiny silver film which can be pried from the walls of the reaction vessel and washed to remove the majority of the catalyst residues. Low temperature (-78 °C) synthesis insures the formation of a film with a high content of cis double bonds in the chain. Heating the film to 150 °C for an hour or so isomerizes the double bonds to the predominantly trans configuration. This isomerization process has been studied in the solid state and is complicated and dependent on the morphology of the film.[224, 225] Also, cis-trans isomerization is believed to occur during doping.[226] The single bonds in the polymer can also adopt a cis or trans configuration (termed cisoid and transoid). The barrier for conversion from cisoid to transoid is much lower than for cis/trans isomerization in which the pi (second) bond must formally be broken and reformed. All four combinations of cis/trans single/double bonds in planar[227] as well as nonplanar structures[228] have been considered and a variety of conformations including helical structures[227, 229, 230] have been proposed. Also, neither *cis-* nor *trans-*polyacetylene contains double bonds of exclusively one type. It has been acknowledged that, even after isomerization, residual cis units exist in *trans*-polyacetylene.[231, 232]

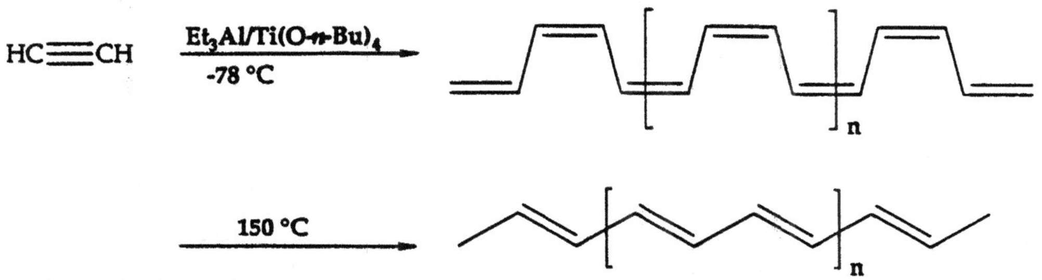

Figure 27. Polymerization of acetylene showing cis and trans isomers.

A number of block and graft copolymers of polyacetylene have been synthesized, generally with the idea of using the second half of the block as a solubilizing tail.[233] In most cases, this modification rendered the polymer soluble and has allowed a variety of studies of the linear[234, 235] and nonlinear[236, 237] optical properties as well as photoinduced absorption studies[238] of polyacetylene segments in solution. There is evidence however, that the polyacetylene segments in these polymers aggregate in solution[239] and these studies have not been made in the absence of interchain effects. Some of these block copolymers were observed to phase separate to produce "microdots" of polyacetylene in an insulating matrix that were potentially small enough to observe quantum dot effects.[240-242]

A very exciting modification of this synthesis has been described by Naarman and others and involves both the use of an additional reducing agent in the catalyst mixture and mechanical alignment of the polymer by stretching by a factor of 6 which is facilitated by synthesizing the polymer in a mixture of the catalyst and silicone oil.[243-245] This new polyacetylene (N-PA) is very well ordered and is believed to have a significantly lower number of non-conjugated (sp^3) defects. The result is extremely high iodine-doped conductivities greater than 100,000 S/cm with values routinely obtained of 10,000-50,000 S/cm. Many have taken this work as evidence that well-ordered organic polymer matrices can be made as conductive as metals.

This type of polymerization has been carried out on numerous acetylene derivatives, often producing high molecular weight, soluble polyacetylene derivatives.[246-248] However, the high-energy visible absorptions and low iodine-doped conductivities[249-251] of these polymers indicate that they are much less conjugated than the parent polyacetylene, most likely due to twisting of the polymer chain due to steric repulsions between the closely spaced side groups (figure 28).[252] Polyacetylene derivatives with side groups spaced farther apart have been synthesized and do not have such significantly reduced conjugation (see below).

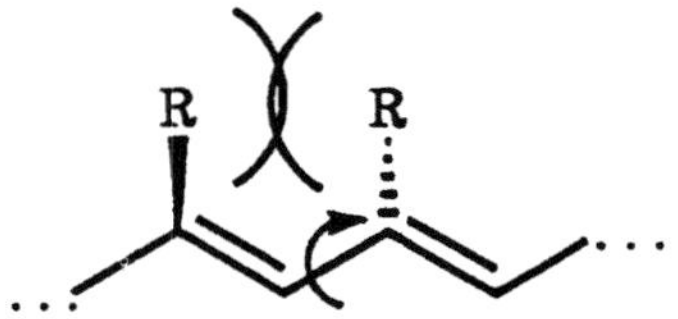

Figure 28. Reduction of conjugation in polyacetylene derivatives due to steric effects.

4.4.2. *Ring-Opening Metathesis Polymerization routes to Polyacetylenes.* Ring-opening metathesis polymerization (ROMP) involves the use of a transition metal carbene complex to cut open a cyclic olefin molecule and "stitch" these molecules back together into a polymer chain (figure 29).[253] This polymerization is of great interest for the synthesis of conjugated polymers because, among other factors, the number of double bonds in the monomer is preserved in the polymer, and because, in principle one can take the repeat unit for any conjugated polymer containing an olefin and cyclize it to form a potential new monomer.

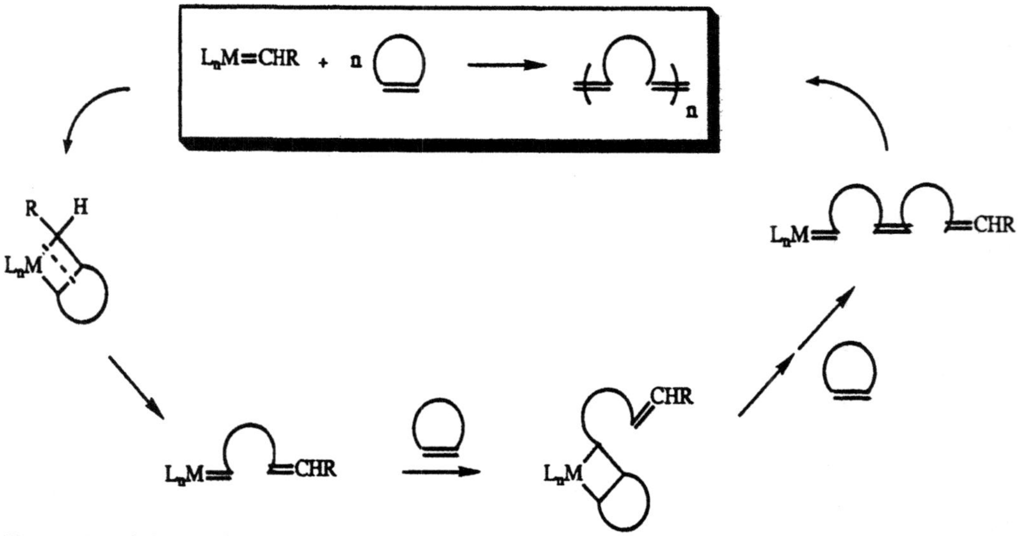

Figure 29. Scheme for ring-opening metathesis polymerization

This polymerization has been very successfully used to form several precursor polymers which will be discussed later. However ROMP has been applied to the direct synthesis of conjugated polymers as well, namely polyacetylene. In this case, the appropriate monomer is a circle of 4 double bonds: cyclooctatetraene (COT). This monomer, prepared by the tetramerization of acetylene with a nickel catalyst, has been polymerized, first using so-called "classical" but also Lewis-acidic catalysts which created a number of saturated defects in the polyacetylene,[254, 255] and more recently using a pre-formed, non-Lewis-acidic tungsten carbene complex to produce polyacetylene of good quality.[256] This method has been extended to the polymerization of mono-substituted COT's which, given the appropriate bulk in the side group, produces soluble, yet highly conjugated polyacetylene derivatives.[257-260] The nonlinear optical properties of some of these polymers have been measured,[261] and these polymers have been exploited in the fabrication of layered structures and efficient solar cells.[262] This procedure avoids the harsh metal vapor conditions typically employed when a thin layer of a metal is desired on a surface.

5. Polymers from precursors

Precursor methods can be very advantageous in the synthesis of insoluble conjugated macromolecules. Most of the insoluble polymers lack adequate characterization data and have low molecular weights. Precursor methods afford a soluble prepolymer that is often of high molecular weight and can be fully characterized. Typically, the best precursors can be converted to the final form under mild conditions that do not disturb the structure of the polymer and with as low a mass loss as possible, although some of the precursors discussed in this section do suffer from the fact that much of the weight of the prepolymer is lost upon conversion.

5.1. POLYPARAPHENYLENE (PPP)

Several step-growth routes to PPP have already been described. Despite some clever synthetic ideas, the end product is still a rigid, insoluble polymer, and all the step-growth methods discussed result in a low degree of polymerization. Thus, precursor routes to PPP have taken on much importance.

Early attempts centered around the polymerization and aromatization of cyclohexadiene (figure 30). This monomer can be polymerized to some degree by Ziegler-type, radical, cationic and anionic initiators.[263-266] In several cases, molecular weights were suspected to be low. However, the biggest problem with this scheme is that dehydrogenation of the polymer (precursor conversion) requires high temperatures and/or strong oxidants and often does not proceed to completion. As an added complication, these polymerizations result in both 1,4- and 1,2-linkages, deviating somewhat from the straight para (1,4) linkages normally drawn. Nevertheless, this precursor route has been employed on several occasions, recently in the synthesis of a soluble polystyrene-PPP block copolymer which was chemically converted from its precursor in solution and could be subsequently doped in solution.[267]

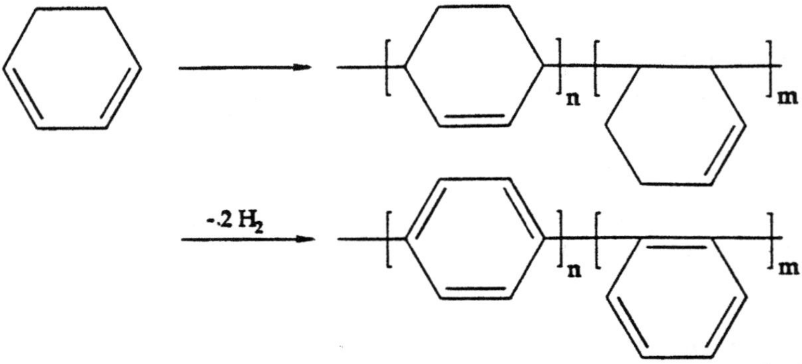

Figure 30. Polymerization and aromatization of cyclohexadiene showing the linkages formed by 1,4- and 1,2-propagation.

More recently, it has been discovered that a cis-dihydrocatechol derivative of cyclohexadiene, conveniently produced from the microbial oxidation of benzene, can be converted to a variety of derivatives and polymerized using radical initiators. These polymers can be converted to PPP thermally at much lower temperatures, eliminating alcohols or acetic acid ($T_{conv} \approx 250\text{-}300$ °C for R = Me, figure 31).[268, 269] The concern over 1,2-linkages in the polymer persists with these precursors. However, they have provided a low temperature route to high molecular weight PPP.

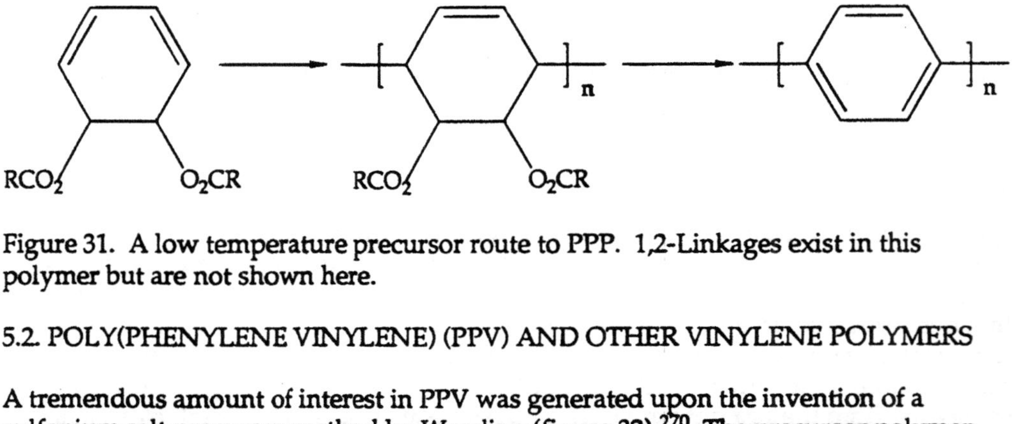

Figure 31. A low temperature precursor route to PPP. 1,2-Linkages exist in this polymer but are not shown here.

5.2. POLY(PHENYLENE VINYLENE) (PPV) AND OTHER VINYLENE POLYMERS

A tremendous amount of interest in PPV was generated upon the invention of a sulfonium salt precursor method by Wessling (figure 32).[270] The precursor polymer, synthesized by adding a base to the *para*-disulfonium salt of the monomer, is thought to form via an anionic chain mechanism.[271] Following the process using X-ray scattering, a reduction in the interchain distances as well as an increase in crystallinity has been observed.[272] This route has allowed for the preparation of a number of derivatives, including alkoxy (-OR), halogen (Cl, Br) and alkyl substituted derivatives, typically at the 2,5-positions of the benzene ring.[273, 274] These substitutions assist in processing of the precursors and also modify the bandgap of the polymer to some extent.[85, 275] A water-soluble 2,5-disubstituted PPV has been reported.[276] Substituents can also be placed on the olefin segment of the polymer, resulting in some twisting of the polymer chain that reduces conjugation, but producing soluble conjugated polymers.[277] Subsequent to the demonstration of this route, further efforts have centered on the optimization of the group to be eliminated[278] and the effect of polymerization conditions on the molecular weight of the precursor polymer.[279] Also, this precursor has been blended with a variety of different types of polymers.[280, 281]

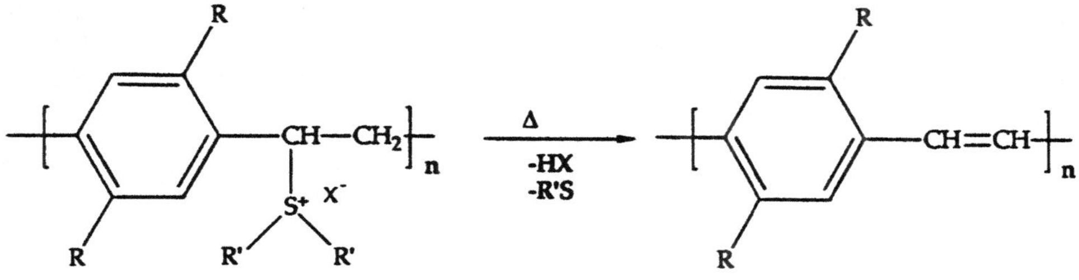

R = H, alkoxy, alkyl, Cl, $O(CH_2)_3SO_3Na$; X = Cl, Br; R' = alkyl

Figure 32. Precursor route to PPV.

Early investigations on this polymer employed it in the formation of graphite intercalation complexes with, for example, potassium[282] and $FeCl_3$.[283] High conductivities were observed in these materials ($\approx 10^5$ S/cm). More recently, however, PPV and its derivatives have been of interest for their third-order nonlinear optical properties,[284-286] and in optical recording devices.[287] PPV has also found application as a light-emitting diode.[288]

The sulfonium salt precursor method has also resulted in the synthesis of a variety of other aromatic and heteroaromatic vinylenes. In particular, poly(napthalene vinylene),[289] poly(thienylene vinylene),[290-292] and poly(furanylene vinylene)[293] and derivatives thereof have been prepared (figure 33).

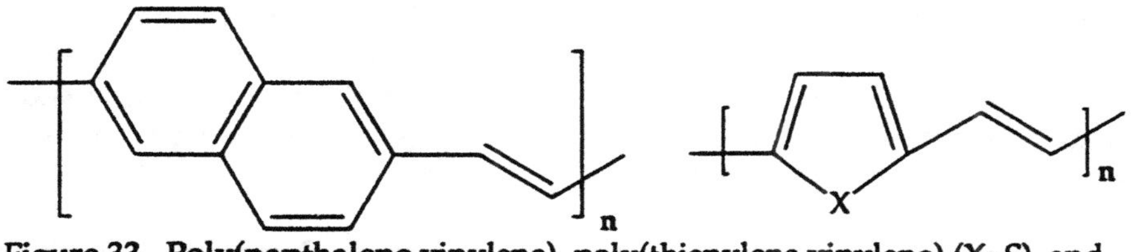

Figure 33. Poly(napthalene vinylene), poly(thienylene vinylene) (X=S), and poly(furanylene vinylene) (X=O).

5.3. PRECURSORS TO POLYACETYLENE

Probably the best known precursor polymer is that developed by Feast and coworkers at the university of Durham.[294, 295] This polymer is typically referred to as "Feast" or "Durham" polyacetylene and involves the ring-opening metathesis polymerization of cyclobutene derivatives followed by a thermal retro-Diels Alder reaction to perform the conversion (figure 34). This polymerization is of chain-growth type and the precursor can be easily rid of catalyst residues which is much more difficult with directly synthesized polyacetylene.[296] Where R = CF_3, this conversion can be performed at a very low temperature (< 50 °C) and results in a high quality polyacetylene with relatively low crystallinity. This low crystallinity can be useful, particularly in nonlinear optical applications where scattering of light off of crystallites becomes a problem. However, crystallinity can be increased, most easily by orienting the material either before[297] or as it is converted, and polarized infrared spectroscopy of oriented material revealed a higher anisotropy than observed for a similarly oriented sample of Shirakawa polyacetylene.[298, 299]

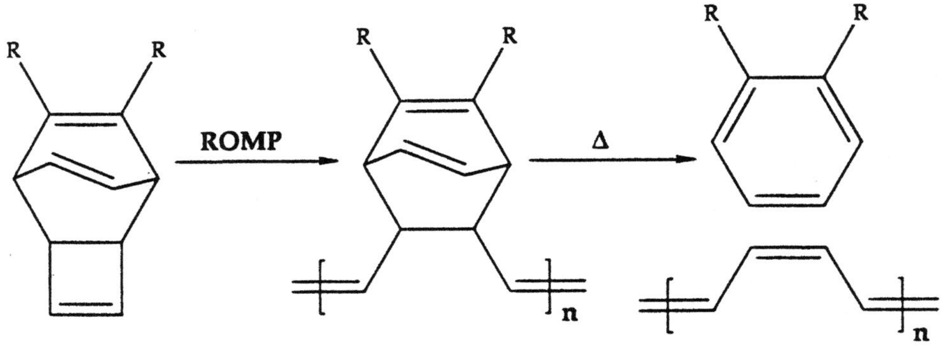

Figure 34. Synthesis and conversion of Durham polyacetylene.

The ring-opening metathesis polymerization (ROMP) employed is subject to a great deal of control, and a controllable polymerization of the Feast monomer has been demonstrated using metathesis catalysts based on tungsten[300] and titanium.[301] Also, by selecting which starting carbene to employ (which now can be easily varied)[302] and taking advantage of the facile end-capping reaction with an aldehyde or ketone,[303] one can individually control both of the end groups on the polymer. Oligomerization of the monomer followed by conversion produced a series of polyenes which were separated and characterized Also, block copolymers were synthesized by polymerizing this monomer and then polymerizing norbornene, a monomer that yields an elastomeric polymer.[300] After conversion, the two sections of this block copolymer were observed to phase separate to form well-defined polyacetylene "microdots" as has been noticed for other polyacetylene block copolymers (section 4.4.1).[304]

In an effort to reduce the amount of mass loss during conversion, Swager and Grubbs ring-opened benzvalene, also using ROMP. The prepolymer possesses considerable ring strain (the polymer is prone to detonation and can be converted to polyacetylene using mercury salts. This precursor is ideal because it loses no mass upon conversion. However, the polyacetylene produced displays low iodine-doped conductivities and possesses a number of saturated (sp^3) defects as determined by ^{13}C CP-MAS NMR, possibly due to cross-linking induced by residual mercury salts.[305]

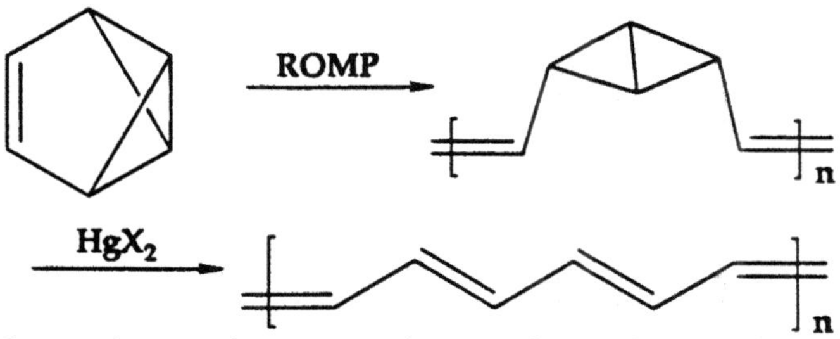

Figure 35. Polybenzvalene and its conversion to polyacetylene.

6. Extensions of these methods in the synthesis of "small-bandgap" polymers

All of this synthetic methodology is designed to establish and exploit the link between structure and properties. In the case of conjugated polymers, one motivation for the synthesis of new structures is the reduction of the bandgap: the energy difference between the highest occupied crystal orbital and the lowest unoccupied crystal orbital orbital of the polymer and, physically, more-or-less the energy of the lowest electronic transition. Briefly, the bandgap of a polymer relates to the ease of generation of carriers. A small bandgap results in easier doping or higher photoconductivities. Ultimately, a polymer with a very small bandgap would be transparent in the visible region of the spectrum and should be easy to dope. A smaller bandgap would also lead to greater mixing between the ground and excited states of the polymer, which could, in principle, lead to larger hyperpolarizabilities (nonlinear optical coefficients).

A very intuitive way of thinking about the bandgaps in conjugated polymers has been advanced by J. L. Bredas.[306] Most conjugated organic systems are drawn using single and double bonds, and, in the molecule, the single bonds are longer than the double bonds: they have bond length alternation. Essentially, one wants to avoid large changes in bond length in going from the ground to the first excited state of the polymer (i.e. jumping its bandgap). In aromatic systems (cyclic arrays of 4n+2 electrons which are especially stable, as in benzene) one wants to avoid losing that aromaticity in the excited state because, in the non-aromatic (quinoid, see figure 36) form, formally long single bonds are found in the aromatic (short bond) part of the polymer, and the long single bonds joining the rings become short double bonds. There is a large change in the bond lengths resulting from this loss of aromaticity, and the large energy cost that is required to go from the aromatic to quinoid form is manifested in a large bandgap (3.2 eV for PPP). If polyacetylene's bonds were all the same length, it would have a zero bandgap and be metallic in nature (elimination of bond length alternation in one-dimensional systems results in a zero bandgap material). However, polyacetylene is bond-length alternate due to a second-order Peierls distortion that is predicted to occur in all strictly one-dimensional systems, and it has a bandgap (1.8 eV for *trans*-polyacetylene).

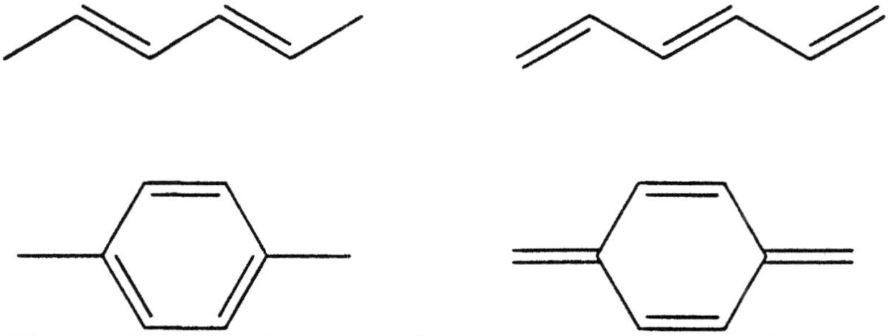

Figure 36. The two bond-length alternate forms of polyacetylene and polyparaphenylene. The aromatic (left) and quinoid (right) forms of PPP are shown.

Wudl cleverly exploited this type of reasoning in the design of a low bandgap polythiophene derivative. As in polyparaphenylene, the quinoid form of polythiophene experiences a loss of aromaticity. By benzannelating polythiophene to give polyisothianapthene (PITN, figure 37), Wudl created a polymer where bond-length alternation was reduced since both forms have some aromatic stabilization.[307-309] In fact, the band gap for this polymer is approx. 1 eV. More recent theoretical studies have suggested that the band gap reduction is better explained by orbital-mixing arguments.[310] However, these very qualitative arguments continue to enjoy success in researcher's design ideas.

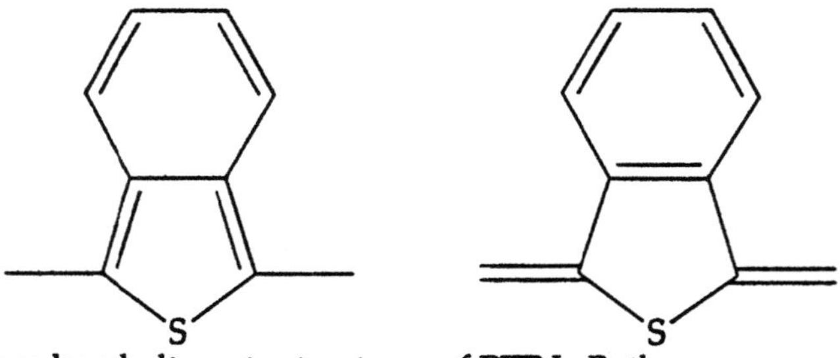

Figure 37. The two bond-alternate structures of PITN. Both possess some degree of aromaticity.

Another approach has been advanced by Jenekhe.[311, 312] The structure of interest contains alternating aromatic and quinoid units (figure 38). Reversal of the bonds leads to an identical structure, and bandgaps as low as 0.75 eV have been claimed. The third-order nonlinear optical properties of this polymer have been examined.[313] However, these types of low bandgap materials have yet to be definitively synthesized. Here a precursor is converted with bromine (figure 38), and it has been shown that this reaction step results in bromination of the material, both covalently and ionically.[314] Since these defects will certainly have an effect on the bandgap of the polymer, further work is warranted. A similar structure was synthesized by reacting 2-pyrrolealdehyde units together using $POCl_3$. The authors note structural irregularities in the polymer such as saturated defects and branched units. Nevertheless, they report a band gap of about 1 eV.[315]

Alternating quinoid/aromatic polymers containing benzene rings[316] (and napthalene and anthracene rings as well)[317] have been synthesized as well, although they have a low degree of polymerization. Moreover, their low AsF_5-doped conductivity ($\approx 10^{-3}$ S/cm), and high-energy optical absorption maxima (3.3 eV) suggest that these polymers contain saturated defects.

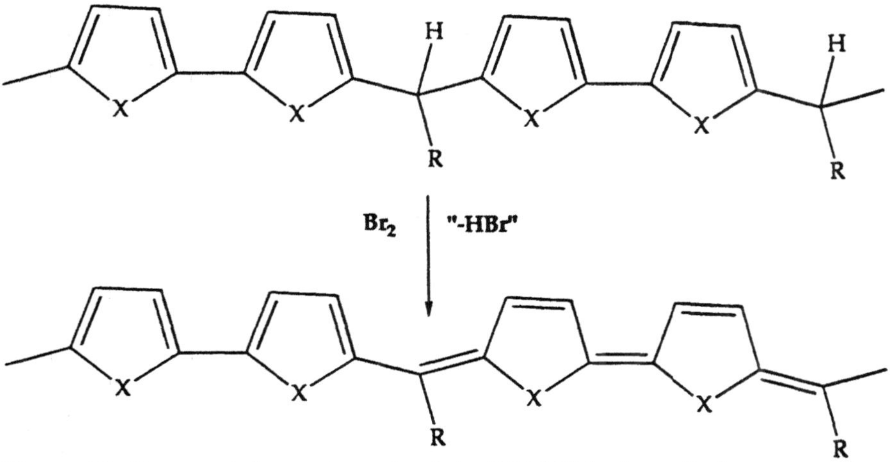

Figure 38. Proposed formation of poly(α-(5,5'-bithiophenediyl)-*p*-acetoxybenzylidene) (PBTAB). X = S; R = *p*-acetyoxyphenyl.

7. Conjugated Polymer Matrices

The previous section focused on tailoring the structure of a conjugated polymer. However, several properties of conjugated polymers including their optical absorptions, but most notably their doped conductivities are dependent on contacts between polymer chains. Stretch aligning a fiber or film of a conjugated polymer helps to increase crystallinity, which can easily lead to increases in the doped conductivity of 3-4 orders of magnitude compared to the unoriented material. This increase in conductivity is primarily due to increased carrier mobility associated with the more conjugated (planar) and more closely packed chains found in a crystalline region. Thus, there have been a number of efforts towards organizing a conjugated polymer matrix as well as efforts towards isolating individual polymer strands to study their properties in the absence of neighbors. Conjugated polymers have also been synthesized in matrices which aid in processing.

The first efforts towards synthesizing a conjugated polymer in a matrix were designed to assist in processing. For example, dispersing polyacetylene as a powder in another polymer melt was not useful because of the high loadings required for reasonable conductivity and the realization that these materials were little or no better than the same matrix containing carbon-black as a filler.[318] More interesting was the method which employed a Ziegler-Natta catalyst impregnated into a low-density polyethylene matrix. Acetylene gas was admitted and polymerized within the matrix. Relatively high iodine-doped conductivities (≈ 1 S/cm) were obtained for relatively low loadings of polyacetylene (4-6 wt %).[319] In contrast, it was also discovered that polyacetylene could be incorporated up to 82% w/w in a polyethylene fiber and drawn and doped to a conductivity of 6000 S/cm.[320] This technique was extended to elastomer composites which were stretched and doped to relatively high conductivities (575 S/cm), [321] and polymerization in aramid gels from which strong, oriented conductive fibers could be pulled.[322]

As mentioned before, composites can be prepared electrochemically, typically by polymerizing the monomer on an electrode that had been pretreated with a polymer. This process has been performed with pyrrole using (with a variety of modifications as well) poly(vinyl alcohol),[323] poly(vinyl chloride),[324, 325] polyurethane,[326] and as a graft onto polystyrene functionalized with benzyl chloride side-chains.[327] Pyrrole has been mixed with Kevlar (an ultra high modulus polymer) and chemically polymerized to form a very strong "alloy" fiber of polypyrrole and Kevlar.[328]

Synthesis in an ordering medium has also resulted in polymer orientation. Aldissi originally discovered that dissolution of a Ziegler-Natta catalyst in a liquid crystalline solvent that was oriented by an external magnetic field followed by polymerization produced an oriented polyacetylene with a very anisotropic doped conductivity of 1235 S/cm ($\rho_{||}/\rho_{\perp} = 4.25$).[329] Later researchers used this technique to achieve conductivities of up to 4800 S/cm.[330] Aldissi has oriented a water-soluble derivative of polythiophene by passing a solution of the polymer through a magnetic field while simultaneously removing the solvent, achieving a conductivity of 10^4 S/cm.[28]

Porous membranes have also been utilized in the synthesis of oriented conjugated polymers. Polypyrrole and polythiophene have been electrochemically grown in a nucleopore membrane[331] to form conjugated polymer microtubules.[332] Polyacetylene, synthesized by allowing catalyst and acetylene to diffuse into the membrane from opposite sides is partially ordered as shown by polarized IR absorption spectra and higher doped conductivities.[333]

Mechanical stretch orientation of polymer films has been one of the most common methods for achieving higher crystallinity and higher doped conductivities in polymer films. It has successfully been applied to polyacetylene,[334-336] polypyrrole,[337] and polyaniline heated and plasticized with solvent[338] to mention only a few of the many investigations that have been reported. Moreover, polyacetylene synthesized according to the modifications of Naarman (section 4.4.2) is also stretch oriented, and this orientation is believed to be one of the reasons that this material displays such a high doped conductivity. A related method involves mechanical orientation during conversion of a precursor polymer. For example, Durham (Feast) polyacetylene has been stretched during conversion by Leising and coworkers and this material yielded reasonably well-resolved X-ray fiber patterns which allowed the geometry of the chain to be investigated.[339] The precursor polymers to poly(phenylene vinylene), its 2,5-

alkoxy derivatives and poly(thienylene vinylene) have been oriented and then converted to the conjugated polymer, producing films and fibers of mechanically strong material that displays doped conductivities up to 1000 times greater than the unoriented material[219, 340-343] as well as enhanced and anisotropic nonlinear optical properties.[344]

Incorporation of conjugated polymeric chains in inorganic host materials has been used to assist polymer ordering and perhaps ultimately provide a matrix that can be analyzed by X-ray diffraction so as to determine the structure of some of these conjugated polymers. Intercalation of pyrrole in FeOCl results in oxidation of the pyrrole monomers and formation of polypyrrole between FeOCl layers.[345] This work has been extended to the intercalation/polymerization of polythiophene in FeOCl,[346] and the intercalation/polymerization of polyaniline[347] and polythiophene[348] in V_2O_5 xerogels. Zeolites have also been used as well for the oxidation, entrapment and subsequent spectral characterization of polyenes[349] and for the oxidative polymerization of thiophene and spectral evaluation of the oligomers produced.[350] It is of fundamental interest to understand how the spectroscopic properties of these polymers evolve as the chain length increases, and a number of spectra of the radical cations and dications (corresponding to polarons and bipolarons in the polymer) have been obtained in this manner.

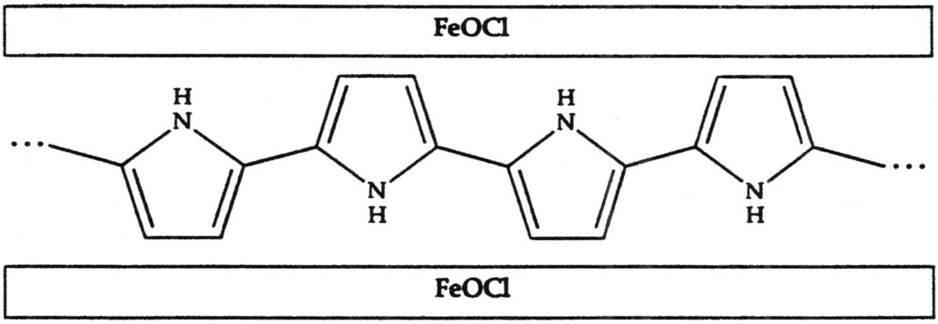

Figure 39. Polypyrrole/FeOCl layered complex.

9. Conclusions and Caveats

This chapter has attempted to give the reader a review of conjugated polymeric materials by outlining the structural variation in both the types of conjugated polymers that have been prepared and in the various methods that have been employed for processing the polymeric matrix. It has categorized this enumeration mostly by the type of synthesis employed for the polymer, making an effort to point out which synthetic methodologies offer the most control over the nature of the resulting material, i.e. how the monomeric units are linked together and what molecular weight and molecular weight distribution can be expected.

Almost all of the research cited here has been performed with the purpose of finding ways to understand and gain control over the properties of these polymeric structures.

Few researchers would be willing to admit that they are content in engaging in a "hunt-and-peck" search for that hunk of material that displays the highest figures of merit for whatever electrical or optical property they might be trying to optimize. Nevertheless, with the basic uncertainty in geometry and stoichiometry in many of these solid-state reactions, there is obvious doubt about the relationship between structure and properties that is trying to be elucidated. Many researchers claim to be concerned about properties and not the structure of their materials. However, dramatic examples of physical properties have already been demonstrated using very simple systems. Controlled pyrolysis of processible resins,[351, 352] and diynes,[353] produce materials that, although not well-defined structures, are intrinsically conducting due to the large number of free-radical charge carriers that are generated. These materials can relatively easily be made to display conductivities of $\approx 10^2$ S/cm which are rivalled by relatively few of the structures presented here. Graphite, a cheap, common organic material can be doped with AsF_5 to a conductivity of approx. 6×10^5 S/cm, comparable with any naturally occuring metal.[354] Potassium-doped graphite superconducts at low temperatures.[355] These examples are simple practical realizations of the properties that drive much of this search and are described here to encourage researchers to be critical of the different synthetic techniques and aware of possible structural variations in these materials rather than accepting the structure that is drawn on paper. More structural control in the synthesis of materials and acknowledgement of structural flaws in materials should result in a quicker and more correct understanding of their physics and chemistry.

Acknowledgements

RHG acknowledges financial support from the Office of Naval Research. CBG thanks the ACS Division of Organic Chemistry and SmithKline & French Laboratories for a fellowship. We thank R. Kaner for helpful comments and manuscripts of work not yet in print.

9. References

(1) Chiang, C. K.; Druy, M. A.; Gau, S. C.; Heeger, A. J.; Louis, E. J.; MacDiarmid, A. G.; Park, Y. W.; Shirakawa, H. *J. Am. Chem. Soc.* **1978,** *100(3),* 1013-1015.
(2) Chiang, C. K.; C. R. Fincher, J.; Park, Y. W.; Heeger, A. J.; Shirakawa, H.; Louis, E. J.; Gau, S. C.; MacDiarmid, A. G. *Phys. Rev. Lett.* **1977,** *39(17),* 1098-1101.
(3) Shirakawa, H.; Louis, E. J.; MacDiarmid, A. G.; Chiang, C. K.; Heeger, A. J. *J. Chem. Soc. Chem. Commun.* **1977,** 578-580.
(4) Heeger, A. J.; Orenstein, J.; Ulrich, D. R., Eds. *Nonlinear Optical Properties of Polymers.* Materials Research Society Symposium Proceedings Vol. 109; Materials Research Society, Pittsburgh, 1988.
(5) Messier, J.; Kajzar, F.; Prasad, P.; Ulrich, D., Eds. *Nonlinear Optical Effects in Organic Polymers.* NATO ASI Series. Series E. Applied Sciences Vol. 162; Kluwer Academic Publishers, Dordrecht, The Netherlands, 1989.

(6) Prasad, P. N.; Ulrich, D. R., Eds. *Nonlinear Optical and Electroactive Polymers.* Plenum Press, New York, 1987.
(7) Carter, F. L., Ed. *Molecular Electronic Devices, Vol I and II.* Marcel Dekker, Inc., New York, 1982 and 1987.
(8) Figure is for low density polyethylene production for 1990 (est. figure) and does not include 8 billion pounds of high density polyethylene produced as well: Kiefer, D. M. *Chem. Eng. News* **1990**, *68(50)*, 26-28.
(9) Karol, F. J. in *Organometallic Polymers*; Carraher, C. E.; Sheats, J. E.;Pittman, C. U. Eds.; Plenum Press, New York, 1978; pp 135-143.
(10) A remarkably detailed procedure for the most common synthesis of polyacetylene is available: Gibson, H. W.; Pochan, J. M. in *Encyclopedia of Polymer Science and Engineering*; Vol. 1; John Wiley and Sons, New York, 1985; pp 87-130.
(11) This principle has been nicely illustrated for a series of soliton model compounds: Tolbert, L. M.; Ogle, M. E. *J. Am. Chem. Soc.* **1990**, *112(26)*, 9519-9527.
(12) Bredas, J. L.; Heeger, A. J. *Macromolecules* **1990**, *23*, 1150-1156.
(13) Rughooputh, S. D. D. V.; Hotta, S.; Heeger, A. J.; Wudl, F. *J. Polym. Sci. Polym. Phys.* **1987**, *25*, 1071-1078.
(14) Inganas, O.; Salaneck, W. R.; Osterholm, J.-E.; Laasko, J. *Synth. Met.* **1988**, *22*, 395-406.
(15) Yoshino, K.; Park, D. H.; Park, B. K.; Onoda, M.; Sugimoto, R. *Jap. Journ. Appl. Phys.* **1988**, *27(9)*, L1612-L1615.
(16) Clough, S.; Tripathy, S.; Sun, X.-F.; Orchard, B. *Makromol. Chem. Rapid Commun.* **1988**, *9*, 535-538.
(17) Schaefer-Siebert, D.; Roth, S.; Budrowski, C.; Kuzmany, H. *Synth. Met.* **1987**, *21*, 285-291.
(18) Yang, X. Q.; Tanner, D. B.; Arbuckle, G.; MacDiarmid, A. G.; Epstein, A. J. *Synth. Met.* **1987**, *17*, 277-282.
(19) Zuo, F.; Epstein, A. J.; Yang, X.-Q.; Tanner, D. B.; Arbuckle, G.; MacDiarmid, A. G. *Synth. Met.* **1987**, *17*, 433-438.
(20) Heeger, A. J.; Kivelson, S.; Schrieffer, J. R.; Su, W.-P. *Rev. Mod. Phys.* **1988**, *60(3)*, 781-850.
(21) Bredas, J. L.; Street, G. B. *Acc. Chem. Res.* **1985**, *18*, 309-315.
(22) Ehinger, K.; Roth, S. in *Electronic Properties of Polymers and Related Compounds*; Kuzmany, H.; Mehring, M.;Roth, S. Eds.; Springer Ser. Solid State Sci. Vol. 63; Springer-Verlag, New York, 1985; pp 67-74.
(23) Abadie, M. J. M.; Boukli, S. M.; Cadene, M.; Rolland, M. *Polymer* **1986**, *27*, 2003-2008.
(24) A recent review of this topic has appeared: Kaner, R. B. in *Electrochemical Science and Technology of Polymers-2*; Linford, R. G. Ed.; Elsevier Applied Science, Essex, England, 1990; pp 97-147.
(25) Diaz, A. F.; Bargon, J. in *Handbook of Conducting Polymers*; Vol. 1; Skotheim, T. A. Ed.; Marcel Dekker, Inc., New York, 1986; pp 81-115.
(26) Genies, E. M.; Bidan, G.; Diaz, A. *J. Electroanal. Chem.* **1983**, *149*, 101-113.
(27) Pfluger, P.; Street, G. B. *J. Chem. Phys.* **1984**, *80(1)*, 544-553.
(28) Kanatzidis, M. G. *Chem. Eng. News* **1990**, *68(49)*, 36-54.
(29) Kaye, B.; Underhill, A. E. *Synth. Met.* **1989**, *28*, C97-C102.

(30) Winter, H.; Gotschy, B.; Dormann, E. *Synth. Met.* **1990**, *38(3)*, 341-352.
(31) Cheung, K. M.; Bloor, D.; Stevens, G. C. *J. Mat. Sci.* **1990**, *25(9)*, 3814-3837.
(32) Kuwabata, S.; Nakamura, J.; Yoneyama, H. *J. Chem. Soc. Chem. Commun.* **1988**, 779-780.
(33) Kuwabata, S.; Okamoto, K.; Yoneyama, H. *J. Chem. Soc. Faraday Trans. I* **1988**, *84*, 2317-2326.
(34) Osaka, T.; Naoi, K.; Ogano, S. *J. Electrochem. Soc.* **1988**, *135(5)*, 1071-1077.
(35) Chao, T. H.; March, J. *J. Polym. Sci. Polym. Chem.* **1988**, *26(3)*, 743-753.
(36) Panero, S.; Prosperi, P.; Scorsati, B. *Synth. Met.* **1989**, *28*, C133-C137.
(37) Yakushi, K.; Lauchlan, L. J.; Clarke, T. C.; Street, G. B. *J. Chem. Phys.* **1983**, *79(10)*, 4774-4778.
(38) Nazzal, A.; Street, G. B. *J. Chem. Soc. Chem. Commun.* **1984**, 83-84.
(39) Berlin, A.; Ferraccioli, R.; Pagani, G. A.; Sannicolo, F. *Synth. Met.* **1987**, *22*, 89-91.
(40) Naitoh, S. *Synth. Met.* **1987**, *18*, 237-240.
(41) Bolognesi, A.; Catellani, M.; Destri, S.; Zamboni, R.; Taliani, C. *J. Chem. Soc. Chem. Commun.* **1988**, 246-247.
(42) Lazzaroni, R.; Dujardin, S.; Riga, J.; Verbist, J.; Bredas, J. L.; Dehalle, J.; Andre, J. M. in reference 22; pp 191-193.
(43) Naarman, H. *Makromol. Chem. Macromol. Symp.* **1987**, *8*, 1-15.
(44) Naarman, H. in *Conjugated Polymeric Materials: Opportunities in Electronics, Optoelectronics, and Molecular Electronics*; Bredas, J. L.;Chance, R. R. Eds.; NATO ASI Series. Series E: Applied Sciences Vol. 182; Kluwer Academic Publishers, Dordrecht, The Netherlands, 1990; pp 11-51.
(45) Fujii, M.; Saeki, Y.; Arii, K.; Yoshino, K. *Jap. Journ. Appl. Phys.* **1990**, *29(11)*, 2501-2505.
(46) Bidan, G. *Mat. Sci. Forum* **1987**, *21*, 21-30.
(47) Wang, H. L.; Toppare, L.; Fernandez, J. E. *Macromolecules* **1990**, *23*, 1053-1059.
(48) Faulkner, L. R. *Chem. Eng. News* **1984**, *28*.
(49) Okabayashi, K.; Ikeda, O.; Tamura, H. *J. Chem. Soc. Chem. Commun.* **1985**, 684.
(50) Noufi, R. *J. Electrochem. Soc.* **1983**, *130*, 2126.
(51) Bull, R. A.; Fan, F.-R.; Bard, A. J. *J. Electrochem. Soc.* **1983**, *130*, 1636.
(52) Ikeda, O.; Okabayashi, K.; Tamura, H. *Chem. Lett.* **1983**, 1821.
(53) A recent example is given for the chemical synthesis of polypyrrole: Machida, S.; Miyata, S.; Techagumpuch, A. *Synth. Met.* **1989**, *31*, 311-318.
(54) Mohammadi, A.; Lundstrom, I.; Inganas, O.; Salaneck, W. R. *Polymer* **1990**, *31*, 395-399.
(55) MacDiarmid, A. G.; Epstein, A. J. *J. Chem. Soc. Faraday Discuss.* **1989**, *88*, 317-332.
(56) Yue, J.; Epstein, A. J. *J. Am. Chem. Soc.* **1990**, *112*, 2800-2801.
(57) Diaz, A. F.; Logan, J. A. *J. Electroanal. Chem.* **1980**, *111*, 111-114.
(58) Wei, Y.; Focke, W. W.; Wnek, G. E.; Ray, A.; MacDiarmid, A. G. *J. Phys. Chem.* **1989**, *93*, 495-499.
(59) Leclerc, M.; Guay, J.; Dao, L. H. *Macromolecules* **1989**, *22*, 649-653.
(60) Chiang, J.-C.; MacDiarmid, A. G. *Synth. Met.* **1986**, *13*, 193-205.
(61) Andreatta, A.; Cao, Y.; Chiang, J. C.; Smith, P.; Heeger, A. J. *Synth. Met.* **1988**, *26*, 383-389.
(62) Cao, Y.; Smith, P.; Heeger, A. J. in reference 44; pp 171-193.

(63) Tang, X.; Scherr, E.; MacDiarmid, A. G.; Epstein, A. J. *Bull. Am. Phys. Soc.* **1989**, *34*, 583.
(64) Vachon, D.; Angus, R. O.; Lu, F. L.; Nowak, M.; Liu, Z. X.; Schaffer, H.; Wudl, F.; Heeger, A. J. *Synth. Met.* **1987**, *18(1-3)*, 297-302.
(65) MacDiarmid, A. G.; Chiang, J.; Halpern, M.; Huang, W. S.; Krawczyk, J. R.; Mammone, R. J.; Mu, S. L.; Somasiri, N. L. D.; Wu, W. *Polym. Prepr.* **1984**, *25(2)*, 248-249.
(66) Anderson, M. R.; Mattes, B. R.; Reiss, H.; Kaner, R. B. *Synth. Met.* **1991**, In Press.
(67) Schoch, K. F. *Polym. Prepr.* **1984**, *25(2)*, 278-279.
(68) Shacklette, L. W.; Elsenbaumer, R. L.; Chance, R. R.; Eckhardt, H.; Frommer, J. E.; Baughman, R. H. *J. Chem. Phys.* **1981**, *75(4)*, 1919-1927.
(69) Short, J. N.; H. W. Hill, J. *CHEMTECH* **1972**, *2*, 481-485.
(70) Lenz, R. W.; Handlovits, C. E.; Smith, H. A. *J. Polym. Sci.* **1962**, *58*, 351-367.
(71) Chance, R. R.; Shacklette, L. W.; Miller, G. G.; Ivory, D. M.; Sowa, J. M.; Elsenbaumer, R. L.; Baughman, R. H. *J. Chem. Soc. Chem. Commun.* **1980**, 348-349.
(72) Novak, B. M.; Hagen, E.; Viswanathan, A.; Magde, L. *Polym. Prepr.* **1990**, *31(2)*, 482-483.
(73) Krische, B.; Zagorska, M. *Synth. Met.* **1989**, *28*, C263-C268.
(74) Yamamoto, T.; Sanechika, K.; Yamamoto, A. *J. Polm. Sci. Polym. Lett. Ed.* **1980**, *18*, 9-12.
(75) Kobayashi, M.; Chen, J.; Chung, T. C.; Moraes, F.; Heeger, A. J.; Wudl, F. *Synth. Met.* **1984**, *9*, 77-86.
(76) Amer, A.; Zimmer, H.; Mulligan, K. J.; Mark, H. B.; Pons, S.; McAleer, J. F. *J. Polym. Sci. Polym. Lett. Ed.* **1984**, *22*, 77-82.
(77) Hotta, S.; Soga, M.; Sonoda, N. *Synth. Met.* **1988**, *26*, 267-279.
(78) Yamamoto, T.; Morita, A.; Maruyama, T.; Zhou, Z.; Kanbara, T.; Sanechika, K. *Polym. J.* **1990**, *22(2)*, 187-190.
(79) Sugimoto, R.; Takeda, S.; Gu, H. B.; Yoshino, K. *Chemistry Express* **1986**, *1(11)*, 635.
(80) Farazdel, A.; Dupuis, M.; Clementi, E.; Aviram, A. *J. Am. Chem. Soc.* **1990**, *112(11)*, 4206-4214.
(81) Aviram, A. *J. Am. Chem. Soc.* **1988**, *110(17)*, 5687-5692.
(82) Tour, J. M.; Wu, R.; Schumm, J. S. *J. Am. Chem. Soc.* **1990**, *112(14)*, 5662-5663.
(83) Elsenbaumer, R. L.; Jen, K. Y.; Oboodi, R. *Synth. Met.* **1986**, *15*, 169-174.
(84) Elsenbaumer, R. L.; Jen, K. Y.; Miller, G. G.; Shacklette, L. W. *Synth. Met.* **1987**, *18*, 277-282.
(85) Elsenbaumer, R. L.; Jen, K. Y.; Miller, G. G.; Eckhardt, H.; Shacklette, L. W.; Jow, R. in *Electronic Properties of Conjugated Polymers*; Kuzmany, H.; Mehring, M.;Roth, S. Eds.; Springer Ser. Solid State Sci. Vol. 76; Springer-Verlag, New York, 1987; pp 400-406.
(86) Yoshino, K.; Nakajima, S.; Gu, H. B.; Sugimoto, R. *Jap. Journ. Appl. Phys.* **1987**, *26(8)*, L1371-L1373.
(87) Kaeriyama, K.; Sato, M.; Tanaka, S. *Synth. Met.* **1987**, *18*, 233-236.
(88) Hotta, S.; Rughooputh, S. D. D. V.; Heeger, A. J.; Wudl, F. *Macromolecules* **1987**, *20*, 212-215.
(89) Osterholm, J.-E.; Laakso, J.; Nyholm, P.; Isotalo, H.; Stubb, H.; Inganas, O.; Salaneck, W. R. *Synth. Met.* **1989**, *28*, C435-C444.
(90) Masuda, H.; Tanaka, S.; Kaeriyama, K. *J. Chem. Soc. Chem. Commun.* **1989**, 725-726.

(91) Higher values of conductivity (up to 720 S/cm) are seen when the film is doped at higher temperatures: Satoh, M.; Imanishi, K.; Yasuda, Y.; Tsushima, R.; Yamasaki, H.; Aoki, S. *Synth. Met.* **1989**, *30*, 33-38.
(92) Prasad, P. in reference 4; pp 271-282.
(93) Prasad, P. N. in reference 5; pp 351-363.
(94) Pang, Y.; Prasad, P. N. *J. Chem. Phys.* **1990**, *93(4)*, 2201-2204.
(95) Singh, B. P.; Samoc, M.; Nalwa, H. S. *J. Chem. Phys.* **1990**, *92(5)*, 2756-2761.
(96) Heffner, G. W.; Rochefort, W. E.; Pearson, D. S. Presented at the 200th National Meeting of the American Chemical Society, Washington, DC.; paper POLY 163
(97) Aime, J. P.; Bargain, F.; Schott, M.; Eckhardt, H.; Miller, G. G.; Elsenbaumer, R. L. *Phys. Rev. Lett.* **1989**, *62(1)*, 55-58.
(98) Nowak, M. J.; Rughooputh, S. D. D. V.; Hotta, S.; Heeger, A. J. *Macromolecules* **1987**, *20*, 965-968.
(99) Feldhues, M.; Kaempf, G.; Litterer, H.; Mecklenburg, T.; Wegener, P. *Synth. Met.* **1989**, *28*, C487-C493.
(100) Kaeriyama, K.; Tanaka, S.; Sato, M.-A.; Hamada, K. *Synth. Met.* **1989**, *28*, C611-C620.
(101) Tanaka, S.; Kaeriyama, K. *Bull. Chem. Soc. Japan* **1989**, *62*, 1908-1912.
(102) Lowen, S. V.; D. MacInnes, J.; Funt, B. L. *J. Polym. Sci. Polym. Chem.* **1989**, *27*, 4087-4097.
(103) Roncali, J.; Marque, P.; Garreau, R.; Garnier, F.; Lemaire, M. *Macromolecules* **1990**, *23*, 1347-1352.
(104) Wudl, F. Presentation at the International Conference on Science and Technology of Synthetic Metals, Tuebingen, FRG, September, 1990.
(105) Ruiz, J. P.; Nayak, K.; Marynick, D. S.; Reynolds, J. R. *Macromolecules* **1989**, *22*, 1231-1238.
(106) Ruiz, J. P.; Gieselman, M. B.; Nayak, K.; Marynick, D. S.; Reynolds, J. R. *Synth. Met.* **1989**, *28*, C481-C486.
(107) Lemaire, M.; Delabouglise, D.; Garreau, R.; Roncali, J. *J. Chim. Phys.* **1989**, *86(1)*, 193-198.
(108) Patil, A. O.; Ikenoue, Y.; Wudl, F.; Heeger, A. J. *J. Am. Chem. Soc.* **1987**, *109(6)*, 1858-1859.
(109) Patil, A. O. *Synth. Met.* **1989**, *28*, C495-C500.
(110) Shu, C.; Wrighton, M. S. in *Electrochemical Surface Science: Molecular Phenomena at Electrode Surfaces*; Soriaga, M. P. Ed.; ACS Symposium Series Vol. 378; American Chemical Society, Washington, DC, 1988; pp 408-430.
(111) Baeuerle, P.; Wuerthner, F.; Heid, S. *Angew. Chem. Int. Ed. Engl.* **1990**, *29(4)*, 419-420.
(112) Yoshino, K.; Nakajima, S.; Fujii, M.; Sugimoto, R.-I. *Polymer Commun.* **1987**, *28*, 309-310.
(113) Yoshino, K.; Nakajima, S.; Sugimoto, R. *Jap. Journ. Appl. Phys.* **1987**, *26(6)*, L1038-L1039.
(114) Laasko, J.; Osterholm, J.; Nyholm, P. *Synth. Met.* **1989**, *28*, C467-C471.
(115) Hotta, S.; Rughooputh, S. D. D. V.; Heeger, A. J. *Synth. Met.* **1987**, 22, 79-87.
(116) Wessling, B. *Synth. Met.* **1988**, *24*, 271-272.

(117) Isotalo, H.; Stubb, H.; Yli-lahti, P.; Kuivalainin, P.; Osterholm, J.; Laasko, J. *Synth. Met.* **1989**, *28*, C461-C466.
(118) Wellinghoff, S.; Deng, Z.; Kedrowski, T.; Dick, S.; Jenekhe, S.; Ishida, H. *Mol. Cryst. Liq. Cryst.* **1984**, *106*, 289.
(119) Jenekhe, S.; Wellinghoff, S.; Reed, J. *Mol. Cryst. Liq. Cryst.* **1984**, *105*, 175.
(120) Hayashida, S.; Sukegawa, K.; Niwa, O. *Synth. Met.* **1990**, *35*, 253-261.
(121) Mort, J. *Science* **1980**, *208(23)*, 819-825.
(122) Ito, S.; Katayama, H.; Yamamoto, M. *Macromolecules* **1988**, *21*, 2456-2462.
(123) Zelent, B.; Messier, P.; Gauthier, S.; Gravel, D.; Durocher, G. *J. Photochem. Photobiol. A.* **1990**, *52*, 165-178.
(124) Tsujii, Y.; Tsuchida, A.; Onogi, Y.; Yamamoto, M. *Macromolecules* **1990**, *23*, 4019-4023.
(125) Ghoshal, S.; Chopra, P.; Singh, B. P.; Swiatkiewicz, J.; Prasad, P. N. *J. Chem. Phys.* **1989**, *90(9)*, 5078-5081.
(126) Ellis, J. R. in reference 25; pp 489-499.
(127) DeMartino, R. N.; Choe, E. W.; Khanarian, G.; Haas, D.; Leslie, T.; Nelson, G.; Stamatoff, J.; Stuetz, D.; Teng, C. C.; Yoon, H. in reference 6; pp 169-187.
(128) Griffin, A. C.; Bhatti, A. M.; Hung, R. S. L. in reference 6; pp 375-391.
(129) Griffin, A. C.; Bhatti, A. M.; Howell, G. A. in reference 4; pp 115-125.
(130) Lee, Y.; Kertesz, M. *J. Chem. Phys.* **1988**, *88(4)*, 2609-2617.
(131) Lowe, J. P.; Kafafi, S. A.; LaFemina, J. P. *J. Phys. Chem.* **1986**, *90(25)*, 6602-6610.
(132) Bevierre, M.; Mercier, F.; Ricard, L.; Mathey, F. *Angew. Chem. Int. Ed. Engl* **1990**, *29(6)*, 655-657.
(133) Aeiyach, S.; Soubiran, P.; Lacaze, P. C.; Froyer, G.; Pelous, Y. *Synth. Met.* **1989**, *32*, 103-112.
(134) Goldenberg, L. M.; Pelekh, A. E.; Krinichnyi, V. L.; Roshchupkina, O. S.; Zueva, A. F.; Lyubovskaya, R. N.; Efimov, O. N. *Synth. Met.* **1990**, *36*, 217-228.
(135) Tsuchida, E.; Yamamoto, K.; Asada, T.; Nishide, Y. *Chem. Lett.* **1987**, 1541-1544.
(136) Ye, J. H.; Chen, Y. Z.; Tian, Z. W. *J. Electroanal. Chem.* **1987**, *229*, 215-222.
(137) Satoh, M.; Tabata, M.; Kaneto, K.; Yoshino, K. *J. Electroanal. Chem.* **1985**, *195*, 203-206.
(138) Kovacic, P.; Kyriakis, A. *J. Am. Chem. Soc.* **1963**, *85(4)*, 454-458.
(139) See references 30-39 in: Elsenbaumer, R. L.; Schacklette, L. W. in reference25; pp 213-263.
(140) Pradere, P.; Boudet, A. *J. Mater. Sci.* **1987**, *22*, 4240-4246.
(141) Rehahn, M.; Schlueter, A.; Wegner, G.; Feast, W. J. *Polymer* **1989**, *30(6)*, 1054-1059.
(142) Rehahn, M.; Schlueter, A.; Wegner, G.; Feast, W. J. *Polymer* **1989**, *30(6)*, 1060-1062.
(143) Rehahn, M.; Schlueter, A.; Wegner, G. *Makrom. Chem.* **1990**, *191(9)*, 1991-2003.
(144) Schiavon, G.; Zotti, G.; Bontempelli, G.; LoCoco, F. *Synth. Met.* **1988**, *25*, 365-373.
(145) Laasko, J.; Osterholm, J.; Lindberg, J. J. *Polym. Bull.* **1987**, *18*, 195-201.
(146) Dezotti, M. W. C.; DePaoli, M. A. *Synth. Met.* **1989**, *29*, E41-E45.
(147) Santos, D. A. D.; Galvao, D. S.; Laks, B.; Dezotti, M. W. C.; DePaoli, M. A. *Chem. Phys.* **1990**, *144*, 103-106.
(148) All of the points covered in this section have been recently reviewed: Miller, R. D.; Michl, J. *Chem. Rev.* **1989**, *89*, 1359-1410.

(149) Harrod, J. F. in *Inorganic and Organometallic Polymers*; Zeldin, M.; Wynne, K. J.;Allcock, H. R. Eds.; ACS Symposium Series Vol. 360; American Chemical Society, Washington, D. C., 1988; pp 89-100.
(150) Woo, H.; Tilley, T. D. *J. Am. Chem. Soc.* **1989**, *111*, 8043-8044.
(151) Woo, H.; Tilley, T. D. *J. Am. Chem. Soc.* **1990**, *112(7)*, 2843.
(152) Matyjaszewski, K.; Chen, Y. L.; Kim, H. K. in reference 149; pp 78-88.
(153) Cypryk, M.; Gupta, Y.; Matyjaszweski, K. *J. Am. Chem. Soc.* **1991**, *113*, 1046-1047.
(154) Nate, K.; Ishikawa, M.; Ni, H.; Watanabe, H.; Saheki, Y. *Organometallics* **1987**, *6*, 1673-1679.
(155) Chicart, P.; Corriu, R. J. P.; Moreau, J. J. E. *Chem. Mater* **1991**, *3(1)*, 8-10.
(156) Hanack, M.; Datz, A.; Fay, R.; Fischer, K.; Keppeler, U.; Koch, J.; Metz, J.; Mezger, M.; Schneider, O.; Schulze, H. in reference 25; pp 133-212.
(157) Snow, A. W.; Griffith, J. R. in reference 10, Vol. 11; pp 212-225.
(158) Datz, A.; Metz, J.; Schneider, O.; Hanack, M. *Synth. Met.* **1984**, *9*, 31-40.
(159) Marks, T. J. *Science* **1985**, *227*, 881-889.
(160) Gaudiello, J. G.; Kellogg, G. E.; Tetrick, S. M.; Marks, T. J. *J. Am. Chem. Soc.* **1989**, *111(14)*, 5259-5271.
(161) Almeida, M.; Gaudiello, J. G.; Kellogg, G. E.; Tetrick, S. M.; Marcy, H. O.; McCarthy, W. J.; Butler, J. C.; Kannewurf, C. R.; Marks, T. J. *J. Am. Chem. Soc.* **1989**, *111(14)*, 5271-5284.
(162) Dirk, C. W.; Inabe, T.; Schoch, K. F.; Marks, T. J. *J. Am. Chem. Soc.* **1983**, *105(6)*, 1539-1550.
(163) Greene, R. L.; Street, G. B. *Science* **1984**, *226*, 651-656.
(164) Wudl, F. *Acc. Chem. Res.* **1984**, *17*, 227-232.
(165) Marks, T. J.; Kalina, D. W. in *Extended Linear Chain Compounds, Vol. 1*; Miller, J. Ed.; Plenum Press, New York, 1983; pp 197-331.
(166) Alcacer, L.; Novais, H. in reference 165, Vol. 3; Plenum Press, New York, 1983; pp 319-351.
(167) Dirk, C. W.; Bosseau, M.; Barrett, P. H.; Moraes, F.; Wudl, F.; Heeger, A. J. *Macromolecules* **1986**, *19*, 266-269.
(168) Vicente, R.; Ribas, J.; Cassoux, P.; Valade, L. *Synth. Met.* **1986**, *13*, 265-280.
(169) Reynolds, J. R.; Lillya, C. P.; Chien, J. C. W. *Macromolecules* **1987**, *20*, 1184-1191.
(170) Pittman, C. U.; Carraher, C. E.; Reynolds, J. R. in reference 10, Vol. 10; pp 541-594.
(171) Wang, F.; Reynolds, J. R. *Macromolecules* **1988**, *21*, 2887-2889.
(172) Wang, F.; Reynolds, J. R. *Macromolecules* **1990**, *23*, 3219-3225.
(173) Bohlmann, F.; Inhoffen, E. *Chem. Ber.* **1956**, *89*, 1276-1281.
(174) Kiji, J.; Iwamoto, M. *J. Polym. Sci. Polym. Lett. Ed.* **1968**, *6*, 53-55.
(175) Tanaka, K.; Ohzeki, K.; Yamabe, T.; Yata, S. *Synth. Met.* **1984**, *9*, 41-52.
(176) Manassen, J.; Wallach, J. *J. Am. Chem. Soc.* **1965**, *87(12)*, 2671-2677.
(177) Grassie, N.; McNeill, I. C. *J. Polym. Sci.* **1958**, *27*, 207-218.
(178) Renschler, C. L.; Sylwester, A. P. *Mat. Sci. Forum* **1989**, *52/53*, 301-322.
(179) Wilbourn, K.; Murray, R. W. *Macromolecules* **1988**, *21(1)*, 89-96.
(180) Blatter, K.; Schlueter, A. *Macromolecules* **1989**, *22(8)*, 3506-3508.
(181) Dalton, L. R. in reference 5; pp 123-141.
(182) Dalton, L. R. in reference 6; pp 243-271.
(183) Feast, W. J. in reference 25; pp 1-44.

(184) Kenny, P. W.; Miller, L. L. *J. Chem. Soc. Chem. Commun.* **1988**, 84-85.
(185) Dietz, T. M.; Stallman, B. J.; Kwan, W. S. V.; Penneau, J. F.; Miller, L. L. *J. Chem. Soc. Chem. Commun.* **1990**, 367-369.
(186) Cammarata, V.; Kolaskie, C. J.; Miller, L. L.; Stallman, B. J. *J. Chem. Soc. Chem. Commun.* **1990**, 1290-1292.
(187) Sum, W.; Kwan, V.; Penneau, J. F.; Miller, L. L. *J. Electroanal. Chem.* **1990**, *291*, 295-299. See also references to the work of Nuzzo and Whitesides contained therein.
(188) DePra, P. A.; Gaudiello, J. G.; Marks, T. J. *Macromolecules* **1988**, *21(7)*, 2295-2297.
(189) Rao, D. N.; Swiatkiewicz, J.; Chopra, P.; Ghoshal, S. K.; Prasad, P. N. *Appl. Phys. Lett.* **1986**, *48(18)*, 1187-1189.
(190) Wolfe, J. F.; Bitler, S. P. in reference 6; pp 401-413.
(191) Goldfarb, I. J.; Medrano, J. in reference 5; pp 93-99.
(192) Zhao, M. T.; Samoc, M.; Prasad, P. N.; Reinhardt, B. A. *Chem. Mater.* **1990**, *2(6)*, 670-678.
(193) Jenekhe, S. A.; Johnson, P. O.; Agrawal, A. K. *Macromolecules* **1989**, *22*, 3216-3222.
(194) For a leading reference, see: Bloor, D.; Chance, R. R., Eds. *Polydiacetylenes.* NATO ASI Series. Series E: Applied Sciences Vol. 102; Martinus Nijhoft Publishers, Boston, 1985.
(195) Wegner, G. *Pure Appl. Chem.* **1977**, *49*, 443-454.
(196) Thakur, M.; Lando, J. B. *Macromolecules* **1983**, *16*, 143-146.
(197) Baughman, R. M. *J. Polym. Sci. Polym. Phys. Ed.* **1974**, *12*, 1511-1535.
(198) Tieke, B.; Lieser, G.; Wegner, G. *J. Polym. Sci. Polym. Chem. Ed.* **1979**, *17*, 1631-1644.
(199) Tieke, B.; Wegner, G. *Makromol. Chem.* **1978**, *179*, 1639.
(200) Day, D. R.; Ringsdorf, H. *Makromol. Chem.* **1979**, *180*, 1059.
(201) Collins, M. A. *J. Polym. Sci. Polym. Phys.* **1988**, *26*, 367-387.
(202) Lopez, E.; O'Brien, D. F.; Whitesides, T. H. *J. Am. Chem. Soc.* **1982**, *104(1)*, 305-307.
(203) Johnston, D. S.; Sanghera, S.; Pons, M.; Chapman, D. *Biochim. Biophys. Acta* **1980**, *602*, 57-69.
(204) Pons, M.; Johnston, D. S.; Chapman, D. *J. Polym. Sci. Polym. Chem. Ed.* **1982**, *20*, 513-520.
(205) Pons, M.; Johnston, D. S.; Chapman, D. *Biochim. Biophys. Acta* **1982**, *693*, 461-465.
(206) Leaver, J.; Alonso, A.; Durrani, A. A.; Chapman, D. *Biochim. Biophys. Acta* **1983**, *732*, 210-218.
(207) Patel, G. N.; Chance, R. R.; Witt, J. D. *J. Chem. Phys.* **1979**, *70(9)*, 4387-4392.
(208) Biegajski, J. E.; Cadenhead, D. A.; Prasad, P. N. *Macromolecules* **1991**, *24(1)*, 298-303.
(209) Wenz, G.; Mueller, M. A.; Schmidt, M.; Wegner, G. *Macromolecules* **1984**, *17*, 837-850.
(210) Lim, K. C.; Fincher, C. R.; Heeger, A. J. *Phys. Rev. Lett.* **1983**, *50(24)*, 1934-1937.
(211) Lim, K. C.; Kapitulnik, A.; Zacher, R.; Heeger, A. J. *J. Chem. Phys.* **1985**, *82(1)*, 516-521.
(212) Lim, K. C.; Heeger, A. J. *J. Chem. Phys.* **1985**, *82(1)*, 522-530.
(213) Nallicheri, R. A.; Rubner, M. F. *Macromolecules* **1991**, *24(2)*, 517-525.
(214) Peiffer, D. G.; Chung, T. C.; Schulz, D. N.; Agarwal, P. K.; Garner, R. T.; Kim, M. W.; Chance, R. R. in reference 6; pp 205-215.

(215) Allegra, G.; Brueckner, S.; Schmidt, M.; Wegner, G. *Macromolecules* **1986**, *19*, 399-405.
(216) Sauteret, C.; Herrmann, J. P.; Frey, R.; Pradiere, F.; Ducuing, J.; Baughman, R. H.; Chance, R. R. *Phys. Rev. Lett.* **1976**, *36*, 956-959.
(217) Thakur, M.; Verbeek, B.; Chi, G. C.; O'Brien, K. J. in reference 4; pp 41-64.
(218) Sandman, D. J.; III, J. W. S.; Jones, M. T. in reference 6; pp 367-374.
(219) Tokito, S. Z.; Smith, P.; Heeger, A. J. *Synth. Met.* **1990**, *36(2)*, 183-194.
(220) Natta, G.; Mazzanti, G. P.; Corrandi, P. *Atti. Naz. Lincei. Cl. Sci. Fis. Mat. Nat. Rend.* **1958**, *25(8)*, 3.
(221) Shirakawa, H.; Ikeda, S. *Polym. J.* **1971**, *2(2)*, 231-244.
(222) Ito, T.; Shirakawa, H.; Ikeda, S. *J. Polym. Sci. Polym. Chem. Ed.* **1974**, *12*, 11-20.
(223) The synthesis and properties this material have been reviewed: Chien, J. C. W. *Polyacetylene. Chemistry, Physics, and Materials Science*; Academic Press: San Diego, 1984.
(224) Gibson, H. W.; Kaplan, S.; Mosher, R. A.; W. M. Prest, J.; Weagley, R. J. *J. Am. Chem. Soc.* **1986**, *108(22)*, 6843-6851.
(225) Robin, P.; Pouget, J. P.; Comes, R.; Gibson, H. W.; Epstein, A. J. *Journ. de Phys.* **1983**, *44(6)*, C3-77 to C3-81.
(226) Hoffman, D. M.; Gibson, H. W.; Epstein, A. J.; Tanner, D. B. *Phys. Rev. B* **1983**, *27(2)*, 1454-1457.
(227) Elert, M. L.; White, C. T. *Phys. Rev. B* **1983**, *28(12)*, 7387-7389.
(228) Cernia, E.; D'Ilario, L. *J. Polym. Sci. Polym. Chem. Ed.* **1983**, *21*, 2163-2176.
(229) Bozovic, I. *Mod. Phys. Lett. B* **1987**, *1(3)*, 81-88.
(230) Rao, B. K.; Darsey, J. A.; Kestner, N. R. *Phys. Rev. B* **1985**, *31(2)*, 1187-1190.
(231) Gibson, H. W.; Weagley, R. J.; Mosher, R. A.; Kaplan, S.; W. M. Prest, J.; Epstein, A. J. *Mol. Cryst. Liq. Cryst.* **1985**, *117*, 315-318.
(232) Gibson, H. W.; Weagley, R. J.; Mosher, R. A.; Kaplan, S.; W. M. Prest, J.; Epstein, A. J. *Phys. Rev. B* **1985**, *31(4)*, 2338-2342.
(233) Due to the vast number of approaches that have been taken, the reader is referred to an excellent review: Stowell, J. A.; Amass, A. J.; Beevers, M. S.; Farren, T. R. *Polymer* **1989**, *30*, 195-201.
(234) Tubino, R.; Dorsinville, R.; Lam, W.; Alfano, R. R.; Birman, J. L.; Bolognesi, A.; Destri, S.; Catellani, M.; Porzio, W. *Phys. Rev. B* **1984**, *30(11)*, 6601-6605.
(235) Piaggio, P.; Cuniberti, C.; Dellepiane, G.; Bolognesi, A.; Catellani, M.; Destri, S.; Porzio, W.; Tubino, R. *Synth. Met.* **1987**, *17*, 337-342.
(236) Dorsinville, R.; Yang, L.; Alfano, R. R.; Tubino, R.; Destri, S. *Solid State Commun.* **1988**, *68(9)*, 875-877.
(237) Pfleger, J.; Kminek, I.; Nespurek, S.; Prasad, P. N. *Synth. Met.* **1990**, *37*, 255-261.
(238) Dorsinville, R.; Tubino, R.; Krimchansky, S.; Alfano, R. R.; Birman, J. L.; Bolognesi, A.; Destri, S.; Castellani, M.; Porzio, W. *Phys. Rev. B* **1985**, *32(6)*, 3377-3380.
(239) VanNice, F. L.; Bates, F. S.; Baker, G. L.; Carroll, P. J.; Patterson, G. D. *Macromolecules* **1984**, *17*, 2626-2629.
(240) Bates, F. S.; Baker, G. L. *Macromolecules* **1983**, *16*, 704-707.
(241) Armes, S. P.; Vincent, B. *Synth. Met.* **1988**, *25*, 171-179.
(242) Lee, K. I.; Jopson, H. *Makromol. Chem. Rapid Commun.* **1983**, *4*, 375.
(243) Naarman, H.; Theophilou, N. *Synth. Met.* **1987**, *22*, 1-8.

(244) Schimmel, T.; Reiss, W.; Gmeiner, J.; Denninger, G.; Schwoerer, M.; Naarman, H.; Theophilou, N. *Solid State Commun*. **1988**, *65(11)*, 1311-1315.
(245) Basescu, N.; Liu, Z.; Moses, D.; Heeger, A. J.; Naarman, H.; Theophilou, N. *Nature* **1987**, *327*, 403-405.
(246) Masuda, T.; Higashimura, T. *Acc. Chem. Res*. **1984**, *17*, 51-56.
(247) Ziegler, J. M. *Polym. Prepr*. **1984**, *25(2)*, 223-224.
(248) Masuda, T.; Higashimura, T. *Adv. Polym. Sci*. **1987**, *81*, 121-165.
(249) Petit, M. A.; Soum, A. H.; Leclerc, M.; Prud'homme, R. E. *J. Polym. Sci. Polym. Phys*. **1987**, *25*, 423-433.
(250) Leclerc, M.; Prud'homme, R. E. *Macromolecules* **1987**, *20*, 2153-2159.
(251) Furlani, A.; Napoletano, C.; Paolesse, R.; Russo, M. V. *Synth. Met*. **1987**, *21*, 337-342.
(252) Leclerc, M.; Prud'homme, R. E. *J. Polym. Sci. Polym. Phys. Ed*. **1985**, *23*, 2021-2030.
(253) Ivin, K. J. *Olefin Metathesis*; Academic Press: San Diego, 1983.
(254) Tlenkopachev, M. A.; Korshak, Y. V.; Orlov, A. V.; Korshak, V. V. *Dokl. Akad. Nauk SSSR (Engl. Trans.)* **1986**, *291*, 1036-1040.
(255) Korshak, Y. V.; Korshak, V. V.; Kanischka, G.; Hoecker, H. *Makromol. Chem. Rapid Commun*. **1985**, *6*, 685-692.
(256) Klavetter, F. L.; Grubbs, R. H. *J. Am. Chem. Soc*. **1988**, *110(23)*, 7807-7813.
(257) Ginsburg, E. J.; Gorman, C. B.; Marder, S. R.; Grubbs, R. H. *J. Am. Chem. Soc*. **1989**, *111*, 7621-7622.
(258) Gorman, C. B.; Ginsburg, E. J.; Marder, S. R.; Grubbs, R. H. *Angew. Chem. Adv. Mater*. **1989**, *101*, 1603-1606.
(259) Gorman, C.; Ginsburg, E.; Marder, S.; Grubbs, R. *Polym. Prepr*. **1990**, *31(1)*, 386-387.
(260) Ginsburg, E. J.; Gorman, C. B.; Grubbs, R. H.; Klavetter, F. L.; Lewis, N. S.; Marder, S. R.; Perry, J. W.; Sailor, M. J. in reference 44; pp 65-81.
(261) Grubbs, R. H.; Gorman, C. B.; Ginsburg, E. J.; Perry, J. W.; Marder, S. R. in *Materials for Nonlinear Optics: Chemical Perspectives*; Marder, S. R.; Sohn, J. E.;Stucky, G. D. Eds.; ACS Symposium Series Vol. 455; American Chemical Society, Washington, DC, 1991; pp 672-682.
(262) Sailor, M. J.; Ginsburg, E. J.; Gorman, C. B.; Kumar, A.; Grubbs, R. H. *Science* **1990**, *249*, 1146-1149.
(263) Marvel, C. S.; Hartzell, G. E. *J. Am. Chem. Soc*. **1959**, *81*, 448-452.
(264) Lefebvre, G.; Dawans, F. *J. Polym. Sci. A* **1964**, *2*, 3277-3295.
(265) Frey, D. A.; Hasegawa, M.; Marvel, C. S. *J. Polym. Sci. A* **1963**, *1*, 2057-2065.
(266) Cassidy, P. E.; Marvel, C. S.; Ray, S. *J. Polym. Sci. A* **1965**, *3*, 1553-1565.
(267) Zhong, X. F.; Francois, B. *Makromol. Chem. Rapid Commun* **1988**, *9*, 411-416.
(268) Ballard, D. G. H.; Courtis, A.; Shirley, I. M.; Taylor, S. C. *Macromolecules* **1988**, *21*, 294-304.
(269) McKean, D. R.; Stille, J. K. *Macromolecules* **1987**, *20*, 1787-1792.
(270) Wessling, R. A. *J. Polym. Sci. Polym. Symp*. **1985**, *72*, 55-66.
(271) Lahti, P. M.; Modarelli, D. A.; Denton, F. R.; Lenz, R. W.; Karasz, F. E. *J. Am. Chem. Soc*. **1988**, *110*, 7258-7259.
(272) Moon, Y. B.; Rughooputh, S. D. D. V.; Heeger, A. J.; Patil, A. O.; Wudl, F. *Synth. Met*. **1989**, *29*, E79-E84.

(273) Hoerhold, H.; Helbig, N. *Makromol. Chem. Macromol. Symp.* **1987**, *12*, 229-258.
(274) Murase, I.; Ohnishi, T.; Noguchi, T.; Hirooka, M. *Synth. Met.* **1987**, *17*, 639-644.
(275) Eckhardt, H.; Shacklette, L. W.; Jen, K. Y.; Elsenbaumer, R. L. *J. Chem. Phys.* **1989**, *91(2)*, 1303-1315.
(276) Shi, S.; Wudl, F. in reference 44; pp 83-89.
(277) Hoerhold, H. Presentation at the International Conference on Science and Technology of Synthetic Metals, Tuebingen, FRG, September, 1990.
(278) Lenz, R. W.; Han, C.; Stenger-smith, J.; Karasz, F. E. *J. Polym. Sci. Polym. Chem.* **1988**, *26*, 3241-3249.
(279) Machado, J. M.; III, F. R. D.; Schlenoff, J. B.; Karasz, F. E.; Lahti, P. M. *J. Polym. Sci. Polym. Phys.* **1989**, *27*, 199-203.
(280) Machado, J. M.; Schlenoff, J. B.; Karasz, F. E. *Macromolecules* **1989**, *22*, 1964-1973.
(281) Machado, J. M.; Karasz, F. E.; Lenz, R. W. *Polymer* **1988**, *29*, 1412-1417.
(282) Uneo, H.; Nogami, K.; Yoshino, K. *Phys. Rev. B* **1987**, *36(15)*, 8142-8146.
(283) Ueno, H.; Yoshino, K. *Phys. Rev. B* **1987**, *36(15)*, 8138-8141.
(284) Singh, B. P.; Prasad, P. N.; Karasz, F. E. *Polymer* **1988**, *29(11)*, 1940-1942.
(285) Kaino, T.; Kobayashi, H.; Kubodera, K.; Kurihara, T.; Saito, S.; Tsutsui, T.; Tokito, S. *Appl. Phys. Lett.* **1989**, *54(17)*, 1619-1621.
(286) McBranch, D.; Sinclair, M.; Heeger, A. J.; Patil, A. O.; Shi, S.; Askari, S.; Wudl, F. *Synth. Met.* **1989**, *29(1)*, 85-90.
(287) Yoshino, K.; Kuwabara, T.; Iwasa, T.; Kawai, T.; Onoda, M. *Jap. Journ. Appl. Phys.* **1990**, *29(8)*, L1514-L1516.
(288) Burroughes, J. H.; Bradley, D. D. C.; Brown, A. R.; Marks, R. N.; Mackay, K.; Friend, R. H.; Burns, P. L.; Holmes, A. B. *Nature* **1990**, *347*, 539-541.
(289) Capistran, J. D.; Gagnon, D. R.; Antoun, S.; Lenz, R. W.; Karasz, F. E. *Polym. Prepr.* **1984**, *25(2)*, 282-283.
(290) Murase, I.; Ohnishi, T.; Noguchi, T.; Hirooka, M. *Polymer J.* **1987**, *28(8)*, 229-231.
(291) Yamada, S.; Tokito, S.; Tsutsui, T.; Saito, S. *J. Chem. Soc. Chem. Commun* **1987**, 1448-1449.
(292) Jen, K.; Eckhardt, H.; Jow, T. R.; Shacklette, L. W.; Elsenbaumer, R. L. *J. Chem. Soc. Chem. Commun.* **1988**, 215-217.
(293) Jen, K.; Jow, T. R.; Elsenbaumer, R. L. *J. Chem. Soc. Chem. Commun.* **1987**, 1113-1115.
(294) Edwards, J. H.; Feast, W. J. *Polymer* **1980**, *21*, 595-596.
(295) Bott, D. C.; Brown, C. S.; Chai, C. K.; Walker, N. S.; Feast, W. J.; Foot, P. J. S.; Calvert, P. D.; Billingham, N. C.; Friend, R. H. *Synth. Met.* **1986**, *14*, 245-269.
(296) Edwards, J. H.; Feast, W. J.; Bott, D. C. *Polymer* **1984**, *25*, 395-398.
(297) Kahlert, H.; Leising, G. *Mol. Cryst. Liq. Cryst.* **1985**, *117*, 1.
(298) Bott, D. C.; Brown, C. S.; Winter, J. N.; Barker, J. *Polymer* **1987**, *28*, 601-616.
(299) Brown, C. S.; Vickers, M. E.; Foot, P. J. S.; Billingham, N. C.; Calvert, P. D. *Polymer* **1986**, *27*, 1719-1724.
(300) Schrock, R. R. *Acc. Chem. Res.* **1990**, *23(5)*, 158-165.
(301) Klavetter, F. L.; Grubbs, R. H. *Synth. Met.* **1988**, *26*, 311-319.
(302) Johnson, L. K.; Virgil, S. C.; Grubbs, R. H.; Ziller, J. W. *J. Am. Chem. Soc.* **1990**, *112(13)*, 5384-5385.

(303) Schrock, R. R.; DePue, R. T.; Feldman, J.; Schaverien, C. J.; Dewan, J. C. *J. Am. Chem. Soc.* **1988**, *110*, 1423-1435.
(304) Steltzer, F.; Leitner, O.; Leising, G.; Pressl, K.; Grubbs, R. H. *Synth. Met.* **1991**, in press.
(305) Swager, T. M.; Dougherty, D. A.; Grubbs, R. H. *J. Am. Chem. Soc.* **1988**, *110(9)*, 2973-2974.
(306) Bredas, J. L. *J. Phys. Chem.* **1985**, *82*, 3808-3811.
(307) Wudl, F.; Kobayashi, M.; Heeger, A. J. *J. Org. Chem.* **1984**, *49*, 3382-3384.
(308) Kobayashi, M.; Colaneri, N.; Boysel, M.; Wudl, F.; Heeger, A. J. *J. Chem. Phys.* **1985**, *82(12)*, 5717-5723.
(309) Bredas, J. L.; Heeger, A. J.; Wudl, F. *J. Chem. Phys.* **1986**, *85(8)*, 4673-4678.
(310) Pranata, J.; Grubbs, R. H.; Dougherty, D. A. *J. Am. Chem. Soc.* **1988**, *110(11)*, 3430-3435.
(311) Jenekhe, S. A. *Nature* **1986**, *322*, 345-347.
(312) Jenekhe, S. A. *Macromolecules* **1986**, *19*, 2663-2664.
(313) Jenekhe, S. A.; Lo, S. K.; Flom, S. R. *Appl. Phys. Lett.* **1989**, *54(25)*, 2524-2526.
(314) Patil, A. O.; Wudl, F. *Macromolecules* **1988**, *21(2)*, 540-542.
(315) Becker, R.; Bloechl, G.; Braeunling, H. in reference 44; pp 133-139.
(316) Al-Jumah, K.; Fernandez, J. E. *Macromolecules* **1987**, *20(6)*, 1177-1180.
(317) Al-Jumah, K.; Fernandez, J. E. *Macromolecules* **1987**, *20(6)*, 1181-1183.
(318) Wnek, G. E. in *Handbook of Conducting Polymers*; Skotheim, T. A. Eds.; Vol. 1; Marcel Dekker, Inc., New York, 1986; pp 205-211.
(319) Galvin, M. E.; Wnek, G. E. *Polym. Commun.* **1982**, *23*, 795.
(320) Chiang, J. C.; Smith, P.; Heeger, A. J.; Wudl, F. *Polymer Commun.* **1988**, *29*, 161-163.
(321) Rubner, M. F.; Tripathy, S. K.; J. Georger, J.; Cholewa, P. *Macromolecules* **1983**, *16*, 870-875.
(322) Smith, P.; Heeger, A. J.; Wudl, F.; Chiang, J. in reference 4; pp 283-290.
(323) Lindsey, S. E.; Street, G. B. *Synth. Met.* **1984/85**, *10*, 67-69.
(324) Niwa, O.; Kakuchi, M.; Tamamura, T. *Macromolecules* **1987**, *20(4)*, 749-753.
(325) DePaoli, M. A.; Waltman, R. J.; Diaz, A. F.; Bargon, J. *J. Polym. Sci.* **1985**, *23*, 1687-1698.
(326) Bi, X.; Pei, Q. *Synth. Met.* **1987**, *22*, 145-156.
(327) Nazzal, A. I.; Street, G. B. *J. Chem. Soc. Chem. Commun.* **1985**, 375-376.
(328) Nemoto, H.; Marks, T. J.; Groot, D. C. D.; Kannewurf, C. R. *Chem. Mater.* **1990**, *2*, 349-351.
(329) Aldissi, M. *J. Polym. Sci. Polym. Lett. Ed.* **1985**, *23*, 167-170.
(330) Akagi, K.; Katayama, S.; Shirakawa, H.; Araya, K.; Mukoh, A.; Narahara, T. *Synth. Met.* **1987**, *17*, 241-246.
(331) Cai, Z.; Martin, C. R. *J. Am. Chem. Soc.* **1989**, *111(11)*, 4138-4139.
(332) Martin, C. R.; Dyke, L. S. V.; Cai, Z.; Liang, W. *J. Am. Chem. Soc.* **1990**, *112(24)*, 8976-8977.
(333) Liang, W.; Martin, C. R. *J. Am. Chem. Soc.* **1990**, *112(26)*, 9666-9668.
(334) Druy, M. A.; Tsang, C.; Brown, N.; Heeger, A. J.; MacDiarmid, A. G. *J. Polym. Sci. Polym. Phys. Ed.* **1980**, *18*, 429-441.
(335) Lugli, G.; Pedretti, U.; Perego, G. *J. Polym. Sci. Polym. Lett. Ed.* **1985**, *23*, 129-135.

(336) Park, Y. W.; Park, C.; Lee, Y. S.; Yoon, C. O.; Shirakawa, H.; Suezaki, Y.; Akagi, K. *Solid State Commun.* **1988**, *65(2)*, 147-150.
(337) Hagiwara, T.; Hirasaka, M.; Sato, K.; Yamaura, M. *Synth. Met.* **1990**, *36*, 241-252.
(338) MacDiarmid, A. G.; Epstein, A. J. in reference 44; pp 53-63.
(339) Kahlart, H.; Leitner, O.; Leising, G. *Synth. Met.* **1987**, *17*, 467-472.
(340) Gagnon, D. R.; Karasz, F. E.; Thomas, E. L.; Lenz, R. W. *Synth. Met.* **1987**, *20*, 85-95.
(341) Jen, K.; Shacklette, L. W.; Elsenbaumer, R. *Synth. Met.* **1987**, *22*, 179-183.
(342) Jin, J.; Park, C. K.; Shim, H.; Park, Y. *J. Chem. Soc. Chem. Commun.* **1989**, 1205-1206.
(343) Machado, J. M.; Masse, M. A.; Karasz, F. E. *Polymer* **1989**, *30(11)*, 1992-1996.
(344) Swiatkiewicz, J.; Prasad, P. N.; Karasz, F. E. *Appl. Phys. Lett.* **1990**, *56(10)*, 892-894.
(345) Kanatzidis, M. G.; Tonge, L. M.; Marks, T. J.; Marcy, H. O.; Kannewurf, C. R. *J. Am. Chem. Soc.* **1987**, *109(12)*, 3797-3799.
(346) Kanatzidis, M. G.; Hubbard, M.; Tonge, L. M.; Marks, T. J.; Marcy, H. O.; Kannewurf, C. R. *Synth. Met.* **1989**, *28*, C89-C95.
(347) Kanatzidis, M. G.; Wu, C.; Marcy, H. O.; Kannewurf, C. R. *J. Am. Chem. Soc.* **1989**, *111(11)*, 4139-4141.
(348) Kanatzidis, M. G.; Wu, C.; Marcy, H. O.; DeGroot, D. C.; Kannewurf, C. R. *Chem. Mater.* **1990**, *2*, 222-224.
(349) Ramamurthy, V.; Caspar, J. V.; Corbin, D. R. *J. Am. Chem. Soc.* **1991**, *113(2)*, 594-600.
(350) Caspar, J. V.; Ramamurthy, V.; Corbin, D. R. *J. Am. Chem. Soc.* **1991**, *113(2)*, 600-610.
(351) Keller, T. M. *CHEMTECH* **1988**, *October*, 635-639.
(352) Keller, T. M.; Gatz, R. F. *Polymer Commun.* **1987**, *28*, 334-336.
(353) Hopf, H.; Kretschmer, O.; Naarman, H. *Angew. Chem. Int. Ed. Engl. Adv. Mater.* **1989**, *28(12)*, 1745-1746.
(354) Foley, C. M. T.; Zeller, C.; Falardeau, E. R.; Vogel, F. L. *Solid State Commun.* **1977**, *24*, 371-375.
(355) Kobayashi, M.; Tsujikawa, I. *J. Phys. Soc. Japan* **1979**, *46(6)*, 1945-1946.

Properties of Highly Conducting Polyacetylene

Th. Schimmel, D. Gläser, M. Schwoerer and H. Naarmann*
Physikalisches Institut and Bayreuther
Institut für Makromolekülforschung (BIMF)
Universität Bayreuth, D-8580 Bayreuth, Germany;
*BASF, D-6700 Ludwigshafen, Germany

J.L. Brédas and R. Silbey (eds.), Conjugated Polymers, 49–111.

1 Introduction

During the past 15 years, the synthesis and characterization of conducting polymers has attracted the attention of numerous groups of scientists /1/, stimulating a close cooperation of experimental and theoretical physicists, chemists, physico-chemists, materials research scientists and engineers.

On the one hand, this widespread interest is due to the high potential of possible technical applications. "Taylor-made polymers" are of increasing technological and economical importance. If chemists succeed in developing chemically stable electrically highly conducting polymers, there will be a wide range of applications ranging from rechargable batteries containing no heavy metals to the production of new antistatic materials and electronic devices.

On the other hand, research on conducting polymers is an intriguing field within solid state physics, as the mechanisms of charge transport in these materials seem to be very different from those of conventional metals and semiconductors. This is partly due to the high degree of disorder in these materials and partly to the - at least locally - high anisotropy of many of their physical properties. New phenomena occur in systems with reduced dimensions: a Peierls transition /2/ is observed in quasi-onedimensional conductors. In many cases, band filling, energy gap and electron density of states can be deliberately varied by variation of dopant concentration. And there are many opportunities for the synthesis of new or modified systems with different band structures.

Within the class of conducting polymers, polyacetylene, $(CH)_x$, due to its simple molecular structure, has become a model system for the investigation of charge transport /3/. Since about 1977, this substance has aroused considerable interest as a material, whose conductivity can be varied by "doping" (i.e. oxidation or reduction) /4,5/ over 16 orders of magnitude, ranging from values of typical insulators to those of typical metals (see fig. 1). In the past decade, numerous groups have investigated its transport properties /6-9/. However, no uniform picture of the electrical conductivity could be derived, partly due to sample imperfections and chemical instability of the substances, partly to different sample preparation techniques in different laboratories. Depending on the exact procedure of preparation, the maximum values of the room temperature conductivity varied between 10^2 and 10^3 S/cm /3/. SEM investigations showed a strongly disordered

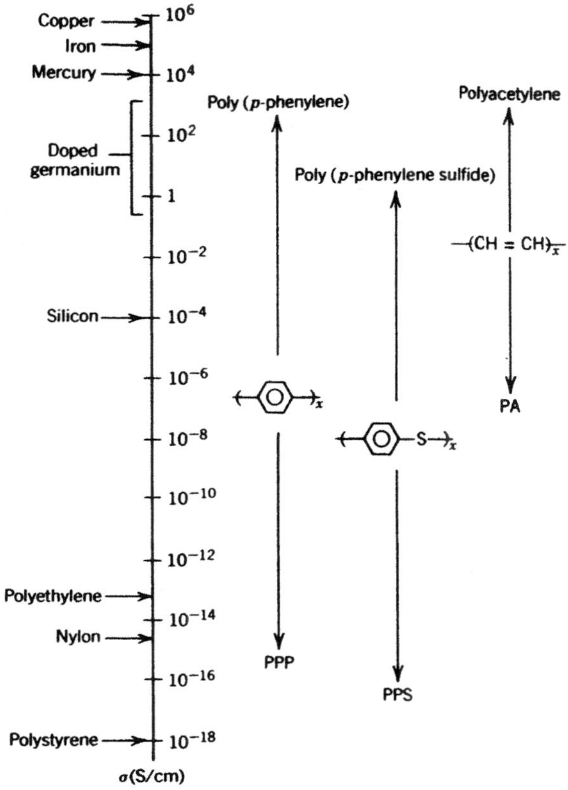

Fig. 1: Comparison of the room temperature values of the conductivities achievable with different conducting polymers with those of typical metals, semiconductors and insulators (from Ref. /10/).

fibrillar morphology. By mechanical stretching of the samples, which led to a partial orientation of the fibrils, conductivities of up to 10^4 S/cm could be achieved, still in contrast to theoretical calculations /11,12/, which predicted values of the order of 10^6 - 10^7 S/cm for the ideal, non-defective polyacetylene sample. This indicated that the conductivity σ was still limited by imperfections, defects and by sample morphology. Surprisingly, the synthesis of non-fibrillar polyacetylene /13,14/ with a high degree of crystallinity and a much higher degree of orientation than obtained before did not lead to significantly higher values of σ.

In this context, a new type of polyacetylene was of special

interest, which yielded room temperature conductivities which were much higher than obtained previously. In 1987, Naarmann and Theophilou /15-18/ presented a route of synthesis which allowed the preparation of free-standing $(CH)_x$ films with room temperature conductivities between 20 000 S/cm and 100 000 S/cm /19-24/ after mechnical stretching and doping with iodine. In the meantime, these high values of σ of the order of 10^5 S/cm could be reproduced by Tsukamoto et al. /25/, who used a modified Shirakawa route for their synthesis.

This progress in sample synthesis led to questions concerning the difference between high-σ samples (σ > 20 000 S/cm) and standard samples of both the fibrillar and the non-fibrillar type. One asks what we can learn from experiments on high-σ polyacetylene on the mechanism of charge transport in conducting polymers and especially about the processes limiting the conductivity. To get an answer, both morphological studies and experiments concerning the conductivity are necessary. As σ increases by stretching, the simultaneous investigation of the conductivities parallel and perpendicular to the stretching direction could provide a key for understanding the charge transport. It is of special interest to get information about the correlation between conductivity on the one hand and morphology and conjugation length on the other. To study the influence of the conjugation length on both the absolute value of σ and the mechanism of charge transport, samples have been investigated both before and after systematic treatment with oxygen.

2 Sample Synthesis, Morphology and Properties

2.1 Standard Routes of Synthesis

A variety of different catalyst systems for the polymerization of acetylene has been described in literature /26-35/. In addition, synthesis of polyacetylene is possible by elimination reactions from PVC /36/ and other polymers /13, 37, 38/. In most cases, the Shirakawa route /35/ and its modifications are used. For this route of synthesis, the catalyst is contained in solution at a concentration of about 0.2 mol/l, while acetylene gas is blown onto the surface of the liquid. As a result, an acetylene film grows on the surface, whose thickness can be varied with the reaction time between several microns and more than one millimeter /39/.

2.2 Naarmann-Type Polyacetylene

Using the Naarmann route /16/, polymerization of the acetylene is performed with a mixture of $Ti(OC_4H_9)_4$ and $Al(C_2H_5)_3$ as catalyst system, which is aged at 120°C in silicon oil as reaction medium. n-buthyl-lithium is added to the catalyst as reducing agent (ARA method: Addition of Reducing Agent).

For polymerization, the catalyst is brought onto a substrate as a thin layer and acetylene is blown over the catalyst at room temperature in contrast to Shirakawa synthesis, were the reaction takes place at -78°C. As a result, rather homogeneous polyacetylene films are obtained, whose thickness can be controlled by the thickness of the layer of catalyst.

When using rigid substrates, the sample can be removed from the substrate after the reaction. For a thickness of 4 μm and more, free-standing and mechanically stable films are obtained, which can be stretched mechanically to get an alignment of the fibrils parallel to the stretching direction. The stretching is more homogeneous and higher degrees of stretching of up to a factor of 6.5 are obtained without macroscopic damage to the sample, if the polyacetylene films are grown on polypropylene as substrate. For removal of the catalyst from the film, the samples are rinsed for three hours in toluene and for 16 hours in a mixture of methanole with 6 % of HCl and finally twice for 60 - 90 minutes in pure methanole. After this procedure, only small amounts of catalyst remain in the sample (titanium and aluminium content below 0.01 % /18/). Doping of the samples is achieved in solution prior to measurement as described in chapter 3.4.

The undoped samples exhibit lower defect concentrations than conventional polyacetylene. Crystallinity is about 80 % and the density is 0.9 g/cm^3 /15, 16, 18/. SEM pictures of strech-aligned samples show a fibrillar morphology with a partial orientation parallel to the stretching direction and typical fibrillar diameters of 50 - 100 nm /40/ (see chapter 6).

3 Conductivity: Experimental

3.1 The Standard Four-Probe and Montgomery Techniques

To determine the value of an electrical resistance, it is necessary to measure both the current passing through the

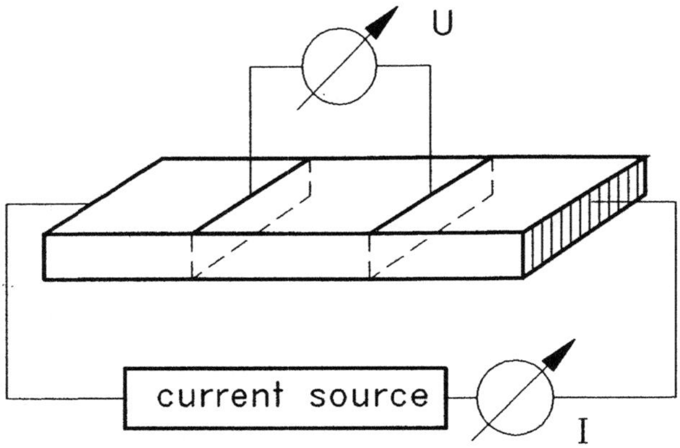

Fig. 2: Resistance measurement using the standard four-probe technique.

resistance and the voltage drop across the resistance due to this current. In order to measure the resistance of the sample itself without that of the contacts and that of the wires leading to the voltmeter, it is necessary to keep the current passing these contacts and wires sufficiently low. For this reason, the current to the sample is supplied by two separate contacts ("four-probe method").

Depending on the contact arrangement and procedure of evaluation, different such methods have been developed. The most well-known, the standard four-probe technique, is illustrated in fig. 2. A current is applied to the outer contacts of a rectangular sample of known geometry, while the voltage drop is measured between the inner contacts.

In contrast to this method, the Montgomery technique /41, 42/ is a procedure which allows the simultaneous determination of two components of the conductivity tensor of the same sample. For this purpose, a sample is prepared in the shape of a rectangular prism one of whose faces has its edges parallel to those principal axes of the conductivity tensor, whose σ values have to be determined. At the four corners of this area, small contacts are attatched (fig. 3). On the two corners of one of the short edges of this area, a current is applied, while the resulting voltage drop is measured between the corners of the opposite edge of this area (fig. 3a). The same procedure is repeated for the two long edges of the same area (fig. 3b). In this way, $R_1=U_1/I_1$ and $R_2=U_2/I_2$, which have the dimension of a resistance, are determined. If the geometry of the sample is known, the two components of the conductivity tensor can be derived by a procedure given by Logan, Rice and Wick /42/ and Montgomery /41/.

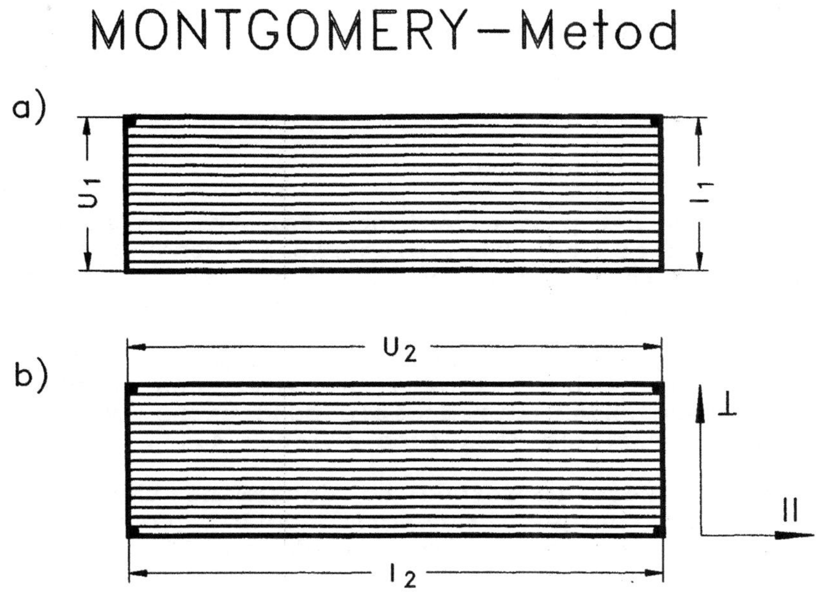

Fig. 3: Simultaneous measurement of the resistivity and its anisotropy using the Montgomery technique. The voltage drop U due to the sample current I is determined for the two contact arrangements shown in (a) and (b). To determine the Montgomery nesting ratio, current and voltage contacts are interchanged.

3.2 Test of Sample Homogeneity

If a current is applied at the corners of one edge of a sample, the resulting voltage drop between the corners of the opposite edge strongly depends on the current distribution within the sample. Conductivity measurements using the Montgomery technique therefore require homogeneous samples and are very sensitive to sample inhomogeneities and microcracks. The reliability of results obtained via the Montgomery technique requires a proof of sufficient homogeneity of the samples not only prior to but also during a series of measurements.

For this reason, the voltage contacts have been regularly interchanged with the current contacts and the resulting differences of the R_1 and R_2 values have been determined /23/. This procedure has been repeated at certain intervals within the whole temperature range. As a limit for tolerance, we required agreement of the values before and after interchanging contacts within 0.1 %, for precision measurements within 0.01 %. Samples with higher nesting differences have been discarded.

3.3 Conductivity Measurement

The conductivity setup was designed for the investigation of the dependence of the electrical conductivity on temperature, current direction relative to the sample, pressure and time in controlled environment (e.g. oxygen, air or vacuum) with high accuracy and reproducibility. In order to meet these requirements, a computer controlled setup was used, which allowed a slow and controlled temperature sweep under reproducible conditions, yielding a high density of acquired data (see chapter 4.1).

The experiments were performed to determine the relevant mechanisms of charge transport both parallel and perpendicular to the stretching direction. Of course, the temperature dependence of σ can give information about the dominating mechanism, but usually the macroscopic sample resistance is due to a superposition of different mechanisms rather than a single process, so that additional experiments are necessary.

Moreover, different mechanisms of conductivity sometimes show only minor differences in $\sigma(T)$. For the case of high-σ polyacetylene σ varies by only typically a factor of 5 in the temperature range between 300 K and 14 mK (see below), a fact which increases the requirements on both conductivity and temperature measurements necessary for a clear discrimination between different mechanisms of charge transport and for a quantitative evaluation of the results (determination of activation energies, barrier parameters etc.). To obtain high measurement accuracy, in addition to fulfilling the standard requirements of low level signal measurements /43/, it is necessary to eliminate potentials superimposed on the signal to be measured; these potentials are due to thermovoltages within the measurement circuit as well as electrochemical potentials. While the level of the first can be reduced by selection of adequate circuit materials and adequate positioning of junctions, the latter is reduced by different precautions like cleaning of soldering contacts within the sample circuit and the use of chemically inert contacts for the iodine doped samples (platinum instead of gold). The remaining potential superimposed to the signal to be measured can be eliminated by determining the voltage drop ΔU across the sample for both directions of the current I:

$$U^{+} = R_{x} \cdot I + \Delta U \qquad (1)$$

$$U^- = -R_x \cdot I + \Delta U \qquad (2)$$

and thus:

$$R_x = \frac{U^+ - U^-}{2 \cdot I} \qquad (3)$$

3.4 Sample Preparation

Prior to measurement, the samples were doped in a saturated solution of iodine in CCl_4. By subsequent rinsing with pure CCl_4, excessive iodine was washed from the sample surface. The samples were dried for approximately 30 minutes in vacuum (10^{-2} mbar) at room temperature.

Contacting was performed by mechanically pressing platinum contacts elastically on the film. By means of small screws, the pressure of the contacts could be adjusted. For the Montgomery method, this adjustment could be done for each of the contacts separately. Different sample holders were used for different sample sizes. For most of the conductivity experiments presented here, we used samples of 19 x 2 mm^2 and 14 x 6 mm^2 in size for the standard four-probe and Montgomery techniques, respectively.

The use of platinum as material for the contacts was important to avoid chemical reactions with the iodine of the doped samples, which would lead to electrochemical potentials. Although these potentials are eliminated by the applied method of measuring with both polarities of the current (see above), the superimposed potentials lead to a significant reduction of the signal-to-noise ratio.

Contacting samples with carbon, gold or silver paints was possible. However, it turned out that the solvents in these paints lead to a change of the properties of the doped samples around the contacts, among others to a reduction of the local conductivities mainly at low temperatures, and different temperature dependences of the conductivities are found if parts of the sample are damaged by the use of conducting paints. For this reason, all results presented in this paper were obtained with mechanical platinum contacts.

Sample preparation including cutting, doping and contacting of the samples was performed without exception under inert gas in a glove box. After contacting, the samples were plugged onto a sample carrier and transferred into the sample chamber of a helium flow cryostat. The sample chamber was evacuated and then floated with inert gas (helium).

4 Conductivity Measurements: Experimental Results

4.1 General Remark

The temperature dependence of the electrical dc conductivity of both freshly-prepared and oxygen aged samples doped to a maximum with iodine has been examined for a variety of different samples of Naarmann polyacetylene. In the following chapter, only a selection of these experimental curves is presented, which is representative for these measurements; the values of the parameters derived from the curves presented are given together with their range of variation from sample to sample. Qualitatively, all the effects observed in these experiments have been reproducible with each of the samples investigated; "conductivity anomalies" did not occur.

To ensure the reliability of the Montgomery technique, neighbouring pieces of the same sample have been examined both with the Montgomery and the standard four-probe technique, yielding comparable results for both the absolute values and the temperature dependencies of the conductivity and its anisotropy. However, as the properties of polyacetylene samples can vary significantly as a function of time and as considerable variations of both conductivity and local stretching ratio are found within the same polyacetylene film, reliable results on both the conductivity anisotropy and especially its dependence on temperature and aging require the application of the Montgomery technique which, in contrast to the standard four-probe method, avoids these problems by allowing the simultaneous determination of σ and $\sigma_{\perp}$ on the same piece of sample. For this reason, the Montgomery technique has been applied in most of our measurements.

Each experimental $\sigma(T)$ curve was determined within a temperature cycle both as cooling-down curve and as warming-up curve to check for a possible thermal hysteresis and to ensure the reproducibility of the results. Typical times used for one temperature cycle vary between 30 and 90 hours; typically 10 000 experimental data points were measured during this time. While the accuracy of the absolute value of the conductivity is limited by the error of the determination of the sample thickness by SEM to about 20 %, the relative accuracy and the reproducibility of the data within a temperature cycle is typically 0.01 %. The nesting difference (for definition see above) had values of typically 0.01␣-␣0.15␣%, indicating a high reliability of the Montgomery results.

Where it is not noted otherwise, the experimental data presented below were derived from freshly-prepared samples, which had been stretched mechanically by a factor of 6.5 and doped to saturation with iodine. The iodine concentration was between 6 and 7 % of J_3^- per CH unit. A part of the results has already been published /20-24/.

4.2 Temperature dependence of σ and $\sigma_\perp$

Fig. 4a gives the temperature dependence of the electrical dc conductivity parallel (σ) and perpendicular ($\sigma_\perp$) to the stretching direction of a highly stretched sample of Naarmann polyacetylene doped to maximum with iodine. The electrical conductivity of these samples varies by a factor of about 1.4 - 2 if measured on different pieces of the same film. Using different films prepared by the same method, the variation is much stronger: for most of the samples examined, the room temperature value of σ was between 20 000 S/cm and 80 000 S/cm. Some samples, however, had even higher or lower values including a sample with σ > 100 000 S/cm and another one with σ = 5 000 S/cm.

In contrast to the absolute values, the temperature dependences of both σ and $\sigma_\perp$ exhibited only a weak sample dependence. For all samples investigated, the conductivity was found to be monotonically decreasing with decreasing temperature ("non-metallic behaviour"). The factor of decrease in conductivity from 300 K to 3 K varied between 2.8 and 4.6 for all freshly-prepared samples for both σ and $\sigma_\perp$. A correlation between this factor $\alpha=\sigma$ (300 K)/σ (3 K) and the absolute value of the conductivity at room temperature has not been observed; the assumption that the σ(T) curves become more and more flat and "metal-like" with increasing values of the room temperature conductivity could not be confirmed. In contrast, both among the samples with the highest and among those with the lowest σ values, steep and flat temperature dependences have been found. So, e.g. a sample with σ (300 K) = 80 000 S/cm had a reduction factor of α = 4, whereas a sample with a room temperature conductivity of only 5 000 S/cm showed α = 2.9.

It is remarkable that σ and $\sigma_\perp$ of freshly-prepared samples show an almost identical temperature dependence. The anisotropy of the conductivity is about 25 and varies only slightly between 300 K and 3 K, i.e. by about 0.5 - 10 %, depending on the sample (fig. 4b).

If different freshly-prepared samples are compared, a decrease of the anisotropy A = σ /$\sigma_\perp$ of highly stretched samples is observed with decreasing room temperature

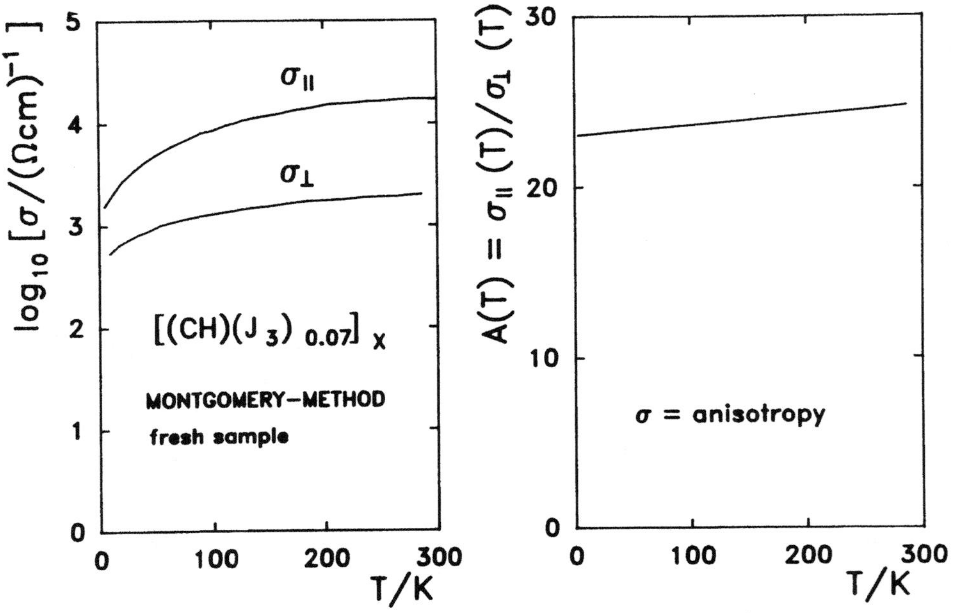

Fig. 4: a) Temperature dependence of the conductivities parallel (σ) and perpendicular ($\sigma_{\perp}$) to the direction of stretch-alignment of Naarmann polyacetylene which was stretched by a factor of 6.5 and doped to a maximum with iodine. The curves in the diagram represent polygons of about 10 000 experimental data points and have been measured as a temperature cycle 290 K - 3 K - 290 K. Experimental curve and fit to Sheng's formula (see chapter 5.3) are identical within less than the graphical linewidth. b) Temperature dependence of $A = \sigma / \sigma_{\perp}$ /22/.

conductivity; for about 70 % of the samples investigated, A had values between 20 and 30; for the less conducting samples, values between 10 and 20 have been found.

4.3 Conductivity at Very Low Temperatures (14 mK - 4.2 K)

Fig. 5 gives the temperature dependence of σ between 14 mK and 4200 mK in a linear and a double logarithmical plot. The conductivity continues to decrease monotonically with decreasing temperature; its value at 14 mK, however, is by only 25 % lower than at 4200 mK. Thus, at 14 mK, σ has still between 18 % and 27 % of its room temperature value. While an almost linear dependence of the conductivity on temperature is observed between 400 mK and 3000 mK, this

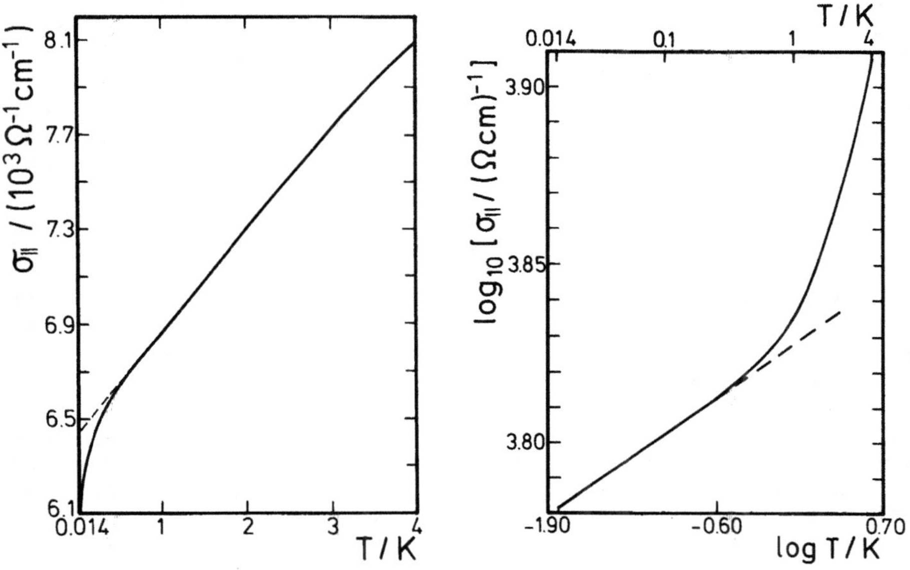

Fig. 5: Temperature dependence of σ between 14 mK and 4 K. The same experimental data are shown both in a linear and in a double logarithmical plot /21/.

behaviour changes drastically towards lower temperatures, showing a downward bending of the σ (T) curve and thus a temperature dependence that is stronger than linear.

4.4 Anisotropy and Stretching Ratio

The investigation of the conductivity anisotropy A as a function of the stretching ratio s is difficult in so far as only the macroscopic stretching ratio of a piece of sample can be determined. Locally - on a length scale of a few millimeters and less - however, inhomogeneities of the local stretching ratio can occur. The measurement of the stretching ratio thus always includes an experimental uncertainty of about 20 %, which results in a corresponding variation of the experimental data of the σ (s) and A(s) dependences.

As during the stretching process the elongation of the inner part of a sample usually is much larger than at its outer ends, equidistant labels have been put onto the sample surface at intervals of about 1 - 2 cm before stretching. Instead of measuring the average stretching ratio of the sample as a whole, the stretching ratio has been determined for each of these intervals individually. In this way, it was possible to get several samples with different stretching ratios from one polyacetylene film. The samples used for measurement have been taken from the area in the middle between two neighbouring labels.

It is found that the conductivity parallel to the stretching direction increases strongly as a function of the stretching ratio for small stretching ratios, but then saturates at values of s between 3 and 4.

4.5 Aging Effects in σ(T)

A systematic treatment of the samples with oxygen leads to a decrease of the conductivity at room temperature. Both the time dependence $d\sigma/dt$, which increases to a multiple of its original value when using oxygen instead of an inert atmosphere, and the different A(T) dependence of aged, highly doped and freshly-prepared, less doped samples indicate that this effect, which has been investigated at ambient pressure, is due to oxygen aging rather than to the evaporation of iodine from the sample. This is also supported by the observation that after exchanging oxygen against helium, the rapid decrease of σ as a function of time comes to an end after a delay time of several hours.

When comparing different polyacetylene films, one finds that the rate of relative decrease of the conductivity during the oxygen treatment varies by up to a factor of 10 from sample to sample, a fact that could be due to differences in sample morphology, which could lead to a different rate of diffusion of the oxygen into the interior of the sample or of the fibrils; on the other hand, a difference in the concentration of reactive sites within the sample (e.g. chemical defects) could also explain this effect. No correlation has been observed between the absolute value of the room temperature conductivity and the rate of oxygen aging.

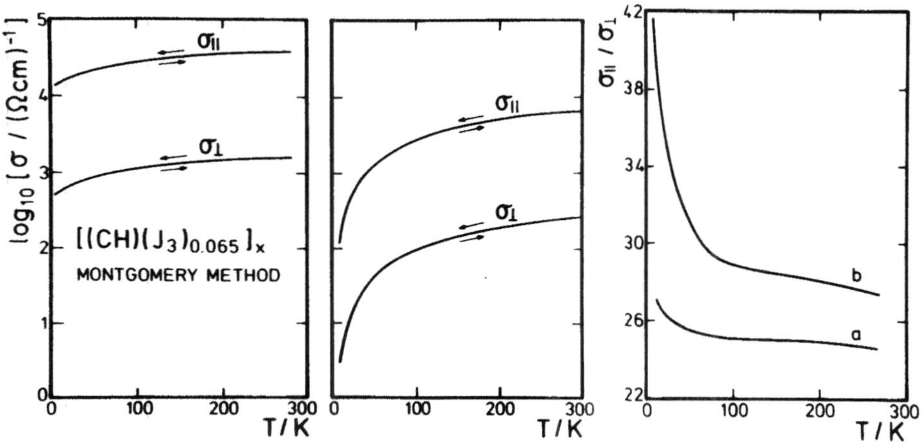

a) b) c)

Fig. 6: σ (T) and $\sigma_{\perp}$(T) of one and the same polyacetylene sample before (a) and after (b) oxygen aging (see text). The curves have been measured via Montgomery technique in a temperature cycle. Fig. 6c gives the corresponding curves of A(T) = σ (T)/$\sigma_{\perp}$(T) /23/.

In addition to the decrease of the conductivity at room temperature, oxygen aging also leads to a drastic change of the temperature dependence of both σ and $\sigma_{\perp}$, which leads to a decrease of conductivity mainly at lower temperatures. The diagrams in fig. 6 show the temperature dependences of σ and $\sigma_{\perp}$ of one and the same sample both before oxygen aging (a) and after treatment with 100 % oxygen at ambient pressure and 310 K for 15 hours (b). While for the freshly-prepared sample, the reduction factor $\alpha=\sigma(300\ K)/\sigma(3\ K)$ is 2.8, a value of $\alpha=100$ is found for the aged sample (fig. 6b).

Fig. 7 shows standard four-probe measurements of σ (T) in the temperature range between 8 K and 250 K. Curve a gives σ (T) of the freshly-prepared sample, while curves b - e correspond to different stages of oxygen aging of the same sample. The aging was performed by floating the cryostat with pure oxygen for different periods at different temperatures. The σ data are also plotted logarithmically versus $T^{-0.25}$, a plot which would give a straight line if the model of three-dimensional variable range hopping could be applied. While the room temperature value of σ decreases by less than a factor of 10, it is found for this sample that the conductivity at 3 K is reduced by more than 5 orders of magnitude (!) from curve a to curve e. Only for the case of strongly aged samples, further aging results in

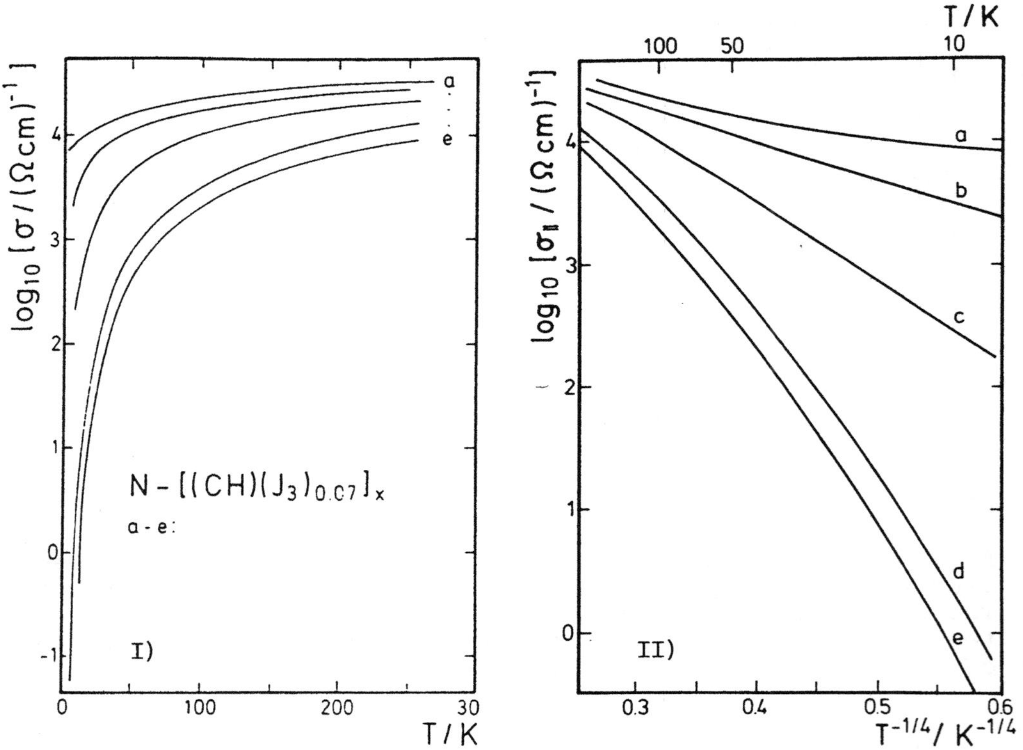

Fig. 7: σ (T) for different degrees of oxygen aging of one and the same sample at one atmosphere of oxygen as determined by the standard four-probe technique: a) freshly-prepared sample, b) after two hours at 260 K, c) after additional 1 hour at 270 K, d) after additional 1 hour at 320 K and e) after additional 2 hours at 320 K. The two diagrams represent two different plots of the same data. The plot of log(σ) versus $T^{-0.25}$ (right diagram) would give a straight line, if the model of three-dimensional variable range hopping could be applied /20/.

an almost constant factor of decrease of the conductivity in the entire temperature range (see e.g. fig. 7, curves c, d and e).

4.6 Anisotropy and Aging

While the absolute value of the room temperature conductivity decreases significantly during the aging process, the conductivity anisotropy shows only small changes. Fig. 8 gives the result of aging a highly

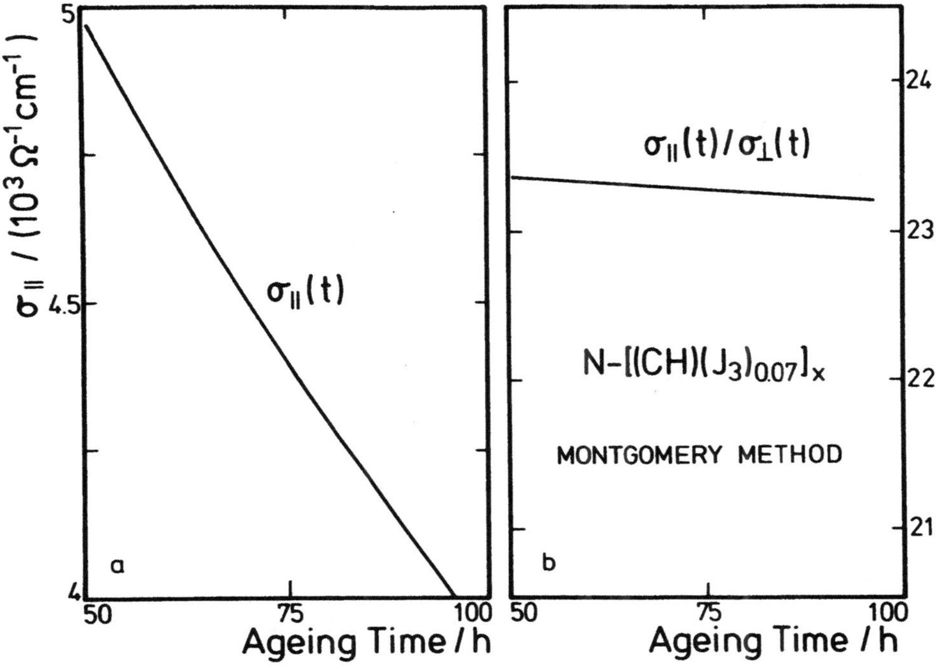

Fig. 8: Time dependence of σ and $A = \sigma /\sigma_{\perp}$ at 310 K and one atmosphere of pure oxygen during a 50 hour time interval /20/.

stretch-aligned sample doped to maximum with iodine at 100% oxygen atmosphere at room temperature and ambient pressure, as determined with the Montgomery technique in the time interval between 50 hours and 100 hours of oxygen exposure. The time interval has been selected with a starting point different from zero, as the aging reaction usually has an induction period of several hours, probably due to the time constant of oxygen diffusion into the interior of the sample and the fibrils.

While the conductivity in the case of fig. 8 decreases by about 20 %, its anisotropy changes by only 0.65 %. However, the aging process leads to a significant change of its temperature dependence. In the case of the sample of fig. 6, an increase of the anisotropy of only about 8 % is observed from 300 K to 3 K; after the aging, this value is more than 50 %. Fig. 9 gives another example of the temperature dependence of the conductivity anisotropy of one sample at different stages of oxygen aging as obtained with the Montgomery technique. The relative increase of the anisotropy, (A(3 K)–A(300 K))/A(3 K), increases from 5 % in the case of curve a (freshly-prepared sample) to 70 % in the case of curve e.

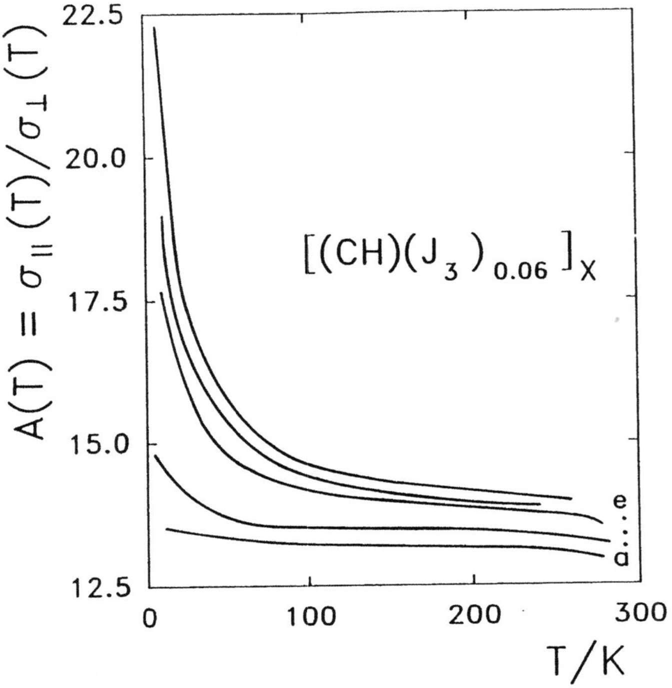

Fig. 9: Temperature dependence of the conductivity anisotropy A for different stages of oxygen aging: a) freshly-prepared sample, b) after one, c) after two, d) after four and e) after five days of oxygen treatment at 300 K and one atmosphere of O_2. The results were determined via Montgomery technique.

4.7 Dependence of $\sigma(T)$ on the Dopant Concentration

With decreasing concentration of iodine, a decrease of the room temperature conductivity as well as an increase of the temperature dependence, i.e. a steeper decrease of σ at lower temperatures is observed, qualitatively similar to the behaviour of oxygen aged samples. Fig. 10 gives results obtained via Montgomery technique with three freshly-prepared neighbouring pieces of the same film and doped to a different degree with iodine. Fig. 10 shows the temperature dependences of σ and $\sigma_{\perp}$ (top) and the corresponding curves of the anisotropy A(T) (bottom). It is observed that in contrast to sample aging with oxygen, a reduced dopant concentration does not result in an increase, but in a decrease of A at low temperatures.

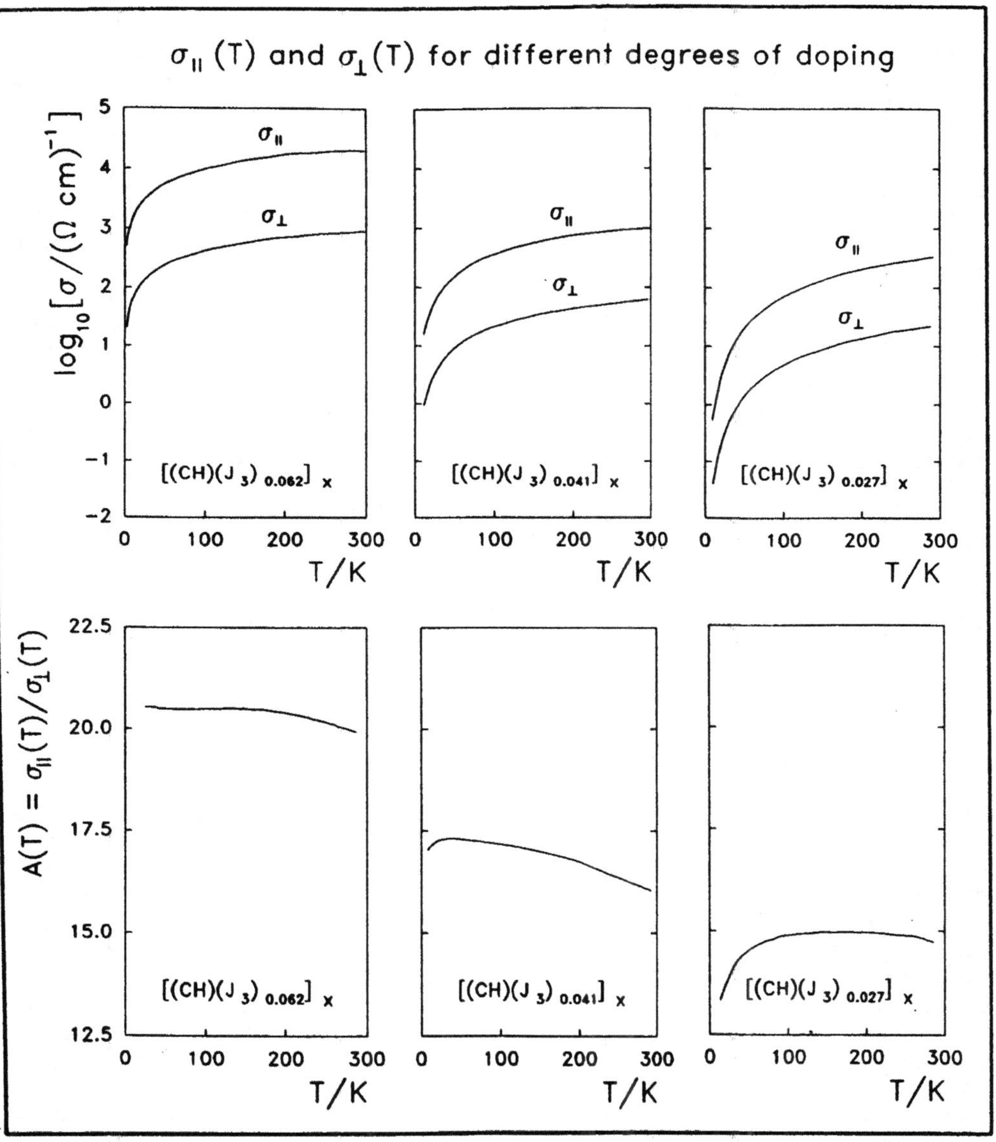

Fig. 10: σ (T) and $\sigma_\perp$(T) (top) and A(T) (bottom) for three different concentrations of iodine doping. I) y = 6.2 mol %, II) y = 4.1 mol % and III) y = 2.7 mol % J_3^- per CH unit /23/.

4.8 Doping with $FeCl_3$

Samples of Naarmann polyacetylene with stretching ratios between 3 and 6.5 have been doped by immersing them into a saturated solution of $FeCl_3$ in acetonitril. The samples obtained in this way have been much more brittle than iodine doped samples, especially at higher stretching ratios. The temperature dependence of σ has been examined between room temperature and 10 K, showing a monotonous decrease with decreasing temperature. Similar to the behaviour of the iodine-doped samples (see below), $\sigma(T)$ can be fitted with the Sheng formula in this temperature range, though the temperature dependence is weaker: σ (300 K)/σ (10 K) is by typically a factor of 2 smaller than for the iodine-doped samples. Though the temperature dependence of σ is smaller, the room temperature conductivity is not larger but by about one order of magnitude smaller than in the case of iodine doping: values between 2000 S/cm and 10000 S/cm have been found. In accordance with the comparison of different iodine doped samples, these results demonstrate that the samples exhibiting the weakest temperature dependence of σ are not necessarily those with the highest absolute values σ(300 K).

4.9 Pressure Dependence

The pressure dependence of σ has been examined in the pressure range between ambient pressure and 3 kbar with a polyacetylene sample doped to maximum with iodine and stretched by a factor of 6.5. Helium gas was used as pressure medium. Within the accuracy of the measurement, which was only 3 % in this experiment, no change of conductivity as a function of pressure was detected, in contrast to measurements of Basescu et al. /19/, who found an increase of σ of about 25 % between 0 and 3 kbars. Though it cannot be excluded that this is due to sample differences, the differing results could also be due to the fact that in our experiments, a gas was used as pressure medium was used: while the gaseous medium fills the interfibrillar spaces and the pressure is applied directly to the individual fibrils, a liquid pressure medium is likely to lead to a compression of the sample as a whole, pressing different fibrils against each other.

5 Discussion of $\sigma(T)$

5.1 Experimental Prerequisites for a Model of Charge Transport for T > 400 mK

A model of charge transport that is able to describe the temperature dependence of the electrical dc conductivity σ (T) of freshly-prepared samples between 400 mK and 300 K has to be compatible with the following experimental results:

1) The conductivity increases monotonically with increasing temperature in this entire temperature range.

2) The slope $d\sigma/dT$ (see fig. 5a) has a finite value for the limit of low temperatures ($T \to 0$), while $d^2\sigma(T)/dT^2$ is zero, i.e. we find an almost linear temperature dependence for temperatures between 400 mK and 3000 mK:

$$\sigma(T) = \sigma_0 + \alpha \cdot T \qquad (4)$$

with $\sigma_0 = 6.45 \cdot 10^3$ S/cm and $\alpha = 4.4 \cdot 10^2$ S/(cm·K) for the data of fig. 5.

3) This means in particular that for $T \to 0$, $\sigma(T)$ approaches a finite value σ_0, which is about 20 % of the room temperature value.

4) For the limit of high temperatures, $\sigma(T)$ asymptotically approaches a saturation value σ_∞, which for most of the samples investigated has values between 30 000 S/cm and 130 000 S/cm. The conductivity $\sigma(T)$ has both an upper boundary (σ_∞) and a lower boundary (σ_0), σ_∞/σ_0 being between 2.8 and 4.6.

For very low temperatures, this temperature dependence changes drastically at about 400 mK: while the linear temperature dependence is observed for T > 400 mK, $\sigma(T)$ is well described by an exponential law of the form:

$$\sigma(T) = \alpha \cdot T^{\beta} \qquad (5)$$

for T < 400 mK with $\alpha = 6.7 \cdot 10^3$ S/(cm·K$^{\beta}$) and $\beta = 0.025$, i.e. $\log[\sigma(T)] = \log(\alpha) + \beta \cdot \log(T)$.

This indicates that between 0.3 K and 0.4 K, a change of the mechanism determining the temperature dependence of σ occurs.

5.2 The Failure of Conventional Models

In the following chapter, only the range T > 0.4 K shall be discussed. The prerequisites listed in the previous chapter for a model of charge transport that could explain the experimental data eliminate most of the conventional models.

So the temperature dependence cannot be explained by phonon scattering alone as in many conventional metals, as the conductivity is increasing with increasing temperature instead of decreasing.

Due to the extrapolation to a finite value of σ for $T \to 0$, all those models of conductivity can be excluded which are based on charge transport between localized states of different energy, in particular all types of one- two- or three-dimensional variable range hopping. In the past, among other models, threedimensional variable range hopping has also been discussed for highly doped polyacetylene (e.g. /44/). Independent of different additional assumptions /45-48/, however, all hopping models have in common that $\sigma(T)$ approaches zero for $T \to 0$.

Particularly due to the possibility of spacially inhomogeneous doping, the model of charge transport via metallic islands is regularly discussed in context with doped polyacetylene /3/. Assuming metallically conducting islands in a dielectric matrix, one obtains:

$$\sigma(T) = \sigma_\infty \cdot \exp[-A'/T^{0.5}] \quad . \qquad (6)$$

This temperature dependence is also expected in the case of granular metals, agglomerates of mesoscopic metallic particles /49/. These metallic particles could possibly be identified with small, highly-conducting regions (e.g. domains of high dopant concentration) in polyacetylene. However, these mechanisms do not fulfil the condition a finite conductivity for $T \to 0$.

On the other hand, it is not necessarily only one mechanism of conductivity which is responsible for the $\sigma(T)$ behaviour above 0.4 K. Therefore it has to be examined if these models are able to describe the temperature dependence of σ at least in certain parts of this temperature range. In this case, the experimental $\sigma(T)$ curve could possibly be described by a superposition of different mechanisms which dominate $\sigma(T)$ at different temperatures. For this reason, the same experimental data are plotted in fig. 11 in five different ways; the plots of figs. 11b-e would give a straight line for the $\sigma(T)$ curve if the corresponding model could be applied, i.e. different models of

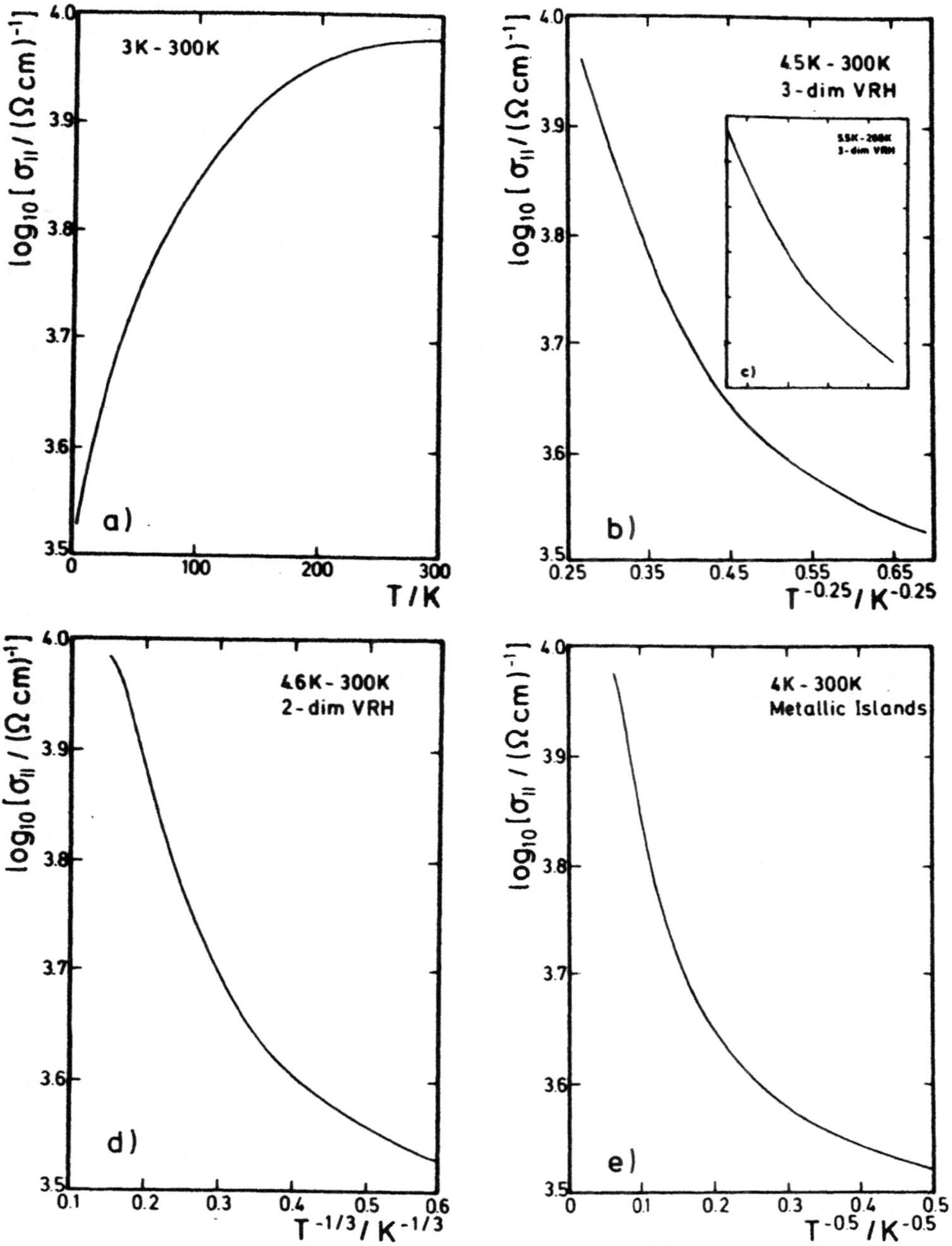

Fig. 11: Different plots of the same experimental data of σ (T) of a "low-σ" sample. For plots (b) - (e), the experimental curves should represent straight lines, if the corresponding conductivity models could be applied (see text) /Sch89/. Even for a fit of $\sigma=\text{const.}\cdot\exp(-A/T^{\alpha})$ with α as free parameter, a description of σ(T) is not possible.

three-dimensional variable range hopping in figs. 11b,c, the model of two-dimensional variable range hopping in fig. 11d and the model of metallic islands, of granular metals or of one-dimensional variable range hopping in fig. 11e. As the $\sigma(T)$ dependence of the highly-doped samples is relatively weak and an arbitrary curve can be fitted with a straight line if only the interval is small enough, a high accuracy and reproducibility of the data is necessary to reliably determine small but systematic deviations from the predictions of a given model even in a smaller temperature interval. The relative accuracy of the data of fig. 11 is well within the graphical linewidth for each of the different plots and the whole temperature range, and the data are reproduced within a temperature cycle. As samples with higher absolute values of the conductivity do not show systematic differences in the temperature dependence as compared with less conducting samples, data derived from a sample with low values of σ (σ (300 K) = 10 000 S/cm) have been selected for fig. 11. It is found that none of the curves gives a straight line in any temperature range, thus indicating that the corresponding models cannot be applied to charge transport in highly doped samples of Naarmann polyacetylene. These different plots have also been drawn for the experimental data of different samples (including highly conducting samples) measured both with the Montgomery and with the standard four-probe technique, leading to the same results.

Even in the low temperature range (14 mK < T < 300 mK) none of the models discussed above could be fitted to the experimental data. Because of the weak temperature dependence, however, the formula $\sigma = \sigma_0 + \sigma_1 \cdot \log(T/1K)$ also gives an agreeable fit to the experimental data.

5.3 Description with the Sheng Formula

In his model /50-52/, Sheng regards a heterogeneously conducting system consisting of large highly conducting regions separated by potential barriers (fig. 12). Contrary to the model of granular metals, however, charging effects of the conducting regions do not have a significant influence on the charge transport due to their larger size. In the case of polyacetylene, the highly conducting regions are sometimes identified with fibrils separated from each other by interfibrillar barriers; but they could also be identified with highly doped regions separated by less doped or undoped regions /23/, or they might finally be crystalline regions separated by amorphous regions or vice versa.

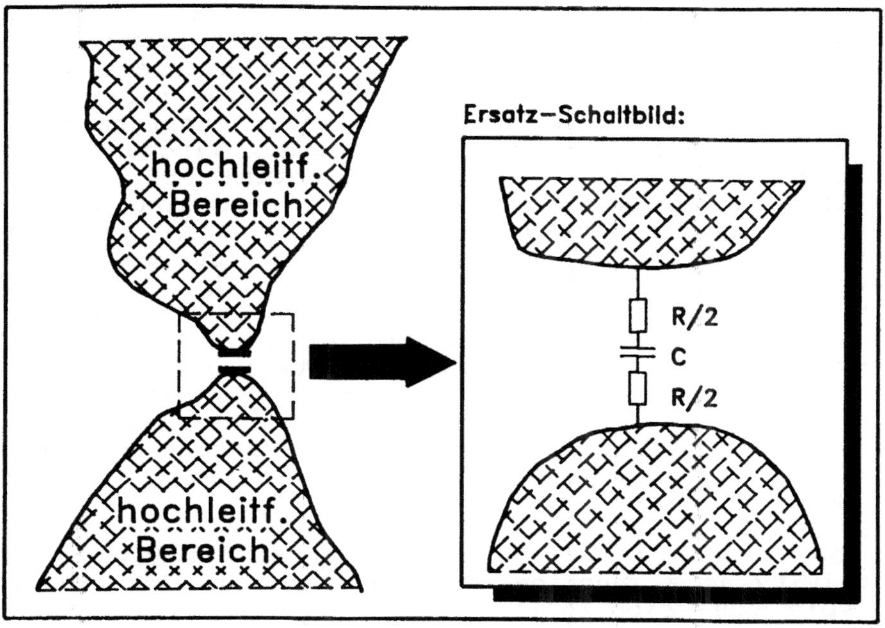

Fig. 12: Sheng's model considers the effect of fluctuation induced tunnelling through potential barriers between extended highly conducting regions. In a model circuit, the barriers can be treated as capacitances with charging resistances /52/.

In contrast to the models discussed above, Sheng's model, which has already been applied to polyacetylene previously /53/, predicts a conductivity that decreases monotonically with decreasing temperature, but still approaches a finite value of σ for $T \to 0$. The basic idea of this model of "fluctuation induced tunnelling" is simple. In the low temperature limit, a crossing of potential barriers is possible only due to quantum mechanical tunnelling, leading to a finite value of the electrical conductivity. For the limit of high temperatures, thermally activated crossing of the barriers is possible, leading to an Arrhenius-type temperature dependence of σ with a saturation value for σ. In addition to these basic processes, Sheng regards the effect of thermal fluctuations on the potential barriers. Fluctuations of the effective barrier height occur in both directions, leading to an increase or a decrease of the probability for a charge carrier to cross the barrier; as, however, this probability is a non-linear function of the barrier height, the net effect of thermal fluctuations on the charge transport through potential barriers by tunnelling is an enhancement of the tunnelling current, leading to an increase of this current with increasing temperature. As the thermal energy of the charge carriers causing the fluctuations in barrier height is of the order kT irrespective of the cross section of the barrier, whereas the capacitance of the barrier is proportional to its cross section, the modulating voltage will depend on the barrier cross section, being larger for smaller cross sections. To summarize, Sheng's mechanism should be important in heterogeneously conducting systems with large highly conducting regions separated by barriers of small cross section.

For the case of parabolic potential barriers, the conductivity as a function of temperature is described by the simple expression /52/:

$$\sigma(T) \quad = \quad \sigma_\infty \cdot \exp[-T_1/(T + T_o)] \qquad (7)$$

leading to the upper and lower boundary values::

$$\sigma(T\to\infty) \quad = \quad \sigma_\infty \qquad (8)$$

$$\sigma(T\to 0) \quad = \quad \sigma_\infty \cdot \exp[-T_1/T_o] =: \sigma_o \qquad (9)$$

The second derivative is:

$$\frac{d^2\sigma(T)}{dT^2} \quad = \quad \frac{T_1}{(T + T_o)^4} \cdot [T_1 - 2\cdot T_o - 2\cdot T] \cdot \sigma(T) \qquad (10)$$

Two of the three parameters of equ. 7 are already determined by σ_o and σ_∞, implying that the shape of the $\sigma(T)$ curve is determined by only one parameter, which, however, is already determined by the slope κ of the curve for $T\to 0$. With:

With the values of σ_o and σ_∞ as well as the slope for $T\to 0$, the whole $\sigma(T)$ curve is determined. The fit of the formula of equ. 7 to an arbitary smooth and monotonically increasing curve therefore is no trivial problem in spite of the existence of three fitting parameters.

It is most remarkable that a fit of the experimental $\sigma(T)$ data is possible with this simple formula for all freshly-prepared samples and a very wide temperature range of almost three orders of magnitude (!) in temperature from 0.4 K to 300 K. The entire experimental curve is reproduced with an error of less than 1 % in most cases.

Figs. 4a and 6a show the experimental $\sigma(T)$ curves together with the fit to Sheng's formula between 300 K and 3 K. Experimental data and fit are identical within less than the graphical linewidth. Fig. 5a gives the experimental $\sigma(T)$ data (full line) together with the fit to Sheng's formula (interrupted line) for the temperature range between 14 mK and 4 K. For T > 400 mK, experiment and fit again are identical within the graphical linewidth; only for still lower temperatures (T < 300 mK), significant deviations are observed. For the fit to these low temperature data, the same parameters have been used for T_o and T_1 as for the fit of the data of fig. 4a in the range T > 4 K, i.e. T_o = 75 K and T_1 = 152 K.

The values of the parameters σ_∞, T_1 and T_o derived from the fit are listed in table 1 for both freshly-prepared and aged samples. For this purpose, the sample with the steepest temperature dependence immediately after preparation (sample I, data of fig. 4a) and the sample with the weakest temperature dependence after preparation,

Table 1: Parameters resulting from the fit of Sheng's formula to the experimental $\sigma(T)$ data of two extreme samples; the T_o and T_1 values of all other freshly-prepared samples have been between the corresponding values of these samples. The evaluation of the barrier height V_o=ΔE and the barrier width w have been performed within the model presented in chapter 5.5.

sample I	σ_o / $\Omega^{-1}cm^{-1}$	T_o/ K	T_1 / K	ΔE/meV	w/nm
fresh: $\sigma_{\parallel}$	81000	75	152	13.0	1.73
$\sigma_{\perp}$	3020	65	122	10.5	1.79
aged: $\sigma_{\parallel}$	12900	54	205	17.7	2.79
$\sigma_{\perp}$	470	49	202	17.4	3.05

sample II	σ_o / $\Omega^{-1}cm^{-1}$	T_o/ K	T_1 / K	ΔE/meV	w/nm
fresh: $\sigma_{\parallel}$	53500	76	112	9.6	1.47
$\sigma_{\perp}$	2180	71	113	9.7	1.58
aged: $\sigma_{\parallel}$	12900	50	240	21	3.26
$\sigma_{\perp}$	519	55	274	24	3.17

(sample II, data of fig. 6) have been chosen to get an estimate of the variation of these parameters from sample to sample. For all the other samples investigated, the values of T_O and T_1 were between these two extremes. A correlation between σ_∞ and the slope of the $\sigma(T)$ curve, i.e. between σ_∞ on the one hand and T_O and T_1 on the other hand could not be observed.

For many freshly-prepared samples, one finds that T_1 is about twice the value of T_O, a fact that explains the almost vanishing curvature of $\sigma(T)$ for $T \to 0$: $d^2\sigma/dT^2$ is zero for $T \to 0$, if $T_1 = 2 \cdot T_O$. Therefore, within the Sheng model the linear $\sigma(T)$ dependence at low temperatures is only due to the values of these parameters rather than being a property essentially correlated with the mechanism of charge transport.

5.4 Limits of the Applicability of Sheng's Model

The calculation that leads to Sheng's formula for $\sigma(T)$ includes several approximations /50-52/, whose validity has to be checked before the model is applied. For the case of the $\sigma(T)$ data presented in chapter 4, different aspects have to be examined in more detail.

5.4.1 Low Temperature Limit

If fluctuation induced tunnelling is a mechanism contributing significantly to the charge transport in a given system, this is only true for a limited temperature range /50-52/. As mentioned above, thermally activated crossing of potential barriers dominates for higher temperatures ($T \gg T_O$), yielding:

$$\sigma_\infty \cdot \exp[-\frac{T_1}{T + T_O}] \xrightarrow{T \gg T_O} \sigma_\infty \cdot \exp[-\frac{T_1}{T}] \quad (11)$$

which corresponds to a thermally activated conductivity with activation energy $k_B \cdot T_1$. On the other hand, for the limit of low temperatures, fluctuations are not important for the crossing of the barriers as well; a temperature independent tunnelling conductivity is obtained in this limit. Whereas the high temperature limit is correctly represented in Sheng's formula, this is not true for the low temperature limit, where the slope of $\sigma(T)$ has a finite value for $T \to 0$ in Sheng's formula.

5.4.2 Image Charge Correction Parameter Λ

In his calculation, Sheng assumes rectangular potential barriers. The shape of the real physical barriers with non-singular potential gradient is generated by image charge correction of these rectangular barriers. The amount of this image charge correction is determined by a parameter Λ, which also determines the shape of the barrier. For the case of $\Lambda = 0.07$, barriers with an almost parabolic potential profile are obtained. Λ obeys the equation:

$$\lambda = 0.795 \cdot e^2/4 \cdot w \cdot \varepsilon \cdot V_o \qquad (12)$$

/52,54/, were ε is the dielectric constant of the insulating barrier and w and V_o are the width and height of this barrier, respectively. Sheng's derivation of equ. 7 assumes parabolic potential barriers, i.e. it assumes a value of $\Lambda = 0.07$. On the other hand, with the equations (see /50-52/):

$$\frac{T_1}{T_o} = \frac{\pi}{4} \cdot \frac{2}{\hbar} \sqrt{2 \cdot m \cdot V_o} \cdot w \qquad (13)$$

and:

$$k_B \cdot T_1 = 8 \cdot \varepsilon_o \cdot \frac{A}{w} \cdot \left(\frac{V_o}{e}\right)^2 \qquad (14)$$

the parameters w and V_o can be determined as a function of the barrier cross section A by using the values of T_o and T_1 from the evaluation of the experimental data. If additional realistic assumptions are made about the value of the dielectric constant, the self consistency of the fitting parameters can be roughly checked within Sheng's model. Table 2 gives the values of Λ for sample I for different barrier cross sections A by assuming $\varepsilon = 1$. The values used for the barrier cross section are based on the widespread assumption that the highly conducting regions within the polyacetylene sample can be identified with the fibrils and the potential barriers with the interfibrillar barriers. In this picture, the barrier cross section could obtain maximum values of of about one fibrillar cross section. It is found that Λ strongly differs from the value of 0.07 that was implicitly used in the calculation by assuming parabolic barriers. Λ even exceeds 0.25; Sheng, however, notes /52/ that $\Lambda > 0.25$ implies the existence of negative barrier heights, a phenomenon that does not make sense in the context of the fluctuation induced tunnelling model.

Table 2: The application of Sheng's model to highly conducting polyacetylene does not lead to reasonable values of the image charge correction parameter Λ.

A	V_o/meV	w/nm	λ for $\varepsilon = 1$
$\pi \cdot (10\ nm)^2$	14	2.1	10
$\pi \cdot (2\ nm)^2$	50	1.1	5
$\pi \cdot (0.5\ nm)^2$	150	0.6	3

The calculations of Voit and Büttner /55/ are based on the experimental $\sigma(T)$ data presented above /20/. They make no assumption on the barrier cross section. By assuming $\varepsilon = 2.5$, they obtain from the data of the freshly-prepared sample I (table 1) values of $V_o = 265$ eV, $w = 0.015$ nm and $A = 6.5 \cdot 10^{-9}$ nm^2, data which do not appear to be physically reasonable.

It could be argued that ε is the dielectric constant within the barrier, a value which cannot be determined directly, and that it cannot be excluded that ε has values which are by far larger than 1. On the other hand, the above values differ from physically reasonable values by orders of magnitude and a value of about 100 would be necessary for ε in order to obtain plausible values for V_o and w. ε values of the order of 100, however can hardly be imagined within a potential barrier which is not even likely to contain strong permanent electrical dipoles.

On the other hand, this problem of contradictions in context with the parameter Λ only concerns the image force correction of the potential barriers, but not the mechanism of charge transport by fluctuation induced tunnelling itself. From contradictions resulting from the application of equ. 12 it can only be concluded that the implementation of the image force correction on the rectangular barriers might not be allowed or needs modification. This, however, does not necessarily have consequences for the applicability of the mechanism of fluctuation induced tunnelling to highly conducting polyacetylene.

5.4.3 Possible Temperature Dependence of σ_∞

A further problem results from Sheng's approximation to regard the function $\sigma_\infty(T)$ as a constant. Sheng assumes that $\sigma_\infty(T)$ is a function, whose temperature dependence is negligible as compared with that of the exponential function $\exp[-T_1/(T+T_0)]$ /52/. For Sheng's most important model systems, PVC composites, this is an agreeable approximation due to the high value of $T_1/T_0 = 17.0$ /50-52/; this , however, is not necessarily the case if the exponential function itself is a function slowly varying with the temperature, like in the case of highly conducting polyacetylene, where $1.4<T_1/T_0<2.0$.

5.4.4 Paasch's Approach

A modified approch on fluctuation induced tunnelling was presented by Paasch et al. /56/. In contrast to other models, Paasch makes the same basic assumptions as Sheng does and considers the thermal fluctuations as vital for the charge transport. However, by a reexamination of Sheng's model and by using a modified calculation, he avoids the contradiction arising from the implementation of Sheng's formulae (Λ-parameter, see above). By applying his model to the data of high-σ polyacetylene shown in figs. 4 and 6 /20-24/, he obtains values between 0.4 eV and 4.5 eV for the barrier heights and between 0.2 nm and 20 nm for the barrier widths. In contrast to his model, the phenomenological model described below does not necessarily assume an important contribution of fluctuations on charge transport. Its applicability does not depend on how strong these thermal fluctuations are in an intermediate temperature range for a given system, as it uses only those temperature limits for evaluation, in which the influence of fluctuations on charge transport is negligible.

5.5 Evaluation within a Phenomenological Model

Due to the reasons mentioned in the above chapter, it is not clear if the model of fluctuation induced tunnelling can be applied to highly conducting polyacetylene in spite of the good agreement with the experimental $\sigma(T)$ data. But independent of the assumptions and calculations of Sheng's model, equ. 7 can be regarded just as a formula which correctly describes the temperature dependence of the conductivity of polyacetylene and which contains two important features of potential barriers in general which are independent of the relevance of Sheng's mechanism:

- the thermally activated crossing of potential barriers for the limit of high temperatures and

- a finite conductivity for $T \rightarrow 0$ due to tunnelling.

Thus, equ. 7 is a simple mathematical expression which correctly describes these physically reasonable limits and which in accordance with the experimental data interpolates between them.

As in this case we neither imply that the tunnelling is fluctuation enhanced nor assume the existence of image force corrected rectangular barriers, equs. 13 and 14 cannot be applied. The effective height V_o of the potential barriers in this case is identical with the activation energy of the electrical conductivity in the limit of high temperatures:

$$V_o = k_B \cdot T_1 . \qquad (15)$$

If the additional assumption is made that the conductivity for $T \rightarrow 0$ is a tunnelling conductivity, the following equation is obtained with $\sigma(T=0) = \sigma_\infty \cdot \exp(-T_1/T_o)$ for rectangular barriers:

$$\exp[-\frac{T_1}{T_o}] = \exp[-\frac{2}{\hbar}\sqrt{2 \cdot m \cdot V_o} \cdot w] . \qquad (16)$$

For other shapes of the potential barrier, corrections are necessary which are typically of the order of 1. If the effective mass m of the mobile charge carriers is approximated by the free electron mass, the width w of the barriers can be calculated.

Within this phenomenological model, the microscopic parameters V_o and w can be determined independent of the relevance of the assumptions which the model of fluctuation induced tunnelling makes. Whereas due to $C \cdot \langle U^2_{fluct} \rangle = k_B \cdot T$ the capacitance C of the barriers is of great relevance for the height of the modulating voltage U_{fluct} and therefore for the charge transport in Sheng's model, the value of C is not necessary for a quantitative evaluation in the phenomenological model suggested above. In contrast to Sheng, this model needs no arbitrary additional assumptions about the average effective cross section of the barriers, a quantity which is not easily accessible to experiments, as it is an effective value and as it is not clear where the barriers are located within the sample. The values of σ_∞, T_o and T_1 as determined with equs. 15 and 16 are given in table 1 (see above); they are derived from the experimental data with the strongest (sample I) and with the weakest (sample II) temperature dependence of σ found in freshly-prepared samples. While the T_0 values of the different (freshly-prepared) samples vary between 65 K and 76 K, the T_1 values of the flattest $\sigma(T)$ curves have been up to 26 % lower than those of the steepest. Barrier heights of about 10 meV and barrier widths of 1 - 2 nm are obtained.

T_0 and T_1 have very similar values for the conductivities parallel and perpendicular to the stretching direction, a fact that results in the observed weak temperature dependence of the conductivity anisotropy A and indicates a **common mechanism responsible for the temperature dependence of both σ and $\sigma_{\perp}$**, exhibiting even similar values of V_0 and w for the parallel and perpendicular directions. If we regard the realistic case of a distribution of barrier heights and widths, the sample can be regarded as a statistical network of the different individual barrier resistances. The conductivity of such a network can be calculated by means of a selfconsistent effective-medium theory as described by Kirkpatrick and Bernasconi /57,58/. The application of this theory is not limited to resistances within Sheng's model, but can also be applied within the phenomenological model described above. In analogy to considerations of Sheng /52/, it can be concluded that the temperature dependence of the conductivity of the network is determined by the behaviour of the individual barrier.

5.6 Influence of the Barriers on σ(300 K)

The slope of $\sigma(T)$ decreases monotonically with increasing temperature; for $T \to \infty$, it asymptotically approaches zero. For $k_B \cdot T >> V_0$, $\sigma(T)$ approaches a finite value σ_∞. Assuming a charge carrier concentration of 6% (100%) (related to the CH-unit), values of σ_∞ = 20 000 - 100 000 S/cm correspond to macroscopic mobilities of the charge carriers of μ = 50 - 250 cm^2/Vs (3 - 15 cm^2/Vs). For organic conductors, which typically show mobilities of the order of 1 cm^2/Vs at room temperature, these are remarkably high values, especially as the sample is not a single crystal, but an only partially crystalline polymer film. Assuming free electron mass for the mobile charge carriers, these values correspond to scattering times of the order of $\tau = 10^{-13}$ s. These macroscopic values of μ and τ, however, are only lower boundaries for the corresponding microscopic values.

At room temperature, the barriers do not play the dominant

role in limiting the conductivity. Using the barrier heights determined from the experiments (see above), one finds that at 300 K, 60 - 70 % of the charge carriers are able to cross the barriers due to their thermal energy. If the effect of the barriers as scattering centres for charge carriers with energy $E > V_0$ is neglected, the conductivity of polyacetylene would be increased by only a factor of about 1.5, if there were no barriers at all in the sample. Thus the assumption that avoiding the presence of barriers, e.g. by an improved technique of synthesis, would lead to a considerable increase of the conductivity by about one order of magnitude, is obviously not correct. **The room temperature value of the conductivity is limited by the resistivity not of the potential barriers, but of the highly conducting regions between these barriers, which which are responsible for typically 60 - 70 % of the macroscopic resistance of the sample.** This, in turn, explains that samples with a very good morphology ("single fibre polyacetylene", /13,14/) do not necessarily exhibit high room temperature conductivities.

5.7 The Influence of Phonon Scattering on the Conductivity

The question remains, which process is limiting the room temperature conductivity within these regions. Potential mechanisms are phonon scattering and defect scattering as well as the interchain resistance due to the finite conjugation length of the polyacetylene chains. Experimentally, a conductivity limited by phonon scattering should lead to an increase of σ with decreasing temperature; this, however, has never been observed at any temperature in Naarmann polyacetylene doped with iodine. On the other hand, a superposition of two effects is also possible: an increase of the conductivity within the intact polyacetylene chains with decreasing temperature could be compensated by a simultaneous increase of the resistance of the barriers. In total, a non-metallic $\sigma(T)$ would be possible in spite of the influence of phonon scattering on the conductivity. In this case, two samples with almost identical barrier height V_0, of which one has a σ value at room temperature, which is by a factor of 10 higher than that of the other sample, should differ in the following way: if one assumes the same probability of phonon scattering in the two samples, the lower conductivity of the second sample is due to its increased concentration of potential barriers; the relative contribution of phonon scattering to the total resistance of the sample is therefore smaller than in the highly conducting sample. Thus, $^{d}/_{dT}$ $(\sigma(T)/\sigma_\infty)$ of the highly conducting sample should be smaller (more "metal-like", showing a stronger influence of phonon scattering) than in the less conducting sample.

The experiment was made with two samples differing by more than one order of magnitude in σ_∞ (5 000 S/cm and 64 000 S/cm, respectively), which exhibited only a small difference in V_o (about 0.6 %). However, the above mentioned effect could not be observed.

This indicates that phonon scattering of the charge carriers does not have a significant effect on $\sigma(T)$ at room temperature. This experimental result is supported by considerations of Pietronero /11/ and of Kivelson and Heeger /12/, who argue that in a onedimensional Fermi gas, only $2k_F$-phonons are relevant for scattering processes. For polyacetylene, the authors made calculations within a linear chain model, which lead to $2k_F$-phonon energies of several $k_B \cdot 1000$ K. For the upper limit of the conductivity given by phonon scattering, they obtain values significantly higher than that of copper (σ_{copper}(300 K) $6 \cdot 10^5$ S/cm).

If σ(300 K) is not limited by phonons, defect scattering or interchain processes due to the finite conjugation length could be responsible for the resistivity of the samples. This is in agreement with the observed $\sigma(T)$, which, in extrapolation for $T \to \infty$, asymptotically approaches a constant σ_∞. Whereas in conventional metals a residual resistance is observed at low temperatures, where scattering by phonons does not play a significant role, polyacetylene shows a residual resistance only at sufficiently high temperatures, where the influence of potential barriers on $\sigma(T)$ is not so important ($T > 250$ K). The influence of phonons on $\sigma(T)$ would only be observed at still higher temperatures, at which, however, there is no chemical stability of the samples.

Kivelson and Heeger /12/ note that in one-dimensional systems only the $2k_F$ fourier component of the scattering potential is relevant for scattering processes. This explains that the spatially slowly varying potential of the J_3^- and J_5^- ions, which are located between the polyacetylene chains, is not so important for the scattering amplitude; spatially strongly varying potentials are expected to be much more important as scattering centres. Whether conductivities in excess of 10^5 S/cm can be achieved by a reduction of the concentration of the scattering centres or by an increase of the average conjugation length of the polyacetylene chains depends on whether σ_∞ is limited by defect scattering or by interchain processes, a question on which more information can be derived from an analysis of the temperature dependence of the conductivity anisotropy /23/, indicating that at least in Naarmann polyacetylene, interchain processes play an important role.

5.8 Low Temperature Behaviour and Aging

5.8.1 Describing $\sigma(T)$ with Sheng's Formula

If one tries to fit equ. 7 to the experimental $\sigma(T)$ data of oxygen aged samples, it is found that an almost perfect fit is still possible, though not in the whole temperature range between 0.4 K and 300 K as for freshly-prepared samples, but only between a temperature T_{loc} and 300 K, where $T_{loc} > 0.4$ K. For $T < T_{loc}$, $\sigma(T)$ decreases much stronger towards lower temperatures than predicted by Sheng's formula. For both freshly-prepared and oxygen aged samples, T_{loc} separates two temperature ranges with different temperature dependence of the conductivity: the range $T > T_{loc}$, where $\sigma(T)$ can be described with Sheng's formula, and the range $T < T_{loc}$, in which the extrapolation of $\sigma(T)$ for $T \rightarrow 0$ is zero. During oxygen aging, a continuous increase of T_{loc} from 0.4 K to about 30 K is observed. This is shown for the case of a strongly oxygen aged sample in fig. 13. For still stronger oxygen aging, the samples are damaged so much that even for higher temperatures a good fit to Sheng's formula is no longer possible: deviations of the order of several percent occur.

Table 1 gives the barrier parameters of two oxygen aged samples in comparison with the parameters of the same samples immediately after preparation; the parameters have been obtained by fitting $\sigma(T)$ to equ. 7 and subsequent evaluation within the phenomenological model suggested in chapter 5.5. It is found that for sample II the aging process led to a decrease of the conductivities σ and $\sigma_\perp$ by a factor of 4.1 and 4.2, respectively, at room temperature, while V_0 and w increased by about a factor of 2. This increase of barrier height and width is plausible: oxygen aging leads to a rupture of double bonds of the conjugated polyacetylene chain, as can be derived from optical experiments /23/, a process that leads to the generation of non-conjugated chain segments, which, in turn, are not electrically conducting. If those reactions occur at the boundaries of a (non-conducting) potential barrier between two highly conducting regions, the corresponding chain segments are transformed from conducting to non-conducting material, and the barrier increases in width.

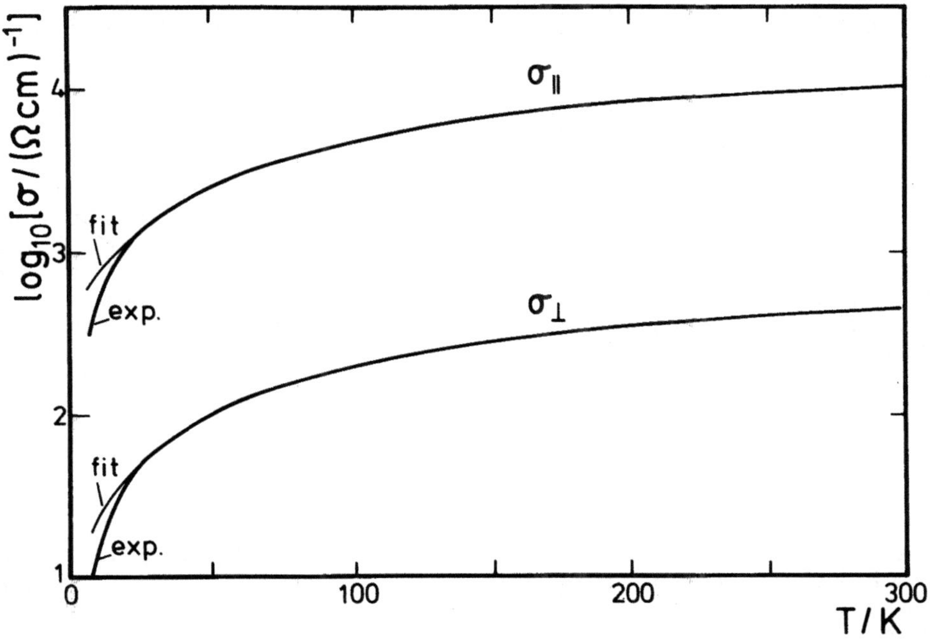

Fig. 13: σ (T) and $\sigma_{\perp}$(T) of an oxygen aged sample (data of sample I in table 2) and the fit to Sheng's formula.

5.8.2 Influence of Finite Chain Lengths

The nature of the mechanism that leads to the low-temperature deviations of the experimental curve from the Sheng fit is yet unknown. One possible explanation is based on the idea that the areas of metallic conductivity are of finite size /23/. As polyacetylene with oriented chains can be regarded as a quasi-onedimensional system, the consequences of this fact on the conductivity are discussed in this and the following chapter for the one-dimensional and the three-dimensional case.

If due to the finite size of the metallic system, only a finite number of states contribute to an energy band, we get a finite level spacing between neighbouring energy levels within a band. If this energy difference ΔE of energetically neighbouring states at the Fermi level with decreasing temperature becomes smaller than $k_B \cdot T$, we no longer have a free mobility of the charge carriers. Below a temperature of

$$T_{loc} = \Delta E/k_B \qquad (17)$$

localization of the charge carriers occurs. In the following, we want to give an estimate if this effect could play a significant role for the conductivity of highly doped polyacetylene at low temperatures. As the band structure of polyacetylene is highly anisotropic, the polyacetylene chain shall be regarded as a one-dimensional system at first. In this estimate we assume that the number of states contributing to a band is identical with the number of CH units of the chain. Then we have:

$$\Delta E = \frac{1}{2 \cdot m} \cdot \left(\frac{\hbar \cdot \pi}{d} \right)^2 \cdot \frac{\alpha}{N} \qquad (18)$$

were the effective mass of the electrons is approximated by the free electron mass and $\alpha = 0.5$ for a half filled band (each CH unit contributes one π electron). The average distance between neighbouring carbon atoms is 0.14 nm. Now, ΔE can be calculated according to equ. 17 from T_{loc}, which, in turn, can be estimated from the $\sigma(T)$ results. We obtain:

$$N = \frac{1}{2 \cdot m \cdot k_B \cdot T_{loc}} \cdot \left(\frac{\hbar \cdot \pi}{d} \right)^2 \qquad (19)$$

For $T_{loc} = 0.4$ K (freshly-prepared sample) conjugation lengths of about 550 000 CH units are calculated, a value that corresponds to a length of about 50 μm of the conjugated $(CH)_x$ chains. For strongly oxygen aged samples we would still have conjugation lengths of 7500 CH units. The effect of oxygen aging qualitatively agrees with what we expect, i.e. we have a rupture of double bonds that decreases the conjugation length. However, the conjugation lengths calculated here appear to be by two or three orders of magnitude too large as compared with estimates derived from Raman results on less conducting samples.

For strictly onedimensional systems, this effect of localization due to level spacing would therefore be too large to explain the experimentally observed low-temperature conductivities with reasonable values of the conjugation lengths. We conclude that the approach of treating polyacetylene as a strictly one-dimensional system is not allowed and that three-dimensional coupling is important. For the three-dimensional case we obtain:

$$T_{loc} = \frac{\hbar^2}{2 \cdot m_o \cdot k_B} \cdot (3 \cdot \pi^2 \cdot n)^{2/3} \frac{1}{N^{2/3}} [(N+2)^{2/3} - N^{2/3}] \qquad (20)$$

n being the microscopic value of the number density of CH groups in the intact material. For N = 1000, 100, and 10 states contributing to the three-dimensional conduction band, T_{loc} values of about 1 mK, 10 mK and 100 mK are obtained. This means that in the case of an isotropic three-dimensional conduction band, localization due to the finite size of the metallic regions cannot explain the experimentally observed values of T_{loc}. Within this localization model, the experimental data support a strongly anisotropic band structure, in which, however, threedimensional coupling plays an important role.

5.9 Conclusions

The experiments described above indicate that in highly conducting polyacetylene, the experimental $\sigma(T)$ dependence can be separated into two temperature regimes: whereas below a temperature T_{loc}, localization effects lead to a temperature dependence that extrapolates to zero conductivity for $T \to 0$, $\sigma(T)$ can be explained as the effect of potential barriers between highly conducting regions for $T > T_{loc}$ for which barrier heights of V_o = 9 meV ... 13 meV and barrier widths of w = 1.5 nm ... 1.7 nm have been derived from the data of freshly-prepared samples within a phenomenological model. Above T_{loc}, σ is limited by these barriers at lower temperatures, whereas at room temperature, it is mainly determined by the resistance within the highly conducting regions. Treating the samples with oxygen leads to the rupture of chemical bonds, thus reducing the conjugation length of the polyacetylene chains. This has two different effects on $\sigma(T)$: on the one hand, the potential barriers between the highly conducting regions are increased in both width and height, and on the other hand, the localization temperature, which is about 0.4 K for freshly-prepared samples, increases monotonically with increasing degree of aging to values of up to 30 K.

To get samples with a significantly higher conductivity, it is necessary to reduce the resistivity of the regions between the barriers. This should be possible, as a comparison with less conducting samples indicates that phonon scattering does not have a strong influence on σ(300 K).

6 Morphology and Charge Transport

6.1 SEM on Freshly-Prepared Samples

6.1.1 Sample Preparation for SEM

The SEM (Scanning Electron Microscopy) studies /40/ have been performed with a JEOL SM 840A microscope, using undoped samples. Sample preparation is illustrated by fig. 14.

The film was transfered from an inert atmosphere (N_2) directly into liquid nitrogen. There the surface was notched with a razor blade to yield a starting point for subsequent treatment (fig. 14a). Using a fine tweezer, a strip of polyacetylene was removed from the bulk sample, applying as little stress as possible, thus uncovering the internal structure (figs 14b and c). This method has been developed since it turned out that otherwise, a modification of the underlying structure is very likely, even under liquid nitrogen, while with the above precautions only negligible artefacts were produced. As the sample exhibits a very low electrical conductivity of typically 10^{-10} S/cm without doping, a layer of either 6 to 7 nm Au or approx. 15 nm C was then sputtered or evaporated, respectively, on the bulk film and the strip, or both methods were combined when necessary for highest resolution micrographs.

6.1.2 Results

a) Surface and Interior of the Films

Due to the catalyst system used in abundance /18/, the surface of the resulting polymer-film is extremly dense /60,61/, showing no structure except for wrinkles and fissures caused by post-polymerization handling. For electron microscopic studies, this is a new situation, since in earlier investigations on the morphology of polyacetylene, the fibrillar structure of the samples was either observable on the surface of the film /59/, or extremely thin layers were polymerized directly onto transmission electron microscope grids /59,62/. We decided not to do the latter, since we wanted to study the morphology of those films which we used for our conductivity measurements (i.e. films of approx. 2-10 μm thickness and lateral dimensions of about 1 by 2 cm). Furthermore we were interested in stretched samples since the procedure of elongation significantly increases the conductivity (/18,23,62/, chapter 4.4). We observed a dependence of the degree of orientation of the fibrils on

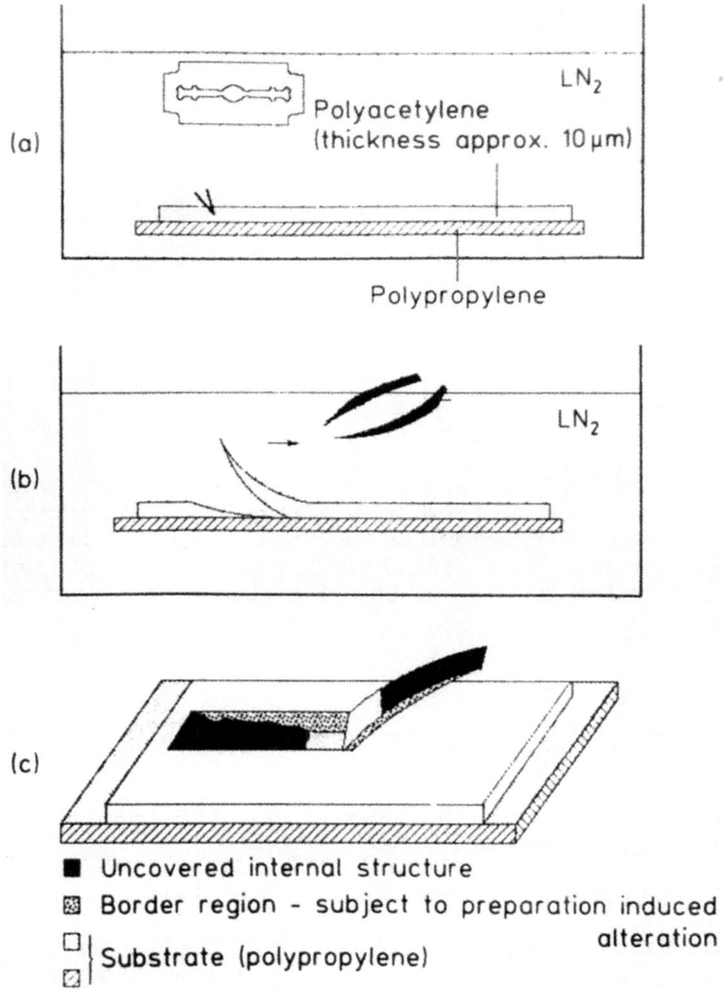

Fig. 14: To study the interior of the polyacetylene-films, the surface was removed by a two-step procedure carried out under liquid nitrogen. a) The surface gets slightly notched with a razor blade. b) The obtained starting point is used to remove a strip of the surface with a tweezer, uncovering the internal morphology. c) Demonstration of the investigated areas. The border regions show artificial structures due to the applied stress and were therefore discarded /40/.

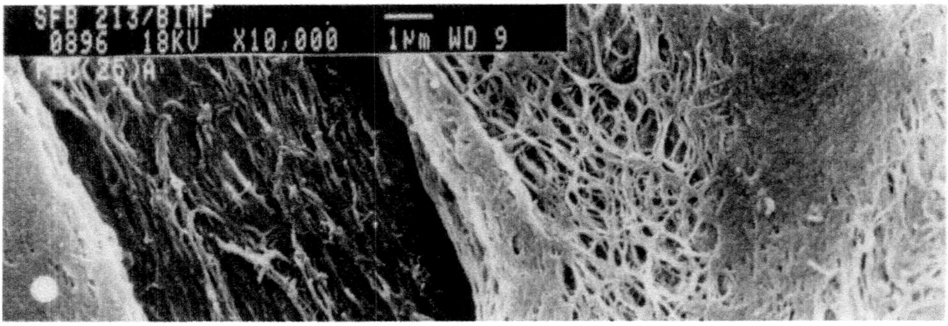

Fig. 15: Film stretched by a factor of 5 and prepared according to fig. 14, thus uncovering three different regions of the sample with different morphologies: the surface of the film is dense and homogeneous whereas immediately below the surface the fibrillar structure of the sample becomes discernable. Fibrils are observed in regions deeper in the bulk of the sample, too, but here a more pronounced orientation of the fibrils is visible,in sharp contrast to the morphology directly beneath the surface /40/.

the depth of the layer, i.e. the distance from the sample surface investigated (fig. 15).

Close to the surface, it appears that the fibrils have almost random directions and are smeared out and molten together, thus yielding a compact and dense surface layer with a rather homogeneous appearance (fig. 15).

The trough in fig. 15 shows a layer of fibrils within the bulk of the sample, which is apparently much more oriented than layers close to the surface, suggesting that alignment by elongation is much effective in interior layers of the film than on the surface, a result which is also supported by a comparison of anisotropic conductivity measurements and polarized optical reflectance experiments /23/ on Naarmann polyacetylene.

b) Stretching

From investigations performed using a stretching stage, we deduce three basic regimes during the process of elongation /23,40/.

- When starting the procedure of stretching, the fibrillar

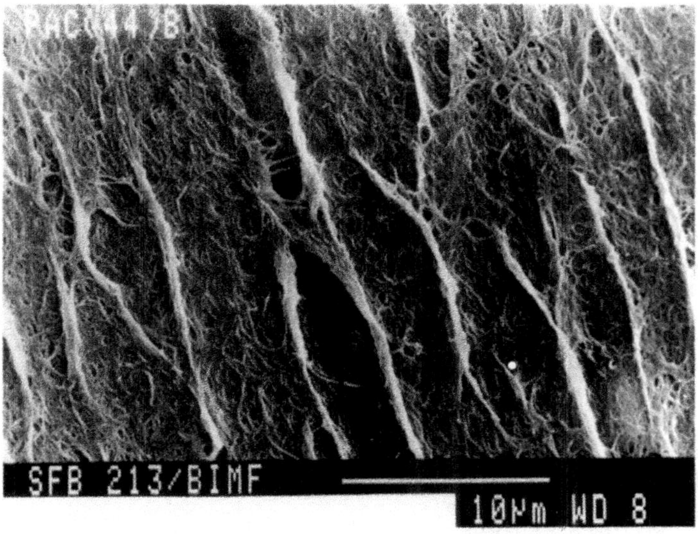

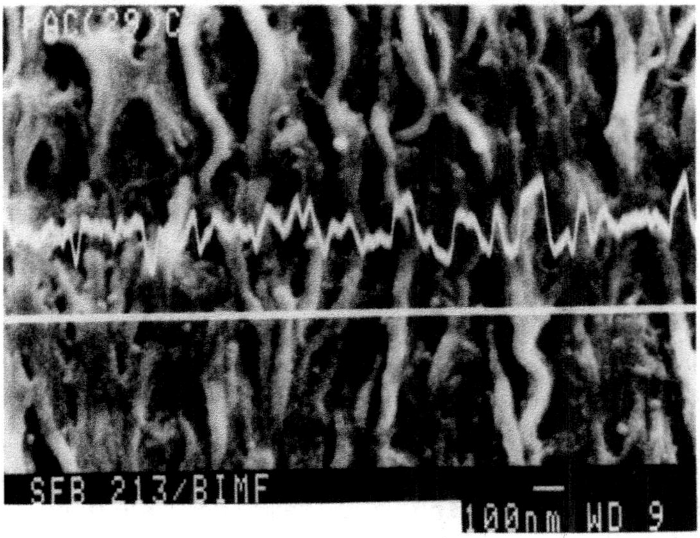

Fig. 16: a) Typical interior of a sample with type I morphology (stretching ratio 5:1). The most prominent feature are thick strings aligned parallel to the direction of elongation. The fibrils forming the tissue between the strings show no significant orientation. b) Fibrillar structure of a type II sample (stretching ratio 6.5:1) as viewed with high magnification. The sample did not show thick bundles of intertwined fibrils. The orientation is much better than in the previous micrograph. The diameter of the fibrils is determined by a y-modulated linescan /40/.

network gets strained, resulting in an orientation of the inner layers of fibrils, thereby causing a plastic deformation of the sample (and the network) up to a stretching ratio of typically 4.

- Continued elongation results in a higher packing density of the fibrils, but no additional fibrillar orientation.

Stretching in this regime is only possible due to sliding of different fibrils against each other.

- When exceeding a stretching ratio of 6 or 7, small cracks form on the surface and the film is torn apart on further stretching. During all stages the typical diameter of the fibrils remains constant within the accuracy of the SEM determination, having values of typically 50 nm to 100 nm.

c) Typical Morphologies

Two different types of morphologies are present in different sets of samples, the morphology being uniform within one set. Fig. 16a shows the typical (internal) morphology of type I samples:

Thick strings are well aligned parallel to the stretching direction (fig. 16a), whereas the tissue between them consists of fibrils showing no significant orientation (fig. 16b). These bundles are a common feature of all investigated samples of this type. Some samples exhibit a different morphology, type II. In these cases, no bundles can be detected. Higher magnification shows well aligned fibrils (fig. 16b) with diameters ranging from 50 nm to 100 nm.

d) Relationship between Morphology and Conductivity

When comparing the conductivity and the morphology of samples stretched by the maximum ratio of 6.5, one feature is especially striking: Samples showing very low conductivities upon doping with iodine (σ = 2 000...3 000 S/cm) exhibit thick bundles of intertwined fibrils after stretching, whereas samples with highest conductivities (σ = 80 000 S/cm) show a much higher degree of orientation. One possible explanation is that the formation of these strings might be due to a considerable amount of interfibrillar crosslinks (fibrils interconnected during polymerization), single crosslinks possibly being the starting point for the formation of the observed strings.

e) Influence of the Stretching Ratio

Both, fibrillar alignment and conductivity (see chapter 4.4) are influenced by the elongation ratio in the same way. The degree of orientation of the fibrils increases as the elongation proceeds, saturating at about a factor of 4.

6.2 Local Density and Bulk Density

The bulk density of Naarmann polyacetylene prepared by the ARA method (chapter 2.2, /16/) is about 0.9 g/cm^3 /20,21/. To find out whether the morphology observed at the cleaved interfaces used for our SEM studies is compatible with the typical bulk properties of the samples, a quantitative comparison of the local density of the sample in these areas with the bulk density is of interest. To get a rough estimate of the local density, we applied line scans(see e.g. fig. 16b) and put the sum L of the line segments l_i occupied by fibrils, in proportion to the total length of the line (fig. 17).

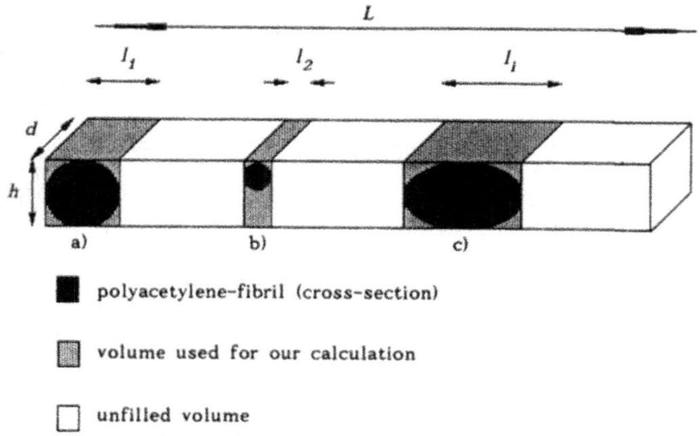

Fig. 17: Model employed for an estimate of the filling factor from our micrographs. Three typical situations are illustrated (from left to right): a) fibril going perpendicular to the scanning direction, b) small fibril from a different layer, contributing to the filling factor and c) non-aligned fibril within the scanned volume.

The result corresponds to a model where the fibrils are taken as ashlars of uniform height h and depth d, and is an upper boundary for the filling factor f_{SEM} of the sample at the regions seen in the SEM studies, since

- the cross section of the fibrils is actually less than the one used, (i.e. $d^2 \cdot \pi/4 < d^2$),
- only the diameters of fibrils perpendicular to the scanning direction are correct, the others being attributed a value which is too high and finally,
- not only fibrils within one layer are seen with this technique, but deeper layers contribute to this estimate of the filling factor as well.

On the other hand, we can calculate the average filling factor, f_{bulk}, of the samples by using a fibre density of 1.2 g/cm^3, which is typical for polyacetylene prepared by Shirakawa and Luttinger catalyst systems /31/; we chose this value for as an upper boundary, since e.g. Durham-route polyacetylene /14/ exhibits a significantly lower density of 1.06 g/cm^3. In this way we obtain $f_{bulk} = 0.75$, thus getting a lower bound for the filling factor, which must be compared with $f_{SEM} = 0.53 - 0.59$ obtained from our micrographs, the lower value being typical for less conducting, the upper value for highly conducting polyacetylene. Using more realistic values for the cross section of the fibrils and the fibre density, we obtain $f_{SEM} = 0.45 - 0.51$ and $f_{bulk} = 0.85$, respectively. The local density of the areas investigated by SEM is therefore lower than the bulk density, the discrepeancy definitely being larger than the error bar of our results, since we made assumptions to minimize the difference (see above).

This means that the mass density is inhomogeneous within the interior of the film. In the areas, which are readily cleaved, it is systematically below the average obtained by bulk measurements, a result, which is not surprising, since less dense areas should more readily be cleaved than more dense regions, as less energy is required for cleaving.

6.3 Geometrical and Electrical Anisotropy

An upper boundary for the mean effective misorientation $\alpha(max)$ of the fibrils can be calculated from the anisotropy of the conductivity. Making the trivial assumption that the molecular chains and crystallites within a fibril have no better alignment to the macroscopic direction of elongation than the fibril itself, the idea is as follows: Let α be the angle between fibril and elongation direction and $\sigma(0)$ the conductivity parallel to the fibril. Charge transport along a fibril then automatically leads to a conductivity $\sigma = \sigma(0) \cdot \cos(\alpha)$ parallel to the stretching direction and

$\sigma_{\perp}=\sigma(0)\cdot\sin(\alpha)$ perpendicular to it. To calculate σ(max), the microscopic anisotropy of the conductivity is - hypothetically - taken as infinity, i.e. the microscopic conductivity perpendicular to the fibril is assumed to be zero, and the finite value of A is regarded as being solely of geometrical origin. In the case of threedimensional misorientation (which is realistic for samples which are thick as compared with the diameter of the fibrils), arbitrary angles of rotation φ of the fibril around the preferential direction given by the angle α are possible, whose probability of ocurrance is independent of φ. Since the diameter of the fibrils typically is between 50 nm and 100 nm, and the thickness of the film was by about a factor of 20 - 200 (typically 2 - 10 mm) higher, one has to consider the threedimensional case. Using A=25 (a typical value obtained for the σ anisotropy of the sample of fig. 16b and similar highly conducting samples) yields $\alpha(\max)<3.6^{\circ}$ in the threedimensional case /40/.

To be able to compare these results with the SEM micrographs, we determined the geometrical anisotropy of the highly conducting samples by digitizing the paths of individual fibrils. The data obtained have been differentiated to yield the angular deviation of the fibrils from the stretching direction.

Samples with a macroscopic misorientation as low as the one predicted above (3.6°) have not been found yet; as demonstrated, good alignment of the fibrils was observed only locally. But this is not necessarily a matter of preparation technique. Rather, this indicates the existence of regions within the sample with a significantly higher orientation than that of the interfaces where the sample is readily cleaved. The fact that the conductivity anisotropy is higher and not lower than the upper boundary for A derived from SEM, shows that the larger part of the electrical current passes through the more highly aligned parts of the sample. The anisotropy could not be explained if the samples were build up homogeneously. This means that not all volume elements contribute to σ in the same way; thus the samples are heterogeneous, containing both regions of high and of low conductivity. One has to conclude that σ is a weighted average, with highly conducting parts having a higher weight in the overall conductivity than the less conducting ones. The comparison of the geometrical and electrical anisotropies indicates that the highly conducting regions are also the highly oriented ones, a conclusion which is also supported by the results of the density measurements (see above, chapter 6.2).

6.4 Influence of Oxygen Aging and Iodine Doping

As described above, it was necessary to apply a special procedure of preparation to conserve the original form of the fibrillar structure. This procedure was applied either before or after O_2 or I_2 oxidation. In addition, it was possible to put the samples in an O_2 atmosphere during SEM investigation. For this purpose a nozzle was built, allowing in situ observation of the influence of O_2 on the fibrillar morphology. In the following, unless explicitly stated, "preparation" refers to preparation for SEM, i.e. uncovering of the internal structure (tweezer technique) and coating with Au.

6.4.1 Oxidation by Oxygen ("Aging")

a) Preparation After Oxidation

The mechanical and morphological properties of polyacetylene depend strongly on the time of exposure to O_2. Aging for approximately one week leads to very brittle samples which are hard to cleave in order to reveal the inner structure. After long periods of aging, the orientation and the morphological information of the former fibrillar structure appears to be lost, whereas intermediate degrees of aging (about 24 h) lead to a "string of pearl" structure (fig. 18).

Even though the individual fibrils appear to have obtained a granular structure, information about the degree of orientation of the fibrils can still discerned. The question is still open where such drastic changes in morphology could come from. To get more information, samples have been prepared according to fig. 14 before applying oxygen.

b) Preparation Before Oxidation

We started by taking a freshly-prepared sample for SEM investigation to make sure the fibrillar structure was intact. We used low acceleration voltages to be able to study undoped and uncoated samples without getting trouble by charging effects. This was necessary because we performed in situ oxidation of the piece of sample under observation via a nozzle (see above). The sample was kept under observation at regular intervals during 8 hours of oxidation. Since no change was observed, we put the sample from vacuum into ambient atmosphere. There it was stored for about two months. Weekly inspections showed no

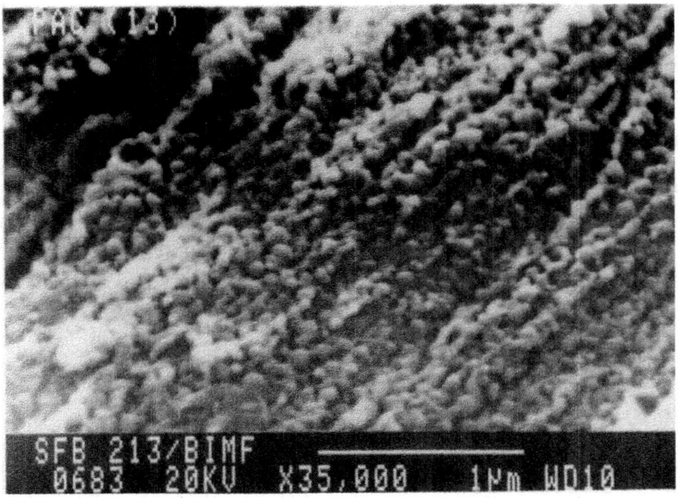

Fig. 18: Shortly (up to 24 h) aged material shows a granular, "string of pearl" like structure. Pristine fibrillar morphology and orientation are conserved. In this case, these features turn out to be artefacts introduced by preparation.

significant change in the fibrillar morphology within the resolution available. This suggests that the morphological changes observed above have been induced due to the preparation of the O_2 aged samples, possibly due to the change in material properties during oxidation.

6.4.2 Oxidation by Iodine ("Doping")

To exclude any possible influences due to the interaction of O_2 with polyacetylene and/or I_2, care was taken to avoid any exposure of the samples to air. No metal coating was applied since the value of the macroscopic conductivity of the sample was sufficient for a direct investigation. To overcome the problems associated with the high vapour pressure of iodine at room-temperature and $p = 2 \cdot 10^{-6}$ mbar (loss of σ and contamination of the microscope), a commercially available cryo-stage was used, where the sample was kept at 80 K.

a) Preparation After Oxidation

During the doping process, which lasted about 30 minutes, no obvious change in material properties like increased brittleness as in the case of O_2 aging was observed. The doped samples showed no change in their fibrillar morphology. Preliminary results indicate that doping, i.e. uptake of I_2, is quite efficient in a relatively small number of fibrils, whereas the dopant concentration is much lower for the major part of the sample volume. Furthermore, doping seems to change the elasticity of the sample on a microscopic scale: the fibrils are more easily broken and torn off.

b) Preparation Before Oxidation

To check for the influence of preparation on the morphology, samples have been investigates, which have been prepared according to fig. 14 before doping. A typical result of this order of preparation is shown in fig. 19. The major difference to the results just described is the clearly discernable "string of pearl" structure, which is similar to that obtained by preparation of the sample after oxidation with O_2. An important difference is the fact that in the latter case, this structure turned out to be induced by preparation.

This morphological change could be due to an increased interaction of the fibrils with the iodine, as, contrary to the experiments described above (a), the absence of the densely packed surface layer of the film could give rise to a larger and faster penetration of the iodine into the interior of the sample. The diameter of individual "pearls" is the same as the diameter of the fibrils where they are contained, but in some cases ellisoidal bumps with lengths of up to 500 nm could be observed. The average diameter of the fibrils and its distribution turn out to be the same as in the former case. To check for the influence of the solvent CCl_4 on the morphology during the 30 minutes solution doping process, undoped polyacetylene was treated for 30 minutes with pure CCl_4 and dried afterwards. No influences on the morphology could be detected in this case.

6.4.3 Charging Effects

Even though pieces of samples with room temperature conductivities between 20 000 and 80 000 S/cm were used, SEM imaging was only possible for low currents and voltages

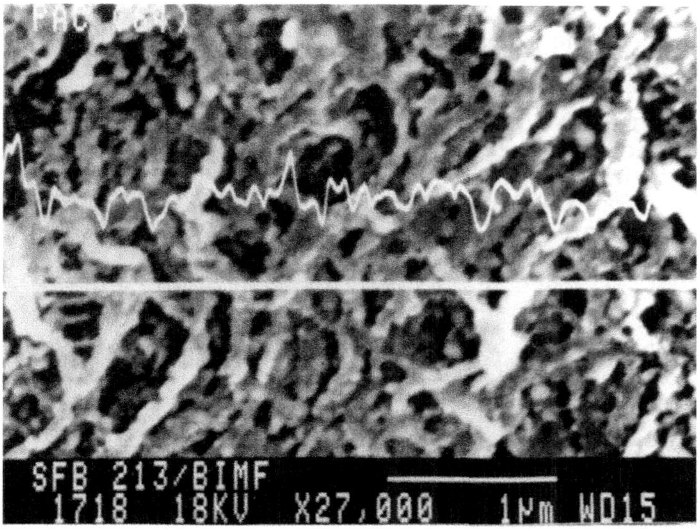

Fig. 19: Polyacetylene film prepared prior to doping with iodine. The granular "string of pearl" structure is clearly visible.

due to charging effects, if no metal coating was present. In general, charging effects are observed in SEM, when the charges applied to a sample spot via the electron probe cannot, due to insufficient electrical conductivity of the sample, be removed fast enough. When returning to such a charged spot scanned in the previous cycle, the electron probe is deflected by the space charges present and the electron multiplier detects an increased yield of secondary electrons. This leads to a distorted picture on the screen, the charged regions being very bright as compared to sufficiently conducting regions of the sample.

Since our preparation ensured maximum doping with iodine leading to macroscopic conductivities between 10^4 and 10^5 S/cm at the time of the SEM investigations, charging effects should not be expected, especially as the probe current of $I_0 = 6 \cdot 10^{-10} - 3 \cdot 10^{-11}$ A was low enough even for much less (approx. 1 000 S/cm) conducting samples. Nevertheless, charging effects have been found for all doped, uncoated samples investigated. Though no correlation between the regions where the charging was observed and their morphology could be established, charging took place at several spots of the sample surface, the fraction of charged versus uncharged surface area varying between approx. 10 to 30% This could not be explained if the macroscopic conductivity were homogeneous within the whole sample. Iodine depletion as a reason is unlikely, as a cryo-stage was used and as conductivity measurements indicate no significant loss in σ at these temperatures even under high vacuum. In addition, a lack of iodine near the surface could be excluded by energy dispersive X-ray analysis of the surface and element distribution images derived from them. Within the lateral resolution of 1 μm of this technique, no inhomogeneity of the iodine distibution in the surface layer (depth approx. 1 μm) could be detected. A more detailed analysis of these effects /64/ supports the model of a heterogeneously conducting material.

6.5 TEM on Individual Polyacetylene Fibrils

6.5.1 Experimental

Again, preparation began with the uncovering of the fibrillar interior of the films. For extraction of single fibrils we chose those films that were actually used for our σ measurements to obtain a direct comparison between theses measurements and the data obtained by inspection of the morphological details. After uncovering the fibrils, the film was put into a nitrogen atmosphere at 170 K. A standard TEM net (material: Cu, width of the meshes: 45 μm) was then put into liquid nitrogen, transfered onto the fibrillar surface and taken off after one or two seconds. The net thus prepared was then transfered in the TEM-chamber and investigated using minimal dosis focussing (see below). For the TEM studies a Zeiss CEM 902 instrument was used. The whole preparation and the sample transfer have been carried out in an inert atmosphere. During the measurement different reasons for a possible damage of the sample have to be considered. Contamination of the sample was not likely to occur since we used a baffle, low beam current intensities and observation times of less than 10 s at one site. With respect to object heating an estimate for the increase of temperature in compact material /64/ shows that critical values are never encountered since the micrographs presented below were checked with probe currents never above 10^{-11} A, yielding a temperature increase below 0.03 K, assuming all beam energy were deposited in one fibril. Thus, thermal effects could be excluded. The influence of radiation damage is crucial in the highly resolution images with regard to the interpretation of the internal structure of the fibrils. Taking 3 mm as the diameter for the illumination of the sample, 3 s exposure time for photography and 10^{-10} A probe current, the charge density is $4.24\cdot10^{-3}$ C/cm^2.

To make sure we observed the undamaged material, we used typically 10^{-11} A probe current and mimimal dosis focussing. This is accomplished by deflecting the electron beam above the sample and leading the beam back to the optical axis beneath the sample. When taking a photograph, the deflection is switched off. Thus it is possible to adjust the instrument at an uninteresting region of the sample and take micrographs at spots not strained beforehand. Furthermore, systematic studies on the deterioration of the samples due to radiation damage were made. The micrographs presented here are selected for showing unaltered, genuine structures of a pristine sample.

For doping, the fibrils collected on a TEM mesh were transfered to a N_2/I_2 atmosphere for periods between one and ten minutes. The oxydized samples were then frozen by shock in liquid propane to stop possible further diffusion of the iodine within the fibrils. This also ensured that water possibly adsorbed on the fibrils was frozen amorphously and could not disturb the results by forming ice crystals.

Fig. 20: Coarse pieces of the fibrillar surface of a polyacetylene-film, collected on a standard TEM net.

6.5.2 Results

a) Undoped Fibrils

The extraction method presented above generally yields coarse pieces of the fibrillar surface and sometimes also individual fibrils (fig. 20) that cling to the Cu-rods of the net by adhesion or are frozen to the net when using cryo-preparation.

A much better yield of individual fibrils is achieved by using commercially available nets coated by a C/formvar film. It is then possible to remove whole layers of fibrils from the sample (fig. 21).

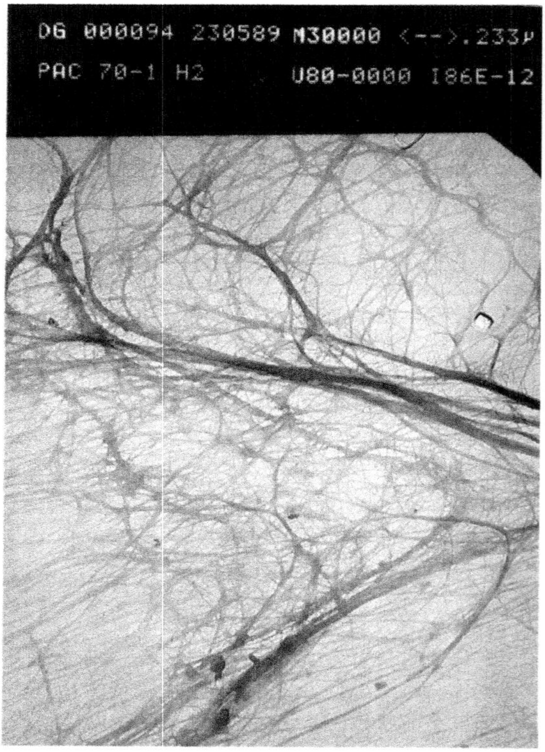

Fig. 21: When using nets covered with a C/formvar film, whole layers can be removed from the fibrillar surface of a polyacetylene film. The magnification in this case is chosen in a way that individial fibrils are visible.

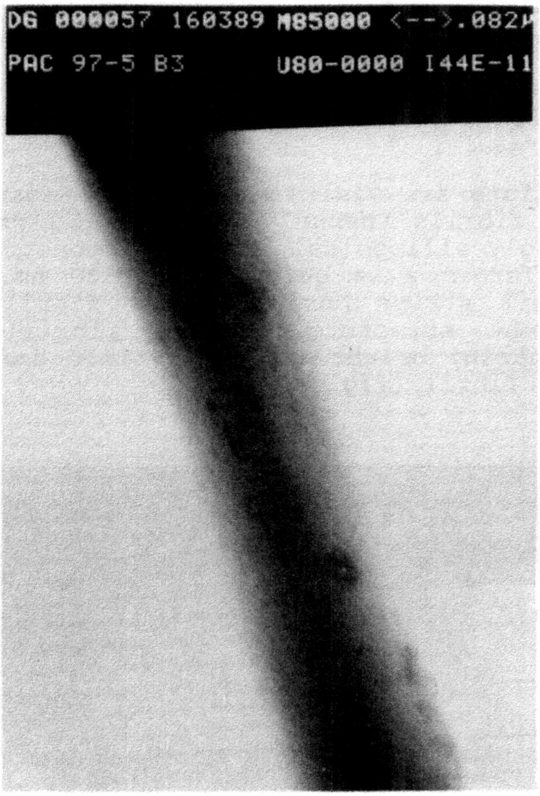

Fig. 22: Part of a fibril with a large diameter (200 nm) prior to doping.

The individual fibrils do not show an internal structure and appear to be homogeneous. Only in some cases inclusions could be detected, indicating minor contaminations of the samples (fig. 22). Oxydation of single fibrils at ambient conditions in air did not lead to any changes in morphology except an increase of the dark spots just described, in accordance with previously performed SEM and TEM investigations of Epstein et al. /67/.

b) Doped Fibrils

Before employing the procedure described above (gas phase doping of fibrils already on a net) we tried to transfer doped fibrils in the same way as the undoped ones by cryo-extraction from a doped sample. This procedure, however, only yields fragments of the sample instead of intact fibrils.

As already pointed out, we solved this problem of fragmentation by doping fibrils already on a net. Short doping times of 1 min to 10 min were necessary, since too much iodine rendered opaque fibrils. During an iodine treatment of approx. 5 min, two morphological details develop (fig. 23a).

Strong constrictions are visible which have about the same diameter as the fibrils themselves (75 - 150 nm). Only in rare cases strongly ellipsoidal bump like constrictions can be seen, their larger axis being about 500 nm. The other feature is a dark stripe bordering the fibril. A closer inspection of the structure of the fibril's edge is possible by comparing bright and dark field images of the same region of a fibril (fig. 23b).

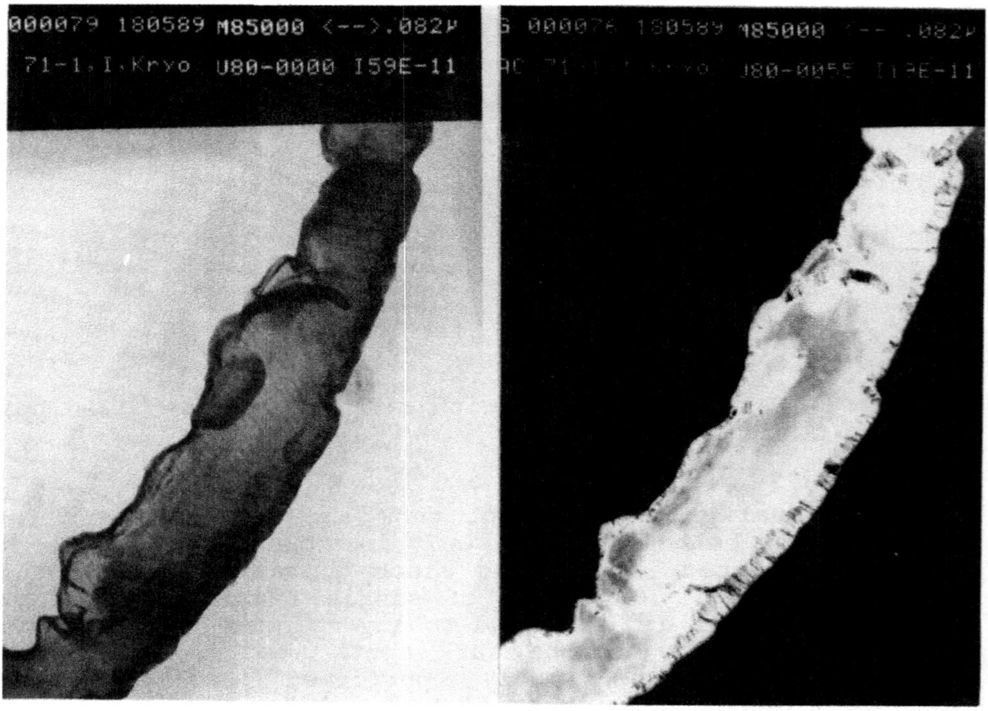

Fig. 23: Left: a fibril doped for approx. 5 min exhibits marked constrictions and a dark border area. Right: dark field image of the same fibril. The stripe-like structure of the border is clearly visible. Dark regions are areas with no iodine present, bright areas are regions with high dopant concentration. The highly doped regions are separated by distances between 1 nm and 3 nm.

The dark field image was obtained by using inelastically scattered electrons for imaging, i.e. electrons with an energy loss greater than 55 eV. This means that regions with high concentrations of I_2 appear bright on the image. One clearly sees that the band like border structure is not confined to the edge of the fibril, but continues across and inside the fibril.

Furthermore, the band exhibits a stripelike structure: regions with low dopant concentration are separated by 1 nm to 3 nm from each other.

7 Conductivity Barriers and Morphology: A Comparison

Different suggestions have been made concerning the nature of possible potential barriers responsible for the temperature dependence of the electrical conductivity, the most common being the assumption that these barriers are due to interfibrillar contacts. However, it has not yet been possible to verify these considerations experimentally.

On the other hand, first information about the nature of these barriers can be derived from the experiments on the temperature dependence of the anisotropy A of the electrical conductivity and its dependence on oxygen aging as described in chapters 4.2 and 4.6. If we assume that the potential barriers determining $\sigma(T)$ act as contacts between different otherwise electrically separated fibrils, we expect that the same barriers have to be crossed for both the electrical conductivities parallel and perpendicular to the stretching direction. This would explain the similar temperature dependence of σ and $\sigma_{\perp}$ (chapter 4.2) and the similar values of the microscopic parameters of the potential barriers (barrier width and height) for different directions (chapter 5.5). In this model, the anisotropy of the conductivity would be of geometrical origin; for the most simple case of negligible intrafibrillar resistances, it would roughly correspond to the length-to-diameter ratio of the fibrils. This, again, in part is in accordance with experimental findings. The anisotropy of σ has only a weak temperature dependence A(T) for freshly-prepared samples, as must be expected if it is of geometrical origin: the fibrillar geometry is largely independent of temperature.

On the other hand, it is difficult to explain the behaviour of A(T) after oxygen aging. While optical experiments /23/ indicate that oxygen treatment results in a reduction of the conjugation length, SEM investigations during the process of oxygen aging (see chapter 6.4.1b) show that this process does not have any influence on the geometry and degree of orientation of the fibrils. In the model of interfibrillar barriers discussed above, we should therefore expect that oxygen aging does not have a significant influence on the conductivity anisotropy and in particular it should not lead to a temperature dependence of A, in contrast to the experimental results, which show a clear temperature dependence of A for aged samples (chapters 4.5 and 4.6, figs. 6c and 9), a behaviour that is more readily explained on the basis of intrafibrillar processes (chapter 5.8, /23/) than as an interfibrillar process.

A different explanation would be the location of barriers between different highly conducting regions within a fibril. This could be due to inhomogeneous doping as well as to the existence of both crystalline and amorphous regions. While the latter needs no further motivation in a partially crystalline polymer, the first is motivated by doping experiments on other types of polyacetylene, which supports a picture of inhomogeneously doped samples /68/.

This idea of heterogeneous doping, in turn, is also supported by the TEM results on Naarmann polyacetylene presented above (fig. 23). In this context, it is interesting that the barrier widths (see table 1) derived from the conductivity experiments (chapters 4 and 5.5), i.e. values between 1 and 3 nm, are in good agreement with the widths of the undoped regions separating the highly doped regions as derived from the TEM results (see chapter 6.5.2). As it is known that undoped polyacetylene is an insulator, whereas doping with iodine leads to high values of σ, an identification of the barriers with the undoped regions and of the highly conducting regions assumed in the barrier models (chapters 5.3, 5.4 and 5.5) with the highly doped regions of fig. 23 is plausible. Whether or not these undoped regions are really identical with the barriers found in $\sigma(T)$, however, needs further experiments and is an object of current investigations.

8 Summary and Outlook

By using a modified technique of synthesis, applying an aged catalyst system at room temperature, polyacetylene samples with room temperature conductivities between 20 000 S/cm and 80 000 S/cm are obtained after mechanical stretching by a factor of 6.5 and subsequent doping with iodine, though some samples exhibit even higher or lower values.

The temperature dependence of the electrical dc conductivity $\sigma(T)$ has been examined between 300 K and 14 mK; for all iodine doped samples, $\sigma(T)$ decreases monotonically with decreasing temperature, freshly-prepared samples still having about 20 % of their room temperature value at 14 mK. Between 300 K and a temperature T_{loc}, $\sigma(T)$ is well described with Sheng's formula. For freshly-prepared samples, T_{loc} has values of about 400 mK; with increasing degree of oxygen aging, it increases monotonically to values of up to 30 K. Below T_{loc}, a different, stronger temperature dependence is found, which leads to a rapid decrease of the conductivity due to oxygen aging mainly at low temperatures.

Sheng assumes fluctuation enhanced tunnelling through potential barriers between highly conducting regions. A critical discussion of Sheng's model shows that its applicability on highly conducting polyacetylene possibly requires a modification of Sheng's calculation. On the other hand, an interpretation of the experimental $\sigma(T)$ data is possible without a detailled knowledge of the influence of the fluctuations and the cross-section of the barriers within a simple phenomenological model which also assumes potential barriers between highly conducting regions. Evaluation of the experimental data yields barrier heights of typically 10 meV and barrier widths of about 1.5 nm for freshly-prepared samples.

Aging with oxygen leads to an increase of both the barrier height and width and to a reduction of the conjugation length of the polyacetylene chains, while no morphological changes are observed.

Although the barriers determine the temperature dependence of σ for $T>T_{loc}$, they are the limiting factor for the conductivity only for lower temperatures ($T_{loc}<T<100$ K), but not at room temperature. Due to the value of the barrier height, the larger part of the electrical resistivity has to be attributed to processes within the highly conducting regions.

The conductivity anisotropy A of highly stretched samples doped to a maximum with iodine is largely temperature independent for freshly-prepared samples and has values between 20 and 30 for highly conducting samples. After oxygen aging, an increase of A towards lower temperatures is observed.

The picture of high-σ polyacetylene as a heterogeneously conducting material, which was derived from the anisotropic conductivity and polarized optical reflectance data, is also supported by scanning electron microscopy studies.

With a special technique, it was possible to extract individual fibrils from the bulk material of the high-σ samples and subsequently study them with transmission electron microscopy at different stages of iodine doping. The micrographs of doped samples show weakly-doped regions of about 1 nm to 3 nm in width between highly doped regions; the values are similar to the widths of the potential barriers as determined from $\sigma(T)$ measurements on doped samples. The question of a possible identification of these TEM structures with the σ-barriers was discussed.

To conclude, high-σ polyacetylene is a highly, but heterogeneously conducting material whose charge transport properties after doping appear to be still determined by sample imperfections and not by the resistance of the intact polyacetylene chains, a situation that allows still higher values of σ on further improvement of sample synthesis.

Acknowledgments

We thank E. Dormann, G. Paasch, G. Denninger and J. Voit for valuable discussions, E. Schliebitz for sample synthesis and J. Gmeiner for doping and sample preparation and W. Schoepe for cooperation in the ultra-low temperature experiments. This work was supported by BASF/BMFT within Project ELP 03 C214-1, by the Deutsche Forschungsgemeinschaft within SFB 213 and by the Fonds der Chemischen Industrie.

Literature

1 A.J. Epstein, "AC Conductivity of Polyacetylene", in: T.A. Skotheim (ed.), "Handbook of Conducting Polymers", Marcel Dekker, New York 1986

2 R.E. Peierls, "Quantum Theory of Solids", Oxford University Press, London, 1955, p. 108

3 J.C.W. Chien, "Polyacetylene - Chemistry, Physics and Material Science", Academic Press, Orlando, 1984

4 H. Shirakawa, E.J. Louis, A.G. MacDiarmid, C.K. Chiang and A.J. Heeger, J. Chem. Soc. Chem. Commum. (1977) 578

5 C.K. Chiang, C.R. Fincher Jr., Y.W. Park, A.J. Heeger, H. Shirakawa, S.C. Gau and A.G. MacDiarmid, Phys. Rev. Lett. 39 (1977) 1098

6 Y.W. Park, A.J. Heeger, M.A. Druy and A.G. MacDiarmid, J. Chem. Phys. 73 (1980) 946

7 Y. Tomkiewicz, T.D. Schultz, H.B. Brom, A.R. Taranko, T.C. Clarke and G.B. Street, Phys. Rev. B 24 (1981) 4348

8 A.J. Epstein, H. Rommelmann, R. Bigelow, H.W. Gibson, D.M. Hofmann and D.B. Tanner, Phys. Rev. Lett. 50 (1983) 1866

9 K. Ehinger and S. Roth, Phil. Mag. B 53 (1986), 301

10 Frommer and Chance, Enc. Polym. Sci. Eng. 5 (1986) 462

11 L. Pietronero, Synth Metals 8 (1983) 225

12 S. Kivelson and A.J. Heeger, Synth. Metals 22 (1988) 371

13 J.H. Edwards and W.J. Feast, Polym. Comm. 21 (1980) 595

14 G. Leising, Polym. Commun. 25 (1984) 201

15 H. Naarmann, Synth. Metals 17 (1987) 223

16 H. Naarmann and N. Theophilou, Synth. Metals 22 (1987) 1

17 N. Theophilou, R. Aznar, A. Munardi, J. Sledz, F. Schue and H. Naarmann, Synth. Metals 16 (1986) 337

18 N. Theophilou and H. Naarmann in: L. Alacer (ed.): "Conducting Polymers: Special Applications; Proceedings of the Workshop held at Sintra, Portugal, July 28-31, 1986", D. Reidel Publishing Company (1987) 65

19 N. Basescu, Z.-X. Liu, D. Moses, A.J. Heeger, H. Naarmann and N. Theophilou, Nature 327 (1987) 403

20 Th. Schimmel, W. Rieß, J. Gmeiner, G. Denninger, M. Schwoerer, H. Naarmann and N. Theophilou, Solid State Commun. 65 (1988) 1311

21 Th. Schimmel, G. Denninger, W. Rieß, J. Voit, M. Schwoerer, W. Schoepe and H. Naarmann, Synth. Metals 28 (1989) 11

22 Th. Schimmel, M. Schwoerer and H. Naarmann, Synth. Metals 37 No.1-3 (1990) 1

23 Th. Schimmel, Dissertation, Univ. Bayreuth (1989)
24 Th. Schimmel, D. Gläser and M. Schwoerer, to be published
25 J. Tsukamoto, A. Takahashi and K. Kawasaki, submitted for publication
26 G. Natta, G. Mazzanti and P. Corradini, Accad. Nazl. Lincei, Sci. Fis. Mat. Nat. 25 (1958) 3
27 W.H. Watson, W.C. McMordie and L.G. Lands, J. Polym. Sci. 55 (1961) 137
28 L.S. Meriwether, E.C. Coltrup, G.W. Kennerly and R.N. Reusch, J. Org. Chem. 26 (1961) 5155
29 M. Hatano, S. Kambara and S. Okamoto, J. Polym. Sci. 51 (1961) 526
30 M. Hatano, Kogyo Kagaku Zasshi 65 (1962) 723, quoted in: [Chi 85]
31 L.B. Luttinger, J. Org. Chem. 27 (1962) 1591
32 W.E. Daniels, J. Org. Chem. 29 (1964) 2963
33 F. Ciardelli, E. Benedetti and O. Pieroni, Makromol. Chem. 103 (1967) 1
34 P.S. Woon and M.F. Farona, J. Polym. Sci. Polym. Chem. Ed. 12 (1974) 1749
35 T. Ito, H. Shirakawa and S. Ikeda, J. Polym. Sci. Polym Chem. Ed. 12 (1974) 11
36 D. Braun and W. Quarg, Angew. Makromol. Chem. 29/30 (1973) 163
37 E.H. Land, J. Opt. Soc. Am. 12 (1951) 957
38 T. Ogawa, R. Cedeno and E.T. Herrera, Makromol. Chem. 180 (1979) 785
39 K. Menke and S. Roth, Chemie in unserer Zeit 20 (1986) 1
40 D. Gläser, Th. Schimmel, M. Schwoerer and H. Naarmann, Makromol. Chem. 190, 3217 (1989)
41 H.C. Montgomery: J. Appl. Phys. 42 (1971) 2971
42 B.F. Logan, S.O. Rice and R.F. Wick, J.Appl.Phys. 42 (1971) 2975
43 J.F. Keithley, J.R. Yeager and R.J. Erdman, "Low Level Measurements", Keithley Instruments, Cleveland, 1984
44 D. Schäfer-Siebert and S. Roth, to be published
45 N.F. Mott, E.A. Davis, "Electronic Process in Non-Crystalline Solids", Clarendon (1979) Oxford
46 W. Brenig, G.H. Döhler and P. Wölfle, Z. Phys. 258 (1973) 381
47 W. Brenig, G.H. Döhler and H. Heyszenau, Phil. Mag. 27 (1973) 1093
48 H. Böttger, V.V. Bryskin, " Hopping Conduction in Solids", VCH (1985)
49 P. Sheng, B. Abeles and Y. Arie, Phys. Rev. Lett. 31 (1973) 44
50 E.K. Sichel, J.I. Gittleman and Ping Sheng, Phys. Rev. B 18 (1978) 5712
51 Ping Sheng, E.K. Sichel and J.I. Gittleman,

Phys. Rev. Lett. 40 (1978) 1197
52 Ping Sheng, Phys. Rev. B 21 (1980) 2180
53 W. Mayr, K. Philipp and K. Seeger, Solid St. Commun. 43 (1982) 117
54 J.G. Simmons, J. Appl. Phys. 34 (1963) 1793
55 J. Voit and H. Büttner, Solid State Commun. 67 (1988) 1233
56 G. Paasch, G. Lehmann and H. Wuckel, [illegible] print.
57 S. Kirkpatrick, Rev. Mod. Phys. 45 ([illegible]) 574
58 J. Bernasconi, Phys. Rev. B 7 (1973) [illegible]52
59 F.E. Karasz, J.C.W. Chien, R. Galkiew[illegible]z, G.E. Wnek, A.J. Heeger and A.G. MacDiarmid, Nature 282 (1979) 286
60 H. Shirakawa, S. Ikeda, Synthetic Metals 1 (1979/80) 175
61 A. Munardi, R. Aznar, N. Theophilou, J. Sledz, F. Schu and H. Naarmann, Eur. Polym. J. 23 (1987) 11
62 G. Lugli, U. Pedretti, G. Perego, Mol. Cryst. Liq. Cryst 117 (1985) 43
63 Y.W. Park, C. Park, Y.S. Lee, C.O. Yoon, H. Shirakawa, Y. Suezaki and K. Akagi, Solid State Comm. 65(2) (1988) 147
64 D. Gläser, Th. Schimmel, M. Schwoerer and H. Naarmann, manuscript in preparation (1990)
65 L. Reimer, Z. Naturforsch. 15a (1960) 405
66 L. Reimer, Z. Naturforsch. 16b (1961) 166
67 A.J. Epstein, H. Rommelmann, R. Fernquist and H.W. Gibson, Polymer 23 (1982) 1211
68 G. Perego, G. Lugli, U. Pedretti and G. Allegra, Macromol. Chem. 189 (1989) 2687

ELECTRONIC PROPERTIES OF HEAVILY DOPED *TRANS*-POLYACETYLENE

S. STAFSTRÖM
Department of Physics, IFM, Linköping University
S-581 83 Linköping, Sweden

ABSTRACT. The structural and electronic properties of heavily doped *trans*-polyacetylene are studied using a total Hamiltonian consisting three terms: 1) the Su-Schrieffer-Heeger Hamiltonian, 2) an external potential term due to electrostatic interactions between electrons of the polymer chain and counterions and charges on neighboring chains, and 3) a term describing the intra-chain electron-electron interactions. The effects of weak disorder in the positions of the counterions are also studied. At all doping levels included in this study, the ground state configuration is a soliton lattice. The energy gap around the Fermi energy and the density of states at the Fermi energy are calculated. A low bandgap system is obatined in the case of weak disorder at doping levels above 8%. The electronic properties of the polaron lattice are also calculated. Previously reported experimental results and theoretical models for the metallic state of heavily doped *trans*-polyacetylene are reviewed and discussed in comparison with the results obtained in this study.

1. Introduction

It is well known today that doping of π-conjugated polymers results in a highly conducting state of the polymer. Due to the one-dimensional character of the polymers, the effects of doping differ considerably from those observed in conventional inorganic semiconductors. The linear chain structure results in a strong electron-phonon coupling, which upon charge transfer from the dopant species to the polymer leads to self-localized particle like defects on the chain [1-2]. Depending on the type of polymer, these defects are termed polarons, bipolarons or solitons. The magnetic, optical and structural characteristics of the defects have been carefully investigated as a function of the doping concentration. The general picture of the effects of doping is the same for all conjugated systems prepared under similar conditions. At low doping concentrations, the conductivity of the polymer increases considerably, additional absorption bands are observed within the optical gap of the undoped polymer as well as in the infrared regime, and except for a reduction of the residual spin

J.L. Brédas and R. Silbey (eds.), Conjugated Polymers, 113–140.

concentration of the undoped polymer, no doping induced spins are observed. These properties persist up to a critical doping concentration (y_c). Above y_c, both the magnetic and the optical properties become metallic. The vibrational properties, however, show the same features both below and above y_c. A number of different models have been proposed in order to explain these intriguing properties of the highly doped state of conjugated polymers. Most of these models are related to *trans*-polyacetylene (*trans*-$(CH)_x$). In this paper, previous studies related to the metallic state of heavily doped *trans*-$(CH)_x$ are reviewed in Section 2. Section 3 describes the theoretical model used by the author in the studies of the metallic state of heavily doped *trans*-$(CH)_x$. The results from these studies are presented in Section 4. Finally, the results are summerized and discussed in Section 5.

2. Models for the Metallic State of Heavily Doped *Trans*-$(CH)_x$.

Trans-$(CH)_x$ exhibits a double degenerate structural ground state. Coexistence of these two states introduces a structural non-linear misfit region between the two phases and an electronic state localized to this misfit region [3]. This excitation is termed soliton [1-2]. Solitons created by means of doping are spinnless and carry a charge ±e. The lack of spin of the soliton is due to the fact that the spin of the charge added to the chain in the doping process (electron or hole) is paired with one spin of the polymer chain. In order for the system to stay spinnless, charged solitons must be created in pairs. The process of creating solitons by means of doping involves, therefore, a breaking of spin-paired electrons in the polymer chain followed by individual pairing of these spins with the spins of the dopant charge.

The electronic state corresponding to the soliton is located in the energy gap between the valence band and the conduction band and is often referred to as the "midgap" state. This terminology is, however, quantitatively not correct since the energy gap in *trans*-$(CH)_x$ is ~1.8 eV and the absorption involving the soliton state appears at ~0.7 eV in the case of solitons created by means of doping [4]. In the case of photogenerated solitons, for which there is no counterion present, the characteristic absorption appears at even lower energies compared to the midgap energy, namely ~0.50 eV [5]. These deviations from the midgap value are due to correlation effects [6] and, in the case of doping, also to the potential of the dopant ions.

It is interesting to study the evolution of the optical spectrum with increasing doping concentration. Up to a concentration of y~4.7 % both the energy gap (π-π^*) and the "midgap" absorption peaks are clearly observed [7]. For the next doping level included in the study, y=7.8 %, the "midgap" absorption shifts slightly towards lower energies and the π-π^* transition disappears. Yang *et al.* [8] have performed frequency dependent conductivity studies in the infrared and optical regime. Also in this case, no signature of the π-π^* transition is observed at high doping levels (y=6.1%). The optical spectrum shows a pseudogap at 0.2 eV and vibrational modes which imply that the carbon-carbon bond length and the charge distribution is not uniform along the polymer chain. Except for the non-uniform distribution of the carbon sites, the features of the optical spectrum for highly doped *trans*-$(CH)_x$ are close to those of a traditional metallic. Metallic properties of highly doped *trans*-$(CH)_x$ are further verified by a high Pauli susceptibility [9-10] and a linear dependence of the thermopower with temperature [11]. In particular the magnetic susceptibility has been studied in great detail. Below y~6%, the susceptibility is essentially zero which is in agreement with the picture doping induced spinless solitons. At the critical doping concentration, $y_c \simeq 6\%$, there occurs an abrupt increase in the Pauli susceptibility. For well ordered samples the transition is sharp and hysteretic and, hence, probably first order [12-13]. For relatively more disordered arrangements of the dopant ions, the transition appears more rounded.

Tanner *et al.* have studied K doped *trans*-$(CH)_x$ [14-15] and found that the oscillator strength of the doping induced IR active (IRAV) modes per dopant molecule is approximately the same both below and above the critical doping concentration. This result indicates that solitons, which are known to be the stable type of defect at low doping levels, remains stable even above the critical doping concentration.

The nature of the metallic state of highly doped conjugated systems is one of the major unresolved problems in the field of conducting polymers. A number of different models have been proposed to rationalize the metallic properties. Mele and Rice [16] have calculated the density of states for *trans*-$(CH)_x$ at different doping levels. Their Hamiltonian includes a potential due to the of counterions, which are randomly distributed. The lattice conformation is that of a soliton lattice with the solitons pinned to the counterions. This conformation is referred to as a disordered incommensurate structure (the regular soliton lattice is referred to as the ideally incommensurate structure). The density of states is calculated with three-dimensional inter-chain couplings included. The inter chain ovelap is adjusted to provide an inter-chain band width of 0.2 to 0.3 eV near the Peierls edges. It is observed that the disorder in the impurity potential and the three-dimensional couplings are of roughly equal importance in surpressing the band gap of the soliton lattice. The disorder effects are shown to lead to a finite density of states at the Fermi energy near y=10%. Including the interchain interactions, a finite density of states at the Fermi energy appears at y~3%. The saturation in the density of states at the Fermi energy, which coincides with the disappearence of dimerization of the plymer chain, occurs in this case around 10% doping. In the model of Mele and Rice, the semiconductor-metal transition is due to a surpression of the incommensurate Peierls gap caused by disorder and three-dimensional couplings.

Another attempt to account for the metallic state by disorder effects is given by Harigaya and Terai. [17]. The Hamiltonian employed for these studies is similar to that used by Mele and Rice. The lattice configuration and the electronic wave functions are determined self-consistently including a complete lattice relaxation of the *trans*-$(CH)_x$ chain. The optimized dimerization order parameter is found to remain large even in the highly doped state. For reasonable values of the impurity potential, the energy gap vanishes around y~12%.

The picture of metallic *trans*-$(CH)_x$ as a disordered incommensurate Peierls insulator is adopted by Yang *et al.* [8] in order to explain their experimental optical conductivity data. Recent *in situ* electron spin resonance experiments on *trans*-$(CH)_x$ during electrochemical doping show, however, that the transition into the metallic state exists and gets sharper for more well ordered and homogeneously doped samples [12-13]. Therefore, even though disorder certainly effects the energy gap and the density of states at the Fermi energy, it seems clear that disorder does not drive the transition into the metallic state.

Kivelson and Heeger [18] proposed that a crossover from a soliton lattice to a polaron lattice could explain the metallic properties of heavily doped *trans*-$(CH)_x$. One electronic band corresponding to the polaron defects is half-filled and therefore contributes the metallic properties of the doped polymer. The mechanism for the crossover from the soliton lattice to the polaron lattice is argued to be due to a lowering of the polaron creation energy with increasing doping whereas the soliton creation energy remains unchanged. Also three-dimensional interchain interactions stabilize the polaron lattice. Furthermore, it is shown that strong attractive interactions exists between electrons in the half-filled narrow polaron subband.

The basic argument against the polaron lattice model is based on the observation that an array of polarons, at a concentration corresponding to y_c, is almost identical to an undimerized lattice [19]. This result is in conflict with the observation that the oscillator strength and the frequency of the doping induced IRAV modes are found to be unchanged during the crossover to the metallic state [14-15]. Tabor and Stafström [20] showed, however, that the polarons become considerably more localized and have a charge density wave similar to that of the soliton lattice if the potential due to the counterions is included. This result is discussed in more detail below.

Another experimental evidence against the polaron lattice model for the metallic state of heavily doped *trans*-$(CH)_x$ comes from Electron-Energy-Loss Spectroscopy (EELS) data [21]. These data show levels spread well across the gap, which is more in agreement with the disordered incommensurate state than with the picture of narrow polaron bands in the gap. Band structure calculations using the Valence Effective Hamiltonian (VEH) technique [22] support this conclusion since it is shown that a large energy gap exists between the polaron bands in the band structure of the polaron lattice. On the other hand, experimental and theoretical results have been presented that support the polaronic metal state for doped polyaniline (emeraldine salt) [23].

Recently, Conwell and co-workers [24-25] have performed calculations at the SSH level with an attractive potential due to the counterions included and found that end-effects can lead to a finite density-of-states at the Fermi level even in the case of a soliton lattice. By not placing any counterions at the chain ends, the strong attractive potential from the counterions (n-type doping) is missing in these regions and the electronic states localized to the ends become strongly destabilized compared to the states localized to the middle of the chain. The energy levels of the occupied end-states lie in this case very close to the energy of the conduction band edge and thereby closing the energy gap and producing a metallic state. The soliton lattice configuration used in these studies is, however, fixed. Recently, it was shown that the particular geometrical structure used by Conwell *et al.* is not that corresponding to the ground state of the system [26]. By allowing for a complete geometry relaxation of the *trans*-$(CH)_x$ chain, the destabilization of the end-states is surpressed and the energy gap for system with the particular arrangement of counterions used in Ref. 25 is in the semiconducting regime.

Within the continuum model of Takayama, Lin-Liu, and Maki (the TLM model) [27], Dinter [28] has shown that a new type of superlattice exists for *trans*-$(CH)_x$. This new type of lattice, which contains considerable dimerization, has a lower total energy than that of the soliton lattice for doping levels above ~8%. The electronic band structure corresponding to this lattice configuration contains a partially filled band, i.e., a gapless metallic state exists for this type of lattice. A crossover from the soliton lattice to this new superlattice could therefore explain the onset of metallic properties in heavily doped *trans*-$(CH)_x$. However, as pointed out by Dinter, calculations using the more realistic discrete SSH model have to be performed in order to check whether or not the new lattice is a mere artifact of the TLM model.

Even though a high concentration of solitons give rise to a quite broad midgap band, band-structure calculations on *trans*-$(CH)_x$ show that the gap around the Fermi level is still too large to explain the metallic properties [22]. These calculations are, however, performed without taking the impurity potential into account. In order to investigate if the dopant impurities are able to cause a closure of the energy gap around the Fermi energy in heavily doped *trans*-$(CH)_x$, we present here results from geometry optimizations on a system including the electrostatic impurity potential as well as inter- and intra-chain Coulomb interactions.

Several attempts to include the effect of the dopant impurities in the calculation of the electronic spectra of doped *trans*-$(CH)_x$ have already been performed. Calculations of Bryant and Glick [29] for a single impurity ion show that the potential of the ion has a strong effect on the distribution of energy levels in both the valence and conduction bands. Furthermore, as a result of the ion potential, several levels are pulled into the gap, and the location of soliton or polaron levels is strongly affected. Bulka [30] calculated the effect of a short range impurity potential limited in interaction to the C-atom closest to the dopant ion. In these calculations, the gap was shown to remain even at high doping levels. Takahashi and Fukutome [31] have calculated the effect of the Coulomb field due to dopant ions on charged solitons on a single *trans*-$(CH)_x$ chain and found that the solitons are strongly pinned to the impurities and their corresponding energy levels located close to the bottom of the conduction band (for p-type doping). Their results show, therefore, a considerable gap around the Fermi level even for very high doping concentrations. Furthermore, the works of Mele and Rice [16], Harigaya and Tarai [17], and Conwell *et al.* [24-25] discussed above are all performed at the SSH level with the potential due to the counterion included. Recently, Stafström extended this type of Hamiltonian to include also electron-electron interaction terms ($H_{el\text{-}el}$) [26]. It will be shown in the following sections that the electron-electron interaction is important as concerns the optimized structure of a doped *trans*-$(CH)_x$ chain. In particular, it is observed that the (mean-field) electron-electron interaction works in favor of a more delocalized soliton, and consequently, a lowering of the electronic gap around the Fermi energy. Sections 3 and 4 below contain a summary of the results obtained in Ref. 26, and in addition some new results on longer polyacetylene chains including calculations of the density of states at the Fermi energy.

3. Methodology

3.1 HAMILTONIAN

The one-dimensional π-electron Hamiltonian employed here for the study of highly doped *trans*-$(CH)_x$ contains the following terms [26]:

$$H = H_{SSH} + H_C + H_{el\text{-}el} \tag{1}$$

H_{SSH} is the Su-Schrieffer-Heeger (SSH) Hamiltonian [1-2] including terms describing the π-electron hopping and the σ-bond repulsion energy.

$$H_{SSH} = -\sum_{n,\sigma} [t_0+\alpha(u_n-u_{n+1})](c^\dagger_{n+1,\sigma}c_{n,\sigma} + c^\dagger_{n,\sigma}c_{n+1,\sigma}) + \frac{K}{2}\sum_n (u_n-u_{n+1}-C)^2 \tag{2}$$

The expression $t_0+\alpha(u_n-u_{n+1})$ is the linearized bond length dependence of the hopping integral, u_n is the displacement of the n-th lattice site, t_0 the hopping integral of adjacent sites when $u_n=0$ for these sites and α is the electron-phonon coupling constant. The compressibility of the σ-bonds is described by the classical, harmonic term in Eq. (2), where K is the effective spring constant. The approximate treatment of the distance dependence in the two terms in Eq. (2) is justified since the displacements, u_n , are small compared to the average C-C bond length.

We perform calculations on chains both with periodic boundary conditions, for which C is set to zero and $u_{N+1}=u_1$, and with fixed-end boundary conditions. In the later case a repulsive term linear in u_n is required to stabilize a finite chain against contraction. The value of C is determined from the condition that when $u_n=0$ for all lattice sites, H_{SSH} yields the ground state energy of an undistorted chain [32]. The values of the parameters appearing in H_{SSH} are the following: K=21 eV/Å^2, t_0=2.5 eV, and α=4.1 eV/Å [1-2].

The doping process includes charge transfer from dopant molecules to polymer chains within an overall neutral system. A proper treatment of the Coulomb interactions of the doped system includes both the intra- and inter-chain electron-electron interactions and the electron-counterion interaction. If, for instance, intra-chain Coulomb interactions are excluded, the system appears as if it has a net positive charge and the stabilization of negatively charged solitons is overestimated. The quantum mechanical treatment is restricted to one single chain. The mean field due to charges on counterions and on surrounding chains give rise to the following one-particle term in the Hamiltonian:

$$H_C = \sum_n (V_n^{imp} + V_n^{sol}) c^\dagger_{n,\sigma} c_{n,\sigma} \tag{3}$$

where V_n^{imp} is the Coulomb potential at site n due to all counterions in the system and V_n^{sol} is the potential at site n due to the charges on surrounding chains. These potentials are described by the following expressions:

$$V_n^{imp} = -e^2 \sum_j \frac{1 - v_{nn_j}(1 - e^{-\gamma|n-n_j|a})}{\varepsilon_1 [a^2 (n-n_j)^2 + d^2]^{1/2}} \tag{4}$$

$$V_n^{sol} = e^2 \sum_{n',\sigma} P^{\sigma}_{n'n'} \frac{1 - v_{nn'}(1 - e^{-\gamma|n-n'|a})}{\varepsilon_2 [a^2 (n-n')^2 + 3d^2]} \tag{5}$$

Here, a=1.222 Å is the average distance between adjacent sites along the chain axis, d=2.4 Å [33-34] is the perpendicular distance of the impurity from the chain, √3d is the perpendicular distance between adjacent chains. The lattice site opposite to an impurity is denoted n_j and a lattice site on an adjacent chain n'. $P_{n'n'}$ denotes the charge density on the n'-th site of an adjacent chain. This quantity is taken to be identical to the self-consistently calculated charge density on the chain treated quantum mechanically, shifted however, in order to match the soliton sites on the adjacent chains (see Fig. 1).

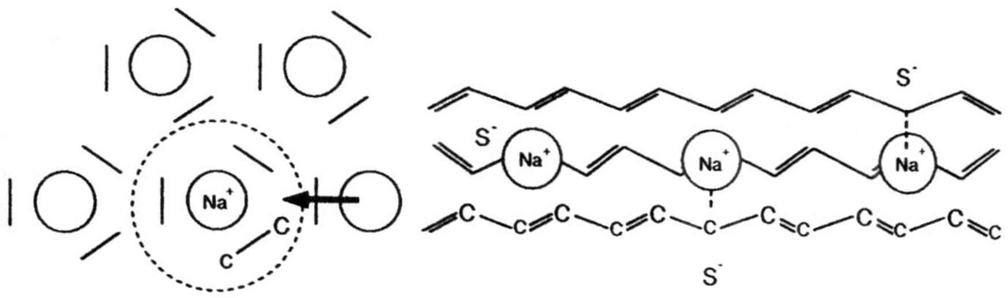

Figure 1. Three-dimensional structure of *trans*-$(CH)_x$ at y=8.33% (see Refs. 33 and 34). The structure is viewed perpendicular (left) and parallel (right) to the direction of the polymer chains. The bottom figure shows the encircled part of the top figure, view as indicated by the arrow. The soliton (S^-)-counterion complexes of each chain are connected with a dashed line. The carbon sites of the chain which is treated quantum mechanically (see text) are labelled with "C".

The expression given for the Coulomb potential due to counterions and charges on surrounding chains are based on X-ray crystallography studies of sodium doped *trans*-$(CH)_x$ [33-34]. In general, for doping of *trans*-$(CH)_x$ with alkali metals, columnar phases are observed. For the formation of these columns, the polymer chains twist around their chain axes to form one-dimensional channels which are filled by a linear arrangement of the dopant ions (see Fig. 1). The electrostatic interactions included in Eqs. (4) and (5) are restricted to one "column" only, i.e., the three-chain system shown in Fig. 1. This restriction is based on the observation that inter-column correlations between alkali-metal channels are found to be weak [35]. Hence, the dipoles consisting of soliton-counterion complexes in neighboring columns have no preferred orientation relative to the *trans*-$(CH)_x$ chain treated quantum mechanically. It is well known that the local net field from an array of such dipoles is approximately zero [36]. The channel surrounded by the polymer chains contains the counterions associated with defects on all three chains. A single chain interacts therefore with $3N_S$ counterions and $2N_S$ singly charged defects on the neighboring chains, N_S being the number of singly charged defects (solitons) on one chain.

Doping with alkali metals is an example of n-type doping (reduction of the polymer chain). Our calculations therefore include negatively charged defects (solitons) only. The impurity potential is in this case attractive (see Eq. (4)). Even though there is no exact electron-hole symmetry in the electronic spectrum in the presence of the extended Hubbard term (see Eq. (6) below), the properties of a p-type doped system are very close to those of a n-type doped system. Therefore, all results presented in the next section also hold for the case of p-type doping.

The dimerization of the lattice as well as other geometrical distortions are neglected in Eqs. (4) and (5) above. This approximation is reasonable since no exact information concerning the positions of the impurities as well as the inter-chain ordering in the doped state is available. Furthermore, the atomic displacements are of the order 0.04 Å, thus very small compared to the average bond length a=1.222 Å and the inter-chain separation $\sqrt{3}d$=4.157 Å, and do not influence the value of the potential in any substantial way.

In order to take into account the screening of the charges, an exponential factor is contained in the numerator in Eqs. (4) and (5). Since only one single column of polymer chains (see Fig. 1) is included in this calculations, all polymer chains appear at the same vertical distance from the impurities and from each other. Therefore, we need not to take into account explicitly any distance dependence of screening perpendicular to the chains. Instead we introduce the static screening parameters ε_1 and ε_2 in Eqs. (4) and (5) above. The values of these parameters, ε_1=1.5 and ε_2=1.8 , are those derived in Ref. 25 for the screening perpendicular to a metallic chain at distances d and $\sqrt{3}d$, respectively. These screening parameters are also shown below to reproduce the external potential used in Ref. 25.

It is observed that the charge distribution on the chains does not exhibit any large spatial variation in the highly doped regime (see Sec. 4.2., i.e., the charge distribution of each individual defect (soliton) is quite extended and overlaps with the charge associated with the adjacent defects. Therefore, the potential due to charges on the surrounding chains is almost site independent. As a consequence of this, a treatment of a soliton on an adjacent chain as an electronic charge (-e) located at the single site opposite to the counterion [25] strongly overestimates the screening of the counterions associated with these chains.

The exponential screening in Eqs. (4) and (5) is controlled by the two parameters γ and v_{nn_j} ($v_{nn'}$). It has been shown that the potential, within the first few bondlengths, show a typical 1/r dependence, whereas, for larger separations the potential become exponentially screened [25]. This situation can be simulated by introducing the factor v_{nn_j} ($v_{nn'}$) which is defined as: $v_{nn_j}=\min(|n-n_j|/4,1)$, i.e., v_{nn_j} is equal to the smallest of the two quantities $|n-n_j|/4$ and 1. For separations larger than three bond lengths, the exponential screening is fully developed, whereas the site opposite to the impurity (for i=j) experiences a pure 1/r potential. These two extremes are smoothly connected by the build up of the exponential screening over three CH-units.

The value of γ=0.373 Å^{-1} is the same as for the exponential screening of the $2p_z$-$2p_z$ appearing in the screened Ohno potential [37] (see Eq. (8) below). Due to the presence of v_{nn_j} ($v_{nn'}$), the exponential screening will be important at large separations only, where a charge on the chain is expected to be identically screened to a charge located in the absolute vicinity of the chain. The value of γ also agrees very well with the value calculated by Conwell *et al.* [25] for the screening parallel to the chains. Takahashi and Fukutome [31] used a similar expression for the screening of impurity ions and charges on adjacent chains but in their expression the full exponential screening starts already at the nearest neighbor site with a screening parameter which is about five times smaller than the one used here.

The electron-electron repulsion term *including diagonal elements only* , often referred to as the extended Hubburd term, is:

$$H_{el\text{-}el}=\frac{1}{2}\sum_{n,m,\sigma,\sigma'} U_{nm}\, c^{\dagger}_{n,\sigma} c_{n,\sigma} c^{\dagger}_{m,\sigma'} c_{m,\sigma'} \tag{6}$$

Here, this interaction is treated within the self-consistent field Hartree-Fock approximation which gives the following mean-field spin-polarized electron-electron interaction Hamiltonian:

$$H^{\sigma}_{el\text{-}el}=\frac{1}{2}\sum_{n,m,\sigma} U_{nm}\,(P^{\sigma'}_{mm}\, c^{\dagger}_{n\sigma} c_{n\sigma} - P^{\sigma}_{nm}\, c^{\dagger}_{n\sigma} c_{n\sigma'}), \qquad \sigma,\sigma'=\uparrow,\downarrow \tag{7}$$

Here, P_{mm} is the charge density at the m-th site and U_{nm}, the effective Coulomb repulsion integrals between $2p_z$ orbitals attached to sites n and m along the polymer chain. We adopt the screened Ohno expression [37] for these integrals:

$$U_{nm} = \frac{U_0 e^{-\gamma |r_{nm}|}}{[1+0.6117 r_{nm}^2]^{1/2}} \tag{8}$$

Here, U_0=11.13 eV and γ=0.373 Å^{-1}. The constant and exponential screening factors were introduced in Ref. 37 in order to produce a soliton charge distribution in agreement with experimental XPS-data. The inter-site distance r_{nm} in Eq. (8) is calculated using the the optimized geometry of the polymer chain.

3.2. SELF-CONSISTENT CALCULATION SCHEME

The ground state of $H=H_{SSH}+H_C+H_{el-el}$ can be derived using a self-consistent, iterative numerical method developed by Stafström and Chao [38]. Assuming the initial values of the lattice configurations and the density matrix, the eigenenergies (ε_i) and the expansion coefficients ($B_{n,i}$) of the molecular orbitals are obtained from the Schrödinger equation :

$$H^\sigma[\sum_n c^\dagger_{n,\sigma} B^\sigma_{n,i}] = \varepsilon^\sigma_i [\sum_n c^\dagger_{n,\sigma} B^\sigma_{n,i}] \qquad \sigma=\uparrow,\downarrow\ ;\quad i=1,2,...N^\sigma_e \tag{9}$$

Here, N_e is the number of electrons of each spin. New values for the dimerization order parameter, $\Delta r_n=-(-1)^n(u_n- u_{n+1})$ are calculated by minimizing the total energy of the system with respect to $v_n= \alpha(u_n-u_{n+1})$. The total energy of the system with the full Hamiltonian (see Eq. (1)) is:

$$\mathcal{E} = \sum_{n,\sigma} -[t_0+\alpha(u_n-u_{n+1})]2P^\sigma_{n,n+1} + [V^{imp}_n + V^{sol}_n]P^\sigma_{n,n} +$$
$$\frac{1}{2}\sum_{n,m,\sigma,\sigma'} U_{nm}\,[P^\sigma_{n,n}P^{\sigma'}_{m,m} - P^\sigma_{n,n}P^{\sigma'}_{m,m}\delta_{n,m}\delta_{\sigma,\sigma'}] \tag{10}$$

where P^σ_{nm} are the elements of the density matrix, which is obtained from the solution of the Schrödinger equation as:

$$P^\sigma_{nm} = \sum_i^{occ} B^\sigma_{n,i} B^\sigma_{m,i} \tag{11}$$

Minimization of the total energy with respect to v_n gives the condition:

$$v_n = \lambda\,[\,C' + 2\sum_\sigma P^\sigma_{n,n+1} + Z\,] \tag{12}$$

where $\lambda=\alpha^2/K$ and $C'=CK/\alpha$. Z contains the derivative of the density matrix terms and the potentials with respect to v_n (see Ref. 38). Since these derivatives can not be evaluated analytically, it is convenient to neglect them during the iteration process. The final self-consistent solution is the same whether Z is neglected or not, because when self-consistency is reached we have Z=0 for the system under equilibrium.

The starting geometry in the iterative optimization process is usually the one with zero dimerization. The stability of the optimized geometry is always tested by using another starting configuration and performing the optimization once again. For all systems included in this study, the same optimal ground state was reached, independent on the starting configuration. The criteria for terminating the optimization are differences less than 10^{-5} Å in the dimerization order parameter and 10^{-5} eV in the total energy, between two succesive iterations.

3.3. Description of the Optimized Systems

Earlier, results from geometry optimizations are performed on chains containing from N=108 sites up to N=132 sites are presented [26]. These studies included periodic as well as fixed-end boundary conditions. No major difference as concerns the structural or electronic properties was, however, found between periodic and terminated systems. Therfore, the additional calcualtions performed for this work all include periodic boundary conditions. The chain length is extended to N=360 sites in order become more realistic. The systems are studied for doping levels $5\%\leq y\leq 12.5\%$. The doping level is equal to the reciprocal of the separation of adjacent defects (solitons or polarons).

Disorder effects are well known to produce tails in the energy bands of regular systems, and in such a way reduce the energy gap between the valence and conduction bands. As mentioned in Sec. 2, models based on disorder effects have been proposed to explain the closure of the energy gap in highly doped *trans*-$(CH)_x$ [16,17]. A study of the disorder effects is also included in this paper. The type of disorder we consider consists of a random displacement of the dopant ions from the regular structure discussed above. The maximum displacement of a dopant ion from the position in the regular lattice is 50 % of a C-C bond-length. This type of disorder is much weaker than those considered in Refs. 16 and 17 and is expected to be found even in the most carefully prepared samples. Properties such as the geometrical structure and the density of states of systems including these types of disorder are compared with those of the regular systems.

3.4. Polaron Lattice

It is shown below that the soliton lattice is always the most stable lattice configuration. This is not surprising since the system is restricted to one chain, i.e., it is one-dimensional. For one-dimensional systems, the Peierls effect results in an opening of a gap in the electronic spectrum around the Fermi level. At present, we have not yet extended the method described above to include three-dimensional effects and we are therefore not able to study if such effects can stabilize the polaron lattice to such an extent that it becomes the ground state configuration as proposed by Kivelson and Hegeer [18] (see Sec. 2 above). Nevertheless, it is interesting to study the electronic properties of the polaron lattice in the one-dimensional model. Calculations are performed using the Hamiltonian described above but with a fixed geometry describing the strucutre of the polaron lattice. The displacement corresponding to the polaron lattice has the form [39]:

$$u_n = -(-1)^n u_0 \prod_{i=1}^{N_P} \left[1 + \frac{1}{\sqrt{2}} \left(\tanh\left(\frac{n-n_i-M}{l_P}\right) - \tanh\left(\frac{n-n_i+M}{l_P}\right)\right)\right] \tag{13}$$

The characteristic width of the polaron is determined by $l_P=\sqrt{2}l_S$ and $M=0.623\ l_P$ [39], where l_S is the soliton width. The soliton width is estimated to $l_S=3$ by fitting a tanh function to the optimized order parameter (see Ref. 26 and below). This length is very close to the value obtained from *ab inito*, extended basis set Hartree-Fock calculations for a single soliton in the presence of a screened sodium counterion [40].

3.5. Density of States

The density of states per site, $N(\varepsilon)$, is calculated following the method used by Mele and Rice [16]:

$$N(\varepsilon) = -\frac{2}{\pi} \operatorname{Im} \operatorname{Tr} G(\varepsilon) \tag{14}$$

where $G(\varepsilon)$ is the resolvant operator of the Hamiltonian given in Eq. (1), $G(\varepsilon)=(\varepsilon-H)^{-1}$. The diagonal matrix elements of G are obtained using a continued fraction expansion [16,41].

$$G_{ii}(\varepsilon) = \cfrac{1}{\varepsilon-\varepsilon_i - \cfrac{t_{i,i+1}^2}{\varepsilon-\varepsilon_{i+1} - \cfrac{t_{i+1,i+2}^2}{\cdots}} - \cfrac{t_{i-1,i}^2}{\varepsilon-\varepsilon_{i-1} - \cfrac{t_{i-2,i-1}^2}{\cdots}}} \tag{15}$$

In order to avoid divergences in the density of states we make the substitution $\varepsilon=\varepsilon+i\tau$ ($\tau=0.03$ eV). This substitution introduces a Lorenzian shaped broadening of the true eigenstates corresponding to the Hamiltonian in Eq. 1.

4. RESULTS AND DISCUSSION

As discussed in Sec. 2 above, Conwell *et al.* [25] have studied the electronic spectrum in the case of an external potential due to dopant ions and charged defects on surrounding chains added to the

SSH-Hamiltonian. With the geometry fixed (the dimerization order parameter is described by a product of tanh functions), they found essentially a zero energy gap for *trans*-$(CH)_x$ above $y \approx 6.67$ %. The first paragraph of this section deals with the same type of Hamiltonian as in their study but instead of performing calculations with the geometry fixed, we allow for a complete geometry relaxation according to the method described above. Paragraph 4.2 contains results of the optimized geometry and the self-consistently calculated charge distribution for the full Hamiltonian $H=H_{SSH}+H_C+H_{el\text{-}el}$ presented in Sec. 3 above. In paragraph 4.3, the effects of disorder in the impurity potential are discussed. Properties of the polaron lattice is presented in paragraph 4.4. The evolution of the energy gap as a function of doping level is presented in paragraph 4.5. Finally, in paragraph 4.6, results from calculations of the density of states at the Fermi energy are presented.

4.1. OPTIMIZED GEOMETRY USING THE CONWELL-MIZES-JEYADEV POTENTIAL.

With the potential due to counterions and charges on adjacent chains described by Eqs. (4) and (5), we obtain an almost perfect fit to the potential presented in Fig. 4 of Ref. 25. Following the procedure of Ref. 25, the distribution of the doping induced charge on the neighboring chains is taken to be that of point charges localized to the sites opposite to the counterions associated with the chain. The chain length in this case is N=108 and the number of solitons on a single chain is $N_S=8$ corresponding to a doping level of y=8.33 %. Using this potential, hereafter referred to as the external potential, we calculate a gap in the electronic spectrum of 0.093 eV, which is almost identical to what is obtained in Ref. 25. The highest occupied states are observed to be localized to the chain ends and considerably destabilized compared to the states in the valence band. It is clear that this situation occurs since the potential is much less attractive in the regions close to the chain ends (see Fig. 2). Consequently, the states localized to these regions appear at lower binding energies compared to the soliton states which are localized to the regions around the minima in the external potential.

Allowing for a full geometry relaxation, shows that the choice of fixed geometrical structure according to Eq. (2) in Ref. 25 is highly unstable. The optimized dimerization order parameter is shown in Fig 2 (top). The quite irregular geometry can be rationalized in the following way. Due to the strong electron-phonon coupling present in conjugated polymers, the lattice distort in a way to avoid electronic charge in the end-regions, since this is unfavorable in the presence of the specific potential used (see Fig. 2, bottom). The somewhat unexpected way for the lattice to distort is by creating positive solitons at both ends of the chain. Compared to the case of fixed geometry, this relaxation increases the positive charge at the ends of the chain. In order to conserve the total charge on the chain, the deficit of negative charge at the end of the chain is compensated for in the regions of the minima in the potential (see Fig. 2, middle). It is this additional charge, approximately two electrons, which cause the deviation from the typical soliton lattice structure of the dimerization order parameter. As clearly demonstrated in Fig. 2, the distribution of electronic charge exhibits eight maxima which coincide with the minima of the potential. Since negative solitons are created for an access of approximately one electronic charge, there are additional distortions created around the minima in the external potential due to the additional charge which is transferred from the ends of the chain.

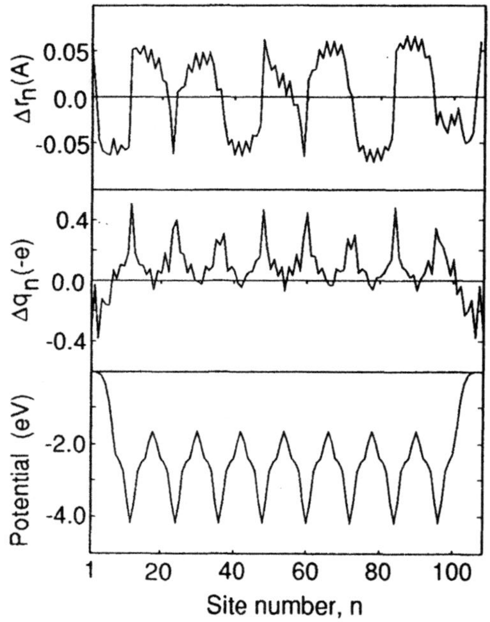

Figure 2. Optimized dimerization order parameter, $\Delta r_n = -(-1)^n(u_n - u_{n+1})$, (top), net charge distribution, Δq_n, (middle), and external potential according to Ref. 25 (bottom) for a system with fixed-end boundary conditions at doping level y=8.33 % (N=108 and N_S=8). No intra-chain Coulomb interactions are included.

After allowing for a full geometry relaxation of the lattice, the electronic spectrum of the *trans*-$(CH)_x$ chain at doping level corresponding to y=8.33 % shows a gap around the Fermi level of 0.461 eV. The highest occupied states of this system are extended over the interior of the chain, whereas the two lowest unoccupied states are localized to the chain ends and are associated with the positive solitons. It should be noted that the wavefunctions corresponding to the highest occupied states in this case do not have the shape typical for soliton states, instead, they resemble the quite irregular geometrical structure which is obtained in this case. The increase in the gap compared to the calculation with fixed geometry is mainly due to a stabilization of the highest occupied states since their spatial distribution is shifted into the region where the external potential is more attractive. Evidently, both the geometrical and the electronic structure of the optimized system looks entirely different from that obtained in the case of fixed geometry [25]. Since the gap in the electronic spectrum of the optimized system lies well into the semiconducting regime, the metallic state of heavily doped *trans*-$(CH)_x$ can not be explained simply by the presence of the impurity potential.

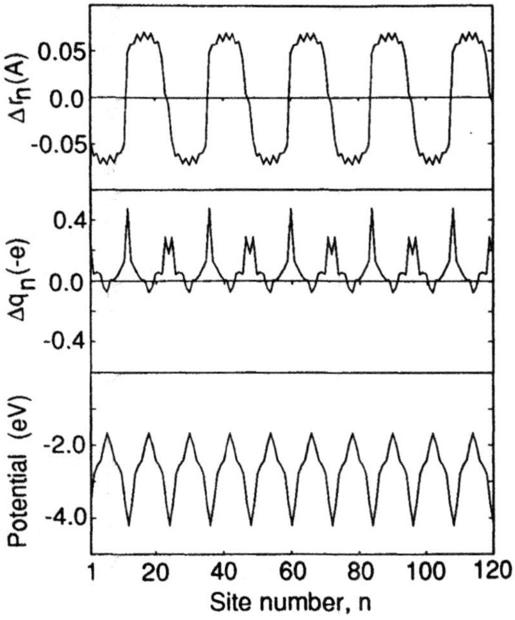

Figure 3. Same as Fig. 2 but with periodic boundary conditions (N=120 and N_S=10). No intra-chain Coulomb interactions are included.

By applying periodic boundary conditions to the polymer chain as well as to the external potential, we avoid the relatively speaking strong repulsive nature of the potential at the chain ends. The optimized geometry of the periodic chain (N=120, y=8.33 %) shows the typical soliton superlattice structure for which the repeat unit is one soliton-antisoliton pair (see Fig. 3). This periodicity is also observed in the charge distribution. Both the order parameter and the charge distribution are commensurate with the external potential. The gap in the electronic spectrum in this case is 0.873 eV, which is considerably larger than for the system with fixed chain ends. This difference is due to the nature of the highest occupied states, which in the case of fixed-end boundary conditions have a quite irregular structure, but in the case of periodic boundary conditions are extended soliton states which are fully commensurate with the external potential and thereby stabilized in energy.

The dramatic difference in the geometrical structure between the terminated and the periodic system is due to the abrupt cut-off in the external potential at the chain boundaries of the terminated system. This is verified by performing a calculation for the two systems without any external potential. Except for a minor increase in the dimerization at the ends of the terminated chain, these systems show very similar geometrical and electronic properties. It is shown below that the dramatic impact of the cut-off in the external potential on the geometrical structure of the system is effectively reduced when intra-chain Coulomb repulsions are included in the model. Furthermore, it is questionable if the metallic screening, which cause the cut off, persist to the very end of the chain. This will be discussed in more detail below in connection with the treatment including the full Hamiltonian.

4.2. THE EFFECT OF INTRA-CHAIN ELECTRON-ELECTRON INTERACTIONS.

It is observed in Figs. 2 and 3 that large net charges reside in regions along the polymer chain where the external potential has its minima. The distribution of the net charges results from a competition between the external potential (H_C), which favorizes a situation where the charge is localized to the lattice site where the potential has a minimum, and the characteristic width of the soliton favored by the SSH Hamiltonian. An additional competitive factor is introduced when the electron-electron repulsion term is included in the Hamiltonian. Due to the strong on-site repulsion included in this term, we expect smaller amplitudes of the charge density wave (CDW) on individual sites in this case.

The treatment of the defect charge on neighboring chains as localized to one site only, which is adopted above as well as in Ref. 25, is not completely adequate. Instead, a proper treatment would be to place the charge distribution, which is calculated at a given step in the iterative self-consistent optimization procedure, on the neighboring chains and then recalculate the external potential for this charge distribution. This procedure is then repeated in each step in the optimization which ensures that, when self-consistency is reached, the charge distribution on the neighboring chains in the three-chain column (see Fig. 1) corresponds to that of the optimized chain. The effect of allowing for a smearing out of the solitons on neighboring chains is to increase the influence of the counterions associated with these chains. Since the attractive potential from the counterions belonging to neighboring chains is increased, notable local minima in the external potential occur on sites opposite to the counterions associated with the neighboring chains (see Fig. 4). Consequently the solitons are less bound to individual counterions in this case. In the search for a minimum in the total energy at very high doping levels, the solitons on one chain start to move from one counterion to the next in the channel in the iterative optimization procedure. This occurs since the barrier between the local minima that exists opposite to each counterion are smaller than the inaccuracy in the iteration procedure (this inaccuracy is mainly due to the neglect of the terms Z in Eq. (12)). Therefore, we are unable to reach a global minimum for the system with the counterions very closely packed. The method described in Sec. 3 is effectively limited to systems with an inter-soliton separation of at least eight CH-units ($y \leq 12.5\%$).

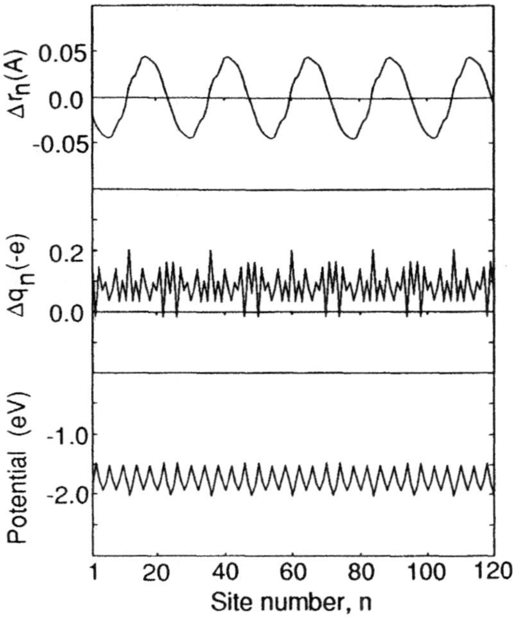

Figure 4. Optimized dimerization order parameter, $\Delta r_n = -(-1)^n(u_n - u_{n+1})$, (top), net charge distribution, Δq_n, (middle), and single particle potential (bottom) for a system with periodic boundary conditions at doping level y=8.33 % (N=120 and N_S=10). Intra-chain Coulomb interactions are included.

The dimerization order parameter and the net charge distribution of the optimized system with the full Hamiltonian is shown in Fig. 4, for the case of periodic boundary conditions (N=120, N_S=10). In this case, the presentation is restricted to a periodic system since the results for a system with fixed-end boundary conditions are qualitatively the same (see discussion below). The bottom graph in Fig. 4 shows the total single particle potential which is the sum of the external potential and the potential energy due to the intra-chain Coulomb interaction. This potential is obtained by summing the terms in Eqs. (4), (5) and (7) at each site, n, along the chain.

In contrast to the results obtained with the intra-chain Coulomb repulsion omitted, the abrupt cut-off in the external potential at the ends of the chain does no longer destroy the soliton lattice structure of the system with fixed end boundary conditions. This is due to a screening of the external potential by the large amount of electronic charge localized to the regions where the external potential has minima. Consequently, the depth of the potential minima becomes strongly reduced relative to the height of the potential wells between two solitons as well as relative to the potential at the chain ends.

The large amount of net positive charge which appeared at the chain ends in the calculations presented in paragraph 4.1, is now reduced since the potential in these regions is, relatively speaking, less repulsive. Therefore, there is no major difference in the charge distribution between the system with fixed-end boundary conditions and the system with periodic boundary conditions. This is indeed in contrast to the result of the calculation presented in paragraph 4.1 above and shows the importance of including the electron-electron repulsion in the calculation.

The optimized dimerization order parameter shows that the ground state of the doped polymer (y=8.33%) corresponds to that of a soliton lattice. Since the minima in the potential are less deep, the width of the solitons is increased compared to the case presented in paragraph 4.1. This is clearly seen by comparing the dimerization order parameter and the charge distribution in Figs. 3 and 4. The soliton length is estimated to $l_S=3$ by fitting a tanh function to the optimized order parameter. This length is very close to the value obtained from *ab inito*, extended basis set Hartree-Fock calculations for a single soliton in the presence of a screened sodium counterion [40].

The overall change in the charge distribution compared to the results presented in Figs. 2 and 3 is, as expected, a reduction of the amplitude of the CDW. The alternating large and small amplitudes of the CDW, typical for a soliton lattice, is present. Due to the strong attractive counterion potential, this feature was lost in the results presented in paragraph 4.1 above. Clearly, the single particle potential including the H_{el-el} is much weaker and has less accentuated local minima. This feature is also reflected in the net charge distribution. The doping induced charge of the polymer chain is non zero not only in the vicinity of the solitons, but is extended over the whole system. The solitons are therefore strongly overlapping. This feature is reflected in the energy level distribution, which exhibits a considerably broader soliton band and a more narrow gap between the soliton band and the conduction band than in the case of H_{el-el} excluded.

At all doping levels included in this study ($5\% \leq y \leq 12.5\%$), the soliton lattice remains stable. This is in agreement with the Peierls theorem which states that a one-dimensional lattice distorts periodically and commensurate with the reciprocal band filling [42] (somewhat misleading, this state is called "ideally incommensurate" in Ref. 16). The Peierls distortion creates a gap around the Fermi level, which in this case lies between the fully occupied soliton band (n-type doping) and the conduction band. If this gap is sufficiently small, residual interactions might close the gap and produce a metallic state even though the lattice still exhibits distortions. The possible metallic properties of the soliton lattice are discussed in paragraphs 4.5 and 4.6 below and compared with the properties of the polaron lattice.

4.3. Disordered System.

Several authors have discussed the role of disorder in obtaining a metallic state of heavily doped *trans*-$(CH)_x$ [16,17]. As discussed in Sec. 2, these studies all involve disorder which is much stronger than the one present in the well ordered samples that show a very abrupt transition into the metallic state [12-13]. The disorder considered here is very weak and is expected to be present even in the most carefully prepared sample. The counterions are randomly displaced from their positions in the regular system. The maximum displacement is limited to half a C-C bond length. The geometry is optimized self-consistently for a number of samples, each with a different random distribution of the counterions. The properties of the different samples are shown to be quite similar (see also paragraph 4.5 below). The optimized dimerization order parameter, net charge distribution and single particle potential for a typical disordered system are shown in Fig. 5, for a periodic system containing 120 CH-units. The doping level is in this case y=8.33%.

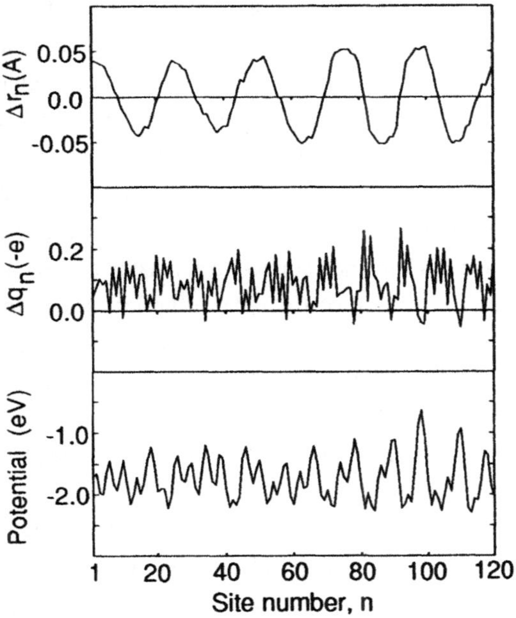

Figure 5. Same as Fig. 4 but for a disordered system (N=120, N_S=10, y=8.33 %). Intra-chain Coulomb interactions are included.

The soliton lattice structure is still stable and not dramatically different from that of the corresponding ordered sample (see Fig. 4). The CDW still exhibits the alternating large and small amplitudes typical for a soliton lattice. The charge distribution corresponding to individual solitons are observed in some regions along the chain, whereas in other regions, the charge is completely delocalized. This disorder enhanced inter-soliton overlap results in an energy level distribution which extends into the energy gap of the regular system.

4.4. POLARON LATTICE

The $H_{el\text{-}el}$ part of the full Hamiltonian used for the calculations of the optimized geometries is based on the Hartree-Fock method. This method is known to be ill-suited for the treatment of metallic systems since it predicts a zero density of states at the Fermi energy for infinite systems [43]. On the one hand, since the polaron lattice is intrinsically metallic, it is not recommended to use the method described above to calculate the electronic properties of the polaron lattice. On the other hand, the system treated here is finite and exhibits therefore a pseudo gap around the Fermi energy even in the case of a polaron lattice. We have performed calculations using both the full Hamiltonian, $H=H_{SSH}+H_C+H_{el\text{-}el}$, and the Hamiltonian with the $H_{el\text{-}el}$ term excluded, i.e., the same Hamiltonian used to calculate the results presented in paragraph 4.1. Consistent with the differences between the results of paragraphs 4.1 and 4.2, it is observed that $H_{el\text{-}el}$ work in favor of a more delocalized defect (polaron or soliton). Kivelson and Heeger [18] have shown that the polaron lattice exhibits a strong attractive interaction between the electrons in the half filled polaron band which lead to a more localized polaron. We believe that this effect is qualitatively obtained if the $H_{el\text{-}el}$ term is neglected. In this way we also avoid the problem of the Hartree-Fock method applied to a metallic system. The most adequate Hamiltonian to be applied to the calculation of the polaron lattice is therefore $H=H_{SSH}+H_C$. With this Hamiltonian, the self-consistent charge distribution is calculated using the expression for the order parameter of the polaron lattice given in Eq. (13). Note that this is not the stable lattice configuration for a one-dimensional system. Therefore, no geometry optimization is performed (such an optimization would converge to the structure presented in Fig. 4). Instead, the dimerization according to Eq. (13) is used in the calculation.

The results for the polaron lattice are shown in Fig. 6. The chain length is N=120, the number of polarons is $N_P=10$, and periodic boundary conditions are applied. The polaron width is set to $l_P=3\sqrt{2}$ which corresponds to the optimized width of the soliton ($l_S=3$) (see discussion in paragraph 3.3).The dimerization order parameter given in Eq. (13) exhibits the valley like structure characteristic for polarons (see Fig. 6). This result is in contrast to the result of a previous work, for which the dimerization order parameter becomes almost completely uniform and close to zero [19]. The difference is due to the fact that we set the polaron width to $l_P=3\sqrt{2}$ and thereby take the localization effect of the counterion potential into account. For $l_P=7\sqrt{2}$, which is the polaron width without the counterion present [39], the dimerization of the polaron lattice above y=7% is very small and almost completely uniform.

In comparison with the dimerization order parameter for the soliton lattice (see Fig. 3), the dimerization of the polaron lattice is about a factor of 2 smaller. Also the CDW has a smaller amplitude than that of the soliton lattice calculated with the same potential (see Fig. 3). The CDW is, however, far from uniform. This indicates that the doping induced IRAV modes of this system, which are due to non-uniformities in the CDW, should have intensities not very different from those of the soliton lattice. In fact, the maximal amplitude of the CDW for the soliton lattice presented in paragraph 4.2, is roughly the same as for the polaron lattice. In order to test the polaron lattice model more carefully, calculations of the frequencies and the oscillator strengths of the IRAV modes of the polaron lattice should be calculated.

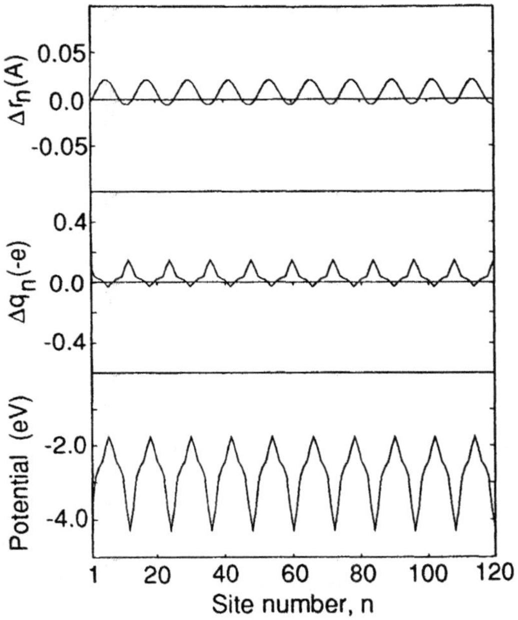

Figure 6. Same as Fig. 3 but for the polaron lattice (N=120, N_P=10, and y=8.33%). Periodic boundary conditions are applied and no intra-chain Coulomb interactions are included.

4.5. Evolution of the Energy Gap as a Function of Doping Level.

The energy gap of a system described by a Hamiltonian including electron-electron interaction terms is adequately described only in a many-particle Configuration Interaction picture. Such a treatment lies beyond the scope of this work. Instead we present values for the single particle gap, which in this case is the part of the electronic gap due to variations in the bond lengths along the chain (the Peierls gap).

The energy gaps are calculated for systems at doping levels y=5.0%, 6.67%, 8.33%, 10%, and 12.5% (see Fig. 7). The structure of the systems is optimized at each doping level. In order to simulate more realistic chain lengths, the size of the systems is set to N=360. This results in a reduction of the energy gap of about 0.01-0.03 eV as compared to previously reported results for shorter chains (N=120 and N=240) [26]. The open circles in Fig. 7 refer to the optimized perfectly ordered systems. The studies of the disordered systems include 10 different samples at each doping level. The value presented in Fig. 7 is the averaged energy gap for these 10 samples. Due to the large size of the system, the fluctuation in the energy gap due to the different distributions of counterions in these samples is small.

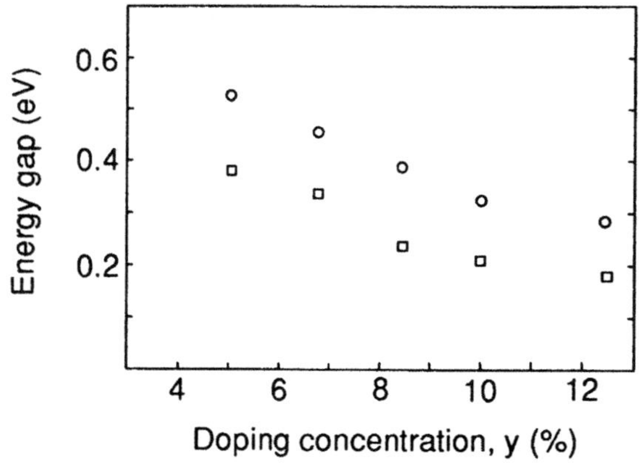

Figure 7. The dimerization gap for a regular (O) and disordered (▫) soliton lattices as a function of doping level.

Even though the disorder is quite weak, it affects the size of the energy gap considerably. At y=8.33%, the gap is of the order of 0.23 eV. The energy gap for the perfectly ordered sample at the same doping level is considerably larger, namely 0.397 eV. At y=6.67%, which is close to the critical doping concentration for which the experimental transition into the metallic state is observed, the energy gaps for the ordered and disordered samples are 0.425 eV and 0.31-0.33 eV, respectively. By including three-dimensional interactions of the same strength as in Ref. 16, the weakly disordered system should exhibit an energy gap close to zero. Therefore, the soliton lattice is expected to give rise to a metallic state at a doping level around 6-7%. However, there is no evidence for a sharp transition into a low energy gap system if the polymer remains in the soliton lattice configuration above the critical doping concentration, unless this transition includes an onset of (weak) disorder. Such an onset is, however, not observed in X-ray scattering data from sodium doped *trans*-$(CH)_x$.

4.6 Density of States

The evolution of the density of state at the Fermi energy, $N(\varepsilon_F)$, as a function of doping concentration is shown in Fig. 8. The three different curves correspond to perfectly ordered soliton lattice, disordered soliton lattice, and polaron lattice, respectively. The systems contain 360 sites.

The values given for the disordered systems are the average values of the $N(\varepsilon_F)$ for 10 different generations of randomly distributed counterions. Apart from the size, the systems are identical to those described above in paragraphs 4.2, 4.3, and 4.4 respectively.

For the soliton lattice, $N(\varepsilon_F)$ is monotonically and smoothly increasing with increasing doping level. Therefore, a phase transition into a metallic state within the soliton lattice, i.e., a genuine electronic phase transition, is not a first order transition. The difference between the ordered and disordered lattices is only quantitative. It is notable that even though calculations are performed up to y=12.5%, the is no tendency for $N(\varepsilon_F)$ to saturate. These results are in sharp contrast to the experimentally determined evolution of the Pauli susceptibility ($\chi_P \propto N(\varepsilon_F)$) which exhibits a sharp transition into a metallic state followed by a saturation at doping levels above y_c [13].

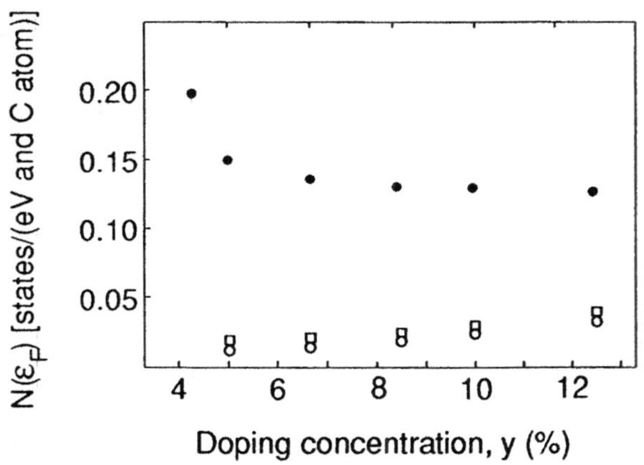

Figure 8. Evolution of the density of states at the Fermi energy, $N(\varepsilon_F)$ as a function of doping concentration, for the perfectly ordered soliton lattice (O), the disordered soliton lattice (▫) and the polaron lattice (•).

The lack of saturation in $N(\varepsilon_F)$ is a direct consequence of a successive reduction in the dimerization amplitude, Δr_{max}, of the polymer chain. As long as the dimerization remains finite, there is always an electronic gap around the Fermi level for the one-dimensional system. However, as discussed above, the energy gap is reduced considerably by three-dimensional interactions. Therefore, $N(\varepsilon_F)$ for the soliton lattice will be considerably higher if inter-chain interactions are taken into account. However, the smooth evolution of $N(\varepsilon_F)$ around y_c=6% will not change unless the strength of the inter-chain interactions change as a function of the doping level around y_c. No change in the inter-chain ordering around y_c is, however, observed in X-ray scattering data from sodium doped *trans*-$(CH)_x$. The abrupt onset of metallic properties is, therefore, difficult to reconcile with the electronic properties of the soliton lattice

The evolution of $N(\varepsilon_F)$ for the polaron lattice is also included in Fig. 8. The polaron lattice exhibits a decrease in $N(\varepsilon_F)$ with increasing doping concentration, i.e., the opposite trend as compared to the soliton lattice. Note that the Fermi energy in the case of a polaron lattice is located in the middle of one polaron band. The decrease in $N(\varepsilon_F)$ is a result of an increase in the dispersion of this band due to the onset of polaron-polaron interactions. We note that $N(\varepsilon_F)$ saturates for doping concentrations around 7%. The saturation in the density of states at the Fermi energy indicates that the polaron band does not change its shape above $y \sim 7\%$, i.e., the slope of the energy versus momentum curve, at the Fermi energy, remains unchanged with increasing doping concentration. Neither disorder, nor changes in the size or localization of the polaron defect, change the evolution of $N(\varepsilon_F)$ in the case of a polaron lattice. The saturation value for $N(\varepsilon_F)$ is 0.13 states/eV and C atom. This value is considerably larger than what is obtained for the soliton lattice at $y=12.5\%$ doping and it is very close to the density of states obtained in the central part of the π-band of an undimerized chain.

We observe that the sharp decrease in $N(\varepsilon_F)$ for the polaron lattice occurs at roughly the same doping concentrations as the experimentally observed onset of metallic properties for doped *trans*-$(CH)_x$. The calculated onset of metallic properties for the soliton lattice occurs, however, at much higher doping concentrations. Therefore, as far as the evolution of the density of states at the Fermi energy is concerned, our results indicate that a transition from a soliton lattice to a polaron lattice at the critical doping concentration is in agreement with the experimental result of a sharp increase in χ_P (or $N(\varepsilon_F)$) at the critical doping concentration, followed by an almost constant $N(\varepsilon_F)$ at higher doping concentration.

It should be noted that the polaron lattice is unstable in the one-dimensional case (see above). Therefore, three-dimensional interactions are necessary in order to, possibly, stabilize the polaron lattice.

5. Summary and Conclusion

The soliton lattice is found to be the ground state configuration of a one-dimensional doped (charged) *trans*-$(CH)_x$ chain. Including the potential due to the counterions results in a very localized soliton. The net charge at the carbon atom opposite to the counterion is approximately 0.5 electronic charges (n-type doping). This localization of the charge is energetically unfavorable when electron-electron interactions are included in the calculation scheme. The ground state configuration is, however, still that of a soliton lattice. The CDW is completely delocalized over the system with alternating large and small amplitudes on consecutive carbon sites. Weak disorder does not change the structural properties of the system dramatically.

The polaron lattice is studied using an expression for the dimerization order parameter proposed by Brazovskii and Kirova [39]. The polaron width is set to $l_P=3\sqrt{2}$ which corresponds to the optimized width of the soliton (l_s=3) in the presence of both counterion potential and electron-electron interactions. The dimerization order parameter for the polaron lattice exhibits in this case a finite dimerization (Δr_{max}=0.02 Å at y=8.33%) between the polarons up to a doping level of y=10%. In contradiction with previously reported results [19], the CDW of the polaron lattice exhibits a considerable modulation along the polymer chain. This modulation leads to an enhancement of the IRAV modes of the system compared to the results presented in Ref. 19. The very weak oscillator strength of these modes obtained in the case of an almost uniform charge distribution has been used as an argument against the polaron lattice. Therefore, it would be very interesting to see if calculations based on the charge distribution obtained here, will produce results that are in better agreement with the experimental infrared data.

The energy gap around the Fermi level is calculated for structurally optimized systems in the doping regime 5%≤y≤12.5%. The energy gap is observed to decrease smoothly with increasing doping concentration. Disorder effects reduce the energy gap by about 0.1-0.2 eV. The energy gap at y=6.67 % for the weakly disordered system is small enough to be closed by three-dimensional interactions. However, since the gap decreases smoothly, the present results for the soliton lattice can not explain the observed experimental evolution of a rapid onset of metallic properties at y_c=6%, followed by a saturation in the density of states above y_c. This is seen very clearly by looking at the evolution of the calculated density of states at the Fermi energy, $N(\varepsilon_F)$, which is shown to increase without saturation up to the maximum doping level included in this study, namely, y=12.5%. The polaron lattice, on the other hand, exhibits a constant $N(\varepsilon_F)$ above y=7% in close agreement with the experimental data.

Unfortunately, neither the results presented in this article, nor any previously reported model are able to present an undisputable explanation for the behavior of heavily doped *trans*-$(CH)_x$ at and above the critical doping concentration. We have pointed out here, that it is essential to include three-dimensional interactions into the Hamiltonian, since calculations with this interaction included will provide information concerning the stability of the polaron lattice vs. the soliton lattice. An extension of the method presented in this article to include inter-chain interactions is under development and we hope to report results from calculations based on this method in the near future.

ACKNOWLEDGMENTS

The author would like to thank Dr. Stanislaw Tabor for stimulating co-operation during the studies of the density of states. This work is financially supported by the Swedish Natural Sciences Research Council (NFR) and the Swedish Board for Technical Development (STU).

REFERENCES

1. Su, W.P., Schrieffer, J. R., and A.J. Heeger (1979) "Solitons in Polyacetylen" Phys. Rev. Lett **42**, 1698-1701.
2. Su, W.P., Schrieffer, J. R., and Heeger, A.J. (1980) "Soliton excitations in Polyacetylene" Phys. Rev. B **22**, 2099-2111.
3. Pople, J. A. and Walmsley, S. H. (1962) "Bond Alternation Defetcs in Long Polyene Molecules" Mol. Phys. **5**, 15-20.
4. Suzuki, N., Ozaki, M., Etemad, S., Heeger, A. J., and MacDiarmid, A. G. (1980) "Solitons in Polyacetylene: Effects of Dilute Doping on Optical Absorption Spectra" Phys. Rev. Lett. **45**, 1209-1213.
5. Orenstein, J., and Baker, G. L. (1982)"Photogenerated Gap States in Polyacetylene" Phys. Rev. Lett. **49**, 1043-1046.
6. Tavan P., and Schulten, K. (1987)"Electronic Excitations in Finite and Infinite Polyenes" Phys. Rev. B **36**, 4337-4358.
7. Feldblum, A., Kaufman, J. H., Etemad, S., Heeger, A. J., Chung, T.-C., and MacDiarmid, A. G. (1982) "Opto-electrochemical Spectroscopy of *Trans*-$(CH)_x$" Phys. Rev. B **26**, 815-826.
8. Yang, X. Q., Tanner, D. B., Rice, M. J., Gibson, H. W., Feldblum, A., and Epstein, A. J. (1987) "Transition to a Gapless Peierls Insulator in Heavily-Doped Polyacetylene" Solid State Commun. **61**, 335-340.
9. Ikehata, S., Kaufer, J., Woerner, T., Pron, A., Druy, M. A., Sivak, A., Heeger, A. J., and MacDiarmid, A. G. (1980) "Solitons in Polyacetylene: Magnetic Susceptibility" Phys Rev. Lett. **45**, 1123-1126.
10. Chung, T.-C., Moraes, F., Flood, J. D., and Heeger, A. J. (1984) "Solitons at High Density in *Trans*-$(CH)_x$: Collective Transport by Mobile Spinless, Charged Solitons" Phys. Rev. B **29**, 2341-2343.
11. Park, Y. W., Denenstein, A., Chiang, C. K., Heeger, A. J., and MacDiarmid, A. G. (1979) "Semiconductor-Metal tTansition in Doped $(CH)_x$: Thermoelectric Power" Solid State Commun. **29**, 747-751.
12. Moraes, F., Chen, J., Chung, T.-C., and Heeger, A. J. (1985) "First-Order Transition to a Novel Metallic State in $[Na_y^+(CH)^{-y}]_x$: In Situ Electron Spin Resonance During Chemical and Electrochemical Doping" Synth. Met. **11**, 271-292.
13. Chen, J., and Heeger, A. J. (1988) "In Situ Electron Spin Resonance Experiments on Polyacetylene during Electrochemical Doping" Synth. Met. **24**, 311-327.
14. Tanner, D. B., Doll, G., Rao, K., Yang, M. H., Eklund, C., Arbuckle, G., and MacDiarmid, A.G. (1987) "Infrared Absorption in K-Doped $(CH)_x$"

Bull. Am. Phys. Soc. **32**, 422.

15. D. B. Tanner, G. L. Doll, A. M. Rao, P. C. Eklund, G. A. Arbuckle, and A.G. MacDiarmid (1989) "Infrared Absorption in K-Doped $(CH)_x$", Synth Met. **28**, D141-146.
16. Mele, E. J. and Rice, M. J. (1981) "Semiconductor-Metal Transition in Doped Polyacetylene" Phys. Rev. B **23**, 5397-5412.
17. Harigaya, K. and Terai, A. (1991) "On a Metal-Insulator Transition of Disordered Conducting Polymers" Solid State Commun., in print.
18. Kivelson, S. and Heeger, A. J. (1985) "First Order Transition to a Metallic State in Polyacetylene: A Strong-coupling Polaronic Metal." Phys. Rev. Lett. **55**, 308-311.
19. Choi , H.-Y. and Mele, E. J. (1986) "Dynamical Conductivity of Soliton Lattice and Polaron lattice in the Continuum Model of Polyacetylene" Phys. Rev. B **34**, 8750-8757.
20. Fink, J., Nücker, N., Scheerer, B., v. Felde, A., and Leising, G. (1987) in H. Kuzmany, M. Mehring, and S. Roth (eds.) "Electronic Properties of Conjugated Polymers, Springer Series in Solid State Sciences No. 76, Springer, Berlin, pp. 84.
21. Tabor, S. and Stafström, S. (1991), submitted for publication.
22. Stafström, S., and Brédas, J.-L. (1988) "Evolution of the Electronic Structure of Polyacetylene and Polythiophene as a Function of Doping Level and Lattice Conformation" Phys. Rev. B **38**, 4180-4191.
23. Stafström, S., Brédas, J. L., Epstein, A. J., Woo, H. S., Tanner, D. B., Huang, W. S., and MacDiarmid A. G. (1987) "Polaron Lattice in Highly Conducting Polyaniline: Theoretical and Optical Studies" Phys. Rev. Lett. **59**, 1464-1467.
24. Conwell, E. M. and Jeyadev, S. (1988) "Insulator-Metal Transition in *Trans*-Polyacetylene" Phys. Rev. Lett **61**, 361-364.
25. Conwell, E. M., Mizes, H. A., and Jeyadev, S. (1989) "Metal-Insulator Transition in *Trans*-Polyacetylene" Phys. Rev. B **40**, 1630-1641.
26. Stafström, S. (1991) "Soliton Lattice in Highly Doped *Trans*-Polyacetylene", Phys. Rev. B, in press.
27. Takayama, H., Lin-Liu, Y. R., Maki, K. (1980) "Continuum Model for Solitons in Polyacertylene" Phys. Rev. B **39**, 2388-2393.
28. Dinter, M. "Semiconductor-Quasimetal Transition of Heavily Doped *Trans*-Polyacetylene." Phys Rev. B **39**, 8423-8433.
29. Bryant, G. W. and Glick, A. J. (1982) "Impurity States in Doped *trans*-Polyacetylene" Phys. Rev. B **26**, 5855-5866.
30. Bulka, B.R. (1988) "Ground State of Doped Polyacetylene" Synth. Met. **24**, 41.
31. Takahashi, A. and Fukutome, H. (1989) "Effects of Dopant Ions on Solitons in Polyacetylene" Synth. Met. **28**, D469-D475.

32. Stafström , S. and Chao, K.A. (1984) "Exact Numerical Investigation of the Polaron-Soliton Generation in Polyacetylene" Phys Rev. B **29**, 7010-7011.
33. Winokur, M., Moon, Y. B., Heeger, A. J., Barker, J., Bott , D. C., and Shirakawa, H. (1987) "X-Ray Scattering from Sodium-Doped Polyacetylene: Incommensurate-Commensurate and Order-Disorder Transformations" Phys. Rev. Lett. **58**, 2329-2332.
34. Murthy, N. S., Shacklette, L. W., and Baughman, R. H. (1989) "Structure of Lithium-Doped Polyacetylene" Phys. Rev. B **40**, 12550-12553.
35. Chen, D., Winokur, M. J., Masse, M. A., and Karasz, F. E. (1990) "Structural Phases of Sodium-Doped Polyparaphenylene vinylene" Phys. Rev. B **41**, 6759-6767.
36. Jackson, J.D. (1975) Classical Electrodynamics, Wiley & Sons, New York.
37. Fukutome H. and Sasai, M. (1983) "Theory of Electronic Structures and Lattice Distortions in Polyacetylene and Itinerant Peierls Systems. II" Prog. Theor. Phys. **69**, 1-19.
38. Stafström , S. and Chao, K.A. (1984) "Polaron-Bipolaron-Soliton Doping in Polyacetylene" Phys Rev. B **30**, 2098-2103.
39. Brazovskii, S. A. and Kirova, N. (1981) "Excitons, Polarons, and Bipolarons in Conducting Polymers" Sov. Phys. JEPT Lett. **33**, 4-8.
40. Fredriksson, C. and Stafström, S. (1991) unpublished results.
41. Haydock, R. (1978) "Study of a Mobility Edge be a New Perturbation Theory" Philos. Mag. B **37**, 97-109.
42. Peierls, R. E. (1955) Quantum Theory of Solids, Clarendon, Oxford, Chap. 5.
43. Delhalle, J. and Calais, J.-L. (1987) "Direct-Space Analysis of the Hartree-Fock Bands and Density of States for Metallic Extended Systems" Phys. Rev. B **35**, 9460-9466.

CONTROLLED MOLECULAR ASSEMBLIES OF ELECTRICALLY CONDUCTIVE POLYMERS

M. F. RUBNER
Department of Materials Science and Engineering, Massachusetts Institute of Technology, Cambridge, MA 02139 USA

T. A. SKOTHEIM
Brookhaven National Laboratory, Upton, NY 11973 USA

ABSTRACT. The Langmuir-Blodgett technique has recently been used to manipulate a variety of electroactive conjugated polymers into multilayer thin films with well defined layered structures and ordered molecular organizations. Multilayer thin films have been successfully fabricated from amphiphilic conjugated polymers and their precursors and from non-surface active conjugated polymers. In most cases, the molecular organizations of these films have been found to be highly anisotropic. The structures of these films have been probed by high energy synchrotron radiation, reflection-absorption infrared and low angle x-ray diffraction techniques. Electrical measurements have also revealed highly anisotropic behavior with the in-plane conductivities of these multilayer structures typically being many orders of magnitude greater than their transverse conductivities. The unique layered structures of these films, in some cases, have also been shown to give rise to unusually large dielectric constants over a wide frequency range. In this chapter, the fabrication, characterization and electrical properties of these new thin film multilayer structures are discussed.

1. Introduction

Over the past few years, the practice of molecular engineering has guided the design and synthesis of new electroactive polymers and has provided a deeper understanding of the structure-property relationships active in this very interesting class of materials. Molecular engineering in the simplest sense simply involves the manipulation and control of the physical and electronic properties of polymeric materials by suitable modifications of their molecular structures. Through this approach, it is possible to systematically alter the electrical, optical, mechanical and processing properties of electroactive polymers with the ultimate goal being the creation of materials with a collection of properties optimized for a particular technological application or scientific endeavor. The logical extension of this molecular manipulation approach is the development of techniques that provide control over the molecular and supermolecular organizations of these newly synthesized molecules. The net result would be an ability to design and fabricate complex molecular assemblies with controllable molecular architectures and well defined molecular arrangements. One could further envision the creation of multicomponent molecular assemblies in which each component serves a specific well defined function and is

J.L. Brédas and R. Silbey (eds.), Conjugated Polymers, 363–403.

1. Doping (through oxidation or reduction) leads to injection of carriers into the π-electron system. Since every monomer is a potential redox site, conjugated polymers can be doped (n-type or p-type) to a relatively high density of charge carriers.
2. The π-bonding leads to π-electron delocalization along the polymer chains, and, thereby, to the possibility of charge carrier mobility, which is extended into three dimensional transport by the interchain electron transfer interactions [1]. In principle, the broad π-electron bandwidths [2] (often several electron-volts) can lead to relatively high carrier mobilities.

As a result of the same intrachain π-bonding and relatively strong interchain electron transfer interactions, the mechanical properties (Young's modulus and tensile strength) of conjugated polymers are potentially superior to those of saturated polymers, such as polyolefins. Moreover, because of these two features, it may be possible to achieve exceptional mechanical properties with aligned conjugated polymers at lower chain lengths than required for their saturated counterparts [3].

Conjugated polymers are, therefore, of special interest because of the potential for a unique combination of electrical and mechanical properties. However, since both the electrical and mechanical properties of such polymers are currently limited by defects and structural disorder, improvement in material quality that will enable the exploration of intrinsic properties and the development of a variety of materials for different applications has become an important goal of conducting polymer research [4].

B. CONDUCTING POLYMERS: APPROACHES TO PROCESSING

Because the interchain electron transfer interactions of conjugated polymers are relatively strong (compared with the Van der Waals and hydrogen bonding interchain interactions typical of saturated polymers) conducting polymers tend to be insoluble and infusible. Thus, there was serious doubt in the early years after the initial discovery that polyacetylene and other π-conjugated polymers could be doped to the metallic state as to whether or not processing methods could be developed that would lead to chain extended and chain aligned materials of sufficient quality to make the intrinsic properties available for study and the materials more generally available in forms useful in applications.

Significant progress has been made. The addition of moderately long side chains onto the monomer units [5] resulted in derivatives of polythiophene, the poly(3-alkylthiophene)'s, P3AT's, and derivatives of poly(phenylene vinylene), the poly(dialkoxy-phelenevinylene)'s or PDalkoxyPV's,[6] as shown in Figures I.1 and I.2. Since the side chains decrease the interchain coupling and increase the entropy these derivatives can be processed either from solution or from the melt. For example, as shown in Figure I.3, within the P3AT system, the melting point decreases monotonically as a function of the length of the side chain [7]. However, the "side chain" route to processible conducting polymers has the disadvantage

a.

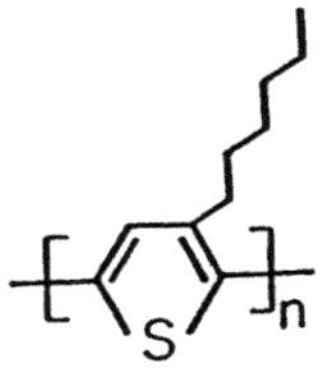

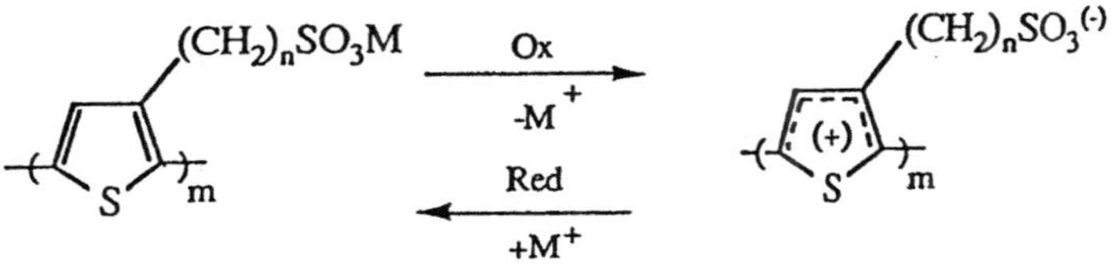

M = H, Li, Na, etc

n = 2, 4

b.

Figure I.1:
a. Molecular structure of the P3AT's
b. Water soluble "self -doped" derivatives of polythiophene

a. PDalkoxyPV's,

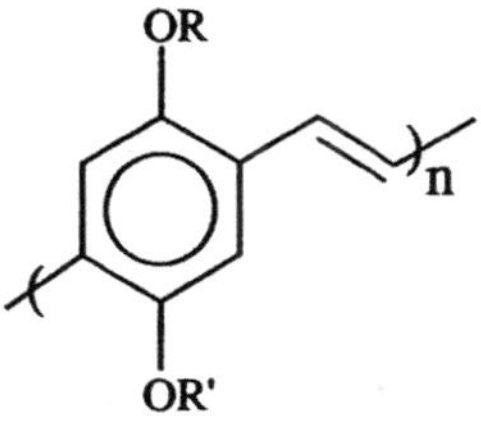

Symmetric: R = R' = C_6H_{13} (DHO-PPV), R = R' = C_8H_{17} (DOO-PPV),

Asymmetric: R = $CH_2CH(C_2H_5)C_4H_9$ and R' = CH_3 (MEH-PPV).

b. PTV

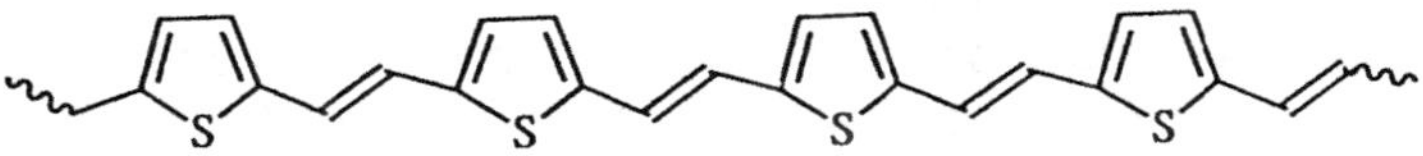

Figure I.2

that the relatively bulky side chains decrease both the π-electron density (and thus the carrier density), and they decrease the interchain coupling, thus making it more difficult to achieve high conductivities and exceptional mechanical properties.

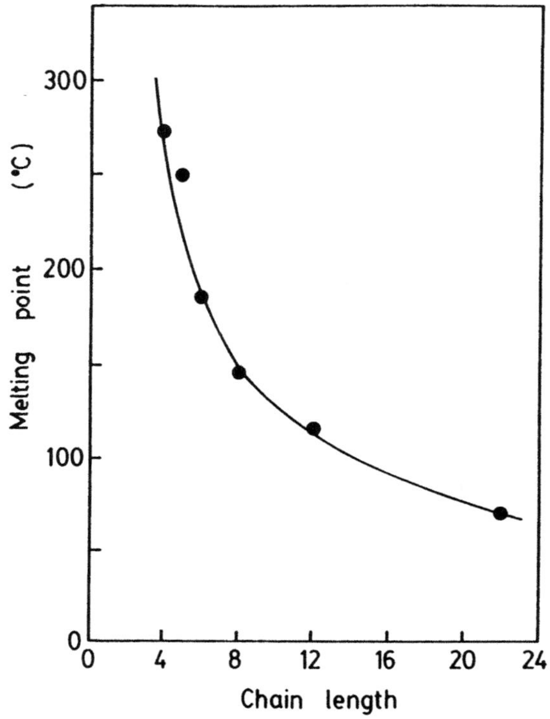

Figure I.3: Melting point vs n from Yoshino et al

By terminating the side chains with $SO_3^-M^+$ (M is either H or any monovalent metal atom, such as Na, etc), as shown in Figure I.1, the conjugated chains are made water soluble [8]. Moreover, since the SO_3^- is covalently bound to the polymer backbone and since it acts as the counterion upon oxidative doping (the M^+ goes off into solution), these derivatives are known as "self-doped" conducting polymers [8].

An alternative strategy is to identify stable conjugated polymer systems that can be processed. Of this class, polyaniline (PANI) is certainly a promising example, for it is soluble in certain organic solvents [9] (e.g N-methylpyrrolidinone, NMP) and in concentrated acids [10]. The molecular structure of the emeraldine base form of polyaniline is shown in Figure I.4. The use of concentrated protonic acids such as H_2SO_4, CH_3SO_3H, and CF_3SO_3H as solvents for PANI has specific advantages in that both the salt and the base form can be completely dissolved at room temperature, with polymer concentrations ranging from extremely dilute to more than 20% (w/w). Perhaps more important is the fact when precipitated from acid solution, PANI comes out in the conducting (protonated) emeraldine salt form.

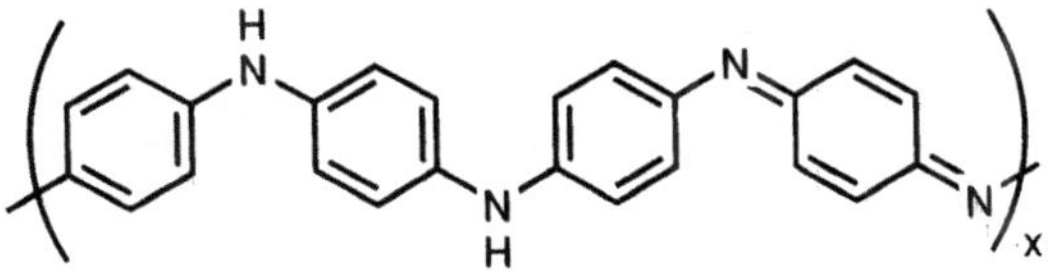

Figure I.4: Molecular structure of the emeraldine base of polyaniline

A third promising strategy for processing conducting polymers utilizes the versatile precursor route which involves the preparation of a processible precursor polymer and subsequent conversion of the precursor polymer to the conjugated polymer [11,12]. The significant advantage of this route is that the saturated precursor polymers can be processed from solution prior to the thermal conversion to the conjugated final product. The precursor polymers may, therefore, be drawn prior to and during the thermal conversion process so as to yield oriented, homogeneous conjugated polymers. Poly(p-phenylenevinylene), PPV, and its PDalkoxyPV derivatives can be prepared from a precursor polymer, a polyelectrolyte, which is soluble in water [12]. The dimethoxy-derivative of PPV, poly(2,5-dimethoxy-p-phenylene vinylene), PDMPV, has been prepared via a similar precursor route to PPV and exhibited relatively high conductivities after doping. However, the commonly used aqueous solutions of the PDMPV precursor polymer tend to form gels, and the gradual elimination of the sulfonium group in the solid precursor cannot be avoided, even at room temperature. Both effects make subsequent processing into highly oriented quality films and fibers difficult. To avoid these difficulties the Kyushu University group [13] succeeded in the preparation of dense PDMPV film from a new precursor polymer which is soluble in common organic solvents, easily processible, and stable even at 100 °C. Similarly, poly(2,5-thienylene vinylene) (PTV), see Figure I.2, is one of the larger class of poly(arylenevinylene) polymers which is attractive as a conjugated polymer and which can be synthesized via the precursor polymer route. As with PDMPV, PTV can be prepared either with a water soluble polyelectrolyte precursor or by conversion to a new precursor polymer which is soluble in common organic solvents and chemically stable.

The side-chain functionalization of PPV derivatives has recently progressed to include polymers which are soluble in the final conjugated form. Three new alkoxy substituted PPV's have been synthesized, all of which are soluble in the fully converted conjugated form (see Figure I.2) [6]. Poly(dihexyloxy phenylenevinylene) (DHO-PPV) is soluble at temperatures above 80° C. Surprisingly, the octyloxy derivative (DOO-PPV) was found to be less soluble in most non- polar organic solvents (probably because of side chain interdigitation and "side chain crystallization"). This observation suggested the design of the asymmetric MEH-PPV, which is readily soluble, at room temperature, in a variety of common organic solvents.

C. BLENDS OF CONDUCTING POLYMERS WITH SATURATED POLYMERS

Once a soluble conducting polymer has been synthesized, polymer blends or composites can be made and fabricated into desired forms (fibers, free standing films, thin films on substrates, etc) from solutions containing the conducting polymer and other compatible insulating polymers such as polyethylene, polystyrene, etc. Since stable solutions of *doped* conducting polymers have been successfully demonstrated, it is also possible to produce conducting blends directly from solution.

Polyaniline provides an example of the opportunity for achieving conducting blends directly from solution; the polyaniline example also serves to illustrate some of the special advantages that can be expected from blends. Although the ability to process conducting polyaniline from solution represents genuine progress, fibers and films made from these solutions have mechanical properties which are not adequate for many applications, due mainly to the low molecular weight of (currently available) polyaniline. Fibers with significantly enhanced mechanical properties have been obtained by blend processing polyaniline with the rigid chain polymer poly-(p-phenylene terephthalamide), PPTA[14]. It is well known that PPTA is processed from solutions in concentrated H_2SO_4 to yield one of the strongest and stiffest fibers commercially available [14]. The process takes advantage of the fact that PPTA is a rod-like polymer that exhibits a liquid crystalline phase at high solution concentrations which facilitates the formation of highly oriented structures. Thus, through blending one can achieve a combination of properties (in this case, the addition of electrical conductivity to the the high performance mechanical properties of the aramids; in other cases, the combination of electrical conductivity and optical transparency, etc.) that might not be possible with a single polymer.

II. Conducting Polymers in Solution

A. ELECTRONIC STRUCTURE (AND CONFORMATION) OF THE NEUTRAL POLYMERS IN SOLUTION

The π-conjugation in these materials is implied by their colors and their electronic spectra. The electronic absorption spectra of thin solid films of the P3ATs (see Figure II.1) indicate that the band edge, the onset of the π-π^* transition occurs at about 2 eV; a value which is typical for the entire P3AT series. Although slight shifts in the edge and in the weak shoulder just above the edge are observed, the spectra in Figure II.1 indicate that the electronic structure of the P3ATs is essentially independent of the alkyl substituent [5b].

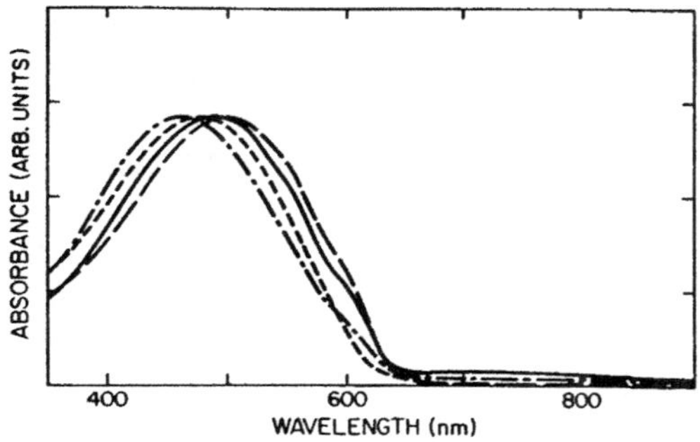

Figure II.1: Absorption spectra of thin solid films of polythiophene and a few of the P3AT derivatives.

Figure II.2 shows the UV-Visible absorption spectrum of a P3HT/THF solution [5b]. On dissolution, a major shift of the π-π^* absorption band is observed (see Figures II.1 and II.2), Dissolution causes localization of the electronic wavefunctions as a result of the disorder brought about by the (random) coil conformation. Based on the similar spectral shifts observed for the P3ATs, it seems that the chains of the yellow P3AT in solution are in a disordered conformation, whereas casting into solid film restores the more ordered crystalline phase with an associated more extensive delocalization of the π-electron wavefunctions.

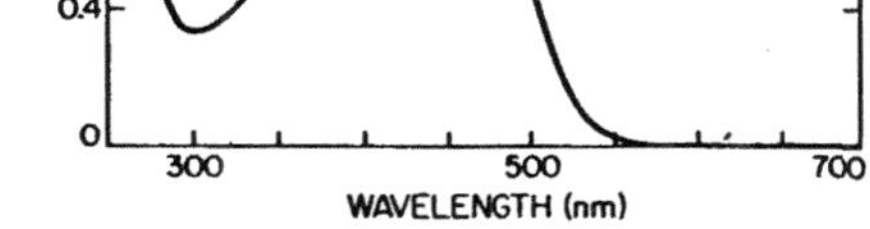

Figure II.2: Absorption spectrum of P3HT in THF

Forced Rayleigh scattering [16] has been used to probe the dynamics of a conducting polymer, poly(3-octylthiophene), in solution. The optical labels which lead to the transient grating are the photoexcited states on the chains themselves (i.e. the photoinduced spectral changes due to the formation of bipolarons); chemical addition of a photochromic labeling attachment was not required. The main feature of the time dependence of the diffracted intensity was an exponential time-decay; from the quadratic dependence of the decay time on the grating wavevector, a translational diffusion coefficient in the dilute regime of $(8\pm1)\times10^{-8}$ cm^2/s was obtained, corresponding to a hydrodynamic radius of $R_h = (130 \pm 15)$ Å. This value is in good agreement with results obtained from light scattering experiments carried out on samples from the same batch of material.

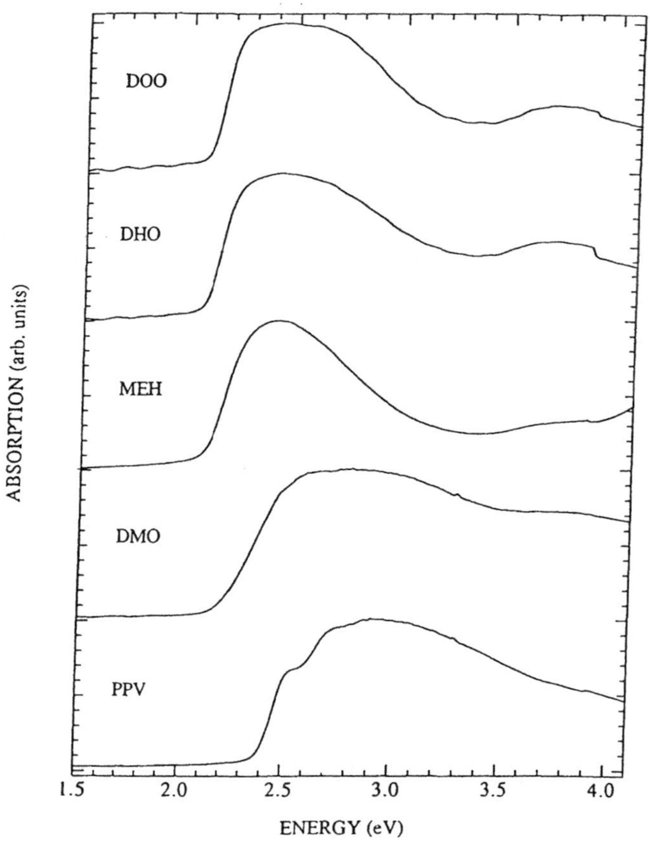

Figure II.3: Absorption spectra for the alkoxy-substituted PPV derivatives.
DOO: R = O-$(CH_2)_7$-CH_3, DHO: R = O-$(CH_2)_5$-CH_3,
MEH: R_1 = O-CH_3 R_2 = O-CH($(CH_2)_5$-CH_2)(CH_3)
DMO: R = O-CH_3

In all cases, the spectral characteristics of the solution-cast films are identical after dissolution and re-casting; the π-π^* absorption edge is neither shifted nor significantly broadened. In addition, the residual absorption below the interband transition is extremely weak for the solution-cast films, indicating the absence of impurity or disorder induced states in the gap.

In addition to the solvatochromism observed on dissolution, thermochromic spectral shifts [17] have been observed for the P3AT's and extensively studied. In this case, the disorder is due to side-chain melting at elevated temperatures (but still well below the melting point of the polymer).

Similar features have been observed for the soluble PDalkoxyPv's. Although in this case, the addition of the alkoxy group reduces the π-π^* gap, adding greater lengths of alkyl chain has only a minor effect (see Figure II.3) [18]. As for the P3AT's, there is a blue shift and broadening on dissolution (although the shift is smaller than in the P3AT's, possibly indicating a more nearly linear conformation and larger localization lengths in solution. Again, the polymers can be dissolved and recast multiple times with no sign of degradation.

The spectroscopic studies in solution and after re-casting demonstrate the following:

1) The electronic structure of the P3ATs and PDalkoxyPV's is unchanged after dissolution and subsequent processing;
2) The solution-cast films have a well-defined electronic structure characteristic of relatively well-ordered material.

These conclusions imply that conjugated polymers can be processed without interfering with or degrading their π-electronic structure. The results imply, therefore, that for conducting polymers in general, solubility (and processibility etc.) and excellent electrical and optical properties are not at all mutually exclusive.

B. ELECTRONIC STRUCTURE (AND CONFORMATION) OF THE DOPED POLYMERS IN SOLUTION

a. Persistence lengths of Conjugated Polymers in Solution [19]. The conjugated polymer in solution illustrates an interesting competition between the conformational entropy, which gives rise to flexibility, and the electronic energy of the π-electrons, which tends to straighten the chains. To bring out these features, Spiegel et al [19] considered a simple model of an undoped conjugated chain in solution as a series of coupled rigid dimers with intra-dimer transfer integral t_o, each containing two π-electrons. The essential feature of the model was the assumption that the inter-site transfer integral is of the form

$$t_{inter} = t_1 \cos\phi \cos\theta \qquad (II.1)$$

where the angles ϕ and θ are defined in Figure II.4

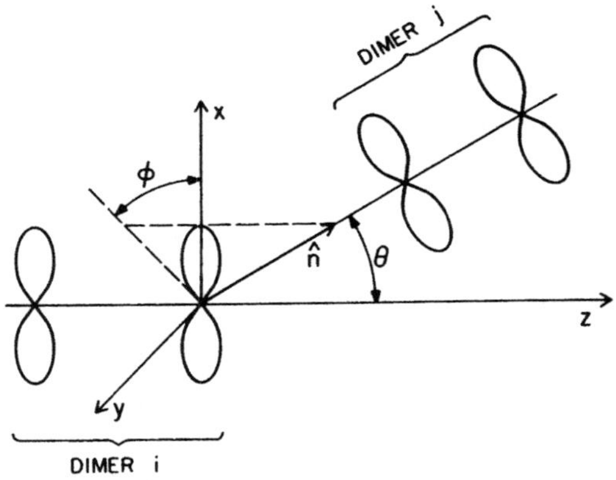

Figure II.4: Relative orientation of two adjacent dimers

Because $t_{inter} < t_o$ and the ground state of the single dimer is nondegenerate, perturbation theory can be used to calculate the effect of the inter-dimer interaction on the ground state energy. The first order matrix element, $\langle g|t_{inter}|g\rangle$, vanishes identically; the second order correction is

$$E_o^{(2)} = (t_1\cos\phi\cos\theta)^2/2t_o \quad \text{(II.2)}$$

Knowing the energy cost due to non-zero (ϕ,θ), one can calculate the associated thermal averages, $\langle\cos\theta\rangle$ and $\langle\cos\phi\rangle$. The result is that

$$\langle\cos\theta\rangle \approx 1\text{-}1/2\alpha \quad \text{(II.3)}$$

where $\alpha=(t_1)^2/2t_ok_BT$, and $\langle\cos\phi\rangle = 1$ (by symmetry). In this model, the dimerized conjugated polymer in solution is considered as being formed from many dimer pairs; the persistence length, b, along an infinite chain may then be found from

$$b/a = [1 + \langle\cos\phi\rangle\langle\cos\theta\rangle]/[1 - \langle\cos\phi\rangle\langle\cos\theta\rangle] \quad \text{(II.4)}$$

$$\approx 4\alpha = 2(t_1)^2/(t_ok_BT) \quad \text{(II.5)}$$

where a is the length of the dimer. For a typical conjugated chain, t_o~1 eV and $t_1 \approx 0.9t_o$, giving a persistence length of 50-80 dimers. Thus, conjugated chains in solution would be expected to be "wormlike".

This model somewhat overestimates the torsional energy for a conjugated chain. Since direct quantum chemical calculation indicate a smaller torsional energy [20], the persistence length may be somewhat smaller. On the other hand, for systems such as the P3AT's, the ratio t_1/t_o may be considerably less than unity. Thus, a persistence length of order 25-50 monomers is reasonable. This result is consistent with the experimental results obtained for P3OT in solution from Forced Rayleigh Scattering and from light scattering (see Section IIA) [16].

On the basis of theoretical arguments, it was expected that when doped away from a half-filled band configuration (one electron per monomeric unit), conjugated polymers would stiffen, i.e. their persistence length would increase. This occurs because the additional carriers may self-localize in rigid segments ("conformons") where the electronic kinetic energy is minimized [21]. In order to test the conformon idea in more detail, simulations have been carried out (hybrid Hartree-Monte Carlo calculations) in which the electronic hamiltonian (including the dependence of the electronic transfer matrix elements on bond rotations) is exactly diagonalized for a given random set of bond angles. The results are in excellent accord with the conformon picture. In particular, the existence of rigid regions containing electrons (i.e. conformons) has been verified; this novel self-localization is unique to conducting polymers in solution. In addition, the dependence of both conjugation and persistence length on temperature were analyzed in detail. These and the corresponding results on the structure of the energy spectrum suggest the existence of joint electronic and conformational changes in qualitative agreement with the thermochromic effects experimentally observed [17]. These effects (i.e. the expected stiffening of the conjugated chains in solution after doping) have not yet been verified experimentally.

b. Electronic Structure: Theoretical Concepts [2] In a non-degenerate ground state conjugated polymer such as the soluble P3AT's, the fundamental nonlinear excitations and the dominant charge storage configurations are polarons and bipolarons [2]. The relative stability of these species depends on the relative strength of the electron-phonon and electron-electron interactions in the polymer. The doubly charged bipolaron is a bound state (resulting from the electron-phonon interaction) between two polarons. However, the repulsive Coulomb interaction and the entropy favors the existence of two separated polarons. Hence, the free energy of creation of two polarons may be comparable to that of a bipolaron.

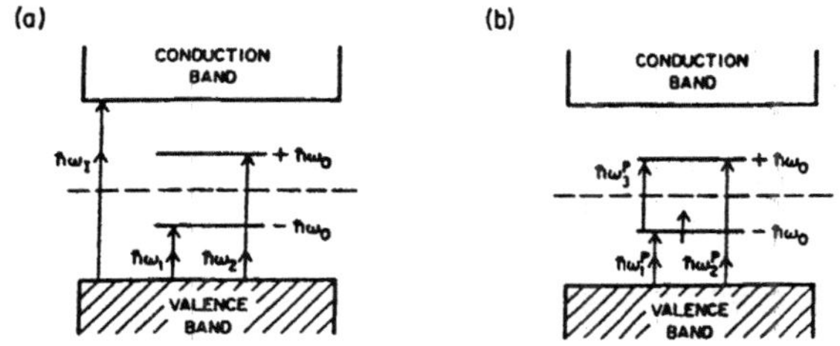

Figure II.5 Band diagrams

a. Bipolaron: The two gap states, the interband ($\hbar\omega_I$) and subgap ($\hbar\omega_1$) and ($\hbar\omega_2$) transitions are shown;

b. Polaron: the gap states are partially filled leading to a third subgap transition.

In Figs. II.5a and II.5b, we show the band diagrams with gap states and allowed optical transitions appropriate for a self-localized bipolaron and polaron, respectively [2]. Clearly, the two have distinguishing optical and magnetic signatures. A bipolaron with its two empty states in the gap carries a double positive charge and is spinless. Two broad electronic transitions from the valence band to the gap states are expected with energies less than the interband transition. On the other hand, a polaron has one unpaired electron in its lower gap state, and, hence, along with one positive charge, it has spin one-half. From Fig. II.5b, a polaron should display an additional sub-gap transition between two localized energy levels.

Using these magnetic and spectroscopic signatures, studies of polythiophene and the poly(3-alkylthiophenes) have provided evidence that upon doping or photoexcitation charge is stored in spinless bipolarons. The initial results were obtained from measurements on the polymer in the form of solid state films, powders and for the polymer in solution [5d,5e] There are reports, however, that polarons are generated as well as bipolarons [22]. This suggests a near degeneracy in the energy of two polarons versus a bipolaron, implying an equilibrium for the reaction

$$P^+ + P^+ \rightleftharpoons B^{2+}. \qquad (II.6)$$

Even in the case where bipolarons have the lowest creation energy, the entropy favors polarons so that a finite concentration of polarons can be expected.

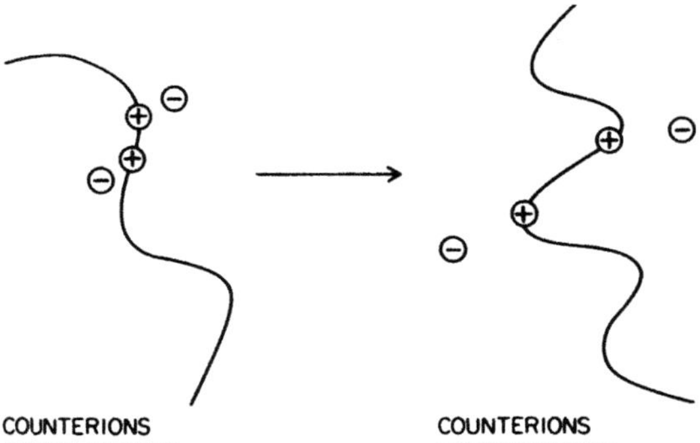

Figure II.6 Two possible limits of ionic screening

By performing the experiments on doped conducting polymer *in solution*, it becomes possible to explore the effect of counterion screening on the interactions in the polymer. The effective on-chain Coulomb interaction between two polarons, although always repulsive, is strongly dependent on the extent to which the "bare" positive charge of the polaron is screened. To first order, this is determined by the proximity of the counterion. Illustrated in Fig. II.6 are two possible limits [5d]. If, as in a non-polar solvent, the counterions are closely bound to the polarons on a chain, the net polaron-polaron interaction due to the electron-phonon interaction can be attractive. In a more ionic environment, such as a polar solvent, the counterions are more loosely bound to the chains. With a larger separation between charges in the π-system and counterion, the net polaron-polaron interaction becomes more repulsive. To some extent, then, it is possible in this way to control the relative energies of the doubly charged bipolaron in comparison to two singly charged polarons.

b. Magnetic and Spectroscopic Measurements of the Doped P3AT's in Solution [5d,5e] In studies of P3HT in solution, the strength of the counterion screening has been altered by varying both the solvent, polymer concentration or the salt concentration of a conducting polymer solution. For example, experimental studies were carried out with P3HT in solution in chloroform (nonpolar) and in methylene chloride (a relatively polar solvent). In methylene chloride, at the most dilute doping levels, the data indicate that charge storage is predominantly in polarons, in marked contrast with the low doping regime results of in chloroform. Only at higher doping levels in methylene chloride is the bipolaron the dominant charge storage configuration. This implies an important solvent dependence to the polaron-bipolaron equilibrium.

The strong dependence of the number of spins per charge on the amount of charge injected is shown in Figure II.7 in which the ratio of spins per charge is plotted as a function of doping level. Because of the fundamentally different spin/charge relationships exhibited by polarons versus bipolarons, the spin to charge ratio plotted in this curve can be thought of as a direct indication of relative fraction of polarons in the material (the remainder of the charge being stored in bipolarons). In this context, note that at the highest doping levels, one spin is present for approximately 20 charges deposited on

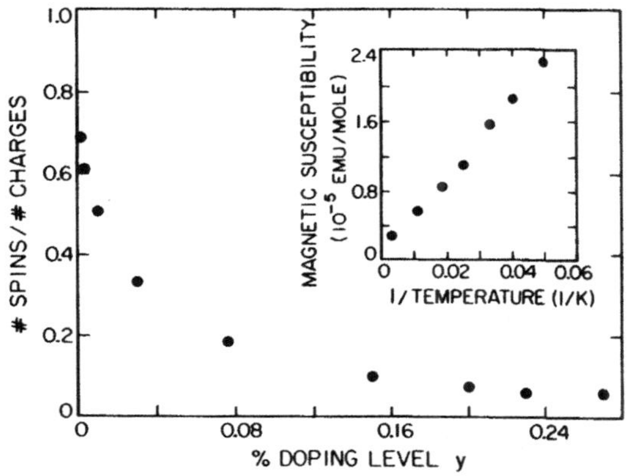

Figure II.7: Ratio of spins per charge vs the dopant concentration at a fixed polymer concentration in solution (in dichloromethane). Inset: Temperature dependence indicative of a Curie Law.

the polymer chains, strongly indicating that bipolarons are the dominant charge storage state in this regime. At low doping levels, however, this is clearly not the case. At the lowest doping level, the ratio approaches the value of 1.0, indicating that charge is stored in the self-localized states associated with polarons. This is consistent with the predictions of various statistical calculations which show that although the bipolaron is energetically favored, entropy favors the polaron, particularly at the most dilute concentrations. In the low y regime, the magnetic susceptibility of the solution, frozen throughout most of the measurement, exhibits a Curie law characteristic of a localized paramagnetic spin.

Most of the optical work was completed using a polymer concentration in solution at which the macromolecules just begin to overlap. The spectra at various doping levels were also measured at lower concentrations; no significant changes in the absorption spectrum were observed. Figure II.8 shows spectra obtained at different doping levels (y). For the neutral polymer, the interband transition has a threshold at approximately 2.2 - 2.3 eV. Upon doping at dilute levels ($y \leq 1\%$), one observes a defined peak at $\hbar\omega = 2.05$ eV, broad but discernable peaks at the energies of 1.2, 1.6 and 2.2 eV along with the trailing edge of an infrared feature. Using Fourier transform infrared interferometry, a broad absorption band was found at about 0.4 - 0.5 eV (although the FTIR results are complicated by strong solvent absorption lines in this spectral region). As the doping level is increased, the sub-gap absorptions grow at the expense of the interband transition. These features broaden and overlap such that at high doping levels ($y \approx 0.20$) there is a broad

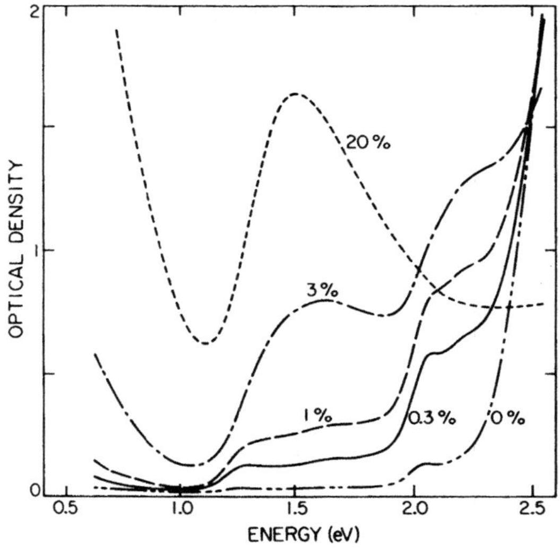

Figure II.8: Absorption spectra of P3HT in solution (in CH_2Cl_2) at various doping levels.

peak at 1.5 eV accompanied by the rising edge of a second absorption in the infrared, peaked at approximately 0.6 eV (as inferred from FTIR studies). Consistent with the optical sum rule, all the oscillator strength associated with the interband transition has been shifted to two broad peaks below the gap.

The higher dopant level data are consistent with optical spectra of heavily doped polythiophenes and their derivatives [6-9]. At y = 20%, oscillator strength has shifted away from the interband transition into the two lower energy gap states. These correspond to the possible absorptions for a bipolaron and suggest that at high doping levels charge is predominantly stored in bipolarons. This is fully consistent with the magnetic resonance results (Figure II.7) which indicate that at $y \approx 0.2$, only approximately 5% of the charge is stored in polarons with the remainder in spinless bipolarons.

It is possible to estimate the strength of the Coulomb interaction in this high doping regime by substituting these values in the relation [23]:

$$\hbar\omega_1 + \hbar\omega_2 = E_g - 2(U_B - E_B) \qquad \text{(II.7)}$$

where $2U_B$ is the difference in Coulomb energy between a singly and a doubly charged state and $2E_B$ is the binding energy of the counterion. The data of Figure II.8, imply $U_B - E_B \approx 0.1$ eV, suggesting that Coulomb energies on a polymer are roughly comparable to the energy of interaction of the self-localized charged states on the polymer with their associated counterions.

The gap states (see Figure II.5) associated with the bipolarons are located at approximately 0.6 eV and 1.5 eV above the valence band. This is in relatively close agreement with the values (0.34 eV and 1.67 eV) obtained by Bredas et al [24] using a generalized SSH model [2] (tight binding band structure plus σ-compressibility). The relatively small shifts between experiment and theory are consistent with relatively weak electron-electron Coulomb interactions, even for P3HT in dilute solution.

At the lowest doing levels (below 1%) the magnetic data (Figure II.7) indicate that charge is stored predominantly in the form of self-localized charged polarons. Thus, the complex spectrum at these low doping levels must be understood in the context of polarons. The corresponding assignments were worked out in detail in ref. 5d.

c. The Statistics of Polarons and Bipolarons [5d] The experimental results thus confirm that competition of the indirect attraction between two polarons via the self-consistent structural distortion and their Coulomb repulsion leads to a near degeneracy in the energy of two polarons versus that of a bipolaron, implying an equilibrium ($P^+ + P^+ \rightleftharpoons B^{2+}$.) in the polaron/bipolaron concentration. Even in the case where bipolarons have the lowest creation energy, the entropy favors polarons (there are more statistically independent configurations for two polarons each with two possible spin degrees of freedom than for a single spinless bipolaron) so that a finite concentration of polarons can be expected; particularly at low doping concentrations and at relatively high polymer concentrations

The configurational entropy of placing P polarons and B bipolarons on a polythiophene chain is an important component of the free energy. Consider, therefore, a simple model of the statistical mechanics of polarons and bipolarons on a chain [25]. Given a chain of N sites, it is possible to derive an expression for the total number of configurations in the continuum limit ($1 \ll B,P < N$) by considering the polarons and bipolarons to be solid beads on a ring (periodic boundary conditions). The number of configurations for P beads of length pa and for B beads of length ba on a ring of circumference Na (where a is the lattice spacing) is the sum (or integral) of the possible configurations of each of the beads, i.e.,

$$\Omega_1 = \int_{\substack{\text{bead}\\P+B}} \cdots \int_{\substack{\text{bead}\\1}} \frac{dx_1}{a} \cdots \frac{dx_{P+B}}{a} \tag{II.8}$$

where x_i is the position of the ith bead. As will become evident, one need not distinguish between polarons and bipolarons at this stage of the calculation. The first bead (a polaron in this case) can be placed from the origin, O, to the end of the second bead x_2 (less one polaronic bead length pa). Then:

$$\int_0^{x_2-pa} \frac{dx_1}{a} = \left(\frac{x_2 - pa}{a}\right) / 1! \tag{II.9}$$

If the next bead is a bipolaron, then

$$\int_{pa}^{x_3-ba} \frac{x_2 - pa}{a}\frac{dx_2}{a} = \frac{\left(\frac{x_3 - pa - ba}{a}\right)^2}{2!} \tag{II.10}$$

and so on:

$$\int_{pa+ba}^{x_4-pa} \frac{\left(\frac{x_3 - pa - ba}{a}\right)^2}{2!}\frac{dx_3}{a} = \frac{\left(\frac{x_4 - 2pa - ba}{a}\right)^3}{3!} \tag{II.11}$$

etc. Hence, the final integral has the following form:

$$\Omega_1 = \int_{Bba+(P-1)pa}^{Na-Pa} \frac{\left(\frac{x_{P+B} - Bba + (P-1)pa}{a}\right)^{(B+P-1)} \frac{dx_{P+B}}{a}}{(B+P-1)!} \tag{II.12}$$

which yields:

$$\Omega_1 = \frac{(N - Bb - Pp)^{(P+B)}}{(P+B)!} \qquad (II.13)$$

This is the number of configurations for P polarons and B bipolarons on a ring of N sites but for only one distinguishable sequence, for example, BBPPBPPP ... etc. To account for all the distinguishable ways to put P polarons and B bipolarons on a ring, a multiplicative factor must be introduced:

$$\Omega = \Omega_1 \binom{P+B}{P} \qquad (II.14)$$

which yields:

$$\Omega = \frac{(N - Bb - Pp)^{(P+B)}}{B!P!}. \qquad (II.15)$$

This is the total number of configurations for P polarons and B bipolarons on a ring of N sites.

Now it is possible to calculate the free energy for P polarons and B bipolarons on a ring corresponding to a polythiophene chain. Denote by E_p the energetic cost of adding a polaron to a chain of polythiophene rings, corresponding to the oxidation reaction:

$$PT^\circ \rightarrow P^+ + e^-. \qquad (II.16)$$

Similarly let E_B denote the energetic cost of adding a bipolaron to the system, corresponding to the reaction:

$$PT^\circ \rightarrow B^{2+} + 2e^-. \qquad (II.17)$$

Given these definitions, the Helmholtz free energy of a system containing P polarons and B bipolarons is:

$$F = PE_p + BE_B - kT \ln \Omega - kTP \ln 2 \qquad (II.18)$$

The last term is due to the spin degree of freedom of the polarons. This free energy determines the equilibrium chemical potentials μ_p and μ_B for polarons and bipolarons, respectively. Making use of Stirling's approximation:

$$\mu_P \equiv \left(\frac{\partial F}{\partial P}\right)_B$$

$$\approx E_P - kT\ln(N - Pp - Bb) + \frac{kT(P + B)p}{N - Pp - Bb} + kT\ln P - kT\ln 2 \quad (II.19)$$

and:

$$\mu_B \equiv \left(\frac{\partial F}{\partial B}\right)_P$$

$$\approx E_B - kT\ln(N - Pp - Bb) + \frac{kT(P + B)b}{N - Pp - Bb} + kT\ln B \quad (II.20)$$

According to the theory of chemical equilibrium, $\mu_B = 2\mu_P$. Hence, a second equation of μ_p can be obtained by writing:

$$\mu_P = \mu_B - \mu_P$$

i.e.

$$\mu_P = (E_B - E_P) + \frac{kT(B + P)(b-p)}{N - Pp - Bb} + kT\ln\left[\frac{2B}{P}\right]. \quad (II.21)$$

An expression for the equilibrium polaron concentration as a function of temperature and the amount of injected charge can be derived by combining the two equations for μ_p and using the law of charge conservation $Q = P + 2B$. Defining Δ as the difference in energy between polarons and bipolarons:

$$\Delta = 2E_p - E_B \quad (II.22)$$

the combined equations yield the relation:

$$kT\ln\left[\frac{4B(N - Pp - Bb)}{P^2}\right] = \Delta + \frac{kT(B + P)(2p - b)}{N - Pp - Bb}. \quad (II.23)$$

The last term in this equation is negligible in the low density limit where the concentration of injected charges is still dilute. Defining the variable, $\eta \equiv e^{\Delta/kT}$, eqn. II.23 becomes

$$\eta = \frac{4B(N - Pp - Bb)}{P^2} . \tag{II.24}$$

With the relation B = (Q-P)/2, this yields a quadratic equation for the polaron concentration P. If we assume 2p ≈ b (see ref. 5d and the discussion in the following paragraphs), this equation is greatly simplified to:

$$P = \frac{1}{\eta}\left[\sqrt{(Qp - 1)^2 + 2\eta Q(1 - Qp)} + (Qp - 1)\right] \tag{II.25}$$

where P = P/N, Q = Q/N.

Figure II.9a shows a series of plots of the calculated polaron concentration as a function of the injected charge Q for different values of the parameter η. For any positive value of the parameter Δ (bipolaron formation energetically favorable over two polarons), this curve exhibits a maximum whose height and position depend on the magnitude of Δ and the temperature. This demonstrates the essential point that the extra entropy of two polarons over a single bipolaron always permits a non-zero concentration of polarons. As bipolaron formation becomes increasingly favorable (Δ becomes arbitrarily large), the maximum polaron concentration becomes very small. At sufficiently high doping levels, the polaron concentration saturates and eventually begins to decline. Finally, if bipolarons are not energetically favored over polarons (Δ zero or negative), the polaron concentration linearly increases as a function of the injected charge for all practical doping levels, i.e., those below the half-filled case.

Figure II.9b is a plot of the calculated polaron concentration as a function of the injected charge Q in direct comparison with the experimental data. A best fit of the data suggests that the polaron width is approximately three thiophene rings (having fixed the bipolaron width to be approximately twice the polaron width) and yields Δ ≈ 0.16 eV, i.e., that bipolarons are indeed energetically favored over polarons. The statistics of the model ignore any interaction between polarons and bipolarons (e.g. polaron-bipolaron Coulomb repulsion) except for recombination. The plateau in the magnetic susceptibility versus doping level above 10% may imply such an inter-species interaction. An excellent fit for the data below 10% can be obtained (Fig. II.9, Inset), yielding approximately the same Δ of 0.15 eV.

The essential difference between the bead model and a previously proposed box model [26,27] is the enumeration of the total number of possible configurations of polarons and bipolarons. In a box model, a polymer chain is broken into an arbitrary collection of boxes into which one can place the charge storage species. In the bead model, a charged species is allowed to move over the chain in a continuous fashion, subject only to the restriction that it not pass through other "beads" on the chain. Simple Huckel calculations [24] (including a σ-compressibility term) have been carried out for

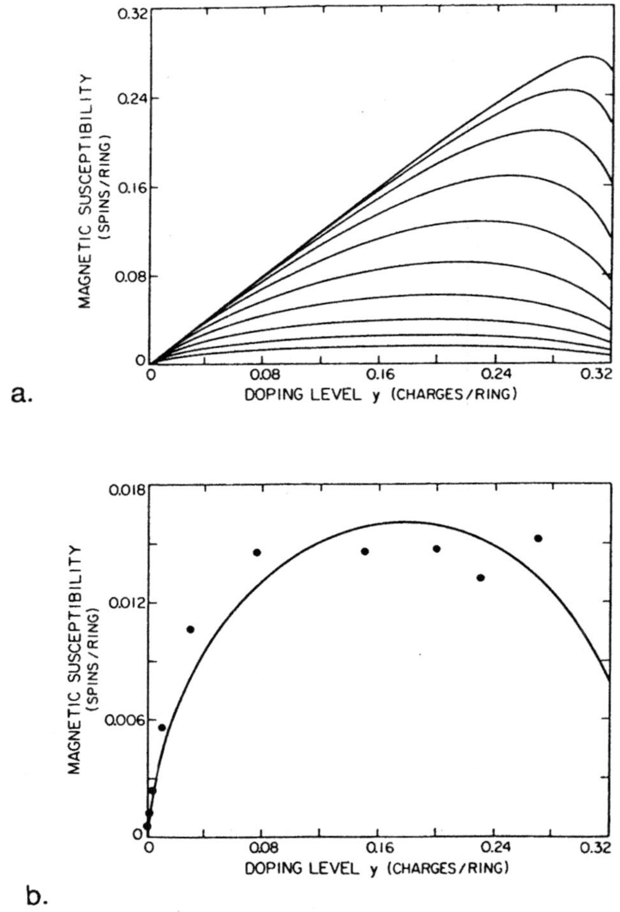

Figure II.9: a. A series of plots of the calculated polaron concentration P as a function of the injected charge Q for different values of the parameter η in the bead model. Beginning with the lowest curve: η = 592, 220, 80, 29.6, 10.8, 4.0, 1.48, 0.56, 0.20, and 0.072.

b. Magnetic susceptibility vs dopant concentration (at a fixed polymer concentration); the solid line is a best fit given by the bead model with η=582 (Δ=0.16 eV).

the energetics of polarons and bipolarons in polythiophene. The results indicate that a bipolaron is favored over two polarons by 0.26 eV. On the basis of the value $\Delta \approx 0.15$-0.16 eV, obtained from analysis of the experimental results, one can estimate that the electron-electron interactions (which are not explicitly included in the Huckel calculations and which favor singly charged polarons with respect to doubly charged bipolarons) reduce the energy difference by about 0.1 eV.

In summary, the statistics of polarons and bipolarons in solution imply an equilibrium, as in equation II.6. Generally, bipolarons have the lowest internal energy, but the entropy favors polarons at low doping levels. These results can therefore be used to generate samples with controlled fractions of either species. By controlling the number of polarons/bipolarons, and by using the implied increase in persistence length resulting from the conformon induced chain stiffness (see Section IIB) , it should be possible to enhance the processing conditions so as to achieve chain aligned polymers with a high degree of structural order.

d. Polyaniline in Solution (Fully and Partially Protonated) Among conducting polymers, polyaniline is unique in that its electronic structure and electrical properties can be reversibly controlled both by charge transfer doping (to vary the oxidation state of main chain) and by protonation [28-30]. Throughout the extensive literature, polyaniline had been generally categorized as an intractable material. Recently, however, methods have been reported to dissolve and process polyaniline (without changing the structure of the polymer) in certain organic solvents [31] or in concentrated sulfuric [10] and other strong acids [10]. In contrast to alternative methods for achieving solubility through preparation of substituted polyaniline or through the synthesis of graft or block copolymers, the resulting films and fibers are highly conductive after processing. The discussion in this section is focused on studies of polyaniline in solution.

The structure of the semiconducting emeraldine base form of polyaniline is (see Figure I.4)

$$[(1A)(2A)]_n \tag{II.26}$$

where

$$(1A) = (B—NH—B—NH—) \tag{II.27a}$$

and

$$(2A) = (B—N{=}Q{=}N—). \tag{II.27b}$$

In the above, B denotes a C_6H_4 ring in the benzenoid form and Q denotes a C_6H_4 ring in the quinoid form. The existence of one quinoid ring (out of four) in the emeraldine base has been well established in the literature [32-35]. The emeraldine base can be fully reduced to leucoemeraldine, $(1A)_n$, a large bandgap insulator with π-π^* transition at nearly 4 eV.

Upon protonation of $[(1A)(2A)]_n$ to the emeraldine salt, there is a structural change (with no change in the number of electrons) leading to a half-filled band and a metallic state (described as a "polaronic" metal) [36-39]

$$[1S]^{\cdot}(A^-)_n = [B\text{—}NH\text{—}B\text{—}NH^+\text{—}]^{\cdot}_n(A^-)_n$$

where A^- is the counter ion (e.g. ClO_4^-, Cl^-, HSO_4^-etc.), and the $[\ ]^{\cdot}$ denotes one unpaired electron per formula unit. Further oxidation of $[1S]^{\cdot}(A^-)_n$ is expected to yield the fully-oxidized copolymer with alternating quinoid and benzonoid monomer units:

$$(B\text{—}NH^+{=}Q{=}NH^+\text{—})_n$$

with an H-atom bonded to each nitrogen, and with one counterion for each charge on the chain. Since the ($-N^+{=}Q{=}N^+-$) unit is a charged bipolaron, the polymer structure is that of a charged bipolaron lattice with a two-fold degenerate ground state.

The solubility of polyaniline opens the way to processing pure (partially crystalline) polyaniline or composites of polyaniline with the other commercial polymers into fibers and films, etc. In addition, this solubility enables extensive characterization of polyaniline as a macromolecular system (e.g. viscosity in solution as a probe of molecular weight, etc.) and of polyaniline as a conducting polymer (e.g. optical studies of spin-cast films as a probe of electronic structure of the salt or base forms). The use of concentrated acids [10] has specific advantages in that both the salt and the base form of polyaniline can be completely dissolved at room temperature, with polymer concentrations ranging from extremely dilute to more than 20% (w/w). Although there has been considerable progress toward the development of soluble conducting polymers (such as the P3AT's and the PDalkoxyPV's), polyaniline in sulfuric acid is the first example of a stable, concentrated, solution of a conjugated polymer in which the material is converted by the solvent to the conducting form and from which the polymer can be processed directly into the metallic salt with no need for subsequent doping.

The inherent viscosity of the polyaniline base (in concentrated sulfuric acid) was used as an indicator of the molecular weight [10,40]. Measurements of the inherent viscosity of the polyaniline emeraldine base (obtained by exposure to aqueous NH_3 solutions and subsequently washed) were carried out using an Ubbelohde viscometer at 25°C using 0.1% w/w polymer solutions in H_2SO_4. By varying the details of the synthesis, the inherent viscosity of various batches of polyaniline was varied from about 1.0 dl/g at 25°C (0.1 wt% of the emeraldine base in H_2SO_4) to about 2.32 dL/g [41]. A given batch could be fractionated to get some finer control on the molecular weight distribution for studies of the molecular weight dependence. For example, samples with low molecular weight were extracted from the batch (i.e. that fraction which was soluble in THF), and samples with higher molecular weight were made from that fraction which was insoluble in THF (after the elimination of THF-soluble part by extraction). Note that the fraction which was insoluble in THF was soluble in concentrated sulfuric acid. Thus, the fraction which is insoluble in common organic solvents is not crosslinked, but simply has a relatively higher molecular weight. A sample with typical viscosity, η_{in}=1.6 dL/g, has a molecular weight (as estimated from the Mark-Houwink relations [42,43]) in the range from about 15,000 (rigid chain limit) to about 60,000 (flexible chain limit).

Studies of polyaniline in dilute solution in concentrated sulfuric acid are thus particularly interesting; in acid solution, the polyaniline chains are fully protonated to the emeraldine salt form. In organic solvents, such as N,N-dimethylacetamide, the degree of protonation can be controlled by the addition of perchloric acid [44]. Thus, spectroscopic and magnetic studies of the electronic structure of the one-dimensional semiconducting and "metallic" chains in solution are particularly interesting. On the other hand, when in solution the interchain interactions are clearly much weaker than in the solid state. As a result, the electronic structure is expected to be more nearly one-dimensional, and the possibility of a Peierls' transition with the formation of an energy gap in the excitation spectrum might be anticipated.

(i) Spectroscopic Studies of the Protonated Emeraldine Salt in Dilute Solution in Sulfuric Acid [40]

The solution spectrum of polyaniline dissolved in sulfuric acid is shown in Figure II.10; the data are presented as a function of wavelength to emphasize the changes found in near infrared region. Based on the above estimate of the molecular weight, the concentration (10^{-4} mole/liter) used for the solution spectroscopic studies is dilute; i.e. involving negligible interchain interaction. Comparing Figure II.10a with the absorption spectrum of a thin film of the emeraldine salt [40], shown in Figure II.10b, one finds that the two near infrared features appear at almost the same positions, 1.0 and 1.5 eV, respectively, while their relative intensities are inverse in the solid and solution spectra. In the dilute solution spectra, the intensity of 1.0 eV feature was always much higher than that of the 1.5 eV absorption; in fact, the 1.5 eV absorption is completely absent in the high molecular weight material, while the 1.0 eV absorption (with a long tail extending farther into the infrared) completely dominates. This result suggests that the 1.0 eV absorption band arises from an <u>intrachain excitation</u>, while the 1.5 eV transition is due to <u>interchain</u> charge separation.

Thus, the absorption spectrum of high molecular weight and fully protonated emeraldine salt, i.e. $[1S]^{\cdot}(A^-)_n$, in solution in sulfuric acid (and in partially crystalline films) the has three important spectral features;

(i) an <u>intra</u>chain absorption band at 1 eV (with a long tail into the infrared)

(ii) an <u>intra</u>chain absorption band at 2.5-3 eV ,

and

(iii) an <u>inter</u>chain absorption at 1.5 eV.

The 1 eV absorption appears to be dominated by the free carrier intraband absorption (with possibly a small gap in the deep infrared; see the following section on magnetic susceptibility). These features of the $[1S]^{\cdot}(A^-)_n$ spectrum are in good agreement with the transitions expected based on an analysis of the VEH electronic band structure calculations [44,45].

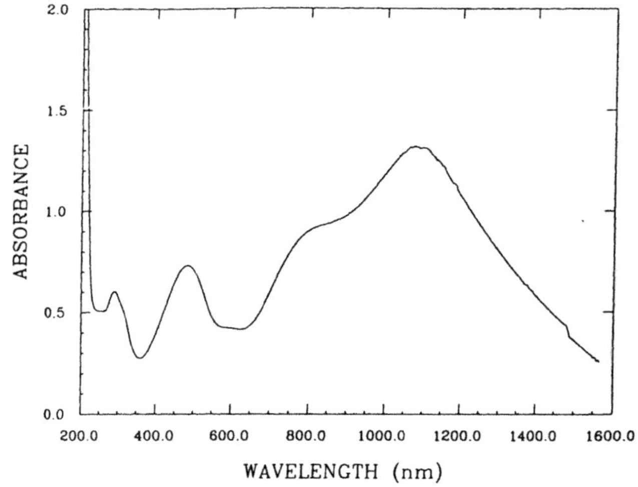

a.

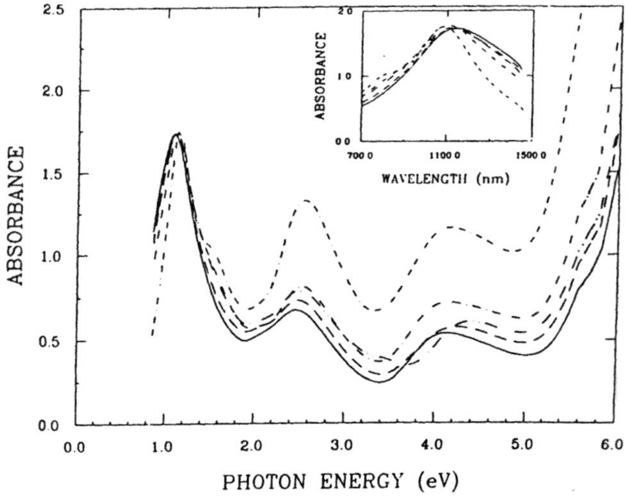

b.

Figure II.10:

a. Absorption spectrum of polyaniline dissolved in sulfuric acid.

b. Optical absorption spectra of fully protonated emeraldine salt, $[1S]^{\cdot}(A^{-})_n$, in sulfuric acid solution (97%); the absorbance is plotted as a function of photon energy. The IR region is shown as a function of wavelength in more detail in the inset. The spectra were obtained from solutions with different molecular weight as inferred from viscosity measurements (see text).

— — · — — · — — η_{in}=0.18 dL/g

— · — · — η_{in}=1.0 dL/g

— ·· — ·· — η_{in}=1.45 dL/g

— — — η_{in}=1.65 dL/g

———— η_{in}=2.27 dL/g

Overoxidation of polyaniline in sulfuric acid leads to conversion to the formation of an array of charged bipolarons on the polyaniline chains. The implied electronic structure was confirmed in detail by UV visible spectroscopy and by ^{13}C NMR spectra.

(ii) Magnetic Susceptibility Studies of the Protonated Emeraldine Salt in Dilute Solution in Sulfuric Acid [46]

Figure II.11 shows the temperature (T) dependence of the susceptibility of a solution with 10 wt% polyaniline in sulfuric acid. The magnetic susceptibility of the $[B\text{-}NH\text{-}B\text{-}NH\text{-}]^{+\cdot}{}_n(X^-)_n$ salt obtained by precipitation from sulfuric acid (and subsequently treated with HCl to ensure complete protonation) are also shown for comparison [47]. As the temperature is lowered, the susceptibility drops from 1.5×10^{-6} at room temperature to 3×10^{-7} emu/mole (2 rings) below 200K, at which point χ is reduced from that of the solid $[B\text{-}NH\text{-}B\text{-}NH^+\text{-}]^{\cdot}{}_n(X^-)_n$ salt by more than a factor of 200.

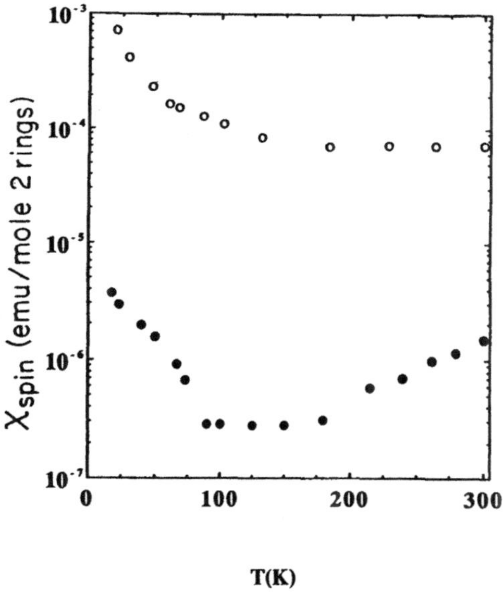

Figure II.11: Magnetic susceptibility (for polyaniline) vs temperature; black points, emeraldine salt in 10% (w/w) solution in sulfuric acid; open circles, solid emeraldine salt.

Because of the reduced interchain electron transfer interaction for conjugated chains in solution, the electronic structure is expected to be more nearly one-dimensional. Thus, the possibility of a Peierls' transition with the formation of an energy gap in the excitation spectrum might be anticipated. The susceptibility results shown in Figure II.11 suggest that this is indeed the case for polyaniline in solution in sulfuric acid. Such a large decrease in χ

cannot be accounted for in the context of the half-filled band expected for $[B\text{-}NH\text{-}B\text{-}NH\text{-}]^{+\cdot}{}_n$. To reduce χ by such a large factor would require an increase in the band width by the same factor; i.e. to a band width of more than 200 eV!

The small density of states, ~0.004 states per eV per formula unit (two rings), indicates the formation of an energy gap at the Fermi surface, consistent with a Peierls' transition. This residual value may arise from a sum of two contributions:

(i) A small fraction of the polyaniline is not in solution, but is suspended as fine particles;

(ii) States in the gap as a result of disorder in the $[B\text{-}NH\text{-}B\text{-}NH\text{-}]^{+\cdot}{}_n$ chains.

The small measured density of states sets an upper limit on the sum of these two contributions.

The magnetic susceptibility data imply, therefore, that polyaniline in sulfuric acid forms a true solution rather than a suspension of fine particles; at least about 99.6% of the polymer is in molecular solution (and that estimate is a lower limit). This is confirmed by the measured esr linewidth [46]. There is no indication of even a small fraction of the material with a linewidth of about 1-2 Gauss; i.e. that of the solid $[B\text{-}NH\text{-}B\text{-}NH\text{-}]^{+\cdot}{}_n(X^-)_n$ salt. Note that this conclusion is valid at low temperatures (e.g. 100K) implying that even in the glass of frozen sulfuric acid, the $[B\text{-}NH\text{-}B\text{-}NH\text{-}]^{+\cdot}{}_n$ chains are molecularly dispersed and in matrix isolation.

Unless the $[B\text{-}NH\text{-}B\text{-}NH\text{-}]^{+\cdot}{}_n$ chains are rigid rods when in solution, disorder would be expected to play a significant role and perhaps to even suppress the Peierls' instability. It is known, however, that for conjugated chains in solution, the conformation is determined by a competition between the increased entropy associated with the many degrees of freedom of a coiled or worm-like chain and the increased electronic energy associated with the breaking of π-bonds [19,20]. Moreover, even if the insulating conjugated chain is coiled or worm-like, the introduction of free carriers leads to the self-consistent formation of straight-chain segments in the vicinity of the carriers, i.e. conformons [21]. Thus, for the $[B\text{-}NH\text{-}B\text{-}NH\text{-}]^{+\cdot}{}_n$ chains, which have one carrier per formula unit, one is perhaps not surprized to find evidence of a rod-like conformation. That this must indeed be the case is implied by the small value of the residual susceptibility for chains in matrix isolation in the frozen glass. The weak disorder is not sufficient to suppress the Peierl's transition, nor is it sufficient to yield a density of localized states within the energy gap greater than about 0.004 states per eV per formula unit (two rings). Moreover, the increase in $\chi(T)$ at low temperatures can be used as an indication of disorder-induced localized states. Even this contribution is extremely small compared with that observed in the partially crystalline material cast from sulfuric acid. Thus, in solution the $[B\text{-}NH\text{-}B\text{-}NH\text{-}]^{+\cdot}{}_n$ chains have a relatively long persistence length (and a correspondingly long localization length) implying that the chain conformation is probably rod-like.

The order of magnitude increase in the ESR linewidth of the $[B\text{-}NH\text{-}B\text{-}NH\text{-}]^{+\cdot}{}_n$ chains in the frozen matrix is consistent with the interpretation that the residual susceptibility arises from localized states in the gap. The onset of narrowing at temperatures above 200 K may imply thermal excitation to more delocalized states, or it may simply reflect the onset of motion of the $[B\text{-}NH\text{-}B\text{-}NH\text{-}]^{+\cdot}{}_n$ chains as the frozen glass begins to melt.

Although the small magnitude of the magnetic susceptibility implies the existence of a gap in the excitation spectrum, it is not possible to directly determine the magnitude of the gap from the susceptibility data. The infrared absorption data (see Figure II.10) do not extend far enough into the IR to detect the gap. The up-turn in $\chi(T)$ at temperatures above 250K (and the corresponding onset of motional narrowing as inferred from the linewidth) may indicate the onset of 1d fluctuations as T approaches the mean field Peierls' temperature (T_c^{MF}). Since the increase in χ continues for temperatures at least up to 325K; i.e. 100K above the solid-liquid transition, the preliminary conclusion is that the increase in χ is not an artifact of the melting transition, suggesting that T_c^{MF} may be somewhat greater than room temperature.

e. Relationship Between the Solid State Electrical Conductivity and the Solution Viscosity of Polyaniline [48] Since polyaniline can be completely dissolved in concentrated sulfuric acid, extensive characterization of the polymer can be carried out using solution viscosity as a probe of the molecular weight. For example, employing solution viscosity as a guide, it was possible to vary the molecular weight of the polymer in a controlled manner by variation of polymerization and/or compensation parameters [41]. This progress made possible a systematic investigation of the dependence of the electrical conductivity of polyaniline on the inherent viscosity (and hence the molecular weight) [48]. In this section, the results of molecular weight dependence are briefly summarized.

(i) Preparation of Polyaniline Samples With Different Molecular Weights and Characterization of the Viscosities and Electrical Conductivities of the Samples

Emeraldine base materials of different inherent viscosities, $1.5<\eta_{in}<1.7$, $1.2<\eta_{in}<1.5$ and $0.7<\eta_{in}<1.2$ dL/g, were obtained using several different methods. Polyaniline base with η_{in}=2.3 dL/g was then prepared by extracting the fraction which was soluble in tetrahydrofuran (THF) from polyaniline base with η_{in}=1.65 dL/g. Oligomeric emeraldine base with η_{in}=0.2 dL/g was derived from the THF-soluble fraction of the extraction procedure. Fully protonated polyaniline samples with different viscosities were then prepared by stirring the various powders in a large excess of 0.5M aqueous HCl solution; complete protonation of polyaniline is achieved in 0.5M HCl solution during two hours, without a significant change in viscosity.

The inherent viscosities of the polyaniline base samples were determined at 25°C using solutions (0.1 wt.%) prepared in 97% sulfuric acid. In addition, a number of polyaniline samples of different inherent viscosities were analyzed by size-exclusion chromatography in concentrated sulfuric acid. Since the conformation and stiffness of polyaniline in sulfuric acid has not yet been determined, it is not possible to make direct correlations between the viscosity and the molecular weight. Instead, upper and lower limits of the molecular weights were estimated from comparison with viscosities measured under identical conditions of nylon 6 and poly-para(phenylene terephthalamide) and the Mark-Houwink relations [42,43] for these polymers.

The dried powder samples were compressed into pellets under a pressure of 160 Kg/cm^2. The compressed pellets were examined by scanning electron microscopy to ensure that under the above molding conditions the original powders materials formed a dense sample that was free of grain boundaries, which would obscure the conductivity measurements. It was established that the porous morphology had disappeared and that the pellets were homogeneous and without defects to a resolution of 50 nm. Electrical conductivities were measured on the pressed pellets by the usual four probe method.

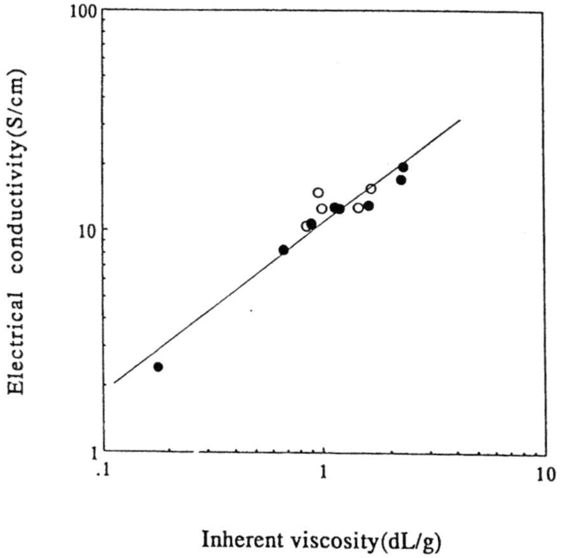

Figure II.12: Electrical conductivity of fully protonated polyaniline (emeraldine salt) vs inherent viscosity of the same polymers in sulfuric acid.

Figure II.12 shows the conductivity of fully protonated emeraldine base vs. the inherent viscosity. These data were obtained from the series of emeraldine base samples prepared by compensation from a single master batch (and subsequently protonated). Moreover, the data demonstrate that samples with identical viscosity, but prepared by different methods, exhibit the same conductivity after protonation. Similar results were obtained from samples precipitated from sulfuric acid, and from samples prepared either in air or in a controlled atmosphere dry box (to exclude moisture). The straight line dependence of the log-log plots (for example as in Figure II.12) indicates that the dependence of the conductivity on viscosity is a power law, $\sigma \propto \eta_{in}{}^{s}$ with $s \approx 0.7\text{-}0.9$. The Mark-Houwink relations between the molecular weight, M, and the viscosity, η, are estimated to be within the following limits: $\eta_{in} \propto M^{1.38}$ for a rigid conformation comparable to that of poly-p(phenylene terephthalamide) and $\eta_{in} \propto {}^{0.7}$ for a flexible conformation comparable to that of nylon 6; both determined in sulfuric acid. With these relations, the upper and lower bounds of the exponent β in the power law dependence, $\sigma \propto M^{\beta}$, over the experimental range studied:

(i) In the rigid chain limit (in solution), $\beta \approx 1.0\text{-}1.2$

(ii) In the flexible chain limit (in solution), $\beta \approx 0.5\text{-}0.6$.

(ii) Theory: Molecular Weight Dependence of the Conductivity

Typically, structural studies of conducting polymers indicate partial crystallinity with crystalline fractions varying from a few percent to greater than eighty percent (e.g. for carefully prepared polyacetylene). Recent X-ray diffraction studies of polyaniline have demonstrated significant crystallinity [49,50], although it is clear that all samples contain a significant volume fraction of amorphous material [51]. It is, therefore, useful to consider two limits; first, a homogeneous glass of Gaussian coils, and second, an oriented polymer with fully extended chains. In each case, a phenomenological description of the charge transport on the chains was developed based on two parameters:

a) the diffusion constant, D_{ch}, describing carrier diffusion along the backbone of the polymer chain;

b) the lifetime, τ_i, for a carrier on a given chain before it hops to a neighboring chain.

If Gaussian coils are assumed and if it is assumed that τ_i is independent of the degree of polymerization of the chains, N, which are taken, for simplicity, to be monodispersed, then the characteristic time τ_c, for a carrier to diffuse along the backbone over the entire length of the chain is given by

$$\tau_c = (Na)^2 / D_{ch}$$

where Na is the contour length of the macromolecule. In order to yield bulk conductivity, the carriers must ultimately perform interchain hopping. Comparing the two time scales, τ_i and τ_c, there are two regimes depending on which time is longer.

For $\tau_i >> \tau_c$, the carriers explore an entire polymer chain before hopping to a neighboring chain. In this limit, one can consider the bulk transport as interchain diffusion with a step-length given by the mean radius of gyration, R_g, of the polymers, which in the Gaussian approximation is given by $R_g = N^{1/2}a$; with a corresponding diffusion constant $D_i=(R_g)^2/\tau_i$. Then, if n is the concentration of carriers the conductivity, σ, may be written as $\sigma = ne\mu$ with

$$\mu = eD_i/k_BT = e(R_g)^2/k_BT\tau_i \qquad (II.28)$$

where k_B is the Boltzmann constant, and T is the absolute temperature. Substitution yields:

$$\sigma = ne^2(R_g)^2/k_BT\tau_i = ne^2Na^2/k_BT\tau_i \qquad (II.29)$$

Thus, for Gaussian chains, this expression yields a conductivity, $\sigma \propto N$. More generally, if ν is the usual Flory exponent defined by $R_g \propto N^\nu$, then $\sigma \propto N^{2\nu}$. For self-avoiding random walks, the coils are expanded such that $\nu \approx 0.6$. Thus, for an amorphous glass of random coils, the conductivity increases with increasing molecular weight as N^β with $1.0<\beta<1.2$.

As N increases, however, eventually τ_c becomes comparable to and then exceeds τ_i. For $\tau_c >> \tau_i$, the distance, s, travelled by the carrier along the polymer contour in a time τ_i is given by $s^2=D_{ch}\tau_i$. Projecting this onto a cartesian frame, the real space distance which the carrier moves, d, is

$$d=a(s/a)^{1/2}=a^{1/2}(D_{ch}\tau_i)^{1/4} \qquad (II.30)$$

and the corresponding conductivity becomes

$$\sigma = (ne^2a/k_BT)(D_{ch}/\tau_i)^{1/2}. \qquad (II.31)$$

Notice that in this regime the conductivity is <u>independent</u> of the molecular weight. Thus, with increasing molecular weight there is a crossover from σ increasing approximately linearly with the molecular weight (eq. II.29) to a saturated behavior at the value (given by eq. II.31) which depends on the geometric mean between the two (intra-chain and inter-chain) characteristic microscopic hopping rates.

A similar analysis can be carried out for an oriented polymer with fully extended chains of length Na. Again, for $\tau_i >> \tau_c$, the carriers explore an entire polymer chain before hopping to a neighboring chain. In this limit, the bulk transport takes place through interchain diffusion with a step-length given by the mean chain length (Na) with a corresponding diffusion constant $D_i=(Na)^2/\tau_i$. Then, if n is the concentration of carriers the conductivity, σ, may be written as

$$\sigma = ne^2(Na)^2/k_BT\tau_i \qquad (II.32)$$

Thus, for chain extended and oriented conducting polymers, $\sigma \propto N^2$.

For $\tau_c >> \tau_i$, the distance travelled by the carrier along the straight polymer chain is determined by D_{ch} so that $\mu = eD_{ch}/k_BT$, and

$$\sigma = ne^2 D_{ch}/k_BT \qquad \text{(II.33)}$$

In this regime the conductivity is once again independent of the molecular weight, and is limited by the intra-chain diffusion constant. For a metal, the intrachain diffusion constant is $D_{ch} = (v_F)^2\tau_{ch}$ where τ_{ch} is the characteristic scattering time for on-chain scattering from defects or phonons etc., and the effective number of carriers is $N(E_F)k_BT$, where $N(E_F)$ is the density of states at the Fermi energy. Thus, using $N(E_F) \approx n/E_F$ and $E_F = mv_F^2/2$, one finds the expected simple expression well-known from the physics of metals;

$$\sigma = ne^2\tau_{ch}/m. \qquad \text{(II.34)}$$

It is of interest to compare the results obtained for the amorphous glassy limit with those obtained for the oriented, chain extended limit.

(i) For $\tau_i >> \tau_c$,

$$\sigma_{orient}/\sigma_{amorph} = N. \qquad \text{(II.35)}$$

(ii) For $\tau_c >> \tau_i$,

$$\sigma_{orient}/\sigma_{amorph} = a^{-1}(D_{ch}\tau_i)^{1/2} = \lambda_d/a \qquad \text{(II.36)}$$

where λ_d is the characteristic diffusion length along a chain prior to interchain hopping. Again, using $D_{ch} = (v_F)^2\tau_{ch}$, one finds

$$\sigma_{orient}/\sigma_{amorph}\lambda_d/a \cong a^{-1}v_F(\tau_i\tau_{ch})^{1/2} >> 1. \qquad \text{(II.37)}$$

The importance of chain extension and alignment is clearly evident; in both limits, the conductivity is greatly enhanced in the oriented, chain aligned case.

(iii) Implications for Polyaniline

The strong molecular weight dependence of the electrical conductivity demonstrated by the experimental results indicates the need for long, uninterrupted chains in order to obtain highly conducting materials. This is consistent with the previous conclusion deduced from the molecular weight dependence of the absorption spectra of polyaniline films cast from sulfuric acid [41]. In these studies, it was observed that upon increasing the molecular weight from $\eta_{in}=1.0$ to $\eta_{in}=1.6$ dL/g the infrared absorption becomes a broad band with a long tail extending into the deep infrared, characteristic of metallic behavior.

For polyaniline, the experimentally observed molecular weight dependence is $\sigma \propto M^{1.0-1.2}$ assuming a rigid rod conformation in solution; or $\sigma \propto M^{0.5-0.6}$ assuming a flexible chain conformation in solution, with the uncertainty due to the lack of knowledge of the chain conformation in solution in sulfuric acid. One can, therefore, infer the following:

(i) In polyaniline the transport is limited by the poor crystalline order;

(ii) Either $\tau_i >> \tau_c$ (the regime of validity of eqn II.29) and the conformation in solution is that of relatively rigid rods, or $\tau_i \sim \tau_c$ such that the data are in the crossover regime (between the linear N-dependence of eqn II.29 and the N-independent limit of eqn II.31) and the conformation in solution is that of a flexible chain.

That transport along the chain is much more facile than interchain hopping is not surprising; as argued in Section I, the large intra-chain π-electron transfer integrals characteristic of conjugated polymers imply that this should be the case. In fact, intra-chain band widths of several electron volts have been estimated for the conduction band of doped polyaniline [44]. The implied relatively large value for τ_i is consistent with relatively weak interchain coupling typical of doped conjugated polymers.

Although a stiff chain conformation of polyaniline in solution in sulfuric acid is consistent with the conclusions inferred from the magnetic susceptibility data [46], the absence of evidence of liquid crystallinity for polyaniline in sulfuric acid suggests a more flexible chain conformation. On the other hand, the strong interchain electron transfer interactions inherent to π-conjugated polymers may lead to crystallization (and precipitation) rather than liquid crystal formation.

Based upon the comparison of the experimental results with theory, it is clear that for polyaniline increased molecular weight will lead to enhanced electrical conductivities. Furthermore, since the transport is currently limited by disorder, it may be possible to achieve a major increase in electrical conductivity (by a factor of order N, the number of monomers in a typical macromolecular chain) with improved chain extension and alignment as a result of post-synthesis processing.

III. Electrical and Mechanical Properties of Oriented Poly(3-alkylthiophenes) Processed from Solution [52]

Since the P3AT's, with side chain lengths greater than 4 carbon atoms, are both fusible and soluble (in a variety of organic solvents), they can be easily processed. The P3AT's exhibit levels of electrical conductivity that are of practical interest; the conductivities can be controlled over a wide range by doping, and they are reasonably stable. To clarify and quantify the relevant materials properties, a systematic study of the effects of both the main chain orientation and the side chain length on the electrical and mechanical properties of these materials has been carried out.

Early work on the poly(3-alkylthiophenes) focused on materials processed from the melt at elevated temperatures [7]. However, because of substantial cross-linking effects at high temperatures, room-temperature solution processing may be preferable as the method to fabricate films and fibers. The solution processing studies concentrated on poly(3-octylthiophene) P3OT as a prototypical system and demonstrated that through a simple solution-processing/drawing route, P3OT can be formed into oriented materials. Upon doping, the highly conducting oriented fibers were found to display significant improvements in mechanical properties [52] with respect to those of the undoped fibers [53a].

A. FIBER SPINNING AND DRAWING

Samples with two different molecular weights were studied. The lower molecular weight sample had a weight average molecular weight ($\bar{M}_w$) of 89,000 and a polydispersity of $\bar{M}_w / \bar{M}_n$=3.3 (vs polystyrene standards); the intrinsic viscosity in chloroform was 0.62 dl/g at 25 ^oC. The higher molecular weight sample had $\bar{M}_w$ of 189,000 and a polydispersity of 4.2; the intrinsic viscosity in chloroform was 1.05 dl/g at 25 ^oC. Filtered P3OT was redissolved in chloroform to produce spinning solutions of 10 wt. % polymer, and the solutions were wet spun into acetone. No draw down was applied during the spinning process. The wet-spun fibers were wound onto bobbins directly from the coagulation bath and subsequently dried under vacuum for 24 hours. Dark red fibers with a linear density of approximately 80 denier were obtained.

Drawing of the dried fibers was carried out continuously in a temperature controlled tube furnace at 105 ^oC. The draw ratio was varied up to a maximum of about 6 by changing the relative speed of the feed and take up spools at either end of the heated tube. The speed of the feed spool varied from 3 to 6 cm/min, while take up varied from 8 to 32 cm/min, depending on the desired draw ratio. The length of the heated zone of the tube furnace was ≅12 cm.

Average cross-sectional areas of fibers were determined from their linear density and specific gravity. The latter was determined by the floatation technique, using water/ethylene glycol solutions for undoped filaments and water/sucrose solutions for doped samples. The densities of undoped fibers ranged from 1.03 to 1.05 g/cc, while those of doped fibers ranged from 1.05 to 1.20 g/cc, depending on the dopant content. The density of an undoped free-standing cast film was 1.05 g/cc.

B. CHARACTERIZATION OF THE DRAWN P3OT FIBERS

In an initial step to design optimal tensile deformation conditions for alignment, solution-cast specimens were drawn over a wide range of temperatures. The maximum draw ratio (l_{max}) versus temperature was determined at an initial elongation rate of 100%/min; a maximum occurred at 105 °C. Dried, wet-spun filaments of were therefore drawn in a continuous fashion through a heated tube furnace at this optimum temperature of 105 °C. The drawing process was carried out both in air and nitrogen atmospheres; no differences were detected in the properties. The higher molecular weight sample could be drawn to slightly higher draw ratios. The maximum values were, however, relatively low for both materials due to their low degrees of polymerization: 139 and 232 respectively.

The modest draw ratios resulted in significantly increased orientation of the P3OT macromolecules. The X-ray fiber pattern of Figure III.1 shows that tensile drawing was moderately effective in increasing the overall order of the macromolecules, as evidenced by the lateral arcs and equatorial diffraction spots. The order was, however, far from perfect as an intense amorphous halo is still present .

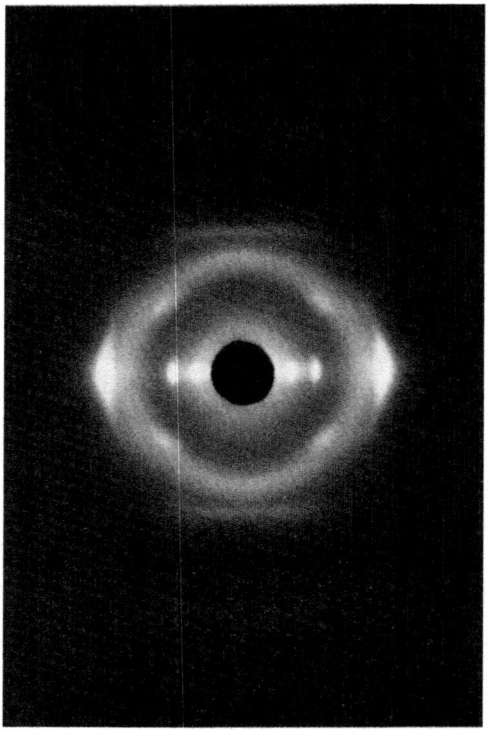

Figure III.1: Wide angle X-ray pattern of drawn P3OT fibers.

The retention of a large portion of disordered material in the drawn samples most likely is the result of two contributing factors. First, the chemical architecture of the poly(3-alkylthiophene) macromolecules is inherently disordered. Nuclear magnetic resonance data suggest [53b] that both chemically and electrochemically prepared poly(3-alkylthiophenes) may contain only 70-80% head-to-tail couplings, the remaining 20-30 % being head-to-head and tail-to-tail defects. Random disruption of the monomer sequence distribution will, of course, significantly affect the ability of the macromolecules to pack, and arrange, in a regular fashion. A second contribution to remnant disorder in drawn P3OT results from chain relaxation and recoil during the drawing process, which is more pronounced as deformation occurs in a non-affine fashion.

Tensile drawing resulted in increases in both the elastic modulus and strength, or tenacity, of the P3OT samples. There was also a commensurate decrease in the strain to break at increasing draw ratio. The mechanical properties achieved were, however, not exemplary. The strength and stiffness of drawn, undoped, P3OT (shown in Figure III.2) were comparable with those of elastomers, and far below those of drawn unsubstituted polythiophene [53a]. Similarly low stiffness has been observed for rigid polymers solubilized also via the attachment of long alkyl side chains[54,55].

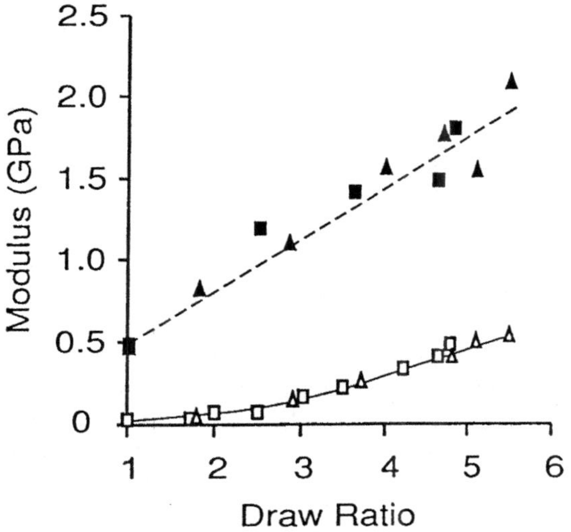

Figure III.2: Young's modulus of undoped P3OT fibers (open squares and triangles for the two different molecular weights) and $FeCl_3 \cdot 6H_2O$ doped P3OT (solid squares and triangles for the two different molecular weights).

Unlike many other conducting polymers, such as polyacetylene (see Section VIII), PTV and PDMPV (see Section VII), the mechanical properties of P3OT did not decrease, but significantly *increased* upon doping to moderate levels (> 18 mole %). Doping with $FeCl_3$ resulted in a 400% (or more) increase in the stiffness, while the strength improved by up to a factor of 3. This is illustrated in Figure III.3.

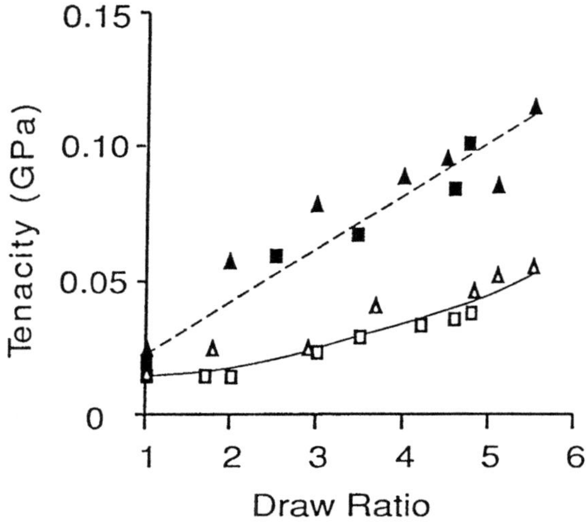

Figure III.3: Tenacity of undoped P3OT fibers (open squares and triangles for the two different molecular weights) and $FeCl_3 \cdot 6H2O$ doped P3OT (solid squares and triangles for the two different molecular weights).

A possible cause for the dramatic increase of the mechanical properties could be the occurrence of considerable cross-linking resulting from doping. To more fully understand the effects of the doping reaction on the mechanical properties of P3OT (and to check for evidence of doping-induced cross-linking), different dopants were utilized. Because the charge transfer doping reaction with iodine is fully reversible ($3I_2+2e \rightleftharpoons 2I_3^-$), iodine doping was investigated in order to clarify the possibility of cross-linking reactions. The effect of the I_3^- dopant-counterion on the mechanical properties was examined in a vapor doping/vacuum dedoping experiment. The gradual loss of dopant molecules from the P3OT fibers, accelerated slightly by subjecting the samples to a moderate vacuum, made it possible to monitor the changes in the mechanical (and electrical) properties as a function of dopant level.

Although the effect of iodine doping on the mechanical properties was not completely reversible, any residual changes were significantly lower than for $FeCl_4^-$, since redissolution was nearly complete after doping and undoping with iodine. The degree of irreversibility with iodine doping was, in fact, so low that the large strain at break of both the isotropic and slightly drawn sample was nearly recovered after dedoping. If there had been substantial levels of crosslinking, a decrease in the strain at break following the doping/dedoping procedure would have anticipated. In contrast to the reversible changes in the modulus, the doping induced increase of the tensile strength appeared to be of a more permanent nature, which is most likely due to a slight amount of cross-linking.

X-ray studies by Winokur et al [56] imply a continuous structural change as a function of the iodine doping level with alkyl side-chain rotation into the plane containing the backbone and with associated interdigitation. Diagrams of the structure of P3OT in the neutral and doped forms are shown in Figures III.4a and III.4b.

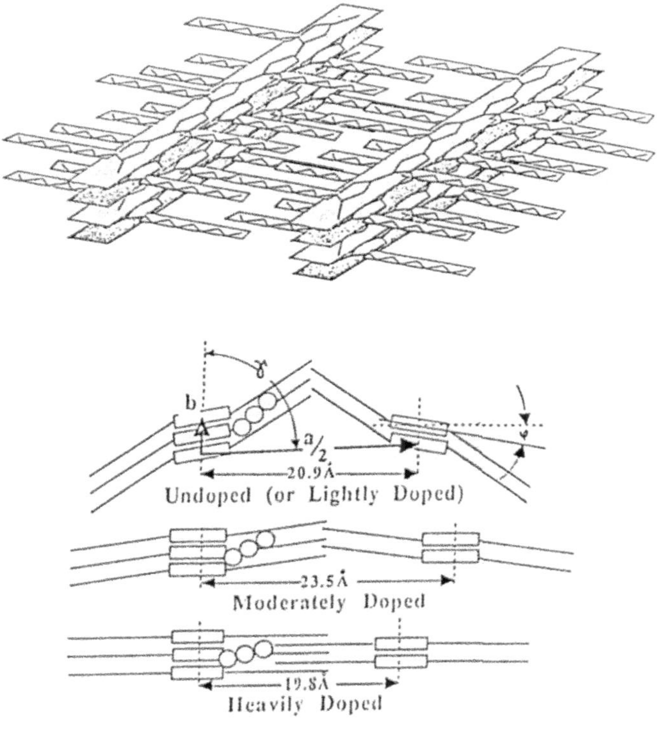

Figure III.4: The structure of P3OT: Upper; lamellar packing of the P3AT's, lower; projected equatorial structure at various I_3^- doping levels

Side-chain interdigitation may, of course, improve the stiffness through increased interactions of the alkyl groups. Moreover, the data indicate that the polythiophene main chains are packed planar structures with relatively good interchain π-π overlap. Thus, in addition to the enhanced side-chain interaction, the observed increase in stiffness may benefit from a more effective secondary (interchain) π-bond overlap in the doped samples. In accord with the mechanical measurements, Winokur et al [56] have also noted that changes in the equatorial d-spacings are reversible during the doping/dedoping process.

Commensurate with the modulus and tenacity, the electrical conductivity doped samples increased with draw ratio (Figure III.5). This increased conductivity with improved chain orientation is consistent with similar observations for many other conducting polymers (see following sections). As anticipated from the conductivity vs. draw ratio data, there were also strong, essentially linear correlations between the conductivity and the mechanical properties, as is illustrated in Figures III.6a and b.

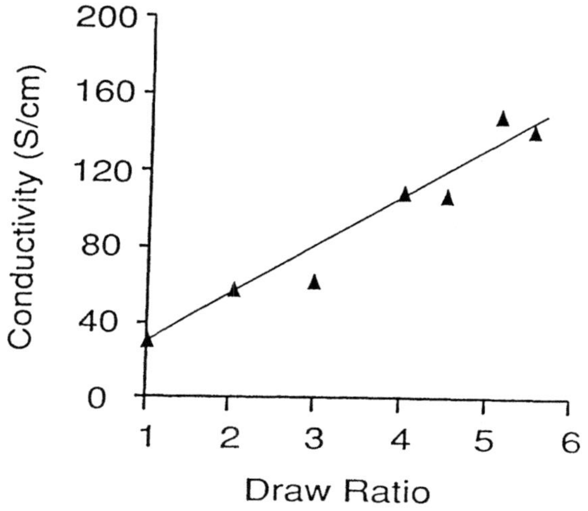

Figure III.5: Electrical conductivity of P3OT (doped with $FeCl_4^-$) as a function of the draw ratio.

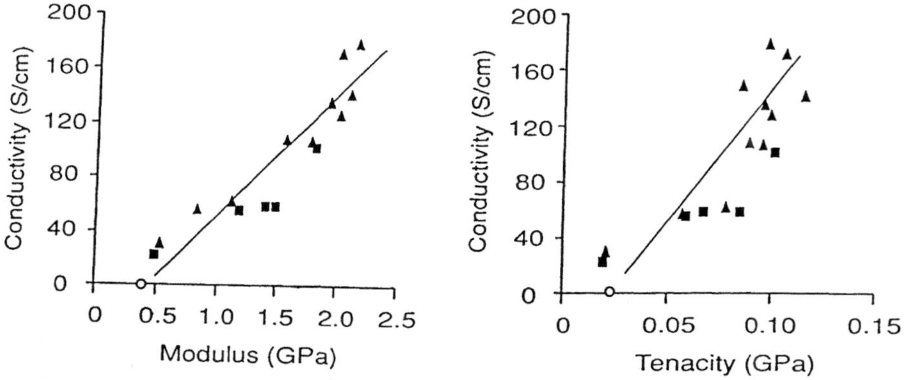

Figure III.6: a. Electrical conductivity vs modulus of P3OT as a function of the draw ratio.
b. Electrical conductivity vs tenacity of P3OT as a function of the draw ratio.

C. EFFECT OF SIDE-CHAIN LENGTH [57]

In order to examine the effects of the side chain length, chemically prepared [2] samples of poly(3-hexylthiophene) (P3HT), poly(3-octylthiophene) (P3OT) and poly(3-dodecylthiophene) (P3DDT) were used. Wide-angle x-ray scattering patterns indicated that all of the samples were oriented, although there appeared to be an increasing level of disorder with increasing side chain length, as evidenced by an increase in the intensity of the amorphous halos.

The side chain length had a significant effect on the relative changes in stiffness with doping, i.e. the increase in doped modulus to undoped modulus at constant draw ratio. P3DDT showed the larger change, while P3HT showed the least. Denoting the modulus as E, a plot of $(dE_{doped}/d\ l)/(dE_{undoped}/dl)$ vs side chain length extrapolates to approximately 1, indicating that there should not be a significant change in stiffness upon doping unsubstituted polythiophene, consistent with the data of Ito et al[53a].

Drawing improved the electrical conductivity by factors of 3-5 for doped samples, ≈ 20 mole% $FeCl_4^-$. As with P3OT, strong correlations between the electrical conductivity and mechanical properties were observed. In particular, there were linear relationships between the conductivity and modulus for all of the doped P3ATs. As would be expected, the P3DDT, with the lowest fraction of main chain per unit cell cross-section, exhibited the lowest modulus and the lowest conductivity, while the highest values were observed for the P3HT. More importantly there was no dependance on side chain length when the conductivities were compared with respect to strength. In fact, data for unsubstituted polythiophene was superimposable on the P3AT data.

In summary, upon doping, the highly conducting oriented P3AT fibers displayed significant improvements in mechanical and electrical properties with respect to those of the undoped fibers. Oriented and doped P3OT fibers showed an elastic modulus and tenacity of 2.1 GPa and 0.12 GPa versus values of 0.5 GPa and 0.05 GPa, respectively, for the undoped filaments; with conductivities as high as 180 S/cm. The increase in the Young's modulus upon doping with iodine was reversible. This (reversible) increase in mechanical properties results from increased interchain π-electron interactions, and from interdigitation of the side-chains in the structure of the doped material. Strong correlations were observed between the mechanical properties and conductivity of doped P3OT; both improved with the degree of chain extension and chain alignment resulting from tensile drawing.

IV. Gels and Blends of the P3AT's Processed From Solution

A. CONDUCTING POLYMER BLENDS OF SOLUBLE POLYTHIOPHENE DERIVATIVES IN POLYSTYRENE [58]

The optical and electronic properties of blends of soluble polythiophene derivatives and polystyrene (PSt) were investigated over the full range of compositions. The polyblend films, made conducting by charge transfer doping (reaction with $NOPF_6$ in acetonitrile) display an insulator-metal transition at the percolation threshold; a volume fraction of ≈ 16% conducting polymer in PSt.

The conducting fraction was a copolymer, poly(3-hexylthienylene-co-3-benzylthienylene, hereafter referred to as P3HBzT, chosen because of excellent solubility in many organic solvents. Infrared and electronic absorption spectroscopy studies verified that the resulting co-polymer has a well-defined chemical structure with negligible crosslinking.

Electronic absorption spectra of thin films of the blend materials are consistent with the spectral details of the parent polymers; the absorption spectra of P3HBzT/PSt blends are identical to those of the P3AT's, both in the neutral state and after doping. These spectroscopic results indicate that the P3HBzT is highly ordered in the PSt matrix and suggest that the P3HBzT and PSt have segregated. The spectral profiles of the fully doped blends are independent of the ratio of P3HBzT to PSt, indicating that the doping level is not affected by dilution of the conducting component. However, DSC scans show that the sharp glass transition of PSt alone is smeared out in the blend so that a model of macroscopic segregation is certainly oversimplified. Evidently the P3HBzT is relatively uniformly dispersed in the PSt in aggregates with a characteristic length scale that is small enough to interfere with the glass transition. Since the latter must be of the order of the radius of gyration of the PSt, the characteristic scale is about 5 nm.

The electrical conductivity (at room temperature) of the doped polyblend films is shown as a function of the volume fraction (in %) of conducting polymer in Figure IV.1. At 100%, corresponding to pure P3HBzT, the conductivity is ≈ 10 S/cm, comparable to that for a solution cast film of P3HT. A relatively sharp conductivity threshold was found at ≈16% volume fraction of conducting polymer; at lower volume fractions, the doped polyblend is essentially an insulator. The conductivity percolation transition covers more than eight orders of magnitude with reasonably high conductivities above approximately 30%. Thus, for volume fractions less than 16%, there are no macroscopically connected paths, whereas at higher volume fractions successively larger numbers of connected paths lead to the relatively high conductivity. Similar results have been reported for polyacetylene/PSt blends with the percolation threshold reported at 16±2% volume fraction of polyacetylene.

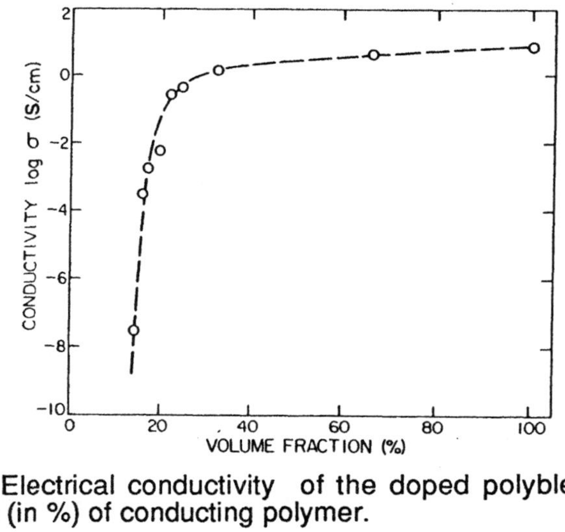

Figure IV.1: Electrical conductivity of the doped polyblend films vs volume fraction (in %) of conducting polymer.

Although, one might expect some sensitivity of the percolation threshold to the specifics of the micro-structure of the aggregates which make up the phase-segregated polyblend, it has been shown [59] that both for site percolation and for continuum percolation, the critical volume fraction, ϕ_c, is quite universal; for percolation in three dimensions $\phi_c \approx 0.16$.

The problem of phase segregation in polymer blends is well-known [60]; because of the large molecular weight of macromolecules with high polymerization index, the entropy of mixing is small. This can be seen directly from examination of the Flory-Huggins free energy for a mixture of two polymer species (A and B) with degrees of polymerization N_A and N_B.

$$F/k_BT = [\phi/N_A]\ln\phi + [(1-\phi)/N_B]\ln(1-\phi) + \chi\phi(1-\phi) \qquad \text{(IV.1)}$$

where ϕ is the volume fraction of species A, $(1-\phi)$ is the volume fraction of species B, and χ is the coefficient of the inter-species interaction term. The first two terms represent the configurational entropy for arranging the A and B chains. For $\chi = 0$, entropy effects are dominant, and they favor mixing. However, since the entropy terms are of order 1/N due to the relatively small number of degrees of freedom of the macromolecules compared to the number of degrees of freedom of the N unconnected monomers, finite $\chi > \chi_c$ (where χ_c is finite, but small) leads to phase separation. The coexistence curve is defined by $\partial F/\partial\phi = 0$, which (for the simple case of $N_A=N_B=N$) leads to

$$\chi_c = 2/N \qquad \text{(IV.2)}$$

As noted by DeGennes [60], this very small value of χ_c is the essential reason for the strong incompatibility of different polymers and the essential reason why phase separation in polymer blends is generally to be expected.

On the other hand, it has been shown that for thin rods (length L and diameter d), in three dimensions, the percolation threshold is determined by the excluded volume per rod, $V_{exc} = \pi L^2 d$ [61,62]. Thus if the conducting polymer fraction could be dispersed as rod-like aggregates, percolation at a much lower volume fraction would be expected, i.e. when $NV_{exc} \sim \phi_c$, where N is the number of rodlike aggregates per unit volume. Thus, by re-organizing the conducting fraction into aggregates with desired microstructure, one can extend the conducing region to lower volume fractions.

B. CONDUCTING POLYMER GELS: A SELF ASSEMBLING CONDUCTING NETWORK WITH REMARKABLY LOW PERCOLATION THRESHOLD [63]

It is of interest to consider percolation of a conducting polymer on the pre-existing network of a polymer gel. Since the conjugated polymer can adsorb onto the mechanically connected gel network, the formation of conducting paths may be guided by (or assembled by) the pre-existing network. In this way, one can envision obtaining conducting gels at volume fractions of conducting polymer far below that required for percolation (gelation and conductivity) of a soluble conducting polymer alone.

An initial study conceived and implemented along these lines has demonstrated that gel processing can provide a facile route to stable blends with low electrical conductivities [63]. Although the volume fraction (f) of P3OT in the gels ranged from below 0.0005 to about 0.03, there were connected conducting pathways even at volume fractions as low as 0.0005, with no indication of a percolation threshold; i.e. more than three orders of magnitude below the volume fraction for percolation for composite conductors in three dimensions [59]. These results have led to a deeper fundamental understanding of how to obtain connected conducting pathways in conducting polymer blends.

The percolative aspect of the conducting network is shown in Figure IV.2 where the conductivity is plotted as a function of the volume fraction of conducting polymer for two frequencies, 1kHz and 1GHz. The 1kHz data characterize the low frequency regime (below ω_c) in which the transport is independent of length scale; the 1 GHz data are in the power law regime (above ω_c) in which the conductivity is strongly sensitive to length scale. Although no specific percolation threshold was observed, for volume fractions of P3OT less than $f \cong 2.5\times10^{-4}$ the 1kHz conductivity was too small to measure with the shorted-coax configuration. The conductivity increases by several orders of magnitude for f in the range from $5\times10^{-4} < f < 3\times10^{-2}$. For f greater than approximately 2×10^{-2}, the conductivity of the network saturates toward values in the range $\sigma \sim 10^{-3}$-10^{-2} S/cm.

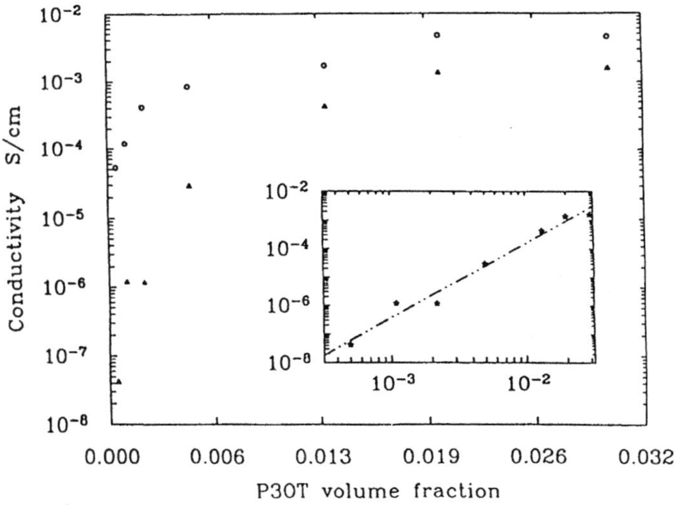

Figure IV.2: The conductivity is plotted as a function of the volume fraction of conducting polymer for two frequencies, 1kHz and 1GHz. Inset: log-log plot of the conductivity (at 1kHz) vs volume fraction of P3OT.

The inset to Figure IV.2 is a log-log plot of the 1kHz conductivity vs volume fraction of P3OT. The results indicate that over a relatively wide range of concentrations, the conductivity follows a simple power law also observed for poly-aniline/poly-p(phenylene terephthalamide) systems [14]

$$\sigma = c_2 f^{\alpha} \tag{IV.3}$$

where c_2 is a constant and $\alpha \approx 2.6$. The dashed curve in 4 shows the power law fit to the 1kHz data with $\alpha \approx 2.6$.

Two observations help clarify the nature of the P3OT/UHMW PE gels. First, the conjugated polymer is not expelled from the gel when stored in a larger volume of decalin, even for the lowest concentrations of P3OT. Since decalin is a good solvent for P3OT, this result is important and implies that the conjugated chains "like" PE better than the solvent; i.e. they are either adsorbed onto or entangled within the PE network. Second, the color of the P3OT/UHMW PE solution changes gradually from yellow to red as the temperature is lowered and gelation occurs. A color change from yellow (the color of P3OT in solution) to red (the color of solid crystalline films) is well-known [17] for the poly(3-alkylthiophenes). Reversible, disorder induced thermochromism (red to yellow color change) has also been observed even in solid films [17,64]; in this case, it appears that the disorder is due to side-chain melting [65]. Thus, the red color of the gel suggests the adsorption of relatively well-ordered P3OT chains onto the PE network, since more complex entanglement of the conjugated polymer would lead to sufficient disorder to cause the P3OT to retain a yellow color.

Percolation theory [59] predicts that at concentrations sufficiently dilute that there are no connected paths, the conductivity is zero. As f is increased above the percolation threshold (f_p), the conductivity becomes finite and increases as the connectivity (i.e. the number of conducting paths) increases. In contrast, the data presented in Figure 1 show no indication of a well-defined percolation threshold. By attempting to fit the data to the form $\sigma = c_2(f - f_p)^{\alpha}$, it was concluded that $f_p < 2.5 \times 10^{-4}$ P3OT. By contrast, classical percolation theory [59] for a three-dimensional network of conducting globular aggregates in an insulating matrix predicts a percolation threshold at a volume fraction $f_p \approx 0.16$, in agreement with results obtained for composites of poly(3-alkylthiophene) in polystyrene [58] and for other conducting polymer composites [66,67].

The extremely low percolation threshhold has, therefore, been understood as arising from decoration of the pre-existing mechanical network of the PE gel by the conducting polymer [68].

The continuum percolation threshold in two dimensions is inversely proportional to the excluded area [69]. Assuming that the conjugated polymer adsorbs onto the PE network either in the form of rods (length l) or pancakes (radius of gyration l), the excluded area in either case is l^2. The area fraction for percolation, Φ_c, is expressed with a coefficient of order unity, α, as

$$\Phi_c \cong \alpha s / l^2 \tag{IV.4}$$

where s denotes the actual surface area occupied by the conducting polymer; for a rod $s \cong al$ and for a pancake $s \cong l^2$, where a denotes the monomer size in the conducting polymer. The volume of the conducting polymer at threshold is expressed in terms of the surface area of the gel network, A,

$$v_c \cong A(\alpha s/l^2)a. \qquad (IV.5)$$

The volume of the gel network is expressed in terms of the typical diameter of the structural features (interconnected lamellar crystals, etc) [70] which make up the network, V=Ad. Thus, the volume ratio of conducting polymer to gel at threshold is $\rho_c = v_c/V$. The critical volume fraction of conducting polymer for percolation in the gel is expressed in terms of the volume fraction of the network in the gel, Φ_o:[68]

$$\Phi_c = \rho_c\Phi_o$$

$$= (a^2/dl)\alpha\Phi_o, \text{ for rods} \qquad (IV.6)$$

$$= (a/d)\alpha\Phi_o, \text{ for pancakes.} \qquad (IV.7)$$

According to the studies by Smith et al [70], d is approximately 100Å. The molecular weight of P3OT and the volume fraction of the PE gel are $M_w \approx 89{,}000$ and $\Phi_o = 0.02$, respectively. Thus, a≈3Å and l≈1000Å for rods and l≈100Å for pancakes. Thus one obtains $\Phi_c \approx 2\times10^{-6}\alpha$ for rods and $\Phi_c \approx 6\times10^{-4}\alpha$ for pancakes, consistent with the experimental results.

These relatively highly conducting gels constitute a novel state of matter with an unusual combination of properties. The use of a pre-existing gel network to guide the formation of conducting paths at volume fractions which are orders of magnitude below that needed for bulk percolation may be a useful method for achieving conducting composites with a small volume fraction of conducting polymer.

V. Electrical and Mechanical Properties of Polyaniline and Blends of Polyaniline with PPTA Processed from Solution in Sulfuric Acid [14]

Although the ability to process conducting polyaniline from solution represents genuine progress, fibers and films made from these solutions have mechanical properties which are not adequate for many applications due mainly to the low molecular weight of the polyaniline used. Fibers with significantly enhanced mechanical properties have been obtained by blend processing polyaniline with the rigid chain polymer poly-(p-phenylene terephthalamide), PPTA [14]. It is well known that PPTA is processed from solutions in concentrated H_2SO_4 to yield one of the strongest and stiffest fibers commercially available [14]. The process takes advantage of the fact that PPTA is a rod-like polymer that exhibits a liquid crystalline phase at high solution concentrations which facilitates the formation of highly oriented structures.

A. PREPARATION OF THE PANI/PPTA BLENDS AND PANI/PPTA FIBER SPINNING [14]

Polyaniline, optimized to yield relatively higher values of the molecular weight, was used for the PANI/PPTA blends. The poly (p-phenylene terephthalamide) with inherent viscosity of 7.43 dl/g was obtained from Du Pont as Kevlar® powder. A stock solution of 2 wt% PPTA in sulfuric acid was prepared by mixing 3.76 g of PPTA in 100 ml of 98% H_2SO_4 (Fisher). The mixture was mechanically stirred overnight to yield a homogeneous solution. This solution was also used to produce the 100% PPTA fibers. Polyaniline was weighed into a vial and concentrated H_2SO_4 was subsequently added. The mixture was stirred until the polyaniline dissolved and the solution was homogeneous. An amount of 2 wt% PPTA stock solution was added to yield a desired specific PANI/PPTA ratio. The mixture was mechanically stirred for 5 to 12 hours at room temperature, and then allowed to stand for 24 hours before spinning, to let the trapped air escape. In all cases, optically homogeneous solutions were obtained. Since the concentration of PPTA in the solutions was maintained at 1.5 wt% (which is below the onset of the formation of the lyotropic phase; typically at 6-8 wt% PPTA), these polyblend fibers were spun from isotropic solutions.

The polymer blend solutions were wet spun into 1N H_2SO_4 using a high precision syringe pump. Monofilaments were collected onto a take-up spool; applied tension was applied to elongate the fiber during coagulation. The draw down ratio (take up speed/extrusion speed) invariably was as high as possible, for the production of continuous fibers; and it was increased from 7 to 20 with increasing PPTA concentration in the solution. Extrusion speed varied from 0.12 to 0.3 m/min and the windup speed from 1.8 to 4.2 m/min.

The fibers were prevented from drying on the bobbin by continually spraying them with deionized water; this procedure impeded the open structure of the wet spun fibers from collapsing and allowed removal of the residual H_2SO_4. Subsequently, the fibers were washed with running deionized water for 48 hr. Half of the bobbin was then submerged in 1.5N HCl for 12 hr. This allowed the acid to penetrate the fibers and homogeneously protonate them to the conducting emeraldine salt form (partial reduction of the emeraldine salt occurs during the washing; the HCl treatment restores PANI to the fully protonated form). Finally, the bobbin was placed in an oven and the fibers dried under vacuum at 50°C while maintained at constant length by the bobbin. The pure PPTA fibers were spun with the same method. The pure polyaniline solution was dry-jet wet spun into a 1N H_2SO_4 solution. Washing, HCl treatment, and drying was carried out as described for the blend fibers.

B. PROPERTIES OF THE PANI/PPTA FIBERS

Wide angle X-ray patterns of the composite fibers consisted of superimposed reflections from the two components. The patterns, therefore, indicate that PANI and PPTA segregated during the coagulation, consistent with the general arguments presented earlier. Scanning electron micrographs of fractured fibers containing 10, 40, 60, 100 wt% PANI show that increased PANI content rendered the blend fibers more brittle. This effect is also quantitatively revealed in the measured values of the strain at break, which decreased at increasing PANI concentration. Quite likely, this effect is due primarily to the low molecular weight of the currently available polyaniline.

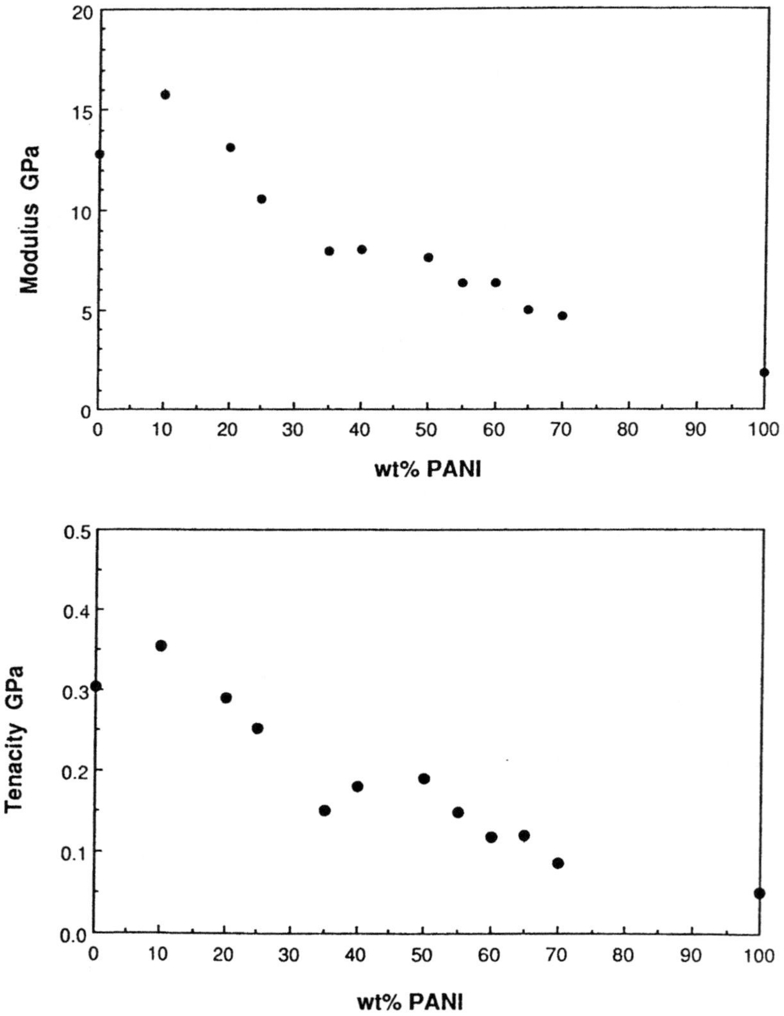

Figure V.1: Mechanical properties of the PANI/PPTA blend fibers as a function of the volume fraction of PANI
a. modulus
b. tensile strength

The mechanical properties (modulus, tensile strength and strain at break) as well as the electrical conductivity of the various monofilaments are shown in Figure V.1. The data clearly indicate that the tenacity and modulus systematically increase with increasing PPTA content. In fact, a simple, nearly linear, relation was observed between the mechanical properties and the fiber composition. This behavior is in accord with that generally found for composites loaded under isostrain conditions [71]. The enhancement of the mechanical properties with increased PPTA (not unexpectedly) came at the expense of the electrical conductivity of the polyblend fibers.

The conductivity vs volume fraction of PANI in the PANI/PPTA blend fibers is shown in Figure V.2. The same data are shown on a log-log plot in the inset; the results indicate that over a relatively wide range of concentrations, the conductivity follows the simple power law of equation IV.3. The solid curve in Figure V.2 shows the power law fit to the experimental data with α=7.9 and σ_o=393; the conductivity varies with fractional concentration of conducting polymer (from 10-70 % PANI) according to this power law equation over a range of values spanning nearly seven orders of magnitude.

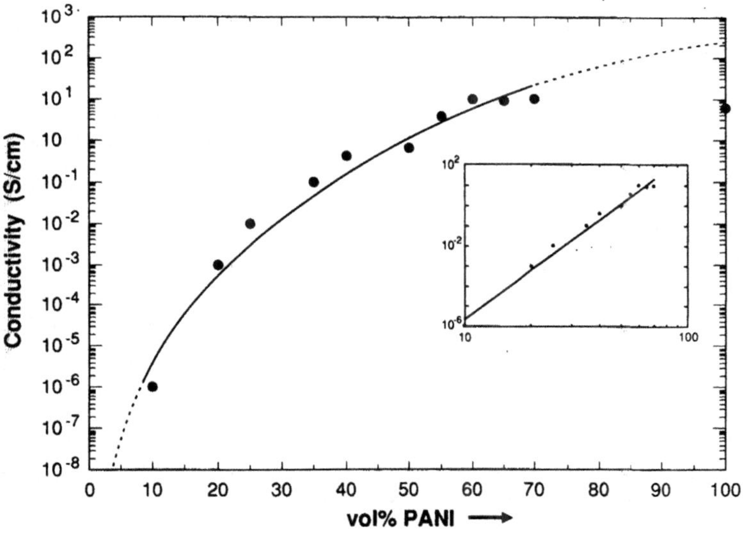

Figure V.2: Conductivity vs volume fraction of PANI in the PANI/PPTA blends; The inset shows the same data on a log-log plot.

As emphasized in the previous sections, in the context of percolation theory, at sufficiently dilute concentrations such that there are no connected paths, the conductivity would be zero. As the concentration of conducting polymer is increased above the percolation threshold, the conductivity would become finite and increase as the connectivity (i.e. the number of conducting paths) increases. In contrast, the data presented in Figure V.2 show no indication of a well-defined percolation threshold. Attempts to estimate a value for the percolation threshold (f_p) by fitting the data to the form $\sigma = \sigma_o(f - f_p)^\alpha$ lead to the conclusion that f_p is below ≅1 %.

The mechanism for the low percolation threshold may be closely related to that involved with the self-assembling PE/P3OT gels (see Section IVB).

In the PANI/PPTA system, PPTA comes out of sulfuric acid at a water content of about 10% and PANI comes out at a water content of about 25%. Thus, when spinning fibers into a water coagulating bath, the PPTA comes out first, forms a mechanically connected network. That structure may serve to assemble the conducting PANI along the pre-existing connected paths, thereby leading to the kind of fibrillar conducting pathways that would decrease the percolation threshold to $f_p << 0.16$.

On the other hand, for rod-like structures with length L and diameter d (e.g. a more fibrillar conducting network), the percolation threshold is determined by the excluded volume per rod ($\approx \pi L^2 d$) [62,69]. With N rods per unit volume, the percolation threshold is given by $f_p \approx N\pi L^2 d = N\pi L d^2 (L/d) = f_{rod}(L/d)$ where f_{rod} is the volume fraction of conducting material in the connected network. Thus, the onset of conductivity would occur at $f_{rod} = f(d/L) << 0.16$.

VI. Electrical and Mechanical Properties of PTV and PDMPV [72,73]

Poly(p-phenylenevinylene), PPV, and its derivatives such as poly(dimethyl-p-phenylenevinylene) can be prepared via a precursor polymer route; either as a polyelectrolyte, which is soluble in water, or as a precursor polymer which is soluble in common organic solvents. The aqueous solutions of the PDMPV precursor polymer tend to form gels, and the gradual elimination of the sulfonium group in the solid precursor cannot be avoided even at room temperature; both effects make subsequent processing into highly oriented films and fibers difficult. Thus, use of the precursor polymer which is soluble in common organic solvents, easily processible, and stable even at 100 °C appears to be preferable. Similarly, poly(2,5-thienylene vinylene) (PTV) is one of the larger class of poly(arylenevinylene) polymers which is attractive as a conjugated polymer and which can be synthesized via the precursor polymer route. As with PDMPV, PTV can be prepared through a new precursor polymer which is soluble in common organic solvents and chemically stable.

A. PREPARATION OF PRECURSOR POLYMERS, FIBER SPINNING, DRAWING AND CONVERSION OF PTV AND PDMPV

The preparation of the precursor polymer and conversion to the conjugated polymer are summarized in Figure VI.1a (PTV) and Figure VI.1b (PDMPV). Details on the synthesis of the precursor polymer and the conversion to the conjugated polymer are presented elsewhere [72,73].

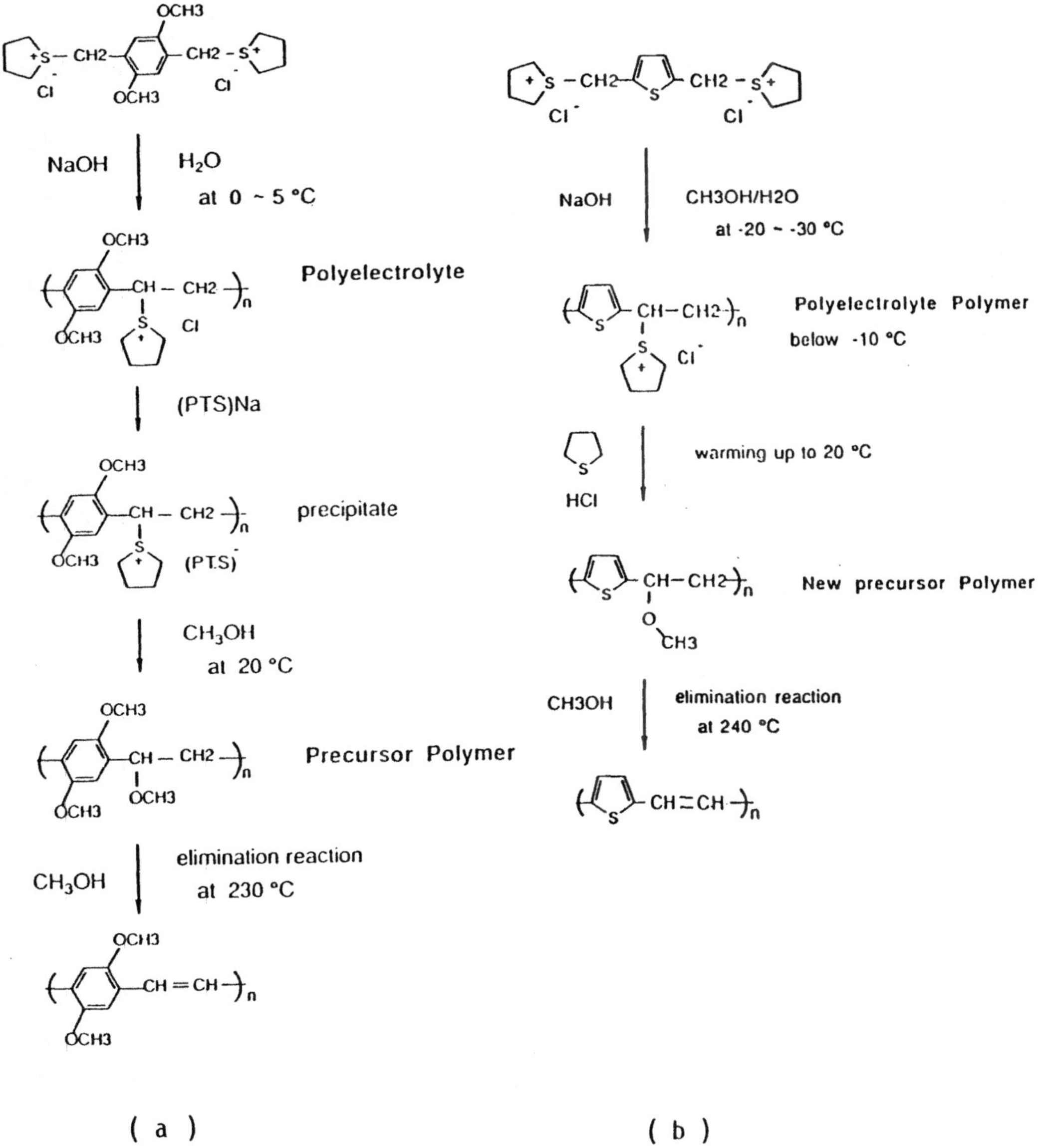

Figure VI.1: Scheme for preparation of the precursor polymer of PTV and PDMPV and conversion to the conjugated polymers
a. PDMPV
b. PTV

The interval between the glass transition temperature of the precursor polymer and the final conversion temperature to the conjugated polymer defines the processing window; the DSC thermogram of the PDMPV precursor polymer indicated a glass transition at 110 °C, well-separated in temperature from that needed (>200 °C) for thermal elimination of the methoxy leaving groups. For the PTV precursor polymer, the glass transition is at 50 °C, again, well-separated in temperature from the elimination of methoxy leaving groups. The precursor polymers were dissolved into chloroform and the solutions were spun using a high precision syringe pump into hexane; the resulting precursor fibers were taken up onto a bobbin at a speed of 30 cm/min. The precursor polymers were dryed in a vacuum oven overnight; uniform pale yellow fibers were obtained with diameters of 50 -100 μm.

The precursor fibers were drawn and converted in a continuous process using the experimental set-up shown in Figure VI.2. The amorphous yellow precursor fibers were converted to oriented crystalline PTV fiber with a metallic golden luster, or to oriented crystalline PDMPV fiber with a metallic greenish luster. Deformation of the precursor fibers was found to occur near the entrance of the tube furnace, where the temperature was above the respective glass transition temperature but below the conversion temperature; i.e within the processing window. Parameters critical to the optimization of this process are the temperature of the tube furnace and the conversion time.

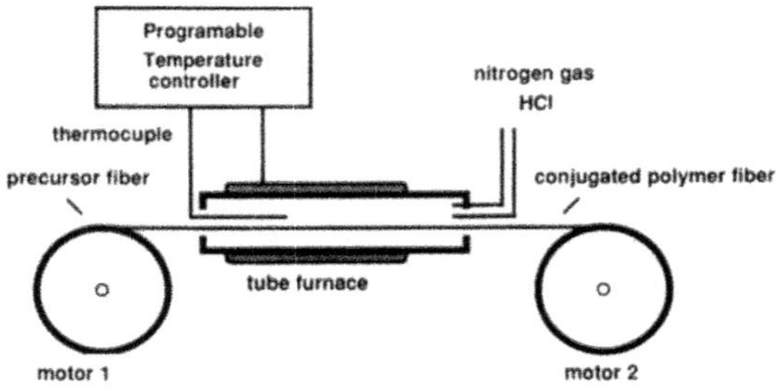

Figure VI.2: Schematic of apparatus for simultaneous drawing and conversion of precursor polymer fiber.

Information on molecular orientation in the drawn fibers was obtained by X-ray diffraction. Figure VI.3 compares diffraction patterns of undrawn (as-spun) PTV fiber and a PTV fiber with draw ratio of 16. Figure VI.4 compares the corresponding diffraction patterns for PDMPV. The undrawn (but converted) fibers exhibited Debye-Scherrer rings in the diffraction pattern indicative of significant crystallinity, but with an isotropic distribution of crystallites. On the other hand, the highly drawn fibers exhibited both equatorial reflections and meridional reflections (draw direction is vertical), confirming that the crystalline regions are aligned with the polymer backbone along the drawing direction. Although both PTV and PDMPV retained significant orientation after tensile drawing, the PDMPV fibers exhibit significantly higher degree of structural order even though the draw tration for PDMPV was less than that of PTV.

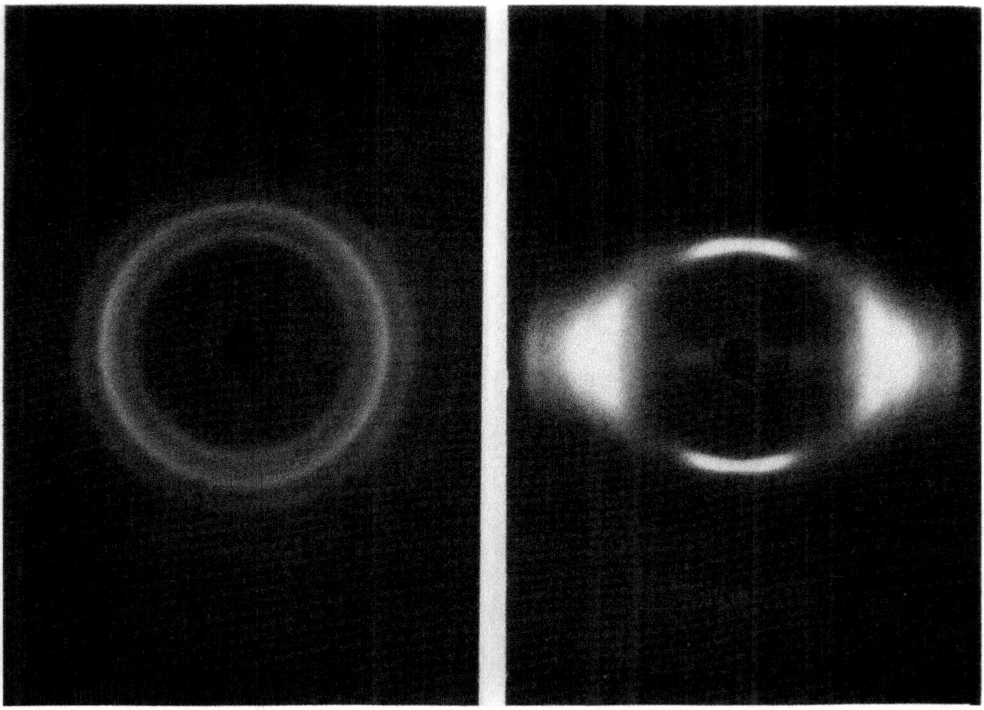

Figure VI 3: X-ray diffraction patterns of PTV:
Left; Undrawn (as-spun) PTV fiber
Right; PTV fiber with draw ratio of 16

The arcing observed in the latter X-ray diffraction pattern is typical of moderately oriented crystalline material. For the PTV fibers, the orientation distribution of crystallites was about 18° (full width at half maximum) with respect to drawing direction; for the PDMPV fibers, the full width at half maximum about 8°. A quantitative measure of the degree of orientation may be obtained from the angular half-width at half height of the intensity distribution. For PDMPV, the 8° angular spread corresponds to an orientation function

$$f= \langle(3\cos^2\theta -1)\rangle/2 \approx 0.96 \quad \text{(VI.1)}$$

where the brackets imply an average over the experimental distribution as indicated in the X-ray pattern. The orientation function for the drawn fibers is, thus, comparable to that reported for oriented PPV films [12].

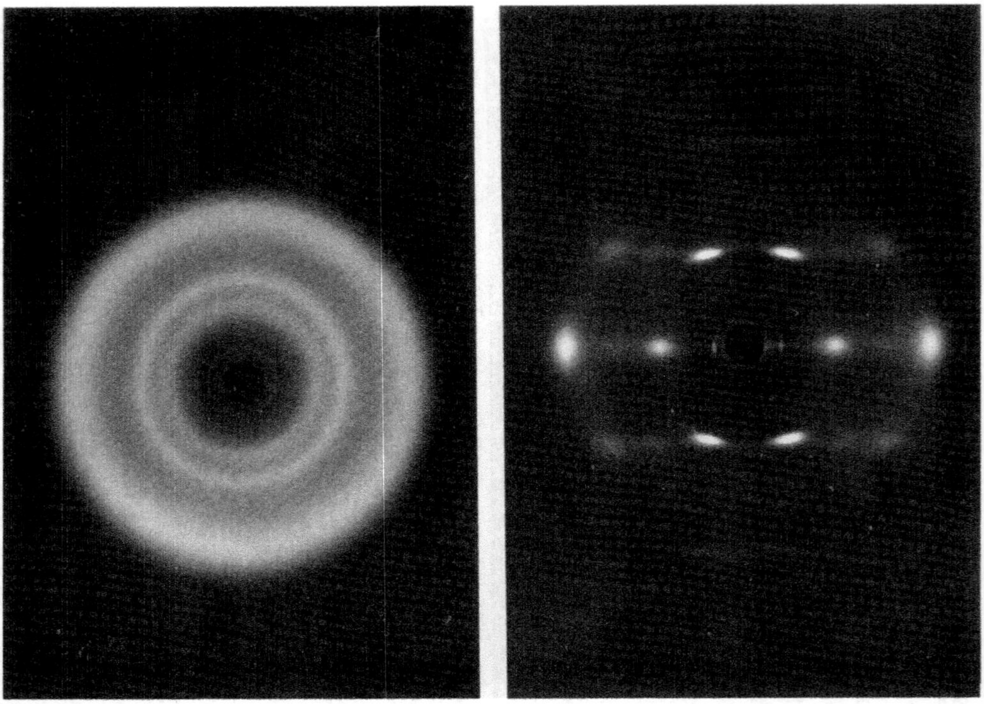

Figure VI.4: X-ray diffraction patterns of PDMPV;
Left; Undrawn (as-spun) PDMPV fiber
Right; PTV fiber with draw ratio of 8

B. ELECTRICAL AND MECHANICAL PROPERTIES OF PTV [72]

The Young's modulus and the tensile strength of the PTV fibers were measured at room temperature: Figure VI.5 shows the Young's modulus plotted against draw ratio; Figure VI.6 displays the tensile strength plotted against draw ratio. The modulus of the undrawn, converted PTV fiber was 1.1 GPa. Values of the Young's modulus of 7 GPa and of the tensile strength of 0.5 GPa were obtained for fibers which had been drawn to 20 times their initial length.

The tensile strength of 0.5 GPa is identical to the value as reported for highly stretched PPV film prepared by a heated roll method [74] and comparable to the values obtained with stretched <u>trans</u>-polyacetylene films [75]. The modulus, however, was lower than the values reported for stretched PPV films [74] and for stretched polyacetylene [75] films.

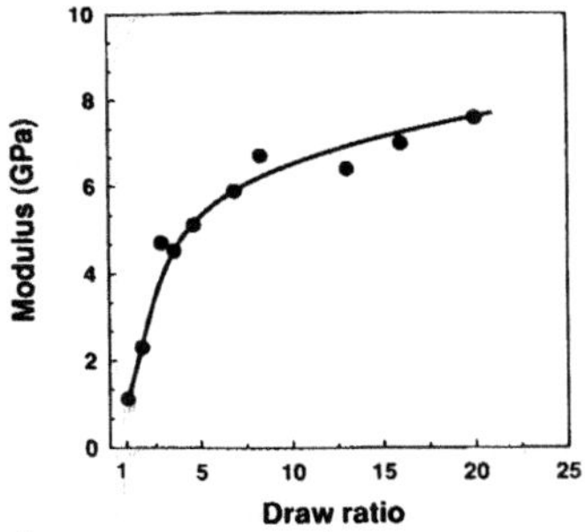

Figure VI.5: Young's modulus of PTV plotted against draw ratio

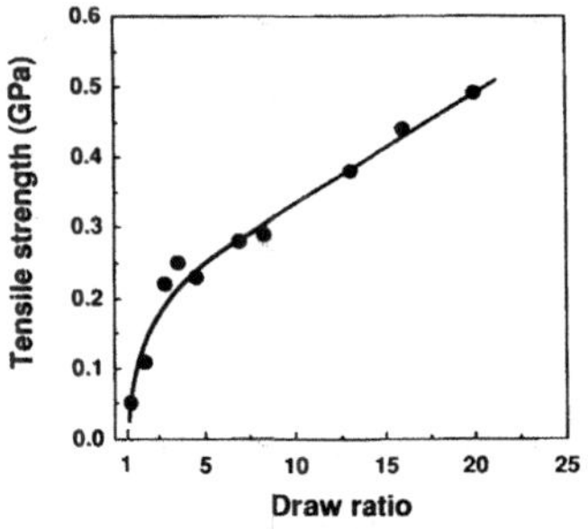

Figure VI.6: Tensile strength of PTV plotted against draw ratio.

The conductivities of the iodine-doped PTV fibers were measured at room temperature. The conductivity rapidly increased after the PTV fibers were exposed on iodine vapor, approaching maximum values after about 1 hour. Figure VI.7 shows the conductivity plotted against the draw ratio. The undrawn PTV fiber exhibited a conductivity of 80 S/cm. As expected, the conductivity was found to increase with the draw ratio; the maximum conductivity of 2000 S/cm was obtained at the maximum draw ratio. This value was 20 times higher than that of an undrawn fiber and comparable to that of an oriented film reported by Murase et al.[12] The relationship between the conductivity and the draw ratio was similar to that for the modulus and tensile strength (and draw ratio), indicating a strong correlation between conductivity and the mechanical properties.

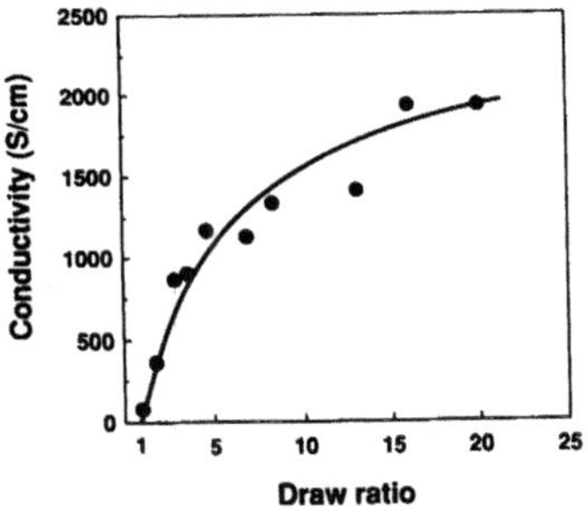

Figure VI.7: Conductivity of PTV plotted against the draw ratio.

This correlation between the structural dependence (as induced by the draw ratio) of the mechanical and electrical properties are illustrated more clearly by Figures VI.8 and VI.9, where the conductivity of drawn/converted PTV samples is plotted against, respectively, the modulus and the tensile strength. In both cases, an essentially linear correlation is observed.

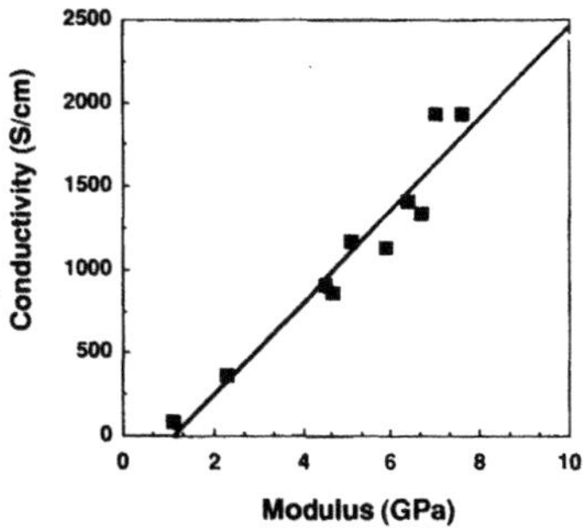

Figures VI.8: Conductivity of drawn/ converted PTV samples is plotted against the modulus

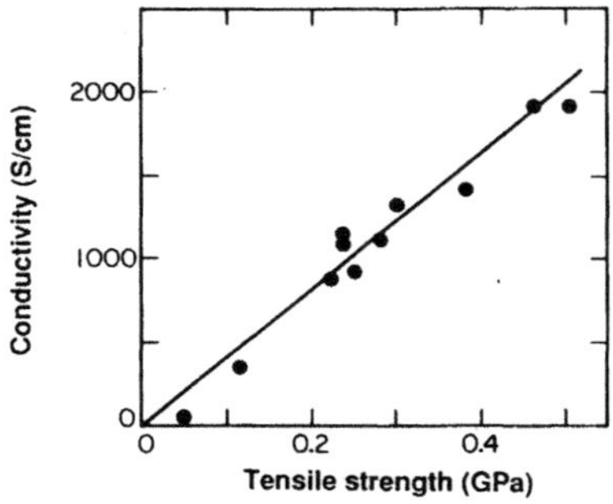

Figures VI.9: Conductivity of drawn/ converted PTV samples is plotted against the tensile strength.

The many complex molecular processes that simultaneously occur in the drawing/conversion process need to be carefully optimized so that their time scales are in harmony. The critical parameters for process optimization are the degree of pre-conversion, deformation rate, process-temperature profile, conversion-catalyst concentration, and residence time. Obviously, in this initial study of PTV, we have not developed a fully balanced set of processing parameters. This is illustrated by the modest orientation indicated by the X-ray patterns. Nevertheless, the current materials have a Young's modulus of 7 GPa, a tenacity of 0.5 GPa and a conductivity of 2000 S/cm; a combination of properties which are adequate for conductive textile applications. It should be clear that further optimization of the inter-related processing variables will undoubtedly result in materials of superior quality.

C. ELECTRICAL AND MECHANICAL PROPERTIES OF PDMPV

The Young's modulus and the tensile strength of PDMPV fibers are shown versus draw ratio in Figures VI.10 and VI.11. The undrawn PDMPV fiber exhibited relatively poor mechanical properties: the modulus and tensile strength were 1.3 GPa and 0.07 GPa, respectively. Fibers which had been drawn to 8 times their initial length had a Young's modulus as high as 35 GPa and a tensile strength of 0.7 GPa. The effects of doping on the mechanical strength are also displayed (filled circles in Figures VI.10 and VI.11). The data presented on these graphs indicate that doping caused only a moderate reduction of the modulus and essentially no loss of tensile strength.

Figure VI.12 shows the conductivity plotted against draw ratio. The undrawn PDMPV fiber exhibited a conductivity of 20 S/cm, with a gradual increase of conductivity seen up to a draw ratio of 5. At draw ratios greater than 5, the conductivity increased dramatically; at a draw ratio of 8, the conductivity was 1200 S/cm, 60 times higher than that of the undrawn material.

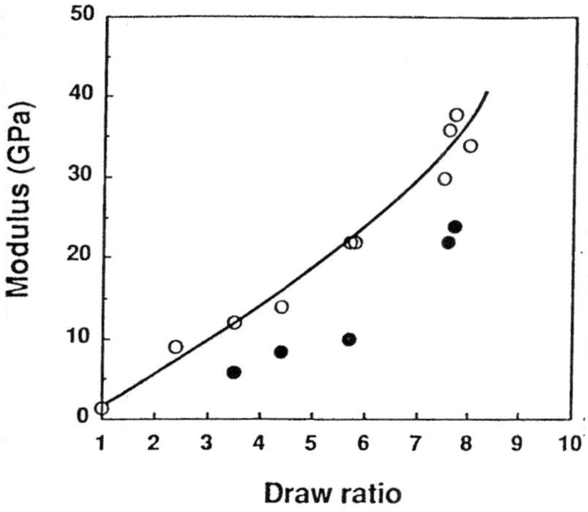

Figure VI.10 : Young's modulus as a function of the draw ratio for PDMPV fiber (open circles) and for PDMPV fiber doped with iodine (closed circles).

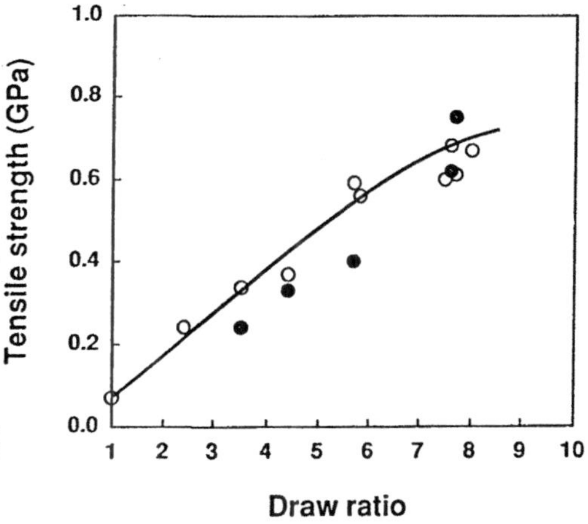

Figure VI.11: Tensile strength as a function of the draw ratio for PDMPV fiber (open circles) and for PDMPV fiber doped with iodine (closed circles).

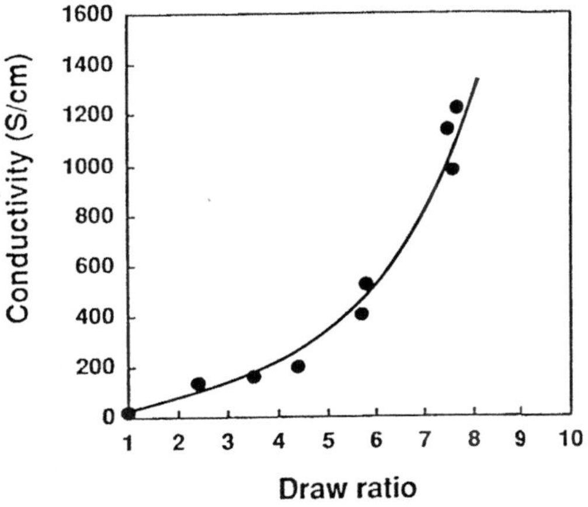

Figure VI.12: Conductivity of PDMPV as a function of the draw ratio.

Figures VI.13 and VI.14 show the relationship between Young's modulus and electrical conductivity and between the tensile strength and electrical conductivity for drawn PDMPV fibers. Once again, in both cases, a strong correlation is observed.

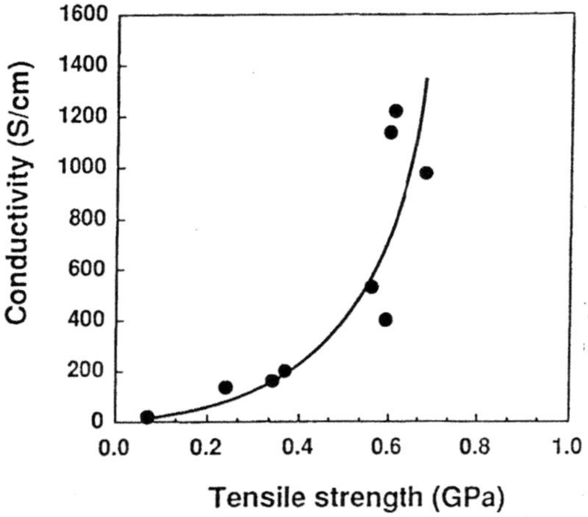

Figures VI.13: Conductivity of drawn/ converted PDMPV samples as a function of the modulus

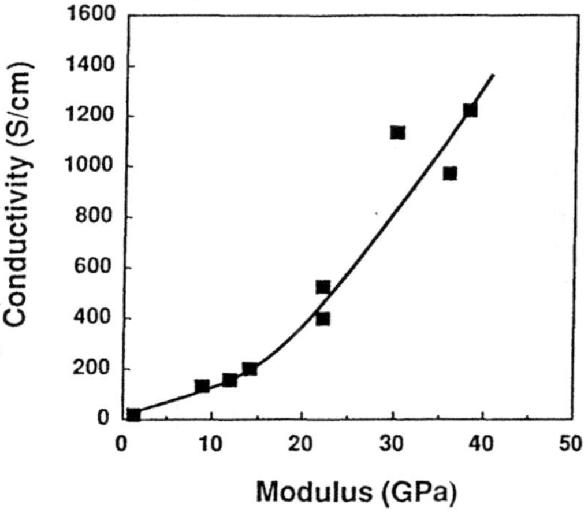

Figures VI.14: Conductivity of drawn/ converted PDMPV samples as a function of the tensile strength.

VII. Mechanical and Electrical Properties of Polyacetylene Films Oriented by Tensile Drawing [75]

Although polyacetylene is often considered an intractable material, there has been significant progress in processing this important conjugated polymer. To date, polyacetylene has exhibited both the highest electrical conductivity and the best mechanical properties (modulus and tensile strength) of any conducting polymer. Recent studies [76-79] demonstrated that major improvements in solid state properties can be achieved with higher quality polyacetylene obtained through improved synthesis. It is, therefore, the prototype example of a High Performance Conducting Polymer to which all other systems are ultimately to be compared.

An improved method for tensile drawing of polyacetylene films (prepared by non-solvent polymerization using a thermally pre-treated variation of the Shirakawa catalyst) has been developed [75]. After equilibration in certain organic liquids, which serve as plasticizers in the deformation process, the films can be stretched to a maximum draw ratio (λ) as high as λ_{max} = 15. The mechanical and electrical properties of such highly drawn polyacetylene films were measured before and after doping. The data demonstrate that the electrical conductivity increases linearly with the modulus and tensile strength, with chain orientation as the implicit variable. For polyacetylene films prepared in this way, the maximum values obtained were 50 GPa and 0.9 GPa for the modulus and tensile strength, respectively. Thin films doped with Iodine, exhibited electrical conductivities along the orientation direction (at room temperature) up to 3×10^4 S/cm, with an anisotropy, $\sigma_{par}/\sigma_{perp}$ > 250. Both Young's modulus and the tenacity decreased upon doping, typically by a factor of about 4-5, indicative of reduced interchain interaction after doping.

A. POLYMERIZATION AND TENSILE DRAWING

Polyacetylene films were prepared using a variation of to the non-solvent method recently described by Akagi et al [78]. After completion of the polymerization, the polyacetylene films were held at -78 °C and washed 20 times with pure toluene trapped into the polymerization vessel from a side flask. After raising the container with polyacetylene films (in toluene) to room temperature, the polymer sample was removed and cut into 15x40 mm^2 size specimens while still wet, in a dry-bag filled with argon. The cut samples were then quickly transferred into a flask containing an appropriate liquid to be used as a plasticiser. After storing in the plasticizing liquid at -78°C overnight, the sample was manually stretched at room temperature using a simple apparatus for tensile drawing. The film was drawn under argon in a dry bag to the desired draw ratio while still wet, and then pre-dried on the drawing apparatus under dynamic vacuum at room temperature overnight. After dismounting from the drawing apparatus, the films were once again dried under dynamic vacuum (less than 1 mtorr) for 48 hours. The thicknesses after drawing to their maximum draw ratio were 25-30 μm for thick films and 3-5 μm for thin films. Isomerization of predominantly cis-polyacetylene films to the trans-$(CH=CH)_n$-configuration was accomplished by heating at 160 °C and dynamic vacuum with the samples under stress (approximately $1x10^3$ Kg/cm^2) for additional stretching and to prevent shrinkage and/or curling.

Tensile drawing of the wet polyacetylene films yielded a maximum stretching ratio of about 15 for the thick films and about 10 for thin films. By contrast, fully dried polyacetylene films could not be extended beyond 6 times their original length under otherwise similar experimental conditions. Some variation in the maximum stretching ratio, λ_{max}, was found, dependent on the characteristics of the liquid used as "plasticizer". Specifically, it was observed that λ_{max} was dependent on the boiling point and the specific composition in the case of mixed liquids. The results seem to indicate that λ_{max} is dependent on the relative content of solvent in the film and on the speed of evaporation of the solvent from bulk of the film. Hexadienes, for example, were found to be excellent plasticizers for polyacetylene, presumably due to the easier penetration of these liquids into the polyacetylene lattice because of the similarity in molecular structures. Pure aromatic solvents appeared to be the least effective; however, a 4:1 mixture of toluene and cumene yielded excellent results.

B. X-RAY DIFFRACTION

The wide-angle X-ray diffraction patterns of drawn polyacetylene films show crystalline order with preferred orientation of the chain axis parallel to the stretching direction (for trans-polyacetylene, see Figure VII.1); the (hk0) reflections are observed only with momentum transfer perpendicular to the stretching direction. The large number of equatorial reflections and off-axis reflections indicate a regular three dimensional arrangement of the polymer chains. The mosaic spread (derived from the half width at half maximum) is found to decrease to a value of about 3° as the draw ratio increases to $\lambda \geq 10$, comparable with that observed for polyacetylene prepared by the Durham method [80,81]. The strong small-angle reflections observed in the equatorial plane are connected with the long range order transverse to the polyacetylene chain direction.

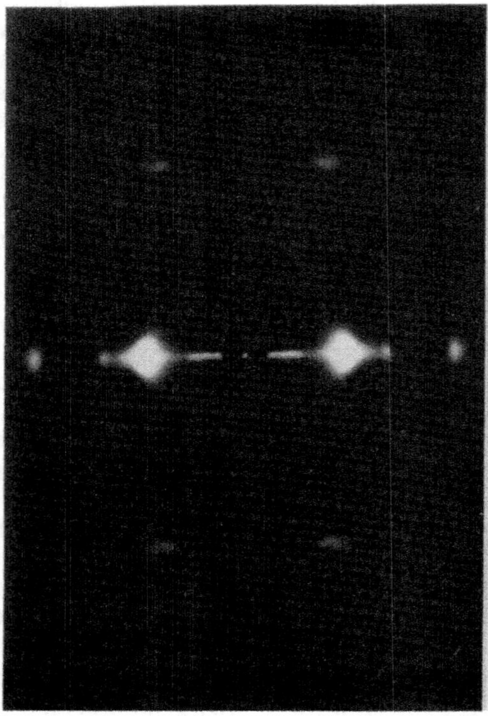

Figure VII.1: X-ray diffraction pattern of a drawn trans-polyacetylene film (neutral)

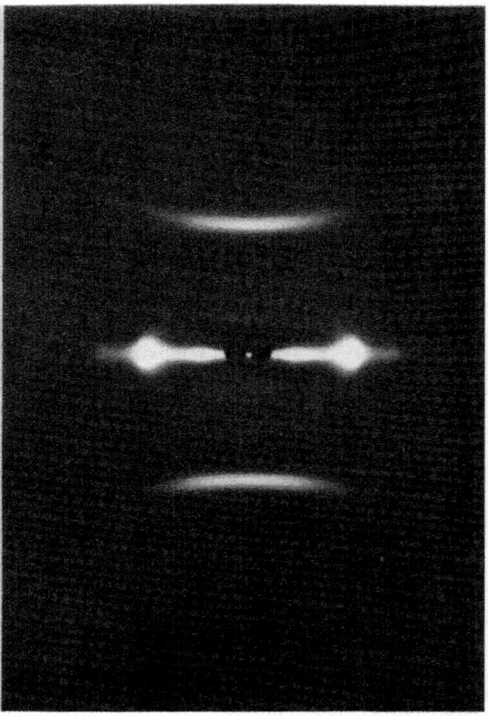

Figure VII.2: X-ray diffraction pattern of a drawn trans-polyacetylene film (Iodine doped)

Assuming that the line broadening results from finite crystallite size, one can calculate the coherence length, ζ, from the Scherrer formula,

$$\zeta = 2\pi\ \lambda / \delta(2\theta) \qquad \text{(VII.1)}$$

where $\delta(2\theta)$ is the full width at half-maximum; for the highly drawn samples, ζ=20nm for the order in the direction perpendicular to the chain direction. This value is larger than that reported previously for trans -$(CH{=}CH)_n$-prepared by the Durham method, by using a liquid crystalline solvent in a magnetic field, or by the usual Shirakawa method (ζ ~5-10 nm). Comparing the diffraction patterns from samples with different draw ratio leads to conclusion that the structural coherence length perpendicular to the draw direction increased by about a factor of two (from approximately 10 nm at $\lambda=4$ to approximately 20 nm at $\lambda = 15$) with increased orientation. This observation is of particular importance, for the improvement in structural caused by tensile drawing demonstrates the positive effect of post-synthesis processing of polyacetylene on the molecular scale (in contrast to simple fibrillar alignment).

The diffraction patterns for the doped samples indicate that the polyacetylene chain orientation remains essentially unchanged after doping. As observed in earlier studies [81], comparison of the fiber diffraction patterns of oriented trans-polyacetylene before and after doping shows that the principal equatorial reflections remain with only modest shifts in their positions as a result of lattice expansion to accommodate the dopant counterions.

For the I_3^- doped samples, the strong layer lines (Figure VII.2) along the fiber orientation direction result from scattering from the counterions. Since these lines in the plane of the photograph indicate sheets of scattering rather than discrete reflections, it can be concluded that the I_3^- counterions are arranged in columns in one-dimensional (1d) channels between the polyacetylene chains, with almost no phase coherence from column to column. For Iodine doping, the diffraction patterns do not change with the doping level either in peak position or in the width of the streaks. Thus, the one-dimensional I_3^- columns form essentially independently of one another. Shimamura et al.[82] observed (by electron diffraction) similar streaks on the meridian of very thin polyacetylene films polymerized epitaxially on dibromobenzene single crystals and lightly doped by Iodine.

C. MECHANICAL PROPERTIES

Figures VII.3 and VII.4 show the experimental values of the Young's modulus and the tensile strength, respectively, for thick films of undoped trans-polyacetylene as a function of draw ratio (all samples were derived from the same polymerization batch). Although there is some scatter in the data, the modulus and tenacity increase approximately linearly with the draw ratio, as is commonly observed for most polymers drawn to moderate draw ratios. The modulus and tensile strength of trans-polyacetylene films stretched up to 15 times are 50 GPa and 0.9 GPa, respectively. These values are essentially equivalent to those observed for ultra-high molecular weight polyethylene [83] drawn to the same draw ratio. Recently, Akagi et al.[78] reported remarkable mechanical properties for drawn polyacetylene films prepared by non-solvent polymerization (100 GPa and 0.9 GPa for the modulus and tensile strength, respectively). The origin of difference in the modulus (in the two studies) is unknown.

Values of the modulus and tensile strength *versus* draw ratio of polyacetylene doped to maximum conductivity are also plotted in Figures VII.3 and VII.4, for comparison with the corresponding properties of the undoped samples. Both the modulus and tenacity consistently decreased by about a factor of 4-5. It is not surprizing to find some degradation of the mechanical properties after doping; this is to be expected when the mechanical properties

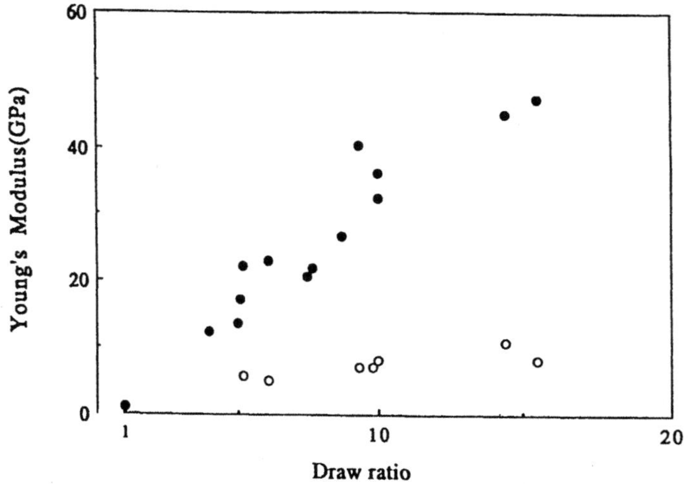

Figure VII.3: Young's modulus for thick films of trans-polyacetylene as a function of draw ratio; open circles represent neutral samples and closed circles represent doped samples.

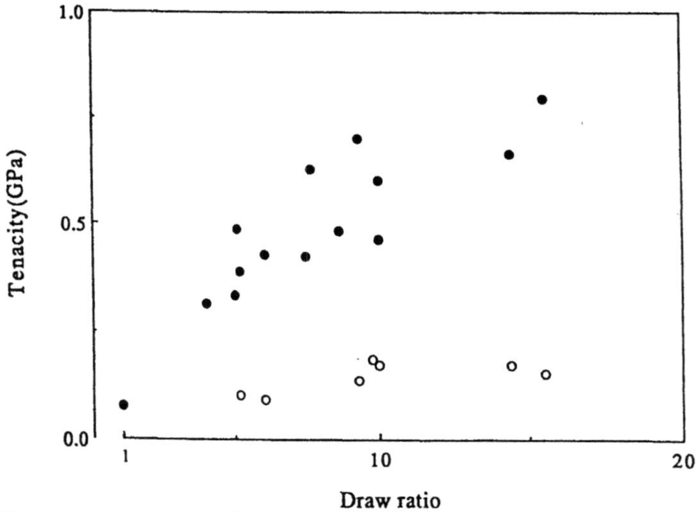

Figure VII.4: Tensile strength for thick films of trans-polyacetylene as a function of draw ratio; open circles represent neutral samples and closed circles represent doped samples.

are limited by the relatively weak secondary interactions (for conjugated polymers, the interchain electron transfer interaction and Van der Waals bonds). The observed decreases in modulus and tenacity after doping imply an associated reduction in the interchain electronic coupling. This reduction presumably arises because of the intercalation of the counterions (columns or sheets, etc) into the structure.

D. ELECTRICAL CONDUCTIVITY

Figure VII.5 shows the correlation between the electrical conductivity and the draw ratio (λ) for both thick (~25-30 μm, after tensile drawing) and thin (~3-5 μm, after drawing) Iodine doped polyacetylene films. The conductivity was found to increase essentially linearly with the draw ratio: the maximum conductivity was $2x10^4$ S/cm at a draw ratio of λ=15 for the thick films, and $3x10^4$ S/cm at a draw ratio of λ=10 for thin films. The slope of σ vs λ for thin films is approximately a factor of two greater than that of thick films. The data imply that, if more homogeneous thin films can be prepared and stretched to higher draw ratios, higher conductivities may be possible.

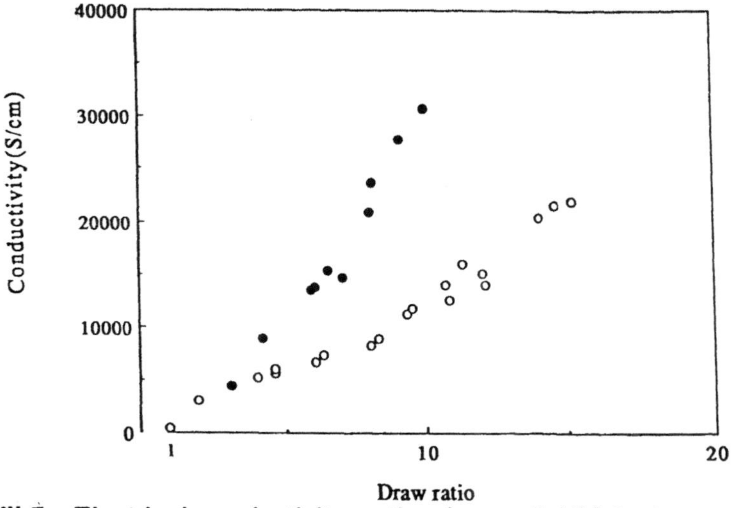

Figure VII.5: Electrical conductivity vs the draw ratio (λ) for both thick (25-30) μm, after tensile drawing) and thin (~3-5 μm, after drawing) Iodine doped polyacetylene films.

Although the dependence on film thickness has been noted, there is no specific microscopic information on the differences between the thin and thick films; evidently the details of the polyacetylene polymerization and/or doping result in higher quality material for the thin films. The conclusion, however, is unambiguous; thin films may have fewer structural defects (e.g. sp^3 carbons), higher molecular weight (or both), or they might simply be more homogeneously doped.

Figure VII.6 shows the anisotropy of electrical conductivity vs. draw ratio. The anisotropy increased with increasing draw ratio reaching $\sigma_{\parallel}/\sigma_{\perp} \approx$ 250 as λ→15. Although the parallel conductivity increased dramatically as a function of the draw ratio, the conductivity perpendicular to the tensile drawing direction was essentially constant with increasing draw ratio (for λ≥2). These data set a lower limit on the anisotropy of the intrinsic conductivity of oriented polyacetylene. The results imply that heavily doped polyacetylene is a quasi-one-dimensional "metal" with relatively weak interchain coupling.

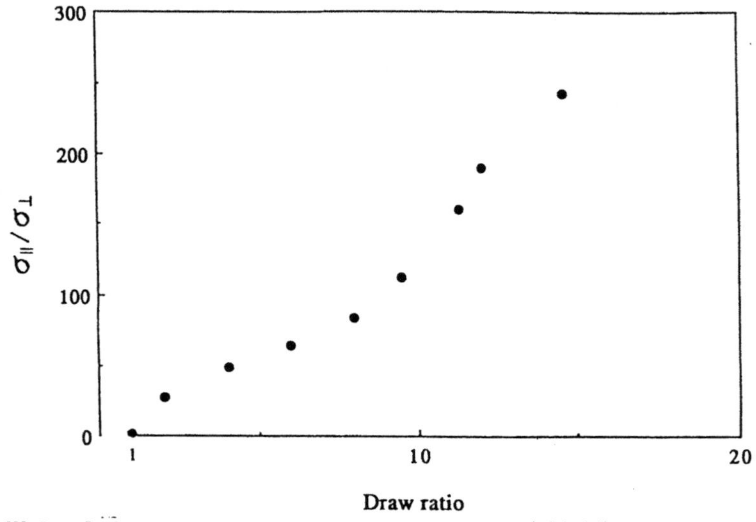

Figure VII.6: Anisotropy of electrical conductivity vs. draw ratio.

Figures VII.7 and VII.8 demonstrate a direct correlation between the electrical conductivity and the mechanical properties (Young's modulus and tenacity, resp.) of polyacetylene. The linear relationship implies that the increase in the electrical conductivity with increasing draw ratio must result from increased uniaxial orientation, improved lateral packing and associated enhanced interchain interaction, as is the case for the mechanical properties.

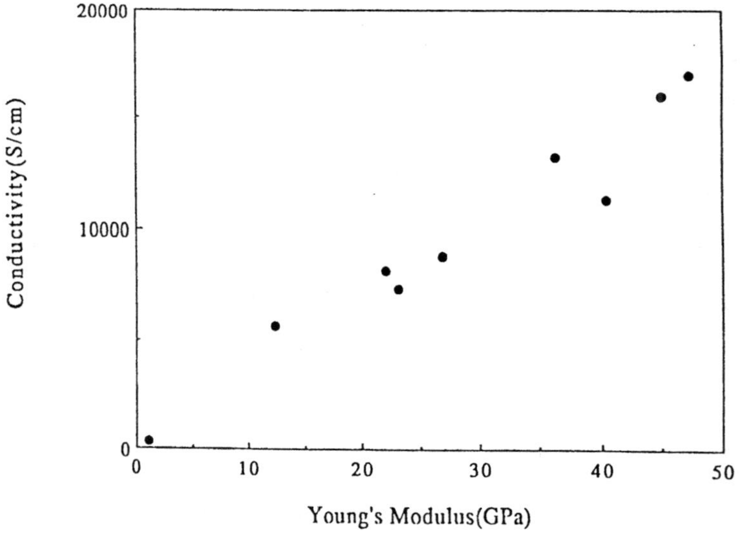

Figures VII.7: Electrical conductivity vs modulus of trans-polyacetylene

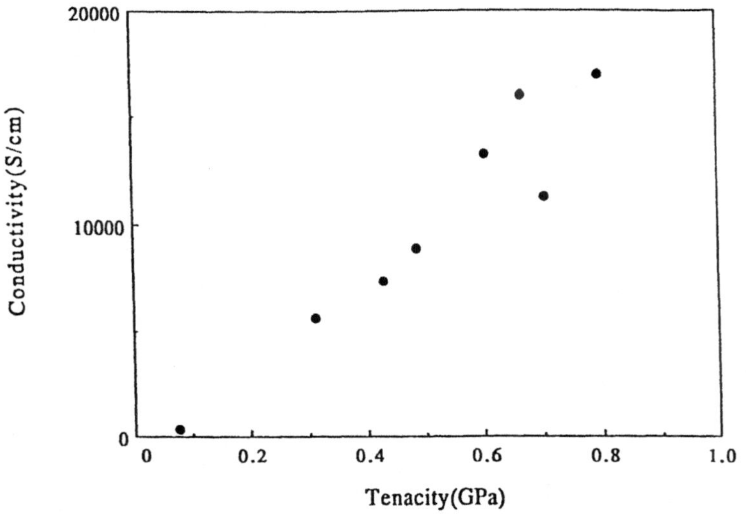

Figures VII.8: Electrical conductivity vs tensile strength of trans-polyacetylene

The reduction of the modulus and tenacity after doping is consistent with the large anisotropy in the electrical conductivity; the quasi-one-dimensionality implies relatively weak interchain coupling. The reduced inter-chain electronic coupling implies an associated reduction in the electronic energy bandwidth in the transverse direction.

VIII. CORRELATION BETWEEN ELECTRICAL CONDUCTIVITY and MECHANICAL PROPERTIES [72,73,75,84]

The correlations observed between the electrical conductivity and the mechanical properties (modulus and tensile strength) are general. In every case studied, the achievement of improved structural leads to increased electrical conductivity and increased mechanical properties. This generality can be understood as a fundamental feature of conjugated polymers.

Although the electrical conductivity is enhanced by the relatively high mobility associated with intra-chain transport, one must have the possibility of inter-chain charge transfer to avoid the localization inherent to one-dimensional electronic systems [1]. The electrical transport becomes three-dimensional (and thereby truly metallic) only if there is a high probability that an electron will have diffused to a neighboring chain between defects on a single chain. For well-ordered crystalline material in which the chains have precise phase order, the interchain diffusion is a coherent process. In this case, the condition for extended anisotropic transport is that [1,4]

$$L/a \gg (t_o/t_{3d}) \qquad \text{(VIII.1)}$$

where L is the coherence length, a is the chain repeat unit length, t_o is the intra-chain π-electron transfer integral, and t_{3d} is the inter-chain π-electron transfer integral. An analogous argument can be constructed for achieving the intrinsic strength of a polymer, i.e. the strength of the main-chain covalent bonds. If E_o is the energy required to break the covalent main-chain bond and E_{3d} is the weaker interchain bonding energy (from Van der Waals forces and hydrogen bonding for saturated polymers), then the requirement is coherence over a length L such that [3,4]

$$L/a >> E_o/E_{3d}. \qquad \text{(VIII.2)}$$

In this limit, the large number (L/a) of weak interchain bonds add coherently such that the polymer fails by breaking the covalent bond. The direct analogy between equations VIII.1 and VIII.2 is evident. In fact, for conjugated polymers, E_o results from a combination of σ and π bonds (the latter being equal to t_o) and E_{3d} is dominated by the interchain transfer integral, t_{3d}. Thus, the inequalities imply that, quite generally, the conductivity and the mechanical properties will improve in a correlated manner as the degree of chain alignment is increased. This prediction is in excellent agreement with all the data summarized in this review; i.e. for the poly(3-alkylthiophenes), the poly(phenylenevinylenes), polythienylenevinylene) and polyacetylene.

Assuming that the linear correlation persists to even higher draw ratios, the extrapolated modulus of 300 GPA for polyacetylene would imply that the electrical conductivity for perfectly oriented polyacetylene would be approximately $2x10^5$ S/cm. However, this extrapolated value may be still limited by structural defects (e.g. sp^3 defects, etc.) rather than by intrinsic phonon scattering [1]. Thus, one can envision a series of curves like those shown in Figures VII.7 and VII.8, but with slopes that increase with sample perfection until the mean free path becomes limited by phonon scattering. A theoretical estimate of the intrinsic conductivity (limited by phonon scattering) at room temperature is $2x10^6$ S/cm an order of magnitude greater than that achieved to date.

IX. CONCLUSION

There is ample evidence, summarized in this review, that conducting polymers can be processed. We conclude that there exists a possibility of two broad categories of heavily doped conducting polymers:

I. *High Performance Materials:* True metals in which the transport is coherent. Both the carrier density and the carrier mobility are large. In such polymers, the molecular weight is sufficiently high, the interchain order is sufficiently good, and the defect density is sufficiently low that delocalization occurs, leading to "free" metallic carriers with mean free paths much greater than the monomer repeat unit. Such highly ordered systems will also have outstanding mechanical properties.

II. *Commodity Materials with Routine Properties:* Disordered conductors in which the electronic wavefunctions are localized (by the disorder) and in which the transport is via (thermally assisted hopping. To achieve such materials (either with pure polymers or in the form of blends) all that is required is a moderately high density of carriers; carrier delocalization is not required.

A simple estimate can be used to define the boundary between these two categories. In free electron theory, the mobility can be written in terms of the mean scattering time (τ) and the electron mass;

$$\mu = e\tau/m. \tag{IX.1}$$

Since the mean free path (λ) is the product of τ and the Fermi velocity (v_F),

$$\mu = e\lambda/v_F\, m. \tag{IX.2}$$

Materials in which λ is large compared to the monomer repeat unit (a) belong to the *High Performance* category; materials with λ~a belong to the *Commodity Materials* category. Combining eqn I.1 and I.3 with λ~a, one obtains an estimate for the crossover conductivity (σ_C) between hopping conductivity and coherent transport

$$\sigma_C \sim ne^2a/v_Fm. \tag{IX.3}$$

With $n \sim 10^{21}$ (typical of heavily doped conducting polymers) a~1Å, and $v_F \sim 10^8$ cm/s, one obtains $\sigma_C \sim 50$ S/cm. For materials with $\sigma < \sigma_C$, the electronic wavefunctions are localized with transport via thermally assisted hopping (σ increases with increasing temperature); whereas, only for materials with $\sigma >> \sigma_C$ will the transport be like that of a metal. Note that even in the latter case, the intrinsic temperature dependence can be masked by series resistance arising from the polymer morphology (amorphous regions, interfibrillar contacts, etc); this is evidently the case even in the highest quality polyacetylene with $\sigma \approx 10^5$ S/cm in which λ>1000Å!

Each of these two categories of conducting polymer materials is interesting from the point of view of potential commercial applications, each offers special challenges in the context of processing, and each offers special opportunities for basic science.

The requirements of chain extension, chain alignment and interchain structural coherence inherent to the high performance category provide a clear challenge and a special opportunity to polymer scientists. Although one might initially guess that the materials science associated with the high performance materials would be the most challenging, that is not necessarily the case. For example, the potential of conducting polymers as commodity materials (for electromagnetic shielding, etc) may never be realized on a large commercial scale unless the materials can be melt-processed in the conducting state.

The achievement of high performance polymers with electrical and mechanical properties that approach intrinsic values will provide the opportunity to explore the intrinsic mechanisms of transport. For example, the study of the temperature dependence, pressure dependence, etc of phonon scattering in quasi-one-dimensional (quasi-1d) metals with sufficient interchain interaction to avoid 1d localization will surely be of interest. On the other hand, in the disordered conductor limit, hopping processes in anisotropic systems are poorly understood. The ability to make polymer blends immediately raises a number of fundamental issues. Can one hope to achieve stable conductivities at low levels (e.g. in the range 10^{-6} and even lower, as might be useful for antistatic applications), or is the onset of conductivity in mixed systems inherently a threshold phenomenon as implied by standard percolation theory? Thus, the ability to process conducting polymers, both as pure materials and as polymer blends, offers special opportunities for science and for technology.

Acknowledgement: This review was prepared under support from the Electric Power Research Institute (EPRI grant RP8007-9).

References:

1. S. Kivelson and A.J. Heeger, Synth. Met. 22, 37, (1988).
2. a. A.J. Heeger, S. Kivelson, J.R. Schrieffer and W.P. Su, Rev.Mod.Phys. 60, 782 (1988).
   b. *Handbook of Conducting Polymers*, Vol. I and II, Ed. by T. Skotheim, (Marcel Dekker, New York, 1986).
   c. J.L. Bredas and G.B. Street, Accounts Chem. Research, 18, 309 (1985).
3. Y. Termonia and P. Smith, in "The Path to High Modulus Polymers with Stiff and Flexible Chains", edited by A.E. Zachariades and R.S. Porter (Marcel Dekker, New York, 1988) p. 321.
4. A.J. Heeger, Faraday Discuss. Chem Soc. 88, 1 (1989).
5. a. K.Y. Jen, R. Oboodi, and R. Elsenbaumer, Polym. Materials: Sci. Eng. 53, 79 (1985).
   b. S. Hotta, S.D.D.V. Rughooputh, A.J. Heeger and F. Wudl Macromolecules, 20, 212 (1987).
   c. S.D.D.V. Rughooputh, S. Hotta, A.J. Heeger and F. Wudl, J. Polym. Sci. Polym. Phys. Ed, 25, 1071 (1987).
   d. M.J. Nowak, D. Spiegel, S. Hotta, A.J. Heeger, and P. Pincus, Macromolecules, 22, 2917(1989).
   e. M.J. Nowak, S.D.D.V. Rughooputh, S. Hotta and A.J. Heeger, Macromolecules, 20, 965 (1987).
   f. M. Sato, S. Tanaka, and K. Kaeriyama, J. Chem Soc. Chem. Commun. 295, 873 ((1986).
6. a. S. Shi and F. Wudl, Macromolecules, 23, 2119 (1990).
   b. S. Shi and F. Wudl, in "Conjugated Polymeric Materials: Opportunities in Electronics, Optoelectronics and Molecular Electronics", Ed. by J.L. Bredas and R.R. Chance, NATO ASI Series, Series E: Applied Sciences - Vol. 82, (Kluwer Academic Publishers, Dordrecht, 1990), p.83.
7. K. Yoshino, S. Nakajima, M. Fujuii and R. Sugimoto, Polymer Commun. 28, 309 (1987).
8. A.O. Patil, Y. Ikenoue, N. Basescu, N. Colaneri, J. Chen, F. Wudl and A.J. Heeger, Synth. Met. 20, 151 (1987); A.O. Patil, Y. Ikenoue, F. Wudl and A.J. Heeger, J. Am. Chem. Soc.109, 1858 (1987); Y. Ikenoue, J. Chiang, A.O. Patil, F. Wudl and A.J. Heeger, J. Am. Chem. Soc. 110, 2983 (1988).

9. M. Angelopoulos, G.E. Asturias, S.P. Ermer, A. Ray, E. M. Scherr, A.G. MacDiarmid, M. Akhtar, Z. Kiss, and A.J. Epstein, Mol. Cryst. Liq. Cryst. 160, 151, (1988).
10. A. Andreatta, Y. Cao, J.C. Chiang, A.J. Heeger and P. Smith, Synth. Met. 26, 383 (1988).
11. J.H. Edwards and E.J. Feast, Polym. Commun. 21, 595 (1980).
12. a. D.R. Gagnon, J.D. Capistran, F.E. Karasz and R.W. Lenz, Polym. Bull. 12, 93 (1984).
    b. I. Murase, T. Ohnishi and M. Hirooka, Polym. Commun. 25, 327 (1984).
13. T. Momii, S. Tokito, T. Tsutsui and S. Saito, Chem. Let. 1201 (1988).
14. A. Andreatta, A.J. Heeger and Paul Smith, Polymer Commun. 31, 275 (1990).
15. S. Hotta, S.D.D.V. Rughooputh, A.J. Heeger and F.Wudl, Macromolecules, 20, 212 (1987).
16. D.R. Spiegel, Macromolecules, 23, 3568, (1990).
17. O. Inganas, G. Gustaffson and W.R. Salaneck, Synth. Met., 28, C377 (1989); W.R. Salaneck, O. Inganas, J.O. Nilsson, J.E. Osterholm, B. Thermans and J.L. Bredas, Synth. Met. 28, C419 (1989); and references therein.
18. K.F. Voss, C.M. Foster, L. Smilowitz, D. Mihailovic, S. Askari, G. Srdanov, Z. Ni, S. Shi, A.J. Heeger and F. Wudl, Phys.Rev. B (in press).
19. D.R. Spiegel, P.A. Pincus and A.J. Heeger, Polymer Commun. 29, 264 (1988).
20. J.-L. Bredas and A.J. Heeger, Macromolecules, 23, 1150 (1990).
21. a. P.Pincus, G. Rossi and M. Cates, Europhys. Let. 4, 41 (1987).
    b. A. Viallat and P. Pincus, Polymer 30, 1997 (1989).
    c. A. Viallat and G. Rossi, J. Chem Phys. 92, 4548 (1990).
    d. A. Viallat, J. Chem. Phys. 92,4557 (1990).
22. G. Harbeke, E. Meier, W. Kobel, M. Egli, H. Kiess and E. Tossatti, Solid St. Commun. 55,419 (1985).
23. Z. Vardeny, E. Ehrenfreund, O. Brafman, M. Nowak, H. Schaffer and A.J. Heeger, Phys. Rev. Let. 56, 671 (1986).
24. J.L. Bredas, F. Wudl and A.J. Heeger, Solid State Commun. 63, 577 (1987).
25. J.H. Kaufman, T.C. Chung, and A.J. Heeger, J. Electrochem. Soc. 131, 2847 (1984).
26. F. Genoud, M. Gugliemi, M. Nechtschein, E. Genies and M. Salmon, Phys. Rev. Let. 55, 118 (1985).
27. F. Devreux, F. Genoud, M. Nechtschein, and P. Villerat, Synth. Met. 18, 89 (1987).
28. Y. Cao, P. Smith and A.J. Heeger, in "Conjugated Polymeric Materials: Opportunities in Electronics, Optoelectronics and Molecular Electronics", Ed. by J.L. Bredas and R.R. Chance, NATO ASI Series, Series E: Applied Sciences - Vol. 82, (Kluwer Academic Publishers, Dordrecht, 1990), p. 171.
29. A.G. MacDiarmid and A.J. Epstein, in "Conjugated Polymeric Materials: Opportunities in Electronics, Optoelectronics and Molecular Electronics",

Ed. by J.L. Bredas and R.R. Chance, NATO ASI Series, Series E: Applied Sciences - Vol. 82, (Kluwer Academic Publishers, Dordrecht, 1990), p.53.
30. A.J. Epstein and A.G. MacDiarmid, in "Conjugated Polymeric Materials: Opportunities in Electronics, Optoelectronics and Molecular Electronics", Ed. by J.L. Bredas and R.R. Chance, NATO ASI Series, Series E: Applied Sciences - Vol. 82, (Kluwer Academic Publishers, Dordrecht, 1990), p. 195.
31. M. Angelopoulos, G.E. Asturias, S.P. Ermer, A. Ray, E. M. Scherr, A.G. MacDiarmid, M. Akhtar, Z. Kiss, and A.J. Epstein, Mol.Cryst.Liq.Cryst. 160, 151, (1988).
32. T. Hjertberg, W.R. Salaneck, I. Lundstrom, N.L.D. Somasiri, and A.G. MacDiarmid, J. Polym. Sci. Polym. Let. Ed. 23, 503 (1985).
33. Y. Cao, S. Li, and D. Guo, Synth. Met. 16, 305 (1986).
34. A. Ray, G.E. Asturias, D.L. Kershner, A.F. Richter, A.G. MacDiarmid and A.J. Epstein, Synth. Met., 29 (1) E141, (1989).
35. S. Kaplan, E.M. Conwell, A.F. Richter and A.G. MacDiarmid, J. Amer. Chem. Soc. 110, 7647 (1988); Macromolecules, 22, 1669 (1989).
36. A.G. MacDiarmid, J.C. Chiang, A.F. Richter, and A.J. Epstein, Synth. Met. 18, 285 (1987).
37. A.G. MacDiarmid, J.C. Chiang, W.S. Huang, B.D. Humphrey, and N.L.D. Somasiri, Mol. Cryst. Liq. Cryst. 121,181 (1985).
38. F. Wudl, R.O. Angus, F.L. Lu, P.M. Allemand, D.J. Vachon, M. Nowak, Z.X. Liu, and A.J. Heeger, J. Am. Chem. Soc. 109, 3677 (1987).
39. J.M. Ginder, A.F. Richter, A.G. MacDiarmid, and A.J.Epstein, Solid State Commun. 63,97 (1987).
40. Y. Cao, P. Smith and A.J. Heeger, Synth. Met. 32, 263 (1989).
41. Y. Cao, A. Andreatta, A.J. Heeger and P. Smith, Polymer, 30, 2305 (1989).
42. D.G. Baird and J.K. Smith, J. Polym. Sci. Polym. Chem. Ed. 16, 61 (1978).
43. K. Kamide and M. Yukio, Kobunshi Ronbunshu 35, 467 (1978).
44. V.M. Geskin, Ya. A. Letuchy, B.Z. Lubentsov, L.M. Grundel, S.M. Tikhomirov and M.L. Khidekel, J. Molec. Electronics 5, 207 (1989).
45. a. S. Stafstrom, J.L. Bredas, A.J. Epstein, H.S. Woo, D.B. Tanner, W.S. Huang, and A.G. MacDiarmid, Phys. Rev. Let. 59, 1464 (1987).
b. D.S. Boudreaux, R.R. Chance, J.F. Wolf, L.W. Shacklette, J. L. Bredas, B. Themans, J.M. Andre and R. Silbey, J. Chem. Phys. 85, 4584 (1986).
46. C. Fite, Y. Cao and A.J. Heeger, Solid State Commun. 73, 607 (1990).
47. C. Fite, Y. Cao and A.J. Heeger, Solid State Commun. 70, 245 (1989).
48. Y. Cao, P. Smith, A.J. Heeger, P. Pincus and T.A. Witten, Macromolecules (in press).
49. Y. Moon, Y. Cao, P. Smith and A.J.Heeger, Polymer Commun. 30, 196 (1989).
50. M.E. Jozefowicz, R. Laversanne, H.H.S. Javadi, A.J. Epstein, J.P. Pouget, X. Tang and A.G. MacDiarmid, Phys. Rev. B 39, 12958 (1989).
51. J. Fischer, Proceedings of ICSM '90, Tübingen (in press).
52. J. Moulton and P. Smith, Synth. Met. (in press).
53. a. M. Ito, A. Tsuruno, S. Ogawa and K. Tanaka, Polymer 29, 1161 (1988).

b. M. Leclerc, F.M. Diaz and G. Wegner, Makromol. Chem. 190, 3105 (1989).
54. S.M. Aharoni, Polymer 22, 418 (1981).
55. A.R. Postema, K. Liou, F. Wudl and P. Smith, Macromolecules 23, 1842 (1990).
56. M.J. Winokur, P. Walmsley, J. Moulton, P. Smith and A.J. Heeger, Phys. Rev. Lett. (in press).
57. J. Moulton and P. Smith, Polymer (in press).
58. S. Hotta, S.D.D.V. Rughooputh and A.J. Heeger, Synth. Met. 22, 79, (1987).
59. R. Zallen, *The Physics of Amorphous Solids*, (Wiley, New York, 1983) Chap. 4.
60. P.G. DeGennes, Scaling Concepts in Polymer Physics, (Cornell University Press, Ithaca and London, 1979) Chap.4.
61. I. Balberg, C.H. Anderson, S. Alexander and N. Wagner, Phys. Rev. B 30, 3933 (1984).
62. S.A. Safran, G.S. Grest, A.L.R. Bug, and I. Webman, Surfactant Sci. Ser. 24, 235 (1987).
63. A. Fizazi, J. Moulton, K. Pakbaz, S.D.D.V. Rughooputh, P. Smith and A.J. Heeger, Phys. Rev. Let., 64, 2180 (1990).
64. S.D.D.V. Rughooputh, S. Hotta, A.J. Heeger and F. Wudl, J. Polym. Sci. Polym. Phys. Ed. 25,1071 (1987).
65. M.J. Winokur, D. Spiegel, Y. Kim, S. Hotta, and A.J. Heeger, Synth. Met. 28, C419 (1989).
66. M. Aldissi, and A.R. Bishop, Polymer, 26, 622 (1985).
67. M. Aldissi, Synth. Met. 13, 87 (1986).
68. Y. Suzuki, P. Pincus and A.J. Heeger, Macromolecules (in press).
69. A.L.R. Bug and S. Safran, Phys. Rev. B 33, 4716 (1986).
70. P. Smith, P.J. Lemstra, J.P.L. Pijpers, and A.M. Kile, Colloid and Polymer Sci. 259, 1070 (1981).
71. C.R. Barret, W.D. Nix, and A.S. Tatelman, The Principles of Engineering Materials, (Prentice-Hall, Englewood Cliffs, NJ, 1973) p. 318.
72 S. Tokito, P. Smith and A.J. Heeger, Synth. Met. 36,183 (1990).
73. S. Tokito, P. Smith and A.J. Heeger, Polymer (in press).
74. J.M. Machado and F.E. Karasz, Polym. Preprints 30, 154 (1989).
75. Y. Cao, P. Smith and A.J. Heeger, Polymer (in press).
76. G. Lugli, U. Pedretti, G. Perego, J. Polym. Sci. Polym. Lett. Ed. 213, 129 (1985).
77. H. Naarman, Synth. Met. 17, 223 (1987); H. Naarman and N. Theophilou, Synth. Met. 22, 1 (1987).
78. K. Akagi, M. Suezaki, H. Shirakawa, H. Kyotani, M. Shimamura, and Y. Tanabe, Synth. Met. 28, D1 (1989).
79. J. Tsukamoto, A. Takahashi, and K. Kawasaki, Japanese J. of Appl. Phys. 29,1 (1990).
80. D.D.C. Bradley, R.H. Friend, T. Hartmann, E.A. Marseglia, M.M. Sokolowski and P.D. Townsend, Synth. Met. 17, 473 (1987).
81. M. Winokur, Y.B. Moon, A.J. Heeger, J. Barker, D.C. Bott, and H. Shirakawa, Phys. Rev. Lett. 58, 2329 (1987).

82. K. Shimamura, Y. Yamashita, H. Kasahara, and K. Monobe, Synth. Met. 17, 485 (1987).
83. P. Smith and P.J. Lemstra, J. Mater. Sci. 15, 505 (1980).
84. A. Andreatta, S. Tokito, P. Smith, and A.J. Heeger, Advanced Organic Solid State Materials, Materials Research Society Symposium Proceedings, Vol. 173 (Materials Research Society, Pittsburgh, 1990), p.269.

THE POLYANILINES: MODEL SYSTEMS FOR DIVERSE ELECTRONIC PHENOMENA

ARTHUR J. EPSTEIN
Department of Physics and Department of Chemistry
The Ohio State University
Columbus, Ohio 43210-1106

ABSTRACT. Polyaniline has three stable insulating forms, leucoemeraldine base (LEB), emeraldine base (EB), and pernigraniline base (PNB). These three forms exhibit differing phenomena ranging from intramonomer luminescing excitons and ring rotation polarons in LEB, to nonluminescing charge transfer excitons and also ring rotation polarons in EB, to a degenerate ground state with bond order parameter solitons and ring order parameter polarons in PNB. The electronic structure and excitations of these three forms are contrasted. The LEB form can be *p*-doped (oxidatively doped), the EB form can be protonic acid doped and the PNB form can be *n*-doped (reductively doped) to form conducting systems. The essential role of crystallinity and coherence in the metallic state of the emeraldine salt form is reviewed. The role of derivatization in controlling the electronic phenomena in the insulating and conducting forms is outlined.

INTRODUCTION

The polyanilines have been known for more than one hundred years [1-5]. Recent progress in the chemistry of these materials has enabled the preparation of pure systems in the form of solutions [6, 7], films [6-11], and fibers [11-13]. Fig. 1 summarizes the repeat unit for the stable insulating forms, leucoemeraldine base (LEB), emeraldine base (EB), and pernigraniline base (PNB) together with the "metallic" emeraldine salt (ES) hydrochloride form. Extensive studies have demonstrated that these materials differ critically from earlier studied conducting polymers such as polyacetylene and polythiophene because of the important role of the C_6 rings, together with the presence of a heteroatom (N) within the conjugation path. Because of this only the pernigraniline base form may be considered a Peierls bandgap polymer analogous to the earlier studied polymer systems [4, 14]. Extensive studies of oriented emeraldine salt [9, 10] and polyaniline derivatives have demonstrated the critical role of three-dimensional charge delocalization for the achievement of high conductivity in this and other polymer systems [15, 16].

J.L. Brédas and R. Silbey (eds.), Conjugated Polymers, 211–228.

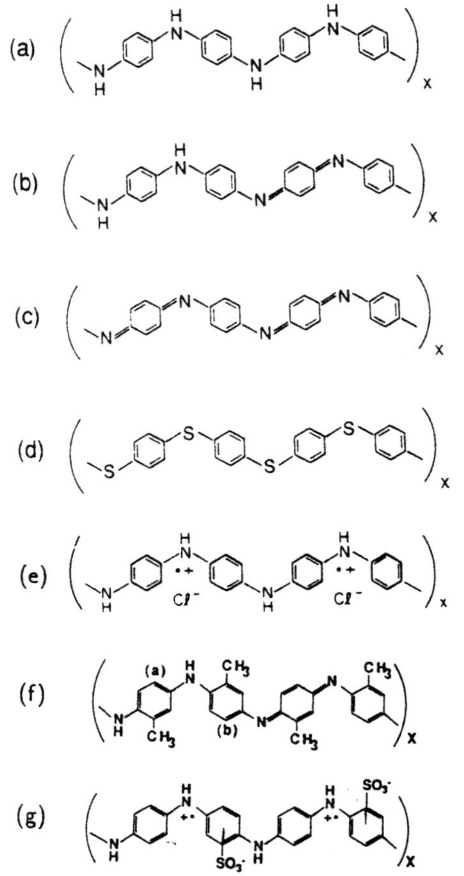

Fig. 1. (a) Leucoemeraldine base; (b) emeraldine base; (c) pernigraniline base; (d) poly(paraphenylene sulfide); (e) emeraldine hydrochloride salt polymer; (f) poly(orthotoluidine) (emeraldine base form); (g) sulfonated polyaniline (self-doped salt form).

In this chapter we will review the essential differences in the physics of the three different polyaniline insulating forms as well as present a summary of the critical aspects controlling conductivity in the emeraldine salt system. The effects of derivatization are briefly indicated.

LEUCOEMERALDINE BASE (LEB)

The optical absorption spectrum of LEB [17] is shown in Fig. 2 together with the results of a Hückel calculation for the electronic band structure of LEB carried out assuming that the repeat unit is a single (C_6H_4NH) unit as the repeat [18, 19]. This simple calculation reproduces most of the band structure features obtained in more extensive calculations [20, 21]. The seven energy bands are derived from the six p_z energy levels of benzene together with the p_z energy level of nitrogen. The lower four energy are completely filled while the upper three are empty. It is noted that the valence band is relatively broad (its breadth depending upon the ring torsion angle) while the conduction band is dispersionless. Hence, charge conjugation symmetry is absent in leucoemeraldine base. The exact width and separation of each of the bands are sensitive to the torsion angle the C_6 rings make with respect to the plane defined by the nitrogen atoms [19, 22].

In polyaniline, the π-electron delocalization energy that favors phenyl rings in the plane of the nitrogen atoms is countered by a substantial steric repulsion between adjacent rings, forcing them out of the plane. The angular dependence of the effective lattice energy is commonly approximated by a ϕ^4-like function of the ring angles, i.e.

$$V\left(\psi_{2m-1},\psi_{2m+1}\right) \simeq -\frac{V_{2,0}}{2}\left(\sin^2\psi_{2m-1}+\sin^2\psi_{2m+1}\right)+\frac{V_{4,0}}{2}\left(\sin^4\psi_{2m-1}+\sin^4\psi_{2m+1}\right)$$

$$+V_{1,1}\left(\sin\psi_{2m+1}-\sin\psi_{2m-1}\right)^2+\text{constant} \qquad (1)$$

with ψ_1 and ψ_2 the torsion angles of adjacent rings and $V_{2,0}$, $V_{4,0}$ and $V_{1,1}$ constants; the equilibrium ring angle ψ_0 is given by $\sin^2\psi_0 = V_{2,0}/(2V_{4,0})$. For example, the interaction of neighboring benzene rings in diphenyl oxide (and thus PPO) can be described by $V_{2,0} \simeq 1.50$ eV and $V_{4,0} \simeq 1.45$ eV. Figure 3 shows the approximate dependence of the lattice energy on the torsion angle for LEB, obtained by modifying the PPO steric parameters (setting $V_{2,0} \simeq 2.0$ eV) to yield $\psi_0 \simeq 56°$ as suggested by the band structure results [19].

Thermochromism temperature dependent blue shifts are observed for LEB in *N*-methylpyrrolidone (NMP) solution [23]. This increase in bandgap (π-π^*) for transition can be accounted for by increasing the most probable ground state ring angle with increasing temperature due to the anharmonic nature of the steric potential [19], as well as a contribution to the thermochromism due to fluctuations around the optimum ring angle [22].

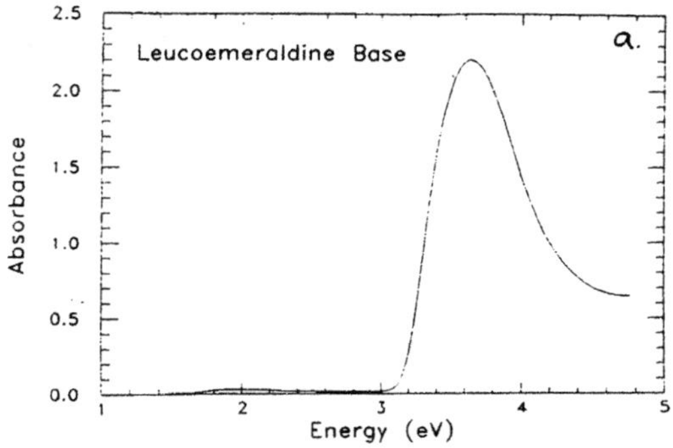

Fig. 2. (a) Direct absorption spectra of leucoemeraldine base in NMP (from Ref. 27); (b) Hückel electronic band structure of leucomeraldine (from Ref. 19).

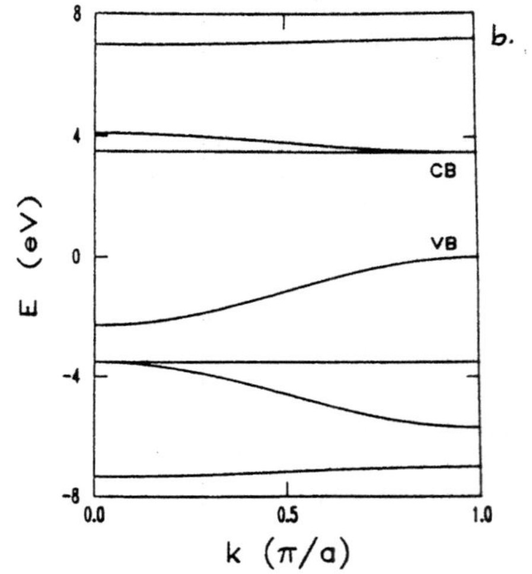

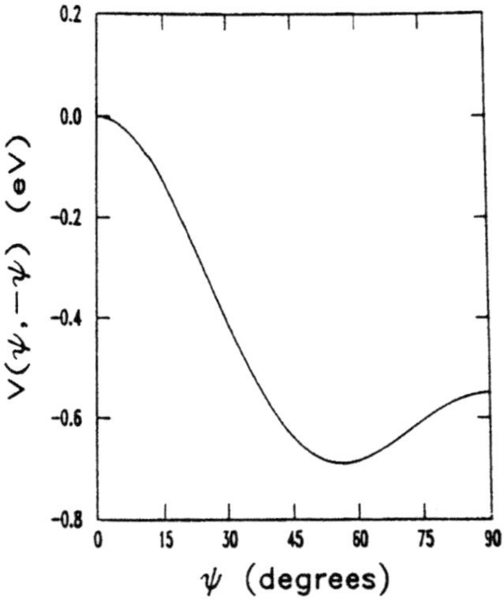

Fig. 3. Approximate total energy per site vs. ring torsion angle ψ for LEB, with equal-magnitude nearest neighbor angles assumed (from Ref. 19).

Excitation of leucoemeraldine solutions in *N*-methylpyrrolidone (NMP) with bandgap light produces an intense luminescence peak at 3.5 eV [24], Fig. 4. The magnitude of the luminescence intensity decreases linearly with increasing concentration of imine groups within the polymer chain. The luminescence is nearly zero when the emeraldine base oxidation state is achieved. The decay time for the luminescence has been measured and found to be ~4ns at room temperature [24]. The presence of a similar luminescence in three-ring oligomers [24] and also in the aniline ($C_6H_5NH_2$) monomer [25] supports that the luminescence derives from radiative decay of an intramonomer polaronic exciton [26].

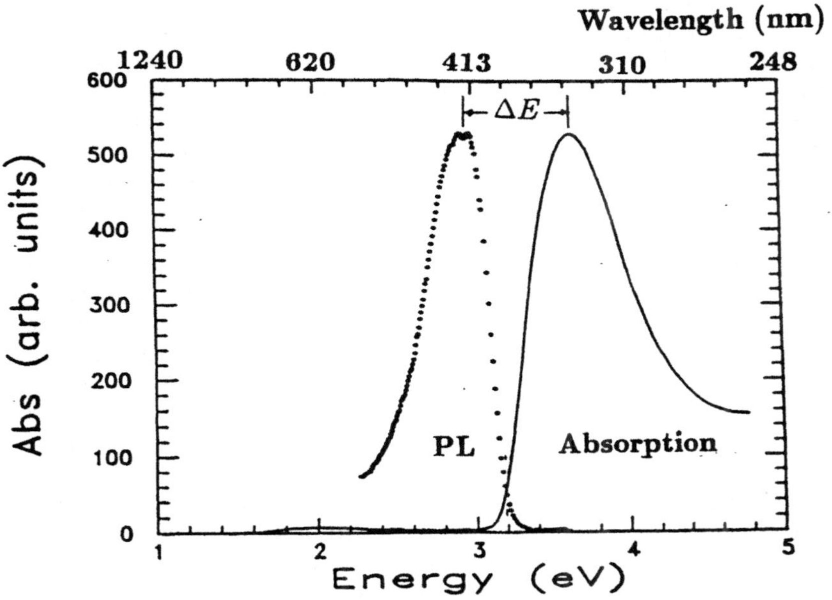

Fig. 4. Absorption and photoluminescence spectra for leucoemeraldine base in NMP, the exciting wavelength for the PL spectrum is ~345 nm (from Ref. 24).

Photoexcitation of LEB films and powders with super bandgap light leads to new intragap absorptions centered at 0.7 eV and 2.9 eV [17, 27]. This photoinduced absorption has been described in terms of formation of a ring rotation polaron [17-19,27]. Because of the large charge conjugation asymmetry in the system there is only one energy level introduced into the gap by the distortion of the ring torsion angles. This polaron differs substantially from earlier studied polarons in polythiophene and polypyrrole and their derivatives in that it has a very large effective mass (~50 m_e) due to the effects of a large moment of inertia involved in rotating the polyaniline rings about the N-N axis. Also observed in the photoexcitation spectrum is a bleaching at ~2.0 eV associated with the presence of some residual imine groups in the otherwise pure LEB. At long times a photoinduced absorption at ~1.4 eV is stable. This absorption is associated with a ring rotation polaron trapped adjacent to a quinoid ring. The short time and long time photoexcitation spectra are summarized in Fig. 5. Comparison of the magnitude of the photoinduced *ir* active vibrations and electronic absorption confirms the large mass of the photoexcited defects.

RING ROTATION POLARONS AND SOLITONS

Several theoretical studies have shown that ring-rotational conformation is important in determining the ground-state electronic structure and geometry of LEB, EB, and PNB [18, 19, 22, 28, 29]. In addition, an excited-state ring rotation was invoked to explain the interring charge-transfer exciton energy in EB [30 - 34]. Despite the experimental evidence for long-lived massive charged defects in polyaniline [17, 27], as well as the demonstration of the importance of ring flipping as a relaxation mechanism [35], the role of ring conformational changes in the self-trapping of charged defect states has received little attention [19].

The removal of an electron from LEB results in the formation of a hole polaron, P_B^+, described as a ring-centered distortion in which the center phenyl ring twists out of its ground state configuration toward the plane formed by the nitrogens [18, 19]. The distortion also causes neighboring rings to distort toward planarity and away from the ground state angle; see Fig. 6. The resulting configuration costs some steric energy, but the π delocalization energy, which favors a planar ring conformation, increases. This model predicts that the energy levels of a P_B^+ split off from the valence band, with little or no splitting from the conduction band. Thus, one would expect two new absorptions in the gap: a transition from the valence band to the new polaron level and one from this new state to the conduction band. Our results for the photoinduced absorption spectrum of LEB [17,27] and the analogous polymer, polyparaphenylene sulfide (PPS) [36], show two absorptions at ~0.75 eV and 2.8 eV that are consistent with this defect level scheme. Both of these absorptions have similar dependences on pump intensity and chopper frequency supporting that they are associated with excitations of the same defect.

Calculations suggest that a negative polaron P_B^- in LEB causes almost no ring-rotational effects, although bond-length effects are expected to produce very small asymmetrical splittings from the conduction and valence bands, perhaps less than a few tenths of an electron-volt. New absorptions arising from this state may be difficult to observe, although changes in the infrared vibrational modes would be expected.

We suggest that the 1.4 eV photoinduced absorption peak may result from a bound state of the hole polaron and a nearby quinoid ring (~4% of the sites in LEB [27]), consistent with calculations by Stafström et al. [34]. If a

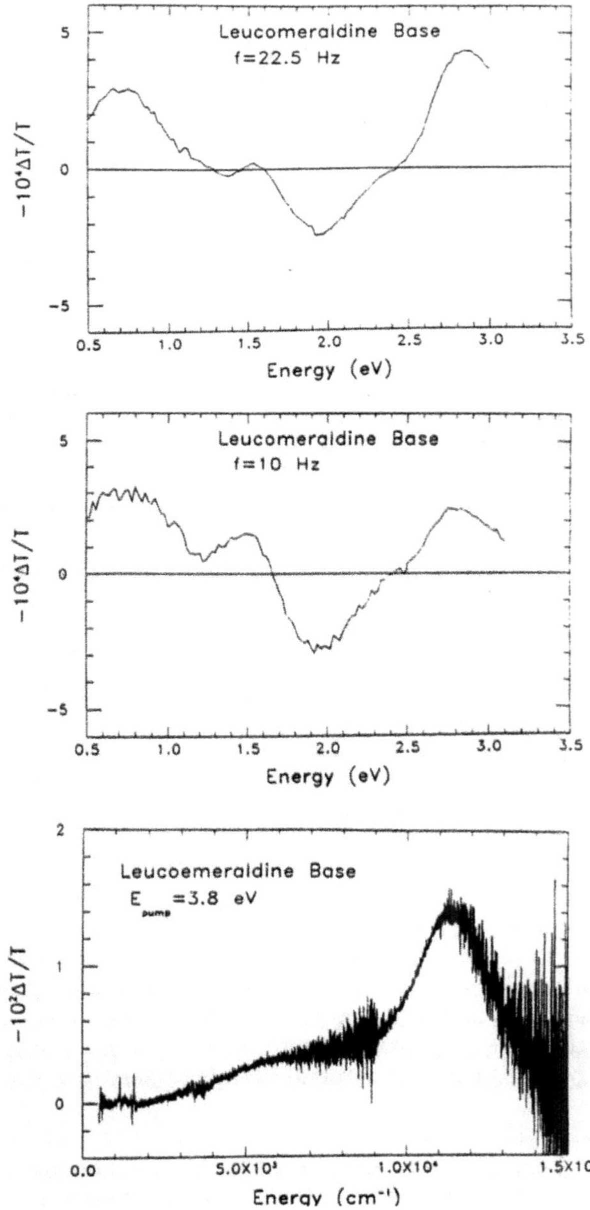

Fig. 5. Near-steady-state photoinduced-absorption spectra of a thin film of LEB on a quartz substrate at 10 K with $E_{pump} = 3.8$ eV, pump intensity of 190 mW/cm^2, at chopper frequency 22.5 Hz (top) and 10 Hz (middle). The long time PA spectrum of LEB in KBr at 80 K for pump energy $E_{pump} = 3.8$ eV is shown in the lower part of the figure (from Ref. 27).

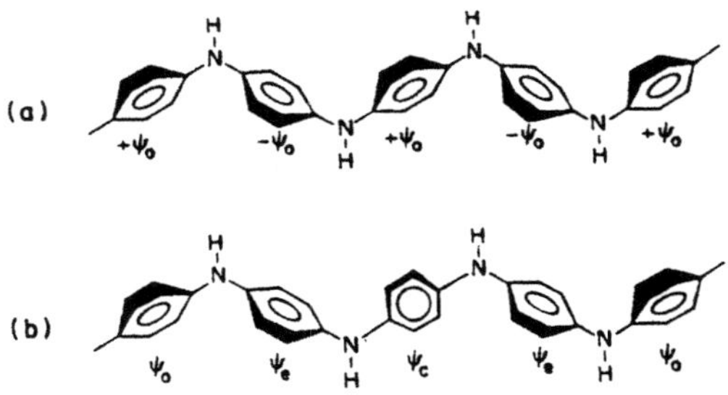

Fig. 6. (a) Sketch of a segment of a leucoemeraldine base chain showing the phenyl rings rotated by $\pm\psi_0$ out of the plane formed by the nitrogens. (b) Representation of a ring-rotational hole polaron in LB where the center phenyl ring and the end phenyl rings rotate toward the nitrogen plane to new ring angles ψ_c and ψ_e.

$P_B{}^+$ is located near a planar (or nearly planar) quinoid ring, π holes will delocalize over several lattice sites without having to drive additional rings toward planarity. The gain in delocalization energy may be enough to bind or trap a $P_B{}^+$ near a quinoid site, forming a $P_{BQ}{}^+$ defect. The $P_{BQ}{}^+$ is expected to take a longer time to form since the ring-rotational polaron must travel to a region containing a quinoid site, and in LEB any quinoid rings should be well-separated. In addition, once formed, the $P_{BQ}{}^+$ would be expected to be long-lived because of the large inertial mass involved for diffusion and thus recombination. The very small 1.4 eV peak at higher chopper frequencies (short times) and the strong low chopper frequency (long time) behavior in the near-steady state experiments, as well as the long-lived behavior in the long-time experiments, support this assignment.

The photoinduced IRAV modes in LEB are quite small compared to the higher-lying electronic PA peaks. This signature indicates that the polarons are massive ($M_{polaron} > 60m_e$) when analyzed in the amplitude mode [37, 38] or bond-order polaron [39] formalisms, in accord with ring rotation polaron predictions [18, 19]. Recent experiments have shown $M_{polaron}$ (as measured in EB) to initially increase with increase of the moment of inertia of the rings through addition of substituents [40], though with increasing substituent size the mass tends to assympote to a (large) value.

The LEB backbone has been a two-fold degeneracy in the phase of its ring order [19]. Hence, soliton defects are possible. Very massive solitons for which the rings near the center of the soliton have nearly zero torsion angle [18] and lighter mass solitons for which the rings near the center of the soliton are at ~90° to the plane defined by the nitrogen atoms [41] have been proposed. Recent studies suggest [42] that solitons may have a more complex variation of ring torsion angle to minimize steric energies.

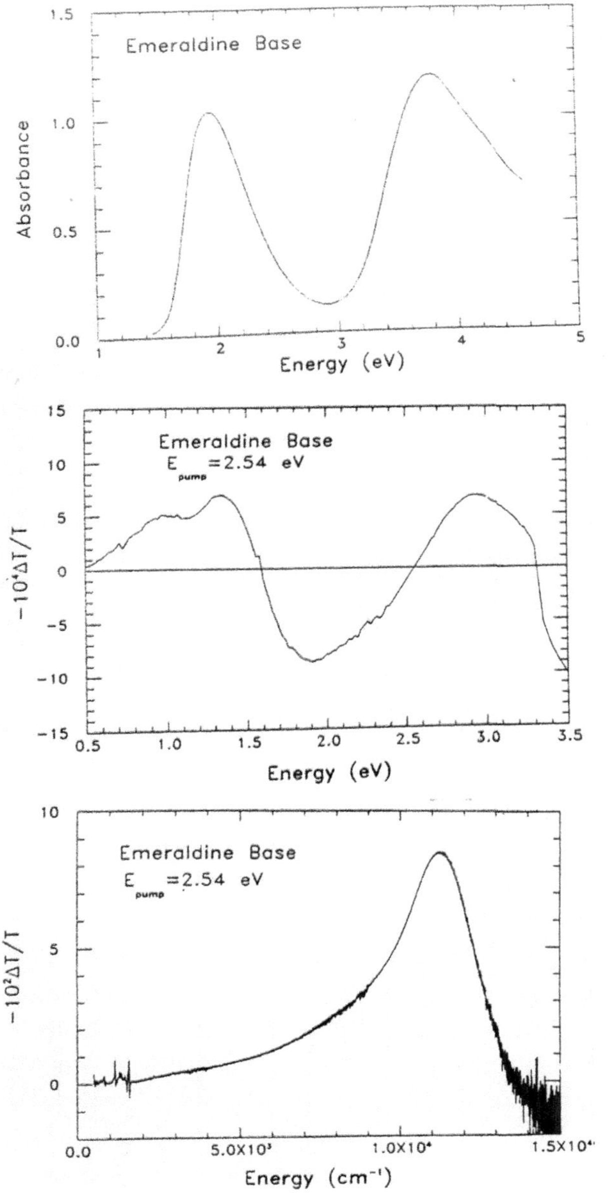

Fig. 7. UV/visible/near-ir absorption spectra of emeraldine base (top) in solutions of NMP; near-steady-state photoinduced-absorption spectra of a thin film of EB on quartz substrate at 15 K for E_{pump} = 2.54 eV, pump intensity of 250 mW/cm^2, and chopper frequency of 21 Hz (middle); and long time PA spectra of EB in KBr at 80 K for E_{pump} = 2.54 eV (bottom) (from Ref. 24).

EMERALDINE BASE

The optical absorption of emeraldine base film, together with selected millisecond photoinduced absorption and long time photoinduced absorption results are shown in Fig. 7. The "steady state" (ms) spectrum of emeraldine base has features indicative of both the ring rotation polaron (at ~0.9 eV and 2.9 eV) and the ring rotation polaron bound to imine groups (1.4 eV and 2.9 eV) as well as bleaching of the π-π^* band (3.7 eV) [43]. At long times only the trapped polaron is observed [17, 27].

The photoinduced bleaching of the 2.0 eV peak is linear in applied power indicative of monomolecular decay kinetics [43]. This supports that this peak is associated with a charge transfer exciton formed by transitions from a benzenoid group to an adjacent quinoid group [30,32]. Picosecond spectroscopy studies [44] of this bleaching show a temperature (T) independent dispersive decay, $\sim t^{-0.11}$, in contrast with a weakly temperature dependent $t^{-.05}$ behavior at millisecond times [43]. Preliminary femtosecond studies [45] of the 2.0 eV peak suggest an initial rapid decay indicative of ring rotation relaxations, followed by a slow dispersive decay. The growth in intensity with constant illumination of the photoinduced ir modes of emeraldine has been shown to follow a stretch-exponential behavior. In contrast, the decay of the photoinduced defects is extremely slow, especially at low temperatures [46]. Light induced electron spin resonance (LESR) experiments show a build-up of spin population which follows a similar stretch exponential or Kohlrausch behavior. The decrease in ESR linewidth (from 25G to 7 G) as a function of increasing photoinduced spin concentration suggest that the photoinduced spins aggregate into regions of high defect density [47]. All of these experiments point to the critical role of steric interaction among polyaniline rings in determining the stability and dynamics of charged defects.

Recently picosecond photoinduced absorption studies of emeraldine base polymer in film and NMP solutions have been reported [48]. They show that for pumping at 2.0 eV and probing ant 2.9 eV a dispersive decay is obtained with more than 50% of the photoinduced signal for the thin film remaining after 5 ns. In contrast, only fast decay components (<100 ps) were found in solution. These differences between solution and film dynamics are attributed to differences in the barriers to ring rotation and differences in interchain processes. In contrast to the luminescence observed from the exciton in LEB, no luminescence is observed for EB [24].

PERNIGRANILINE BASE

The optical absorbance of pernigraniline base, Fig. 8, though similar to that of EB is different in origin. In particular, the 2.3 eV absorption in PNB has been assigned to the excitation across a Peierls energy gap. This Peierls gap has been proposed to arise from two independent contributions, variation in bond length order parameter and variation in ring torsion angle order parameter, [19, 28, 29] Fig. 9. The effects of these two order parameters recently have been shown to vary additively [22].

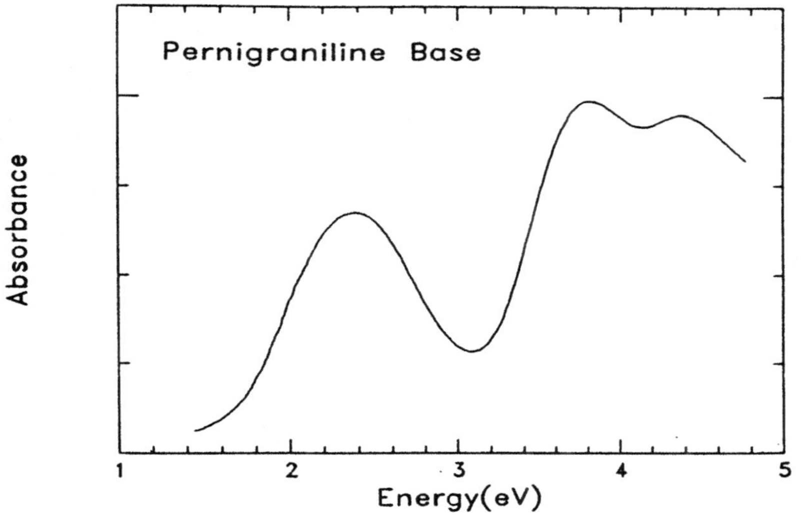

Fig. 8. Direct absorption spectrum of pernigraniline base in NMP solution (from Ref. 49).

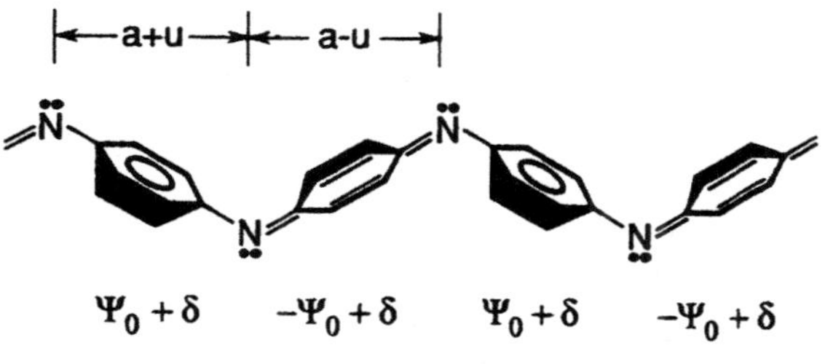

Fig. 9. Schematic structure of bond length dimerized and ring torsion angle dimerized pernigraniline base (from Ref. 50).

Steady state photoinduced absorption experiments show the presence of both a "low energy" (LE) pair of absorptions (at ~1.1 eV) that is short lived [49, 50], and a middle energy (ME) absorption (at ~1.5 eV) that is long lived [40, 49, 50] Fig. 10, 11. Different photoinduced infrared active vibrations are observed for the LE and ME photoinduced absorptions. By studying the time dynamics of the photoinduced infrared active vibrations, these photoexcitations can be unambiguously assigned to the low energy and middle energy peaks. Comparison of the oscillator strength in the infrared active vibrations with the corresponding electronic absorption indicates that the low energy peak is associated with a photoinduced charge defect of effective mass ~5 m_e while the middle energy peak is associated with a photoinduced charge defect of effective mass ~300 m_e. Because of the enormous difference in their effective mass and their lifetimes, the low energy peak is associated with a solitonic defect in the bond length order parameter while the middle energy peak is associated with a polaronic defect in the ring torsion angle order parameter. Light induced electron spin resonance studies support these assignments [47]. This is the first known material where two order parameters contribute independently to the bandgap and sustain separate excitations.

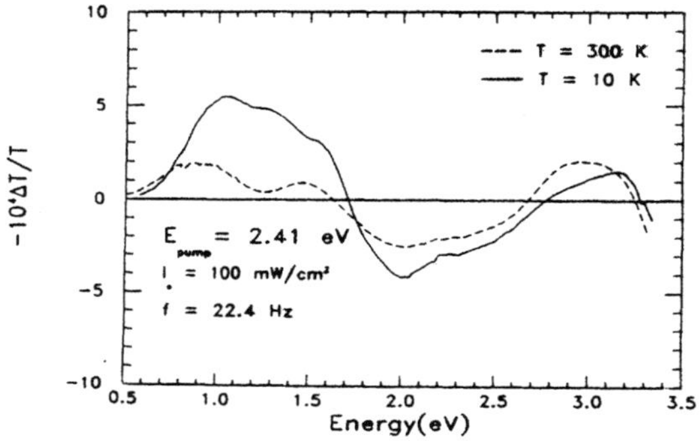

Fig. 10. Near-steady-state photoinduced absorption spectra of pernigraniline base cast on quartz substrate (from Ref. 50).

"METALLIC" POLYANILINE

The emeraldine salt state can be obtained by equilibration of emeraldine base samples with protonic acid of appropriate pH or by *p*-doping (oxidation) of leucoemeraldine base [5]. When prepared in powder form, the resulting emeraldine salt is up to 50% crystalline [51, 52], though when prepared as films [6, 8] or as stretch oriented films [9, 10] the resulting salt may be of lower crystallinity [11, 52].

Correlated transport studies (including temperature dependent dc conductivity, audio-frequency (10Hz-10^5 Hz) conductivity and dielectric constant, electric field dependence of conductivity, thermopower, electron spin resonance, microwave conductivity and dielectric constant) have been carried out on oriented emeraldine hydrochloride films [15], the

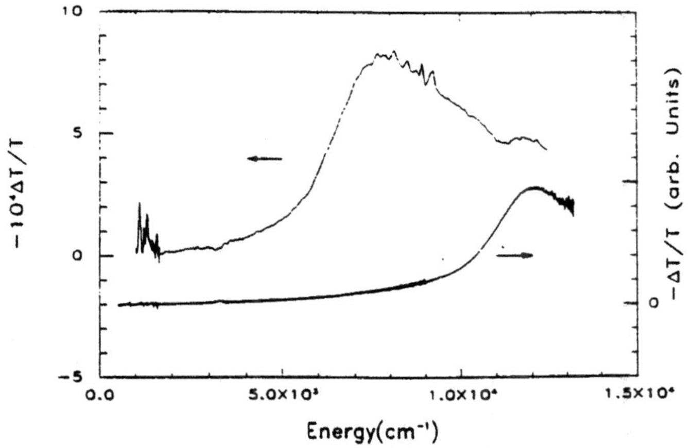

Fig. 11. Photoinduced absorption spectra of PNB in powder form mixed with KBr. Left hand scale: near-steady-state (22 Hz, ~15 K); right hand scale: long-lived photoinduced absorption (80K) (from Ref. 50).

monomethyl derivative (poly(orthotoluidine hydrochloride) [16] and the sulfonic acid self-doped derivative[53, 54]. These results demonstrate the important role of crystallinity and coherence in achieving a highly conducting state in the material. In particular, it is shown that within the crystalline regions of emeraldine hydrochloride the conducting electrons are three-dimensionally delocalized leading to large dielectric constants and "metallic-like" thermopower. The dc conductivity, however, is limited by charge motion in the amorphous material present between crystalline regions. For poly(orthotoluidine) hydrochloride and sulfonated polyaniline the conduction electrons are localized to states or individual chains and charge motion is well accounted for by quasi-one-dimensional variable range hopping.

EFFECTS OF DERIVITIZATION

One of the most attractive aspects of the polyanilines is the ease of preparation of numerous derivatives at ring and nitrogen positions [5]. This provides an opportunity to systematically probe a number of critical issues. In general, the effects of the C_6 ring substituents can be discussed in terms of three broad categories, changes in chemical properties, changes in moment of inertia and steric potentials, and changes in interchain separation, disorder and crystallinity.

While replacement of an H on the C_6 rings with a methyl, ethyl or other alkyl group has only modest effects on the chemical properties, use of alkoxy (e.g. methoxy, ethoxy, etc.) causes considerable solubility changes [5]. Even more dramatic is the replacement of a ring hydrogen with sulfonic acid $-SO_3H$ [53, 54]. In this case the sulfonic acid acts to protonate the imine sites resulting in a self-doped polymer, as well as increased solubility and charged pH equilibrium.

Replacement of C_6 hydrogen with a more massive and bulky group increases the moment of inertia for rotation of the ring about the N-N axis as well as increases the steric repulsion. This causes increases in the effective mass of ring-rotation polarons and changes their lifetimes [40]. For large substituent groups the polaron effective mass tends to asymptote to a constant value.

For the salt forms of polyaniline the C_6 rings tend to lie nearly in the plane of the nitrogen atoms [51, 52]. The primary effect of large substituent groups is to increase the interchain spacing and disorder, even within the crystalline regions [55, 56]. This reduces the interchain transition rate for conduction electrons, leading to charge localization and lower conductivity [16]. Improved processing with resulting greater crystallinity and order may lead to better chain packing, higher coherence and resulting higher conductivities, even for the derivatives of polyaniline.

SUMMARY

The polyaniline family of polymers is a versatile system that enables studies of excitations and charge motion in three distinctly different insulating ground states, leucoemeraldine base, emeraldine base, and pernigraniline base. Formation of the conducting emeraldine salt forms allows study of the mechanisms for insulator-to-metal transition and localization. The ability to derivatize, process, and control the molecular weight and crystallinity, orient, and dope enables probing a wide range of phenomena in ring containing polymer systems.

ACKNOWLEDGEMENT

The author acknowledges the outstanding collaboration with Alan G. MacDiarmid and the active contributions and discussions of the colleagues, postdocs, and graduate students. This work is supported in part by DARPA through a contract monitored by ONR.

REFERENCES

1. Letheby, H. (1862) *J. Chem. Soc.* **15**, 161.

2. Green, A.G., Woodhead, A.E. (1910) *J. Chem. Soc. Trans.* **97**, 2388.

3. Green, A.G., Woodhead, A.E. (1912) *J. Chem. Soc. Trans.* **101**, 1117.

4. For recent progress, see Proc. Int. Conf. on Science and Technology of Synthetic Metals, Kyoto, Japan, June 1986 (*Synth. Met.* **17-19**

(1987)); Sante Fe, NM, June 1988 (*Synth Met.* **27-29** (1989)); Tübingen, Germany, September 1990 (*Synth. Met.* xx (1991)).

5. MacDiarmid, A.G., Epstein, A.J. (1989) *Faraday Discuss. Chem. Soc.* 88, 317.

6. Angelopoulos, M., Ermer, S.P., Manohar, S.K., MacDiarmid, A.G., and Epstein, A.J. (1988) *Mol. Cryst. Liq. Cryst.* **160**, 223.

7. Angelopoulos, M., Asturias, G.E., Ermer, S.P., Ray, A., Scherr, E.M., MacDiarmid, A.G., Akhtar, M., Kiss, Z., and Epstein, A.J. (1988) *Mol. Cryst. Liq. Cryst.* **160**, 151.

8. Angelopoulos, M., Ray, A., MacDiarmid, A.G., and Epstein, A.J. (1987) *Synth. Met.* 21, 21.

9. Cromack, K.R., Jozefowicz, M.E., Ginder, J.M., McCall, R.P., Epstein, A.J., Scherr, E., MacDiarmid, A.G. (1989) *Bull. Am. Phys. Soc.* 34, 583.

10. Cromack, K.R., Jozefowicz, M.E., Ginder, J.M., McCall, R.P., Du. G., Leng, J.M., K. Kim, Li, C., Wang, Z., Epstein, A.J., Druy, M.A., Glatkowski, P.J., Scherr, E.M., MacDiarmid, A.G. (1991) *Macromolecules* xx, xxx.

11. Scherr, E.M., MacDiarmid, A.G., Manohar, S.K., Masters, J.G., Sun Y., Tang, X., Druy, M.A., Glatkowski, P.J., Cromack, K.R., Jozefowicz, M.E., Ginder, J.M., McCall, R.P., Epstein, A.J. (1991) *Synth. Met.* xx, xxx.

12. Tang, X., Scherr, E., MacDiarmid, A.G., Epstein, A.J. (1989) *Bull. Am. Phys. Soc.* **34**, 583.

13. Andreatta, A., Y. Cao, Chiang, J.C., Heeger, A.J., Smith, P., Synth. Met. 26, 383 (1988); Andreatta, A., Tokito, S., Smith, P., Heeger, A.J. (1990) *Mat. Res. Soc. Symp. Proc.* **173**, 269.

14. For a review, see Heeger, A.J., Kivelson, S., Schrieffer, J.R., and Su, W.P. (1988) *Rev. Mod. Phys.* **60**, 781.

15. Wang, Z.H., Li, C., Scherr, E.M., MacDiarmid, A.G., Epstein, A.J. (1991) *Phys. Rev. Lett.* **55**, 1745.

16. Wang, Z.H., Javadi, H.H.S., Ray, A., MacDiarmid, A.G., Epstein, A.J. (1990) *Phys. Rev. B* **42**, 5411; Wang, Z.H., Ray, A., MacDiarmid, A.G., Epstein, A.J. (1991) *Phys. Rev. B* **43**, 4373.

17. McCall, R.P., Ginder, J.M., Roe, M.G., Asturias, G.E., Scherr, E.M., MacDiarmid, A.G., Epstein, A.J. (1989) *Phys. Rev. B* **39**, 10,174.

18. Ginder, J.M., Epstein, A.J., and MacDiarmid, A.G. (1989) *Solid State Commun.* **72**, 987.

19. Ginder, J.M. and Epstein, A.J. (1990) *Phys. Rev. B* **41**, 10674.

20. Stafstrom, S. and Brédas, J.L. (1986) *Synth. Met.* 14, 297.

21. Boudreaux, D.S., Chance, R.R., Wolf, J.F., Shacklette, L.W., Brédas, J.L., Thémans, B., André, J.M., and Silbey, R. (1986) *J. Chem. Phys.* **85**, 4584.

22. Brédas, J.L., Quattrocchi, C., Libert, J., MacDiarmid, A.G., Ginder, J.M., and Epstein, A.J. (1991) *Phys. Rev. B* **44**, xxx.

23. Masters, J.G., Ginder, J.M., MacDiarmid, A.G., and Epstein, A.J., to be published.

24. Masters, J.G., MacDiarmid, A.G., Kim, K., Ginder, J.M., and Epstein, A.J. (1991) *Bull. Am. Phys. Soc.* **36**, 377.

25. Berlman, I.B. *Handbook of Fluorescence Spectra of Aromatic Molecules*, 2nd ed., Academic Press (1971).

26. Colaneri, N.F., Bradley, D.D.C., Friend, R.H., Burn, P.L., Holmes, A.B., Spangler, C.W. (1990) *Phys. Rev. B* **42**, 11,670.

27. McCall, R.P., Ginder, J.M., Leng, J.M., Ye, H.J. Manohar, S.K., Masters, J.G., Asturias, G.E., MacDiarmid, A.G., Epstein, A.J. (1990) *Phys. Rev. B* 41, 5202.

28. Ginder, J.M. and Epstein, A.J. (1990) *Phys. Rev. Lett.* **64**, 1184.

29. dos Santos, M.C. and Brédas, J.L. (1990) *Phys. Rev. Lett.* **64**, 1185.

30. Epstein, A.J., Ginder, J.M., Zuo, F., Bigelow, R.W., Woo, H.S., Tanner, D.B, Richter, A.F.,. Huang, W.S., and MacDiarmid, A.G. (1987) *Synth. Met.* **18**, 303.

31. Duke, C.B., Conwell, E.M., and Paton, A. (1986) *Chem. Phys. Lett.* **131**, 82.

32. Conwell, E.M., Duke, C.B., Paton, A., and Jeyadev, S. (1988) *J. Chem. Phys.* **88**, 3331.

33. Sjögren, B. and Stafstrom, S. (1988) *J. Chem. Phys.* **88**, 3840.

34. Stafstrom, S., Sjögren, B., and Brédas, J.L. (1989) *Synth. Met.* **29**, E219.

35. Kaplan, S., Conwell, E.M., Richter, A.F., and MacDiarmid, A.G., (1989) *Macromolecules* **22**, 1669 ; Kaplan, S., Conwell, E.M., Richter, A.F., and MacDiarmid, A.G. (1988) *Polymer Preprints* **29**, 212.

36. Ginder, J.M., Epstein, A.J., and MacDiarmid, A.G. (1991) *Synth. Met.* xx, xxx .

37. Ehrenfreund, E., Vardeny, Z., Brafman, O., and Horovitz, B., (1987) *Phys. Rev. B* **36**, 1535; Vardeny, Z., Ehrenfreund, E., Brafman, O., Heeger, A.J., and Wudl, F. (1987) *Synth. Met.* **18**, 183.

38. Horovitz, B. (1982) *Solid State Commun.* 41, 729.

39. Rice, M.J. and Phillpot, S.R. (1987) *Phys. Rev. Lett.* **58**, 937.

40. McCall, R.P., Ginder, J.M., Leng, J.M., K.A. Coplin, Ye, H.J. Epstein, A.J., Asturias, G.E., Manohar, S.K., J.G. Masters, Scherr, E.M., Sun Y., and MacDiarmid, A.G. (1991) *Synth. Met.* xx, xxx.

41. MacKenzie, R., Ginder, J.M., and Epstein, A.J. (1991) *Phys. Rev. B* xx, xxx.

42. Brédas, J.L. et al, to be published.

43. Roe, M.G., Ginder, J.M., Wigen, P.E., Epstein, A.J., Angelopoulos, M., and MacDiarmid, A.G. (1988) *Phys. Rev. Lett.* **60**, 2789.

44. Roe, M.G., Ginder, J.M., Gustafson, T.L., Angelopoulos, M., MacDiarmid, A.G., and Epstein, A.J. (1989) *Phys. Rev. B* **40**, 4187.

45. Vardeny, Z., Epstein, A.J., and MacDiarmid, A.G., to be published.

46. Jozefowicz, M.E., McCall, R.P., Epstein, A.J., and MacDiarmid, A.G., to be published.

47. Cromack, K.R., Epstein, A.J., Masters, J.G., Sun Y., and MacDiarmid, A.G. (1991) *Synth. Met.* xx, xxx ; Cromack, K.R. et al., to be published.

48. Lin, L.B., Iwata, K., Ginder, J.M., McCall, R.P., Gustafson, T.L., Epstein, A.J., and MacDiarmid, A.G. (1991) *Bull. Am. Phys. Soc.* **36**, 377.

49. Leng, J.M., Ginder, J.M., McCall, R.P., Ye, H.J. Epstein, A.J., Sun Y., Manohar, S.K., and MacDiarmid, A.G. (1991) *Synth. Met.* xx, xxx.

50. Leng, J.M., Ginder, J.M., McCall, R.P., Ye, H.J. Sun Y., Manohar, S.K., MacDiarmid, A.G., and Epstein, A.J., to be published.

51. Jozefowicz, M.E., R. Laversanne, Javadi, H.H.S., Epstein, A.J., Pouget, J.P., Tang, X., and MacDiarmid, A.G. (1989) *Phys. Rev. B* **39**, 12,958.

52. Pouget, J.P., Jozefowicz, M.E., Epstein, A.J., Tang, X., and MacDiarmid, A.G. (1991) *Macromolecules* **24**, 779.

53. Yue, J. and Epstein, A.J. (1990) *J. Am. Chem. Soc.* **112**, 2800.

54. Yue, J. Wang, Z.H.,Cromack, K.R., Epstein, A.J. and MacDiarmid, A.G. (1991) *J. Am. Chem. Soc.* **113**, 2665.

55. Jozefowicz, M.E., Epstein, A.J., Pouget, J.P., Masters, J.G., Ray, A., MacDiarmid, A.G., (1991) *Macromolecules* **xx**, **xxx**.

56. Jozefowicz, M.E., Epstein, A.J., Pouget, J.P., Masters, J.G., Ray, A., Sun, Y., Tang, X. and MacDiarmid, A.G., (1991) *Synth. Met.* **41**, 723.

STRUCTURAL CHARACTERIZATION OF CONJUGATED POLYMER SOLUTIONS IN THE UNDOPED AND DOPED STATE

By JEAN-PIERRE AIME

Groupe de Physique des Solides
Université PARIS VII-Tour 23
2, Place Jussieu
75251 PARIS CEDEX 05, France

1 INTRODUCTION

This part is mainly devoted to the study of soluble conjugated polymers in doped or neutral state in solution. Electrically conducting polymers have been the subject of growing interest since the middle seventies, partly because a great variety of potential applications have been proposed. Besides, from a more fundamental point of view, they exhibit new and complex phenomena which motivate intensive research in a large domain of scientific field.

The development on these systems in fundamental or applied science belongs to the material research field. One aim of materials research is to design a well characterized material combining several desired properties. In the case of conjugated polymers this corresponds to developing electronic or optical properties with keeping the elastic behavior of a polymer. Indeed, polymers offer, through a variety of synthetic routes or chemical units, many possibilities (more in fact than with inorganic materials) for obtaining selected properties.

Polymers are complex systems. They have been extensively studied first from a chemical and physical chemistry point of view, (P. Flory, "Principles of Polymer Chemistry" (Cornell University Press, Ithaca) (1970)). Later, the analogy with critical phenomena was successfully applied, (P.G. de Gennes, "Scaling concepts in Polymer Physics" (Cornell University Press,Ithaca) (1979)), the use of scaling laws becoming a powerful approach for describing chains in melt and solution, or predicting their behavior in a great variety of situations, adsorption, micelles, star-like structures etc...

For conducting polymers it is well established that the pure substances are electrical insulator, only after doping (oxidation or reduction) do they become conductors. For a polyacetylene sample, the simplest and most studied system, a "good" conductivity is about 100 S/cm, while a conductivity of 10^5 S/cm has been obtained for oriented samples. This insulator to metal transition was in itself a phenomenon attracting original theoretical and experimental works. Studies on polyacetylene: morphology, crystalline structure, synthetic routes, models coming mainly from solid state physics, have lead to a number of publications which counts in thousands. Here we do not intend to discuss any properties related to the crystalline state of polyacetylene, numerous reviews of the subject have been published (see for instance S. Roth and H. Bleier, Adv. Phys., 36, p385-462

J.L. Brédas and R. Silbey (eds.), Conjugated Polymers, 229–314.

(1987) and A.J. Heeger, S. Kivelson, J.R. Schrieffer and W.P. Su, Rev. Mod. Phys., 60, 781 (1988)).

The substituted polydiacetylenes are another system forming an unique class of conjugated polymers. These macromolecules are obtained through a solid-state polymerization in the monomer crystal. The topochemical reaction is initiated either with X, γ, U.V. or electron irradiation, or thermally. In some cases, the mixed crystals of monomer and polymer form perfect solid solutions and lead to a perfect single crystal of polymer after the polymerization has been completed. The specific properties of such assemblies of perfectly ordered conjugated macromolecules have been widely studied, the system providing a large third order nonlinear optical susceptibility. A review can be found in "Nonlinear Optical properties of organic molecules and crystals" Vol 2, edited by D.S. Chemla and J.Zyss (Acad. Press) 1987. Moreover substituted polydiacetylenes were the first system providing large molecular weight conjugated soluble macromolecules.

Since the end of the seventies, increasing efforts have been done for obtaining soluble conjugated polymers, mainly by adding appropriate substituent side-groups. It now happens that the volume of work devoted to the synthesis of new soluble materials may have reached or even overpassed that devoted to the control of the solid phase through the polymerization. Soluble macromolocules like substituted polyacetylenes or polythiophenes often have a rather low molecular weight, in most cases in the order of a few 10^4 corresponding to a chain of about 100 monomers. In that context, polydiacetylenes again occupy a special place, not only because their molecular weight can be larger than 10^6 (several thousand monomer units), but also because the topochemical polymerization hinders parasitic reactions, producing an almost perfect conjugated backbone, free from chemical defects like sp^3 defects, branching etc....

Solubility provides the opportunity to process the material in an easier way with the aim to control the chain packing in the solid state. Even more important is the fact that we now have access to the physical properties of a single chain state. This is a real breakthrough, implying that concepts taken from polymer physics should be very helpful for describing the chain statistics, then providing a new approach for understanding the solid state phase. For example, a recurrent problem is the respective roles of the chain conformation and chain packing in the transport properties. The relation between the way the charges are stored and the chain conformation is not understood either.

We then focus the attention on the physical properties of polymers in solution. Basic concepts related to polymer physics in solution are given together with a description of the small angle scattering technique. The later, in spite of the loss of information at the atomic scale, is a powerful tool for having access to the chain structure. Numerous experimental studies have been performed on "conducting" polymer solutions from which a clear picture of the chain behavior emerges. For the aggregation process, which can be described in terms of transition from good to poor solvent, the situation is less clear. This is due in part to the difficulty in describing the role of the π-electrons on chain-chain interaction, and to the lack of experimental methods allowing to get accurate informations on the chain conformation within an aggregate.

Another way for obtaining soluble conjugated polymers consists to synthesize a copolymer with alternating saturated and conjugated blocks. Only few studies have already been performed, in many cases it appears that elementary characterizations at the synthetic level have still to be done. Nevertheless this route seems very promising, up to now many possibilities have not been investigated and there is an increasing effort for synthesizing new systems.

2 POLYMER SOLUTION

The physics of saturated polymers is now well understood. We can find a considerable amount of work in the literature which form a well defined theoretical and experimental framework. In this chapter, we wish to emphasize the basic concepts which we believe are relevant for the understanding of the statistics of conjugated polymers in solution. For a more complete description, it is recommended to read the books written by P. Flory (1) and "Scaling concepts in polymer physics", P.G de Gennes (2). Also useful and exhaustive informations are found in "Les polymeres en solution..", J. des Cloizeaux and G. Jannink (3).

2-1. Models for single chains

2-1-1. Ideal chain

A flexible long macromolecule is an object with a large number of degrees of freedom. To describe completely the state of an isolated chain is a formidable and hopeless task. The attempt to define an elementary unit is a good example of the complexity of these objects. We can be easily convinced that for each different polymer the size and local structure of the unit is different. Even for the same polymer, we can expect changes in the description of the unit with changes of the effective medium surrounding the chain. An infinity of elementary units is thus created depending of the chemical structure of the monomer and the interaction of the chain with the medium.

To avoid such a complication, the mathematical models describing the behavior for a single chain simply do not take into account the structure at a microscopic scale. These idealized models becomes realistic for very long chains. For a flexible polymer the most simple model is the random walk.

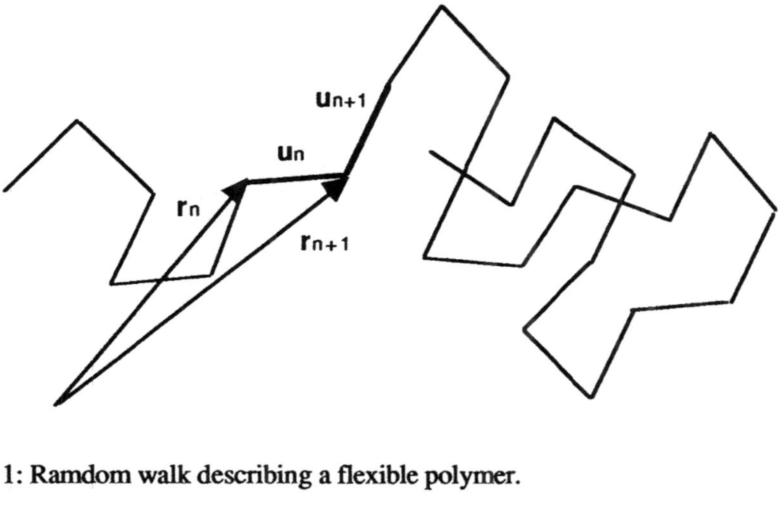

Figure 1: Ramdom walk describing a flexible polymer.

Let $\mathbf{u}_n$ be the vector joining the end points of two consecutive segments located at $\mathbf{r}_n$ and $\mathbf{r}_{n+1}$ (figure 2-1), at position n the next vector $\mathbf{u}_{n+1}$ may have any direction. Each vector has a completely independent orientation hence the relations

$<\mathbf{u}_n> = 0$

$<\mathbf{u}_n.\mathbf{u}_m> = a^2\ \delta_{nm}$

with a the segment length and $\delta_{nm} = 0$ for $n \neq m$ and 1 for $n = m$

the end to end distance is given by

$$< R_0^2 > = <(\mathbf{r}_N - \mathbf{r}_0)^2> = \Sigma\ \mathbf{u}_n.\mathbf{u}_m = N\ a^2 \qquad (2\text{-}1)$$

for very large N the end to end distance follows a Gaussian probability distribution with a variance given by equation 2-1

$$P_N(\mathbf{r}) = \frac{1}{(2\pi R_0^2)^{3/2}} \exp\left(-\frac{3\mathbf{r}^2}{2R_0^2}\right) \qquad (2\text{-}2)$$

this relation is typical of scaling law because the probability depends on the variable r only through the ratio r^2/R_0^2. We do not have to consider the specificity of the macromolecule. This provides an universal description of the chain behavior.
At fixed elongation, equation 2-2 gives the entropy of the chain

$$S(r) = S(0) - \frac{3\mathbf{r}^2}{2R_0^2} \qquad (2\text{-}3)$$

which leads to the free energy

$$F(r) = F(0) + \frac{3T\mathbf{r}^2}{2R_0^2} \qquad (2\text{-}4)$$

where T/R_0^2 has the dimensionality of a spring constant.

2-1-2. Real chain in good solvent

In the random walk model, there is no restriction about the direction of a step, for example we can return to the same position as the previous one or anyone already investigated between the first and the current step. In other words, a segment representing the unit length of the polymer can overlap another segment. Because the interactions between monomer units are ignored, it is easily understandable that such a model will be unable to describe a real chain in solution. A more realistic description must take into account the so called excluded volume effect corresponding to the steric repulsion between units. Intuitively we expect a more extended macromolecule, for large N we write :

$$< R^2 > \approx N^{2\nu}\ a^2 \qquad (2\text{-}5)$$

where $\nu = 1/2$ for a gaussian coil. Since for $\nu = 1/2$ the interpenetration between units is possible and for $\nu = 1$ the chain is fully extended corresponding to the rod conformation, the behavior of a real polymer swollen by the solvent is described with an exponent such that $1/2 < \nu < 1$. The dimensionality for the chain in good solvent is between 2 (ideal chain) and 1 (rod).

The value of ν was first calculated by Flory. The fundamental assumption is to consider an homogeneous monomer concentration within the volume of the coil. In a space of dimension d, the internal concentration is given by

$$c = \frac{N}{R^d} \tag{2-6}$$

The repulsive interaction is proportional to the number of contacts between monomer units, and since we consider a pair interaction the energy per unit volume is proportional to c^2. The total repulsive energy for a given conformation is

$$F_{rep} = \frac{1}{2} T v c^2 R^d = \frac{1}{2} T v \frac{N^2}{R^d} \tag{2-7}$$

where v is the excluded volume parameter
Minimizing versus R the entropy (equation 2-4) and repulsive contribution (equation 2-7), Flory obtains (1)

$$R^{d+2} \approx v N^3 a^2 \tag{2-8}$$

$$\text{and } \nu = \frac{3}{d+2} \tag{2-9}$$

with $\nu = \frac{3}{5}$ for d=3.

The relation (2-9) is in remarkable agreement with experiment in spite of the crude approximation involved. It is surprising that considering an average concentration within the coil and an entropy term given by an ideal behavior gives such a good result. A large amount of work has been devoted trying to understand the success of the mean field picture. The reader interested by this problem will find an exhaustive discussion in ref (2) page 43 and in ref (3) page 275.

2-1-3. Chain with local stiffness: Kratky-Porod model.

In a real polymer the positions between neighboring monomer units are strongly correlated. A simple model proposed by Kratky and Porod (4) describes the properties of a chain with local rigidity. The constraint applied to the monomer units, with a monomer unit length equal to a, is

$$\langle \mathbf{u}_i . \mathbf{u}_{i+1} \rangle = a^2 \cos\alpha \tag{2-10}$$

the correlation between successive monomer units induces a correlation at longer distance and we have

$$\langle \mathbf{u}_n . \mathbf{u}_m \rangle = a^2 (\cos\alpha)^{|n-m|} \tag{2-11}$$

writing the sum of the projections on the first unit, we have for large N (4)

$$l_p = \Sigma\, a\, (\cos\alpha)^n = \frac{a}{1-\cos\alpha} \qquad (2\text{-}12)$$

where l_p is a length characteristic of the local rigidity.

A continuous description is obtained by keeping the total contour length L = Na and the ratio a/α^2 constant, with N going to infinity and a and α going to zero. Setting $l_p = a/(\alpha^2/2)$ and $N = L/l_p$ we have

$$(\cos\alpha)^N \approx (1 - \frac{\alpha^2}{2})^N \approx e^{-N} = e^{-\left(\frac{L}{l_p}\right)} \qquad (2\text{-}13)$$

For the continuous curve the angular correlation is related to the relative orientation between the unit vector tangent to the curve

$$\mathbf{t}\,(l) = \frac{\delta \mathbf{r}(l)}{|\mathbf{r}|\,\delta l} \qquad (2\text{-}14)$$

with l the curvilinear coordinate.

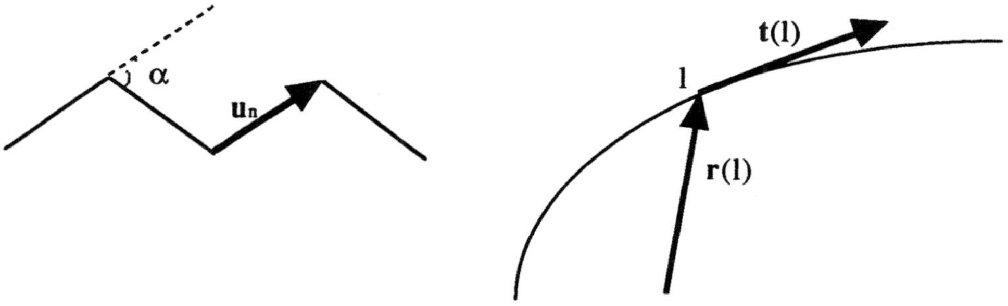

Figure 2 : Description of the local correlations for a macromolecule : Discrete and continuous models

The local rigidity described by the equations (2-11) and (2-12) is now written

$$< \mathbf{t}(l).\mathbf{t}(l')> = \exp\left(-\frac{|l-l'|}{l_p}\right) \qquad (2\text{-}15)$$

Equation (2-15) allows to calculate the mean square end to end distance of the chain:

$$\langle R^2 \rangle = \int_0^L t(l_1)\, t(l_2)\, dl_1\, dl_2 = 2Ll_p \left(\frac{L}{l_p} - 1 + e^{\left(-\frac{L}{l_p}\right)} \right) \qquad (2\text{-}16)$$

For a total chain contour length smaller than l_p

$$\langle R^2 \rangle \approx N^2 l_p^2 \qquad (2\text{-}17)$$

the local rigidity leads to a rod behavior for a short chain with respect to the persistence length.
For N large or $L >> l_p$, we obtain a result similar to the case of an ideal chain with an effective length equal to l_p (5):

$$\langle R^2 \rangle \approx 2\, N\, l_p^2 = L\, 2\, l_p = L\, b \qquad (2\text{-}18)$$

the latter equation gives the definition of the Kuhn length or statistical length as a function of the persitence length $b = 2\, l_p$.
Equation 2-18 clearly shows that the Kratky- Porod model does not take into account the repulsion between units. Nevertheless, equation (2-16) is of practical interest for chains with finite length as it happens in experimental situation.

2-2. Notion of theta and good solvent for linear saturated polymer.

2-2-1. Mean field picture

The lattice model (1,6) is a convenient way for describing the thermodynamics of polymers in solution. Let $\phi = ca^3$, the fraction of sites occupied by a monomer unit, with a^3 the volume of the unit cell in the cubic lattice, c the concentration of monomer units (number of monomer units per cm^3).

The free energy has two terms: the entropy and the interaction terms. For a mixture of several components, the entropy term per site is given by $S = -\sum p_i \ln p_i$, where p_i is the probability distribution for species i. The probabilities to find a monomer unit or solvent molecule at a given site are ϕ and $1-\phi$ respectively. The corresponding probability for the polymer we have ϕ/N, where N is the number of monomer units in the chain. This is the translational entropy, that is the internal degrees of freedom of polymer and solvent units are not considered.

$$-S|_{site} = \frac{\phi}{N} \ln \frac{\phi}{N} + (1-\phi)\ln(1-\phi) \qquad (2\text{-}19)$$

substracting the entropies of the polymer and solvent alone $\phi\, S(1)$ and $(1-\phi)\, S(0)$, yields the entropy of mixing :

$$-S_{mix}|_{site} = \frac{\phi}{N} \ln\phi + (1-\phi) \ln(1-\phi) \qquad (2\text{-}20)$$

The second term in the free energy corresponds to the interaction between monomer units, between solvent units and between monomer and solvent units. The interaction between species i and j is written

$$U_{ij} = \chi_{ij}\, \phi_i \phi_j \tag{2-21}$$

the χ_{ij} term is construct with van der Waals interaction. Solvent and monomer unit have the same kind of interaction and additive rule may be applied leading to a simple dependence as a function of the concentration. The total interaction is obtained by using the average concentration without considering local fluctuations as it is typically done with a mean field approach. The sum term is given by:

$$\chi_{ms}\, \phi_m \phi_s + \frac{1}{2}\chi_{mm}\, \phi_m \phi_m + \frac{1}{2}\chi_{ss}\, \phi_s \phi_s \tag{2-22}$$

again substracting the contribution coming from polymer and from solvent alone, and substituting ϕ_m and ϕ_s by ϕ and $1\text{-}\phi$ yields:

$$\chi\, \phi\, (1-\phi) \tag{2-23}$$

with χ the Flory parameter equal to $\chi = \chi_{ms} - 1/2\, (\chi_{mm}+\chi_{ss})$

With these approximations, the free energy in the lattice model is given by (setting $k_B=1$):

$$\frac{F_{mix}}{T}\text{site} = \frac{\phi}{N} \ln\phi + (1-\phi)\ln(1-\phi) + \chi\, \phi(1-\phi) \tag{2-24}$$

The basic equation 2-24 gives us the thermodynamic quantities related to the polymer and solvent. Expanding equation 2-24 at small ϕ we find

$$\frac{F_{mix}}{T}\text{site} = \frac{\phi}{N} \ln\phi + \frac{1}{2}\phi^2(1-2\chi) + \frac{1}{6}\phi^3 \ldots. \tag{2-25}$$

It is interesting to compare equation 2-25 to an expression for the free energy reminiscent of the one for non ideal gases:

$$\frac{F_{mix}}{T}(\text{cm}^3) = \frac{c}{N} \ln c + \frac{1}{2} v c^2 + \frac{1}{6} w^2 c^3 \tag{2-26}$$

where v is the excluded volume and c the concentration of monomer units. From the definition of ϕ and equation 2–25, the excluded volume v is obtained as a function of the two particle interaction between monomer units:

$$v = a^3\, (1-2\chi) \tag{2–27}$$

The value of χ is dependent of the temperature and the pressure. For small χ the solvent is good, in particular, $\chi = 0$ corresponds to an athermal solvent. The free energy only comes from the conformational entropy of the polymer. The statistical conformation of the chain is governed by the steric repulsion between monomer units.

With increasing χ the excluded volume term decreases, for $\chi = 1/2$, the two particle interaction is cancelled. In such a case we have a theta solvent with a nearly ideal behavior for the polymer. Actually, there is a difference between the definition of the theta solvent, that is the cancellation of the ϕ^2 term, and an ideal behavior of the polymer. The remaining third order term becomes now relevant for the description of the chain conformation (2).

For $\chi > 1/2$, the excluded volume becomes negative, the attractive interaction between monomers is larger than the attractive interaction between monomer and solvent, in that case the chain collapses.

2-2-2. Osmotic pressure

The osmotic pressure of the macromolecules is obtained by considering a change in the solution volume keeping the number of monomer units n_m constant.

$$\Pi = -\frac{\partial F_{tot}}{\partial V}\Big|_{n_m} \tag{2-28}$$

the total free energy is related in the lattice model to the free energy per site by $F_{tot} = F_{site} V_{tot}/a^3$, and $n_m = \phi V_{tot}/a^3$. Inserting this relation in equation 2-28, and using the Flory-Huggins approach (eq.2-24), we get:

$$a^3\Pi = \phi^2 \frac{\partial}{\partial\phi}(F_{site}/\phi) \tag{2-29}$$

$$\frac{a^3\Pi}{T} = \frac{\phi}{N} - \ln(1-\phi) - \phi - \chi\phi^2 \tag{2-30}$$

Expanding equation 2-30 for small ϕ we have:

$$\frac{a^3\Pi}{T} = \frac{\phi}{N} \tag{2-31}$$

For $1/N << \phi << 1$ we may still expand Π in power of ϕ

$$\frac{a^3\Pi}{T} = \frac{\phi}{N} + \frac{1}{2}(1-2\chi)\phi^2 + ... \tag{2-32}$$

a leading term in ϕ^2, that is a quadratic dependence on the concentration, is predicted within the mean field picture. This behavior has been shown to be wrong by using scaling law arguments. For strongly interacting system at concentration where the chains overlap, the osmotic pressure is predicted to be $\Pi = T \phi^{9/4}$ (7).

For very dilute solutions, equation (2-31) corresponds to the equation of state for an ideal gas. We can obtain a more general formulation, without the help of the lattice model by expanding the free energy (per cm^3) as a function of the concentration.

$$\frac{\Pi}{T} = \frac{c}{N} + B\, c^2 + \ldots \tag{2-33}$$

where B is the second virial coefficient. An expansion in term of polymer concentration gives

$$\frac{\Pi}{T} = \frac{c}{N} + N^2\, B\, (c/N)^2 + \ldots\ldots \tag{2-34}$$

the direct comparison with the mean field equation 2-14 gives a virial coefficient of the polymer proportional to the square of the degree of polymerization N, the proportionality constant being the excluded volume v.

In three dimensions, two polymers which occupy the same volume of radius R have N^2v/R^3 contact points. The free energy of the two interpenetrating polymers is proportional to the number of contacts. Setting as usual the Boltzmann constant equal to 1,

$$F_{rep} = T\, v \frac{N^2}{R^3} \approx T\, v\, R^{2d_v - 3} \tag{2-35}$$

where $d_v = v^{-1}$

whenever $d_v < 3/2$ the repulsive energy goes asymptotically to zero and the mean field theory may be used to evaluate the second virial coefficient. When $d_v > 3/2$, the repulsive energy is much larger than T. The two polymers interpenetrate only slightly and a hard sphere picture is appropriate. The virial coefficient is proportionnal to the volume occupied by the polymer and equation 2-16 is rewritten as

$$\frac{\Pi}{T} = \frac{c}{N} + \text{constant}\; R^3 \left(\frac{c}{N}\right)^2 + \ldots\ldots \tag{2-36}$$

In good solvent the polymer dimension of the coil being $d_v = 5/3$, equation 2-36 applies, while for a rod the dimensionality being equal to 1, the Onsager result can be used and the virial coefficient is proportional to the square of the rod length (8), i.e the square of the degree of polymerization.

An interesting question is the behavior of two interacting polymers when the sum of their dimension is equal to 3. It happens with the interaction of a coil and a rod. For the flexible chain in good solvent the sum is equal to 8/3 and the mean field result applies. The case of the interaction with the coil in theta solvent is less straightforward since the sum is equal to 3. The question has been solved by J.F. Joanny (9), the second virial coefficient is equal to $LR^2/\text{Log}N_c$, i.e $N_r N_c/\text{Log}Nc$. That is, it remains proportional to the product of the degree of polymerization of the two macromolecules with a proportionality constant equal to the inverse of the logarithm of the degree of polymerisation of the flexible polymer in theta solvent instead of the monomer-monomer interaction between the coil and the rod .

In this short introduction, the only interaction taken into account between two monomers was the so called excluded volume interaction. This corresponds to the case where the repulsive part is much bigger than the attractive part. Usually the monomer-monomer interaction can change by varying either the temperature or pressure, and the concept of theta solvent corresponds to the temperature at which the excluded volume interaction cancels. Below the Θ temperature the dominant contribution is the attractive part and the chain collapses.

In chapter 7 the transition between good and poor solvent is discussed in relation with the gel formation and the precipitation observed for soluble polydiacetylenes. Without anticipating this discussion, it is interesting to note that theta behavior has not yet been found for conjugated polymers. Only structure of such macromolecules in good solvent or aggregation processes have been studied. While the study of aggregation phenomena may be of some help for improving the processability of conjugated polymers, for example the folding process, the study of the chain conformation in good solvent provides quantitative information on the origin of the solubility and description of the chain statistics. In good solvent, the chain statistics can not differ drastically from the conformation of saturated polymers. Therefore only at short scale (a few monomer units) we can expect changes on the intrachain correlations in which the effective influence of the π-electrons can be evaluated.

References

1 P. Flory, "Principles of Polymer Chemistry" (Cornell University Press, Ithaca) (1970).

2 P.G. de Gennes, "Scaling concepts in Polymer Physics" (Cornell University Press,Ithaca) (1979).

3 J. des Cloizeaux, G. Jannink, "Les Polymères en Solution, leur modélisation et leur structure" (Editions de Physique Les Ulis) (1987).

4 O. Kratky, G. Porod, Rec. Trav. Chim. Pays-Bas, **68**, 1106 (1949).

5 W. Kuhn, Kolloid Z., **76**, 258 (1936) and **87**, 3 (1939).

6 M. Huggins, J. Phys. Chem., **46**, 151 (1942). M. Huggins, Ann. N.Y. Acad. Sci., **41**, 1 (1942)

7 J. des Cloizeaux, J. Phys. (France), **36**, 281, (1975).

8 L. Onsager, Ann. N.Y. Acad. Sci., **51**, 627 (1949).

9 J.F. Joanny, J. Phys. (France), **49**, 1981 (1988).

3. STRUCTURAL STUDIES WITH SMALL ANGLE SCATTERING

3-1. Basic Principles

Small angle scattering is a technique for measuring spatial correlations at large distance. The resolution is lost at the atomic and even at the monomer unit scale, typically we can not expect to have an accurate structural information for length smaller than 10 Å. The experiment can be achieved either by using light, X-ray or neutrons.

In a scattering experiment, a beam of electromagnetic radiation (light, X-rays) or neutrons is directed onto a sample and the scattered intensity is measured as a function of the angle θ to the incident direction. The incident and scattered beams are characterized by their wavenumbers $\mathbf{k_i}$ and $\mathbf{k_s}$ and the scattering event by the momentum transfer $\mathbf{q}$:

$$\mathbf{q} = \mathbf{k_i} - \mathbf{k_s} \tag{3-1}$$

At small scattering angles θ, inelastic contributions can be neglected and the quasi elastic scattering is dominant. The scattered intensity is mainly described by a process where the magnitude of the wavenumber remains constant: $k_i = k_s = k$ and consequently the Bragg relation applies:

$$q = 2k \sin\left(\frac{\theta}{2}\right) \tag{3-2}$$

$$\text{with } k = \frac{2\pi}{\lambda} = \frac{2\pi n}{\lambda_0} \tag{3-3}$$

Table 3-1

Type of radiation	wavelength (Å)	q range ($Å^{-1}$)	Spatial resolution (Å)
Light	4000-6000	$5\ 10^{-5}$ - $3\ 10^{-3}$	2000 - 100 000
X-rays	1-4	10^{-2} - 1	7 - 500
neutrons	1-30	10^{-3} - 1	7 - 5000

λ is the wavelength of the incident beam in the sample, λ_0 the wavelength in vacuum and $n = \lambda/\lambda_0$ the refractive index, which is close to and generally slightly smaller than 1 for X-Rays and neutron scattering. The momentum transfer $\mathbf{q}$ behaves like the reciprocal of the wavelength giving the range of the spatial scale studied. Table 1 gives the length scales available with the different wavelengths. Therefore the use of a particular technique depends on the size of the particle and the type of information sought for, average shape or local conformation. The second aspect concerns the interaction between the beam and the particle, central diffusion occurs with density-density fluctuations in an homogeneous liquid or solid medium. In fact, one does not directly observe density fluctuations, but rather variations in scattering intensity due to differences between the magnitudes of the interactions between the incident radiation and the scatterers and surrounding medium (solvent) respectively. For light scattering the change of refractive index is the relevant factor and the scattering intensity I(q) is proportional to

$$I(q) \approx (n_e - n_0)^2 \approx n^2\left(\frac{dn}{dc}\right)^2 \quad (3\text{-}4)$$

with n_e the refractive index of the particle and $\left(\frac{dn}{dc}\right)$ the index change as a function of the concentration of particles.

X-rays interact with electrons, so that the intensity scattered by a particle of average electron density Z_1 in a medium of electron density Z_2 will scale as $e^2\,(Z_1 - Z_2)^2$: there is no isotopic selectivity and the contrast between particle and medium can be varied by chemical modification only, and X-rays are vastly more sensitive to heavy atoms, uncommon in usual polymers.

For thermal neutrons, scattering is the superposition of potential scattering, essentially by a hard sphere having the nuclear radius $R \sim 1.45\ 10^{-5}\ A^{1/3}$ angströms, where A is the atomic weight, and terms associated to resonances of the neutron-nucleus compound state. The latter term can be of either sign and comparable in magnitude to first one, so that total scattering length may vary widely and can be positive or negative (Figure1). Therefore, neutron scattering is not dominated by heavy atoms, and isotopes can scatter very different: contrast variation can be achieved by mere isotopic substitution. Fortunately, the hydrogen isotopes have very different coherent scattering lengths: b(H) = $-3.74\ 10^{-5}$ Å, b(D) = $+6.67\ 10^{-5}$ Å, which is about the greatest possible contrast (1,2).

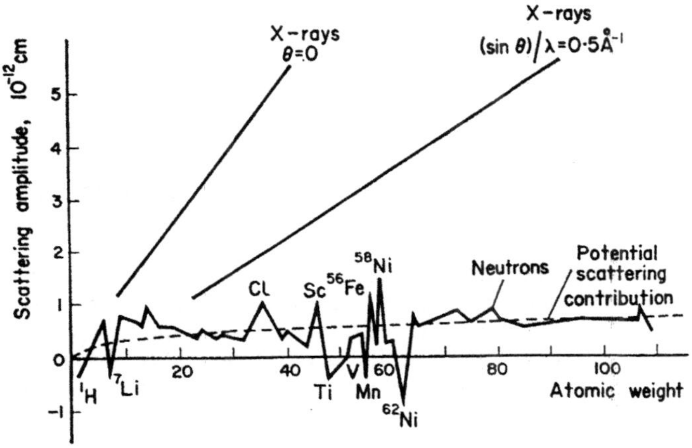

Figure 1 : Neutron and X-ray cross-section as a function of the atomic number. (from reference 1).

3-2. Scattering at small angle.

Let us start with a collection of homogeneous particles diluted in a volume V_0, we do not consider the medium and consider the case where the concentration is low. The meaning of low concentration is given below. Setting $\mathbf{R_k}$, the vector position of the center of gravity of particle k, and $\mathbf{r}_{kl}$ the vector position of the atoms belonging to the particle, such that any atom l has a position defined by $\mathbf{r}_l = \mathbf{R}_k + \mathbf{r}_{kl}$, the scattering function of the assembly of particles is given by the sum:

$$S(q) = < \sum_{kl} \sum_{jm} \exp\left(i\mathbf{q}(\mathbf{R}_k-\mathbf{R}_j)\right) \exp\left(i\mathbf{q}(\mathbf{r}_{kl}-\mathbf{r}_{jm})\right)> \qquad (3\text{-}5)$$

where the bracket means an average upon the different orientations accessible to the particles. Let us define the form factor of the isolated particle k (depending of the equilibrium shape of the particle) as:

$$F_k(q) = \sum_{l} \exp i\mathbf{q}\mathbf{r}_{kl} \qquad (3\text{-}6)$$

and the scattering function for the particle

$$S_1(\mathbf{q}) = < F^2(\mathbf{q}) > \qquad (3\text{-}7)$$

With definitions 3-6 and 3-7, equation 3-5 is rewritten as:

$$S(\mathbf{q}) = < \sum_{k\,j} F_k(\mathbf{q})\, F^*_j(\mathbf{q}) \exp\left(i\mathbf{q}(\mathbf{R}_k-\mathbf{R}_j)\right) > \qquad (3\text{-}8)$$

$$S(\mathbf{q}) = N\, S_1(\mathbf{q}) + < F(\mathbf{q}) >^2 < \sum_{k\, j\neq k} \exp\left(i\mathbf{q}(\mathbf{R}_k-\mathbf{R}_j)\right) > \qquad (3\text{-}9)$$

the first term corresponds to k = j and is the scattering function for N isolated homogeneous particles. The second term is the interparticle term describing the coherent scattering coming from the existence of the interaction between particles. At low concentration the second term can be expressed as a function of the pair interaction potential between the particles.

Let us consider first that the particles behave like isotropic object and the probability to find a particle in an elementary volume dv_j located at j together with a particle located at k in the volume dv_k is only function of the distance between the two points: R_{kj}. In order to describe the spatial correlation between the particles we add first that the probability to locate at the same time the center of gravity of the two particles in the same elementary volume ($R_{kj} = 0$) is zero , while for R_{kj} going to infinity the probability is equal to 1. Replacing the sum in the second term of the equation 3-9 by an integration over the irradiated volume V, with v_1 the volume of a particle, we obtain:

$$< \int_V \int_V \exp(i\mathbf{q}\mathbf{R}_{kj})\;\; P(R_{kj})\, \frac{dv_k}{v_1}\frac{dv_j}{v_1} > \qquad (3\text{-}10)$$

It is convenient to split equation 3-10 in two terms

$$< \int_V \int_V \exp(i\mathbf{q}\mathbf{R}_{kj}) \frac{dv_k}{v_1}\frac{dv_i}{v_1} > + < \int_V \int_V \exp(i\mathbf{q}\mathbf{R}_{kj}) \, (P(R_{kj}) - 1) \frac{dv_k}{v_1}\frac{dv_i}{v_1} > \quad (3\text{-}11)$$

the first term corresponds to the scattering at large scale, such that for R_{kj} large we have $P(R_{kj}) = 1$. Only the values of q close to 0 contribute in the integral (for an infinite volume this term corresponds to $\delta(q)$). This means that in a scattering experiment, with the particle scattering function $S_1(q)$ only the second term in 3-11 is important.

Since $P(R_{kj}) - 1$ goes to 0 for large R_{kj}, we can now perform the integration to an infinite volume. Noting that $V = Nv_1$, where N is the number of particles, and performing one integration, equation 3-9 is rewritten:

$$S(\mathbf{q}) = N \{ (S_1(\mathbf{q}) + \frac{< F(\mathbf{q}) >^2}{v_1} < \int_V \exp(i\mathbf{q}\mathbf{R}) \, (P(R) - 1) \, dv \, >\} \quad (3\text{-}12)$$

the second term is the Fourier transform of a function characteristic of the interaction between particles. The interparticle term is called the Fourier transform of the excluded volume (3, 6, 7, 24). As a consequence of the interference between particles, we do not have a direct observation of the isolated particle form factor. At low q values, such as the ones reached by light scattering, the interparticle contribution is eliminated through a study at different concentrations, putting:

$$v(q) = - \int \exp(i\mathbf{q}\mathbf{R}) \, (P(R) - 1) \, dv \quad (3\text{-}13)$$

we obtain from 3-12:

$$S(\mathbf{q}) = N \{ S_1(\mathbf{q}) - \frac{c < F(\mathbf{q}) >^2}{m} v(q)\} \quad (3\text{-}14)$$

where c and m are the concentration and the mass of the particle.

3-3. Small angle scattering from polymers in solution.

3-3-1. Incompressibility and contrast factor.

Let us consider a binary mixture of species noted A and B. The incompressibility rule implies that the density is a constant (see figure 2):

$$\sum_{i \in A} \rho_i + \sum_{j \in B} \rho_j = cst$$

or

$$\sum_{i \in A} \delta\rho_i + \sum_{j \in B} \delta\rho_j = 0 \tag{3-15}$$

since the scattering function is the Fourier transform of the density fluctuation, that is $S_{ij}(q) = \langle\rho_i(q)\rho_j(-q)\rangle - \langle\rho_i(q)\rangle\langle\rho_j(q)\rangle$, equation 3-15 gives the relation:

$$S_{AA}(q) = S_{BB}(q) = -S_{AB}(q) \tag{3-16}$$

setting b_A and b_B the scattering length of the species A and B, the scattering function observed is given by (4):

$$S(q) = \sum_{i \in A\, j \in B} b_i\, S_{ij}(q)\, b_j \tag{3-17}$$

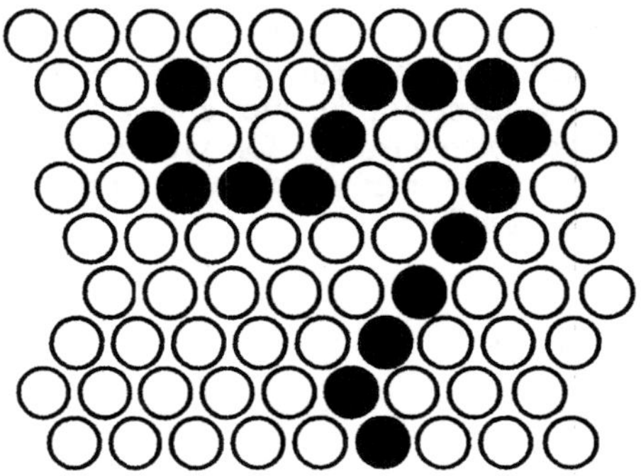

Figure 2: Sketch of polymer chains (black circles) and deuterated solvent (white circles). Contrast is due to protons along the chain. Selective deuteration either on polymer or solvent change the magnitude of the contrast.

If $b_A = b_B$, $S(q) = 0$, we do not observe any scattering, the binary mixture appears homogeneous. In general, for a polymer in solution, both a solvent and an elementary unit of the polymer (the repeat unit) are polyatomic entities, which in addition usually have different molar volumes v_s and v_m respectively. Furthermore the phase changes between the atoms belonging to the same monomer unit are negligible. Within a monomer or solvent unit, the atomic distances are in the order of 1 Å, while the q range is typically between 10^{-3} and 10^{-2} Å^{-1}, then a phase change $qr \ll 1$. Therefore the average scattering length densities ρ_m and ρ_s may be defined as:

$$\rho_m = \sum_i \frac{b_i}{v_m} \quad \text{and} \quad \rho_s = \sum_j \frac{b_j}{v_s} \tag{3-18}$$

Note that in equation 3-18, the scattering length density can differ markedly from the nuclear density. In figure 3 are reported the scattering length density of common solvents and monomer units of conjugated polymers. For a chain containing only one kind of monomer unit the contrast factor is given by:

$$K = v_m (\rho_m - \rho_s) \tag{3-19}$$

then the measured scattering function is directly related to the structure of the chain , and for an isolated chain:

$$S(q) = K^2 S_{AA}(q) \tag{3-20}$$

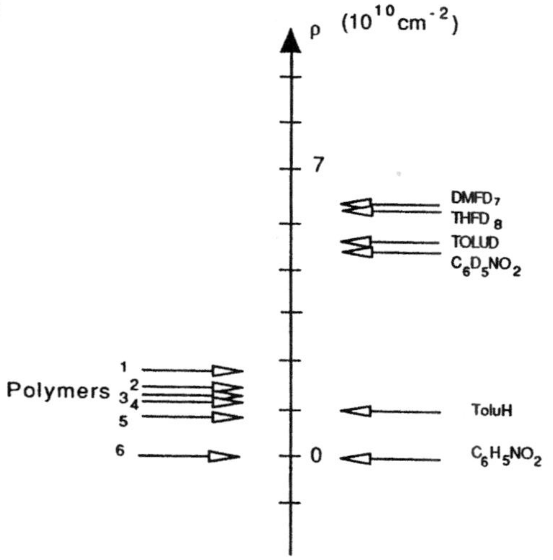

Figure 3:Scattering length densities of polymer and solvent units. Polyphenylacetylene (1), polydiacetylene PTS12(2), polydiacetylene p3BCMU(3), polydiacetylene p4BCMU(4), poly-3butylthiophene (5), poly-2octyne (6).

3-3-2. Form factor

In a scattering experiment, the measured intensities do not give any information on the phase relation between the elementary parts. On the other hand polymers are non homogeneous particles, most of the volume occupied by a chain is filled with the solvent molecules. Therefore the smooth scattering curve observed as a function of q is not very hepful if we do not have robust theoretical models describing the chain statistics. The form factor for the chain is

$$S(q) = K^2 < \sum_{k\,l} \exp (iq\mathbf{r}_k - q\mathbf{r}_l) > = K^2 S_1 (q) \tag{3-21}$$

for a chain with an isotropic behavior $S_1(q)$ can be written

$$S_1(q) = \left\langle \sum_{k\,l} \frac{\sin(qr_{kl})}{qr_{kl}} \right\rangle \tag{3-22}$$

where the brackets mean an average upon all the conformations available to the macromolecules. The average r_{kl} can be appropriately done if the distribution function $P(r_{kl})$ is known. With the assumption that the mean distance between segments k and l is a function of $|k - l|$ only, i.e only r_{kl}^2 enters into the function $P(r_{kl})$, we have $r_{kl} = r_{lk}$. In addition we consider that the function $P(r_{kl})$ is the same whatever the number of segments involved between the segments k and l. In other words, $P(r_{kl})$ is molecular weight independent. The later assumption being exact only for the case of ideal chain. The scattering function of the isolated chain can be now expressed as the Fourier transform of the distribution function $P_n(r)$, with r the mean distance corresponding to n segments:

$$S_1(q) = N_p + 2 \int_0^{\infty} \sum_n (N_p - n)\, P_n(r)\, 4\pi r^2 \frac{\sin(qr)}{qr} dr \tag{3-23}$$

3-4. Models for polymer chains

3-4-1. Ideal chains.

For an ideal chain we can use the Gaussian distribution (equation 2-2) :

$$P_n(r) = \frac{1}{(2\pi \langle r^2 \rangle /3)^{3/2}} \exp\left(-\frac{3r^2}{2\langle r^2 \rangle}\right) \tag{3-24}$$

with $\langle r^2 \rangle = n\, a^2$, inserting equation 3-24 in equation 3-23 we obtain, with $N_p << N_p^2$, the Debye function (5).

$$S_1(q) = \frac{2\, N_p^2 (e^{-x} + x - 1)}{x^2} \tag{3-25}$$

with $x = q^2 R_G^2$ and R_G the radius of gyration of the chain related to the mean square end to end distance by the relation

$$R_G^2 = \frac{1}{N_p + 1} \sum_0^{N_p} s_k^2 = \frac{1}{(N_p + 1)^2} \cdot \sum_{k \leq l \leq N_p} r_{kl}^2 = \frac{\langle R^2 \rangle}{6} \tag{3-26}$$

with s_k the distance between segment k and the centre of gravity of the chain. A characteristic feature of this function is its asymptotic behavior

$$\lim_{q \to \infty} S_1(q) = \frac{2 N_p^2}{q^2 R_G^2}\left(1 - \frac{1}{q^2 R_G^2}\right) \tag{3-27}$$

for large q we have a q^{-2} behavior characteristic of a chain described with a random walk. In other words at scale R, the chain fills the three-dimensional space as a two-dimensional fractal object (see equation 2-1).

3-4-2. Chain in good solvent, flexible chain with interactions.

The interactions between repeat units of a chain are of two types, with different physical consequences. Interactions at a local scale, depends on the microscopic structure and influences the local order. The spatial range along the chain of these local correlations correponds to the persistence length, i.e. the distance at which correlations along the backbone are lost. In addition, two repeat units separated by a large number of monomers repel each other. This excluded volume interaction increases the volume occupied by the macromolecule compared to that of the brownien conformation: the chain is swollen in good solvent. A new statistical description is needed, but the key point is that the scaling still applies (6, p294). We shall discuss this point first, and the effects of local interactions in the next paragraph 4-c.
The probability distribution $P_n(r)$ can be rewritten:

$$P_n(r) = \frac{1}{C(d) n^{d\nu} a^d} f\left(\frac{r}{n^{\nu} a}\right) d_{\nu} \tag{3-28}$$

where C is a constant, d the dimensionality of the space, ν an exponent equal to .588 ($\nu = 1/2$ for the brownian chain) and a the unit segment length. The mean square end to end distance being the second moment of the distribution Pn(r), the end to end distance scales like N^{ν}. As an example, this result can be obtained by using a Gaussian distribution with a variance equal at $n^{2\nu}$. As for most of the computations related to the description of a chain in good solvent, the value of ν has been the subject of numerous calculations which we do not intend to discuss in this chapter. An illuminating and exhaustive discussion is given by J. Des Cloiseaux and G. Jannink (ref 7 chapter XIII and XV).

From an experimental point of view, it is not obvious to obtain the value of ν given by the theory. With respect to the brownian chain, the swelling is dependent on the size of the macromolecule and the strength of the interaction. A good example of the effect of the solvent quality is given by the case of the polystyrene in cyclohexane. For this system, the theta temperature is equal to 35°C, the measurement of the radius of gyration as a function of the temperature shows a continuous decrease of the volume occupied by the polymer down to the theta behavior at T = 35°C. This shows the importance of the magnitude of the interaction upon the statistical conformation of the macromolecule. But even in the case of a good solvent with a large enough repulsion, the asymptotic behavior is only observed for a very large polymer. Typically for polystyrene in benzene or CS_2, $M_W \sim 10^7$ is needed (7).

An usual way for measuring the swelling of a macromolecule, is to measure the radius of gyration as a function of molecular weight. However, a large number of different samples must be studied, an accurate correction for the effect of molecular weight distribution is needed for each sample, and obviously all measurements must be performed at the same temperature. A critical review of many measurements on polystyrene in benzene yielded $\nu = 0.586$ (8), while for polystyrene in solvent CS_2 the measured exponent was $\nu = 0.588$ (7).

Another method is to measure the scattering function in the asymptotic regime: $qR \gg 1$, in which equation 3-27 reads

$$S(q) = \frac{2\,N_p^2}{q^2\,R_G^2} \tag{3-29}$$

equation 3-29 can be expressed in a more general way by using the normalized scattering function P(q) such that we have P(q=0) = 1, and substituting the exponent $\nu_b = 1/2$ for the ideal chain.

$$P(q) = \frac{2}{(q\,R_G)^{1/\nu_b}} \tag{3-30}$$

For a chain in good solvent reaching the asymptotic state we write

$$P(q) = \frac{C_\infty}{(q\,R_G)^{1/\nu}} \tag{3-31}$$

where the calculated value of $C_\infty = 1.04$ (9). Equation 3-31 tells us that the swelling of a macromolecule can be determined using a single sample, as soon as the chains are long enough for the asymptotic behavior to obtain. Equation 3-31 is characteristic of the scattering laws obtained from fractal objects. But as for any fractal objects, the scattering range where the self-similarity is preserved must be specified. The conditions are given by :

$$q \gg \frac{1}{R} \text{ and } q \ll \frac{1}{\xi} \tag{3-33}$$

At low q we must be below the Guinier range $qR \gg 1$: where the end-to-end distance plays the role of a cut off length at large scale. At short scale the magnitude of ξ is more difficult to define. For a common fractal object, the lower limit at which the self-similarity is lost is simply given by the length of the segment unit. This also happens for a polymer in brownian state. But in most cases, the segment unit is not related in a simple way to the length of the monomer unit.

As was first point out by Kuhn (10), we can use an equivalent unit length describing the effect of the local rigidity. The statistical length b (twice the persistence length) gives a first lower limit such that for an ideal chain behavior equation 3-31 applies for $q \gg b^{-1}$, where b is the magnitude of the statistical length. In good solvent the macromolecule requires a more specific treatment. The asymptotic behavior is difficult to reach. An experiment performed on short chains will yield an exponent close to, or equal to, the one given by the brownian model, so that the observed statistical conformation is dependent of the length scale : there is a limiting size below which the chain is brownian (7, chapter XIII), and that size is dependent on the interaction and not on the overall size of the polymer. The lower limit length deduced, here noted ξ, is called the thermal length. A careful experiment has been done on the polystyrene with $M_w = 6\ 10^6$, the constant $C_\infty = 1.19$ has been found in good agreement with the computed value (11).

At this stage it is useful to make some remarks concerning the soluble conjugated polymers. Up to now the samples available have a fairly small molecular weight, typically 10^4 to 10^5, as for substituted polythiophenes and polyacetylenes. Some polydiacetylenes where M_W can be larger than 10^6 are exception, but these are semi-rigide macromolecules (see chapter V and VII). The statistical lengths are at least an order of magnitude larger than for saturated polymers. Without anticipating the discussion given in chapter V, we can expect a conformation of the conjugated polymers in good solvent described by a brownian statistics.

3-4-3. Scattering function of chain with persistence length.

In this paragraph we focus our attention on the study of the conformation at a local scale, in other words we are interested with the asymptotic regime, $q \to \infty$, of the scattering function. Following Kuhn (10), a chain exhibiting a local stiffness can be described as an ideal chain by substituting the monomer unit length by an equivalent unit length related to the local rigidity or the bending elastic modulus. Here neglecting the excluded volume effect, the scattering function should approach the scattering of a Gaussian coil for $q \to 0$ and the contour length $L \to \infty$ (equation 3-27). It should approach the scattering of an infinitely thin and long rod for $q \to \infty$.

The structure factor for a rod of length L, (that is the normalized scattering function with $P(q{=}0) = 1$) has been computed a long time ago by Neugebauer (12):

$$P(q) = \frac{2}{qL}\left(Si(qL) - \frac{1-\cos(qL)}{qL}\right) \qquad (3\text{-}33)$$

$$\text{where } Si(x) = \int_0^x \frac{\sin t}{t}\,dt$$

with the relations (13)

$$Si(\infty) = \frac{\pi}{2}$$

$$\lim_{q\to\infty} \frac{Si(qL)}{qL} = \frac{\pi}{2qL} - \frac{\cos(qL)}{(qL)^2}$$

the scattering function 3-33 has an asymptotic law in q^{-1} given by

$$\lim_{q \to \infty} P(q) = \frac{\pi}{qL} - \frac{2}{q^2L^2} \tag{3-34}$$

it is often convenient to normalize the scattering function to the molecular weight or degree of polymerization of the chain. Introducing the function $g(q) = \frac{M_w}{m} P(q)$, we rewrite the equation 3-34

$$\lim_{q \to \infty} g(q) = \frac{\pi M_L}{mq} - \frac{2M_L}{mq^2L} \tag{3-35}$$

with m the molar mass of the monomer unit and M_L the mass per unit length, for long chain we have

$$\lim_{q \to \infty} g(q) \approx \frac{\pi M_L}{mq} \approx \frac{\pi}{qa} \tag{3-36}$$

For a chain with persistence length, the end to end distance is given for the two limiting cases by :

$R^2 = N^2 b^2$ for L small or $\frac{L}{b} < 1$

$R^2 = Lb$ for $\frac{L}{b} > 1$

We have now two asymptotic laws : a q^{-2} law for the gaussian coil and a q^{-1} for the rod. For actual chain exhibiting local rigidity, we now have two scales at which the two asymptotic laws are observed.

$qR >> 1$ and $qb < 1$ $g(q) \propto q^{-2}$

$qb >> 1$ $g(q) \propto q^{-1}$

The two regimes can be easily seen with a Kratky plot, $q^2g(q) = f(q)$, suggested by Kratky and Porod (14) see figure 4. The plot shows the transition between the gaussian coil behavior and the rod conformation. The location of the intersection between the two regimes depends on the local rigidity of the chain. With equation 3-29 and 3-36 we obtain the point intersection between the two asymptotes

$$q^2g(q) \approx \frac{12M_L}{mb} = q\frac{\pi M_L}{m} \qquad q^*b \approx 3.8 \tag{3-37}$$

This method does not require the knowledge of the absolute intensity and seems to give sufficient information on the chain conformation. Nevertheless, for most practical cases, absolute measurements are needed. Examples are given in chapter V, they show that measuring the absolute intensity is the unique way to an unambiguous description of the chain conformation. As often, a good structural study is only achieved if analytical calculations are available. One of the main results is the absence of a plateau in a Kratky plot for the chain with persistence length. A lot of analytical calculations have been devoted to the structure factor of a worm-like chain (15,16,17,18). For an infinite chain an exact computation has been performed by J Des Cloiseaux (18). The result gives the asymptotic behavior (18):

$$\lim_{q\to\infty} q^2 g(q) = q\pi + \frac{4}{3b} \qquad (3\text{-}38)$$

where we set the unit length equal to 1.

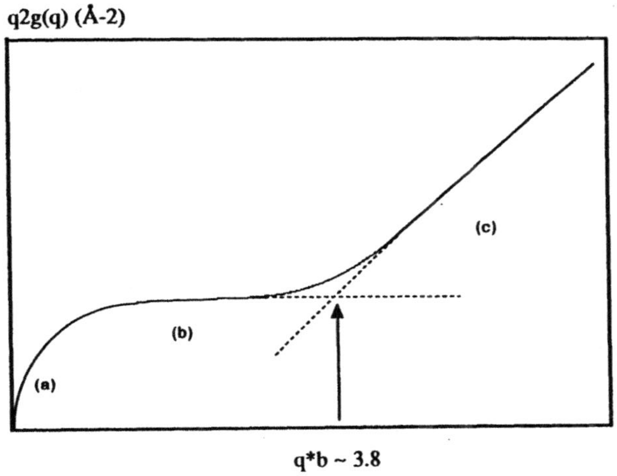

Figure 4: Kratky plot. (a) Guinier range. The arrow indicates the intersection between the two asymptotic regimes : Gaussian coil (b) and rod (c).

Different approaches have been used for chain with a finite length, see for instance Sharp and Bloomfield (15) and the serie of papers written during the last twenty years by Yamakawa and his collaborators, which give the structure factor for the widest q range (16,17). We now have computed values of the structure factor in the whole q range accessible by the experiment. In a continuous limit, the asymptotic behavior is formally given by:

$$\lim_{q\to\infty} P(q) = \pi q + \frac{C(L)}{b} \qquad (3\text{-}39)$$

we have $C(L) = -2\,L^{-1}$ for the rod (equation 3-34) and $C(\infty) = 4/3$ for the infinite worm-like chain (equation 3-38). It is impossible to evaluate C(L) analytically for chain with finite size L, nevertheless the computation for $L = 10^4$ b almost agree with the des Cloiseaux result in the range of $qb > 0.5$.

Since for an infinite chain the value at q = 0 corresponds to a random coil behavior : $q^2g(q) = 12/b$, the result obtained by Yoshizaki and Yamakawa emphazises that $q^2g(q)$ increases monotically with increasing q for all values of L. In other words the chain with persistence length does not exhibit a plateau in the transition range between the random coil and the rod as it has been previously proposed (see figure 4). Yoshizaki and Yamakawa show that a plateau can be only observed if the chain has a persistence in curvature, i.e some helicity remains along the chain. Indeed the chain conformation is neither a pure random coil, nor a chain with only curvature fluctuation, nor a pure helical chain but mostly a mixture of these three basic conformations. Thus for avoiding any misinterpretation of the scattering curve we need a calculation of the form factor for an helical worm-like chain.

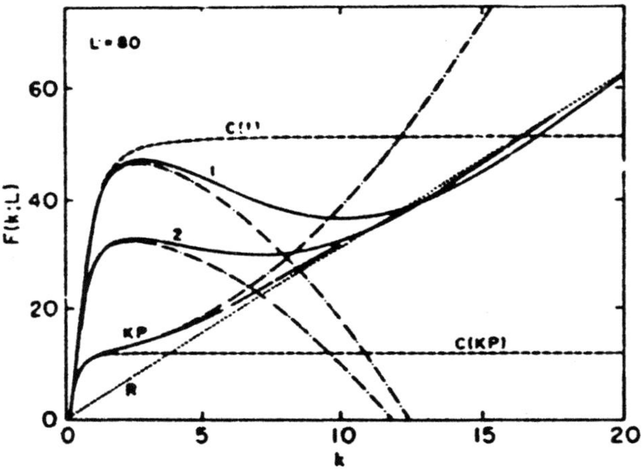

Figure 5: Scattering function calculated with the Yoshizaki and Yamakawa method. Reduced coordinates are used : $k = qb$ and $F(k;L) = k^2 (L/b) P(q)$. The chain contour length is L = 80b. KP is the Kratky-Porod or wormlike chain. 1 and 2 refer to the helical wormlike chain model with $\mu = 0.2$, $\nu = 4$ and $\mu = 0.2$, $\nu = 3$ respectively. The dotted curve R represents the corresponding values of F for the rod, the broken curves C(1) and C(KP) the values of F for the random coils having the same mean-square radii of gyration as those of code 1 and the KP chain, respectively. Note that in contrast to the Kratky plot (figure 4), the exact computation does not show a plateau except for an helical worm-like chain.

The helical conformation has been described by Bugl and Fujita (19). Let us note the curvature and torsion at the curvilinear abscissa l along the line: $\rho(l)$ and $\tau(l)$. With κ and c the force constants for the curvature and the torsion, the free energy of an infinite thin thread is written (19):

$$F = \int_0^L \frac{1}{2}\kappa\, [\rho(l) - \rho_0(l)]^2 + \frac{1}{2}c\, [\tau(l) - \tau_0(l)]^2\, dl \qquad (3\text{-}40)$$

where ρ_0 and τ_0 correspond to the stable state. The radius r_h and pitch P of the helix are given by the relations

$$r_h = \frac{\rho_0}{(\rho_0^2 + \tau_0^2)} \qquad P = \frac{2\pi\tau_0}{(\rho_0^2 + \tau_0^2)} \tag{3-41}$$

we now have three parameters $b = 2\kappa T^{-1}$ (see chapter VII), r_h and P. Yoshizaki and Yamakawa used two parameters μ and ν defined by:

$$\mu = \frac{\tau_0}{(\rho_0^2 + \tau_0^2)^{1/2}} \qquad \nu = (\rho_0^2 + \tau_0^2)^{1/2} \tag{3-42}$$

μ gives the shape of the chain, i.e the ratio P/r_h while ν b gives the number of helical turns per statistical length. Yoshizaki and Yamakawa use reduced coordinates (in unit of statistical length), such that the structure factors reported in figure 5 for $\nu = 4$ and 3 correspond to ν/π helical turn per statistical length. The scattering functions show smooth oscillations and a plateau at low q (case $\nu = 3$) which does not correspond to a random coil behavior (see figure 5).

3-5. Scattering measurements in real polymer/solvent systems

3-5-1. Polydispersity effect.

The molar mass dependence of the radius of gyration of polymer coils is widely used to characterize the properties of a polymer in solution (see chapter II). Moreover, if the molecular weight and the structure of the monomer unit are known, the radius of gyration also provides information about the internal structure of the coil. For a given total contour length of a polymer, the magnitude of the radius of gyration depends on the local correlations between monomer units. Equation 2-16 gives the mean square end to end distance of a chain with persistence length hence the radius of gyration (20):

$$\langle R_G^2 \rangle = \frac{L_w b}{6}\left\{1 - \frac{3b}{2L_w}\left[1 - \frac{b}{L_w}\right] - \frac{3}{4}\left(\frac{b}{L_w}\right)^3\left[1 - \exp\left(\frac{-2L_w}{b}\right)\right]\right\} \tag{3-43}$$

Equation 3-43 is valid for monodisperse polymers. However, in most cases the polymers and soluble conjugated polymers are more or less polydisperse. What is then measured is a z-average of the radius of gyration (21).

$$\langle R_G^2(M_w) \rangle_z = \frac{1}{M_w} \Sigma \frac{N_i M_i^3 R^2(M_i)}{\Sigma N_i M_i^2} \tag{3-44}$$

$$\text{with } M_w = \sum \frac{N_i M_i^2}{\sum N_i M_i} \tag{3-45}$$

with N_i the number of macromolecules with mass M_i.
The contribution of the highest molecular weight will be predominant for polydisperse samples. Therefore the mass distribution must be included in equation 3-43. R.C. Oberthur (22) has obtained the following expression for chains with persistence length:

$$\langle R_G^2 \rangle = \frac{1+2U}{1+U}\frac{L_w b}{6}\left\{1 - \frac{1+U}{1+2U}\frac{3b}{2L_w}\left(1-\frac{b}{L_w}\right) - \frac{(1+U)^2}{1+2U}\frac{3}{4}\left(\frac{b}{L_w}\right)^3\left(1-\left(1+\frac{U}{1+U}\frac{2L_w}{b}\right)^{-\frac{1}{U}}\right)\right\} \tag{3-46}$$

where $U = M_w/M_n - 1$.

Figure 6 illustrates the change of the statistical length as a function of the ratio M_w/M_n for a given radius of gyration. The polydispersity can be an important factor, and would yield a fairly inaccurate value of b if ignored.

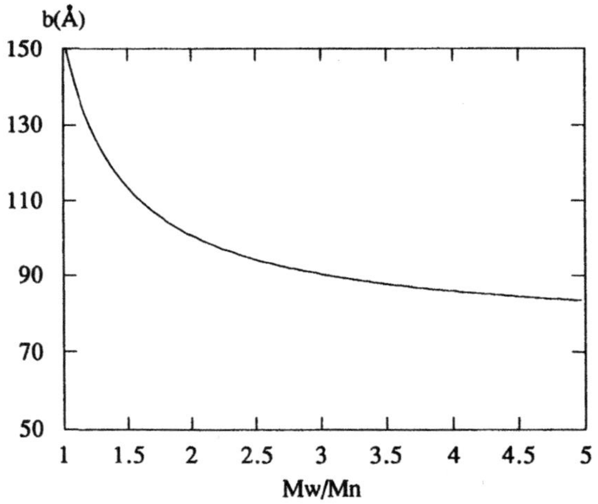

Figure 6: Variation of the statistical length as a function of M_w/M_n for a given radius of gyration. With $M_w = 540000$, $R_g = 530$ Å.

3-5-2. Models for chain cross section.

Whatever the statistical conformation of a polymer, the structure factor has been up to now calculated with the assumption that the chain is an infinite thin thread. However, the finite thickness of the chain often changes the scattering curves. It can happens at rather low q values, $q \sim 0.05$ Å^{-1} or less (11,23). The chain structure factor is then written as (24):

$$g(q) = g_0(q)\,\Phi(q) \tag{3-47}$$

where now, $g_0(q)$ is the normalized structure factor of the infinite thin thread given in paragraph 3-4 and $\Phi(q)$, that of its cross section. At large enough scale with respect to the size of the substituent, the effect of the cross section can be simply modelled with the Guinier approximation for $\Phi(q)$, and equation 3-47 is replaced by (24)

$$S(q) = S_0(q)\exp\left(-\frac{q^2R_c^2}{2}\right) \tag{3-48}$$

with R_c^2 the mean square radial radius of gyration given by:

$$R_c^2 = \frac{\int r^2\left(\rho_{SG} - \rho_s\right) dr}{\int\left(\rho_{SG} - \rho_s\right) dr} \tag{3-49}$$

where ρ_{SG} and ρ_s are the scattering length densities for the side groups and solvent respectively. If the scattering length density has a radial dependence identical to the one of the nuclear density, the mean radial radius of gyration is a direct measurement of the approximate geometrical extent of the side groups. For side groups extended in the solvent without large torsion between neighboring monomer units, a good approximation is to use the relation for the rod:

$$R_c^2 = \frac{L^2}{12}$$

For a cylinder, the relation is simply $R_c^2 = R^2/2$.

Equation 3-48 holds for $qR_c < 1$. At larger q a more explicite expression for $\Phi(q)$ is needed. If the substituents form a thin plane, the cross section leads to the characteristic q^{-2} scattering behavior of a flat particle. The normalized structure factor $\Phi(q)$ is (24) :

$$\Phi(q) = \frac{2\pi}{Sq^2} \tag{3-50}$$

where S is the surface of the homogeneous planar monomer unit.

For an homogeneous cylinder of radius R, $\Phi(q)$ is given by (24):

$$\Phi(q) = \left(2\,\frac{J_1(qR)}{qR}\right)^2 \tag{3-53}$$

with $J_1(qR)$ the first order Bessel function.

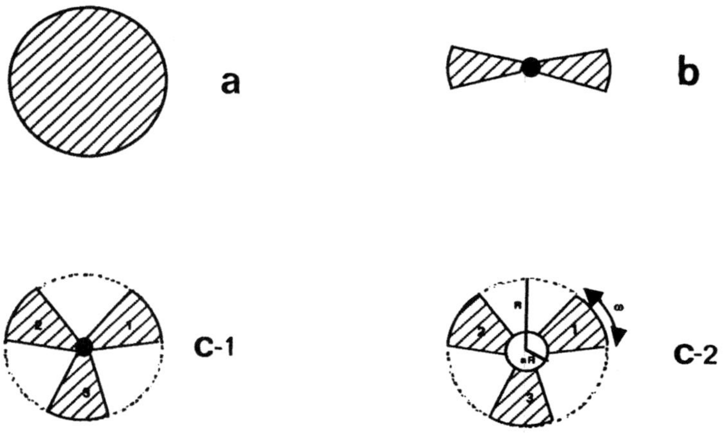

Figure 7: Geometrical parameters of different cross sections: a-Cylinder, b-planar ribbon structure, c-1 helix, c-2 helix with central hollow part.

For locally ordered chains, for instance with a non-random average torsion between successive monomer units, a more accurate description is again needed. The case of local helical conformation is of particular interest for polymers with substituent side-groups, since their steric interaction.can dictate a sens of rotation, providing a regularity in structure. The theoretical structure factors for helices as on figure 7 were calculated by O. Pringle et al (25) and are shown on figure 8 : a local helical order should show up oscillations on the structure factor q-dependence, but at large q (for instance a pitch length equal to 20Å corresponds to $q_p = 2\pi/P \sim 0.3Å^{-1}$) the poor signal to noise ratio often prevents accurate enough measurements. Nevertheless, one can see that for an helical conformation with an hollow central part, the decrease of the scattering intensity is more pronounced at large q. Moreover the scattering length density increases, hence a scattering intensity at low q values larger than for a cylinder or ribbon like structure (figure 8).

To summarize, the small angle scattering techniques provides smooth scattering curves which are, with the help of powerful mathematical tools, able to give quantitative information at many length scales. Indeed, most of the questions concerning the conformational statistics of conjugated polymers in solution can be solved unambiguously. With suitable models we are even able, at least qualitatively, to describe the structure at the monomer unit scale. Examples are given in chapter V and VII with the scattering curves obtained for substituted polydiacetylene, polyacetylene, polythiophene, poly-1octene and 1decene.

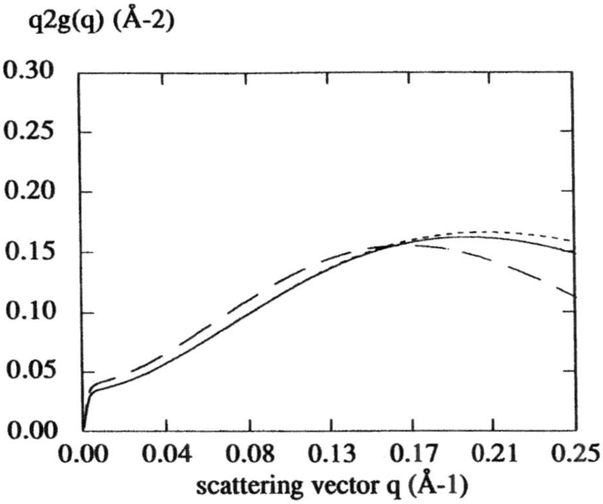

Figure 8: Scattering functions for different cross sections. The dotted curve corresponds to the Guinier approximation with R_c = 5.09Å, the continuous curve to the cylinder R = 7.2Å (R_c = 5.09Å), the dashed curve represents an helical conformation with central hollow part with P = 20Å, R = 8.5Å and a = 0.15, i.e. an internal radius aR = 1.275Å.

References

1 G.E. Bacon, Neutron Diffraction, first edition (Pergamon Pres, Oxford) (1966).

2 V.F. Sears, AECL-8490, Chalk River Nuclear Laboratories, Ontario (1984).

3 L.D. Landau, E.M. Lifshitz., Statistical Physics, 2nd edition (Pergamon Press, London) 1959.

4 P.G. de Gennes, J. Phys. France, **31**, 235 (1970).

5 P. Debye, J. phys. Colloid. Chem., **51**, 18 (1947).

6 P.G. de Gennes, Scaling concepts in Polymer Physics (Cornell University Press, Ithaca) (1979).

7 J. des Cloizeaux, G. Jannink, Les Polymères en Solution, leur modélisation et leur structure (Editions de Physique Les Ulis) (1987).

8 J.P. Cotton, J. Phys. Lett., **41**, L-231 (1980).

9 J. des Cloizeaux, B. Duplantier, J. phys. Lett., **46**, L-457 (1985).

10 W. Kuhn, Kolloid Z., **76**, 258 (1936) and **87**, 3 (1939).

11 M. Rawiso, R. Duplessix, C. Picot, Macromolecules, **20**, 630 (1987).

12 T. Neugebauer, Ann. Physik, **42**, 509 (1943).

13 Handbook of Math. Functions, M. Abramovitz and J.A. Stegun eds, Dover Pub., Inc., New York (1970).

14 O. Kratky, G. Porod, Rec. Trav. Chim. Netherlands, **68**, 1106 (1949).

15 P. Sharp, V.A. Bloomfield, Biopolymers, **6**, 1201 (1968)

16 H. Yamakawa, M. Fujii, Macromolecules, **7**, 649 (1974).

17 T. Yoshizaki, H. Yamakawa, Macromolecules, **13**, 1518 (1980).

18 J. des Cloizeaux, Macromolecules, **6**, 403 (1973).

19 P. Bugl, S. Fujita, J. Chem. Phys., **50**, 3137 (1969).

20 H. Benoit, P. Doty, J. Phys. Chem., **57**, 958 (1953).

21 B.H. Zimm, J. Chem. Phys. , **16**, 1099 (1948).

22 R.C. Oberthür, Makromol. Chem., **179**, 2693 (1978).

23 D.Y. Yoon, P.J. Flory, Polymer, **16**, 645 (1975).

24 G. Porod, in "Small Angle X-ray Scattering", O. Glatter and O. Kratky eds (Acad. Press, New York) (1982).

25 O.A. Pringle, P.W. Schmidt, J. Appl. Cryst. **4**, 290 (1971).

4 SOLUBLE CONJUGATED POLYMERS

4-1 Conjugated polymers with substituents.

Since the discovery that conjugated polymers can exhibit high conductivity when doped, a tremendous amount of work has been done on these materials. Solid conducting polymers form inhomogeneous porous media, with an ill-defined structure. In many case, complex structures are obtained, with coexisting amorphous and crystalline, more or less fibrillar, parts. The average density is low, typically for Shirakawa polyacetylene, $\rho \sim 0.4$ g/cm^3. The chain packing may be described at a local scale, let say within 100Å, but with no long range order. The structural study of solid polyacetylene (1), polythiophene (2), or polyaniline (3) materials are good examples of the difficulties encountered for understanding the physical properties of these polymers.

Moreover, except may be in the polyacetylene case, a clear picture of the structural change accompanying the doping process is not available. The doping process involves an intricate interplay between chain conformational changes and packing changes. Different models have been suggested, mainly by analogy with graphite intercalation compounds. Various stages are proposed as a function of the doping level but the experimental results do not form a clear unambiguous picture (4). The major impediment is the lack of information at an elementary level, elementary characterization such as the knowledge of the molecular weight were not available complicating any attempt to model the relation between the chain conformation and the electronic properties.

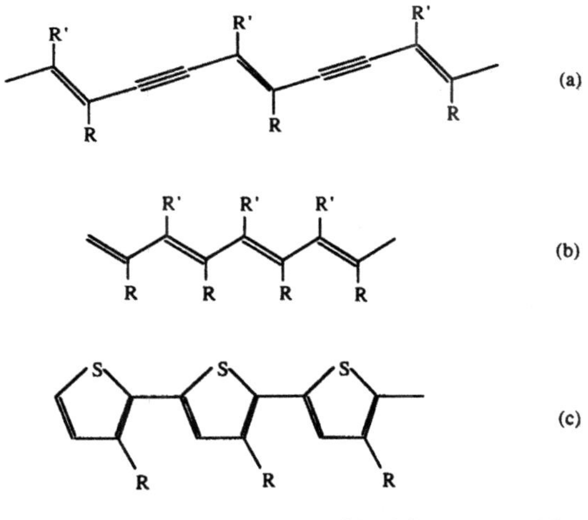

Figure 1 : Example of conjugated polymers with side-groups. Most of the substituents are aromatic or alkyl groups : (a) Polydiacetylene, (b) polyacetylene, (c) polythiophene. Side-groups of some soluble polydiacetylene are given in table 7-1.

The discovery that conjugated polymers can be solubilized with appropriate substituents has given access to single chain properties. The volume of work devoted to the synthesis of new soluble materials has may be reached or even overpassed that devoted to the control of the solid phase through the polymerization. In this part we focus the attention on substituted polyacetylene, polythiophene and polydiacetylene, which have been the most investigated.

4-1-1 Substituted polyacetylenes.

Since the middle seventies soluble polyacetylenes with not too small molecular weight were obtained. Using for the first time tungsten hexachloride (WCl_6) and molybdenum pentachloride ($MoCl_5$) catalysts, Masuda et al (5) were able to polymerize phenylacetylene, hereafter noted PPA, with a molecular weight larger than 10^4.

With these two kinds of catalysts, polymerization of soluble polyacetylenes with numerous substituents was achieved. The list of the various substituted polyacetylene is too long to be given here, a detailed description will be found in ref 5.

Numerous parameters govern the polymer properties : the catalyst, cocatalyst and even the solvent used for the polymerization. N.M.R spectrum, Resonant Raman Scattering and visible optical absorption show changes function of these parameters. Figure 2 illustrates the change on the visible absorption with the catalyst used. Moreover the synthesis made by S.Ramakrishnan, following the same route as Masuda with WCl_6 but with a different cocatalyst, gives a large red-shift in the visible absorption (6). The solution is red for the case WCl_6-Me_4Sn while the solution of PPA previously obtained by Masuda et al either by using $MoCl_5$ or WCl_6 are more or less yellow. The solution of PPA synthesized with WCl_6-Me_4Sn shows an absorption spectrum close to the one observed for the substituted Ortho-Methyl polyacetylene. In addition, the polyacetylene chain may have two different conformers cis and trans, so the various synthetic routes may favor either conformation.

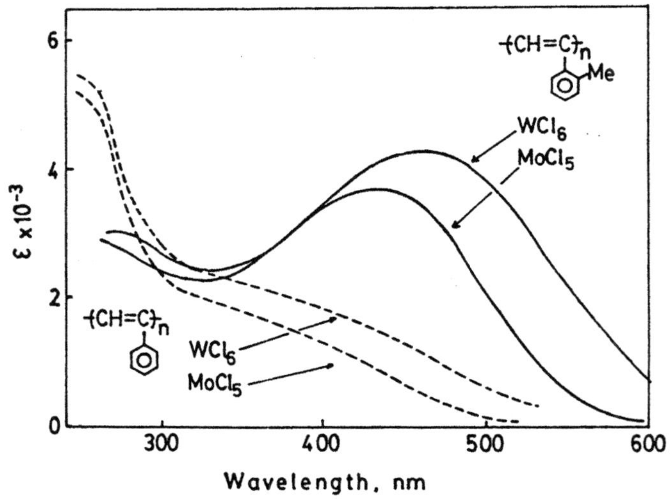

Figure 2 : UV-visible absorption of poly-phenylacetylene and poly(o-MePA).(from reference 10)

One can consider that the central part of the overall work made on the synthesis concerns the description of the chain in term of the two conformers cis and trans. Two decades ago, a Resonant Raman Scattering experiment concluded to a trans conformation for PPA (7). Since then, extensive work has shown a more complex situation (8-10). Sanford et al (8) obtained C_{13} N.M.R spectra similar to those of 3-phenylbenzene, then concluded to a cis-cisoid conformation. Masuda et al reached the same conclusion for the PPA made with $MoCl_5$ while for WCl_6 they concluded to a trans-rich conformation (5).

In most cases a mixing of cis and trans conformations is considered, for example in poly-Ortho-MethylPA (10). Obviously a mixture of the two possible conformers increases the number of available configurations, rendering the description of the statistical conformation much more difficult. Preliminary SANS studies have been performed on PPA synthesized with $MoCl_5$ and WCl_6. The scattering curves for the two samples in solution are almost identical. They correspond to a coil behavior without any indication on the existence of a persistence of curvature or a partial rod conformation due to an helical local order (6).

An important feature concerns the stability of those samples. Masuda et al (5) indicated that PPA films show some degradation when they are kept at room temperature and exposed to air for a few weeks. The degradation is still more effective for the polymer in solution. After a few hours, the yellow and red solutions, i.e. the PPA synthetisized with $MoCl_5$ and WCl_6 respectively, become colorless. In addition, the scattering intensity at small angles decreases within one hour. The latter result can be easily interpreted as a decrease in molecular weight (chain scission). To prevent the degradation, the solution is cooled at 0°C, where no change is observed on the scattering curve over a period of one day, while the solvent remains a good one even at this temperature. Therefore, in contrast to polydiacetylene or polythiophene, substituted polyacetylenes like PPA require special care for avoiding degradation effects.

A lot of work, mainly synthesis and elementary characterizations, have been performed on substituted polyacetylenes. It is somewhat surprising that almost no companion structural studies are available. The lack of structural data does not allow an accurate knowledge of the internal conformation of the polymer. Nevertheless these systems appear very interesting, the existence of a possible cis-cisoid conformation may involve a well defined helical structure such that we can speculate a relation between the specific elastic behavior of an helical-spring like structure and the optical properties of the conjugated backbone.

Polyacetylene with n-alkyl substituents form another class, the tensile properties are markedly different from those of the aromatic substituted polyacetylenes : the Young modulus being 600 to 800 MPa and 2500 MPa respectively (5), that is polyacetylene with alkyl side chains exhibit a behavior similar to the one of the saturated polymers. On the other hand they are colorless in solution. For example, poly-2octyne in toluene has a maximum absorption located around 280 nm with no additional structure at longer wavelength. In term of electronic delocalization, the location of the maximum absorption in the near U.V indicates a π-electron localization at the monomer unit scale.

Therefore poly-2octyne apparently does not differ much from saturated polymers. In other words the conjugated backbone structure must play a minor role on the chain conformation in the isolated state. SANS studies have been performed on p-2octyne in toluene. The scattering curves of poly-2octyne in good solvent show a worm-like chain conformation with a statistical length equal to 130Å (11).That is the statistical length is about three times larger than for the poly-3butylthiophene (chapter 6 and 7), while the backbone does not show much conjugation. The physical origin of the local rigidity can be simply understood with the location of the bulky substituent. The methyl unit is close to the backbone inducing a large torsion between monomer units, then increasing the local rigidity (chapter VII).

4-1-2 Poly-n-alkylthiophenes.

Polythiophene is a conjugated polyheterocycle in which backbone structure differs markedly from those of PDA or Polyacetylene. In contrast to PDA these polymers are easily dopable. A common way for describing the doped state is to use the concept of polaron or bipolaron (12). Since the polymer ground state structure is non-degenerate we can not expect a soliton like excitation and therefore the possibility of charged soliton as it has been predicted for polyacetylene (13). In contrast to the polyacetylene case, an important and useful property of the polythiophenes is their stability, then allowing numerous experiments without being constrained to a low temperature or inert atmosphere.

The introduction of alkyl-group at the 3-position of the thiophene ring yields poly-nalkylthiophenes which are both soluble (14) and melt processable (15). The Linköping group has studied these materials in depth: investigation of the properties of the blend formed with poly-3 octylthiophene and the random copolymer poly(ethylenevinylacetate), or extensive study on the physical properties of poly-nalkylthiophene films(15). Experiments and discussion are reported elsewhere in this book by Salanek et al. Here we only emphasize some aspects concerning the thermochromic properties of the films: They show a color transition and the transition temperature T_c decreases as the size of the alkyl chain increases suggesting a direct relation between the color transition and the melting process of the alkyl chains (15). This behavior is reminiscent of the variation of the glass transition temperature T_g of substituted n-alkylpolyacetylenes (5) : Tg is decreasing when the size of the substituents is increased. Incidently the variation of T_c as a function of the side-group size together with a temperature transition for the films occuring at higher temperature than in solution (for example for the case of poly-nBCMU films and solutions) have been understood as an experimental evidence of the screening effect of the side-groups (ref 16 and chapter 7).

For slightly different reasons, the study of soluble poly-n alkylthiophene also appears fruitful. In the case of films, the stability and the opportunity to process easily the conjugated polymer are the leading properties. For the solution studies, the existence of a conjugated backbone structure differing markedly from the polydiacetylenes or polyacetylenes gives the opportunity to improve our understanding of the statistical conformation in the single chain state. A discussion on the chain conformation of poly-3butylthiophene is given in details in chapter VI and VII. Here we note that the poly-3butylthiophene exhibits a very flexible behavior, typically b ~ 50 Å, corresponding to a behavior similar to the one of saturated polymers. At first glance, this result is very surprising and appears in contradiction with the common idea on the statistics of the isolated chain state for conjugated polymers. The results obtained with the poly-butylthiophene clearly show that conjugated polymer do not have necessarily a stiff behavior.

4-1-3 Polydiacetylenes

Substituted polydiacetylenes form a unique class of conjugated polymers. A decade ago substituted polydiacetylenes have been solubilized and shown various color transition either by changing the solvent quality or the temperature. This has been the origin of a growing interest on the solution of conjugated polymers (17).

A very specific property of PDA is the polymerization route: it is performed in solid phase starting from monomer crystals. The polymerization is initiated either with X, γ, U.V. or electron irradiation, or thermally. In many case the mixed crystals of monomer and polymer form perfect solid solutions and lead to a perfect single crystal of polymer after the polymerization has been completed. Depending of the substituents, the PDA chains are or not soluble. For example PDA para-toluene sulfonate or parafluorobenzene sulfonate which contains only one CH_2 unit between the backbone and the end of the side group are insoluble, while pTS12, the PnBCMU or the polymers made by Plachetta et Schulz which contain $CH_2)_n$ with $n \geq 3$ are soluble (18). In contrast with most other soluble conjugated polymers, large molecular weights are available, the molecular weight being controlled by the the topochemical reaction rules.

4-2 Diblock Copolymers and Graft Copolymers.

Other routes to soluble conjugated polymer consist to graft the conjugated sequence to a saturated one. The synthesis of diblock copolymer and or block/graft copolymer have for primary purpose to impart the desired physical properties of the carrier (the saturated polymer) to the copolymer. These properties may be the elasticity , or, for an insoluble block, solubility. The swelling state of the saturated part controlling the packing of the conjugated polymer. In contrast with substituted conjugated polymers, such an attempt can give the opportunity for studying the single chain behavior of a "pure" conducting chain.

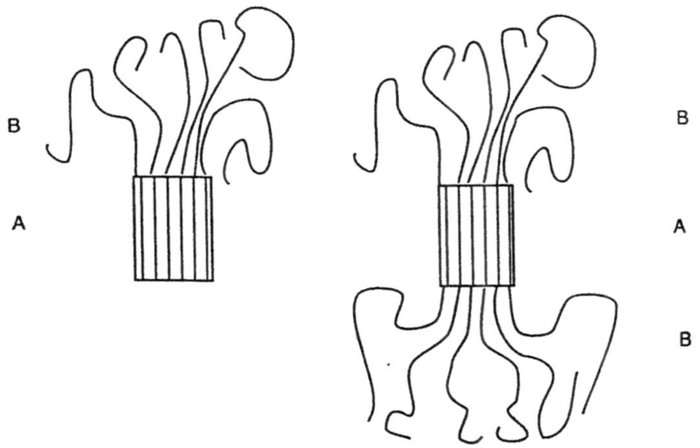

Figure 3 : A schematic cross section of di-block and tri-block copolymers. The core consists of immiscible aligned conjugated polymers (A). The saturated parts are swollen in the solvent and adopt a star like structure (B).

Common organic solvents act very selectively on saturated and conducting polymers. With the couple saturated-conducting polymer, we can speculate that block copolymers incorporate rigid, rod-like blocks (figure 3). The saturated part is highly soluble and swollen in the solvent while the conjugated immiscible part form a core with a rod-like structure. The balance between the two conformations of the respective polymers should produce a great variety of structure. By varying the ratio of the molecular weight of the soluble and insoluble polymers and the selectivity of the solvent we obtain, at least theoretically (19), different structures: micelles with a rod-like cylindrical structure for the core, lamellar phases and single chain state.

Several attempts have been made with the couples polystyrene-polyacetylene (PS-PA) or polyisoprene-polyacetylene (PI-PA) (20,21,22). The choice of the PA as the companion polymer is quite understandable. PA has been the most studied and is the simplest polymer providing a high conductivity while it is doped. A huge amount of work has been devoted to PA films, but PA films are inhomogeneous, insoluble, and poorly characterized.

The synthesis of copolymer should be a way for solving these difficulties, either through the solubilisation of the copolymer giving the opportunity to study the single-chain state, or by developing methods for preparing well-defined forms of PA crystals.

4-2-1. Graft Copolymer.

We shall focus the attention on the Block/Graft Copolymer synthesized by Baker et al (20,23). They have employed polyisoprene and polystyrene as carrier polymers. These polymers form monodisperse carriers $M_w/M_n = 1.06$ with $M_n \sim 2\ 10^5$. The grafting process is a process in which the growing polyacetylene chain initiated by a Ziegler-Natta catalyst is intercepted by a graft site on a carrier polymer. A typical number of graft sites for a 200 000 molecular weight polyisoprene carrier is 15-50 per chain. Multiple grafting of growing polyacetylene to a single carrier polymer was avoided by carrying the grafting reaction to a low conversion (>95% of the carrier chains have no graft).

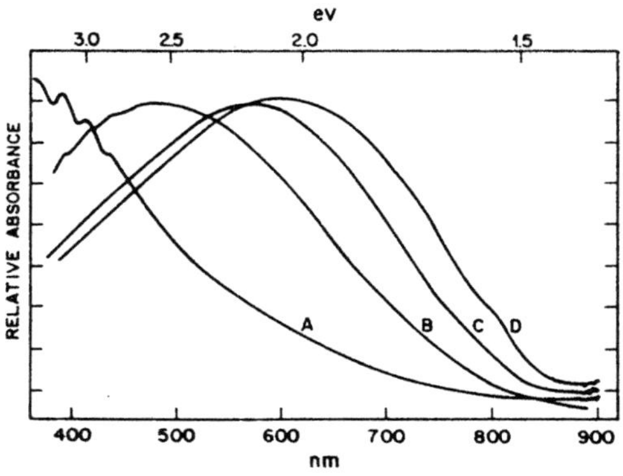

Figure 4 : UV/vis spectra of polyacetylene/polystyrene block copolymers. The ratio of acetylene to graft sites was increased from A to D. (from reference 23)

Baker et al were able to solubilize the graft copolymer. Unfortunately, they also emphasized the impossibility to separate the graft copolymer from the homopolymer. Therefore their definition of "solubilized polyacetylene" corresponds to the capability for the copolymer to pass through filters with pore sizes as small as 2000Å with a large amount of homopolymers. For oligomers of thiophene grafted on polystyrene carrier, the same difficulty occurs. In that case the graft system has rubber properties but it is still impossible to separate the graft copolymer from the homopolymer (6).

A proper characterization of the graft copolymer is not obvious. The size of the conducting chain is mainly deduced from spectroscopic studies either with IR providing an elementary characterization, i.e. the effectiveness of the graft process, or visible optical absorption providing a more or less accurate information on the polyacetylene chain length grafted. Figure 4 illustrates the products obtained for four different copolymers. Baker et al (23) suggest that for curve A, the polyene segment corresponds to a chain with 6 double bonds. For the other curves the red-shift is analysed as an increase of the chain length, but can not provide an estimation upon the molecular weight. As the authors noted, the absorption shift toward lower energy can be due to an increase of the conjugation length or an increase in the coupling between chains, so a crystallisation process improved by a chain folding due to an increase of the chain length.

The microscopic state of the solubilized polyacetylene may have various structure. Small amorphous isotropic aggregates, microcrystals with fibrous structure, or single chains are all possible states for the polyacetylene chains. Light scattering experiments have been performed for the sample corresponding to curve C (figure 4) (24). The light is noticeably depolarized suggesting an anistropic behavior with a strong angular correlation. This implies that microcrystals with aligned chains are formed. A radius of gyration of 350 ± 70Å is measured, much larger than for the homopolymer alone. Moreover the hydrodynamic radius for the homopolymer was 160Å, while for the Block/graft copolymer an hydrodynamic radius of 1050 ± 200 Å was measured. The authors concluded that the molecules aggregate in solution to form a micellar particle with an average radius near 1050Å, the core of the micelle being made of highly oriented polyacetylene chain.

4-2-2. Sequence of Block-Copolymer.

The diblock copolymer of saturated-conjugated polymer form a very fascinating system. The solvent selectivity may lead to a flexible behavior for the saturated sequence and a rigid one for the conjugated part. The problem is of interest to liquid crystal research as well as to polymer or surface science. Accordingly, rod-coil block copolymers are the macromolecular analogs for monomeric smectogens. The major impediment for the couple saturated-conjugated polymer is a too large solvent selectivity, so that the core formed with the conducting polymer may precipitate without being influenced by the swollen state of the saturated part.

The couples PS-PA and PI-PA have been synthetizised by different groups (20,21,22). Baker et al (20) have studied the copolymers PS_{100}-PA and PS_{200}-PA, where the number means the molecular weight for a monodisperse polystyrene: 10^5 and 2 10^5 respectively. The diblock copolymer PS_{100}-PA contains 35-40% by weight of polyacetylene, i.e. a molecular weight about 6 10^4 for the polyacetylene part. The authors indicated a solubilisation of the copolymers PS_{200}-PA and PI_{200}-PA, while the couple PS_{100}-PA undergoes a phase separation from solution to produce lamellar single crystal. With an argument concerning the interfacial area occupied by the polystyrene chain (S/N_p) of a diblock copolymer containing a lamellar structure: $S/N_p = 800$ Å^2, with N_p the degree of polymerization corresponding to $M_n = 10^5$, the authors proposed a chain folding of the polyacetylene with a fold length estimated to be ~ 100Å. They obtain a lozenge-shaped crystal of typical size 0.5µm and a thickness in accordance with the chain folding length. The analysis of the X-ray pattern is interpreted as an hexagonal structure providing a cis-cisoid helix conformation of the polyacetylene. In the case of the Shirakawa synthesis, polymerization and crystallization occur at the same time providing a structure based upon the cis-transoid conformation. Therefore, if for the couple PS_{100}-PA the solubilisation can not be readily obtained, the saturated part has still an effect controlling the packing of the conducting polymers. For lower molecular weight, typically PI_8-PA_2, L. Dai et al (21) have shown that the polyacetylene goes at the solvent (toluene)-air interface, the polyisoprene tail being swollen in the solvent. From neutron scattering reflectivity the authors estimate a layer thickness of the "soluble" copolymer equal to 50 Å.

It appears clearly difficult to obtain a well-defined soluble copolymer, even forming a micelle structure. Besides, copolymer made with PA sequence often exhibits degradation implying that the material must be kept under inert atmosphere (20,21,25).

Another attempt has been made, using a poly-paraphenylene (PPP) sequence, more stable than polyacetylene (26). The copolymer PS-PPP is obtained through the aromatisation of the copolymer PS-PCHD, where PCHD is the poly-cyclohexadiene. The respective molecular weights are for the PS sequences 25000 and 30000 and the PCHD sequences 3000 and 10000. The aromatisation is complete but with the creation of chemical defects such that sequence of PPP are not greater than about 10 units (26). Light scattering experiment on the sample with a molecular weight of 10^4 for the PPP gives a radius of gyration about 400 Å. The M_w is found to be equal to 10^7 meaning that aggregates are formed in THF. The GPC of the sample with low molecular weight for the PPP sequence shows two peaks: one is very close to the value of the polystyrene M_w = 25000, the other corresponds at $M_w = 10^7$ with a polydispersity equal to 1.3. The authors concluded that solubility is well improved for copolymer with low molecular weight of PPP and show that aggregation can be reduced by heating the solution. In contrast to the couple PS-PA the PS-PPP aggregates induce a weak depolarization factor suggesting a micellisation with a core more spherical than the one for the Block/Graft copolymers PS-PA.

The study of saturated-conjugated copolymer shows a complex behavior. It appears that elementary characterizations at a synthetic level have still to be done in many cases. An accurate understanding of the physical properties is obviously very dependent of this first step. Nevertheless these routes are very promising, up to now many possibilities have not been investigated. For example, larger molecular weight for the saturated sequences or triblock systems, i.e. systems with saturated-conjugated-saturated sequences (figure 2-b), would improve the solubility therefore the tractability of the materials. Also tentative with block containing sequences of substituted and conjugated polymers should allow to play with the solvent selectivity, managing various kind of micelle shapes and phases.

References:

1 J.P. Pouget, P. Robin, R. Comes, H.W. Gibson, A.J. Epstein, D. Billaud, Physica, **127 B**, 158 (1984).

2 F. Garnier, G. Tourillon, J.Y. Barraud, H. Dexpert, J. Mat. Sci., **29**, 2687 (1986)

3 M.E. Jozefowicz, R. Laversanne, H.H.S. Javadi, A.J. Epstein, J.P. Pouget, X. Tang, A.G. MacDiarmid, Phys. Rev. B, **39**, 12958 (1989).

4 R.H. Baughman, N.S. Murthy, G.G. Miller, L.W. Schacklette, J. Chem. Phys., **79**, 1065 (1983). H.W. Haesslin, C. Riekel, K. Menke, S. Roth, Makromol. Chem., **185**, 397 (1983).

5 T. Masuda, T Higashimura, Adv. Polym. Sci. , **81**, 121 (1986).

6 S. Ramakrishnan unpublished results.

7 D. Bloor, Chem. Phys. Lett., **43**, 270 (1976).

8 T.J. Sanford, R.D. Allendoerfer, E.T. Kang, P. Ehrlich, J. Polym. Sci. Polym. Phys. Ed., **18**, 2277 (1980).

9 T. Masuda, N. Sasaki, T Higashimura, Macromolecules, **8**, 717 (1975).

10 Y. Abe, T. Masuda, T Higashimura, J. Polym. Sci. Polym. Chem. Ed., **27**, 4267 (1989).

11 J.P. Aimé, P. Garrin, G. Baker, S. Ramakrishnan, P. Timmins, ICSM'90 Tübingen.

12 J.L.Brédas, R.R. Chance, R. Silbey, Phys. Rev. B, **26**, 5843 (1982).

13 W.P. Su, J.R. Schrieffer, A.J. Heeger, Phys. Rev. B, **22**, 2099 (1980).

14 R.L. Elsenbaumer, K.Y. Jen, R. Oboodi, Synth. Met. **15**, 164 (1986).

15 W.R. Salaneck, O. Inganäs, B. Thémans, J.O. Nilsson, B. Sjögren, J.E. Österholm, J.L. Brédas, S. Svenson, J. Chem. Phys., **89**, 4613 (1988).

16 K.S. Schweizer, J. Chem. Phys., **85**, 1156 (1986).

17 G.N. Patel, R. Chance, J. Witt J. Chem. Phys., **70**, 4387 (1979) and ref. chapter 5.

18 C Plachetta, R.C. Schulz, Makromol. Chem., Rapid. Commun., **3**, 815 (1982).

19 A. Halperin, Macromolecules, **23**, 2724 (1990).

20 F.S. Bates, G. L. Baker, Macromolecules, **16**, 1013 (1983).

21 L. Dai, J.W. White, J. Kerr, R.K. Thomas, J. Penfold, M. Aldissi, **28**, D69 (1989).

22 M. Aldissi, Chem. Com., **20**, 1387 (1984).

23 G.L. Baker,F.S. Bates, Macromolecules, **17**, 2619 (1984).

24 F.L. VanNice, F.S. Bates, G.L Baker, P.J. Carroll, Macromolecules, **17**, 2626 (1984).

25 C. Mathis, R Nuffer ICSM'90 Tübigen (1990).

26 X.F. Zhong, B. François, Makromol. Chem. Rapid Commun., **9**, 411 (1988). X.F. Zhong, B. François, Synt. Met., **29**, E35 (1989).

5. POLYDIACETYLENES

5-1. Introduction

Polydiacetylene derivatives -hereafter noted PDA- have the generic formula : $\leftarrow$CR-C≡C-CR'$\rightarrow_n$, with R and R' the substituent side groups. They have been available since 1969, the polymer chain being obtained by a solid-state 1-4-addition polymerization in the monomer crystal (1). As for common conjugated polymers, most PDA are insoluble even in exotic solvents, and infusible.

A decade ago, the discovery that with a proper choice of side-groups, some PDA can be dissolved in common organic solvents (2) opened the way for extensive experimental studies in various areas and provided access to basic and fundamental characteristics of conjugated polymers.

Most of the attention was focussed at first on the optical properties. Thermochromism of single crystals and films has been thoroughly investigated (3), while the dramatic color change exhibited by PDA solutions (yellow to red or blue) when the solvent/non-solvent ratio or the temperature is varied were the origin of extensive studies by several groups (4-11). Patel et al (3) first argued that the color change signals a single chain planar-non planar conformational transition with a strong coupling of the delocalized π electrons to the polymer conformation. Numerous hypothesis were proposed for the statistical conformation of the chain and the physical origin of the color change. In this chapter we report experimental results mainly devoted to P3BCMU, P4BCMU and PTS12 which are the most extensively studied PDA's (see table 7-1). The properties of these macromolecules should apply to other soluble ones whatever the substituents used.

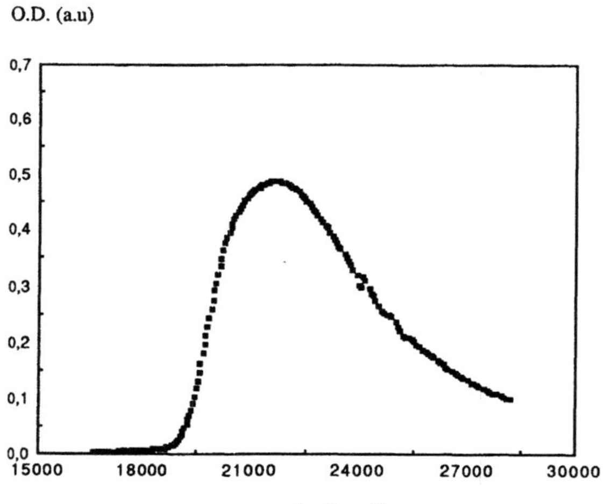

Fgure 1: Typical visible absorption spectrum of a polydiacetylene in the yellow state : P4BCMU in THF.

Depending on the nature of the substituents, PDA can be dissolved in many solvents. For example the above mentioned PDAs are soluble in $CHCl_3$, DMF, nitrobenzene, THF, toluene (at high temperature). The solutions are yellow and correspond to true solutions of isolated chains with a characteristic visible absorption spectrum shown in figure 1. Because only the backbone is conjugated, the visible optical absorption can be seen as a probe of the backbone structure. Therefore broadening of the absorption and shift of the peak position should be the first evidences of chain conformational change for any kind of substituted PDA. We have already emphasized the importance of the side groups for soluble conjugated polymers, here more than 90% of the atoms of the monomer unit belong to the side-group. Therefore an accurate knowledge of the substituent conformation is required in order to describe the effect of the side-group upon the overal chain conformation.

An important and helpful feature observed with P3 and P4BCMU is that through the CONH groups, H-bonds can be formed between neighboring side-groups on the same side of the chain. The H-bonds formation can be easily measured by infra-red spectroscopy (Figure 2). Free NH groups absorb near 3435 cm-1 whereas NH groups engaged in H-bonds absorb much more strongly and at lower frequency: 3320 cm^{-1} (3,7). Incidently, Patel et al used the infra-red results to suggest the planar- nonplanar change at the color transition (3). The formation of an H-bonds network being observed at the color change.

Through visible and infra-red spectroscopy, light and small angle neutron scattering, we do have many tools providing a more accurate picture of the statistical conformation of PDA in solution.

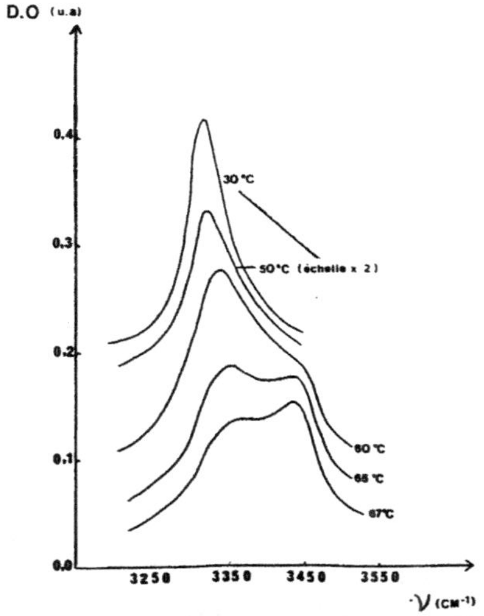

Figure 2 : Infra-red absorption of the NH groups in the yellow (T > 65°C) and red phase of P4BCMU in toluene. H-bonds formation (in the red phase) induces a spectral shift of about 100 cm^{-1} and a large increase of the absorption intensity.

5-2. Statistical conformation in good solvent: Yellow solution.

Most of the works performed in the early or middle 80's described the isolated chain in yellow solution as a coil (4,5,6,11). This was shown either by measuring the hydrodynamic radius of the polymer (quasi elastic light scattering) or the integrated light scattering giving the radius of gyration. On the other hand infra-red spectroscopy showed that the side-groups were not linked together, suggesting a random position of the side-groups with regard to the backbone.

In these first studies the samples were very polydisperse: $M_w/M_n = 5$ to 8, and the discussion of the physical properties such as the internal conformation of the chain or the threshold for the gel formation were made without considering the polydispersity effect. The molecular weight and mass distribution depends drastically from the way the sample is prepared. Because the polymer grows in monomer single crystal, a Shulz-Flory distribution does not necessarily occur. The size and the polydispersity are very dependent of the γ irradiation dose. The polydispersity is decreased by using a small amount of irradiation. With a 90 krad γ dose, a 30% of polymer conversion in crystals of 4BCMU monomer is obtained. The molecular weight is about $5\ 10^6$ and the polydispersity smaller than 3 (13). The mass distribution is asymmetric with a sharp edge at the large molecular weight and a tail for the short chains (figure 3).

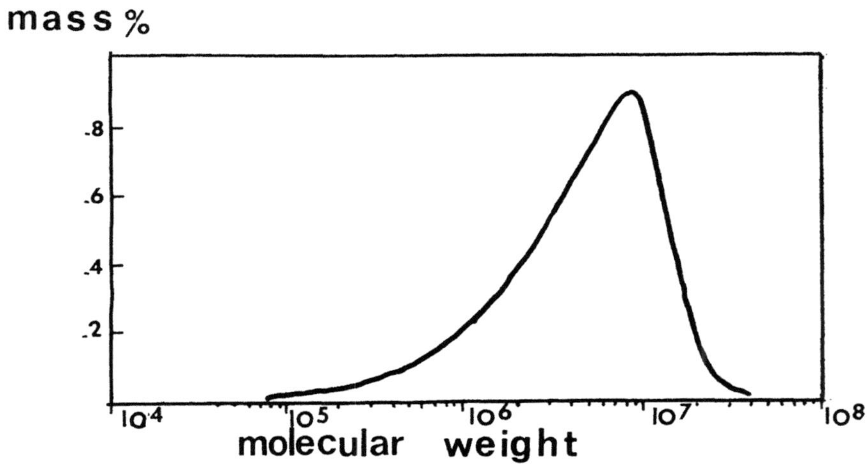

Figure 3 : Molecular weight distribution of p-4BCMU obtained with a 90 krad γ irradiation. $M_w/M_n = 2.8$, $M_w = 5\ 10^6$.

In 1984, Wenz et al (11) made a systematic study on the chain conformation of PTS12 in good solvent with the light scattering technique. Integrated light scattering (ILS) does not allow to reach the appropriate range of transfer momentum for describing the internal conformation of the coil. Nevertheless, information can still be obtained on the local correlation inside the chain by measuring the average size of the coil as a function of the molecular weight (§ 3-5). The different sizes of the chain were obtained by photodegradation of the PTS12 in solution. Wenz et al (11) showed that the photodegradation is a random chain scission process providing a normal distribution of chain lengths. Therefore a Shulz-Flory distribution can be used for describing the final polydispersity. The range of M_w was 10^5 to $2\ 10^6$. Wenz et al have measured a dependence of the radius of gyration on the molecular weight given by the relation:

$$< R_G^2 >_z = 0.45\ M_w^{0.55} \tag{5-1}$$

The value of the exponent is close to the one for a pure random coil. This can be easily understood with the fact that the molar mass of the monomer unit is about 5 times larger than for usual polymers and that PDA chains exhibit a large local rigidity (see below). Therefore the equivalent unit length is much larger than the monomer unit length. The asymptotic behavior for flexible polymers in good solvent being reached for Mw larger than 10^6 (see chapter 2), the asymptotic behavior for PDA chain in good solvent should be reached for molecular weight equal to or larger than 10^8 which is an order of magnitude above the largest PDA chains ever made.

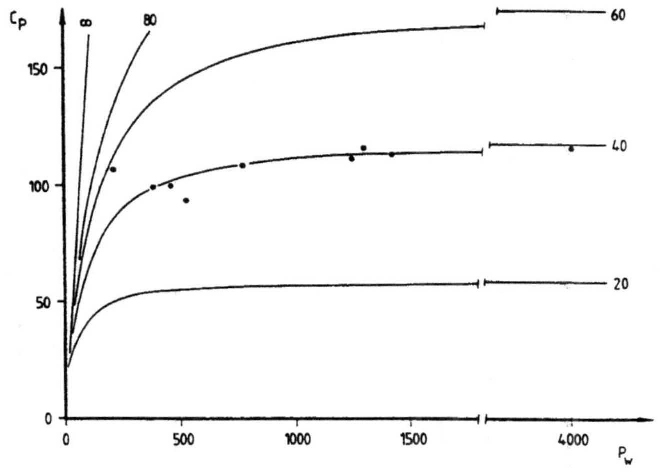

Figure 4 : Characteristic ratio C_p (equation 5-2) vs. degree of polymerization N_w: solid lines, calculated for a wormlike chain with the persistence length (in number of monomer units) as the parameter; points, results from ILS. (from reference 11)

The value of the statistical length is obtained following the method described in § 3-5. The relation between the statistical length and the radius of gyration for a monodisperse system is given by equation 3-43, while using the Shulz-Flory distribution, the z-average is given by the equation 3-46. Wenz et al reported the ratio between the radius of gyration $<R_G^2>_z$ and the radius of gyration of the equivalent perfect random coil: $R_{GC}^2 = N_W a^2/6$.

With $N_W = M_W/m$, the number of monomer units for a chain with mass M_W, and a the monomer unit length. Figure 4 shows the variation of the ratio (equation 5-2) as a function of the degree of polymerization:

$$C_p = \frac{<R_G^2>_z}{\frac{N_w a^2}{6}} \qquad 5\text{-}2$$

The varying parameter is the statistical length, or equivalently, the number of monomer units within a statistical length b/a = n. The best fit was obtained with n = 80 corresponding to b= 400 Å. This value was the first non ambiguous result describing the chain statistics as a worm like chain and giving the good order of magnitude for the statistical length.

Because they measured approximatively the same radius of gyration (for a given M_w) for P3BCMU and PTS12, the authors also concluded that side groups have no or little influence on the chain conformation. In other words the semi-rigid behavior of the PDA chains would come from the conjugated structure of the backbone (14,15). Following this aim, attempts have been made to relate the magnitude of the statistical length to the position of the maximum absorption (16,17). These approaches for understanding the physical origin of the chain conformation were later found to be incorrect (see below).

Small angle neutron scattering (SANS) gives the appropriate q range for studying correlation at local scale. SANS was performed in a q range between 5 10^{-3} and 0.25 $Å^{-1}$ (18). Obviously XRay scattering also provides such a domain but could induce chain degradation while neutron does not cause any damage. Besides, absolute measurements are possible, allowing to measure the contour length of the monomer unit (§ 3-3-1). The later point is of particular interest for PDA solutions, since it has been proposed that the coil in the yellow phase consists predominantly of a cis structure, assumed to be more flexible than the trans structure (19). A coil with a cis structure for the backbone would be a strong support for the hypothesis of a rod to coil transition, while if not the hypothesis of a pure single chain process for the sol-gel transition becomes much more speculative.

ILS experiments indicate a semi-rigid behavior for the PDA chains. Therefore we can expect to observe the form factor of the Porod-Kratky chain. More precisely, the q^{-1} scattering behavior of a rod like molecule should be measured since the normalized form factor P(q) of an infinitely long worm-like chain has the asymptotic form:

$$\lim_{\substack{L\to\infty \\ q\to\infty}} \frac{L}{b} P(q) = \frac{\pi}{qb} + \frac{4}{3(qb)^2} \qquad (5\text{-}3)$$

and for long enough chains ($\frac{L}{b} > 2$), equation 5-3 holds for $qb > 7$ (chapter 3-4).

At larger q values, when the spatial resolution is increased, this model is no longer accurate. We know as a characteristic of soluble conducting polymers that bulky side-groups are needed. The larger the side-groups the more important will be their contribution to the form factor. Depending of the spatial scale, the effect of the side-group can be treated as a simple thickness. In the q range where $qR_c < 1$ (§ 3-5), the chain structure factor is rewritten:

$$P(q) = P_0(q) \exp\left(- \frac{q^2 R_c^2}{2} \right) \qquad (5\text{-}4)$$

where R_c is related to the extension of the side-group in the solvent.

In figure 5 is reported the structure factor normalized to the degree of polymerization N_w:

$$g(q) = \frac{M_W}{m} P(q) \qquad (5\text{-}5)$$

The qg(q) versus q plot shows directly the effect of the local rigidity through the existence of a q^{-1} behavior. The statistical length deduced is equal to 300Å. According to equation 5-4 and 5-5, the height of the plateau at which the q^{-1} behavior is reached gives the mass per unit length or equivalently the monomer unit length : $A = \pi M_L/m = \pi/a$. So $M_L = 94$ g mole^{-1} Å^{-1} or the length of the monomer unit is about 5 Å. In PDA single crystal, such as PTS6 or PFBS, the trans conformation is observed with a monomer unit length equal to 4.91 Å (20,21), while a length of 3.5 Å is expected for a cis conformation. The difference between cis and trans is much larger than the experimental error (less than 10%). These results rule out a cis conformation for the coil structure in the yellow solution.

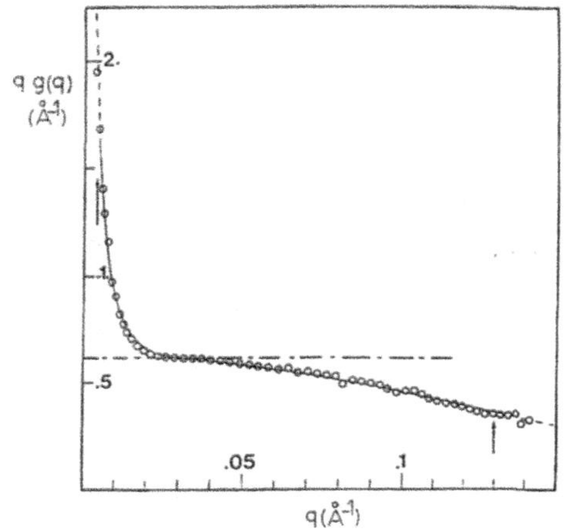

Figure 5 : qg(q) plot of p-3BCMU in good solvent. The full line represents the best fit by the theoretical structure factor $g(q) = g_0(q) \exp(-q^2R^2_c/2)$. The worm-like chain model is assumed for $g_0(q)$ (see § 3-4). The dashed line corresponds to the height of the q^{-1} plateau which yields the mass per unit contour length M_L. (from reference 18)

The experimental curve was compared to the theoretical curve in the domain 5 10^{-3} to 0.13 Å^{-1}. The spatial resolution within this q range gives $2\pi/q_{max} = 50$ Å, such that the phase change between the different parts of the side-group can be neglected. In other words we can consider an average scattering length density for the monomer unit. Fully extended side-groups correspond to a width of the order of 35Å. For P3 and P4BCMU, R_c^2 equal respectively to 64 and 60 Å^2 were found. Using the relation for a rod: $R_c^2 = l_c^2/12$ the respective extension l_c are 28 and 27 Å with an estimated error ± 2 Å. Alternatively the scattering behavior can be ascribed to a ribbon like structure with a q^{-2} dependence :

$$P(q) = \frac{2\pi}{Sq^2} \qquad (5\text{-}6)$$

for $1/W << q < 1/l_e$, where W and l_e are the width and the thickness of the ribbon and S the surface. The height of the plateau in a $q^2g(q)$ plot is proportional to the mass per unit area : $B = 2\pi\ Ms/m = 2\pi/S$ giving the width $W = S/a$ (figure 6). The method gives a respective width for P3 and P4BCMU 27 and 25 Å with an estimated error ± 1.5 Å.

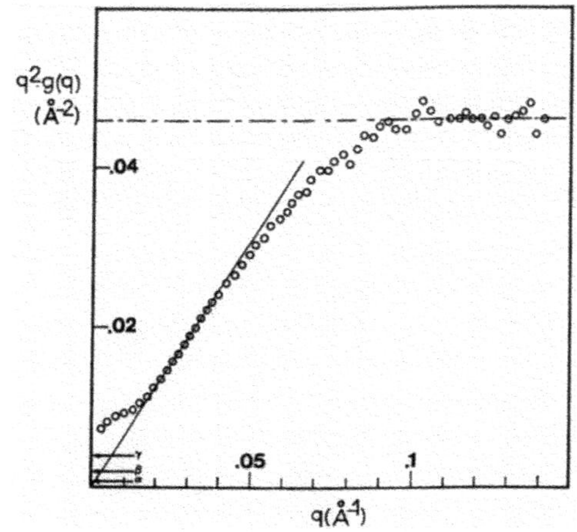

Figure 6 : $q^2g(q)$ plot of p-3BCMU in DMF. The q^{-2} behavior at large q values is characteristic of the ribbon like structure. (from reference 18)

Even if they are not fully extended the side groups exhibit a rather large extension with regard to the central part of the chain. The picture is quite reasonable, the side-groups repel each other in good solvent such that a coil conformation for the substituents can not be built along the chain.

The scattering curves cannot be described neither with broken segments nor with a distribution of broken segments (18), equation 2-12 gives an average misorientation between monomer unit equal to 14° for a statistical length equal to 300 Å. The neutron scattering data provide a definite proof that the soluble PDA are much better described as a chain with a continuous curvature.

5-3. Origin of the blue and red shifts in good solvent : chain conformation and solvatochromism

The worm-like chain picture immediately emphasizes the importance of the fluctuation in curvature. Whatever the solvent used, the yellow solutions exhibit a slight blue shift when the temperature is increased. On the other hand the model describing the fluctuations in curvature of the macromolecule predicted a T^{-1} dependence for the statistical length (ref. 22 and chapter VII). With increasing the temperature the local curvature increases, therefore suggests to relate the flexibility increase, then a localization of the π-electron, to the blue shift of the visible absorption (16,17).

Neverthelèss the shift of the maximum absorption is not in itself a proof that the backbone conformation has been changed. With increasing temperature, thermal expansion of the solvent occurs and a simple solvatochromism effect should be enough for understanding the spectral shift. Within the framework of a simple theory, solvent effécts on the excitation energy of a transition is proportional to the ratio $(n^2-1)/(n^2+2)$ with n the refractive index of the solvent (23):

$$h(\nu-\nu_0) = a^{-3}\frac{n^2-1}{n^2+2}\frac{E}{E'+E}[M_{eg}^2 - E(\alpha_g - \alpha_e)] \quad (5\text{-}7)$$

where a is an effective solute cavity radius, E and E' are mean energies of the solvent and solute, M_{eg} is the electric dipole transition moment for the solute, and the α's are polarizabilities of the solute in the ground and excited states, respectively. The quantity $(n^2-1)/(n^2+2)$ is proportional to the solvent density ρ_s and the polarizability of the solvent molecule α_s.

$$\frac{n^2-1}{n^2+2} \propto \rho_s\,\alpha_s \quad (5\text{-}8)$$

Between 25°C and 80°C, the solvent polarisability can be considered as temperature independent, and a linear dependence with the solvent density is thus expected for a pure solvatochromism effect.

Sklar et al have shown that the absorption maximum of an alkyl-substituted tetraene shifts in response to temperature in exactly the same fashion as a shift in response to solvent change (24). An additionnal proof of the solvent effect is obtained by measuring the change of the optical absorption for a solution of P4BCMU in THF as a function of the pressure (25). THF is a good solvent for PDA chains, remaining fluid even at pressure as high as 20 kbars. In contrast to the temperature, the available pressure cell has the advantage to offer a larger change on the solvent density.

The experiments were performed with long macromolecules but also with the short oligomers of PDA (4 monomer units) without bulky side-groups synthesized by F. Wudl and S.P. Bitler (26). Short oligomers being rod-like molecules which cannot have any conformational change, they form a good probe for testing the solvent effect. The spectral shift (a red shift with pressure) exhibits a linear dependence with the solvent density suggesting that the macromolecule is only sensitive to solvent change. In addition the slopes of the spectral shift (12500 cm^{-1} ± 1000) as a function of the ratio $(n^2-1)/(n^2+2)$ are almost identical for the P4BCMU chains and the PDA oligomers (figure 7).

Therefore a continuous change of the visible absorption can be simply understood by considering the variation of the solvent properties. If any conformational changes occur in good solvent for the long chains, in most cases they may have almost no influence on the optical properties. The preceeding results raise a problem within the worm-like chain model. The structural data showing that the macromolecules are semi-rigid, the flexibility must increase with increasing the temperature (see chapter VII). Since a pure solvatochromism effect is enough for understanding a blue or red shift, this would imply that the fluctuations in curvature do not have any influence on the π electron structure or at least on the optical properties.

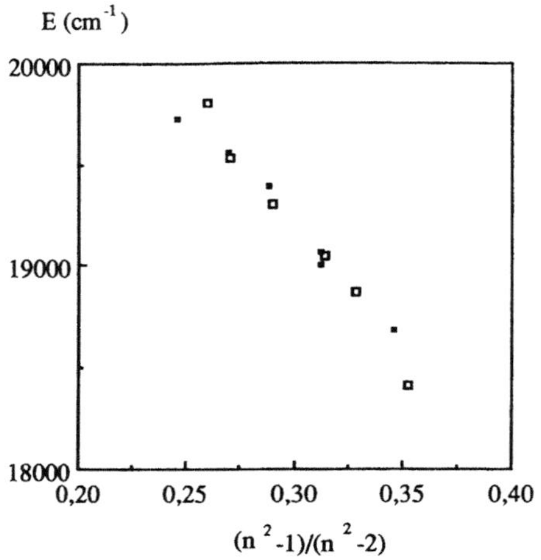

Figure 7 : Absorption spectral shifts of the oligomer and the macromolecule p-4BCMU. For a direct comparison the absorption energy of the oligomer has been shifted by 6250 cm^{-1}.

One way for solving this question is to study the evolution of the chain conformation as a function of the temperature. SANS experiments were performed on the PDA PTS12 in the good solvent DMF. PTS12 has been chosen because there is no possibility to form H-bonds between side-groups (Table 7-1). Therefore additionnal effects due to link formation between side-groups like those observed for P3BCMU in bromoform were avoided (12).

For a statistical length equal to 300 Å at RT a value equal to 255 Å at 75°C is expected. Figure 7-1 shows the scattering curves measured at RT and 75°C, no change on the local correlation was detected. Such absence of variation can be understood if we take into account the size of the side-groups and the average torsion between monomer units. This point is of particular importance for understanding the physical properties of conducting polymers in solution and will be described in details in chapter VII. Here it is worth mentioning that if the chain statistics is well described as a worm-like chain, the physical origin of the semi-rigid behavior is not so obvious to understand. The π electron structure does not seem to play a major role on the chain conformation. Moreover a continuous shift of the maximum of the visible optical absorption, does not appear to be related to change of π electron delocalization but rather to change of the surrounding medium.

5-4. Color transition : aggregation versus single chain process.

The study of the sol-gel transition is of a great interest for many reasons : a good understanding of the gel formation should provide a better insight on the crystallization process, and more generally on the packing rules for conducting polymers. On the other hand if there is evidence for a rod-like conformation of the macromolecules, this will be also of particular interest because rod-like structure can lead to liquid-crystal structure.

Drastic color change with large hysteresis can be observed by varying either the solvent/non-solvent ratio or the temperature (figure 8). Here we focus our attention on solution of PDA P4BCMU in toluene as a function of the temperature, on cooling the transition occurs around 66°C while on the heating run the transition occurs at 75°C, the values of the temperature transition being dependent on the M_w. The use of the temperature as the relevant variable offers advantages. It can be precisely controlled and continuously varied (both of these are not easily achieved with changes in solvent/non-solvent ratio).

The sol-gel transition has been studied extensively by several groups. The interpretation on the physical origin of the red phase was controversial. For example Patel et al (3) or Lim et al (8,9) argued that the yellow to red change was due to a single chain process i.e a coil to rod transition, while Wegner and colleagues (10,11) hold the opposing view that the color change in the polymer solutions results from the onset of aggregation. At the beginning of the 80's, the rod to coil interpretation was mainly supported with spectroscopic experiments in the visible or infra-red (3) which, however could not prove alone the absence of aggregation. Later quasi elastic light scattering measurement showing the existence of a single rod-like macromolecule in the red phase was still not convincing. Indeed up to now, there are no robust experimental results showing a rod conformation in solution. It has been argued that field induced birefringence experiment in red solution at low concentration ($5\ 10^{-6}$ to $5\ 10^{-5}$ gcm^{-1}) was a good proof of the rod-like conformation (9). We shall first discuss these experiments, then in details the structural data (ILS and SANS) available on the sol-gel transition (11,18,27).

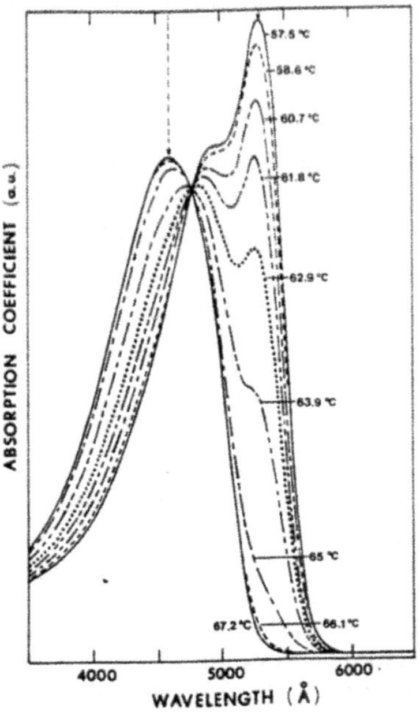

Figure 8 : Change of the absorption spectra of p-4BCMU in toluene under cooling. The yellow solution corresponds to a maximum absorption located at 4600Å, the red phase at 5300Å.

The aim of the field induced birefringence experiment is to relate the magnitude of the observed birefringence to the chain conformation. The light intensity detected due to the field induced orientation is proportional to $(\Delta n)^2$, where Δn is the birefringence.

$$\Delta n = n_{//} - n_{\perp} \quad (5\text{-}8)$$

and $n_{//}$ and $n_{\perp}$ are the refractive indices parallel and perpendicular to the applied field, respectively. For polymer rods, the birefringence is written:

$$\Delta n(\text{rod}) = \frac{2\pi N_p}{15n} (\alpha_{//} - \alpha_{\perp})^2 \frac{E^2}{k_B T} \quad (5\text{-}9)$$

where $\alpha_{//}$ and $\alpha_{\perp}$ are the polarizabilities of the polymer parallel and perpendicular to the chain axis, n the refractive index of the isotropic solution, E the applied field and N_p the degree of polymerisation. If a polymer exists in the two forms: flexible and rod, the ratio between the two birefringences should be given by:

$$\frac{\Delta n(\text{rod})}{\Delta n(\text{coil})} \propto N_p \quad (5\text{-}10)$$

A large ratio $\Delta n(\text{rod})/\Delta n(\text{coil}) > 300$ was found. As discussed by the authors (9), the result cannot be considered as a definite proof of a single rod-like chain structure in the red solution. The field induced birefringence in the red solution could result from orientation by the field either of individual rod-like molecules or of large aggregates in which the individual chains are more or less aligned. To differentiate between these two cases, the authors carried out measurements of the transient decay of the birefringence. Since the rotational diffusion constant for a rod varies as L^{-3}, the relaxation time is very sensitive to the particle dimension. They found that the relaxation time is concentration independent in the range $5\ 10^{-6} < c < 5\ 10^{-5}$. Using a Gaussian distribution for the polydispersity, from the measure of the time relaxation, they found a length of 2μm in rather good accordance with the expected contour length of the P4BCMU. Moreover, even when aggregates are formed the relaxation time remains constant within the experimental accuracy. They concluded that the aggregates do not contribute to the field induced birefringence.

Such a conclusion requires an additionnal hypothesis on the aggregate structure. For aggregates which would be formed prior to gelation they proposed a completely random coil structure of P4BCMU inside the aggregate, thus an isotropic particle which could not contribute to the birefringence. But as it is shown below (chapter VII), we do not have a proper knowledge on the internal structure of the aggregates formed with conducting polymers. Indeed, a description with a fibrillar structure rather than a random or self-similar one seems to be more appropriate.

Wenz et al (11) reported a study of the blue solution of P3BCMU (in 1,2-dichloroethane) with light scattering. The molecular weight is $1.36\ 10^6$ and the concentration used was 10^{-2} mg cm-3. Whatever the incident wavelength used, the integrated intensity is at least an order of magnitude larger than for the case of the yellow solution in $CHCl_3$. To relate the intensity increase to the mass of the particle one must know the index increment, i.e. the change of the refractive indice of the solution as a function of the polymer concentration. This is very difficult and probably impossible to measure in the blue form of P3BCMU because for concentrations allowing an accurate dn/dc measurement the polymer precipitates and the solution becomes inhomogeneous. In order to have a measure of the mass of the particle, they used the refractive index increment of the yellow form in $CHCl_3$. A $M_w = 2\ 10^8$ is then calculated. Compared to the M_w of the macromolecule the result indicates an aggregate structure with 150 to 200 chains. This experiment is not completely conclusive, as it was noted by K. C. Lim et A. J. Heeger (8), the large increase of the intensity might come from the change of the refractive index. Because there is a large red shift for the blue and red solution as well, the contribution of the imaginary part should be important (the typical wavelength used for light scattering are between 436 and 633 nm), therefore the number of chains within the aggregates and may be even the existence of aggregates becomes questionable. Moreover, the concentration is not small enough for avoiding a percolation of rods, assuming that rod-like structure occurs prior to aggregation.

Absolute measurement with light scattering may be more convincing with the red solution formed with P4BCMU in toluene. Muller et al have estimated dn/dc for the case of P4BCMU in toluene at RT (27). In contrast to P3BCMU, P4BCMU does not form micro crystallite but a well defined homogeneous gel. ILS were performed at 20°C, on P4BCMU with $M_w = 1.67\ 10^6$ and $M_w/M_n = 2.5$. The overlap concentration for polydisperse rods is given by :

$$c_p(U) = \frac{6\ M_L^3}{\mathcal{N}\ \pi\ M_w^2} \frac{(1+U^2)}{(1+2U)(1+3U)} \qquad 5\text{-}11$$

where $U = M_w/M_n - 1$, and M_L the mass per unit length is known from the study in the yellow solution. The concentration used was $0.47\ 10^{-6}$ g cm^{-1} slightly larger then c_p (U) (see table) but at a concentration at which the rod-like chain were proposed in the red form. The Zimm extrapolation method yields an average molecular weight M_w of the order of $50\ 10^6$ g mole^{-1}. Again, comparing with the M_w of the isolated macromolecule, the result proves an aggregate structure, each particle containing about 30 macromolecules.

M_w (g mole^{-1})	$U = (M_w/M_n)$-1	$c_p(U)$ (g l^{-1})
1 670 000	1.5	0.34 x 10^{-3}
~ 43200	1.5	~0.51
	3	~0.41

Table 5-1 : Critical concentrations related to the percolation threshold of rods. The critical concentration are obtained from equation 5-11. For $M_w = 43200$, U = 3 is given in order to show the influence of polydispersity on the critical concentration (from reference 18).

In order to solve the problem of the absolute measurement, SANS experiments have been done (18). The neutron scattering cross section being weak, a polymer concentration of about 10^{-4} g cm^{-3} is required. With the molecular weight of PDA as prepared, this is well above the overlap concentration for rod (equation 5-11), and an increase of c_p(U) requires a drastic decrease of M_w. This was achieved with a sample having $M_w = 43200$, for which, with a reasonable polydispersity, c_p is 0.47 10^{-3} g cm^{-3} (see table 5-1).

The solution exhibits on cooling a yellow to red transition similar to the one observed for the long chains. It begins at 54°C, in a temperature range down to 44°C. The dependence of the transition temperature and of the temperature range on M_w is reminiscent of what was observed for the polypeptides (28). The analogy with the helix-coil transition of polypeptides (already noted by A.J Berlinsky et al (19) and Patel et al (3)) is quite helpful and will be discussed in chapter VII. In figure 9 are shown the scattering functions measured in the yellow and red phases. The large increase of the intensity definitely proved an aggregation process.

The experimental results show an aggregation process whatever the color transition considered : yellow to blue (p3BCMU) or yellow to red (p4BCMU). In addition, aggregation is also observed for the short chains having a contour length no longer than the statistical length. Nevertheless they do not provide a clear picture about the physical origin of the color transition. Up to now the fact that we are unable to observe a rod-like conformation in the red or blue phase does not rule out an aggregation process triggered by a single chain phenomenon.

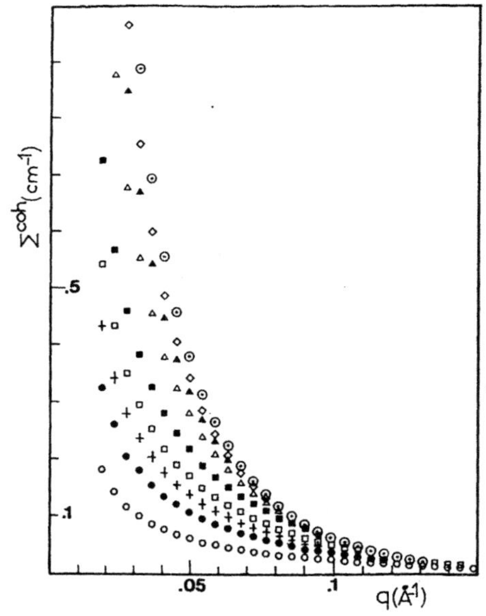

Figure 9 : Change of the scattering intensity with the gel formation. P4BCMU in toluene at different temperatures. T = 70°C (o), 65°C (•), 64°C (+), 63°C (□,), 61°(■), 58°C (Δ), 55°C (▲), 51°C (◇), 38°C (⊙). (from reference 18)

A way to address this question is to study the chain conformation close to the transition temperature in the yellow phase. If an intra-chain process occurs, a change of the side group conformation or an increase of the local rigidity might be observable.

The infra-red spectroscopy do not show any change allowing to propose a precursor effect due to the formation of an H-bonds network. Moreover, the scattering function at large q (.05 to .15 Å^{-1}) for a temperature as close as 0.5°K above the temperature transition is identical to the scattering function observed for the yellow solution. Within the experimental accuracy, these results imply a side-group conformation identical at all temperatures in the yellow solution range. As far as the side-groups are concerned with an intra-chain process, there is no evidence of conformational change prior to aggregation.

Another method consits to measure a possible increase of the chain stiffness through the study of the q^{-1} behavior of the scattering function. In the yellow solution, the q^{-1} behavior occurs at $q \geq 0.023$ Å^{-1} corresponding to a statistical length equal to 300 Å. If the statistical length increases, a q^{-1} behavior must be observed at lower q values. Here again, experiments have been done down to half a degree above the temperature of the transition. The curves remain identical providing an upper limit for the statistical length change, i.e an increase of the rigidity must not lead to a statistical length larger than 330Å. In other words careful SANS experiments do not give any evidence of a pure intra-chain process prior to aggregation.

References:

1 G. Wegner, Z. Naturforsch. (1969), **24b**, 824

2 G.N. Patel, J. Polym. Sci. Letters Ed., (1978), **16**, 607.
G.N. Patel, E.K. Walsh, J. Polym. Sci. Letters Ed., (1979), **17**, 203 (1979)

3 G.N. Patel, R.R. Chance, J.D. Witt, J. Chem. Phys. **70**, 4387 (1979).

4 G.N. Patel, J.D. Witt, Y.P. Khanna, J. Polym. Sci. Polym. Phys. Ed., **18**, 1383 (1980).

5 R.R. Chance, G.N. Patel, J.D.. Witt, J. Chem. Phys. **71**, 206 (1979)

6 R.R. Chance, M.L. Shand, C.Hogg, R.Silbey, Phys. Rev. B, **22**, 3540 (1980).

7 M.L. Shand, R.R. Chance, M. Le Postollec, M. Schott, Phys. Rev. B, **25**, 4431 (1982). R.R. Chance, J.M. Sowa, H. Eckhardt, M. Schott, J. of Physical Chemistry **90**, 3031 (1986).

8 K.C. Lim, A.J. Heeger, J. Chem. Phys., **82**, 522 (1985).

9 K.C. Lim, A. Kapitulnik, R. Zacher, A.J. Heeger, J. Chem. Phys., **82**, 516 (1985).

10 G. Wenz, G. Wegner, Makromol. Chem. Rapid. Comm., (1982), **3**, 206

11 G. Wenz, G. Wegner, M.A. Müller, M. Schmidt, Macromolecules, (1984), **17**, 37

12 J.P. Aimé, F. Bargain, J.L. Fave, M. Rawiso, M. Schott, J. Chem. Phys., **89**, 6477 (1988).

13 J. Berrehar, S. Ramakrishnan, J.P. Aimé unpublished results.

14 G. Allegra, S. Brückner, G. Wegner, M. Schmidt, Macromolecules, **19**, 399 (1985).

15 P.A. Pincus, G. Rossi, M.E. Cates, Europhys. Lett., **4**, 41 (1987).

16 Z.G. Soos, K.S. Schweizer, Chem. Phys. Let., **139**, 196 (1987).

17 J.P. Aimé, M. Rawiso, M. Schott, International Winter School on " Electronic prope of polymer " p 58, (1987), Kirchberg, Springer-Verlag, F.R.G.

18 M. Rawiso, J.P. Aimé, J.L. Fave, M.Schott, M.A. Müller, M. Schmidt, H. Baumgartl, G. Wegner, J. Phys. France, **49**, 861 (1988).

19 A.J. Berlinsky, F. Wudl, K.C. Lim, C.R. Fincher, A.J. Heeger, J. Polym. Sci. Polym. Phys. Ed., **22**, 847 (1984).

20 D. Kobelt, E.F. Paulus, Acta Cryst., **B30**, 232 (1974).

21 J.P. Aimé, M. Schott, M. Bertault, L. Toupet, Acta Cryst., **B44**, 617 (1988).

22 L.D. Landau, E.M. Lifshitz., Statistical Physics, 2nd edition (Pergamon Press, London) 1959, p. 478.

23 B.S. Hudson, B.E. Kohler, K. Schulten, in "Excited States", **6**, E.C. Lim Ed.(Acad. Press, N.Y (1982).

24 L.A. Sklar, B. Hudson, M. Peterson, J. Diamond, Biochemistry, **16**, 813 (1977).

25 J.P. Aimé, H. King, M.W. Kim, R.R. Chance, ICSM'90 Tübingen.

26 F. Wudl, S. Bitler, J. Amer. Chem. Soc., **108**, 4685 (1988).

27 M.A. Müller, M. Schmidt, G. Wegner, Makromol. Chem.., Rapid Commun., **5**, 83 (1984).

28 B.H. Zimm, J.K. Bragg, J. Chem. Phys., **28**,1246 (1958).

6 STUDY OF DOPABLE POLYMERS : PANI AND POLY-n-ALKYLTHIOPHENES.

Conjugated polymers are often modelled as low-dimensional semiconductors, with an intrinsic dimensionality lying between 1 and 2. The typical "conductivity" in the neutral state is between 10^{-14} and 10^{-6} S/cm (1,2). Conjugated polymers become conducting polymers when they are doped, with a conductivity depending on the doping level and the type of materials.

The conductivity for a doped polyacetylene film can reach a value higher than 10^{5} s/cm (2) at room temperature, comparable to the conductivity of metal. This high level is somewhat surprising if we keep in mind that the polymer shows a high degree of disorder by comparison with usual solids or even with other organic systems exhibiting high conductivity such as organic charge-transfer crystals.

As for semi-conductors the conductivity is achieved by doping, but with main differences due to the type of doping process, interstitial rather than substitutional hence a mobility of the dopant, and to the softness of the host matrix. The softness of the polymer necessarily imply a dopant insertion different than the one considered for a rigid well-defined lattice. The symetry, if we can define it, of the host matrix can change but also rotation of the whole chain and torsion between monomer units have to be considered. Due to the large deformation imposed, the electronic states responsible for the conduction are believed to be self-localized bound states. Except for the case of polyacetylene where the degenerate state in the trans conformation allows a charged soliton state, two kinds of states are expected : polaron which bears one charge and one spin, and bipolaron which bears two charges and no spin (3). In this chapter we give a brief, and not at all exhaustive, review, first of the most recent problems concerning the physical properties of the doped polymers in solid phase, then of the recent investigations on the doped polymers in solution. More detailed description for the solid phase are found elsewhere in this book (see Epstein et al for polyaniline in "the chemistry and physics of polyanilines", and Salanek at al for poly-n alkylThiophene in "Processable conducting polymers and their physical characterization").

6-1. Doped polymers in the solid phase

The transport properties in the solid phase of an infusible polymer raise a great variety of problems. We first face to the need of an accurate description or at least an unambiguous definition of the effective medium : homogeneity, local structure order ..., second the kind of charges stored and their relative stability versus the chain conformation and chain packing. In order to describe the properties of the electronic states responsible for the conductivity, N.M.R. and E.S.R. are powerful tools for probing the spin motion (4). It has been shown that nuclear relaxation can be directly related to the electronic spin motion providing an access to the electron spin mobility parallel and perpendicular to the chain without being obscured by trapped spins at defect sites (5). A fundamental hypothesis consists to consider that spin and charge are closely related such that the study of the spin dynamic gives access to the charge motion. Moreover a change of the spin density, for example as a function of the doping level, is directly understood as a change of the charge storage on the chain, i.e. polaron to bipolaron (4-9).

Several properties are deduced from the knowledge of the spin dynamics, either upon the doping state or the way the macroscopic dc conductivity is achieved. As we have already noted, the relation between the chain structure, the compactness of the surrounding medium and the macroscopic dc conductivity is far from being understood. Nevertheless the spin motion (or the charge motion) and phenomenological models on transport properties are hepful even if they are unable to provide a description at a microscopic scale.

As an example of such an approach, the behavior of the conducting polymer, polypyrrole-perchlorate (Ppy-ClO_4), show interesting features. Through the measurement of the frequency dependence of the nuclear relaxation times, F. Devreux and H. Lecavelier were able to determine an effective dimensionality $\tilde{d}$ of the spin motion (6). For a given doping level, they note an unidimensional spin motion between 50 and 150 K and conclude that in this temperature range the polarons are confined along the polymer chain. At higher temperature, interchain hopping occurs changing the dimensionality such that $\tilde{d}$ varies from 1.2 at 200 K up to 2 at room temperature. In other words, at room temperature , PPy films does not behave like a compact structure. The effective dimensionality expresses the number of sites visited by a random walker as a function of time $N(t) \propto t^{\tilde{d}/2}$, it is a measure of the compactness of the medium with respect to spin motion. Such an analysis demonstrates that the polymer film is a porous material whatever the physical origin of the porosity, i.e. either controlled by different electronic states giving the best hopping rate between chains, or a purely structural effect due to the way the polymeric chains fill the space.

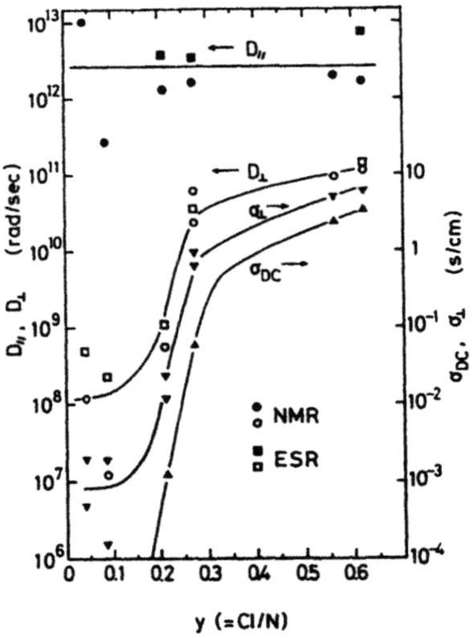

Figure 1 : Room-temperature spin-diffusion rates, as obtained from NMR (o,•) and from ESR (□, ■), and conductivity in polyaniline as a function of the protonation level. σ_{dc} is the measured conductivity and $\sigma_{\perp}$ has been calculated from $D_{\perp}$ (the interchain diffusion rate), using $\sigma_{\perp} = ne^2 D_{\perp}/kT$, with n, the carrier concentration, proportional to y.(from reference 4).

From that point of view, the polyaniline film seems to have a completely different behavior. The polyaniline is known for its unusual insulator-to-conductor transition as a function of protonation. Upon protonation of the emeraldine form of polyaniline the conductivity increases by 10 orders of magnitude while the number of electrons on the polymer chains remains constant (4,7). The electronic spin motion shows an highly anisotropic diffusion $D_{//} >> D_{\perp}$. K Mizoguchi et al (5) show that the diffusion rate along the chain, $D_{//}$ does not depend on the protonation level. Figure 1 illustrates the anisotropic behavior for a protonation level lower than $y_p = 0.2\text{-}0.3$. At $y = y_p$, the anisotropy drops suddenly by more than two orders of magnitude. The authors interpret this behavior as the consequence of a percolation process between conducting islands formed at value lower than y_p. In contrast to doped polypyrrole, the unidimensional spin motion becomes a three-dimensional one at a rather well defined level of protonation. There are no particular reason for expecting a more compact structure for polyaniline films with respect to polypyrrole films, however they exhibit a completely different spin motion at room temperature.

On the other hand, because a drastic change of the anisotropic behavior occurs (figure 1), K Mizoguchi et al concluded that the isolated conducting islands are only formed by a single conducting chain only, while F. Zuau et al proposed that the conducting islands contain packed chains (9).

In order to discriminate between these two proposals, SANS experiments were performed (10). The central aim was to investigate the structure of the cluster in the polyaniline amorphous matrix. The use of neutrons as a probe is particulary suitable because the conducting chains are protonated and water molecules are selectively fixed on the acidic protons of the conducting chains (11). The measurement can be described as a structural study of the water chains. Moreover, since the porosity of the sample adds an additional density fluctuation which can be very large and obscure the density fluctuation coming from the conducting chains, a possible index matching can be obtained by using deuterated solvent to fill the voids in the film. Several attempts were done and the most significant was obtained by using deuterated cyclohexane, which scattering length density is very close to that of PANI. In addition it was the best solvent for filling efficiently the neutral matrix. The scattering intensity due to the void was significantly reduced, but still large enough and hardly reproducible, so that the scattering intensity coming from the conducting chain was not accuratly measured. The difficulty for performing elementary characterization, such as to determine whether or not the clusters contain several chains, is a good example of the kind of problems encountered with these materials. Up to now, no clear answer is obtained on the structure of the conducting islands.

M.E Jozefowicz et al (12) have performed a structural study of polyaniline by wide angle scattering. Both insulating and doped state were measured, the detail of this structural study is given by A.J. Epstein et al (this book). Here we will note that this first complete structural measurement shows unambiguously an increase of the ordering with the doping, the insulating state being amorphous while the doped state has a crystallinity of about 50% for $y = 0.5$. On the other hand the authors indicate that the protonated amorphous region contains bipolarons and not polarons. The latter result is just the opposite of the recent result obtained by M. Nechtschein et al (13). In a thin film made layer per layer they observe only few spins $\sim 10^{-4}$ per monomer, while with thick film polarons are observed. The authors concluded that the higher disorder expected in thin films is responsible for polarons formation. A main question is the stability of the polaron state versus the bipolaron state and their relation with the average structure of the matrix. It is obvious that at the present time no clear picture emerges.

6-2. Doped polymers in solution : poly-n-alkylthiophene.

6-2-1. Structure of Poly-3-butylthiophene in the neutral state.

The impetus to discover the solution phase for conducting polymers is due to the aim to elucidate the molecular conformation and structure in the active conducting state. For a long time, attempts to solubilize doped polymers have been unsuccessful. One can rationalize the insolubity with the presence of the species responsible of the conductivity, i.e. free radicals, radical ions and ions. These reactive species participate to side reactions, i.e. cross-linking, decreasing the tractability of the material.

The first system which has been dissolved in the doped state is AsF_5^- doped poly(p-phenylene sulfide) in AsF_3 (14). The solutions in AsF_3 can be diluted or concentrated in a large range of concentration without precipitation. The solution is stable suggesting that cross-linking is not occuring to any significant degree. Even with that unusual solvent, this system is the first to show that it is possible to dissolve conducting polymers.

Since the discovery that Poly-n-alkylthiophenes can be solubilized (15), they have been the subject of intensive studies. The stability of the polymer together with the possibility to study the doped polymer in solution are the origins of the interest. Besides, poly-n alkylthiophene films exhibit drastic change as shown with the thermochromism which varies with the size of the substituent (Salaneck et al, this book). Moreover poly-n alkylthiophene provides an unique opportunity to investigate the chain conformation of a conjugated polymer with a conjugated backbone markedly different from the substituted polydiacetylene or polyacetylene.

Before investigating the properties of the doped polymer solutions, a good knowledge of the chain statistics in the neutral state in solution is needed. SANS experiments were first performed on poly-butylthiophene in deuterated nitrobenzene (16), then in deuterated THF(17). Two types of samples were used, one with $M_w = 49000$ and a weight distribution $M_w/M_n = 1.5$ used for the neutral solutions in nitrobenzene, and another with $M_w = 22000$ and $M_w/M_n = 5$ for the doped solutions and the neutral solutions in THF.

For the solution in deuterated nitrobenzene, even at a concentration as low as 0.1 mg/cm^3 a significant attractive interaction between chains is observed at room temperature This strong attractive interaction prevents one from getting quantitative information on the single chain conformation. Therefore experiments have been performed at T = 65°C, with the aim to decrease the interchain interactions through an increase of the thermal fluctuations. The scattering curves obtained at three different temperatures are reported in Figure 2.

Inset in figure 2 illustrates the difference between the scattering curve measured at 65°C and the calculated structure factor. A residual interchain interaction is still observed in a range of scattering vector q lower than 0.035 $Å^{-1}$. Nevertheless, the single chain conformation can be obtained with the help of the constraints given by the contour length $L \cong 1400$ Å ($M_w = 49000$) and the mass per unit length $M_l = 35.4$ g mole^{-1} $Å^{-1}$. Incidentally the absolute measurements made with SANS do not allow to discriminate between the aromatic and quinoïd structure. A M.N.D.O calculation gives a monomer unit length a = 3.91 Å for the aromatic structure and a = 3.89 Å for the quinoïd structure. This difference is too small for the accuracy of SANS (typically 5 to 10%). The best fit obtained with the wormlike chain model (see chapter 3) gives $M_l = 35$ g mole^{-1} $Å^{-1}$, a statistical length b = 55 Å, and a mean square radial radius of gyration $R_c = 3.5$ Å. The latter value together with a measure of the height of the plateau in a q^2 g(q) plot in the range 0.18 to 0.25 $Å^{-1}$, give a total geometric extent for the lateral groups L_T = 11-12 Å. That is, the side groups are extended and the polythiophene is in a trans conformation.

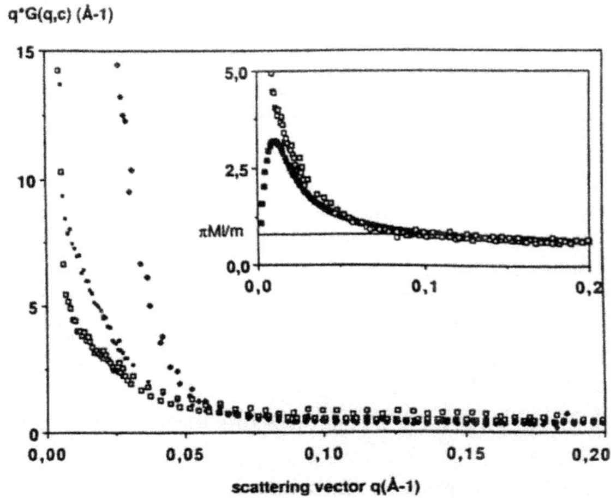

Figure 2 : qg(q) plot of p-3BT neutral chains in nitrobenzene : T = 25°C (•), T = 55°C (■), T = 65°C (□). Inset : qg(q) plot of p-3BT at T = 65°C (□) and theoretical curve (■) computed with the constraints given in the text. (from reference 16)

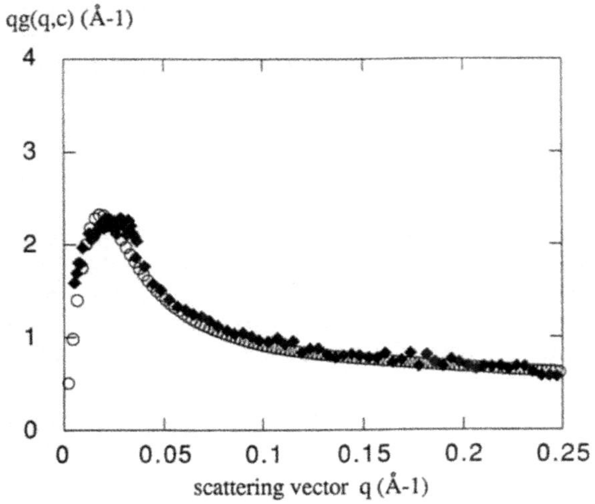

Figure 3: qg(q) plot of poly-3butylthiophene in the good solvent THF. Comparison between the worm-like chain model (§ 3-4) (o) and experimental data (♦) gives a statistical length b = 45Å.

The conformation of poly-3-butylthiophene in nitrobenzene at 65°C shows a fairly flexible behavior. The value is not much larger than for a typical flexible polymer, like polystyrene ($b \approx 20$ Å). In contrast to the semirigid conjugated polymer polydiacetylene ($b = 300$ Å), the conformation of poly-3-butylthiophene shows that a polymer with a conjugated backbone structure can be flexible. One might argue that interchain interaction prevents an accurate measurement of the statistical length. The typical q^{-1} behavior could be obscured at small q by the intensity increase due to this interaction, so that the true statistical length could be larger than found above.

This result is clearly important enough to be confirmed by another experiment. It has been done with the $M_w = 22000$ sample in deuterated THF. The solvent is a good one for poly-3-butylthiophene, allowing to measure a single chain conformation at room temperature (this is not the case with the nitrobenzene even for this molecular weight). Figure 3 shows the scattering curve obtained. The results are : a mass per unit length identical to the previously measured one, an extension of the side group slightly lower $R_c = 3$ Å, giving $L_T = 10$ Å, and a statistical length $b = 45$ Å. In the good solvent THF the polymer appears even more flexible. One can speculate that the more flexible conformation is due to a slightly smaller extension of the side groups (see § 7-1) or a decrease, in good solvent, of the attractive interaction between chain. In any case, the important result is the existence of at least one case of flexible conjugated polymer.

6-2-2. Charged poly-n-alkylthiophene in solution

As discussed in part 6-1, when we are considering charge storage on the polymer, the question of the respective stability of polarons or spinless bipolarons arises. The stability of the charge configuration on the chains depends on numerous parameters. We can mention at least three factors: the electrostatic repulsion between charges, the chain conformation, and the stucture of the medium. To discriminate between these different factors is not an easy task, suffice to remember that in the case of polyaniline films two different experimental works lead to opposite conclusions concerning the effect of the disorder on the charge storage (§ 6-1-1).

In this part, we are dealing with the properties of doped polymers in solution. Here we report experiments measuring the screening effect on the electrostatic repulsion between polarons, then SANS experiments giving the chain conformation in the doped state.

It is worth recalling first some useful concepts applied to polyelectrolyte solutions. These are solutions of macroions with counter-ions insuring the electroneutrality. The macroions consist of polymer chains with fixed ionized groups as polystyrene sulfonate. The counter-ions are more or less condensed around the ionized groups of the chain depending on the dielectric constant of the solvent and the linear density of charge along the chain.

A length characteristic of the solvent screening effect is the Bjerrum length:

$$Q = \frac{e^2}{\varepsilon k_B T} \tag{6-1}$$

with e the elementary charge, ε the dielectric constant of the solvent, k_B the Boltzman constant, and T the temperature. For nitrobenzene $\varepsilon = 35$ gives $Q = 15$ Å.

An important parameter is the ratio between the Bjerrum length and the average distance A between the charges along the chain : $\xi = Q/A$. G.S. Manning demonstrated that for a one-dimensional distribution of charges, in the strong coupling limit corresponding to an average distance A lower than Q, $\xi > 1$, a partial condensation of the counter-ions occurs (18). A fraction of the charges $1 - \xi^{-1}$ is neutralized, and the segment of length A bears an effective linear charge density $\rho\xi^{-1}$, with $\rho = q/A$. In other words the chains bearing the charges q separated by the length A are equivalent to chains with charges q separated by the length Q. The mobile counter-ions can be in first approximation analysed with the Debye-Hückel theory. The screening of the electrostatic potential is characterized for the case of monovalent counter-ions by the Debye length κ^{-1}:

$$\kappa^2 = 4\pi Q n_c \qquad (6\text{-}2)$$

the screening effect being due essentially to the uncondensed counter-ions. For the case of partial condensation the total counter-ions concentration n_c is replaced by $\xi^{-1}n_c$. Incidentally the Debye-Hückel theory strictly applies if the Coulomb interaction between two ions is smaller than k_BT, $e^2/\varepsilon r \ll k_BT$, where r the mean distance between ions scales like $n^{-1/3}$ (19). Following this criterion, the concentration of counter-ions or dopants must be :

$$n \ll \left(\frac{\varepsilon kT}{e^2}\right)^3 = Q^{-3} \qquad (6\text{-}3)$$

At the local scale, the Coulomb interaction between charges increases the local rigidity of the chain. T. Odijk and A.C. Houwart calculated that the persistence length L_p of a polyion near the rod limit can be given by the sum of two contributions (20) : $L_p = L_0 + L_e$, with L_0 the intrinsic rigidity and L_e the rigidity due to the electrostatic repulsions. The magnitude of L_e depends on the counterion condensation or effective charge of the polyion.

In the strong coupling limit Q > A, Odijk gives the expression for the electrostatic persistence length (21) :

$$L_e = \frac{1}{4\kappa^2 Q} \qquad (6\text{-}4)$$

Without condensation, even partial, of counterions, the electrostatic persistence length is (21)

$$L_e = \frac{Q}{4A^2\kappa^2} \qquad (6\text{-}5)$$

This description of the macroion conformation applies to isolated chains in dilute solution, that is to non-interacting chains. For large enough concentration, with an overlap of the chains, different models have been proposed describing the structure of the solution (22-23). It is beyond the scope of this chapter to discuss these models. We only note that the osmotic pressure for polyelectrolyte solutions is larger than for neutral polymer solution. The osmotic forces decrease the density fluctuations at large scale (at low scattering vectors q), which implies a scattering intensity going to zero at q = 0, while at the local scale a rodlike structure providing a q^{-1} behavior is expected. Therefore a peak in the scattering curves must be observed at the overlap concentration. The definition of the peak depends on the screening, by adding salt in the solution, the peak disappears (24).

For doped polymer there is an additional effect, the charges along the chains are mobiles, and the way the charges are stored depends on the screening strength. This is a major difference with polyelectrolyte which have fixed ionized groups on the chains.

The screening effect upon the charge storage has been thoroughly investigated by M.J Nowak et al (8,25). The polymer is poly-3-hexylthiophene with a molecular weight about 41000. A first experiment has been performed in chloroform ($\varepsilon = 4.73$). In the dilute regime, for $c < c_0 = 1$mg/cc (with c_0 the overlap concentration), the ESR measurements do not show the existence of spins. The authors concluded that the charges are stored as spinless bipolarons. By increasing the polymer concentration for a given doping level ($y = 0{,}08$), i.e. 8% of charge per thiophene ring, they observed an increase of the number of spins per thiophene ring (see figure 4). In term of polarons and bipolarons, this result suggested that part of bipolarons are broken to form polarons.

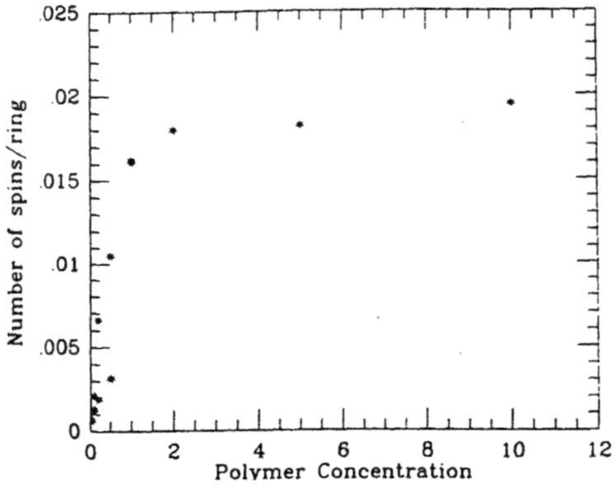

Figure 4 : Number of spins per thiophene ring as a function of doped polymer concentration (in chloroform) at a fixed dopant level ($y = 0.08$). The polymer concentration is given in units of c_0, where $c_0 \sim 6\ 10^{-3}$ M. (From reference 25).

As noted above, the chain conformation is very dependent on the screening of the Coulomb interaction between charges. In the same way the recombination of two charges to form a bipolarons is drastically dependent on the electrostatic interaction between polarons. To rationalize the change of charge storage, the authors consider, to first order, that the polaron is screened by the proximity of its counterion, a second-order effect including the solvent itself. By increasing the polymer concentration, at the same doping level, the counterion becomes more loosely bound to its charge on the polymer because there are more nearby ionic species : the other polyions and counterions. The electrostatic interaction between charges along the chain becomes more efficient, stabilizing a polaron structure.

In order to test the above proposition two other experiments were performed (25). One was to use a more polar solvent, methylene chloride ($\varepsilon = 9.73$), such that interaction between the charge and its counterion in the dilute regime is weaker. In contrast to P3HT in chloroform, spins are observed suggesting that part of the charge is stored in a polaron structure in the dilute regime. The other concerns a common test used for polyelectrolyte solutions. The screening effect can be modulated by adding salt, κ^2 increases linearly

increases linearly with the concentration of salt, then again a weaker interaction between charge and counterion. The measurements were performed with P3HT in chloroform at concentration $c << c_0$, where in the salt free solution no spin are observed. By adding salt (tetrabutylammonium perchlorate) a significant increase of the number of spin per charge is observed (see figure 5).

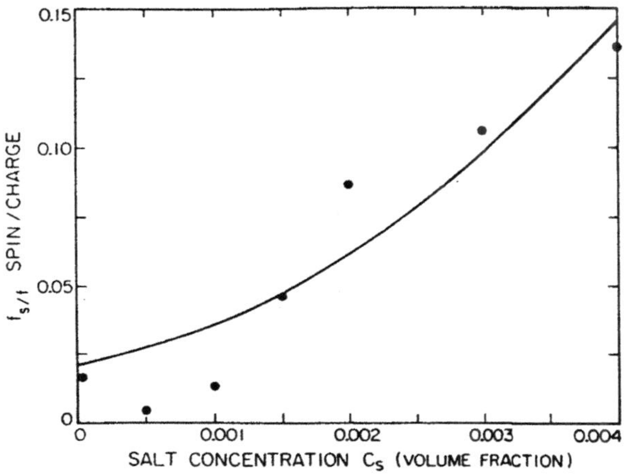

Figure 5 : Magnetic susceptibility versus concentration of added salt solution at a fixed polymer concentration, $c = 0.1c_0$, and fixed dopant level, y = 2%, in chloroform. A 0.1 M salt concentration is equivalent to a volume fraction of 2 10^{-3} on the abcissa. (From reference 25).

These experiments show the effect of the counterion on the structure of the charges stored on the polymer chain. On the other hand the electrostatic interaction between charges, as for polyelectrolyte, must change the chain conformation. Then, one can ask the influence of the chain conformation in the neutral state on the structure of the charges and the way the chains are sensitive to the different kinds of charges.

The structural study at small angle scattering were performed on doped P3BT with $M_w = 22000$ (16, 26). Figure 6 shows the scattering curves obtained for two polymer concentrations 0.5 mg/cc and 2.4 mg/cc. The dopant molecules $NOSbF_6$ were dissolved in deuterated nitrobenzene and added to the polymer solution at room temperature. In the experiment reported in figure 6, the nominal dopant concentration is one dopant molecule per monomer unit. The neutrality of the solution is insured by the SbF_6^- counterions. Except the near-infrared absorption spectrum showing a maximum absorption at 830 nm and a broad peak around 1800 nm, we do not have quantitative information on the doping process. In the following, the doping level refers to the nominal dopant concentration, but this does not necessarily mean that a complete doping is achieved.

The structure of the doped chains in solution leads to completely different scattering curves depending on the neutral-polymer concentration. A q^{-2} behavior is observed for $c_p = 2.4$ mg/cm^3, while for $c_p = 0.5$ mg/cm^3 a q^{-1} behavior is observed in the investigated q range, i.e. down to $q = 8\ 10^{-3}$ Å^{-1}. Keeping in mind that the P3BT is a flexible coil with a statistical length equal to 55 Å in nitrobenzene and 45 Å in THF, the doped chain in the dilute regime is much more rigid, showing in the q range measured a rodlike behavior.

Using the relation qb = 7 (see § 3-4), with $q_{min} = 8\ 10^{-3}$ Å^{-1} one can give a lower limit for the statistical length : b = 850 Å.

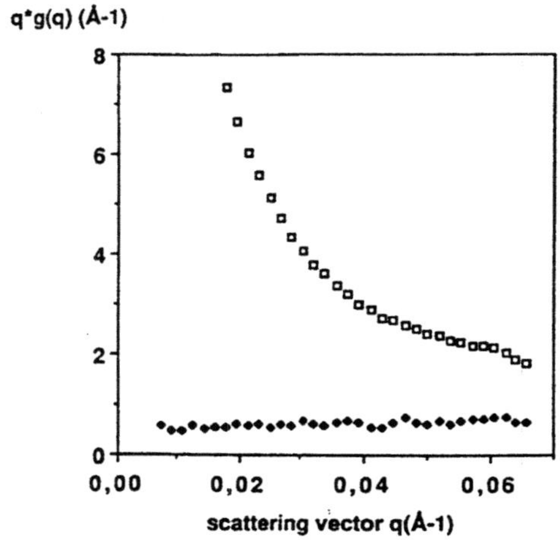

figure 6 : qg(q) plot of the chain p-3BT doped with $NOSbF_6$: $c_p = 0.5$ mg cm^{-1} and c_d = 100% (♦), $c_p = 2.4$ mg cm^{-1} and c_d = 100% (□). (from reference 16).

In the framework of polyelectrolyte theory, the doped solution corresponds to the strong-coupling limit : A < Q. From optical spectroscopy studies (27), it is estimated that the effective doping level does not exceed 50%, then A ~ 8 Å. Equation 6-4 with $\kappa^2 = 4\pi A n_c$, where n_c is the nominal dopant concentration, gives a statistical length b ~ 150 Å, markedly lower than the measured one. Even if we assume a 100% doping level, the calculated statistical length b ~ 300 Å is still far too low.

An alternative is to consider the rod-limit case without condensation, even partial of counterions. Using equation 6-5 with $\kappa^2 = 4\pi Q n_c$ we obtain for one charge per thiophene ring a statistical length b ~ 1200 Å, while for 50% doping the length is reduced to b ~300 Å.

Except if we consider a complete doping process at a high level without any condensation of the counter-ions, the theory of polyelectrolyte solutions is unable to describe the rod like behavior of the doped polymer. One can emphasize a major difference between polyelectrolyte and doped polymer. The total persistence length is the sum of the electrostatic persistence length L_e, and the intrinsic persistence length L_0. For the case of polyelectrolyte there is no reason for expecting a change of L_0 when ionized groups are added. For conjugated polymer the situation can be completely different, the flexible behavior of P3BT in the neutral state can be drastically changed with the addition of charges. In other words together with a Coulomb repulsion stretching the polymer, a change of the electronic configuration from aromatic to quinonic structure and a change on the average torsion between monomer units can increase the "intrinsic" statistical length L_0.

The differences with conventional polyelectrolytes may be also found from the behavior of the upper curve in figure 6. If a correlation hole is present it should be observed at a concentration $c_p = 2.4$ mg/cc, since the overlap concentration will be equal to $c^* \sim 6M_w/\pi N L^3 = 0.3$ mg/cc for the chain rodlike state. Instead, a large scattering intensity is observed with a q^{-2} behavior. It follows that the density fluctuations are much larger than the average density, which is inconsistent with an increase of the osmotic pressure. Nevertheless, we must add that due to the strong attractive interaction in the neutral state in the solvent nitrobenzene, at concentration as high as 2.4 mg/cc, the doping is performed in an aggregate structure, then the homogeneity of the doping remains questionable. To insure an identical doping on the chains, the dopant should be added to an isolated chain state.

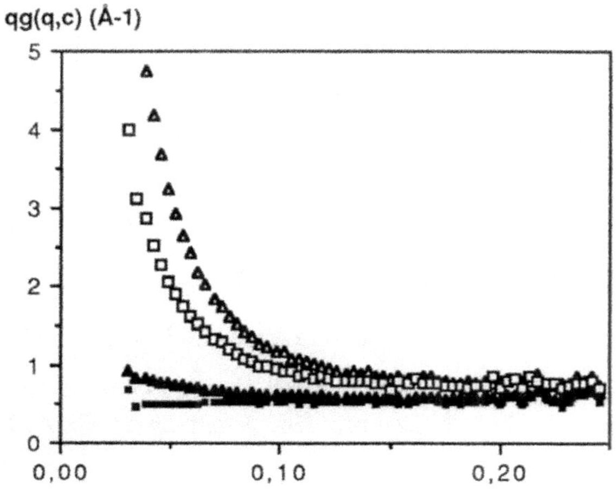

figure 7 : $c_d = 5\%$ doped at 25°C (Δ), $c_d = 5\%$ doped at 75°C (□); $c_d = 50\%$ doped at 25°C (Δ), $c_d = 50\%$ doped at 75°C (■)).

In order to investigate possible consequences of doping chains either in the aggregate structure, or in the isolated chain state, solutions have been doped in two different ways (26) :

-Type 1: as before, dopant molecules are added to the neutral solution of P3BT at room temperature.

-Type 2: chains are dissolved at T = 80°C in a deuterated nitrobenzene containing the desired number of dopant molecules $NOSbF_6$.

In both case the measurements are then performed at room temperature. A Zimm plot made on the P3BT (with $M_w = 22000$) neutral solution at 70°C shows that the interaction between chains becomes negligible (down to 6 10^{-3} Å^{-1}). Scattering curves measured for concentrations up to 2mg/cc are very close to the one extrapolated for $c_p = 0$ (26). Therefore the second method can be considered as a good one for doping chain in the isolated state in the solvent nitrobenzene.

Figure 7 shows the scattering curves obtained for two different doping levels and for the two doping processes. With increasing doping level the scattering intensity decreases leading to a more rigid behavior. But also the way the doping is achieved drastically changes the scattering curves. The most obvious change concerns the chains doped at 50%. In contrast to the doping at room temperature, a q^{-1} behavior is observed in the overall q range. At least down to 0.035 Å^{-1}, a rodlike conformation is obtained at high doping level even for a polymer concentration as high as 2 mg/cc.

The polymers doped in solution exhibit a more rigid behavior leading to a rodlike structure at high doping level. Moreover the study of the chain conformation in solution indicates a marked change depending on the method used for doping the chain. The later result emphasizes the importance of the structure of the chain in the neutral state. One can rationalize the difference observed as an evidence of a non homogeneous doping along the chain in the aggregate structure, while in the isolated chain state the rigid structure measured is understood as a doping which is much better achieved. This result indicates that the compactness of the polymer chains is a key factor for controlling the doping of the macromolecules.

References

1 R.R. Chance, J.E. Frommer, in "Electrically Conductive Polymers", Encycl. Polym. Sci. and Eng.J. Wiley & Sons, **5**, 462 (1986).

2 S. Stafström, J.L. Brédas, J. of Molecular Structure, **188**, 393 (1989).

3 J.L.Brédas, R.R. Chance, R. Silbey, Phys. Rev. B, **26**, 5843 (1982).

4 K. Misoguchi, M. Nechtschein, J.P. Traverse, C. Menardo, Phys. Rev. Let., **63**, 66 (1989).

5 K. Holczer, J.P. Boucher, F. Devreux, M. Nechtschein, Phys. Rev. B, **23**, 1051 (1981).

6 F. Devreux, H. Lecavelier, Phys. Rev. Let., **56**, 2585, (1987).

7 S. Stafström, J.L. Brédas, A.J. Epstein, A.G. H.S. Woo, D.B. Tanner, W.S. Huang, A.G. MacDiarmid, Phys. Rev. Lett., **59**, 1464 (1987).

8 M. Nowak, S.D.D.V. Rughooputh, S. Hotta, A.J. Heeger, Macromolecules, **20,** 965 (1987).

9 F. Zuau, M. Angelopolous, A.G. MacDiarmid, A.J. Epstein, Phys. Rev. B, **36,** 3475 (1987).

10 J.P. Aimé, M. Nechtschein, I.L.L. report 9-11-202.

11 A. Lemoine, M. Nechtschein, J.P Traverse, Synth. Met., **29**, E457 (1989).

12 M.E. Jozefowicz, R. Laversanne, H.H.S. Javadi, A.J. Epstein, J.P. Pouget, X. Tang, A.G. MacDiarmid, Phys. Rev. B, **39**, 12958 (1989).

13 Y. Nicolau, M. Nechtschein et al, ICSM'90 Tübingen.

14 J.E. Frommer, R.L. Elsenbaumer, R.R. Chance, Org. Coat. Appl. Polym. Sci. Proc., **48**, 552 (1983).

15 R.L. Elsenbaumer, K.Y. Jen, R. Oboodi, Synth. Met., **15**, 164 (1986).

16 J.P. Aimé, F. Bargain, M. Schott, H. Eckhardt, G.G. Miller, R.L. Elsenbaumer, Phys. Rev. Let., **62**, 55 (1989).

17 J.P. Aimé, P. Garrin, G. Baker, S. Ramakrishnan, P. Timmins, ICSM'90 Tübingen.

18 G.S. Manning, J. Chem. Phys., **51**, 924 (1969).

19 L.D. Landau, E.M. Lifshitz., Statistical Physics, 2nd edition (Pergamon Press, London) 1959.

20 T. Odijk, Macromolecules, **12**, 688 (1979).

21 T. Odijk, A.C. Houwart, J. Polym. Sci. Polym. Phys. Ed., **16**, 627 (1978).

22 M. Nierlich, C.E. Williams, F. Boué, J.P. Cotton, M. Daoud, B. Farnoux, G. Jannink, C. Picot, M. Moan, C. Wolf, M. Rinaudo, P.G. de Gennes, J. Phys. , **40**, 701 (1979).

23 P.G. de Gennes, P. Pincus, R.M. Velasco, F. Brochard, J. Phys. , **37**, 1461 (1976).

24 M. Drifford, J.P. Dalbiez, J. Phys. Chem. , **88**, 5368 (1984).

25 M.J. Nowak, D. Spiegel, S. Hotta, A.J. Heeger, P.A. Pincus, Macromolecules, **22**, 2917 (1989).

26 J.P. Aimé, F. Bargain, M. Schott, H. Eckhardt, G.G. Miller, R.L. Elsenbaumer, Synth. Metals, **28**, C407 (1989).

27 S. Hotta, S.D.D.V. Rughooputh, A.J. Heeger, F. Wudl, Macromolecules, **20**, 212 (1987).

7. DOES CONJUGATED POLYMER BEHAVE LIKE SATURATED ONE.

7-1. Conformation of soluble conjugated polymers : Origin of the local rigidity.

A major interest in the study of conjugated polymers in solution is the opportunity to investigate the relation between electronic properties and conformational disorder in low dimensional materials. Because it is an extremely difficult task, color transition and more generally any color change exhibited by soluble conjugated polymers were considered as very promizing features. The experimental results suggest that these materials should be model systems for exhibiting the influence of any kind of disorder. On the other hand our ability to describe the physical properties of the polymer at any scale, either by using scaling law for describing the overall chain structure or using simple free energy expansion for describing the local correlations, provides powerful tools.

Different measurements suggest a strong interplay between the chain conformation and the π electrons. For example electronic properties of conjugated polymers are very dependent of the number of monomer units in the conjugated sequence. Experiments on short oligomers show a large change of the position of the maximum absorption as a function of the number of monomer units, the infinite chain state being approached asymptotically in $1/n$ (1). On the contrary for saturated polymers, there is basically no difference between the absorption spectrum of an isolated monomer unit, in solution phase for example, and a monomer unit belonging to a macromolecule.

For saturated polymers we do not expect any change in the electronic excitation spectrum as a function of conformational disorder, while for conjugated polymers large changes are expected with a strongly modified local conformation, in other words an increase of the local rigidity.

At first glance, the statistical conformation measured for the PDA systems strongly supports a direct influence of the π-electron on the chain structure. The polymer is semi-rigid in good solvent suggesting an effect of the conjugation at the local scale. The shift and broadening of the optical absorption in good solvent has been interpreted in that way (2,3), i.e as evidence that the electronic wave functions had localized to a distribution of short conjugation lengths. Resonance Raman results were subsequently interpreted from a similar point of view (4). The "conjugation length" can be understood as a parameter describing the π electron delocalization along the conjugated backbone structure. This length is related to the average overlap between the π electron orbitals. The overlap being directly dependent of the geometry of the macromolecule, the curvature and torsional fluctuations must decrease the conjugation length (5-7) providing a direct link between the chain conformation and the electronic properties (6-10).

An immediate consequence is to consider the π electronic structure as the leading term of the local rigidity. Attempts have been made to describe the electronic contribution to the statistical conformation quantitatively and qualitatively (9,10). The fundamental hypothesis is that the local correlation between monomer units is enhanced by the electron delocalization. This hypothesis appears rather reasonable but is unable to explain some experimental results.

A completely different approach consists in taking into account explicitly the geometry of the side group. We know that substituents are needed for having soluble conjugated polymers. For the polydiacetylenes, the side groups are so long that we can consider them as short chains. In other words most of the atoms of the monomer unit belong to the pendant groups, reaching in many cases a proportion of about 90% (table 7-1). Therefore it is somewhat a crude approximation to neglect their influence upon the chain statistics.

The physical properties of the polymer are now well understood in term of scaling laws. Obviously, scaling laws do not apply at a microscopic scale and will not be very hepful for understanding the effect of the side groups. On the other hand a purely microscopic approach does not allow to predict a general behavior and could be a formidable work giving only some informations for very specific cases. Therefore one may try to reduce the number of variables, still describing the effect of the monomer unit structure. Such approach is relevant only if it is strongly supported by experimental measurements. We give first the experimental results that we believe show unambiguously the effect of the side group upon the chain statistics, then with the help of a simple model, discuss the consequence of the use of the side group for the physical properties of the polymer in good solvent and the aggregation process as well.

7-1-1. Thermal behavior of the PDA PTS12 in good solvent

The fluctuations of the atomic positions within a macromolecule are typically about 0.01 and 0.1 Å and correspond to high frequency modes. On the other hand at finite temperature the equilibrium straight shape of a long macromolecule exhibits large fluctuations and can be described as a macroscopic linear system in which statistical methods can apply. The fluctuation of the rod shape, i.e the existence of a local curvature can be obtained by using a quadratic expansion of the free energy in term of curvature (11), meaning that the straight line is the stable state at T = 0 °K :

$$F = \frac{1}{2} a \rho^2 + O(\rho^4) \tag{7-1}$$

where ρ is the curvature vector equal in magnitude to the reciprocal of the radius of curvature. a is a coefficient characterizing the chain rigidity.

The total change of the free energy due to the curvature is obtained by integration along the total contour length L:

$$\Delta F = \frac{1}{2} \int_L a\rho^2 dl \tag{7-2}$$

where l is a coordinate along the chain. Here we note that equations 7-1 and 7-2 apply for an infinitely thin thread, that is the chain is considered without any thickness.

Let θ (l) be the angle between the unit vectors along the tangents $\mathbf{t_a}$ and $\mathbf{t_b}$ at two points a and b separated by a length l along the thread (see chapter 2). The curvature vector is related to θ (l) by $\rho = d\theta(l)/dl$ and equation 7-2 may be written:

$$\Delta F = \frac{1}{2} \int_L a \left(\frac{d\theta(l)}{dl}\right)^2 dl \tag{7-3}$$

An elegant argument due to Landau and Lifshitz (11) allows to get the mean angular deviation between the tangents separated by the contour length l :

$$<(\theta(l))^2> = \frac{lT}{a} \quad (7\text{-}4)$$

and using equation 2-15 we obtain the expression for the statistical length :

$$b = \frac{2a}{T} \quad (7\text{-}5)$$

whatever the physical origin of the stiffness and the magnitude of the radius of curvature, equation 7-5 predicts a T^{-1} behavior for the statistical length of an infinitely thin thread.

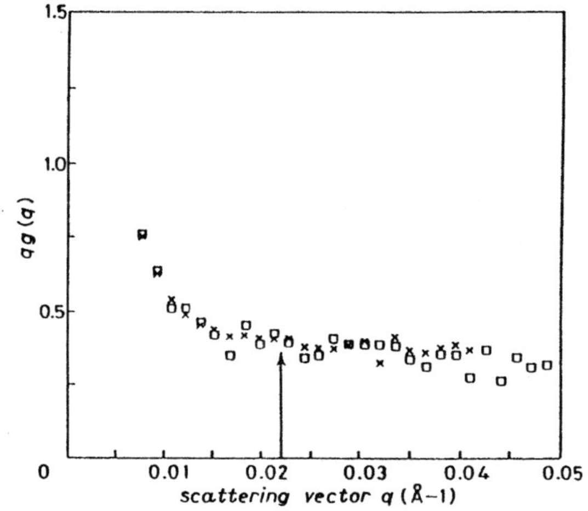

Figure 1: qg(q) plots vs q of pTS12 in deuterated nitrobenzene : (x) T = 25°C; (□,) T = 75°C.(from ref 12).

SANS gives the appropriate range of transfer momentum q for studying the change of the local stiffness. Experiments were performed on PTS12 in good solvent DMF and nitrobenzene (12). For any decrease (increase) of the chain stiffness by heating the solution, we should observe that the characteristic q^{-1} behavior of the rod conformation extends to larger (smaller) q values with respect to the one observed at room temperature : $q = 0.024$ Å^{-1}. Figure 1 shows the scattering curves measured at RT and 75°C, no variation is detected. For a statistical length equal to 300 Å at room temperature, a value equal to 255 Å is expected at 75°C while if we consider that at infinite temperature the statistical length does not go to zero, (let for instance $b_{T=\infty}$ =50 Å i.e 10 monomer units), a T^{-1} behavior gives a value equal to 265 Å at T = 75°C. Therefore we face the contradiction that the chain statistics is well described by a worm-like chain model but does not have the predicted thermal behavior.

To rationalize this apparent contradiction, a first step is to look at what happens if the macromolecule is no longer described by an infinitely thin thread (Figure 2). The finite lateral size of the polymer can be described in a very simple way : The backbone and lateral parts of the monomer form an homogeneous unit and simple geometrical forms are considered. With a circular cross section the size of the side group, or more accurately their extension in the solvent, is the unique parameter. With a rectangular cross section, we can define two additional lengths for the polymer chain, one corresponds to the extension of the side group and the other to the thickness of the monomer unit (figure 3). The latter geometrical form is able to give the essential features : the effect of the side group size and the influence of the average torsion between monomer units.

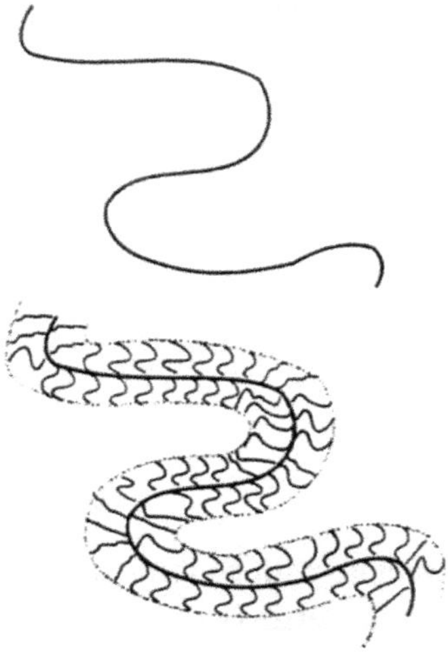

Figure 2: Sketch of polymer chains with and without side-groups. Soluble conjugated polymers have pendant groups.

An additional assumption concerns the elastic constant, which is considered independent of the side group size and also temperature independent at least within the experimentally accessible temperature range.

For a thread with finite lateral extension, equation 7-3 is rewritten (13):

$$\Delta F = \frac{1}{2} \int_L EI \left(\frac{d\theta(l)}{dl}\right)^2 dl \qquad (7\text{-}6)$$

where E is the elastic modulus, I the moment of inertia. For circular cross section $I = \pi r^4/4$, for a rectangular section the principal moment of inertia are :

$$I_1 = \frac{(a_1^3 a_2)}{12} \text{ and } I_2 = \frac{(a_2^3 a_1)}{12}$$

The expression for the statistical length becomes:

$$b = 2\frac{EI}{T} \qquad (7\text{-}7)$$

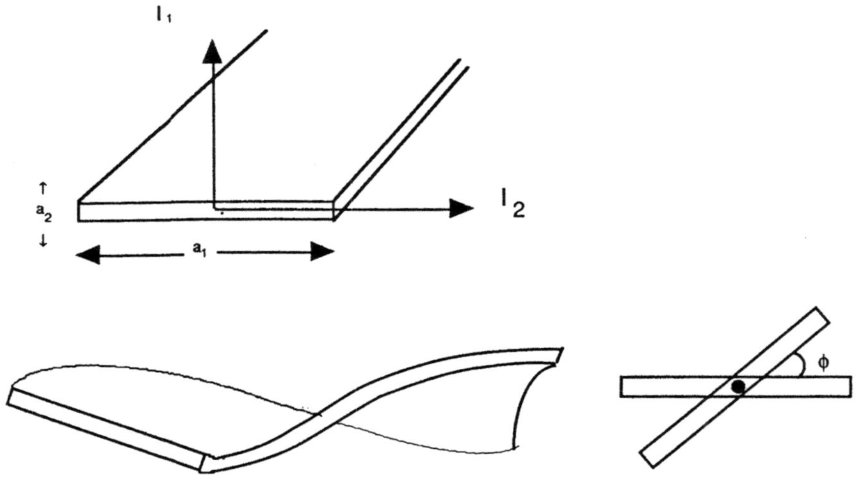

Figure 3: Model for ribbon-like structure with continuous angular deviation Rectangular cross-section with principal moment of inertia I_1 and I_2 and rotation ϕ between monomer units leads to an average effective moment of inertia.

As noted above, the rectangular section is the most suitable for describing the effect of torsion and side group conformation. The smaller dimension corresponds to the thickness of the monomer unit and we set $a_2 = 4Å$, while a_1 corresponds to the extension of the side group in the solvent. With α the angle between the plane of the applied force and the principal bending plane, the effective moment of inertia is given by:

$$I_e = \frac{I_1 I_2}{\sqrt{I_2^2 \cos^2(\alpha) + I_1^2 \sin^2(\alpha)}} \tag{7-8}$$

but looking at the elastic susceptibility of a polymer, the chain consists of randomly oriented parts with respect to the direction of the applied force. To take this into account we have to calculate an average effective moment of inertia. With β the angle between the bending plane and the principal bending plane given by $tg\beta = I_1/I_2\, tg\alpha$ (the choice of the principal bending plane corresponding either to I_1 or I_2 being unimportant), the average is obtained by summing over all possible orientations within the mean angular deviation between monomer units. In a continuous limit we obtain an average effective moment of inertia governing the curvature with the integral (12)

$$\langle I_e \rangle = \frac{1}{\phi} I_1 \int_0^{\phi} \sqrt{1 - k^2 \cos^2(\beta)}\, d\beta \tag{7-9}$$

where $k^2 = 1 - (I_2/I_1)^2$. This is an incomplete elliptic integral of the second kind for $0 < \phi < \pi/2$ and a complete one for $\phi = \pi/2$.

The two limiting cases are given by $\phi = 0$ (the planar ribbon structure) and $\phi = \pi/2$ corresponding to a $\pi/2$ rotation between neighboring monomer units.

For $\phi = 0$, the curvature is governed by the lowest moment of inertia, here noted I_2 and the statistical length is a linear function of the lateral extension a_1. But for any value of ϕ different from zero, a contribution due to the largest moment of inertia I_1 enters and the larger the magnitude of ϕ, the larger the average effective moment of inertia.

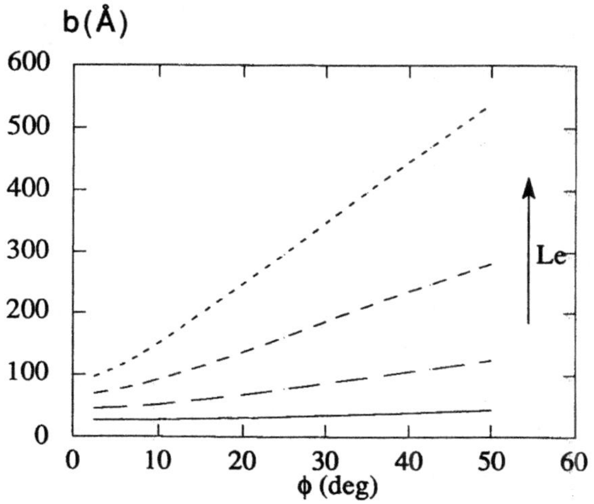

Figure 4: Variation of the statistical length as a function of the size of side groups and torsion between monomer units with the lateral extensions : —— $L_e = 8Å$, — — $L_e = 12Å$, - - $L_e = 16Å$, - - $L_e = 20Å$.

Equation 7-9 predicts a non linear dependence with ϕ, an increase of the torsion between monomer units increases the statistical length. Figure 4 shows the effect of the mean angular deviation between monomer units together with the effect of the lateral extension. Equation 7-7 is now rewritten

$$b = \frac{2E\langle I_e \rangle}{T} \tag{7-10}$$

Equation 7-10 emphasizes two points: the side groups behave like short chains swollen in the solvent. Neighboring lateral short chains repel each other, straightening the side groups. On the other hand, the packing along the backbone induces a torsion between side groups rather than a planar ribbon structure.

Only a microscopic computation would give a quantitative information on the magnitude of the torsion. Nevertheless equation 7-10 predicts a very interesting feature: an increase of the angular deviation between monomer units does not cost much energy as it has been computed for the polydiacetylene backbone (14), the calculated barrier height for a $\pi/2$ rotation is O.6 kcal mole^{-1}, i.e an energy barrier of the order of T. A direct consequence is that heating the solution increases the torsional fluctuations then the local rigidity of the macromolecule (figure 4). We now have two opposite effects with the temperature change : the curvature fluctuations behave like T^{-1} while the torsional fluctuations through $<I_e>$ increase the statistical length when the solution is heated. Therefore it becomes very difficult to observe any change on the scattering curves since the two effects can cancel each other.

7-1-2 Experimental evidence of the influence of the side-group extension.

These experimental results do not provide a definite proof showing that the side groups are the relevant parameters for understanding the local stiffness. We can still object that the magnitude of the statistical length is due mainly to the conjugated backbone structure providing a statistical length in the order of the one observed whatever the temperature used. Considering the π-electron delocalization, the contribution of the transfer integral between neighboring sites, the band width, is of the order of one eV, that is almost two orders of magnitude larger than T. Even if we consider torsional fluctuations, reducing the transfer integral by a factor $\cos\phi$, a value of 10 or 20 degrees still gives an order of magnitude. In other words, the temperature effect becoming negligible within the temperature range available for studying soluble conjugated polymers in good solvent, we should not expect to measure any change.

Additional measurements are needed in order to assess whether the π electrons have a major influence upon the local correlations or not. The first ones were obtained on poly-n-alkylthiophenes, where it was found that the statistical length of an isolated poly-3butylthiophene chain is equal to 45Å (see chapter 6). Obviously the conjugated backbone of polythiophene is markedly different from polydiacetylene or polyacetylene, so a direct comparison between the different statistical conformations may not be meaningful. Nevertheless, it remains that the observed flexible behavior of the poly-3butylthiophene is a first and direct evidence that a conjugated backbone structure does not necessarily mean a stiff chain behavior.

Table 7-1 : Some examples of substituents for soluble polydiacetylenes.

Name	Ref.
$(CH_2)_3OCONHCH_2CO_2C_4H_9$	P3BCMU
$(CH_2)_4OCONHCH_2CO_2C_4H_9$	P4BCMU
$(CH_2)_4OSO_2C_6H_4CH_3$	PTS12
$(CH_2)_9OCOCH(CH_3)_2$	1
$(CH_2)_9OCOCH_2(CH_3)_2$	2
$(CH_2)_9OCOCH_2$(1-naphthyl)	3

Polydiacetylene have provided additionnal experimental results showing the influence of the side groups. Plachetta et Schulz (15) synthesized PDA with very long substituents (see table 7-1) and studied them by light scattering. No structural information on the side group conformation can be obtained by that method, only a relative change of the radius of gyration as a function of the nature of the substituent can be observed. For contour lengths roughly equal to those of P3BCMU or PTS12, the PDAs made by Plachetta and Schulz give a much larger radius of gyration (see table 7-2). Using equation 3-46, statistical lengths respectively equal to 550Å and 660Å are deduced. The backbone structure being identical in the series, the magnitude of the statistical length is seen to be strongly dependent of the size of the side group. Moreover the visible optical absorption is similar to the one obtained with the others PDA while the statistical length is about twice larger. This suggests that the optical properties due to the π electrons are unsensitive to the magnitude of the statistical length measured. In other words, the magnitude of the statistical length is not very hepful for understanding the optical properties of the conjugated polymers in solution, and even could be confusing if we wish to use the concept of curvature fluctuations.

Table 7-2 : Conformation of polymers with substituent in good solvent. (*) Statistical lengths measured with small angle neutron scattering.

Polymer	L_w (nm)	R_g (nm)	b (Å)	b* (Å)
PTS12	614	70.8 (2.4)	380	300
P3BCMU	1 386	94.0 (2.4)	260	310
P4BCMU				310
p-3BT(nitrobenzene)				55
p-3BT(THF)				45
1	832	110.0 (5)	550	
2	1 123	131.0 (5)	550	
3	882	126.0 (5)	660	
poly-1octene			65	45
poly-1decene			85	55

The above results are strong evidence of the influence of the finite lateral size of the monomer unit. This is also shown by structural study of saturated polymers. The electronic properties of these systems are independent of the size of the chain : they are very local properties, i.e at the atomic or monomer unit scale. Therefore saturated polymers are model systems for measuring the effect of the lateral extension upon the local rigidity. This has been achieved with the most simple systems available: poly-1octene and poly-1decene. The backbones are the same while the side groups differ by two CH_2 units.

Light scattering measurements have been done on samples with different molecular weights. For a given contour length, poly-1decene has a larger radius of gyration than poly-1octene, hence a larger statistical length (16). This result was confirmed by SANS. The scattering curves obtained give the best fit for poly-1octene and poly-1decene with $\pi M_l/m = 0.4$, $b = 45$Å, $R_c = 4$Å and $b = 55$Å, $R_c = 5$Å respectively (figure 5). The measurements show a direct relation between the lateral extension of the pendant group and the magnitude of the statistical length. Moreover the typical statistical length for saturated polymers with no or short side-chains is about half that observed for poly-1decene and poly-1octene. The conformation of the side-group in the solvent implies that the substituents are alternatively on either side of the backbone. The extended state of the substituents can be seen as an effect of the repulsion between side groups meaning that the monomer unit length becomes a relevant parameter. For saturated polymers the unit length is about 2.5Å, for the polythiophene 3.9Å and for PDA 5Å. Therefore we expect an almost fully extended state of the side-group for poly-1octene and poly-1decene which is in good agreement with the experimental results : 14Å and 17Å , while 18 and 22 Å are the theoretical total extensions respectively for poly-1octene and poly-1decene.

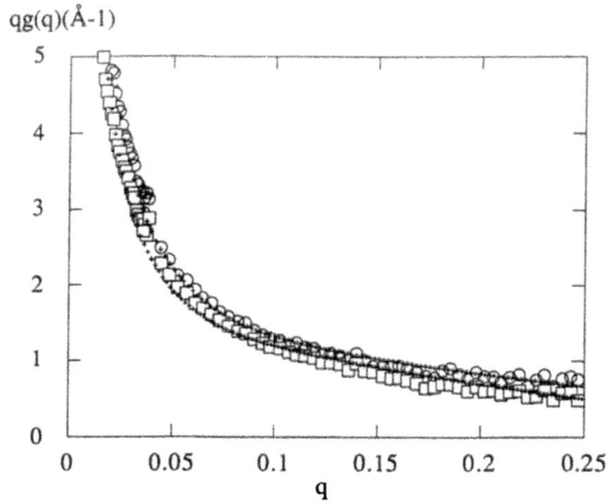

Figure 5 : qg(q) plots of poly-1octene: o, poly-1decene: □,. Theoretical curves calculated with the parameters given in the text : +.

In figure 6 are reported the statistical lengths as a function of the lateral extension of the side groups in good solvent. The extension of the side group are either measured with SANS or calculated from the chemical structure. We also report computed value using equations 7-9 and 7-10 with $E = 10^{10}$ dyne cm^{-2} and $\phi = 12°$. The general behavior is surprisingly well reproduced. The model demonstrates that macromolecules with substituent side-groups of significant length are fairly well described as a ribbon like structure with torsional fluctuations. The physical picture corresponds to an angular deviation between monomer units randomly distributed along the backbone leading to an effective moment of inertia $<I_e>$. The main result is a statistical conformation dependent of the side group size and structure. Moreover, optical properties and conjugation length should not be very sensitive to the magnitude of the statistical length as it is shown with the series of PDA, then suggesting that the torsion between monomer units is the pertinent parameter for describing the optical properties of conjugated polymers in solution.

Angular deviations may not be randomly distributed along the backbone. Various structures can be obtained (§ 3-4 and discussion below). From that point of view, the substituted polyacetylene poly-2octyne shows an effect of the side-group slightly different from the one described above. The methyl unit is very close to the backbone for poly-2octyne while it is not for poly-1octene. Therefore it is very reasonable to expect a large torsion between monomer units, hence a rigid behavior not related to the π electronic structure. On the other hand, poly-2octyne has a maximum of the optical absorption located at 2800Å. As far as the position of maximum absorption measures the π electron delocalization, we could say that the π electrons are localized at the monomer unit scale, a behavior close to the one of saturated polymers. Poly-2octyne has a statistical length equal to 130Å. This is much larger than for the poly-1octene or even poly-1decene, but far from being that of a rigid rod as can expected for a polyacetylene system.

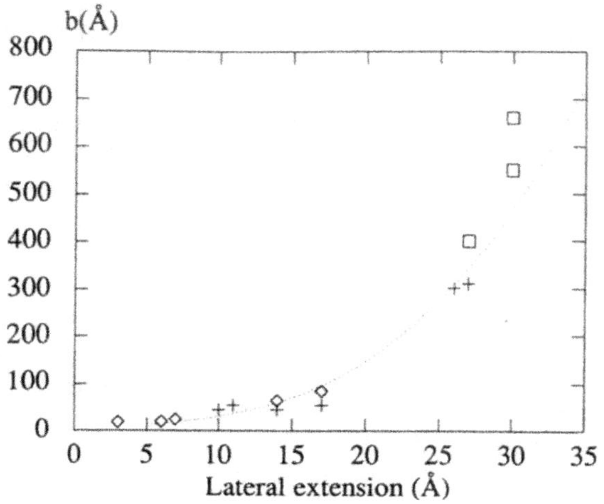

Figure 6 : Variation of the statistical length as a function of the lateral extension. Light scattering data : Saturated polymers (◊), conjugated polymers (□); Neutron scattering data for conjugated polymers (+); dotted curve corresponds to calculated values with equations 7-9 and 7-10.

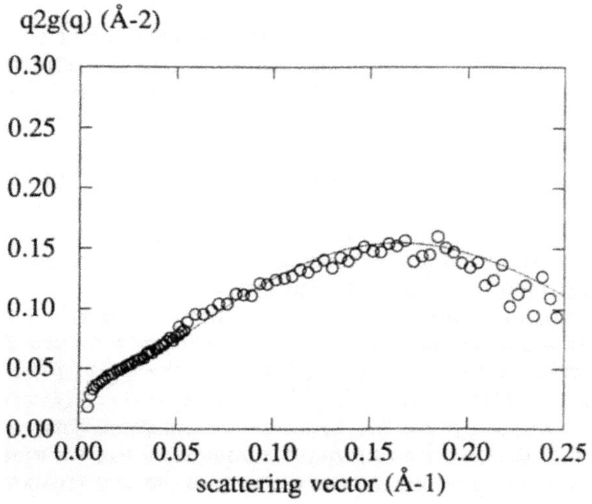

Figure 7 : $q^2g(q)$ plot of poly-2octyne : o. The theoretical curve corresponds to a worm-like chain form factor with a cross section given by an helical structure with a central hollow part (see text). The pitch is equal to 20Å corresponding to 8 monomer units.

The scattering curve at large q values, between 0.15 and $0.25Å^{-1}$, shows a pronounced decrease (figure 7). The decrease is better described with a cylindrical cross section, an even better fit is obtained considering a local helical order with a central hollow part (ref 17 and § 3-5). This means that equation 7-9 is not appropriate for describing the average effective moment of inertia and the chain stiffness. The limiting case is given by a cylinder cross-section providing a stiff behavior, for example using a radius R = 7Å and E = 10^{10} dyne cm^{-2}, $<I_e> = \pi R^4/4$ gives b = 900Å. An helical structure with an external radius R_e = 8.5Å and an internal radius R_i = 1.28Å exhibits a less compact structure, therefore a more flexible behavior. The present conformation can be seen as an intermediate structure between the ribbon like and cylinder like structure providing a coherent picture for describing the chain stiffness together with the angular correlation at local scale.

7-2. Aggregation process for conjugated polymers.

Consider the dissolution of a system of macromolecules in rigid, rod-like or helical like conformation initially packed in crystalline array. The process usually can be separated, hypothetically at least, into two steps. One involves randomization of the chain configuration, the rigid, rod-like (or helical) conformation being replaced by the so-called random coil. The other consists in the actual mixing of solute with the solvent. In actual situations the two usually are very difficult to distinguish and often take place simultaneously.

Similarly, aggregation from solution may involve the reverse of the two above-mentioned processes. Understanding the packing of chains in the structures thus formed (cast films, possibly gels, etc..) requires an accurate understanding of both processes.

7-2-1. Observed conformations for polymers in good solvent

A major question is the stability of the chain conformation in solution or in the amorphous state. In other words does it exist a stable rigid or helical conformation in solution implying a mixing with solvent without randomization of the chain conformation of the crystalline state. For saturated polymers, the steric repulsion of the pendant groups governs the chain conformation. If the polymer is isotactic, i.e., if successive asymmetric carbons possess the same tetrahedral configuration, then the trans state is excluded by interferences between neighboring side-groups. The resulting conformation is the helix proposed by Natta et al for various crystalline isotactic polymers on the basis of X-ray diffraction (18). The steric interactions dictate the same sign of rotation for each bond pair between substituted carbons, providing the regularity.

In solution or in the amorphous state, no such regularity is imposed. Helical arrangement is preserved if the angular deviation remains of the same sign. For observing such a conformation in solution, we do not need that the sign of the angular deviation propagates throughout the total contour length, but need only to observe a persistence in curvature. In other words a preferred torsional orientation between monomer units imposes a prefered orientation of the curvature fluctuation. For saturated polymers, there is no evidence of such a long range angular correlation, the experimental results show a very flexible behavior for most polymers, the difference between a completely random angular deviation and some short ordered parts containing a few monomer units being unimportant. The chain could be described as a succession of helical segments alternating in sign of rotation and varying in length, but in average consisting of only a few bond pairs between points of reversal of rotation direction. In both cases, the average chain conformation can be described as a random coil. An interesting system is given by the syndiotactic PMMA chain, syndiotactic PMMA exhibits a persistence in curvature reminiscent of an helical conformation (19) (figure 8). The scattering curve suggests a conformation different from the usual one observed for saturated polymers, the oscillations are characteristic of an helical worm-like chain structure (ref 19,20 and § 3-4).

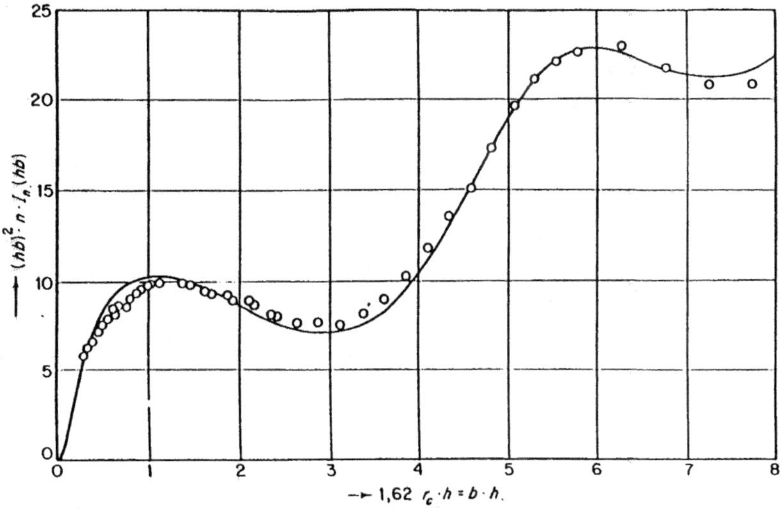

Figure 8 : Kratky-plot of the experimental Small Angle Xray Scattering curve of s-PMMA. The calculated scattering curve corresponds to a chain with a persistence in curvature (from ref 19).

Therefore in most cases polymers are random coil in solution, but exceptions exist in biological systems. Those are macromolecules which exhibit helical conformation in solution. They can form simple double or triple helices (21). For polypeptides, the H-bonds formation is constitutive of the helical structure. Whatever the number of polymers forming the helical structure, these various helices derive their stability from an unusually strong intramolecular interaction (H-bonds) between neighboring units. Polypeptides exhibit sharp transition suggesting a cooperative process to form an helical conformation from a random one. The Zimm-Bragg model (22) was an attempt to describe this cooperative process. With a parameter σ measuring the coexistence of helical part and random coil, $-RT\ln\sigma$ represents the free energy for each junction and can be taken as a measure of the reciprocal of the cooperative interaction favoring perpetuation of the helical state. The smaller σ, the larger is the energy of the boundary (domain wall) between coil and helical sequences and the more abrupt is the transition.

The approximate breadth for the transition is given by (22) :

$$\Delta T = \frac{2RT_c^2\,\sigma^{1/2}}{\Delta H_\upsilon} \qquad 7\text{-}11$$

where ΔH_υ is the enthalpy change per monomer unit. With an appropriate value of σ and the help of equation 7-11, the Zimm and Bragg model is able to relate the observed enthalpy change and temperature width of the transition. For example, for T = 40°C and ΔH_υ = 2 kcal/mole, ΔT = 2°K, while for ΔH_υ = 0.9 kcal/mole, the width is ΔT = 5°K. These are exactly the values observed by Doty et Al. (23).

The analogy with biological systems was widely used for describing the dramatic color change for the soluble PDA (2,24). Various experimental results support such a comparison : The DSC thermograms, the infra-red absorption showing H-bond formation, and the sharpness of the color transition.

The DSC thermograms give an enthalpy change $\Delta H_{\upsilon} = 2.5$ kcal/mole (25), the measured breadth is equal to 2 K. Moreover a second step has been observed (25) showing an enthalpy change about 0.6 kcal/mole with $\Delta T > 5$ K, then a behavior similar to the one observed for the so-called helix-coil transition of polypeptide (21,22). This does not mean that the Zimm and Bragg model would apply straightforwardly to the color change of the PDA. H-bonds play a major role for the helix structure while for P3 or P4BCMU it is an additional parameter which is not needed for making a rod. As a matter of fact there is a PDA (PTS12) without H-bonds providing a color transition and DSC thermograms showing the same features as the ones observed for P4BCMU (15).

In conclusion most polymers in good solvent exhibit flexible behavior. If some more rigid conformations occur, they correspond to an unusually strong intramolecular interaction between monomer units. In other words, if drastic conformational changes happen either by varying the temperature or the good solvent/poor solvent ratio, we must identify the reason why strong intramolecular interaction becomes efficient.

7-2-2. Model for conjugated polymers

We have shown that soluble conjugated polymers in good solvent behave like saturated polymers. That is we do not need to invoke π electron delocalization for understanding the statistical conformation of the macromolecules. Such a claim does not imply that the π electrons have no influence but rules out their role as the leading parameter for predicting the chain statistics for soluble conjugated polymers.

The understanding of the actual influence of the π electron still remains, we can not avoid their influence upon the thermodynamic properties of the solution. One conclusion is that side-groups have a screening effect upon the electronic interactions between monomer and solvent units or between monomer units. A quantitative and even a qualitative analysis of the influence of the π electrons is a difficult task, mainly because the range of the interaction and the magnitude should differ from the usual ones.

K S. Schweizer has attempted to treat the effect of the π electrons (26). The aim was to study the solvatation effect for conjugated polymers. For large, complex molecules, a site-site interaction is usually considered such that a simple additivity rule applies. For saturated polymers, σ electrons only are involved in the dispersion interaction providing an attractive interaction between two chains containing N units proportional to N^2/R^6 (27). For conjugated polymers, the spatial extension of π - electron system virtual transitions is large, therefore a breakdown of the additivity rule is expected leading to a different attractive interaction.

In his theoretical study of the order-disorder transition, K.S. Schweizer focuses the attention on the polymer-solvent interaction. A R^{-6} dependence is assumed and the spatial extension of the π-electrons is introduced through an effective dependence of the backbone polarizability as a function of the conjugation length. The part of the interaction between the macromolecule and the solvent units as a function of the electron delocalization is given by (26):

$$\langle \Delta V \rangle = - C_{\lambda} \sum^{N_s} \sum^{\xi_{\lambda}} \frac{3}{2} \frac{\Delta\alpha_p(\lambda)}{\xi_{\lambda}} \qquad 7\text{-}12$$

with $C_\lambda = \frac{I_s I_\lambda}{I_s + I_\lambda} \rho_s \alpha_s \langle R^{-6} \rangle_\lambda$ 7-13

where I_s and I_λ are the ionization potential of the solvent unit and submolecule λ respectively. N_λ is the number of submolecules with the conjugation length ξ_λ. $\Delta\alpha_p(\lambda)$ the varying part of the polarizability depending on the size of the submolecule. $\langle R^{-6} \rangle_\chi$ is the average geometrical factor giving the distance dependence of the dispersion interaction between the subunits of the backbone and the surrounding medium. The prefactor in the right hand part of equation 7-13 can be simply expressed as a function of the refractive index of the solvent:

$$\rho_s \alpha_s = \frac{3}{4\pi} \frac{n^2 - 1}{n^2 + 1} \qquad 7\text{-}14$$

K.S. Schweizer uses the rod conformation as the reference state, that is for ξ_λ infinite the attractive interaction with the solvent is maximum. The coil structure is modelled as a defect like structure, with a competition between an attractive polymer-solvent interaction and the existence of defects along the chain implying a randomization of the macromolecule. The length of the subunits is controlled by the number of defects N_D : $N_\lambda = N/(N_D+1)$. Defining an energy ε for the creation of one defect, the coil structure and the order-disorder transition, thus the color change becomes dependent of the ratio V_i/ε, where V_i is the magnitude of the attractive part for the rod conformation (equation 7-12 with $\Delta\alpha_p$(rod)).

The model has encountered some success, being able to give a coherent picture on the behavior of chains in good solvent. First it emphasizes the effect of the side-groups. Following K.S. Schweizer, we can separate, at least theoretically, the backbone and the side-groups, so that for the dispersion interaction the substituents are considered as solvent molecules. A direct consequence is that the equation 7-12 can now express the interaction between the backbone and the side-groups by substituing the side-group polarisability α_{sd} to the solvent polarisability α_s. Depending on the size and the conformation of the substituents with respect to the backbone, one may find systems which are weakly sensitive to the solvent refractive index.

The screening effect being more or less efficient, in most case we should have to consider both the interaction with the solvent and side-group units (26). Up to now, there is no clear evidence of a screening effect. The red shift as a function of the pressure for P4BCMU exhibits a linear dependence with the solvent density suggesting a weak screening effect (Chapter 5). To get quantitative informations, a measurement of the visible absorption as a function of the side group structure for the same conjugated backbone in a given good solvent would be more appropriate. But even such an experiment remains questionable, since as we noted before, the angular deviation between units is dependent of the geometry of the side-groups. Two effects, at least, are involved : One is the screening effect acting on the magnitude of the solvent term. The other is the change on the π-electron delocalization due to change in the average torsion and acts on the term $\Delta\alpha_p(\lambda)$. These two terms may either enhance or cancel the side-group influence in a such way that the sign and the magnitude of the absorption shift can not be easily predicted. This means that spectrocopic studies do not provide unambiguous information on the chain structure. Even with the help of SANS experiments we are unable to give accurate quantitative results on the magnitude of the torsion. In any case for such a complex system the conclusions remain very model dependent.

An interesting attempt is the prediction of the color transition temperature T_c. K.S. Schweizer gives T_c as a function of the attractive interaction for the rod and the energy of the defect creation, V_i and ε. Neglecting the defect part we have : $T_c \sim V_i^{0.8}$. The results allow to explain the differences observed for various materials : the serie of polydiacetylenes (28) or poly alkyl-silanes (29). Moreover the ratio of the temperature transitions observed in solution and films is in reasonable agreement with the theoretical predictions.

The above model is an attempt to describe explicitly the solvatation of conjugated polymer, then a possible conformational change of the macromolecule in good solvent. In other words, the analysis of the color transition in term of order-disorder transition suppose a stable rod conformation for long chains. We have shown that such a statement is far from proved (chapter 5). Moreover except for some particular cases of biological systems, a random coil conformation is usually observed for long polymer chains in good solvent. Therefore the structure of the "ordered" phase and the concept used for describing the color transition are questionable.

The key point is that the monomer-monomer interaction is not considered in K.S. Schweizer's model, so the analysis in terms of the transition from good to poor solvent is avoided. The collapse of an isolated chain has been studied long ago by Flory (30). Using the Flory theory de Gennes has shown the analogy existing between the theta transition and the tricritical point for a magnetic system. The free energy F for an isolated chain is written (31):

$$\frac{F}{k_B T} = 3(\alpha^2/2 - \ln\alpha) + \frac{N}{2}\left(\rho\ V_1(T) + \rho^2\ V_2(T) + \ldots\right) \tag{7-15}$$

where $\alpha = R/R_0$ is the expansion factor and R_0 is the radius of the ideal coil: $R_0 = N^{1/2}$ a. N is the number of monomer units. The average density in the coil ρ is

$$\rho = kN/R^3 \tag{7-16}$$

with k a numerical constant. V_1 and V_2 are the second and third virial coefficients which are now dependent on the temperature in contrast with the athermal solvent case (see discussion in chapter 2). Inserting equation 7-16 in equation 7-15, and minimizing the free energy one obtains the Flory equation:

$$\alpha^5 - \alpha^3 - \frac{y}{\alpha^3} = k\ N^{1/2}\frac{V_1}{a^3} = x \tag{7-17}$$

where $y = k^2\ V_2\ a^{-6}$ is a dimensionless parameter corresponding to three-body interaction. The theta temperature is defined by $x(\theta) = 0$, for $T<\theta$ the second virial is negative $(x < 0)$ and the swelling of the chain becomes very dependent on the magnitude of y. For large enough y the coil contracts smoothly while for y close to zero a collapse is expected (31).

Whatever the sign of the interaction, the virial coefficients correspond to short range interactions (even if the parts considered are separated by a large number of segments), $V_1(T)$ for the attractive case is considered to have the typical dependence R^{-6}, while the repulsive part is described as an hard core repulsion. For conjugated polymers the situation can be noticeably changed. One can speculate that the balance between $V_1(T)$ and $V_2(T)$ controlling the chain conformation below θ can be completely modified leading to a collapse whatever the magnitude of y.

To our knowledge, up to now no such study has been done. This requires first an accurate estimation of the $\pi\pi$ interaction, then a description of the chain conformation allowing the $\pi\pi$ interaction to be efficient and finally an ability to describe the resulting packing.

Even within the framework of the Schweizer's model, i.e. considering a dispersion interaction going like R^{-6}, we can expect a chain-chain interaction, or equivalently an attractive interaction between different parts of the chain, very dependent of the conjugation length. In other words, the main conclusion of this model which predicts an increase of the rod conformation due to an enhancement of the attractive polymer-solvent interaction, should rather apply to the interaction between subunits of the macromolecule. We have to consider the balance between a term $V_{ms} \propto \alpha_p(\lambda)\, \alpha_s$ and a term $V_{mm} \propto \alpha_p(\lambda)\, \alpha_p(\lambda)$, the resulting behavior being far fromobvious.

The total interaction energy of conjugated molecules includes the contribution coming from the π-electrons and σ-electrons, and one can write formally :

$$E_{int} = E_{\pi\pi} + E_{\pi\sigma} + E_{\sigma\sigma} \qquad 7\text{-}18$$

The term $E_{\sigma\sigma}$ corresponds to the interaction usually taken for saturated polymers. For saturated polymer solutions, the balance between monomer-monomer, monomer-solvent and solvent-solvent interactions is given by the Flory parameter χ (see chapter II). The behavior of $E_{\pi\sigma}$ is dependent on the length of the submolecule and in Schweizer's model is described by equation 7-12. For conjugated polymer in good solvent it corresponds either to polymer-solvent or polymer-polymer interactions. The interaction term $E_{\pi\pi}$ is a polymer-polymer or submolecule-submolecule interaction and is still more dependent of an effective conjugation length.

A study of the asymptotic behavior of the terms involved in the expression 7-12 for polyenes with N >> 1 leads to the following qualitative relationships at R >> L (27):

$$E_{\pi\pi} \propto \frac{L^5}{R^6}, \quad E_{\pi\sigma} \propto \frac{L^3}{R^6}, \quad E_{\sigma\sigma} \propto \frac{L^2}{R^6}, \qquad 7\text{-}19$$

It appears first that the following relations hold:

$$E_{\pi\pi} > E_{\pi\sigma} > E_{\sigma\sigma}$$

As the submolecule size or the conjugation length increases, the contribution of the $E_{\pi\pi}$-interaction becomes dominant (32). Moreover Coulson and Davies have calculated that important deviations from the law R^{-6} occur at distances R << L. For example the effective power dependence $E_{disp} \sim R^{-n}$ at distances R = 10Å is for a polyene with 10 and 20 carbon atoms n = 4.4 and 3.5 respectively (33).

These remarks suggest that precipitation must occur as soon as $E_{\pi\pi}$ becomes dominant, the macromolecule folds forming parallel vertical parts with an attractive interaction (figure 9). For large enough concentration, say at the overlap concentration, bundles can be formed, or the so-called fringed micelle could be the relevant picture (34).

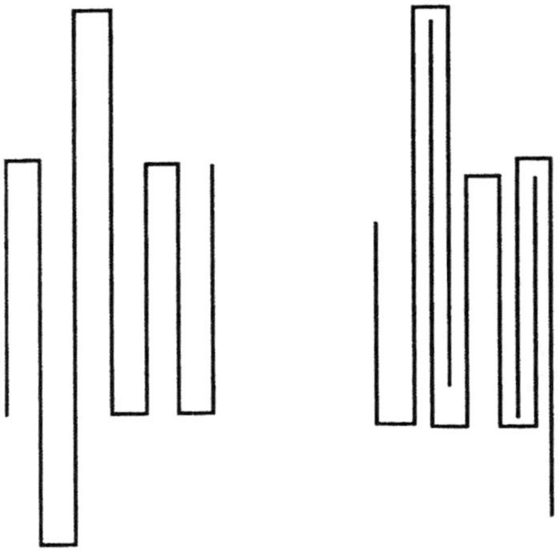

Figure 9: Schematic representation of the chain folding at infinite dilution (collapse of a single chain) and at the overlap concentration.

From that point of view one can understand the effect of the substituents for solubilizing conjugated polymers. First, they screen the π-electrons interaction, allowing to balance the interaction between solvent and monomer units. Second, the average conjugation length or size of an effective submolecule is reduced through the torsion induced by steric interference then hindering a chain folding due to the interchain π-π interaction.

These conjectures should not be taken as proofs, but allow a conclusion which appears at first glance paradoxical : Side-groups have minor effect while they form planar ribbon structures, in which case the torsion between monomer units is zero allowing a larger conjugated sequence (6). In addition the screening effect should be less efficient, since the backbone is not surrounded by the side-groups. The planar ribbon structure corresponds to the lowest effective moment of inertia (§ 7-1-1), that is a more flexible behavior of the macromolecule. Therefore an increase of the conjugation length implying a red shift of the maximum absorption in the visible can be related to an increase of the chain flexibility in good solvent. On the other hand the color transition (the sol-gel transition for P4BCMU) can have for physical origin the reduction of the side-groups influence, i.e. as a first step an increase of the local flexibility allowing a more efficient attractive interaction between chains or part of the chain due to the π-electrons.

ACKNOWLEDGMENT

It is a great pleasure to thank Michel Schott for a critical reading of this paper and fruitful suggestions.

References :

1 B.S. Hudson, B.E. Kohler, K. Schulten, in "Excited States", Vol 6, E.C. Lim Ed.(Acad. Press, N.Y (1982).

2 G.N. Patel, R.R. Chance and J.D. Witt, J. Polym. Sci. Polym. Lett. Ed., **16**, 607 (1978).

3 G.N. Patel, R.R. Chance and J.D. Witt, J. Chem. Phys., **70**, 4387 (1979).

4 M.L. Shand, R.R. Chance, M. Le Postollec, M. Schott, Phys. Rev. B, **25**, 4431 (1982).

5 P.R. Surjan, H. Kuzmany, Phys. Rev. B, **33**, 2615 (1986).

6 G. Rossi, R.R. Chance, R. Silbey, J. Chem. Phys., **90**, 7594 (1989).

7 J.M Ginder, A.J. Epstein, Phys. Rev. B, **41**, 10674 (1990).

8 G. Allegra, S. Bruckner, M. Schmidt, G. Wegner, Macromolecules, **19**, 399 (1986)

9 P.A. Pincus, G. Rossi, M.E. Cates, Europhys. Lett., **4**, 41 (1987).

10 Z.G. Soos, K.S. Schweizer, Chem. Phys. Lett., **139**, 196 (1987).

11 L.D. Landau, E.M. Lifshitz., Statistical Physics, 2nd edition (Pergamon Press, London) (1959), p. 478.

12 J.P Aimé, F. Bargain, Europhys. Lett., **9**, 35 (1989).

13 L.D. Landau, E.M. Lifshitz., Theory of Elasticity (Pergamon Press, London) (1959).

14 J.L. Bredas, A.J. Heeger, Macromolecules, **23**, 1150 (1990).

15 C. Plachetta, R.C. Schulz, Makromol. Chem. Rapid. Commun. , **3**, 815 (1982).

16 J.P. Aimé, S. Ramakrishnan, R.R. Chance, M.W. Kim, J. Phys. France, **51**, 963 (1990).

17 J.P. Aimé, P. Garrin, G. Baker, S. Ramakrishnan, P. Timmins ICSM'90 Tübingen.(1990).

18 G. Natta, P. Corradini, J. Polym. Sci., **39**, 29 (1959).

19 R.G. Kirste, R.C. Oberthur, in "Small Angle X-ray Scattering", edited by O. Glatter and O. Kratky (Acad. Press, New York) (1982).

20 H. Yamaka, in "Molecular Conformation and Dynamics of Macromolecules in Condensed Systems"; M. Nagasawa, Ed.; Elsevier : Amsterdam, p21 (1988).

21 T.M. Birshtein and O.B. Ptitsyn, in "Conformations of Macromolecules", High Polymers, vol. XXII, Interscience Publishers, N.Y., (1966),.chapt. 9 and 11.

22 B.H. Zimm, J.K. Bragg, J. Chem. Phys., **28**,1246 (1958).

23 P. Doty, J.T. Yung, J. Am. Chem. Phys., **78**, 498, (1956).

24 A.J. Berlinsky, F. Wudl, K.C. Lim, C.R. Fincher, A.J. Heeger, J. Polym. Sci. Polym. Phys. Ed., **22**, 847 (1984).

25 J.P. Aimé, J.L. Fave, M. Schott, Europhys. Lett., **1**, 505 (1986).

26 K.S. Schweizer, J. Chem. Phys., **85**, 1156 (1986).

27 I.G. Kaplan, "Theory of Molecular Interactions", Elsevier (Netherlands), (1986).

28 R.R. Chance, G.N. Patel, J.D. Witt, J. Chem. Phys., **71**, 206 (1979).

29 R.D. Miller, D. Hofer, J. Rabolt, G.N. Fickes, J. Am. Chem. Soc., **107**, 2172 (1985).

30 P. Flory, "Principles of Polymer Chemistry" (Cornell University Press, Ithaca) (1970).

31 P.G. de Gennes, J. Phys. Lett., **36**, L55 (1975).

32 E.F. Haugh, J.O. Hirschfelder, J. Chem. Phys., **23**, 1778 (1955).

33 C.A. Coulson, P.L. Davies, Trans. Faraday Soc., **48**, 777 (1952).

34 G. Wenz, G. Wegner, M.A. Müller, M. Schmidt, Macromolecules, (1984), **17**, 37

PROCESSABLE CONDUCTING POLY(3-ALKYLTHIOPENES)

G. GUSTAFSSON, O. INGANÄS, W.R. SALANECK
Department of Physics, IFM, Linköping University
S-581 83 Linköping, Sweden

J. LAAKSO, M. LOPONEN*, T. TAKA, J.-E. ÖSTERHOLM
Neste Oy, Research R&D, POB 310,
SF-06101 Porvoo, Finland

H. STUBB
Semiconductor Laboratory, Technical Research Centre of Finland,
SF-02150 Espoo, Finland

T. HJERTBERG
Department of Polymer Technology, Chalmers University of Technology,
S-412 96 Göteborg, Sweden

ABSTRACT. The poly(3-alkylthiophenes) represent a new generation of conducting polymers, that exhibit both solution and melt processability. These properties open new possibilities for the utilization of conducting polymers in practical applications. Furthermore, the solubility of these polymers enables the characterization of the physical and chemical structure in greater detail than hitherto have been possible. In this chapter we give a brief review of the synthetic routes to the poly(3-alkylthiophenes) as well as an overview of some of the recent results on the physical and chemical characterization of this class of materials. In this overview we include results on stability properties, transport properties, thermo- and solvatochromism. We also discuss some aspects of the utilization of the poly(3-alkylthiophenes) in applications.

1. Introduction

One of the ultimate goals of conducting polymer research is to synthesize a material with the electrical and optical properties of a semiconductor or a metal and the mechanical and unique processing properties of an organic polymer. In general, however, conducting polymers are intractable materials which neither can be melt nor solution processed. This is because of the extended conjugation in these materials which makes the polymer chains

*Deceased, July 1990

J.L. Brédas and R. Silbey (eds.), Conjugated Polymers, 315–362.

very stiff. One way to circumvent this problem is to use solution and/or melt processable precursor polymers which then can be converted into a conjugated polymer by a simple thermal elimination reaction. The most well-known precursor method is the Durham route to polyacetylene developed by Edwards and Feast [1]. The precursor polymer (poly({5,6-bis(trifluoromethyl)bicyclo-[2.2.2]octa-5,7-diene-2,3-diyl}-1,2-ethendiyl)), which is solution processable, is converted under mild heating (less than 100°C) by elimination of hexa-fluoro-*o*-xylen to polyacetylene. Several other conjugated polymers can also be prepared via precursor routes; examples are poly(*p*-phenylene vinylene) [2] and poly(2,5-thienylenevinylene) [3]. In the former case the full conversion requires temperatures above 300°C. Conversion at lower temperatures can be used to obtain copolymers with variable lengths of the conjugated segments.

Another route to confer processability to conjugated polymers is to add suitable side-groups to the conjugated main chain. A variety of side groups can be used and it is possible to taylor processability properties like the melting point, solubility etc. by a suitable choice of the side group. In particular, the introduction of hydrocarbon chains to the 3-position of thiophene yields high molecular weight poly(3-alkylthiophenes) (P3ATs) (Figure 1) which are melt processable and soluble in common organic solvents [4-6]. The melting point of these materials decreases with increasing length of the hydrocarbon chain. Moreover, the polymers can be made water soluble by adding sulfonate groups at the end of the side chain [7-9]. The same strategy can be applied to pyrrole to yield poly(3-alkylpyrrole) which also is soluble in common organic solvents. Only a few studies of poly(3-alkylpyrroles) [10] have been carried out, however, mainly because the monomer is not readily available. Other examples of side group substituted conjugated polymers include poly(diphenylenediphenylvinylene) [11] and poly(*para*-2,5-di-n-alkylphenylene) [12].

One of the main advantages of conjugated polymers which are processable in their conjugated form, is the possibility to form blends with common matrix polymers to give a new generation of electrically conducting polymer blends. This enables the combination of the mechanical properties offered by the matrix polymer with the electrical properties of the conducting polymer. Such blends will also be cost competitive since a relatively low fraction of conducting polymer is needed to achieve conductivities comparable to that of the pure conducting polymer.

Inherently processable conjugated polymers is not only technologically important. There is also a considerable scientific interest in these materials, since they enable the study of conjugated systems in solution and in the melt. Such studies have lead to the discovery of phenomena such as thermochromism and solvatochromism in conductive polymers.

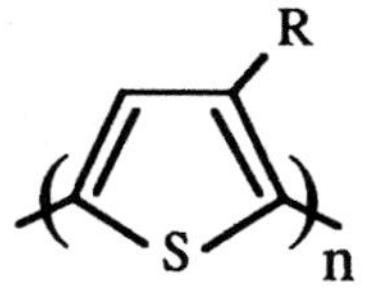

FIGURE 1. Poly(3-alkylthiophene). R=alkyl group.

Another advantage of the inherently processable conjugated polymers is that conventional polymer physics can be applied to these materials. This is a significant advantage, since many experimental techniques and theoretical models for characterization of conventional polymers are available.

In this chapter we give an overview of the chemical and physical characterization of the P3ATs and their blends. In addition some aspects of the applicability of these materials will be discussed. The intention of this overview is not to give a proper review of all work done on the P3ATs, since this would be far to extensive. Rather, we will focus on the work done within the Nordic group and relate this to work done by others in the same areas.

2. Synthesis

2.1. MONOMER SYNTHESIS

The most common route to synthesize 3-alkylthiophenes is to use 3-bromothiophene as a starting material. Via Grignard coupling and using 1-bromoalkanes it is possible to synthesize different 3-alkylthiophenes [13]. Using the dichloro-1.3-bis(diphenylphosphino)propane nickel(II) ($NiDPPPCl_2$) catalyzed crosscoupling method, 3-alkylthiophenes can be prepared with excellent yields [14]:

$$\text{R-Br} + \text{Mg} \xrightarrow{\text{ether}} \text{R-Mg-Br} \qquad (1)$$

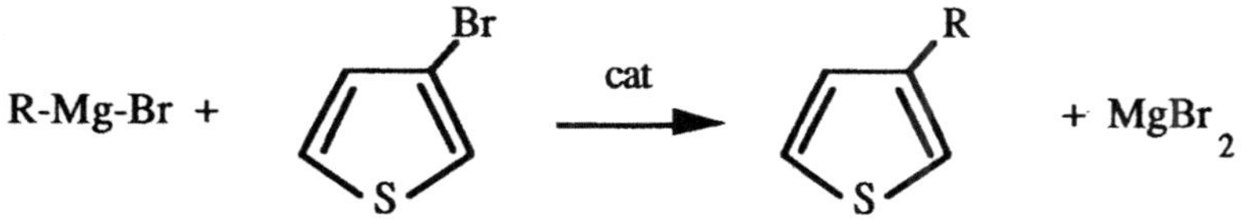

(2)

where R is an alkyl group. The yield of the reaction is 80% and the purity of the product is 92%. After passing the distilled product through silica the purity of the monomer can be raised up to 99%. The formation of the Grignard reagent, of course, affects the yield [15,16]. Table 1 shows the byproducts formed during the reaction.

TABLE 1. Typical byproducts formed during Grignard coupling of (3-octylthiophene).

Product	Amount (%)
2-octylthiophene	0.68
C_9-thiophene	0.78
n-hexadecane	4.42
3.3-dithiophene	0.49

According to Pham *et al.* [17] (table 2) the substituent (R) affects the yield of the synthesis.

TABLE 2. Effect of the substituents on the yield of the Grignard reaction.

Substituent (R)	Yield (%)
ethyl	67
n-propyl	79
n-butyl	71
cyclopentyl	66
n-hexyl	71
benzyl	75

Another convenient way to synthesize this kind of monomer is to brominate 3-methylthiophene and then use Grignard coupling [18]:

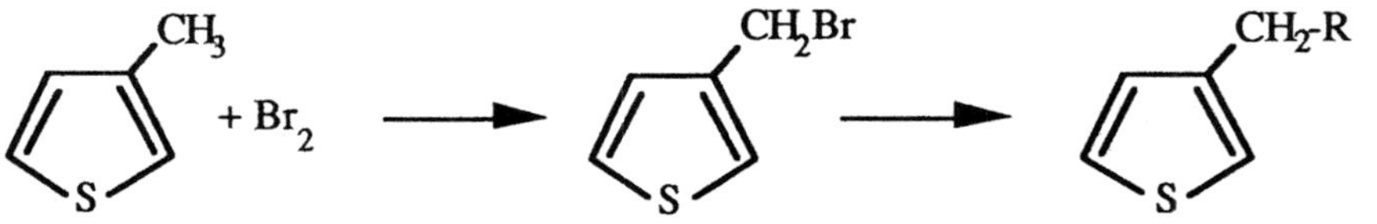

(3)

It is also possible to synthesize alkylthiophene via 3-lithiumthiophene, which is prepared by lithation of 3-bromothiophene:

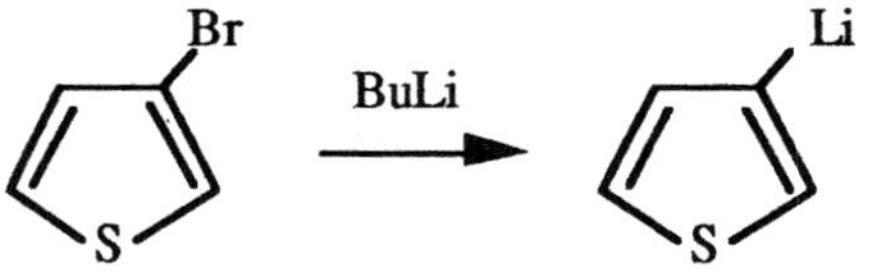

(4)

This reaction can be carried out at -70°C to prevent exchange of 3-thienyllithium into 2-thienyllithium [19]. Further reaction with suitable halogenated alkanes gives the desired alkylthiophene.

2.2. POLYMERIZATION

3-alkylthiophene monomers can be polymerized both electrochemically [20-23] or chemically. In this short review only the chemical polymerization of 3-alkylsubstituted thiophenes is described.

Yamamoto *et al.* [24] have studied the polymerization of 2.5-dibromothiophene by debromination with magnesium, catalyzed by nickel compounds:

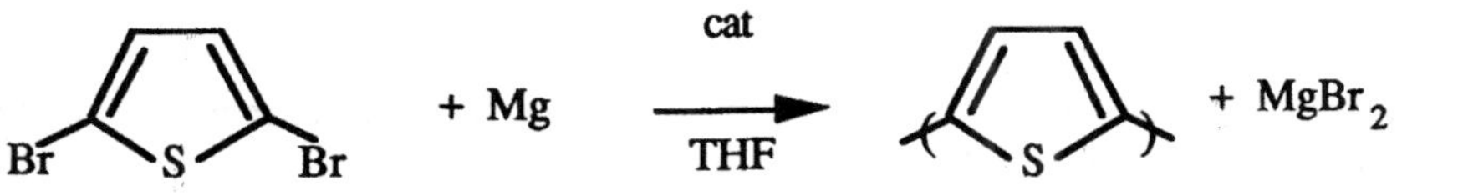

(5)

The resulting polymer is very stable. The yield depends on the catalyst used. This method has also been used for the synthesis of copolymers of dibromothiophenes [25].

Grignard coupling can also be used to prepare poly(3-alkylthiophene). 3-substituted thiophene is diiodinated according to the method of Barker *et al.* [26] in which a 1:1 mixture of HNO_3:H_2O is slowly added to the reactor containing the monomer and finely powdered iodine in dichloromethane.

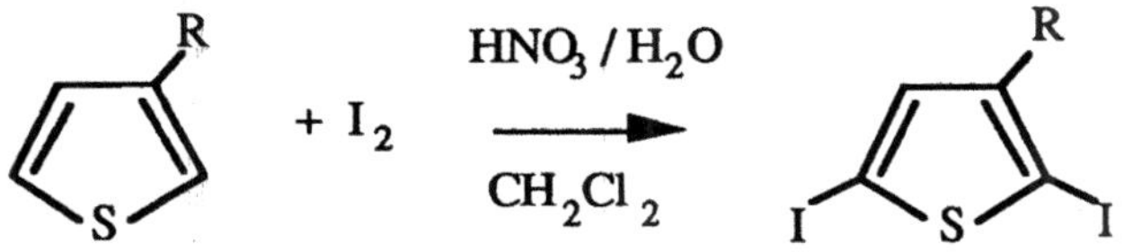

(6)

The isolated crude product is purified chromatographically to give 3-alkyl-2.5-diiodothiophene [27]. Figure 2 shows the formation of the diiodinated monomer as a function of reaction time [28]. Polymerization of this monomer can be carried out by reacting it with magnesium in dry THF or 2-methyltetrafurane (2MeTHF):

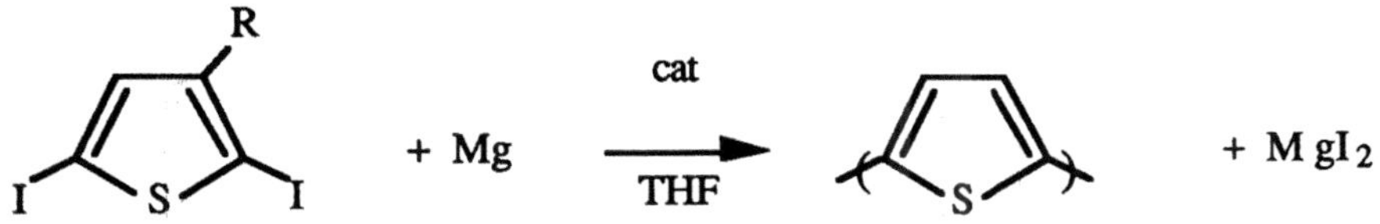

(7)

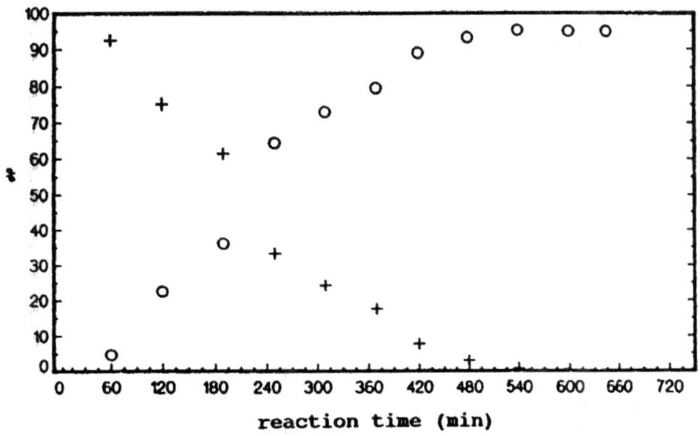

FIGURE 2. The formation of diiodinated 3-alkylthiophene as a function of the reaction time (+=monoiodinated, 0=diiodinated) [28].

The polymer is washed by Soxhlet extractions with methanol and acetone and dried under vacuum [27,29].

Sugimoto *et al.* [30] have studied the direct oxidation of 3-alkylthiophene by using transition metal halides ($FeCl_3$, $MoCl_5$, etc.) as catalysts/oxidants:

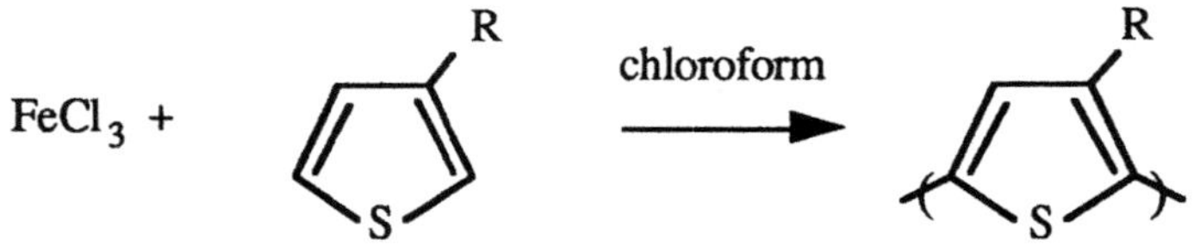

(8)

The crude polymer is precipitated in methanol and carefully washed with methanol, water and finally dried under vacuum.

3. Characterization

3.1. INFRARED SPECTROSCOPY

Figure 3 shows the infrared spectrum of poly(3-octylthiophene), (P3OT). Assignments are 3058 cm^{-1}: C_β-H of the thiophene ring; 2853-2954 cm^{-1}: the octyl group; 1509 cm^{-1}: C = C of the thiophene ring; 1440 cm^{-1}: - CH_2 of the octyl group; 819 cm^{-1}: trisubstituted thiophene; 725 cm^{-1}: -CH_3 of the octyl group [6,14,27].

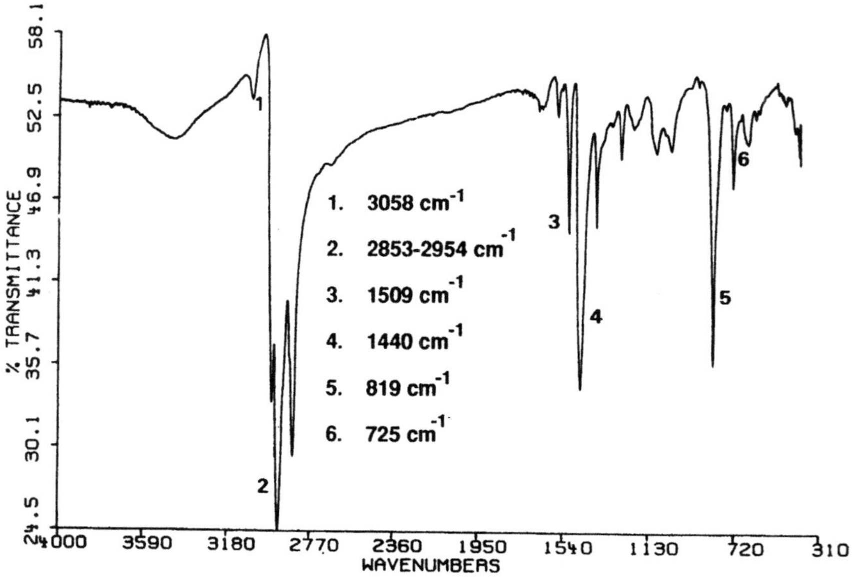

FIGURE 3. IR-spectrum of poly(3-octylthiophene) [27].

3.2. NMR

NMR has become the dominate technique for the characterization of the structure of polymers. The development of new 2D-NMR pulse sequences is continuously increasing the capabilities offered. Compared to most of the earlier known conducting polymers, the solubility of P3ATs, thus should allow a much more detailed structural analysis. Surprisingly, there are, however, only a few reports presenting NMR results on P3ATs [31-34].

The ^1H NMR spectrum of P3OT, synthesized by Grignard coupling, is shown in Figure 4 [34]. This spectrum is similar to those of other P3ATs, published by others [31-33]. The methylene protons next to the thiophene ring are seen as two resonances at 2.8 and 2.55 ppm. The downfield peak has been assigned to head-to-tail couplings [31], which has been confirmed by comparison with spectra of poly(3,4-dihexylthiophene) [33]. The few available data indicate that the content of head-to-tail coupling is about 70-90%, and depends on the polymerization technique, increasing in the order; electrochemical oxidation < chemical oxidation < Grignard coupling.

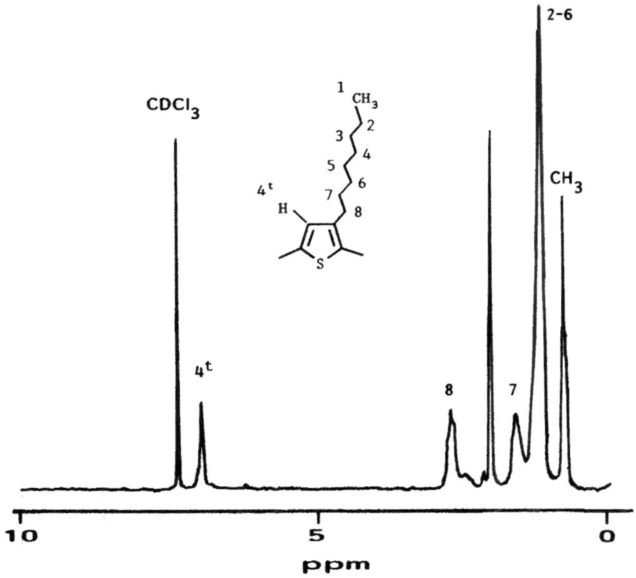

FIGURE 4. ^1H NMR spectrum of P3OT, synthesized by Grignard coupling [34].

The assignments of the ^{13}C NMR spectrum of P3OT, synthesized by Grignard coupling, which is shown in Figure 5, is supported by ^1H-^{13}C correlated spectroscopy [34]. Besides the four major peaks of the carbons in the thiophene ring, several minor peaks are seen in the aromatic region. In the case of P3OT synthesized by Grignard coupling these peaks have been associated with different end groups [34]. Although the molecular weight of P3OT obtained by oxidation is considerably higher (see below), the contribution of minor peaks (with other chemical shifts) in the ^{13}C NMR spectra is even more pronounced [33]. For the latter polymers, these peaks have been related instead to 2,4-couplings.

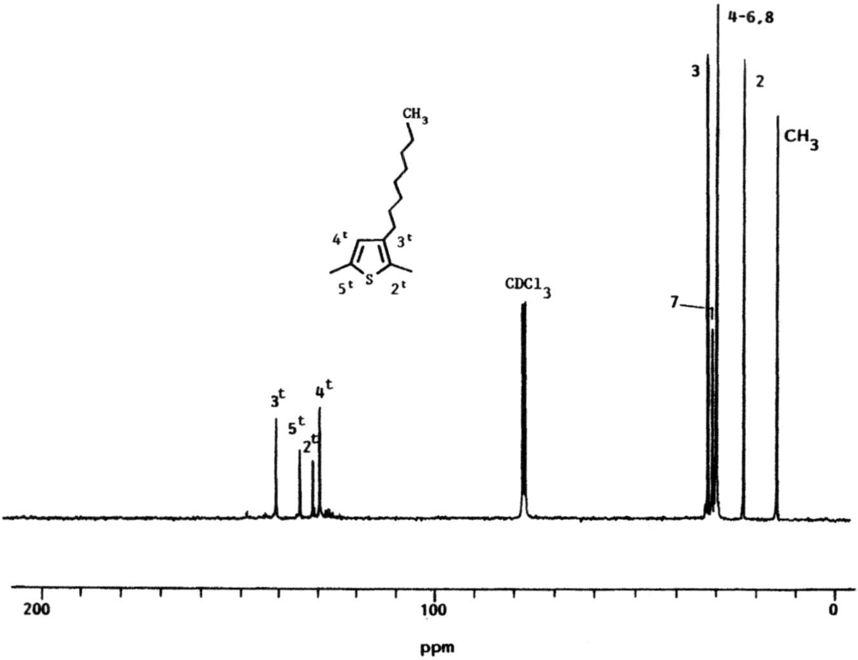

FIGURE 5. ^{13}C NMR spectrum of P3OT, synthesized by Grignard coupling [34].

3.3. ELEMENTAL ANALYSIS

Typical results of the elemental analysis of P3OT, synthesized by Grignard coupling, are shown in table 3. The iodine content can be reduced to less than 0.05% by heat treatment at about 120°C in nitrogen or vacuum [27,35]. P3OT obtained by oxidation contains significant amounts of Fe and Cl (2-4%), if not washed with extreme care.

TABLE 3. Elemental analysis of poly(3-octylthiophene) synthesized by the Grignard reaction.

	found (%)	theory (%)
S	16.10	16.51
C	76.03	74.19
H	9.15	9.30
I	1.35	-

3.4. THERMAL ANALYSIS

Figure 6 shows the result of thermogravimetric analysis, TGA, of P3OT, synthesized by the Grignard reaction. As can be seen this polymer exhibit excellent stability in nitrogen up to at least 300°C. We have found, however, that crosslinking occurs during the heat treatment. In oxygen, P3OT, synthesized by Grignard coupling, starts to decompose at about 250°C, loosing 40% of its weight by 440°C. Figure 7 shows the effect of Fe and Cl impurities on the heat TGA stability of P3OT synthesized by oxidation with $FeCl_3$. The polymer decomposes rapidly in oxygen at about 260°C [27].

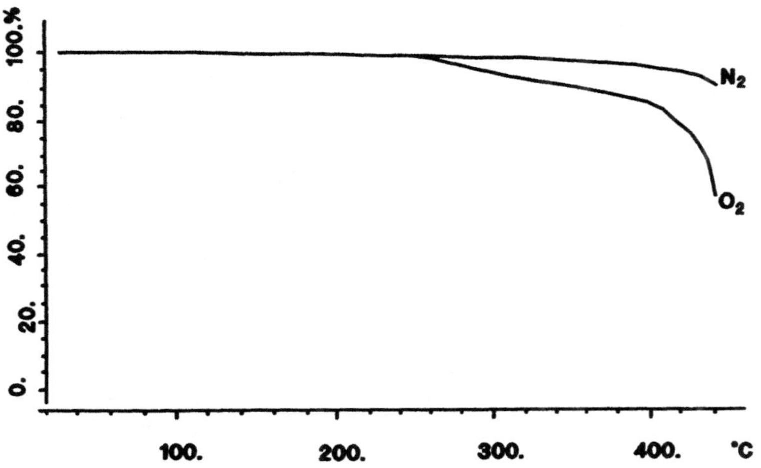

FIGURE 6. TGA curve of poly(3-octylthiophene) synthesized by the Grignard reaction [27].

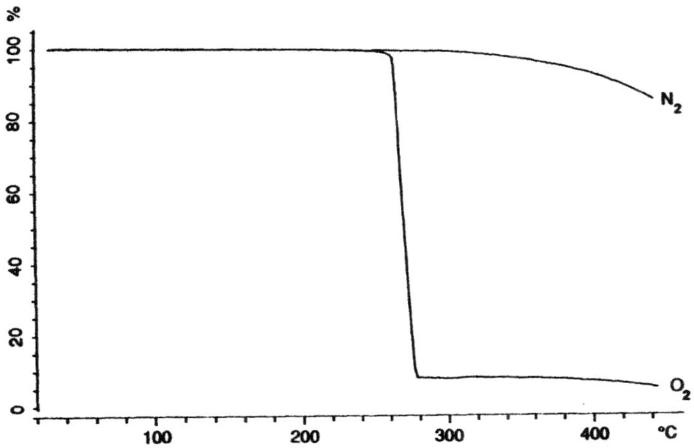

FIGURE 7. The effect of Fe and Cl impurities (both around 1%) on the heat TGA stability of poly(3-octylthiophene).

Differential scanning calometry, DSC, curves of P3OT show one low temperature transition at -20°C corresponding to the glass transition temperature, T_g, and one high temperature transition at 165°C corresponding to melting of the polymer.

3.5. MOLECULAR WEIGHT

The solubility of P3ATs also implies that the molecular weight and the molecular weight distribution (MWD) can be determined by the techniques used in polymer science, e.g., viscometry, light scattering (LS) and size exclusion chromatography (SEC). In principle, SEC can be used to determine the MWD, but the column set must be calibrated with standards of known molecular weight. Most often polystyrene standards with narrow MWD are used, and the calibration curve for the polymer in question is obtained via the "universal calibration" concept [36,37] using appropriate Mark-Houwink relations ($[\eta] = K \times M^{\alpha}$ where $[\eta]$ is the intrinsic viscosity and M the molecular weight). As the Mark-Houwink relations of the P3ATs have not been known, the published molecular weight data have been given in polystyrene calibration [6,27,32,35,38].

Hjertberg *et al.* [39] have used a SEC setup that, besides the ordinary concentration detector, contain a viscometer and a low angle laser light scattering (LALLS) detector on-line. This allows the determination of the Mark-Houwink relation. $[\eta]=10^{-4}M^{0.56}$ was obtained for P3OT. It turns out that the calibration curve for P3OT is similar to that for polystyrene. The latter can thus be used as a rather good approximation. The LALLS recordings show that P3OT is in true solution in tetrahydrofuran (THF). It is further obvious that the higher molecular weight fractions of P3OT obtained by chemical oxidation contain branched material. This has also been demonstrated by Leclerc *et al.* [33].

In the case of P3OT, synthesized by Grignard coupling, the number-average molecular weight, $\langle M_n \rangle$, is typically around 8300, the weight-average molecular weight, $\langle M_w \rangle$ around 19000 , the polydispersity index, $\langle M_w/M_n \rangle = 2.3$ and DP=40. If the polymer is heated, the end-group iodine disappears and the chain length increases: $\langle M_n \rangle = 9700$, $\langle M_w \rangle = 29000$, $\langle M_w/M_n \rangle = 3.0$ and DP=50. In the case of P3OT obtained by chemical oxidation, $\langle M_n \rangle$ and $\langle M_w \rangle$ are typically 30000 and 150000 respectively, depending on the polymerization time and temperature. As a consequence of the higher molecular weight, films made of $FeCl_3$ polymerized P3OT have superior mechanical properties compared to films made of Grignard coupled P3OT [27].

3.6. OPTICAL SPECTRA

An optical absorption spectrum of a chloroform solution of P3OT is shown in Figure 8. Identical spectra are obtained for P3OT synthesized by Grignard coupling and P3OT obtained by chemical oxidation as well as for melt processed P3OT, redissolved in chloroform. The absorption maximum at 440 nm and the absence of absorptions in the 600-800 nm region suggest a rather high purity of the polymers and that neither the method of chemical polymerization nor melt-processing of as synthesized powder in air at 180°C significantly affects the electronic structure.

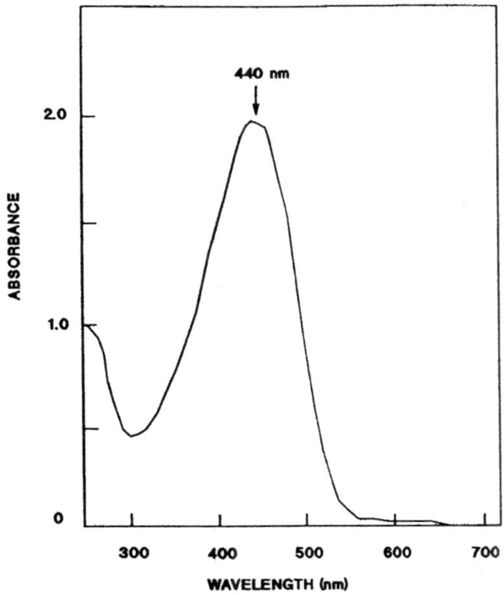

FIGURE 8. Optical absorption spectrum of a chloroform solution of poly(3-octylthiophene) [27].

3.7 X-RAY DIFFRACTION

The P3ATs are semicrystalline in the neutral state and amorphous in the doped state. The degree of crystallinity, approximated by the ratio of the area of the crystalline (sharp) peaks to the total area of a diffractogram of P3AT (corrected for background scattering), varies between 15-40 %, depending on the film preparation technique, i.e., solution casting or melt processing. Figure 9 shows a diffraction scan of a neutral, melt processed film of P3OT measured in the reflection mode. The diffractogram is characterized by a sharp reflection at low scattering angle (4.4°), together with its higher orders (8.8° and 13.1° respectively). This is typical for comb-shaped polymers in general and indicates a layered structure, where the main chains are stacked on top of each other, forming parallel planes which are separated by the alkyl side chains. This interpretation is supported by the fact that the interplanar distance corresponding to the low angle reflection increases with increasing number of methylene groups in the alkyl side chains, as can be seen in Figure 10 [40]. A similar observation has also been made for the poly(3-alkylpyrroles) [10]. As can be seen in Figure 10, the peak at 23.4°, which corresponds to an interplanar distance of 3.8 Å, does not vary with the length of the side chain.

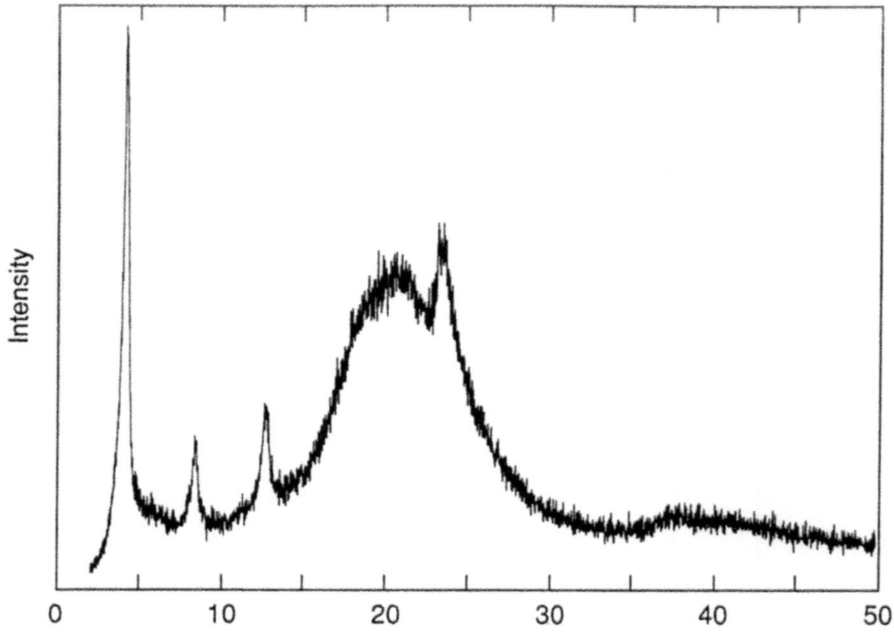

FIGURE 9. X-ray diffraction scan of a neutral, melt processed film of P3OT measured in the reflection mode.

This peak can probably be attributed to a regular packing of the stacked main chains and their side chains in the direction of stacking. The crystal structure is not known. Different models have been proposed [40,41], but more experiments are needed to determine the exact crystal structure.

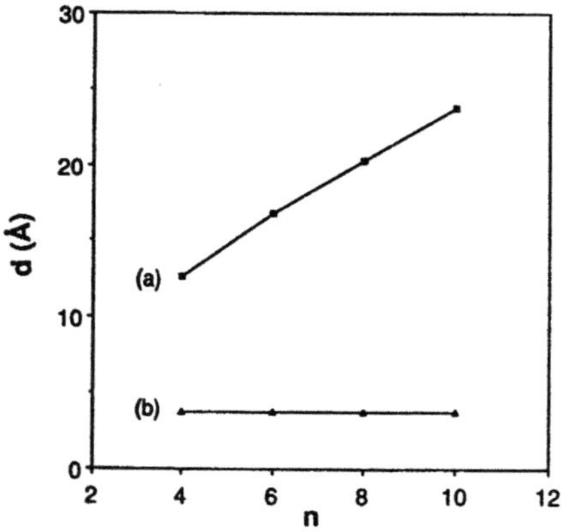

FIGURE 10. Interplanar distances, d, corresponding to the first (a) and the fourth (b) maxima in the diffractogram versus the number of carbon atoms in the alkyl chain, n. [40].

4. Processability-polymer blends

4.1 PROCESSABILITY

Poly(3-alkylthiophenes), with alkyl side chains longer than butyl, are both solution and melt processable. The solubility depends on the solvent, the temperature and the alkyl chain length [42,43]. P3ATs which have alkyl groups longer than 4 exhibit good solubility at room temperature in common organic solvents such as chloroform, toluene, decalin and tetrahydrofuran (THF). The P3ATs can easily be processed into uniform, flexible films by solution casting or spinning. These films can be made conducting by exposing them to doping agents, either from solution or from the vapor phase. The P3ATs also show solution processability in the doped state. Solutions of doped P3ATs can be prepared by mixing a solution of the neutral polymer with a solution of the dopant [23,38]. Films of doped polymer can then be cast directly from the solution. This is a significant advantage, since this enables full control of the doping level and yields uniformly doped films. Furthermore, problems associated with the dimensional changes upon doping pre-casted films can be avoided.

As mentioned above, P3ATs are also melt-processable. The melting/softening point of P3ATs depends on the chain length of the alkyl group [44] (table 4.)

TABLE 4. Effect of the alkyl chain length on the melting point of poly(3-alkylthiophene) [44].

Alkyl chain	Melting point (°C)
butyl	275
hexyl	190
octyl	150
decyl	125
dodecyl	80

This implies that conventional melt processing techniques such as compression molding, extrusion etc. can be used to process these materials. The preparation of conducting fibers of P3ATs by melt spinning (extrusion) has successfully been demonstrated by Yoshino *et al.* [44]. It is also possible to prepare flexible, free-standing films of P3OT by compression molding at 180°C in air without significant decomposition [27]. These films are readily doped to conductivities of 20-30 S/cm using I_2 or $FeCl_3$. However, it is not possible to melt process P3ATs in the doped state, due to the rapid undoping that occurs at higher temperatures (see section 6).

4.2 POLYMER BLENDS

The processability of the P3ATs implies that it is possible to obtain a wide variety of polymer blends or composites with conventional technology polymers such as polyethylene (PE), polystyrene (PS), poly(vinyl chloride) (PVC) and poly(ethylene vinyl acetate) (EVA). The mixing of the polymers can be done either in solution or in the molten state. Mixing in solution limits the number of matrix polymers that can be used to those which are soluble in the same solvents as the P3ATs. Hotta *et al.* [45] have studied the optical properties of polystyrene blends of soluble polythiophene derivatives over the full range of composition.

P3ATs and the matrix material can also be mixed in the molten state using a Brabender Plasticorder. The mixing temperature depends on P3AT and the melting point of the matrix material. In the case of P3OT, a convenient mixing temperature is 180°C, the mixing speed 30 rpm and the mixing time 5-10 min. Besides this, blow molding and extrusion molding can be used. Again, the mixing conditions are chosen according to the combination of P3AT and the matrix polymer. Films and sheets of polymer blends are made using compression molding [27,46].

Very little is known about the degree of compatibility in these blends. SEM micrographs of P3OT-polystyrene blends show larger domains surrounded by thin layers of another phase indicating phase separation [47]. The thin layers have been associated with mainly P3OT, as the x-ray fluorescence mode of the SEM microscope allows elemental assignments. These results have been confirmed by a preliminary study [48] on blends of P3OT with different matrix polymers using DSC and FTIR to study different interactions.

For the amorphous polymers PS and PVC, the T_g of the matrix polymer decreased when the content of P3OT increased up to 10% by weight. At higher levels no further change in T_g is observed. This shows that a certain amount of P3OT can be dissolved in these polymers, and that higher proportions lead to phase separation. To date, it has not been possible to detect any transitions of P3OT in the blends. However, FTIR has been used to study the interaction on a molecular level. The out-of-plane vibrations of C-H in the thiophene ring at 820 cm^{-1} are not obscured by any absorption of the matrix polymers. Although somewhat depending on the mode of preparing the blend, it can in general be said that this absorbance shifted towards shorter wavelengths in the blends. The shift is comparable to that observed when increasing the temperature of pure P3ATs [41], and may be taken as an indication of increased mobility of the polythiophene chains. This is in turn an indirect evidence that some matrix polymer is dissolved in the P3OT-phase as well.

The alkyl group of P3ATs should give a certain compatibility with polyethylene. This was also indicated in the DSC experiments by a decreased T_m of the PE crystals. Obviously the presence of P3OT had disturbed the crystallization process. Somewhat surprisingly, the effect was clearly larger in the case of high density PE compared to low density PE, the decrease was around 8 and 4°C, respectively. Again, a shift towards shorter wavelengths for the 820 cm^{-1} absorbance indicated increased mobility for the major part of the polythiophene chains. Taken together the results thus indicate that P3OT has a limited compatibility with several matrix polymers leading to phase separation at higher contents of P3OT in the blends.

P3AT blends can be made conducting by doping. The conductivity of the blends can be varied from less than 10^{-10} to 1 S/cm by varying the P3OT content, the doping time, the matrix polymer or a combination of all of those parameters. In the case of EVA/P3OT blends, a rapid increase in the conductivity of the films is observed at P3OT concentrations between 5-10 weight-% [27,46]. Within this concentration region the conductivity increases from 10^{-10} - 10^{-1} S/cm. Although conductivities in the range 10^{-6} - 10^{-4} S/cm have been measured in films containing less than 5 weight-% P3OT, the percolation threshold is reproducible observed between 5-10 weight-% P3OT. Using solution-doping, a rapid diffusion of the dopant molecules to the interior of the blends requires that the matrix polymer is swellable in the chosen solvent. This is clearly illustrated when comparing the doping profiles of EVA/P3OT and PE/P3OT blends, doped with $FeCl_3$ in nitromethane. In the case of EVA/P3OT, a uniform dopant distribution throughout the 150 μm thick film was observed [46]. In contrast, due to the chemical inertness of PE towards organic solvents, only a surface doping was achieved in the case of PE/P3OT.

5. Electronic structure and conformational excitations

5.1 ELECTRONIC STRUCTURE

By the terminology "*electronic structure*" of a molecular system is meant the energy (and momentum) distribution of the valence electrons, given the chemical (and hence the geometrical) structure of the molecule, or, in the present case, of the polymer chains. In the context of the P3ATs, the recent developments in the processability of these conjugated polymers has lead to better science being done on these materials. In the case of the polythiophenes, electrochemically made polythiophene (PT) [49] and poly(3-methylhiophene) [50] have been studied some years ago. More recently, the P3ATs, and in particular poly(3-hexylthiophene) (P3HT), has been studied more extensively by the same methods, employing high-quality, ultra-thin, solution-cast films, enabling much better spectra to be obtained, as discussed below.

Three of the most direct and readily comprehensible measures of the electronic structure of a molecular system are optical absorption spectroscopy and the two branches of photoelectron spectroscopy, known as ultraviolet and X-ray photoelectron spectroscopy, or UPS and XPS. The XPS of polymers [51], the UPS of polymers [52], as well as both the XPS and UPS of *conducting* polymers [53] have been reviewed recently. The reader is referred to the literature for further details. It should be relatively obvious from the figures presented below, however, how the calculated band structure, the density-of-valence-states (DOVs) derived therefrom, and the corresponding experimental valence band spectra can be compared with one another.

The molecular geometry of PT has been determined to be planar [54]. That is to say, the thiophene rings in the polymer backbone all lie approximately flat within a plane. There are discussions in the literature, however, concerning the possibilities that helical geometries might be possible [55, 56]. Indeed, total energy calculations of the geometry of some P3ATs predict only small energy differences between the lowest energy planar configuration and helical conformations. At the present, however, there are no definitive measurements indicating anything other than a planar geometry at low temperatures for all of the polythiophenes.

For a given geometry of the conjugated backbone, the presence of the alkyl groups does not directly affect the π-system of the thiophene-ring backbone [57-59], since the alkyl groups themselves contain no π-electrons. The presence of the alkyl groups, however, leads to intermolecular effects, which, in turn, do induce changes in the π-electron system as a function of temperature.

The electronic band structure of planar PT is shown in Figure 11 [49]. The electronic structure is presented as the usual energy-versus-momentum diagram (one dimensional, in this case), as calculated using the Valence Effective Hamiltonian (VEH) method [59]. The π-band portion of this band structure is identical to that of P3ATs at low temperatures [59], since the conjugated ring backbone of the substituted polythiophenes is also believed to be planar at low temperatures [59]. The delocalized π-bands of lowest binding energy in Figure 11 are labelled. In a simple single-electron model with the occupied atomic orbitals of the constituent atoms as the basis set, these bands are derived mostly out of atomic p_z-orbitals of carbon, oriented perpendicular to the planar thiophene backbone. S(4p) atomic orbitals do not contribute significantly to the lowest energy occupied band, labelled π_3 in the figure. The delocalized π-band structure results in a much smaller electronic energy

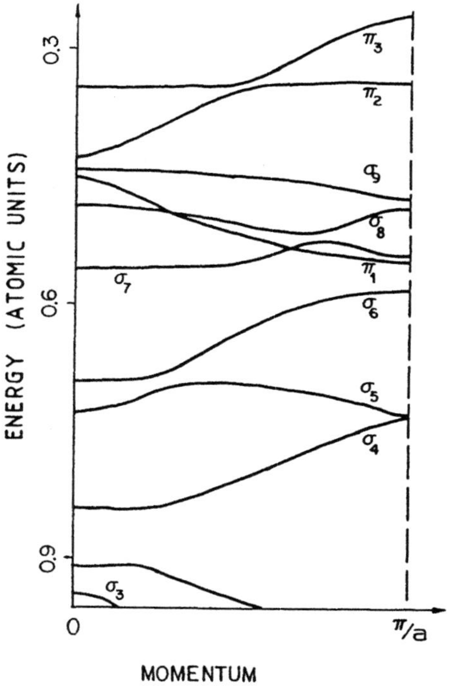

FIGURE 11. The band structure (E_k vs k) for planar polythiophene [49].

band gap, E_g, than the lowest energy π-π^* energy difference in an isolated thiophene molecule. The optical absorption spectrum corresponds to that of a semiconductor with an electronic band gap of $E_g = 2$ eV , as can be seen in Figure 12 [60]. The geometry of the optically excited state of P3AT is not well known. The presence of vibronic structure in the optical absorption spectra of the P3ATs points to a strong Franck-Condon modulation of the oscillator strength, and thus to large excited state molecular relaxation. The vibronic structure appears to correspond to 0-n transitions, where n is the index of the vibrational state. These vibronic peaks are separated by approximately 0.2 eV, and can be seen as shoulders on the optical absorption spectra of the polymers in the solid state at low temperatures [60], in solution in poor solvents [35,61], and in polymer blends in the solid state [47,62]. Depending upon conditions, up to 5 shoulders can be distinguished. Such vibrational structure has *also* been observed in the optical absorption spectra in other polymers when they are well ordered [60]. In any case, the appearance of this vibrational structure is an indication of geometrical relaxation in the excited state, which, in turn, indicates the presence of electronic localization in the optically excited state of the P3ATs.

The processability of the P3ATs has also contributed to studies of doping processes in P3HT using UPS and XPS [63,64]. Extremely thin (~few hundred Ångströms) films, solution cast from chloroform in a saturated chloroform atmosphere, were cleaned in UHV simply by heating to almost 200°C in order to remove solvent and small adsorbed hydrocarbons, just as in the case of the studies of the thermochromism discussed below. After studying the undoped films spectroscopically, doping was carried out *in situ* (in a flowing N_2 atmosphere) using $NOPF_6$ in acetonitrile. By following the degree of doping with XPS core-level spectroscopy, and recording the corresponding UPS spectra, an unexpected evolution of the dimensionality of the electronic structure of P3HT was observed [63,64].

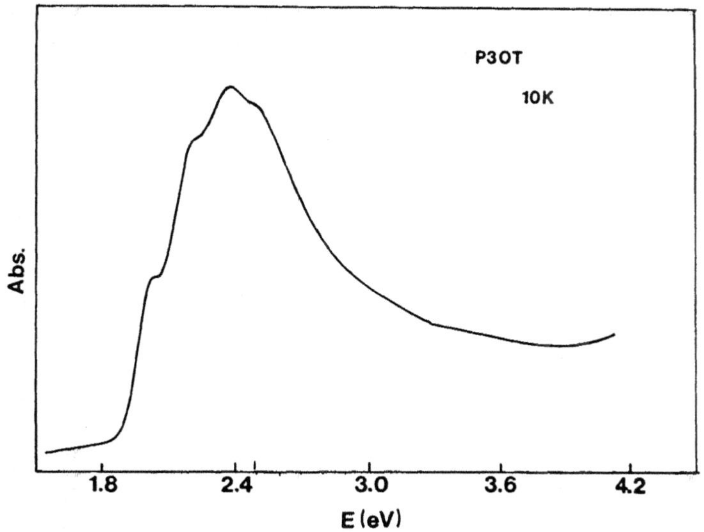

FIGURE 12. Optical absorption spectra of a thin solid film of P3OT [60].

Undoped P3HT exhibits a valence band electronic structure consistent with the one-dimensional nature of isolated (1-D) polymer chains [59]. Upon doping from $NOPF_6$ to the saturation level (i.e., repeated treatment does not lead to an increase in the doping concentration in the material), two important changes in the electronic structure were observed: the Fermi level moved by 0.8 eV, towards higher binding energy; and a finite density-of-states appeared at the Fermi level. Based upon an analysis of the spectra using the results of VEH calculations [65], these two features are indicative of a transition to a polaron lattice state in highly doped P3HT. The polaron lattice, however, is expected to be unstable in one dimension. Thus the existence of higher dimensional effects, plausibly through interactions of the polymer chains through the dopant counter ions, is indicated. Subsequent VEH level calculations [66] verify the higher dimensionality of the electronic structure of highly doped P3HT.

5.2 CONFORMATIONAL EXCITATIONS: THERMOCHROMISM AND SOLVATOCHROMISM

A very interesting and potentially useful thermochromic effect occurs in the (undoped) alkyl-substituted polythiophenes, for alkyl groups larger than methyl [35]. The thermochromic effect is a (reversible) thermally-induced change in color. In order to characterize the observed thermochromic behavior of the P3ATs, a series of studies of the temperature-dependence of the electronic structure of P3HT have been carried out using optical absorption spectroscopy, UPS and XPS. The central results, as well as a model of the thermochromism which is consistent with all of the experimental data, are reviewed below.

The temperature dependence of the features in the optical absorption spectra of P3HT indicate a shift of the absorption intensity maximum of about 0.5 eV between room temperature and 180°C, as shown in Figure 13 [67]. In addition, there is a loss of optical absorption at low photon energies, as well as an increase in the optical absorption at higher photon energies, as can be seen in the figure. These temperature-dependent features in the optical absorption spectra are indicative of an increase of the energy gap, E_g, and, with the help of the UPS information to be discussed below, a decrease of the width of the highest energy occupied π-band, π_3, with increasing temperature.

In Figure 14 is shown the UPS spectra of P3HT at two different temperatures, along with a comparison with the density-of-valence-states calculated from the VEH band structure shown in Figure 11. The portions of the spectra within the dashed-line box are of interest, since they correspond to the lowest binding energy electron energy bands, labelled π_3 and π_2 in Figure 11 [57-59]. In Figure 15 is shown the UPS spectra of these same π-bands, on an expanded scale, and compared with the π-band portion of the band structure from Figure 11, which has been turned upon its side in order to enable a better comparison with the UPS spectra. The correspondence between the shape of the UPS spectrum and the curvature in the band structure can be seen in the figure. The UPS data for T = -60° C clearly shows structure (peaks) corresponding to the π_3-band edge and the π_2-π_3 flat-band

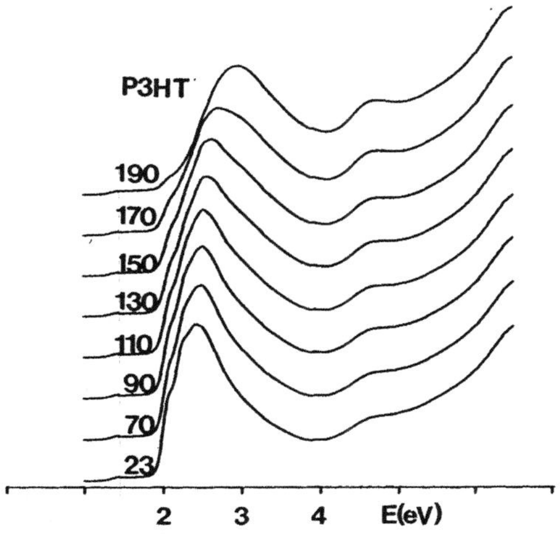

FIGURE 13. Optical absorption spectra of a thin solid film P3HT at temperatures between 20 and 200°C [67].

region, in agreement with the calculated band structure for the isolated, planar polymer chain. At high temperature (190° C in this case), the UPS spectrum indicates a collapse of the π_3-band into essentially one stronger peak (intensity is conserved), which shows up as an increase in the density-of-states as well as a slight shift in energy of the peak in the vicinity of the π_2-π_3 flat-band region.

The collapse of the π_3-band edge at high temperatures, along with the shift observed in the optical absorption spectrum (equivalent to a change in the electron energy gap, E_g) are accounted for *simultaneously* by any of several molecular models of the thermochromism in the P3ATs [59]. These molecular models are to be considered as special cases which are convenient for computational studies. The electronic state of the polymer system might then be represented by a distribution function based upon the model molecular states. The π-band portion of a *dimer model* of the geometrical structure of P3HT (Figure 16) at the elevated temperatures, as calculated using the VEH method [59], is included at the bottom of Figure 15. The correspondence between the shape of the UPS spectrum and the π-bands in the calculated structure can be seen. The optical band gap calculated for the dimer model is consistent with the optical absorption spectrum of P3HT at high temperature, shown in Figure 13. In addition, subtle details in the temperature-dependent XPS spectra [68] are also consistent with the models mentioned above, all of which correspond to various degrees of electronic localization effects as the origin of the thermochromic effect.

The *dimer model*, i.e. two coplanar thiophene rings, twisted together through a torsion angle of $\theta = \pi/4$ relative to the nearest neighbors (Figure 16), is only one of several models of "soft conformation defects" which are theoretically equivalent in accounting for the thermochromic behavior of P3HT; monomers (with $\theta \approx \pi/5$) and trimers (with $\theta \approx \pi/3$) are equally valid within the context of the VEH calculations [59]. The case of monomers is actually identical to the helical geometry discussed in the literature, also

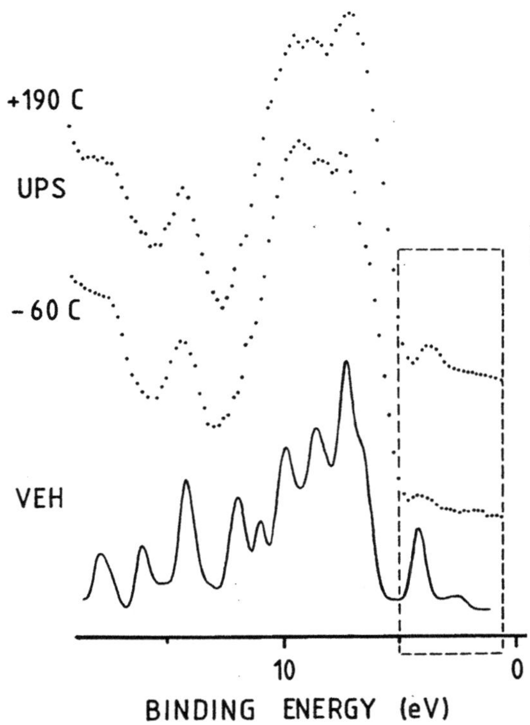

FIGURE 14. Two UPS spectra of P3HT.

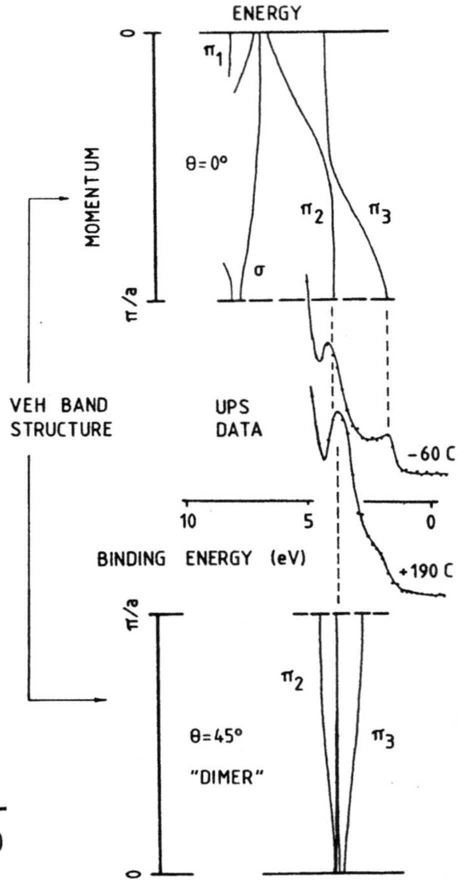

FIGURE 15. The UPS π-bands at two temperatures [57].

known as a "worm". These discrete soft conformational (geometrical) defect models simply are (idealized) special cases, which represent the basis for actual polymer geometry at high temperature.

The significance of the rotations associated with each defect, however, is that each rotation represents a reduced degree of conjugation between neighboring thiophene rings, i.e., a decrease in overlap of the atomic p_z orbitals of neighboring carbon atoms (on either side of the torsion angles θ indicated in Figure 16. For $\theta = 0$, there is maximum atomic overlap. For $\theta = \pi/2$, however, the overlap of the neighboring p_z atomic orbitals is zero, and the π-conjugation is completely broken at that point. Values of $0 < \theta < \pi/2$ result in intermediate values of the p_z overlap and a corresponding weakening of the π-conjugation. Thus, these soft conformation defects are geometrical models, which, within the context of VEH level calculations, reproduce the observed effects on the π-electronic structure, as

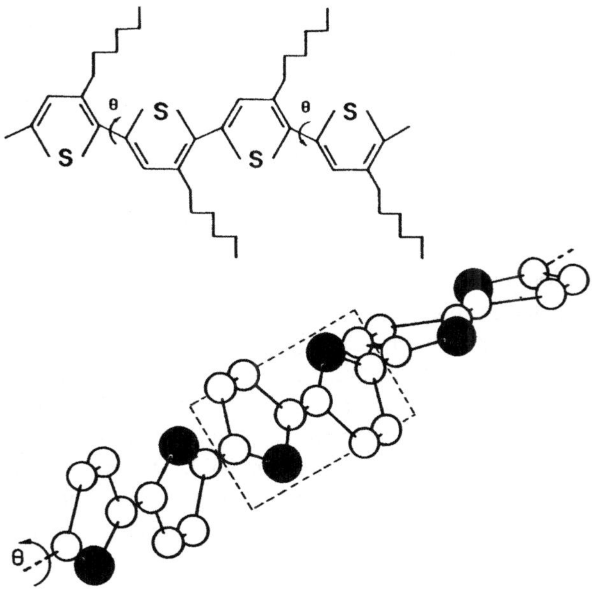

FIGURE 16. The chemical (geometric) structure of P3HT (Top), and the Dimer Model (bottom).

seen in the temperature-dependent optical absorption and UPS spectra, and which are consistent with subtle features in the temperature-dependent XPS spectra [68].

Thus far the characterization of the electronic properties in terms of the change in the π-electronic structure has been discussed. This structure, for a given geometry of the polymer backbone, is independent of the presence of the alkyl groups. On the other hand, the alkyl side groups are responsible for, actually "mechanically" induce, the geometrical changes in the backbone discussed above. The results of studies of the molecular vibrational features associated with the thermochromism using infrared and Raman spectroscopies [69-71] provide evidence for a picture of torsional deformation of the polymer backbone caused by temperature-dependent side-chain coiling. These measurements indicate an increase in the number of coiled side chains with increasing temperature, in parallel with a decrease in the average effective conjugation length. The model that emerges from these studies is one in which at low temperatures the alkyl side chains are in an all-*trans* conformation. With increasing temperature, defects modes are excited on the side chains, and that at highest temperatures the side chains are in random coils rather that straight "rods". The steric interactions among the polymer chains are thereby altered with increasing temperatures, forcing the backbone out of the planar configuration. These results are consistent with the fact that the thermochromic transitions of P3ATs with longer side chains occur at lower temperatures [67]. Apparently longer side chains are more effective in ultimately forcing the polymer backbone out of planarity.

The analogous phenomenon of solvatochromism [35] shows optical absorption features in the poor solvent to be identical to those at low temperatures in solid films, as well as a blue shift of the absorption features when going to a good solvent, via chemical composition or changed temperature. In addition, the poor solvent induces precipitation of the polymer. NMR studies have shown [34] that the shape of the ^{1}H NMR resonance of the thienylene proton is quite sensitive to the solvent quality. In a good solvent at relatively high temperatures the resonance at 7 ppm is homogeneous but rather broad. A decrease in solvent quality, either by an increased fraction of non-solvent or by a decreased temperature, gives a two-component resonance with one more narrow part simultaneously with a clearly observed color change (yellow to magenta). For still lower solvent qualities the sharper signal is decreased in intensity. This is illustrated in Figure 17 with variable temperature spectra of P3OT dissolved in $CDCl_3$:acetone-d_6 (9:1). The changes in the spectrum can be related to the conformational changes associated with the chromatic effects [34]. The broad signal observed at high temperature is thus representative of segments with varying torsion angles between adjacent rings. At a decreased temperature optical spectroscopy indicates an increased conjugation length. [61]. The narrower resonance can therefore be related to planar sequences. At lower temperatures the planar and stiff sequences aggregate, which explains the decreased intensity of the narrow resonance.

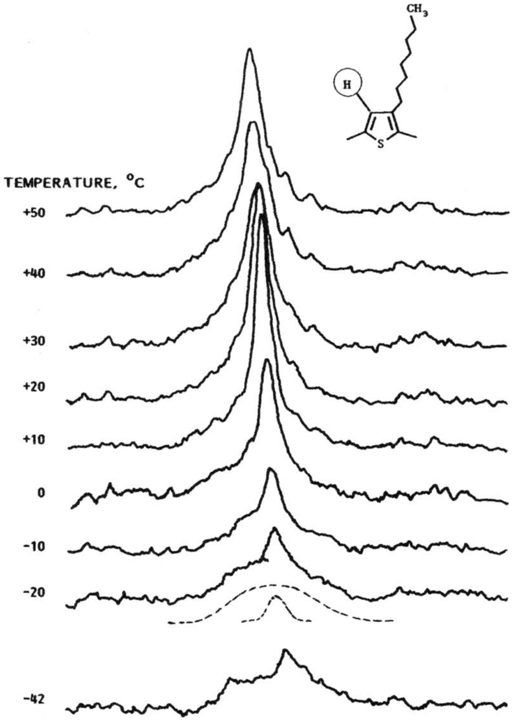

FIGURE 17. Variable-temperature ^{1}H NMR spectra of the thiophene proton of P3OT. The concentration of P3OT was 1mg/ml in $CDCl_3$:acetone-d_6 (9:1) [34].

We have also made a preliminary variable temperature CP MAS ^{13}C NMR investigation of solid P3HT. Increased temperature seems to induce *trans* to *gauche* transformations in the side chain ends below the chromatic transition temperature, indicating that "melting" of the alkyl chains might drive the disordering, leading to decreased planarity in the main chain.

The picture that emerges from studies of thermochromism and solvatochromism is thus consistent with a torsional deformation of the polymer main chain via the disordering of the alkyl side chains. This mechanism is sufficient to account for all the features of the chromatic transitions. Another mechanism, however, invoking the interaction between the polymer and its surrounding medium through the coupling of polarizability and conformational structure, as proposed by Schweizer [72-74], has not yet been ruled out. It may indeed be another manner in which the transition can be explained.

6. Doping and stability

6.1 METHODS OF DOPING

Poly(3-alkylthiophenes), like conducting polymers in general, are not conductive without doping. In the doping process charge carriers, polarons and bipolarons, are introduced into the conjugated chain. For each charge transferred into the thiophene chain, there must be a counterion in the vicinity of the polymer chain to guarantee overall charge neutrality. The available doping methods include electrochemical doping, vapor phase doping and solution doping with Lewis acids and nitrosyl compounds.

In electrochemical doping the reduction or oxidation of the polymer is caused by electric current, and the counterion is drawn into the system from the electrolyte solution with an electric field. This method is applicable to a large number of doping ions. Even anions of protonic acids can be used, because in electrochemical doping protonation of the thiophene chain does not occur [75,76], since the positive ions (e.g. H^+, Na^+) move towards the counter-electrode in the electrochemical cell. Tourillon [55] discusses the use of $CF_3SO_3^-$ anions in polythiophenes, and Nilsson *et al.* [77] have has used dodecylbenzene sulfonate anions for the doping of P3HT.

Doping with Lewis acids is another way to drive a charge transfer reaction with the polymer. An example is doping with ferrichloride, $FeCl_3$. In the doping process negative $FeCl_4^-$ ions are formed as the counter ions for the positively charged polymer chains. The reaction balance is maintained by the formation of $FeCl_2$. The doping is performed in a solution of ferrichloride in e.g. nitromethane. Both hydrous and anhydrous ferrichloride can be used. The main difference is that with anhydrous ferrichloride strong solutions and long doping times have to be avoided, whereas with hydrous ferrichloride $FeCl_3 \cdot 6H_2O$ one can use even 1 M solutions and over 24 h doping times without severe degradation of the polymer.

In vapor phase doping, the polymer is exposed to a gaseous dopant, typically a Lewis acid halide, that can take part in the charge transfer reaction. For instance iodine, I_2, forms

I_3^- and I_5^- ions in the doping process. The obtainable conductivity depends on the vapor pressure and doping time, as demonstrated by Satoh *et al.* [78]. Other dopants used in vapor phase doping with good results include BF_3, AsF_5 and SbF_5.

Doping can also be accomplished by using nitrosyl compounds. In the doping solution a NO molecule is formed when the NO^+ cation captures an electron from the polymer. The anionic part of the molecule then acts as a negative counterion, and gaseous NO escapes from the solution. In the literature NO compounds like $NOBF_4$, $NOPF_6$ and $NOSbF_6$ are mentioned in connection with P3ATs [38], (see also ref. [79] for a description of the NO doping technique).

Self-doping has also been studied with poly(3-alkylsulfonate-thiophenes) [7-9]. The self-doped polythiophenes have a SO_3H group at the end of the alkyl side chain. However, in order to obtain good conductivities, the charge transfer reaction has to be driven electrochemically to remove the cation (H^+ or Na^+) of the sulfonic acid out of the polymer. Otherwise this kind of self-doped polythiophenes have a very low conductivity in the dried state due to the protonation of the thiophene rings [75].

As discussed by Han *et al.* [76], protonic acids are not suitable dopants for PTs or P3ATs, although the optical and mechanical properties of P3ATs, mixed with protonic acids, to some extent resemble those of doped PTs. The reason is the protonation of the thiophene rings, which prevents the conduction of the electric current, in analogy with the case of the self-doped polymers.

6.2 CONDUCTIVITY

The best conductivities reported for polythiophenes is 1 kS/cm for drawn films [78]. These results were obtained with iodine doping at very high iodine concentrations. Conductivities of 190, 450 to 510, and 270 S/cm have been measured in as-synthesized polythiophene, poly(3-methylthiophene) and poly(3-ethylthiophene) films, respectively [80]. So far the conductivities measured for P3ATs with longer side chains have remained smaller even in the best cases. Sato *et al.*[20] have measured conductivities of 95, 78, 67, 17, and 11 S/cm for poly(3-hexylthiophene), poly(3-octylthiophene), poly(3-dodecylthiophene), poly(3-octadecylthiophene) and poly(3-eicosylthiophene), respectively. The samples were obtained in the doped state by electrochemical synthesis and in all cases the anion were PF_6^-. Doping of neutral P3AT films with PF_6^- or I_2 hardly produces conductivities above 10 S/cm [38]. For ferrichloride doped P3ATs conductivities of about 30-50 S/cm have been measured [5,81]. In stretched films values above 100 S/cm have been measured. In electrochemical doping with sulfonic acids the best conductivities have been in the range of 10 mS/cm [77]. Probably the reason is that a low doping level is achieved.

The conductivity of P3AT is dependent on the side chain length. The conductivity decreases with increasing number of methylene groups in the side chains. Wegner *et al.* [82] have studied the conductivity of ferrichloride doped poly(3-methylthiophene) and several poly(3,4- cycloalkylthiophenes) as a function of temperature and size of the cycloaliphatic ring. According to their results, the conductivity is proportional to exp(-2αR), where α is a constant ($1/\alpha = 0.38 \pm 0.16$ nm for polythiophenes) and R is the distance between the polymer backbones. However, the dependence on the length of the cycloaliphatic ring on conductivity observed by Wegner *et al.* is orders of magnitude larger than the effect of the length of the side chain in P3ATs [20], and thus it is uncertain whether these results are applicable to P3ATs.

6.3 DEDOPING

Polythiophene and poly(3-methylthiophene) appear remarkably stable when doped with $SO_3CF_3^-$ ions or ferrichloride [55,83], although they are unstable when doped with e.g. BF_4^- ions. The instability of PT has been studied by Billingham *et al.* [84] in the case of BF_4^-, ClO_4^- and PF_6^-. They noted the irreversibility of the dedoping, the sensitivity to the anion and the importance of moisture to the dedoping rate. On the other hand, P3ATs doped by any of the techniques described above are not completely stable. Especially when heated the thermal instability occurs as a loss of conductivity, change of mechanical properties, and, in some cases, emission of degeneration products. The degraded material is brittle, black, and insoluble, it does not melt and subsequent redoping produces results inferior to the doping of undegraded material.

Jen *et al.* [5] have studied the stability properties of several P3AT polymers and copolymers. In $FeCl_3$ doped poly(3-butylthiophene-co-3-methylthiophene) (50:50) the conductivity decays in ambient air at room temperature from 67 S/cm to 16 S/cm in 10 months, and at 100 °C by a factor of 2 in 24 h. The order of stability for some dopants is $SbF_6^- > FeCl_4^- > PF_6^- \approx BF_4^- > I_3^-$. The order of stability is based on TGA measurements and thus may not be relevant for the stability concept discussed here, since the result depends on the formation of evaporating components in the dedoping process. Especially the half-life of the conductivity in $SbF6^-$ doped polymer is only 60 days at room temperature, which is well below the corresponding stability of $FeCl_3$ doping, (humidity may affect the results, however).

The degradation of the conductivity of thin films (1000 Å) of P3HT at 110 °C in laboratory air is very rapid [85,86]. In one hour the conductivity decreases by several orders of magnitude (Figure 18). Also the importance of humidity has been observed, and actually in dry argon atmosphere the rate of the conductivity degradation is much smaller, as can be seen in Figure 18. The degradation mechanism has been studied through the use of in-situ optical spectroscopy [85,86]. The results clearly show that the degradation of the conductivity is due to dedoping of the polymer. This conclusion has also been fortified by photoemission spectroscopy as well as infrared spectra. The influence of water was clear in the undoping mechanism, but thermal undoping took place also in the ultra high vacuum chamber used for photoemission spectroscopy, although at higher temperatures and longer duration.

Wang et.al. [87,88] have measured the conductivity during dedoping of several P3ATs (butyl, hexyl and octyl) at various temperatures. The dopants used were $FeCl_3$ and $NOPF_6$. They observe a strong temperature dependence. The stability also decreases with increasing length of the alkyl side chains.

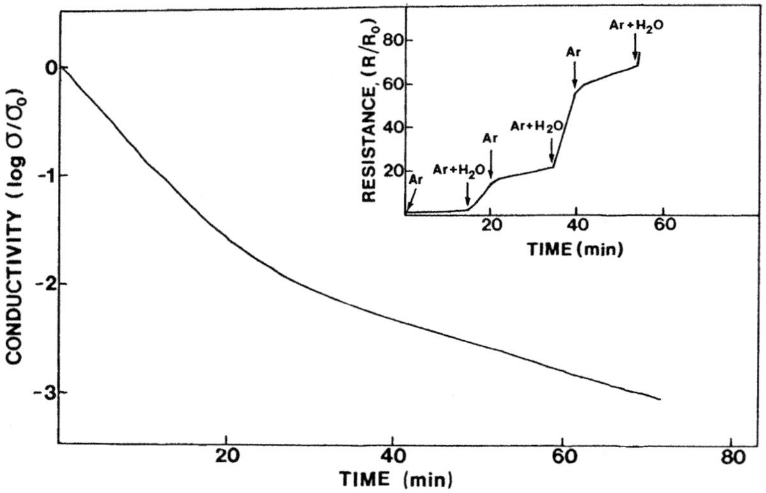

FIGURE 18. Normalized conductivity *vs.* time of a $FeCl_3$-doped P3HT film at 110 °C in laboratory atmosphere. The film thickness was around 1000 Å. The inset shows the resistance *vs.* time of a $FeCl_3$-doped P3HT film at 110 °C in argon with and without water. The film thickness was around 1000 Å [85].

The dedoping process, as studied by following the conductivity at constant and elevated temperature, can be described with an activation energy [87-89]. The activation energy E_A is determined using the temperature dependence of the time constant of the conductivity decay, τ, (the Arrhenius law)

$$\tau = \sigma/(d\sigma/dt) \sim \exp(E_A/RT)$$

A typical result is shown in Figure 19. According to Loponen *et al* two typical activation energies can be identified. One is approximately 15000 K·R (~130 kJ·mol^{-1}). It belongs to such dopants as hydrous and anhydrous ferrichloride and various organic acids including sulfonic acids of very different sizes. Within the experimental accuracy, the activation energy is similar for all these dopants. An activation energy in the range 5000 to 10000 K·R (~ 40 to ~ 90 kJ·mol^{-1}) is typical to such dopants as I_2 and PF_6^-. It is not yet certain whether there is just one or several values of E_A in this category. In addition to the temperature dependence, the dedoping process also depends on the dopant concentration, which can be seen as the dependence of τ on the conductivity. Above about 1 mS/cm τ in ferrichloride doped P3OTs is nearly independent of the conductivity, but at lower conductivities τ increases with decreasing conductivity. The organic acids for which only values below 1 mS/cm have been studied so far also show the $\tau \propto \sigma^{-1}$ behavior in that conductivity range . For I_2 the conductivity dependence, $\tau \propto \sigma^{-1/2}$, is observed.

Comparable results have been published by Hagiwara *et al.* [90] for the thermal dedoping of ClO_4 doped poly(3,4-dimethoxythiophene). Although the given data do not

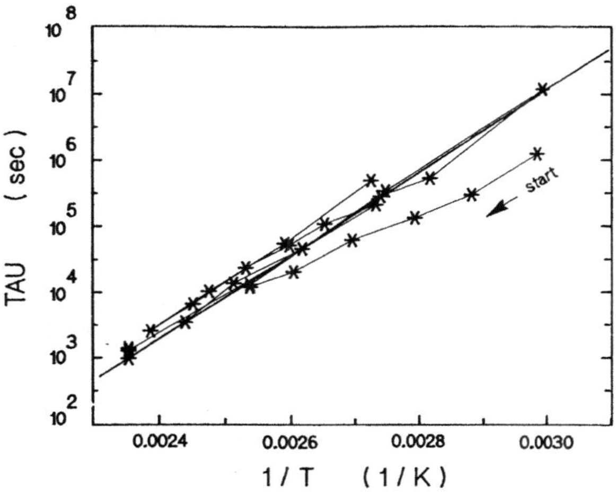

FIGURE 19. The dedoping time constant $\tau = k^{-1} = \sigma/(d\sigma/dt)$ of conductivity σ, of poly(3-octylthiophene) doped with hydrous ferrichloride, as a function of inverse temperature, 1/T. The measurement has been performed by changing the temperature stepwise upwards and downwards, repeatedly, measuring the time constant at each temperature. The thick line is fitted to a subset of points measured consecutively. It corresponds to an activation energy E_a, of 14500 K·R. The first points in the measurement (labelled "start") deviate from the general trend. [89]

allow an accurate interpretation in terms of an activation energy, an estimate $E_A < 10000$ K can be made.

The observed instability of doped P3ATs is a serious problem in practical applications. In the case of ferrichloride, which so far is the most stable dopant in room temperature, the dedoping time constant is shortened from 10 .. 100 years at room temperature in dry atmosphere to minutes at 180 °C. The iodine doped P3ATs have a less severe temperature dependence, but the room temperature value of the dedoping time constant is already much shorter, of the order of one day at 1 S/cm for sample thicknesses of 0.1 mm. Wang et.al. [87,88] also notice that $FeCl_3$ doped samples is much more stable against thermal dedoping than PF_6^- doped samples.

The mechanism responsible for the thermal dedoping is not fully understood yet. One hypothesis is that the undoping process is driven by conformational changes of the polymer backbone [85,86]. This hypothesis is based on the physics of the thermochromic transition in the solid state. In brief, it is suggested that the torsional deformation of the polymer backbone occurs also in the doped state, and that such torsional deformation would lead to an increase of the oxidation potential of the polymer, causing an oxidation of some species and a reduction of the polymer. A conversion of the dopant anion mediated by the postulated chemical species is necessary to form a consistent explanation of the thermal undoping effects. It is thus conjectured that the role of the alkyl side chains, giving molecular mobility and fusibility in the solid/liquid state, gives rise to thermal undoping in the doped state. This theory has not yet been adequately tested, but has received support from recent studies by Wang *et al* [87,88] of undoping in P3ATs .

The chemical processes occurring during the thermal dedoping have not been sufficiently studied so far. It is mentioned by Jen *et al.* [5] that in the dedoping process neutral Lewis acid halides are liberated (BF_3, PF_5, I_2 etc.). A possible dedoping reaction is:

$$P3AT^{+} + PF_6^{-} \rightarrow P3AT\text{-}F + PF_5 .$$

The dedoping mechanism seems to be associated with the fluorination of the polymer chain in these cases. In ferrichloride doped P3OT it has been observed that the anion ($FeCl_4^-$) is converted to $FeCl_2$, possibly via the formation of $FeCl_3$, during thermal dedoping [85,86]. In addition, a loss of chlorine was observed. Recent results have shown that the chlorine leaves the polymer in the form of HCl [89]. In that case the conducting polymer probably looses one hydrogen in the dedoping process. It is also plausible that, when P3ATs doped with organic acid anions are thermally dedoped, a hydrogen is captured by the acid anion. This suggests that the dedoping mechanism of ferrichloride and organic acid anion doped P3ATs is related to the loss of hydrogen from the polymer, forming a possible point for crosslinking.

The observed dedoping activation energies of various dopants of very different types and sizes seem to appear rather similar. For instance, the same dedoping activation energy of 15000 K·R is found both for hydrous and anhydrous ferrichloride and for organic acids of various size. It may be of importance that the dopants for which stability has been reported in the case of PT are just $FeCl_3$ and $F_3CSO_3^-$. As noted above, both of these also absorb hydrogens in the dedoping process. Furthermore, with these dopants there is no dependence on sample thickness (if the thermal undoping is performed in inert atmosphere). Thus, the dedoping process cannot be controlled by diffusion of the dopant ions in these cases. A tentative conclusion is that the rate limiting step in the dedoping process must be related to the conductive polymer itself, not to the dopant ion. However, for iodine the dedoping time constant strongly depends on the sample thickness and thus on the diffusion of I_2 molecules. It is not yet certain whether diffusion is of importance for all Lewis acid halides.

Thermally dedoped P3AT have been studied with many techniques. From FTIR spectra of dedoped films of P3AT no evidence for an irreversible chemical attack on the polymer can be obtained [85,86,88]. Several cycles of doping and thermal undoping did not give rise to accumulated defects high enough in concentration to be detected in the FTIR spectra. The possibilities of a low level of chemical reactions such as crosslinking or chlorination of the alkyl chains or the backbone cannot, however, be excluded with these experiments. Rheological measurements at temperatures above 180°C, indicate that crosslinking actually is involved in the thermal degradation of P3ATs [91]. There have been very few attempts to quantify how this type of reaction occurs. However, Wang *et al* [88] claim that similar crosslinking occurs also in the neutral P3AT, i.e., that temperature is sufficient and that dopants cause no major new crosslinking mechanism. Yoshino *et al.*

[92] have studied crosslinking in solution. They conclude that the amount was not very extensive and well below the analytical testing methods available to them.

Taka *et al.* [93,94] have compared the soluble and insoluble polymerization products of P3OT. The insoluble part is assumed to resemble the thermally dedoped, degenerated polymer but no significantly large differences between the soluble and the insoluble parts can be seen in the NMR and XPS spectra and the interpretation of the results is still open. In crystallinity only a small difference is found.

It is clear that much more work is needed to fully understand the processes in thermal dedoping and, perhaps, in the stabilization of the doped melt processable polythiophene derivatives against it. The absence of clear spectroscopic evidence for chemical degradation of the polymer after thermal undoping does not exclude the possibility that this is actually what occurs. Indeed, the source of protons implicated in hydrogen chloride evolution of the P3AT($FeCl_4^-$) is most probably the polymer. The possibility of reversible charge transfer from the polymer is demonstrated in the reversible thermal undoping observed in solutions of P3AT($FeCl_4^-$) [86]. However, in the solid state irreversible thermal degradation occurs. If reversible charge transfer should be possible both in the polymer solid and polymer solution, fine tuning of charge transfer processes may be necessary.

7. Transport properties

Among the transport properties of P3ATs we shall here primarily be concerned with the electrical conductivity, which is of importance from theoretical as well as practical points of view. Even the prototype polyacetylene which today exhibits a high degree of perfection leaves open questions in its solitonic framework. Much more so do the less perfect polyheterocycles, which, lacking the degenerate ground state, end up with the polaron/bipolaron concepts. These fundamental questions have been thoroughly addressed in a recent review [95]. Entering the real world with imperfect material we also have to consider the homogeneity of the material, i.e., models including conducting particles separated by insulating barriers. The question of homogeneity becomes even more intriguing for the polymer blends based on P3ATs. The unique processability of the materials, on the other hand, gives new means for dealing with the problem. Some of these efforts will be reported in this section.

7.1 FIELD EFFECT TRANSISTORS FOR TRANSPORT PROPERTY STUDIES.

In a survey of conduction properties it is of interest to divide the conductivity into its two components: charge carrier concentration, n or p for electrons and holes, respectively, and the corresponding carrier mobilities μ_n and μ_p. In this connection field effect transistor (FET) structures are useful. They, however, only cover the low conductivity range, typically $\sigma \approx 10^{-6}$ S/cm, i.e. essentially undoped material. A central problem in the fabrication are the thin (some ten nanometers) and still homogeneous layers that are needed. The first series of FET experiments addressing the mobility properties of conducting polymers were made on electrochemically grown PT films [96]. Sophisticated studies on polyacetylene FETs demonstrated the soliton formation in situ [97]. The solubility of the P3ATs facilitates easy preparation of high quality thin films e.g. by spin-coating, giving rise to a new line of transport studies using FETs.

P3ATs with a range of side chain lengths have been studied in FET structures [98,99]. The unintentionally doped materials, synthesized by Grignard coupling of 3-hexyl-2,5-diiodothiophene, have room temperature conductivities in the range 10^{-8} to 10^{-5} S/cm. The corresponding charge carrier concentration is 10^{17} to 10^{18} cm^{-3} resulting from spurious acceptor type impurities. These may at least partly be due to residual iodine from the polymerization process, which is detected at a level of 10^{18} atoms/cm^3 using Rutherford Backscattering Spectroscopy (RBS). The room temperature mobilities were 2×10^{-4}, 1×10^{-4}, 3×10^{-6} and 6×10^{-7} cm^2/Vs in P3BT, P3HT, P3OT and P3DT, respectively [98]. Looking for a systematic trend in these values one finds a decrease in μ with increasing side chain length, qualitatively in accordance with the decreased probability of hopping between chains.

In the case of P3HT a more thorough study on the temperature behavior in the range 130-430 K has been made [98]. The low temperature behavior can be fitted to a hopping type conduction, the nature of which (activated, Kivelson type or variable range hopping) cannot be determined on the basis of these measurements alone. An essential feature is that the mobility roughly shows the same temperature dependence (Figure 20), the charge carrier concentration $p = 3 \times 10^{17}$ cm^{-3} being almost constant with temperature. At higher

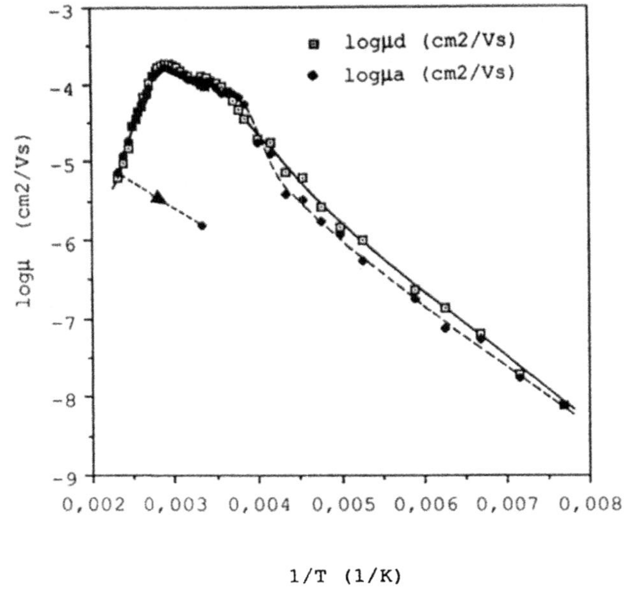

FIGURE 20. Mobilities of charge carriers in a spinned P3HT-FET device, in accumulation (μ_a) and depletion (μ_d) mode, during a temperature cycle, RT → 130K → 430K → RT [98].

temperatures, T>340 K, σ, p and μ decrease because of the decrease of active dopants and/or the formation of conformational excitations. The latter explanation has been further elaborated in a separate study of the mobility during high temperature cycling up to 350 K [99].

FET structures including P3ATs have also been prepared using the Langmuir-Blodgett (LB) technique [100]. Thin films of P3HT mixed with 40 mol % arachidic acid to facilitate LB deposition, exhibit a mobility of 7 x 10^{-7} cm^2/Vs. These LB-structures may be considered as models - on a molecular level and in a highly ordered form - of blend materials to be discussed below.

7.2 POLY(3-ALKYLTHIOPHENE) BLENDS.

The melt processability of P3ATs has facilitated their mixing with a wide variety of technology polymers such as EVA, PS, PE etc. [46]. The blends can further be doped by conventional methods resulting in electrical conductivity. These materials give new possibilities for tailoring e.g. structural and processing properties. In the widely studied EVA/P3OT blends an amount less than 20 weight-% of P3OT is enough to achieve a certain saturation in conductivity [27]. At the same time, however, the questions concerning the homogeneity of the conductivity of the material becomes more pronounced in comparison with the pure P3ATs.

For the P3AT based blends EVA has been the most studied matrix material as its structure seems to allow an efficient blending and doping. The conductivity of EVA/P3OT blends containing 20 weight-% P3OT has been systematically studied in a wide temperature range 10 to 300 K (Figure 21) [101]. The room temperature conductivity of the most highly doped sample reached 1 S/cm. The temperature dependence of the same material followed the charging energy limited tunneling (CELT) model dependence ($\sigma \sim \exp(T^{-0.5})$) indicative of a granular metal structure with metal island sizes of around 10 nm. A comparison with the domains found in the SEM studies on blends (section 4.2), shows that the metal islands are much smaller. This indicates that the conducting islands are formed within an insulating P3OT framework. The modelling of lightly doped material cannot be accomplished on the basis of conductivity behavior alone. Complementary thermoelectric power (TEP) measurements (Figure 22) showed for the lowest conductivity samples a temperature dependence ($S \sim T^{-1}$) typical of band conduction in disordered semiconductors, here for the first time seen in conducting polymers. The most highly doped material showed a metallic TEP connected with the conducting particles.

The instability in the conductivity of P3ATs and their blends at higher temperatures (T>100K) is well documented (see section 6.3 and [102]). In an effort to reduce this problem in blends, doped EVA/P3OT has been crosslinked using an electron beam [103]. The result of this treatment was an increase in the conducting particle size from some ten to some hundred nanometers. This effect, which increased with the amount of P3OT and crosslinking, was deduced from the conductivity model changing from the CELT behavior of the non-crosslinked material to a temperature fluctuation induced tunneling (TFIT) behavior, typical of the mentioned particle sizes, respectively. The dc conductivity

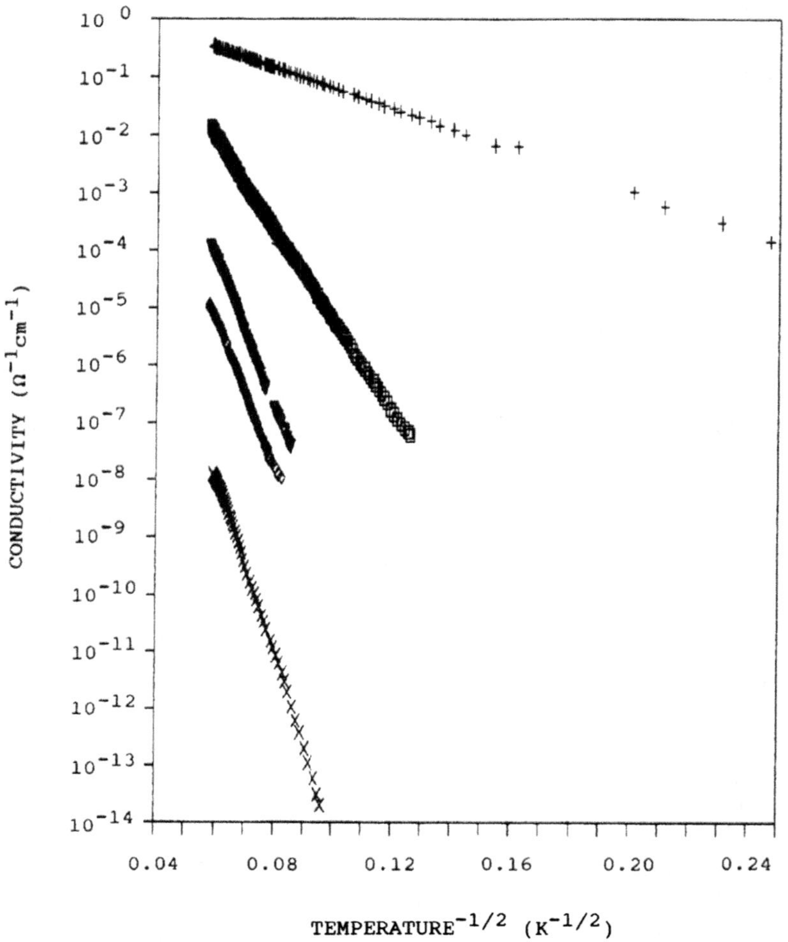

FIGURE 21. dc conductivity *vs*. inverse temperature, $T^{-1/2}$, in $FeCl_3$ doped EVA/P3OT samples [101].

behavior was here confirmed using complementary high electric field measurements. The original effort, i.e. to stabilize the material in a systematic way was not achieved. However, interesting connections between the P3AT polymerization route and thus its defects/impurities and crosslinking behavior became apparent, which emphasizes the importance of the crosslinking mechanisms for the material degradation.

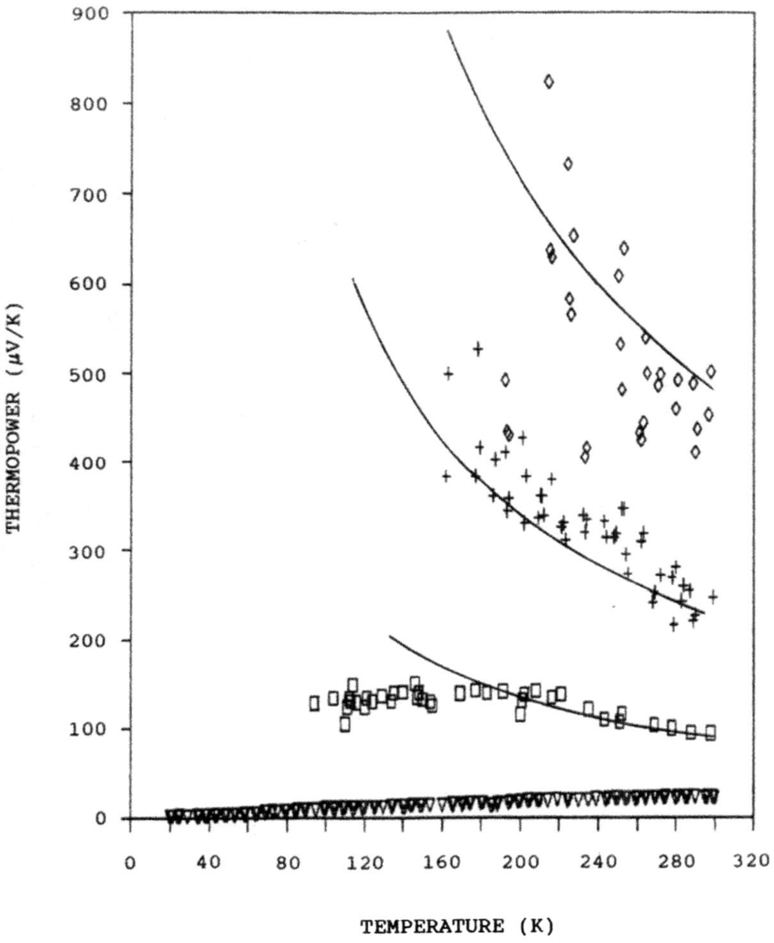

FIGURE 22. Thermoelectric power (TEP) *vs*. temperature in $FeCl_3$ doped EVA/P3OT samples [101].

8. Stretch orientation of poly(3-alkylthiophenes)

The inherent processability of the P3ATs and the good stability properties (at least in the undoped state) make these materials very interesting candidates for stretch orientation. Elongations up to 5 times the initial length upon stretching have been reported for P3HT [104] and P3OT [40]. Furthermore, Yoshino *et al.* [105] have shown that it is possible to stretch gels of P3OT up to 5 times the initial length. These gels showed highly anisotropic and reversible size changes upon change of temperature and solvent composition and also doping. The ability of the P3ATs to form polymer blends with common technology polymers and elastomers gives new possibilities of achieving high extension ratios. This has successfully been demonstrated by Yoshino *et al.* [106] who stretched a P3AT/ethylene-propylene elastomer (EPR) composite to a stretch ratio of 12.

Stretch-oriented P3ATs have been studied by means of x-ray diffraction [40], infrared [40,104] and optical spectroscopy [106,107]. The infrared dichroic spectra of a stretched, free-standing film of P3OT is shown in Figure 23. In both spectra well defined interference fringes appear which indicate the macroscopic uniformity of the film. The difference of the period of these fringes reflects the anisotropy of the refractive index in the material. The dichroic features are evident in this figure and there are several bands that are highly anisotropic. The band at 1510 cm^{-1}, which is due to the ν(ring) mode show the strongest dichroism. According to Hotta *et al* [104] the transition moment of the corresponding mode in PT is polarized in the thiophene ring plane and perpendicular to the symmetry axis of the thiophene ring, i.e., along the chain axis. Assuming the same polarization of the 1510 cm^{-1} band in P3OT, we can conclude that the polythiophene main chains are highly oriented along the stretch direction. This is supported by the dichroic characteristic of the band around 820-830 cm^{-1}. As Hotta *et al.* [104] has pointed out, this band consists of two peaks, one at 824 cm^{-1} and another which appears as a shoulder at 830-835 cm^{-1}. The former of these can be assigned to the CH out-of-plane mode which is

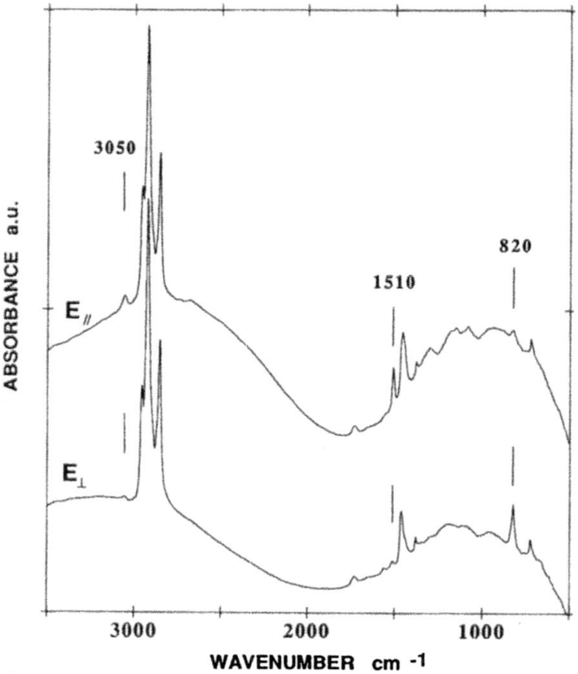

FIGURE 23. Infrared spectra of a stretched, free-standing film of P3OT for light polarized with the electric vector parallel ($E_{||}$) and perpendicular ($E_{\perp}$) to the stretch direction.

polarized perpendicularly to the thiophene plane. The fact that this band is polarized perpendicular to the stretch direction shows that the chains are oriented parallel to the stretching direction.

The bands at 1306, 1184, 1150 and 1084 cm^{-1} show quite strong dichroism. These bands are doping induced bands and the relatively high intensity of these bands show that the polymer is lightly doped. This doping is probably due to remaining $FeCl_3$ from the polymerization process. As can be seen all the doping induced bands are polarized along the chain axis. This has also been observed in P3HT and PT [104].

The degree of orientation in the stretched P3ATs can be expressed by Hermans orientation function, which is defined as [108]:

$$f = [3\langle\cos^2\sigma\rangle - 1]/2$$

where $\langle\cos^2\sigma\rangle$ is the average square cosine of the angle σ between the chain axis and the stretch direction. The value of this orientation function and can be determined by means of the dichroic ratios, i.e., the ratio of absorbance parallel and perpendicular to the stretch direction, of the infrared spectra. By using the dichroic ratios of the peaks at 1510 cm^{-1} and 824 cm^{-1} of an infrared spectra of a film stretched 5 times the initial length at 100°C we obtain the value $f = 0.54$ [40]. This is a surprisingly low value compared to the value 0.80 [40] predicted by the Kratky model [109,110] for the same stretch ratio. The reason for the discrepancy is twofold. Firstly, the Kratky model is known to give an upper level for the orientation in semicrystalline polymers since it is based on the assumption that the units building the polymer are rigid entities that rotate upon orientation. Secondly, the stretching was performed at an elevated temperature where relaxation processes are likely to decrease the degree of orientation. This relaxation is probably primarily occurring in the amorphous phases. This is supported by x-ray diffraction measurements which probes the orientation of the crystalline regions. Figure 24 show the azimuthal scan of the first diffraction peak in the diffractogram of a stretch-oriented film of P3OT. Calculating the value of the Hermans orientation function from the azimuthal variation of the intensity the value 0.72 is obtained [40]. This is considerably higher than the value obtained from the infrared spectra. This shows that the crystalline regions are better oriented than the amorphous parts.

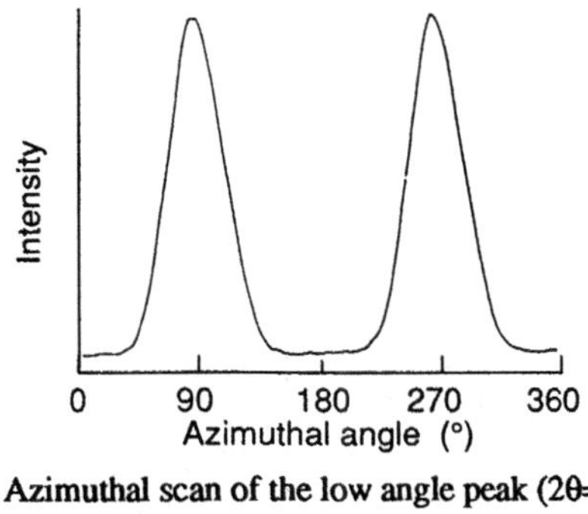

FIGURE 24. Azimuthal scan of the low angle peak (2θ=4.4 °) in the diffractogram of P3OT [40].

A more well oriented material can be obtained if the P3OT is spun on a stretchable substrate, e.g., polyethylene. These kind of structures also have the advantage that films thin enough for optical transmission measurements can be obtained [107]. Figure 25 shows the dichroism of the infrared 1510 cm^{-1} peak of a stretched P3OT/PE structure [107]. As can be seen the dichroic ratio is very high. A calculation of the Hermans orientation function from these peaks gives the value $f = 0.82$. This is considerably higher than the orientation obtained in free-standing films of P3OT, stretched to approximately the same stretch ratio, $l/l_0 = 5$. At present we do not have any explanation for this difference in orientation but it can possibly be due to differences in the actual stretch ratio since the measurements of this ratio always is uncertain. Moreover, the P3OT/PE structure was stretched at room temperature where relaxation process are less likely to occur.

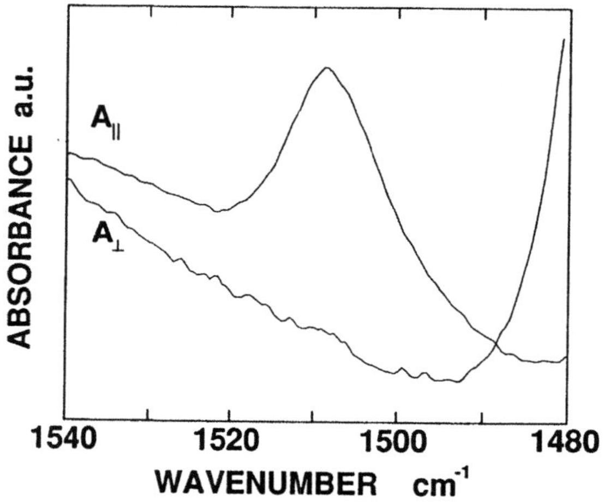

FIGURE 25. The 1510 cm^{-1} infra-red absorption band (C=C stretch) of a stretched P3OT/PE structure for light polarized parallel ($A_{||}$) and perpendicular ($A_{\perp}$) to the stretch direction [107].

9. Applications

9.1 APPLICATIONS THROUGH PROCESSABILITY.

The development of truly processable conductive polymers has been heralded as the crucial turning point for industrial applications. It is also now demonstrated that many of the promises held out are currently being realized. The realization of polymer blends by solution or melt processing enables construction of materials with a considerable conductivity at low volume fraction of the conducting polymer in the blend. We are now able to stretch align the polymers, attaining quite high degrees of orientation. Anisotropic conductivity should be expected. And the preparation of thin polymer films by casting or spinning from solution has enabled studies of transport properties, using field effect transistors, that were very difficult to perform only some years back. The higher level of characterization that is possible with these materials bodes well for further development. And the serious problems of thermal instability certainly requires much progress in the route towards processable and stable conductive polymers.

9.2 ELECTRONIC DEVICES: TRANSISTORS AND DIODES

Our FET structures have been used solely for materials characterization as described in section 7. At the same time they represent a new family of Molecular Electronics devices. Typical current-voltage characteristics of P3HT-FET is shown in Figure 26 [98]. One primary limitation to the applicability of these FETs is the low carrier mobility of the hopping type conduction. A new degree of order, as found in our LB-FETs, does not diminish this problem. On the contrary, especially the thinnest films, consisting of a few molecular layers, show decreasing mobility and conductivity, which has been interpreted as a result of the dimensionality being reduced from 3D to 2D, with a corresponding increase in hopping distances [100]. The order of these fatty acid containing LB-films, on the other hand is not very good. It may be significantly higher in a mechanically oriented material.

Attempts to increase the mobility by introducing impurities as hopping centers have been made. Non-doping or compensating species are then used in order not to increase the charge carrier concentration and correspondingly the conductivity, as the latter limits the performance of the FETs. Thus an increase by a factor of 100 in the mobility of poly(3-

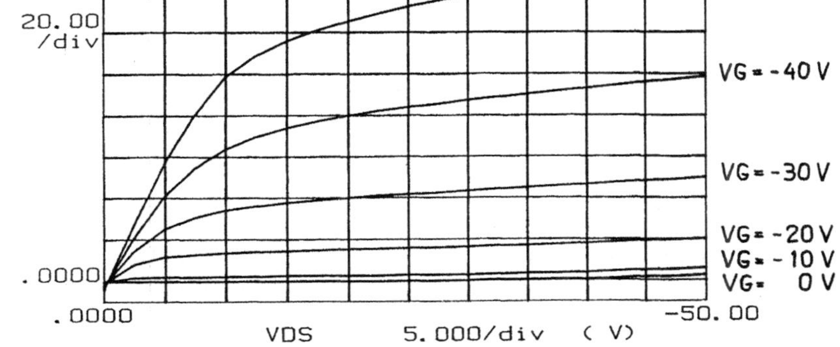

FIGURE 26. Current-voltage characteristics of P3HT-FET [98]

methylthiophene), by using copper, has been reported [111]. However, the exposure of a P3HT -FET structure to ammonia resulted in an almost reversible *decrease* in the mobility, whereas the carrier concentration remained constant [112]. Using potentially doping $SbCl_5$, we have increased the mobility of P3HT by more than an order of magnitude to 4 x 10^{-3} cm^2/Vs without a corresponding increase in carrier concentration. The value is still 1-2 orders of magnitude lower than the one for amorphous silicon, which is applied in, e.g., transistors for displays. The stability problems of conducting polymers are less critical in the case of FETs as undoped material is being used.

P3AT forms rectifying contacts to metals of low work function (Al, In, Cr, Ti) as well as the metal oxide ITO (indium tin oxide). These junctions have been studied in detail [113-115] and it has been shown that charge transport through the metal-polymer energy barrier is not well described by conventional thermionic emission theory. One probable cause of this behavior may be that tunneling through the barrier makes a major contribution to the charge transport. Capacitance-voltage analysis has also shown that the dopant distribution profile is inhomogeneous, and that dopants (in the concentration range of 10^{17} - 10^{18} cm^{-3}) are accumulated close to the rectifying interface [115]. This accumulation makes for thin barriers, and thus enables tunneling through the barrier. Furthermore, it has been shown how an applied electric field can change the dopant distribution profile. Under backward bias, dopants are accumulated close to the blocking interface. This mechanism is also operative in diodes under open circuit, causing the observed inhomogeneous dopant distribution. This mechanism of dopant diffusion can also explain many of the problems of reproducibility of these devices.

The demonstration of melt processed rectifying metal-polymer junctions using P3ATs [116] shows that some of the advantages of polymer processing can, in principle, be combined with the materials quality needed for electronic devices. Such melt processed devices have been formed at temperatures close to the softening point of the P3ATs used. The development of melt processable electronics (meltronics) may not be of immense practical importance, but it certainly demonstrates that new processing aspects will be relevant for polymer electronics.

9.3 NONLINEAR OPTICAL PROPERTIES

The nonlinearity of conjugated polymers was discovered and studied both experimentally and theoretically already in the 1970's [117-119]. Because of the lack of stable and processable materials the research activity then decreased to rise again in recent years with the advent of soluble polydiacetylenes and polythiophene derivatives.

The third-order optical nonlinearity of the conjugated polymers is based on the electron delocalization, which makes the electron cloud easily polarizable. Because of this nonresonant origin, the nonlinearity is very fast with subpicosecond response times [120]. Fukaya *et al* [121] have measured the intensity-dependent index of refraction of P3OT using a surface plasmon resonance method. The result was $n_{NL} = 5 \cdot 10^{-16}$ m^2/W at the wavelength 1.06 μm, which is of the same order of magnitude as the values obtained for polydiacetylenes.

Logsdon *et al* [122] studied the resonantly enhanced nonlinearity in poly(3-dodecylthiophene) Langmuir-Blodgett films. Using a degenerate four wave-mixing method they measured a third-order susceptibility $\chi^{(3)} = 10^{-15}$ m^2/W at 602 nm. In situ iodine doping decreased the value to one tenth of the original one. For a new polythiophene derivative containing aromatic and quinoid moieties Jenekhe *et al* [123] measured a resonant $\chi^{(3)} = 7 \cdot 10^{-15}$ m^2/W at 532 nm. The reported nonlinearity magnitudes indicate that conjugated polymers may be applicable in integrated optics [124-126].

10. Conclusions

The development of processable conductive polymers has now led to a class of materials that show both melt and solution processability, while still being capable of high electronic conductivity. High molecular weight materials of sufficient purity and regularity for fundamental physical studies are routinely obtained. In these conductive polymers, new physics has been discovered in the form of thermochromism and solvatochromism. The mechanisms behind these new effects have been investigated leading to a full and self-consistent account of the geometric and electronic aspects of the chromatic effects. The processing aspects of the polymers have led to a series of new polymer blends, closing into the practical application of these materials for technical purposes. New applications and properties have become feasible through thin film formation by spinning. The main remaining problem of these materials, the thermal instability of the doped polymer, still awaits solution.

ACKNOWLEDGEMENTS

We are grateful to Dr Yading Wang and Prof. Michael F. Rubner for kindly providing their paper prior to publication. Furthermore, we are indebted to all our colleagues at our institutes for their contributions. Financial support from Neste Oy, the Nordic Fund for Industrial and Technological Development and the Swedish Natural Science Research Council is gratefully acknowledged.

REFERENCES

1 Edwards, J.H. and Feast, W.J. (1980) "A new synthesis of polyacetylene", *Polymer* **21**, 595.

2. Wessling, R.A. and Zimmerman, R.G. (1968) US Patent No. 3 401 152 and (1972) US patent No. 3 706 677.

3. Murase, I., Ohnishi, T., Noguchi, T. and Hirooka, M. (1987) " Highly conducting poly(2,5-thienylene vinylene) via a precursor route", *Polymer Commun.* **28**, 229-231.

4. Sato, M., Tanaka , S. and Kaeriyama K. (1986) "Soluble conducting polythiophenes.", *J. Chem. Soc. Chem. Commun.*, 873-874

5. Jen, K.Y., Miller, G.G. and Elsenbaumer, R.L. (1986) "Highly conducting, soluble and environmentally-stable poly(3-alkylthiophenes)", *J. Chem. Soc. Chem. Commun.*, 1346-1347.

6 Hotta, S., Rughooputh, S.D.D.V., Heeger, A.J. and Wudl, F. (1987) "Spectroscopic studies of soluble poly(3-alkylthienylenes).", *Macromolecules* **20**, 212-215.

7. Patil, A. O., Ikenoue, Y., Basescu, N., Colaneri, N., Chen, J., Wudl, F. and Heeger, A. J. (1987) "Self-doped conducting polymers", *Synth. Met.*, **20**, 151-159.

8. Patil, A. O., Ikenoue, Y., Wudl, F. and Heeger, A. J. (1987) "Water-soluble conducting polymers", *J. Am. Chem. Soc.*, **109**, 1858-1859.

9. Patil, A. O. (1989) "Polyalkylsulfonatethiophene: electrically conducting, soluble polythiophene", *Synth. Met.* **28** C495-C500, and references therein.

10. Rühe, J., Ezquerra, T.A., and Wegner, G. (1989) "New conducting polymers from 3-alkylpyrroles" *Synth. Met.* **28**, C177-C181.

11. Feast, W.J., Millichamp, I.S., Friend, R.H., Horton, M.E., Phillips, D.,Rughooputh, S.D.D.V. and Rumbles, G. (1985) "Optical absorption and luminescence in poly(4,4'-diphenylenediphenylvinylene).", *Synth. Met.*, **10**, 181-191

12. Rehahn, M., Schlüter, A.-D. Wegner, G. and Feast, W.J. (1989) "Soluble poly(*para*-phenylene)s. 1. Extension of the Yamamoto synthesis to dibromobenzenes substituted with flexible side chains.", *Polymer*, **30**, 1054-1059.

13. Tamao, K., Kodama, S., Nakajima, I., Kumada, M., Minato, A. and Suzuki, S. (1982) "Nickel-phosphine complex catalyzed Grignard coupling-II - Grignard coupling of heterocyclic compounds" *Tetrahedron* **38**, 3347-3354.

14. Laakso, J. (1988) "Synthesis and analysis of some sulphur and nitrogen containing conducting polymers including conducting polymer blends of poly(3-

octylthiophene).", *Acta Polytechnica Scandinavica, Chemical technology and Metallurgy Series* **184,** 1-48.

15. Gandha, B. and Sugden, J.K. (1979) "Effect of light on the formation of Grignard reagents" *Synth. Commun.* **9**, 845-849.
16. Lai, Y.-H. (1981) "Grignard reagents from chemically activated magnesium" *Synthesis*, 585-604.
17. Van Pham, C., Mark, H.B., Jr and Zimmer, H. (1986) "A convenient synthesis of 3-alkylthiophenes" *Synth. Commun.* **16**, 689-696.
18. Waltman, R.J. and Bargon, J. (1986) "Electrically conducting polymers: a review of the electropolymerization reaction, of the effects of chemical structure on polymer film properties, and of applications towards technology" *Can. J. Chem.* **64**, 76-95.
19. Hurd, C.D. (1969) "Thiophene chemistry 1962-1968" *Quaterly reports on sulfur chemistry* **4**, 187.
20. Sato, M.-A., Tanaka, S. and Kaeriyama, K. (1987) "Soluble conducting polymers by electrochemical polymerization of thiophenes having long alkyl substituents" *Synth. Met.* **18**, 229-232.
21. Kaeriyama, K., Sato, M. and Tanaka, S. (1987) "Electrochemical preparation of conducting polyalkylthiophene films" *Synth. Met.* **18**, 233-236.
22. Bryce, M.R., Chissel, A., Kathirgamanathan, P., Parker, D. and Smith, N.R.M. (1987) "Soluble, conducting polymers from 3-substituted thiophenes and pyrroles" *J. Chem. Soc., Chem. Commun*, 466-467.
23. Hotta, S. (1987) "Electrochemical synthesis and spectroscopy study of poly(3-alkylthienylenes)" *Synth. Met.* **22**, 103-113.
24. Yamamoto, T., Sanechika, K.-I. and Yamamoto, A. (1980) "Preparation of thermostable and electric-conducting poly(2.5-thienylene).", *J. Polym. Sci, Polym. Lett. Ed.* **18**, 9-12.
25. Yamamoto, T., Sanechika, K.-I. and Yamamoto, A. (1983) "Preparation and characterization of poly(thienylene)s" *Bull. Chem. Soc. Jpn.* **56**, 1497-1502.
26. Barker, J.M., Huddleston, P.R. and Wood, M.L. (1975) "Manich reactions of methoxythiophenes" *Synth. Commun.* **22**, 59-64.
27. Österholm, J.-E., Laakso, J., Nyholm, P., Isotalo, H., Stubb, H., Inganäs, O. and Salaneck, W.R. "Melt and solution processable poly(3-alkylthiophenes) and their melts" *Synth. Met.* **28**, C435-C444.
28. Nyholm, P., to be published.
29. Allied Corporation (1986) "Solution processible forms of neutral and electrically conductive poy(substituted heterocycles)" European patent application 0 203 438.

30. Sugimoto, R.-I., Takeda, S., Gu, H.B. and Yoshino, K. (1986) "Preparation of soluble polythiophene derivatives utilizing transition metal halides as catalysts and their property" *Chem. Exp.* **1**, 635-638.

31. Elsenbaumer, R.L., Jen, K.Y., Miller, G.G., Eckhardt, H., Schaklette, L.W. and Jow, R. (1987) "Poly(alkyl thiophenes) and poly(substituted heteroaromatic vinylenes): Versatile, highly conductive, processible polymers with tunable properties" *Springer Series in Solid State Science* **76**, 400-406

32. Hotta, S., Soga, M. and Sonoda, N. (1988) "Novel organosynthetic routes to polythiophene and its derivatives." *Synth. Met.* **26**, 267-279

33. Leclerc, M., Diaz, F.M. and Wegner, G. (1989) "Structural Analysis of Poly(3-alkylthiophene)s" *Makromol. Chem.* **190**, 3105-3116

34. Hjertberg, T and Lagerstedt-Eidrup, M.-L., " A solution and solid state NMR investigation of chromism in poly(3-alkylthiophenes)", submitted.

35. Inganäs, O., Salaneck, W.R., Österholm, J.-E. and Laakso, J. (1988) "Thermochromic and solvatochromic effects in poly(3-hexylthiophene)" *Synth. Met.* **22**, 395-406.

36. Drott, E.E. and Mendelson R.A. (1970) "Determination of polymer branching with gel-permeation chromatography. 1. Theory." *J. Polym. Sci., Polym. Phys. Ed.* **8**, 1361

37. Drott, E.E. and Mendelson R.A. (1970) "Determination of polymer branching with gel-permeation chromatography. 2. Experimental results for polyethylene." *J. Polym. Sci., Polym. Phys. Ed.* **8**, 1373.

38. Elsenbaumer, R.L., Jen, K.Y., Miller, G.G., and Shaklette, L.W. (1987) "Processible, environmentally stable, highly conductive forms of polythiophene" *Synth. Met.* **18**, 277-282.

39. Hjertberg, T. and Lagerstedt-Eidrup, M.-L. "Molecular weight distribution and thermal degradation of poly(3-alkylthiophene)", submitted.

40. Gustafsson, G., Inganäs, O. Österholm, H. and Laakso, J. "X-ray diffraction and infrared spectroscopy study of oriented poly(3-octylthiophene)." to be published in *Polymer*.

41. Winokur, M.J., Spiegel, D., Kim, Y., Hotta, S. and Heeger, H.J. (1989) "Structural and absorption studies of the thermochromic transition in poly(3-hexylthiophene) " *Synth. Met.* **28**, C419-C426

42. Yoshino, K., Nakajima, S., Gu, H.B. and Sugimoto, R.-i. (1987) "Optical properties of solution of polythiophene derivates as functions of alkyl chain length, concentration and temperature" *Jap. J. Appl. Phys.* **26**, L1371-L1373.

43. Yoshino, K., Nakajima, S. Gu, H.B. and Sugimoto, R.-I. (1987) "Absorption and emission spectral changes in a poly(3-alkylthiophene) solution with solvent and temperature" *Jap. J. Appl. Phys.* **26,** L2046-L2048.

44. Yoshino, K., Nakajima, S., Fujii, M. and Sugimoto, R.- I. (1987) "Conducting polymer fibre prepared by melt-spinning method from fusible polythiophene derivative" *Polym. Commun.* **28**, 309-314.

45. Hotta, S., Rughooputh, S.D.D.V. and Heeger, A.J. (1987) "Conducting polymer composites of soluble polythiophenes in polystyrene" *Synth. Met.* **22**, 79-87.

46. Laakso, J., Österholm, J.-Ö. and Nyholm, P. (1989) "Conducting polymer blends" *Synth. Met.* **28**, C467-C471.

47. Inganäs, O. and Gustafsson, G. (1990) "Thermochromism in the poly(3-alkylthiophenes and their polymer blends" *Synth. Met.* **37**, 195

48. Hjertberg, T. and Lagerstedt-Eidrup, M.-L., to be published.

49. Wu, C. R., Nilsson, J. O., Inganäs, O., Salaneck, W. R., Österholm, J.-E. and Brédas, J. -L. (1987) "Electronic Structure of Polythiophene" *Synth. Met.* **21**, 197-202.

50. Jugnet, Y., Tourillion G., and Duc, T.M. (1986) "Evidence of intrinsic extended π-bonding and metallic-like behavior in undoped and doped, electropolymerized poly(3-methylthiophene) films.", *Phys. Rev. Lett.* **56**, 1862-1865.

51. Clark, D. T., (1981) "ESCA Studies of Polymer Surfaces", in D. W. Dwight, T. J. Fabish and H. R. Thomas, (eds.), *Photon, Electron and Ion Probes of Polymer Structure and Properties*, American Chemical Society, Washington, DC, Chap. 17.

52. Salaneck, W. R. (1986) "Photoelectron spectroscopy of the valence electron structure of polymers.",*CRC Rev. Sol. State. Mat. Sci.* **12**, 267.

53. Salaneck, W. R. (1986) "Photoelectron spectroscopy of electrically conducting organic polymers" in Skotheim, T., (Ed.), *Handbook of Conducting Polymers,* Marcel Dekker, New York, Vol. 2, Chap. 37.

54. Mo, Z., Lee, K. B., Moon, Y. B., Kobayashi, M., Heeger, A. J. and Wudl, F.(1985) "X-ray scattering from polythiophene-crystallinity and crystallographic structure.", *Macromolecules* **18**, 1972 -1977.

55. Tourillon, G. (1986) "Polythiophene and its derivates" in in Skotheim, T., (Ed.), *Handbook of Conducting Polymers,* Marcel Dekker, New York, Vol. 1, Chap. 9.

56. Cui, C.X. and Kertesz, M. (1987) "The effect of helical conformation on the electronic structures of σ- and π-conjugated polymers" *Springer Series in Solid State Science* **76**, 400-406

57. Salaneck, W.R., Inganäs, O., Thémans, B., Nilsson, J.O., Sjögren, B., Österholm, J. -E., Brédas, J. -L. and Svensson, S. (1988) "Thermochromism in poly(3-hexylthiophene) in the solid state: A spectroscopic study of temperature-dependent conformational defects."*J. Chem. Phys.* **89**, 4613-4619.

58. Salaneck, W.R., Inganäs, O., Nilsson, J.O., Österholm, J.-E., Thémans, B and Brédas, J -L. (1989) "Thermochromism in the poly(3-alkylthiophenes): a study of

conformational defects by photoelectron spectroscopy "*Synth. Met.* **28**, C451-C460 .

59. Thémans, B., Salaneck, W.R. and Brédas, J. L. (1989) "Theoretical study of the influence of thermochromic effects on the electronic structure of poly(3-hexylthiophene)"*Synth. Met.* **28**, C359-364.

60 Sundberg, M., Inganäs, O., Stafström, S., Gustafsson, G. and Sjögren, B. (1989) "Optical absorption of poly(3-alkylthiophenes) at low temperatures", *Solid State Commun.* **71**, 435-439, and references therein.

61. Rughooputh, S. D.D.V., Hotta, S., Heeger, A.J. and Wudl, F. (1987) "Chromism of soluble polythienylenes.", *J. Polym. Sci. B, Polym. Phys.* **25**, 1071-1078

62. Inganäs, O., Sundberg, M., Gustafsson, G., Nilsson, J.O., Stafström, S. and Sjögren, B. (1989) "Chromism in the poly(3-alkylthiophenes)" *Springer Series in Solid State Sciences* **91**, 336-342.

63. Lögdlund, M., Lazzaroni, R., Salaneck, W. R., Stafström, S., Nilsson, J. -O., Shuang, X., Österholm, J. -E. and Brédas J. -L. (1989), "π-electronic structure of poly(3-hexylthiophene) studied by photoelectron spectroscopy" *Springer Series in Solid State Sciences* **91**, 367-371.

64. Lögdlund, M., Lazzaroni, R., Stafström, S., Salaneck, W. R. and Brédas J. -L. (1989) "Direct observation of charge-induced π-electronic structural changes in a conjugated polymer " *Phys. Rev. Lett.* **63**, 1841-1844 .

65. Stafström, S. and Brédas, J. -L. (1988) "Evolution of the electronic structure of polyacetylene and polythiophene as a function of doping level and lattice conformation" *Phys. Rev. B.* **38**, 4180-4191 .

66. Stafström, S., to be published

67. Inganäs, O., Gustafsson, G., and Salaneck, W.R. (1989) "Thermochromism in thin films of poly(3-alkylthiophenes)" *Synth. Met.* **28**, C377-C384.

68. Salaneck, W. R., Lazzaroni, R., Sato, N., Lögdlund, M., Sjögren, B., Keane, M. P., Svensson, S., Naves de Brito, A., Correia, N. and Lunell, S. (1990) "The electronic and chemical structure of poly(3-hexylthiophene) studied by photoelectron spectroscopy.", NATO ASI Series E, **182**.

69. Zerbi, G., Chierichetti, B. and Inganäs, O. "Vibrational spectra of oligothiophenes as model of polythiophene" submitted, *J. Chem. Phys.*

70. Zerbi, G., Chierichetti, B. and Inganäs, O. "Thermochromism in poly(alkylthiophenes): Molecular aspects from vibrational spectroscopy", submitted, *J. Chem. Phys.*

71. Zerbi, G., Chierichetti, B. and Inganäs, O. "Non equilibrium structure and geometry relaxation in polyoctylthiophene", submitted to *Chem. Phys. Lett.*

72. Schweizer, K.S. (1986) "Order-disorder transitions of π-conjugated polymers in condensed phases. I. General theory" , *J. Chem. Phys.* **85**, 1156-1175

73. Schweizer, K.S. (1986) "Order-disorder transitions of π-conjugated polymers in condensed phases. II. Model calculations" , *J. Chem. Phys.* **85**, 1176-1183

74. Schweizer, K.S. (1986) "Configurational dependence of the optical properties of π-conjugated polymers. I. Strong disorder limit." , *J. Chem. Phys.* **85**, 4181-4193

75. Ikenoue, Y., Uotani, N., Patil, A. O., Wudl, F., and Heeger, A. J. (1987) "Electrochemical studies of self-doped conducting polymers: verification of the 'cation-popping' doping mechanism", *J. Am. Chem. Soc.* **109**, 1858-1859.

76. Han, C. C. and Elsenbaumer, R. L. (1989) "Protonic acids: generally applicable dopants for conducting polymers", *Synth. Met.* **30**, 123-131.

77. Nilsson, J. O. and Inganäs, O (1989) "Alkyl benzene sulfonate doping of poly(alkylthiophenes)", *Synth. Met.* **31**, 359-367.

78. Satoh, M., Imanishi, K. Yasuda, Y., Tsushima, R. Yamasaki, H., Aoki, S. and K. Yoshino (1989) "Electrical conductivity of drawn polythiophene films", *Synth. Met.* **30**, 33-38.

79. Kim, Y. H., Spiegel, D., Hotta, S. and Heeger, A. J. (1988) "Photoexitation and doping studies of poly(3-hexylthiophene)", *Phys. Rev B* **38**, 5490-5495.

80. Sato, M.-A., Tanaka, S. and Kaeriyama, K. (1985) "Electrochemical preparation of highly conducting polythiophene films", *J. Chem. Soc., Chem. Commun.*, 713-714.

81. Väkiparta, K., Moulton, J., Heeger, A. J., Smith, P., Isotalo, H., Stubb, H., and Loponen, M. T. (1990), "Temperature dependence of DC-conductivity in poly(3-alkylthiophenes) in temperature regime 20-430 K", to be published in Proc. ICSM 1990.

82. Wegner, G. and Rühe, J. (1989) "The structural background of charge-carrier motion in conducting polymers", to be published in *Faraday Discuss. Chem. Soc.*

83. Österholm, J.-E., Passiniemi, P., Isotalo, H. and Stubb, H. (1987) "Synthesis and properties of FeCl4-doped polythiophene", *Synth. Met.* **18**, 213-218.

84. Billingham, N.C., Calvert, P.D., Foot, P.J.S. and Mohammad, F. (1987) "Stability and degradation of some electrically conducting polymers", *Polymer Degradation and Stabilituy* **19**, 323-341.

85 Gustafsson, G., Inganäs, O., Nilsson, J. O. and Liedberg (1988) "Thermal undoping in poly(3-alkylthiophenes)", *Synth. Met.* **26**, 297-309.

86. Gustafsson, G., Inganäs, O., and Nilsson J.O. (1989) " Thermal instability of doped poly(3-alkylthiophenes)", *Synth. Met.* **28**, C427-C434.

87. Wang, Yading and Rubner, M.F. (1990) "Thermal stability of the electrical conductivity of poly(3-alkylthiophenes)", *Polymer Preprints* **31**, 402-403.

88. Wang, Yading and Rubner, M.F. "Stability studies of electrical conductivity of various poly(3-alkylthiophenes)" submitted.

89 Loponen, M. T., Taka, T., Laakso, J., Väkiparta, K., Suuronen, K., Valkeinen, P. and Österholm, J.-E. (1990) "Doping and dedoping processes in poly(3-alkylthiophenes)", to be published in Proc. ICSM 1990.

90. Hagivara, T., Yamamura,, M., Sato, K., Hirasaka, M. and Iwata, K. (1989)"Synthesis and properties of poly(3,4-dimethoxythiophene", *Synth. Met.* **32**, 367-379.

91. Pearson, D. (1990), private communication.

92. Yoshino, K., Nakano, K., Onoda, M., and Sugimoto, Ryu-ichi (1989) "Instability of conducting polymer, poly(3-alkylthiophene) gel with solvent, temperature and doping and effect of alkyl chain length", *Solid State Commun.* **70**, 609-613.

93. Taka, T., Loponen, M. T., Laakso, J. Suuronen, K., Valkeinen, P. and Österholm, J.-E., (1990) "Degradation studies of doped and dedoped P3OT", to be published in Proc. ICSM 1990.

94. Taka, T., Nyholm, P., Laakso, J., Loponen, M. T. and Österholm, J.-E. (1990) "Determination of impurity effects on solubility and processability of P3OT, to be published in Proc. ICSM 1990.

95 Heeger, A.J., Kievelson, S., Schrieffer, J.R. and Su, W.-P. (1988) "Solitons in conducting polymers", *Reviews of Modern Physics*, **60**, 781-850.

96. Tsumura, A. Koezuka, H. and Ando, T. (1988) "Polythiophene field-effect transistor: Its characteristics and operation mechanism", *Synth. Met.* **25**, 11-23, and references therein.

97. Burroughes, J.H., Jones, C.A. and Friend, R.H. (1988) "New semiconductor device physics in polymer diodes and transistors.", *Nature*, **335**, 137-141.

98. Paloheimo, J., Punkka, E., Stubb, H and Kuivalainen, P. "Polymer field-effect transistors for transport property studies.", submitted for publication in Proc. of NATO ASI "Lower Dimensional Systems and Molecular Electronics", Spetses, Greece 12-23 June 1989. Ed. R M Metzger. Plenum Press 1990.

99. Assadi, A., Svensson , C., Willander, M. and Inganäs, O. (1988) "Field effect mobility of poly(3-hexylthiophene)", *Applied Physics Letters* **53** (1988), 195-197.

100. Paloheimo, J., Kuivalainen, P., Stubb, H., Vuorimaa E. and Yli-Lahti, P. (1990) "Molecular field-effect transistor using conducting polymer Langmuir-Blodgett films.", *Applied Physics Letters*, **56**, 1157-1159.

101. Isotalo, H., Laakso, J., Kuivalainen, P., Stubb, H., Österholm J-E and Yli-Lahti, P. (1989) "Charge transport in $FeCl_3$-doped polyethylenevinylacetate-polyoctylthiophene blends." *physica status solidi (b)* **154** , 305-313.

102. Yoshino, K., Nakajima, S., Onoda, M and Sugimoto, R. (1989) "Electrical and optical properties of poly(3-alkythiophene)", *Synth. Met.* **28** , C349-C357.

103. Punkka, E., Laakso, J., Stubb H and Kuivalainen, P. (1990) "The effect of polymer crosslinking on the electrical properties of ethylenevinylacetate-poly(3-octylthiophene) polymer blends.", *Phys. Rev. B*, **41**, 5914-5918.

104. Hotta, S., Soga, M and Sonoda, N. (1989) "Infrared dichroic studies of polythiophenes", *J. Phys. Chem.*, **93**, 4994-4998

105. Yoshino, K., Nakao, K. and Onoda, M. (1989) "Anisotropic characteristics of drawn poly(3-alkylthiophene) gel." *Jpn. J. Appl. Phys.*, **28**, L1032-L1034.

106. Yoshino, K., Onoda, M. and Sugimoto, R. "Effect of polymer elongation on the absorption spectrum of poly(3-alkylthiophene).", *Jpn. J. Appl. Phys.*, **27**, L2034-L2036.

107. Gustafsson, G., Inganäs, O. and Stafström, S. "Optical anisotropy of neutral and doped poly(3-octylthiophene).", to be published in *Solid State Commun.*

108. Read, B.E., (1975) " Ultra-violet, visible and infra-red dichroism.", in Ward, I.M. (ed.) *Structure and properties of oriented polymers*, Applied Science Publishers, London, pp. 159-163

109. Holliday, L. and Ward I.M., (1975) "A general introduction to the structure and properties of oriented polymers", in Ward, I.M. (ed.) *Structure and properties of oriented polymers*, Applied Science Publishers, London, pp. 32-35.

110. Kratky, O. (1933) "Zum Deformationsmechanismus der Faserstoffe, I.", *Kolloid-Z.*, **64**, 213-222.

111. Nishikitani, Y., Fichou, D., Horowitz, G. and Garnier, F. (1989) "Carrier mobility enhancement in semiconducting poly(3-methylthiophene) by electrochemical inclusion of copper.", *Synth. Met.*, **31**, 267-273.

112. Assadi, A., Gustafsson, G., Willander, M., Svensson, C. and Inganäs, O. (1990) "Determination of field effect mobility of poly(3-hexylthiophene) under ammonia exposure", to be published in *Synth. Met.*

113. Tomozawa, H., Braun, D. Phillips, S., Heeger, A.J. and Kroemer, H. (1987) "Metal-polymer Schottky barriers on cast films of soluble poly(3-alkylthiophenes).", *Synth. Met.*, **22**, 63-69

114. Tomozawa, H., Braun, D. Phillips, S., Worland,R., Heeger, A.J. and Kroemer, H. (1989) "Metal-polymer Schottky barriers on processible polymers", *Synth. Met.*, **28**, C687-C690.

115. Gustafsson, G. Sundberg, M., Inganäs, O. and Svensson, C. "The nature of rectifying contacts between poly(3-hexylthiophene) and indium-tin oxide or aluminum", to be published in *J. Mol. Electron.*

116. Sundberg, M., Gustafsson, G. and Inganäs, O. "Rectifying metal-polymer contacts formed by melt processing.", to be published in *Appl. Phys. Lett.*

117. Hermann, J.P., Ducuing, J. (1974) "Third-order polarizabilities of long-chain molecules." *J. Appl. Phys.* **45**, 5100-5102.

118. Sauteret, C., Hermann, J.P., Frey, R., Pradere, F., Ducuing, J., Chance, R.R. and Baughman, R.H. (1976) "Optical nonlinearities in one-dimensional-conjugated polymers.", *Phys Rev. Lett.* **36**, 956-959.

119. Agrawal, G.P., Cojan, C. and Flytzanis, C. (1987) "Nonlinear optical properties of one-dimensional semiconductors and conjugated polymers.", *Phys. Rev. B* **17**, 776-789.

120. Sinclair, M., McBranch, D., Moses, D., Heeger, A.J.(1989) "Nonlinear optics of conjugated polymers", *Synth. Met.* **28**, D645-D653.

121. Fukaya, T., Heinämäki, A., Stubb, H. (1989) "Measurement of non-linear refractive index of polyalkylthiophene films" *J. Mol. Electron.* **5**, 187-191.

122. Logsdon, P.B., Pfleger, J., Prasad, P.N. (1988) "Conductive and optically non-linear polymeric Langmuir-Blodgett films of poly(3-dodecylthiophene).", *Synth. Met.* **26**, 369-281

123. Jenekhe, S.A., Lo, S.K., Flom, S.R. (1989) "Third-order nonlinear optical properties of a soluble conjugated polythiophene derivative.", *Appl. Phys. Lett.* **54**, 2524-2526.

124. Heinämäki, A., Honkanen, S. (1989) "Non-linear waves in a multilayer waveguide application to SiO_xN_y with organic overlayer.", *Opt. Quant. Electr.* **21**, 137-146.

125. Stegeman, G.I., Seaton, C.T., Zanoni, R. (1987) "Organic films in non-linear integrated optics structures.", *Thin Solid Films* **152**, 231-263.

126. Carter, G.M., Thakur, M.K., Chen, Y.J., Hryniewicz, J.V. (1985) "Time and wavelength resolved nonlinear optical spectroscopy of a polydiacetylene in the solid state using picosecond dye laser pulses.", *Appl. Phys. Lett.* **47**, 457-459.

CONTROLLED MOLECULAR ASSEMBLIES OF ELECTRICALLY CONDUCTIVE POLYMERS

M. F. RUBNER
Department of Materials Science and Engineering, Massachusetts Institute of Technology, Cambridge, MA 02139 USA

T. A. SKOTHEIM
Brookhaven National Laboratory, Upton, NY 11973 USA

ABSTRACT. The Langmuir-Blodgett technique has recently been used to manipulate a variety of electroactive conjugated polymers into multilayer thin films with well defined layered structures and ordered molecular organizations. Multilayer thin films have been successfully fabricated from amphiphilic conjugated polymers and their precursors and from non-surface active conjugated polymers. In most cases, the molecular organizations of these films have been found to be highly anisotropic. The structures of these films have been probed by high energy synchrotron radiation, reflection-absorption infrared and low angle x-ray diffraction techniques. Electrical measurements have also revealed highly anisotropic behavior with the in-plane conductivities of these multilayer structures typically being many orders of magnitude greater than their transverse conductivities. The unique layered structures of these films, in some cases, have also been shown to give rise to unusually large dielectric constants over a wide frequency range. In this chapter, the fabrication, characterization and electrical properties of these new thin film multilayer structures are discussed.

1. Introduction

Over the past few years, the practice of molecular engineering has guided the design and synthesis of new electroactive polymers and has provided a deeper understanding of the structure-property relationships active in this very interesting class of materials. Molecular engineering in the simplest sense simply involves the manipulation and control of the physical and electronic properties of polymeric materials by suitable modifications of their molecular structures. Through this approach, it is possible to systematically alter the electrical, optical, mechanical and processing properties of electroactive polymers with the ultimate goal being the creation of materials with a collection of properties optimized for a particular technological application or scientific endeavor. The logical extension of this molecular manipulation approach is the development of techniques that provide control over the molecular and supermolecular organizations of these newly synthesized molecules. The net result would be an ability to design and fabricate complex molecular assemblies with controllable molecular architectures and well defined molecular arrangements. One could further envision the creation of multicomponent molecular assemblies in which each component serves a specific well defined function and is

J.L. Brédas and R. Silbey (eds.), Conjugated Polymers, 363–403.

molecularly positioned in the structure to achieve an optimum electrical, optical or chemical response.

The Langmiur-Blodgett (LB) technique is currently one of the few methods available that provides molecular level control over the organizations of organic and polymeric materials. With this technique, it is possible to fabricate highly ordered, complex multilayer structures in which functionally distinct molecules are contained within different molecular layers (heterostructures) or within the same layer (mixed monolayers). Since the LB process involves the direct manipulation of a single monolayer, it is further possible to control the thickness of the large area thin films created by this technique at the monolayer level. Thus, the LB technique can be used to fabricate unusual molecular assemblies with optical and electrical properties simply not found in the parent molecules used to construct the multilayer films. A particularly interesting challenge in this area is the design and successful LB manipulation of monolayer forming systems based on electroactive and optically nonlinear organic materials [1]. The extension of the LB technique to such materials is expected to generate a variety of new thin film structures with electrical and optical properties that can be tailored at the molecular level.

In this chapter, the application of this technique to the thin film processing of electroactive polymers is discussed. The fabrication of controlled molecular assemblies of electroactive polymers provides a unique opportunity to investigate the structure-property and structure-function relationships of multilayer thin films containing electrically conductive, optically nonlinear and redox active polymers. These assemblies can also be coupled to appropriate solid supports for application in such molecular electronic devices as ultrathin electrodes and transistors, optical waveguides and switches, and chemical and biochemical microelectronic sensors.

2. Fabrication of Monolayer and Multilayer Thin Films of Electrically Conductive Polymers

The fabrication of well ordered molecular assemblies of electrically conductive polymers requires the use of a technique that provides control over both the molecular and supermolecular organization of the final assembly. There are a number of possible ways to approach this problem, however, only the Langmuir-Blodgett technique currently offers the versatility and molecular level control needed to fabricate a wide variety of novel and technologically relevant molecular organizations. With this technique, it is possible to control both the thickness and architecture of a large area thin film at the molecular level. It also affords one of the few means available to assemble a collection of functionally different molecules into thin film superlattices with well defined and controllable layer structures.

Traditionally, the LB technique has been used to manipulate surface active molecules such as fatty acids and phospholipids into monolayer and multilayer thin films[1]. This is usually accomplished by spreading a dilute solution of the surface active molecules onto a purified water subphase and subsequently reducing the area available to the surface confined molecules in order to force them into a highly ordered condensed monolayer. The condensed monolayer is then transferred, while held at a constant surface pressure, onto a rigid substrate by passing the substrate vertically through the monolayer. If the substrate is hydrophobic, a single monolayer will be transferred on both the down-stroke and up-stroke producing a head-to-head/tail-to-tail arrangement known as a Y-type multilayer film. It is also possible to have transfer of the monolayer only on the up-stroke (or down stroke) thereby producing a Z-type (or X-type) multilayer in which the molecular layers stack with a head-to-tail arrangement. This latter mode of transfer is very common in rigid monolayers such as those formed from polymers. These three basic transfer processes are illustrated in Figure 1.

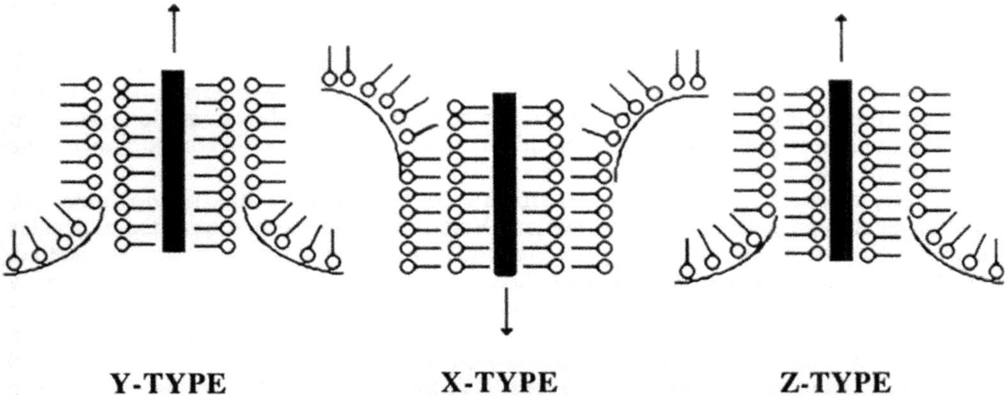

Figure 1. An illustration of the three basic modes of transfer possible for surface active monolayers.

For excessively rigid films that are not easily transferred using the vertical dipping technique, it is necessary to use the horizontal film lifting technique illustrated in Figure 2. In this case, a substrate oriented horizontally to the LB film is lowered until it just touches the monolayer and is then raised; the net result being the transfer of a single monolayer during each dip. Such a process usually delivers a X-type film since most substrates are pretreated to make their surfaces hydrophobic.

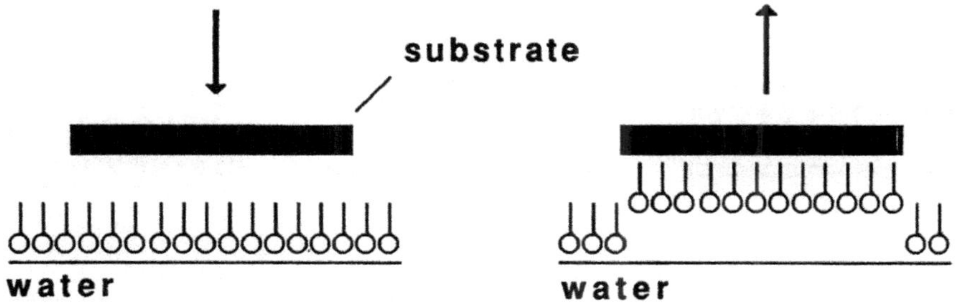

Figure 2. The horizontal lifting technique

It immediately becomes apparent that this technique can be used to precisely control the thickness of a multilayer film by simply controlling the number of monolayers transferred to the substrate (a typical fatty acid monolayer is about 25-30 Å thick). In addition, since it is possible to transfer different monolayers onto the same film (for example by alternately passing the substrate through two different condensed monolayers) it is also possible to create multilayer structures comprised of functionally different layers stacked in a controlled sequence.

A typical surface active molecule is usually comprised of two key molecular features: a hydrophilic head group and a long hydrophobic tail. The tail group prevents the molecule from dissolving in the water subphase and provides control over the spatial orientation of the molecule at the air-water interface whereas the head group encourages spreading at the water surface and hence the formation of a true monolayer and not larger scale aggregates. Thus, in principle, in order to manipulate electroactive polymers with the LB technique, it is necessary to utilize monomers or polymers whose structures have been modified so as to render them surface active. As will become apparent shortly, however, this restriction can be overcome through the use of mixed monolayers in which only one of the components need be a surface active molecule.

To date, a number of conducting polymers have been successfully manipulated into multilayer structures using the LB technique. Variations in the structures of the molecules manipulated, the types of chemistry used to render the molecules electrically conductive, and the techniques used to fabricate monolayers have resulted in a number of structurally distinct conducting polymer based LB films. Figure 3 illustrates examples of the three basic molecular architectures that have been created from conducting polymer based systems. In each case, the films are comprised of electrically conductive regions and electrically insulating regions. Films A and B represent molecular organizations whose electrical properties are highly anisotropic whereas film C would be expected to exhibit a significantly lower level of conductivity anisotropy. Clearly, the electrical and optical properties of these films will be strongly influenced by the nature and form of their molecular organizations.

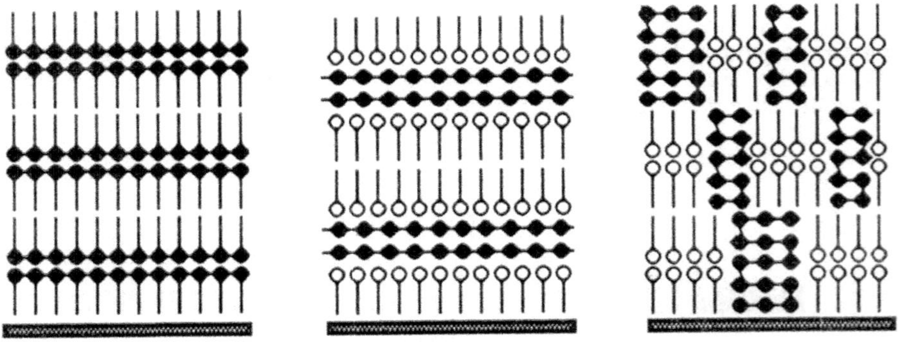

A B C

Figure 3. Examples of three basic types of molecular assemblies possible with conducting polymers. Closed circles represent conducting polymer chains and open circles represent insulating surface active molecules.

The molecular assembly illustrated in Figure 3A resembles a classical Y-type multilayer thin film where, in this case, the head groups have been linked together to form conducting polymer chains. Shimidzu and coworkers [2] have shown that such a multilayer structure can be created by the LB manipulation of mixed monolayers comprised of a surface active pyrrole monomer and a nonsurface active hydrocarbon spacer molecule. The spacer molecule, which is octadecane, is apparently needed to stabilize the resultant condensed monolayer of the surface active pyrrole derivative thereby allowing its successful transfer into Y-type multilayer thin films. Two types of amphiphilic pyrrole derivatives have been successfully manipulated using this approach: 4-methyl-1-octadecylpyrrole-3-carboxylic acid and octadecyl 4-methylpyrrole-3-carboxylate. The structures of the molecules used to form monolayers of the latter pyrrole derivative are shown below in Scheme 1.

Scheme 1

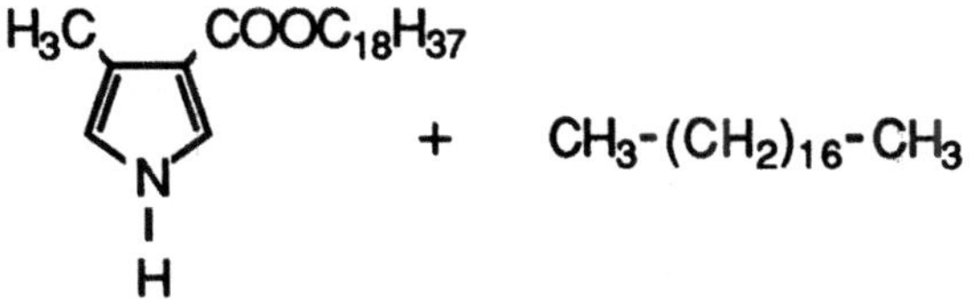

The pyrrole head groups of the surface active molecules are converted into electrically conductive polypyrrole chains by a rather unique electrochemical polymerization process that occurs without significant disruption of the film's molecular organization. This amazing process is illustrated in Figure 4.

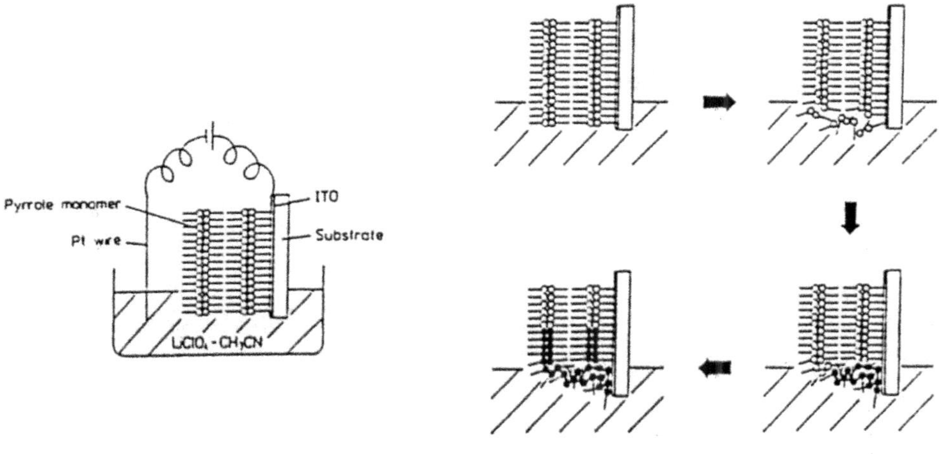

Figure 4. Experimental set-up for the electrolytic polymerization of a surface active pyrrole-based LB film and the proposed mechanism of polymerization. (Ref. 2 and 3)

As indicated in this figure, the lower portion of the multilayer film is immersed into a $LiClO_4$ electrolyte solution and subsequently electrolytically oxidized. The substrate in this case is coated with an electrically conductive film of indium tin oxide and therefore serves as one of the electrodes in the electrochemical cell. The molecular organization of the portion of the multilayer film that is immersed in the electrolyte solution is completely disrupted during the electrochemical polymerization. The remaining pyrrole head groups of the multilayer film, however, are converted into electrically conductive polypyrrole chains without significantly altering the layer structure of the film. The polymerization process can be visually observed as a color front (due to the creation of polypyrrole chains that absorb strongly in the visible region) moving up the substrate during the electrochemical reaction. Apparently, the electrolyte diffuses into the hydrophilic planes of the multilayer structure and facilitates the electrochemical polymerization of the head groups.

Shimidzu's group has also reported [3] that X-type LB films of conducting polymers can be fabricated by the direct manipulation of soluble surface active conducting polymer derivatives of the polyanilines and polypyrroles. For example, using this approach, they have demonstrated that condensed monolayers of poly(2-octadecoxyaniline) (A; scheme 2) can be transferred into multilayer structures containing over 200 layers by using a horizontal lifting method. Soluble complexes of poly(3,4-dibutylpyrrole) doped with perfluorooctanoic acid ($CF_3(CF_2)_6COOH$) have also been directly manipulated into LB films in this manner (B; scheme 2). The perfluoroalkylcarboxylate counterion of the doped polymer in this case imparts sufficient surface activity to the polymer to allow its manipulation into multilayer thin films. In both cases, the films were rendered electrically conductive by subsequent treatment with iodine vapor. The chemical structures of these monolayer forming systems are shown below in Scheme 2.

Scheme 2

(A)

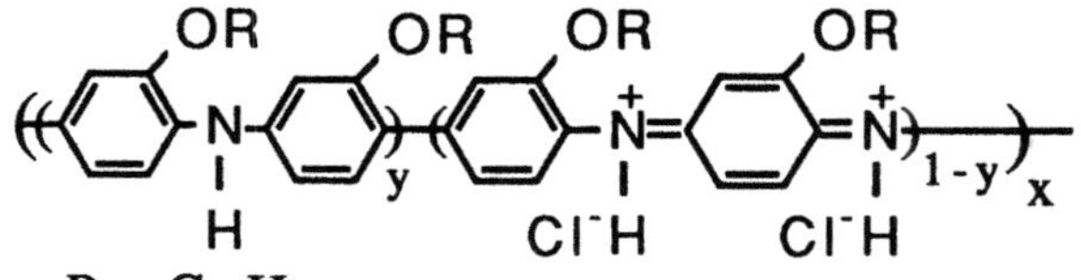

$R = C_{18}H_{37}$

(B)

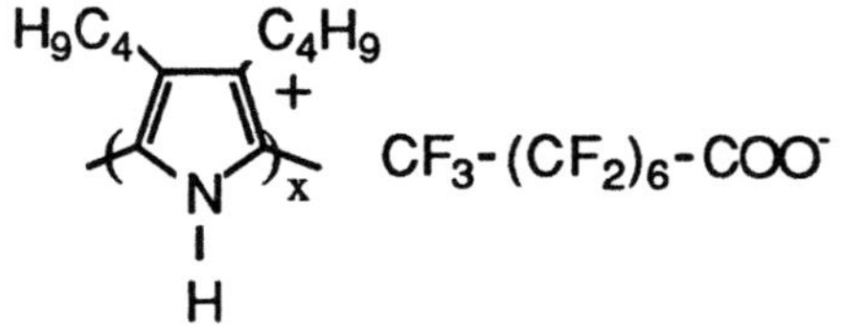

The type of molecular organization illustrated in Figure 3B, in which layers of an electrically conductive polymer alternate with bilayers of an insulating material, has also been successfully fabricated. For example, by adsorbing a nonsurface active conducting polymer onto the monolayer of a surface active material, Rubner and colleagues [4] have fabricated electrically conductive LB films based on polypyrrole. This was accomplished by spreading a solution containing a surface active pyrrole derivative and a large excess of pyrrole monomer onto a oxidizing $FeCl_3$ subphase. The ferric chloride acts to both polymerize the mixture at the air-water interface and oxidize the resultant polymer to its electrically conductive state. The pyrrole monomer is needed to facilitate the polymerization as neither the surface active pyrrole monomer nor pure pyrrole will polymerize independently under the conditions used to prepare the mixed monolayers. A large excess of pyrrole monomer was used due to the high degree of water solubility exhibited by this material.

The specific type of surface active pyrrole derivative used to form the monolayer was found to strongly influence the chemistry that is initiated at the air-water interface. For example, when solutions containing a 5000/1 mole ratio of pyrrole/3-octadecylpyrrole were spread onto the oxidizing subphase, copolymerization of the two monomers occurred as well as homopolymerization of the unsubstituted pyrrole monomer. The net result was a relatively thick surface film (100-200Å) that was very difficult to transfer into multilayers via a conventional vertical lifting technique. With 3-octadecanoylpyrrole as the surface active component, on the other hand, copolymerization of the two monomers was suppressed and only electrically conducting polypyrrole chains were formed. In this latter case, uniform monolayer films about 40 Å thick were formed and these could be readily transferred into multilayer structures. The chemical structures of the molecules used in this particular system are presented in Scheme 3.

Scheme 3

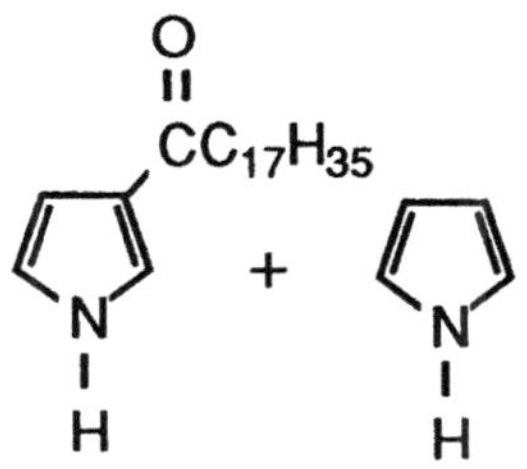

Figure 5 illustrates the reaction that takes place at the air-water interface when 3-octadecanoylpyrrole is used as the surface active component. In this case, the surface active pyrrole simply provides an environment at the air-water interface that is suitable for the homopolymerization of pyrrole and, in addition, enhances the spreading of the resultant mixture into a uniform monolayer film. Copolymerization with the surface active pyrrole monomer is prevented due to the electron-withdrawing nature of the carbonyl group attached to the pyrrole ring which effectively inhibits the chemical polymerization of this material. In the condensed state, the surface film is comprised of a well ordered, tightly packed monolayer of 3-octadecanoylpyrrole onto which is physically absorbed electrically conducting chains of polypyrrole. It is also likely that some of the polypyrrole chains penetrate into the condensed monolayer.

The mixed monolayers containing the physically adsorbed electrically conductive polypyrrole chains were found to transfer in a Y-type fashion thereby creating the structure illustrated in Figure 3B in which the polypyrrole chains are, for the most part, sandwiched between the pyrrole head groups of the surface active octadecanoylpyrrole molecules. This system represents the first example of the direct LB manipulation of a conjugated polymer in a highly conducting form.

The notion of physically adsorbing a polymer onto the condensed monolayer of a surface active molecule and subsequently transferring the entire molecular assembly into a multilayer structure can be exploited to manipulate a number of different non-surface active conducting polymers systems. The LB manipulation of non-surface active molecules, in fact, is frequently accomplished by dissolving the nonsurface active material into the water subphase and adsorbing it onto a monolayer of a more traditional surface active species. The nature of the forces acting between the surface active molecules poised at the air-water interface and the non-surface molecules dissolved in the subphase can range from those based on weak nonbonded interactions to those based on stronger ionic interactions or, in the extreme, primary bonds formed via chemical reactions. The LB manipulation of water soluble precursor polymers provides a good example of this particular approach.

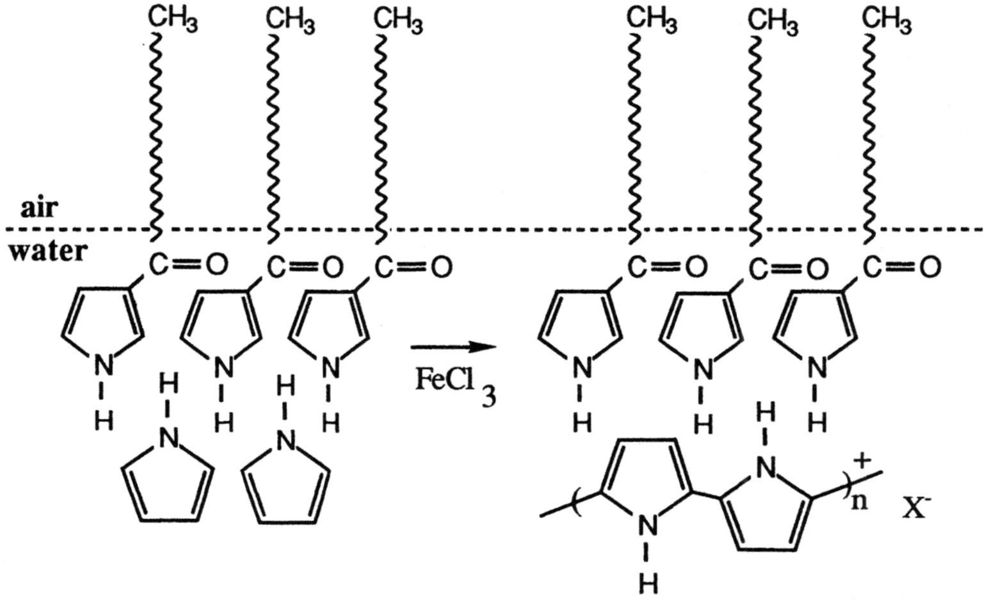

Figure 5. Homopolymerization of pyrrole at the air-water interface in the presence of the surface active pyrrole derivative 3-octadecanoylpyrrole

Saito and associates have shown [5] that it is possible to construct multilayer LB films of poly(*p*-phenylenevinylene) or poly(*p*-thienylenevinylene) based precursor polymers by spreading negatively charged amphiphiles onto aqueous solutions of the positively charged precursor derivatives (A and B; scheme 4). Upon standing, a polyion complex forms between the precursor polymer dissolved in the subphase and the monolayer of the charged amphiphile. The formation of the polyion complex is driven by the ionic attraction established between the positively charged sulfonium ion present on each polymer repeat unit and the negatively charged sulfate ion present in the head groups of the amphiphile. The polyion complexes of the precursor polymers were found to readily transfer into multilayer films by a conventional LB dipping process. In order to convert the precursor polymer into an electroactive polymer, the multilayer films were heated at 100-200°C for one hour under nitrogen. Although a detailed description of the multilayer structure before and after thermal treatment has not yet been provided, infrared analysis indicated that the amphiphile was not removed from the film during heat treatment. Spectroscopic evidence also suggested that, in some cases, the conjugated polymer chains formed during thermal treatment were preferentially oriented within the plane of the LB film.

The precursor approach to the formation of LB films based on poly(*p*-phenylenevinylene) and its derivatives has been utilized by a number of researchers over the past few years [6,7]. In all cases, a monolayer of the precursor polymer is initially manipulated into a multilayer film which, in turn, is subsequently heat treated to convert the precursor polymer to its electroactive conjugated analog. The first reports of this approach were actually based on the LB manipulation of preformed amphiphile-polyion complexes [6,7] as opposed to the previously described adsorption of the precursor polymer onto the monolayer of an amphiphile poised at the air-water interface (for an example of a typical monolayer forming system, see scheme 5). The thermal treatment needed to create the conjugated polymer is usually accompanied by a volatilization of the amphiphilic counterion and hence a drastic change in the structure of the multilayer. It is interesting to note, however, that the LB films obtained after heat treatment generally do not exhibit the kinds of gross defects such as cracks and voids that might be expected from an elimination process involving the loss of volatile reaction products.

Scheme 4

(A)

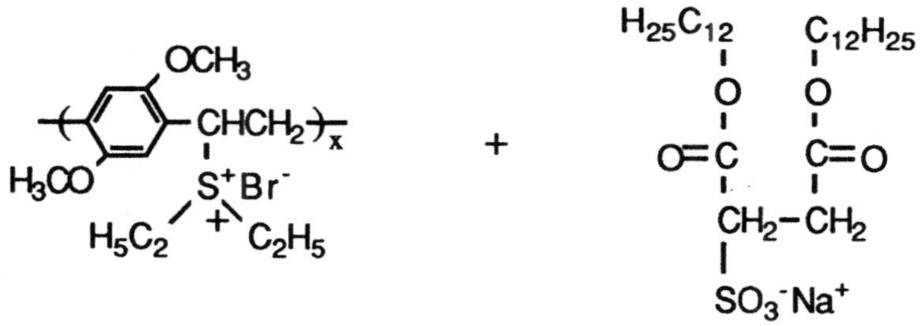

(B)

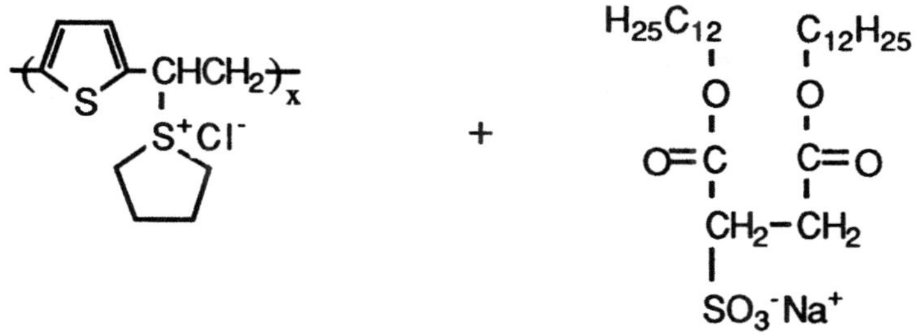

Scheme 5

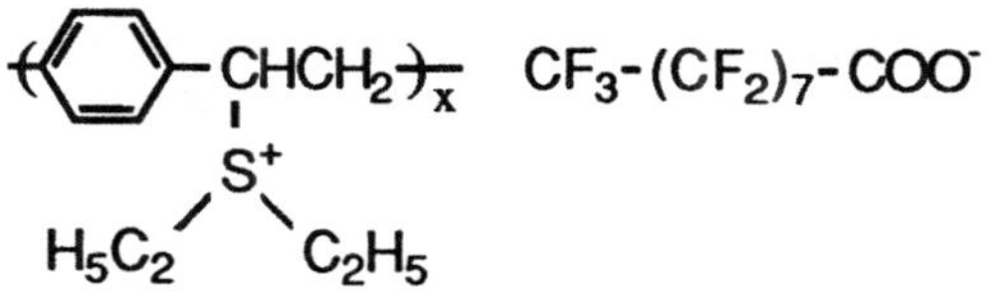

The process of adsorbing a water soluble polyelectrolyte onto the monolayer of a surface active molecule has also been recently utilized [8] to form LB films containing derivatized polythiophene molecules. In this case, carboxylate derivatized polythiophene chains (formed from poly(thiophene-3-acetic acid)) were adsorbed onto a monolayer formed from dimethyl-distearyl-ammonium chloride. This process, which involves the direct manipulation of a non-surface active conjugated polymer, is shown schematically in Figure 6. Transfer of the as-formed polyion complex into multilayer thin films was accomplished with the vertical dipping technique and the resultant LB films were rendered electrically conductive by doping with suitable oxidizing agents. This approach is particularly attractive since it is possible to control the conformational state of the conjugated polymer backbone by adjusting the PH and ionic strength of the subphase. Thus, it should be possible through variations in the nature of the subphase to encourage the conjugated polyelectrolyte chains to adopt conformations ranging from random coils to fully extended backbones. Manipulation of the conformational state of the polymer backbone thereby provides additional molecular level control over the structure of the final assembly.

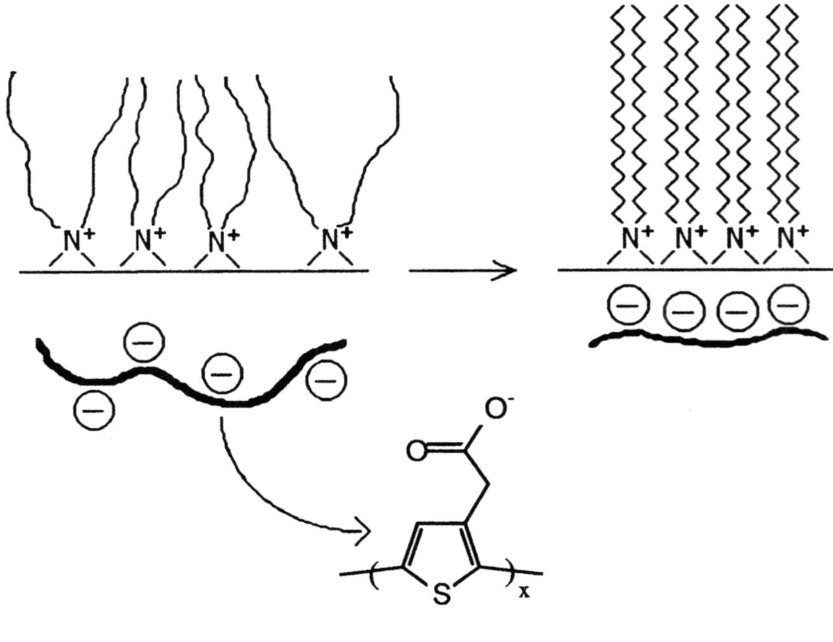

Figure 6. Monolayer formation of a polyion complex based on carboxylate derivatized polythiophene molecules.

An alternative approach to the LB manipulation of nonsurface active conjugated polymers has also been described by Rubner's research group [9]. They have shown, for example, that it is possible to form stable mixed monolayers of alkyl substituted polythiophenes by simply mixing the conjugated polymer with suitable portions of a surface active molecule. Using this approach, they were able to form high quality Y-type multilayer LB films containing as much as 80 mole % of the conjugated polymer. It was also shown that polythiophene chains substituted with alkyl groups ranging from butyl to octadecyl could be successfully manipulated into LB films as mixed monolayers.

The use of mixed monolayers of the poly(3-alkyl thiophenes) is necessary in this case since the thiophene rings of the polymer backbone are not sufficiently hydrophilic to render the polymer molecules surface active. As a result of this, after spreading at the air-water interface, the polymer molecules tend to aggregate into large scale clusters or islands that are not stable to film collapse when held at moderate surface pressures. Thus, regardless of the length of the pendent alkyl group, the poly(3-alkyl thiophenes) by themselves are not easily manipulated via the LB technique. The surface active molecules serve two primary functions. First, they help disperse the polymer molecules into much smaller domains that become uniformly distributed throughout the condensed monolayer and second, they prevent the polymer molecules from aggregating into larger clusters during film manipulation; particularly with low polymer content monolayers. The net result is a flexible two phase monolayer film that is stable enough to be transferred into multilayer structures using a standard vertical lifting technique. Thus, the use of mixed monolayers provides a means to transfer conjugated polymer molecules that would otherwise be difficult to manipulate with the LB technique. This approach is also generally applicable to any soluble, nonsurface active conjugated polymer.

Mixed monolayer films of the poly(3-alkyl thiophenes) were first prepared using stearic acid as the surface active component [9] and a number of research groups have since utilized this amphiphile to fabricate multilayer films with this particular combination of materials [10,11]. More recently, mixed LB films of the poly(3-alkyl thiophenes) have been fabricated with poly(isobutyl methacrylate) [12] and a diacetylene based amphiphile [13] as the dispersing agents. For multilayers fabricated with stearic acid as the surface active component, the molecular organization of the resultant LB film is best represented by the structure illustrated in Figure 3C. Thus, in this case, the polymer molecules are dispersed throughout the bilayer matrix of the surface active molecules. More will be said about the details of this particular molecular organization in the section on structural characterization. A summary of the various monolayer forming systems used to fabricate polythiophene based LB films is presented in Figure 7. Although not shown in this figure, it has been reported that mixed monolayer films containing oligothiophenes can also be sucessfully fabricated into LB films [14,15].

It should be noted that Prasad and coworkers [16] have reported that LB films of poly(3-dodecylthiophene) can be prepared by the direct manipulation of neat surface films of this material. The transfer of these surface confined polymer molecules was accomplished with some difficultly via a horizontal film lifting technique. The current method of choice for fabricating LB films of the poly(3-alkyl thiophenes), however, is based on the manipulation of mixed monolayers. This latter method is extremely versatile, can be used to manipulate a wide variety of nonsurface active polythiophene derivatives, and delivers uniform thin films with high optical quality.

The LB technique is well suited for the preparation of complex molecular assemblies of functionally distinct electroactive molecules and it is to be expected that the creation of such structures will be pursued as this area matures. The first steps toward the realization of functionalized molecular assemblies of conducting polymers have been taken by Skotheim's research group. They have demonstrated, for example, that it is possible to manipulate mixed monolayer films containing an electroactive ferrocene functionalized pyrrole derivative[17]. The ferrocene-derivatized pyrrole (Fc-Py) molecule consists of a pyrrole head group at one end of a 12 carbon aliphatic chain and a hydroxyl ferrocene group at the opposite end as shown in Scheme 6.

This particular pyrrole derivative is not a surface active molecule and therefore must be mixed with a surface active pyrrole derivative to form stable monolayers. Figure 8 shows the pressure-area isotherms of the pyrrole derivatives and their mixtures. The lack of a distinct upturn in the pressure-area isotherm of Fc-Py (see Figure 8b) confirms that this material does not form a stable monolayer at the air-water interface. This is in sharp contrast to 3-hexadecyl pyrrole (3HDP) (a; scheme 6) [17] which displays a steep rise in surface pressure at the point that the molecules become compressed into their condensed phase. By mixing these two pyrrole derivatives together, stable, well behaved monolayer forming systems can be prepared.

(A)
(B)
(C)
(D)
(E)

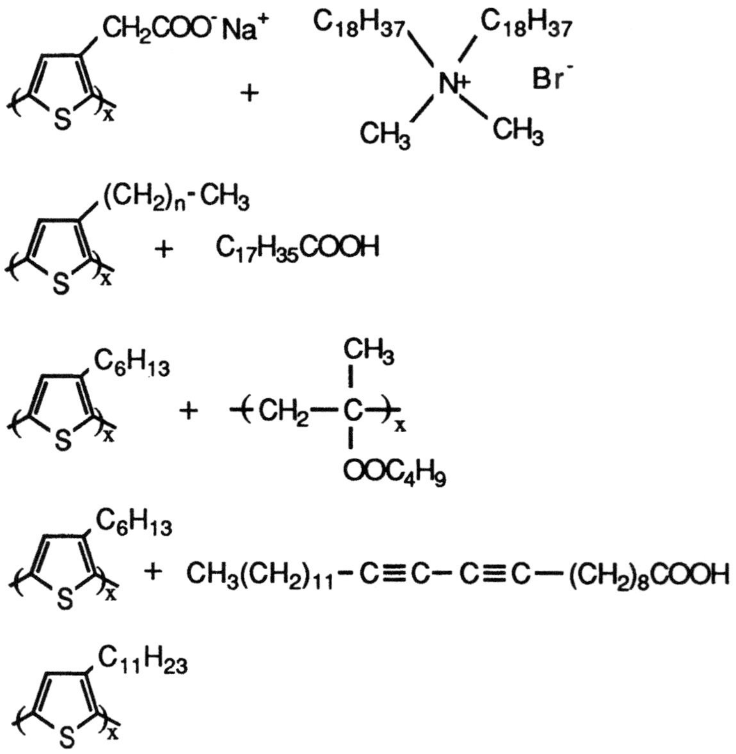

Figure 7. Monolayer forming systems that have been successfully used to form polythiophene based LB films.

The pressure-area isotherms of mixtures of Fc-Py and 3HDP in various mole ratios are also shown in Figure 8. As can be seen, the addition of 3HDP has a pronounced effect on the ability of this material to form a monolayer. With the addition of increasing amounts of 3HDP to the mixture, multiple transition regions develop which reflect a change in the orientation of the Fc-py molecules with increasing surface pressure. These isotherms are highly reproducible and indicate that the 3HDP is optimizing the vertical orientation of the Fc-Py as has been determined by Near Edge X-ray Absorption Fine Structure Spectroscopy (NEXAFS) [18]. Thus, by varying the surface pressure and monolayer composition, it is possible to exert control over the spatial orientation of the electroactive ferrocene groups.

As mentioned earlier, one of the great virtues of using the LB technique to process conjugated polymers is the ability to fabricate mixed layer heterostructures comprised of layers of functionally distinct molecules. Currently the only reported example of such a structure is based on LB films comprised of layers of electrically conductive polypyrrole alternating with bilayers of an insulating pyrrole amphiphile [4,19]. This novel molecular architecture was fabricated by alternately transferring bilayers of the conducting polypyrrole/3-octadecanoyl pyrrole system described earlier and insulating bilayers of 3-octadecyl pyrrole onto the LB film. The insulating bilayers of the surface active pyrrole derivative were used to further isolate the already electrically anisotropic bilayers of the conducting polypyrrole thereby creating a highly anisotropic film with high in-plane conductivities (10^{-1} S/cm) and extremely low conductivities across the film thickness (< 10^{-12} S/cm). Clearly, as this area of research matures, it will be possible to fabricate a wide variety of these mixed layer superlattices each exhibiting a unique collection of electrical and optical properties not found in the parent materials used to construct them. Indeed, the fabrication of LB heterostructures containing mixed layers of conjugated polymers with different electronic states and different redox potentials looks to be a very promising endeavor.

Scheme 6

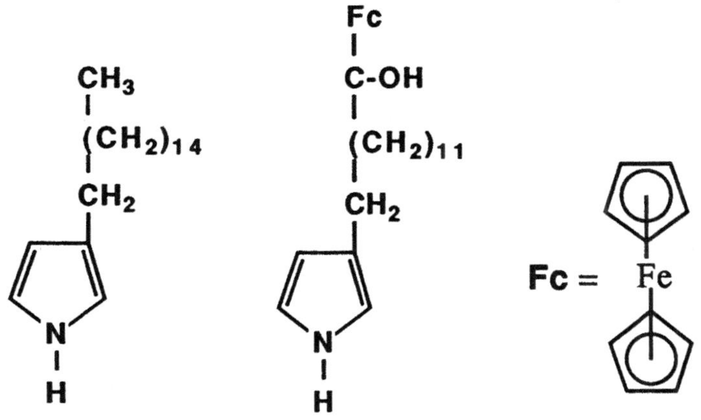

(a) (b)

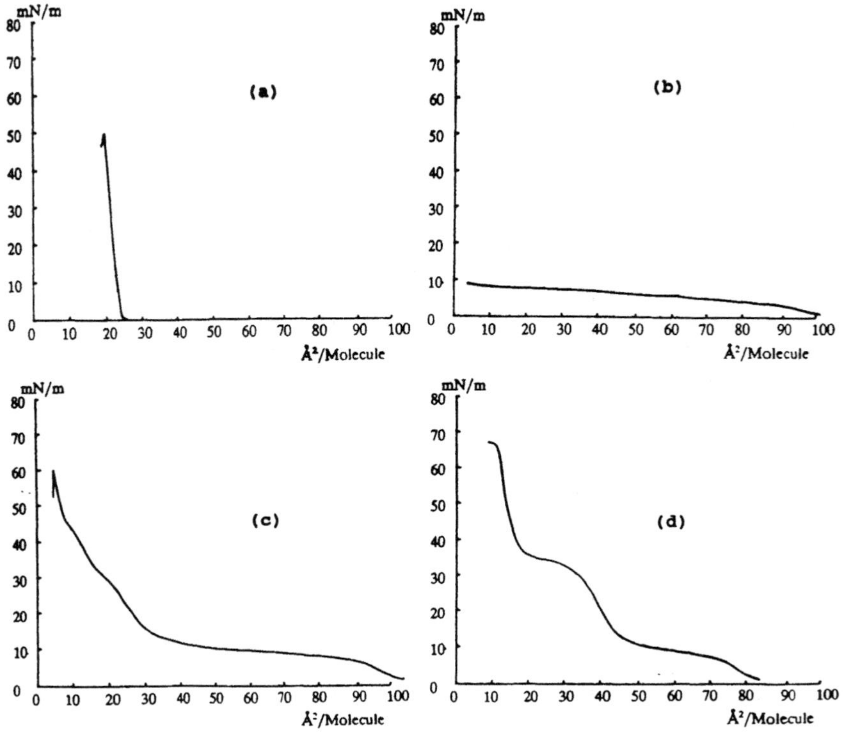

Figure 8. Pressure-area isotherm of (a) 3-hexadecyl pyrrole and mixtures of ferrocene-derivatized alkyl pyrrole and 3-hexadecyl pyrrole for three different ratios: (b) 1:0, (c) 1:2, (d) 1:5 (subphase is purified water).

3. Molecular and Supermolecular Organizations of LB Films Containing Conducting Polymers

The well defined periodic molecular organizations illustrated in Figure 3 are clearly only idealized representations of the actual structures fabricated by the LB technique. Multilayer LB films are also subject to the rules of thermodynamics and therefore contain their share of grain boundaries, layer misregistries, in-plane disorder and other such defects. At this point in time, it is simply not possible to stack monolayers into single crystal-like multilayer structures. The determination of the actual structure of an LB film is usually carried out by examining a number of different parameters including the level and type of molecular orientation present in the film, the layer thickness and type of layer stacking present in the film and the level and type of in-plane order present in the film. Thus, in order to gain an appreciation of the molecular origin of the electrical and optical properties exhibited by these novel molecular assemblies, it is necessary to examine their structures at both the molecular and supermolecular level. This section reviews some of the results obtained from studies aimed at determining the molecular organizations of LB films containing conjugated polymers or their precursors.

3.1. X-RAY DIFFRACTION STUDIES

The process of building a multilayer structure one layer at a time coupled with the amphiphilic nature of the surface active molecules used to construct these films inevitably results in highly anisotropic molecular organizations comprised of well defined molecular layers. A structural parameter of particular interest in multilayer LB films is therefore the out-of-plane repeat distance of the layers deposited during the transfer process. A direct way to probe this layer structure is by low angle x-ray diffraction which provides information about the monolayer and/or bilayer repeat distances of the molecularly stacked layers. Low angle x-ray techniques are needed since the typical bilayer *d*-spacings of Y-type LB films are on the order of 30-70 Å. Generally, a well ordered multilayer LB film will exhibit a number of clearly defined (001) Bragg reflections from which the *d*-spacing of the layers can be extracted. A listing of some of the *d*-spacings and monolayer thicknesses reported for LB films containing electrically conducting polymer chains or the precursor monomers and polymers used to generate conducting chains can be found in Table 1. The monolayer thickness values were estimated from the x-ray data or, in some cases, by independent thickness measurements using an ellipsometer or profilometer.

Table 1. *d*-Spacings and layer thicknesses for multilayer films of conjugated polymers and their precursors

Multilayer System	*d*-spacing (Å)	Average thickness per monolayer (Å)	Ref.
octadecyl 4-methylpyrrole -3-carboxylate/octadecane	57[1]	28.5	2
Poly(octadecyl 4-methyl pyrrole-3-carboxylate) /octadecane	58[1]	29	2
poly(2-octadecoxyaniline)	37	37	3
poly(3,4-dibutylpyrrole) perfluorooctanoate	-	16	3
3-octadecylpyrrole	55	27.5	4
3-octadecanoylpyrrole	28	14	23
poly(3-Alkylthiophene)/CdSA[2]	50	25-50[3]	9
poly(*p*-xylylene-a-diethyl sulfonium) perfluorononanoate	-	10	6
poly(*p*-phenylenevinylene)	-	3.4	6

1 monolayers transfered at 30 mN/m
2 values obtained for butyl, hexyl, octyl and octadecyl substituted poly(3-alkylthiophenes)
3 measured thickness depends on film composition

In general, it can be seen that many of the multilayer thin films containing electroactive polymers or their precursors are sufficiently well ordered to extract meaningful *d*-spacings from their x-ray patterns. The *d*-spacings are also in the range expected for Y-, X-, or Z-type LB films of their respective surface active components. The first two entries in this table represent the Y-type pyrrole based multilayer system that is rendered electrically conductive by the electropolymerization process described earlier. As indicated by the values presented in the table, the electrochemical process used to link the pyrrole head groups into electrically conductive polypyrrole chains does not significantly alter the layer stacking of the molecules. The layered structure of this particular system has also been directly observed by high resolution transmission electron microscopy [2,20] which gave similar values for the bilayer repeat distance. The *d*-spacing of the third entry in the table, (poly(2-octadecoxyaniline)), is close to the value calculated for the thickness of a single monolayer of this material, thereby supporting the notion that it forms an X-type multilayer film. The significance of the *d*-spacings of the multilayer films fabricated from 3-octadecylpyrrole, 3-octadecanoylpyrrole and the poly(3-alkyl thiophenes) will be discussed in some detail in subsequent sections. The relatively small thickness per monolayer indicated for the LB films fabricated from poly(3,4-dibutylpyrrole)/perfluorooctanoate and poly(*p*-xylylene-α-diethyl sulfonium) /perfluorononanoate simply reflects the fact that the surface active counterions of these materials are fitted with shorter than normal tail groups. The average thickness per layer of the multilayer thin film of the latter material is seen to decrease from 10 Å to a value of about 3.4 Å after thermal treatment to its conjugated poly(*p*-phenylenevinylene) (PPV) form. This is due to the volatilization of the perfluoroalkyl group during thermal treatment to PPV.

In order to unambiguously establish the molecular origin of the x-ray diffraction patterns generated from LB films of mixed monolayers, it is necessary to determine whether the film is comprised of a single or two phase morphology. Indeed, it is not uncommon for phase separation of the two components to occur during either the spreading or monolayer transfer process. A good example of this type of behavior is provided by the LB films fabricated from mixed monolayers of the poly(3-alkylthiophenes) and stearic acid. In this case, it has been found [9] that the bilayer *d*-spacings of the multilayer structures are all about 50 Å irrespective of the composition of the film or the type of poly(3-alkylthiophene) used to form the mixed monolayers. Multilayer thin films fabricated with poly(3-butylthiophene), poly(3-hexylthiophene), poly(3-octylthiophene) and poly(3-octadecylthiophene) all exhibit a set of at least seven (00l) Bragg reflections; the intensity and breadth of which were found to depend on the mole ratio of P3AT/SA used in the film and the specific polymer. The LB films containing P3HT displayed the most intense, well defined diffraction peaks whereas films containing P3ODT exhibited reflections that were broader and significantly weaker in intensity. Since all of the *d*-spacings calculated from these reflections were identical to that obtained from a multilayer of pure cadmium stearate (the cadmium salt of stearic acid is formed during the monolayer spreading step from the cadmium chloride present in the water subphase), it was concluded that the mixed films are two phase systems comprised of well ordered domains of cadmium stearate and relatively disordered domains of the poly(3-alkylthiophene) molecules. Thus, the mixed monolayers reorganize during film transfer to form a multilayer structure in which polythiophene molecules are randomly dispersed throughout a matrix of well ordered molecular stacks of cadmium stearate molecules. As will be discussed shortly, this type of molecular organization has also been confirmed by FTIR reflection/absorption studies.

For LB films containing a low percentage of the conjugated polymer, the average thickness per layer is essentially determined by the dimensions of a monolayer of cadmium stearate (similar to the structure illustrated in Figure 3c). At higher polymer contents (>50 mole%), however, the average thickness per layer increases as the film structure is now dominated by the spatial requirements of the polymer molecules. Thus, although the *d*-spacings of the various poly(3-alkylthiophene) mixed LB films are all about 50 Å (suggesting a 25 Å per layer thickness), the average thickness per monolayer increases from 25 to 50 Å as the mole fraction of polymer in the film increases.

X-ray studies utilizing synchrotron radiation have also been carried out on mixed LB films containing poly(3-hexylthiophene) and cadmium stearate [21]. In these studies, the high intensity x-rays from a synchrotron source were utilized to examine the molecular organization of this system at different temperatures. The results of this work are found in Figure 9 which shows the out-of-plane x-ray diffraction patterns for a 34 layer film of poly(3-hexyl thiophene)/Cd-stearate (1:2 mole ratio) at different temperatures. At 25°C, the diffraction pattern displays a set of well defined (001) Bragg reflections identical to those observed in a multilayer thin film of pure cadmium stearate. The *d*-spacing calculated from these reflections is 49 Å. At 50°C, the diffraction peaks move slightly toward higher angles, indicating a small decrease in the *d*-spacing to 48.95 Å. At 75°C, however, a completely new set of peaks appears with a *d*-spacing of 40 Å, whereas the *d*-spacing corresponding to the original set of peaks continues to decrease to 48.4 Å. Upon cooling the sample back to 25°C, the original *d*-spacing moves back to 48.95 Å and the new set continues to contract slightly, to 39.9 Å.

Figure 10 shows the diffraction pattern for the same film reheated to higher temperatures. At 90°C, the *d*-spacing of the new set of peaks decreased further to 37 Å while the original set contracted to 48.3 Å. At 110°C, the peaks corresponding to the new set disappear, whereas all the peaks disappear at 130°C, indicating melting of the film and a complete disruption of the layer structure.

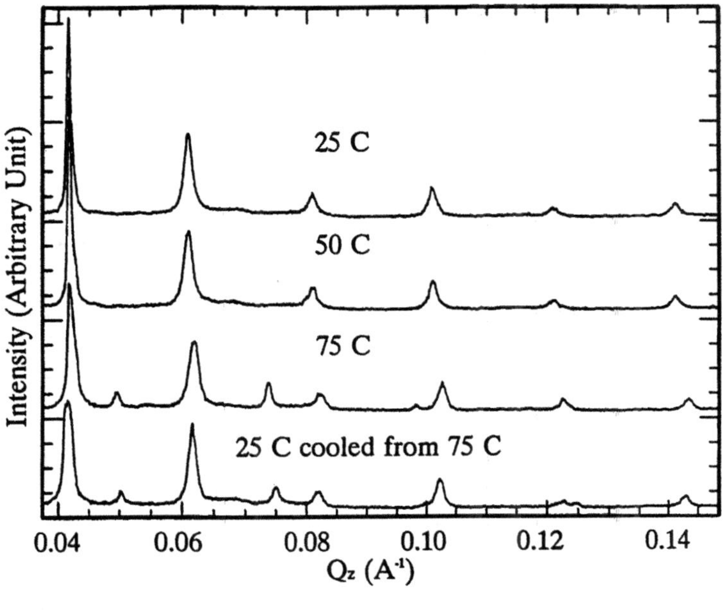

Figure 9. Out-of-plane x-ray diffraction pattern for a 34 layer film of cadmium stearate and poly(3-hexylthiophene) (2:1 mole ratio) recorded at different temperatures.

Since the original set of diffraction peaks has the same *d*-spacing and thermal behavior as the diffraction peaks observed from a multilayer thin film of pure cadmium stearate, it can be concluded that they represent reflections from the cadmium planes associated with the isolated domains of pure cadmium stearate. The new set of peaks, on the other hand, is a feature that is quite unique to the mixed multilayer system. In addition, the thermal behavior of the new set of peaks is completely different from that corresponding to the peaks associated with the pure cadmium stearate domains. As indicated earlier, the multilayer films containing poly(3-hexyl thiophene) and cadmium stearate are two phase systems made up of domains of pure cadmium stearate and domains consisting primarily of poly(3-hexyl thiophene). In the pure stearate domains, as the temperature increases, the hydrocarbon chains of the stearate molecules undergo tilting and twisting through a conformational change (most likely via the introduction of gauche conformers), causing the *d*-spacing to decrease. At sufficiently high temperatures (75°C), it appears that mixing or intercalation of the hydrocarbon chains of the polythiophene molecules with the tails of the cadmium stearate molecules is taking place in some of the cadmium stearate domains. This penetration, in turn, alters the molecular organization of some of the cadmium stearate domains thereby giving rise to the second set of diffraction peaks. At about 110°C, there appears to be complete mixing of the poly(3-alkyl thiophene) molecules with the cadmium stearate molecules in the mixed regions, resulting in a disruption of these domains and the disappearance of the second set of diffraction peaks altogether. The details of the structural changes induced by temperature in this system are still under investigation, however, these preliminary results clearly show that interesting interactions are possible between the different phases present in mixed LB films.

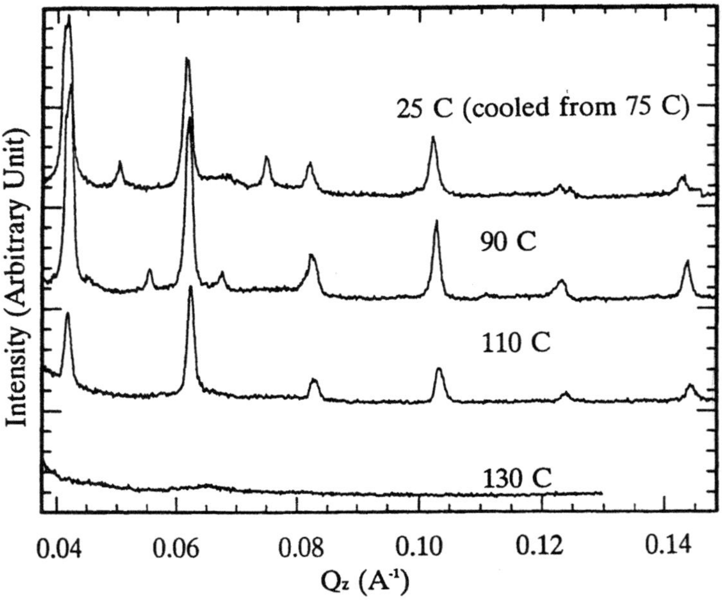

Figure 10. Out-of-plane x-ray diffraction pattern for a 34 layer film of cadmium stearate and poly(3-hexylthiophene) (2:1 mole ratio) recorded at different temperatures (in this case, heated up to the melting temperature).

3.2. ORIENTATION STUDIES BY FTIR

Infrared reflection-absorption spectroscopy (IRRAS) is a powerful method for determining the structure and molecular orientation of thin films [22]. This technique typically involves measuring the spectrum of a film in both the transmission and reflection mode and then comparing the intensities of the absorption bands of key functional groups. When light is reflected from a flat metal surface, the component polarized parallel to the reflecting surface undergoes a phase shift of 180 degrees, and hence, the reflected beam will interfere destructively with the incident beam, producing a node at the reflecting surface. For grazing incident angles, the component polarized perpendicular to the reflecting surface undergoes no phase shift. The net result is an electric field vector polarized normal to the reflecting metal surface. Hence, in the reflectance mode, only the vibrational bands with a component of the transition dipole moment oriented normal to the reflecting surface will be excited. In the transmission mode, on the other hand, only the vibrational bands with a component of the dipole moment parallel to the substrate will be excited. Thus, by comparing the transmission and reflection spectra, it is possible to determine the orientation of well characterized functional groups present in the film. These surface selection rules, and the sensitivity to study single monolayers on metal surfaces make FTIR an exceptionally useful tool for investigating the orientational effects of LB films.

Detailed studies of the level and type of molecular orientation present in LB films of conducting polymers and their precursor monomer assemblies have been carried out for both polypyrrole and polythiophene based systems [23-27] . The results obtained from these studies are summarized below.

3.2.1. *ORIENTATION STUDIES OF LB FILMS FABRICATED WITH THE POLY(3-ALKYLTHIOPHENES).* IRRAS studies have been carried out on multilayer LB films fabricated from mixed monolayers of the poly(3-alkylthiophenes) and deuterated stearic acid [25]. By using deuterated stearic acid, the orientation of the alkyl groups of both types of molecules present in the LB film could be independently examined. This is possible since the vibrational bands of the hydrocarbon tails of the deuterated stearic acid molecules are shifted to lower wavenumbers relative to the hydrocarbon sidegroups of the polythiophene molecules. The specific molecular organization of these heterogeneous multilayer films was found to depend strongly on the length of the alkyl sidechain pendent to the polymer backbone and the amount of polymer present in the film. Two basic types of behavior were observed.

For LB films fabricated from poly(3-alkyl thiophenes) with short alkyl groups (eight or fewer carbon atoms), the cadmium stearate molecules were found to exhibit a strong tendency to assemble into well ordered molecular stacks in which the molecules assume a molecular organization very similar to that found in a homogeneous LB film of cadmium stearate. Specifically, it was found that the fully extended hydrocarbon tails of the cadmium stearate molecules were oriented nearly normal to the substrate surface with a small tilt angle of about 10° from the surface normal. The magnitude of this tilt angle is consistent with the 50 Å bilayer repeat distance observed by x-ray diffraction (see previous section), thus further confirming the existence of a two phase system in which the cadmium stearate molecules form segregated, well ordered molecular stacks. In general, the most ordered cadmium stearate domains were found in films containing a low polymer content, although in the case of films fabricated from poly(3-hexyl thiophene), a high level of molecular orientation of the cadmium stearate molecules was observed even in films containing as much as 80 mole % of the polymer. Another common feature of these films was the absence of any significant molecular orientation of the alkyl groups pendent to the polymer backbone over the entire compositional range of the mixed films.

Quite different results were obtained for LB films fabricated from poly(3-alkylthiophenes) with long alkyl sidechains. For example, when the mixed LB films were fabricated from poly(3-octadecyl thiophene), a polymer which has an alkyl sidechain equivalent in length to the hydrocarbon tail group of cadmium stearate, the formation of well ordered, separate domains of cadmium stearate molecules was found to be strongly inhibited. This was attributed to the greater level of molecular mixing experienced by the two components of the film due to the stronger non-bonded interactions established between the long alkyl groups of each species. In this case, the level of molecular orientation of the polymer sidechains was actually found to be greater than that achieved by the cadmium stearate molecules and was found to increase significantly with increasing polymer content. At higher polymer loadings, some of the cadmium stearate molecules apparently become dissolved in the polymer phase thereby facilitating the orientation and order of the the hydrocarbon sidegroups of the polymer backbone. LB films containing greater than 50 mole % of poly(3-octadecylthiophene) molecules are therefore best described as heterogeneous organizations comprised of small, weakly ordered domains of cadmium stearate molecules (responsible for the weak Bragg reflections observed in the x-ray studies) and mixed polymer rich domains containing highly oriented hydrocarbon tails contributed from both the polymer and the cadmium stearate molecules. Tilt angle calculations indicated that the hydrocarbon tails in the most ordered films (2/1 mole ratio of P3ODT/CdSA) are oriented nearly perpendicular to the substrate surface with a tilt angle of about 12° from the surface normal.

Recently, the influence of doping on the molecular organization of poly(3-hexyl thiophene)/*d*-cadmium stearate (P3HT/*d*-CdSA) multilayer thin films has been examined [26]. Some of the results of this work are displayed in Figure 11 which shows the reflection and transmission FTIR spectra of films containing a 2/1 mole ratio of P3HT/*d*-CdSA recorded before and after gas phase doping with I_2 (a) and $SbCl_5$ (b). The asymmetric and symmetric CD_2 and CD_3 stretching vibrations of the hydrocarbon tail groups of the *d*-cadmium stearate molecules are found in the 2000-2300 cm^{-1} region while the asymmetric and symmetric CH_2 and CH_3 stretching vibrations of the hydrocarbon tail groups of the P3HT molecules are found in the 2800-3000 cm^{-1} region.

Looking first at the spectra of the films prior to doping, it can be seen that the intensities of the CH_2 and CH_3 stretching vibrations of the polymer alkyl groups are not strongly dependent on the mode of polarization (transmission versus reflectance). In sharp contrast, the intensities of the CD_2 and CD_3 stretching vibrations of the hydrocarbon tail groups of the *d*-cadmium stearate molecules display a dramatic polarization dependence. Specifically, the asymmetric (2193 cm^{-1}) and symmetric (2088 cm^{-1}) CD_2 stretching bands observed in the transmission spectra exhibit significantly stronger intensities than the asymmetric (2214 cm^{-1}) and symmetric (2066 cm^{-1}) CD_3 stretching bands whereas in the reflectance spectra the asymmetric and symmetric CD_2 stretching bands are comparable in intensity to their CD_3 counterparts. This type of polarization dependence is characteristic [22] of highly ordered hydrocarbon tail groups oriented with their fully extended chains nearly perpendicular to the substrate surface. Thus, before doping, the multilayer films are comprised of well ordered domains of cadmium stearate molecules and relatively disordered domains of the conjugated polymer (as evidenced by the lack of polarization dependence of the CH_2 / CH_3 stretching bands).

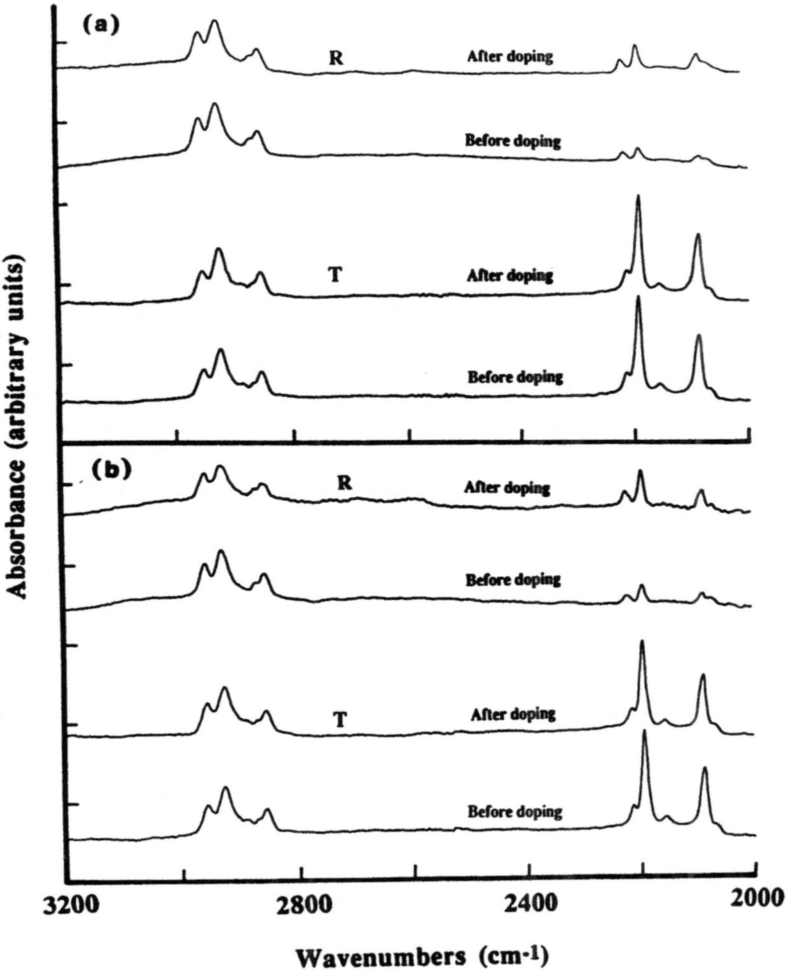

Figure 11. Transmission (T) and reflectance (R) FTIR spectra of P3HT/CdSA (2/1) LB films recorded before and after doping with I_2 and $SbCl_5$.

Exposure of the P3HT based multilayer films to I_2 or $SbCl_5$ gas creates electrically conductive LB films with the I_2 doped film in this case achieving a maximum conductivity of about $2x10^{-2}$ S/cm and the $SbCl_5$ doped film a conductivity of about $1x10^{-3}$ S/cm. The FTIR spectra of the multilayer films before and after doping are almost identical (see Figure 11) indicating that, in both cases, the molecular organization of the film has not been significantly altered by the doping process. The slight increase in intensity of the asymmetric and symmetric CD_2 stretching vibrations relative to the asymmetric and symmetric CD_3 stretching vibrations that is observed in the reflectance spectra does, however, suggest that doping has introduced a small amount of disorder into the cadmium stearate domains (or less likely, there has been a slight modification of the tilt angle). It is nevertheless clear that the polythiophene molecules present in the film can be successfully doped from the gas phase without disrupting the unique heterogeneous molecular organization established by LB technique.

This is not necessarily true for films doped from the solution phase. For example, the FTIR spectra of similar mixed LB films of P3HT/*d*-CdSA change dramatically after exposure to acetonitrile solutions of $NOPF_6$ (spectra not shown). Most noticeable is the complete absence of any vibrational bands associated with the cadmium stearate molecules. Thus, during doping, the original molecular organization of the film is completely disrupted and a portion of the cadmium stearate molecules are extracted into the doping solution. Interestingly, the final doped film is still of high optical quality and displays a conductivity of about 1 S/cm which is close to the value obtained when free standing films of poly(3-hexyl thiophene) are doped from solution with $NOPF_6$.

As indicated above, doping from solution with $NOPF_6$ creates the most highly conductive films but also results in a complete disordering of the original molecular organization. Doping from the gas phase with $SbCl_5$ or I_2, on the other hand, takes place without a significant disruption of the molecular organization. X-ray [28] studies have also confirmed these conclusions. For example, the diffraction patterns of multilayer films doped from solution with $NOPF_6$ are completely devoid of any observable reflections indicating that the doping process has significantly disordered the layer stacking of the cadmium stearate molecules. The x-ray diffraction patterns of films doped with I_2 from the gas phase, however, are the same as those generated from undoped films thereby confirming the fact that doping does not alter the layer structure of the cadmium stearate domains. In the case of $SbCl_5$ doping, x-ray studies have shown that the dopant molecules penetrate into the multilayer structure and interact with both the polymer and cadmium stearate molecules. The interaction of the dopant with the polymer molecules creates electrically conductive polymer chains whereas the interaction with the cadmium stearate molecules causes the layer spacing of the cadmium stearate domains to expand most likely by the insertion of dopant molecules into the hydrophilic planes. As suggested by the FTIR results, this expansion of the layer spacing occurs without any significant changes in the molecular orientation and packing of the cadmium stearate domains. Thus, the gas phase doped films retain the two-phase molecular organization originally present in the film.

The lower electrical conductivities exhibited by the gas phase doped films reflect the fact that insulating domains of the cadmium stearate molecules still impose constraints on the transport of charge carriers through the heterogeneous structure. Conduction in the LB films doped from the gas phase is therefore still strongly influenced by the two phase molecular organization originally established in the film. This is not the case for solution doped films with a high polymer content (> 50 mole %). Such films exhibit electrical properties very similar to the bulk conducting polymer since the polymer chains are now free to adopt a more continuous phase throughout the LB film due the disordering effects of the solvent.

3.2.2. *ORIENTATION STUDIES OF LB FILMS FABRICATED WITH SURFACE ACTIVE PYRROLES AND POLYPYRROLE.* The molecular organizations of multilayer thin films of the surface active pyrrole derivatives used to formed mixed monolayers with electrically conductive polypyrrole chains and ferrocene derivatized pyrroles have also been examined by FTIR reflection/absorption techniques. The materials investigated in these studies included 3-hexadecyl pyrrole, 3-octadecyl pyrrole and 3-octadecanoyl pyrrole.

FTIR studies of LB films fabricated from alkyl substituted pyrroles were first carried out on Y-type films of 3-hexadecyl pyrrole [27]. These studies revealed that the pyrrole rings in films containing five monolayers are oriented vertically with respect to the plane of the substrate and the hydrocarbon tail groups are oriented approximately normal to the substrate. In the case of a single monolayer, however, the hydrocarbon tails were found to be tilted away from the substrate normal. It was speculated that this later effect is due to an interaction between the π-electron cloud of the heterocyclic ring of the head group and the Pt substrate. The nature of this interaction is further discussed in the section on NEXAFS .

Monolayers of the surface active pyrrole derivatives 3-octadecyl pyrrole and 3-octadecanoyl pyrrole have also been successfully transferred into Y-type multilayer thin films. FTIR measurements have shown [23,24] that these materials both form highly ordered molecular organizations, each with its own specific type of molecular orientation. To illustrate these differences, the FTIR reflection and transmission spectra of multilayer films of 3-octadecyl pyrrole (3-ODP) and 3-octadecanoyl pyrrole (3-ODOP) are displayed in Figure 12. Also included in this figure are the reflection and transmission spectra of a multilayer film of electrically conductive polypyrrole made with a 5000/1 mole ratio of pyrrole/3-octadecanoyl pyrrole using the procedure described earlier.

For 3-octadecyl pyrrole, the hydrocarbon tail groups organize in a manner similar to that of 3-hexadecyl pyrrole, that is, they are oriented nearly perpendicular to the substrate surface. Tilt angle calculations [24] based on the work of Umemura et al. [29] indicate that the tail groups tilt at an angle of about 8° away from the surface normal. As indicated in Table 1, x-ray diffraction analysis of multilayer thin films of 3-octadecyl pyrrole has revealed that the molecular layers stack with a bilayer repeating distance of about 55 Å (27.5 Å/layer). Given that a fully extended molecule of 3-octadecyl pyrrole is about 28.4 Å from end to end (based on CPK models) it is reasonable to conclude from the x-ray and FTIR results that the molecules stand nearly perpendicular to the substrate surface and stack in a classical Y-type fashion. Further evidence for this type of structure has been provided by capacitance measurements made on stepped thickness Pt/LB film/Al sandwich capacitors [30]. Such measurements have shown that the dielectric constant of multilayer thin films of 3-octadecyl pyrrole is about 2.9 and is essentially independent of frequency between 10^2 -10^6 Hz. This value is consistent with the head-to-head type of molecular orientation characteristic of a Y-type LB film since any contribution to the polarization by the permanent dipoles of the pyrrole head groups is cancelled in such an arrangement.

The multilayer thin films fabricated from 3-octadecanoyl pyrrole, which is the surface active pyrrole preferred for making electrically conducting monolayers of polypyrrole, are also highly ordered Y-type films. In this case, however, strong hydrogen bonding interactions between the carbonyl and N-H moieties of the pyrrole head groups force the hydrocarbon tails to tilt away from the surface normal with a much larger tilt angle than that observed in the multilayer films of 3-hexadecyl pyrrole or 3-octadecyl pyrrole. As can be seen in Figure 12, this difference in tilt angle has a dramatic effect on the polarization behavior of the hydrocarbon stretching vibrations. In contrast to what is observed in the spectra of the 3-octadecyl pyrrole multilayer, the intensity of the asymmetric CH_2 stretching vibration (2920 cm^{-1}) is actually greater in the reflectance spectrum than in the transmission spectrum whereas the symmetric CH_2 stretching vibration (2851 cm^{-1}) is significantly weaker in reflectance. Calculations based on these FTIR measurements [24] have revealed that the hydrocarbon tail groups are tilted away from the surface normal with an angle of about 30°. In addition, the pyrrole head groups exhibit a tendency to orient parallel to the plane of the substrate as indicated by the presence of a stronger N-H stretching vibration (3200 cm^{-1}) in the transmission spectrum as compared to the reflectance spectrum.

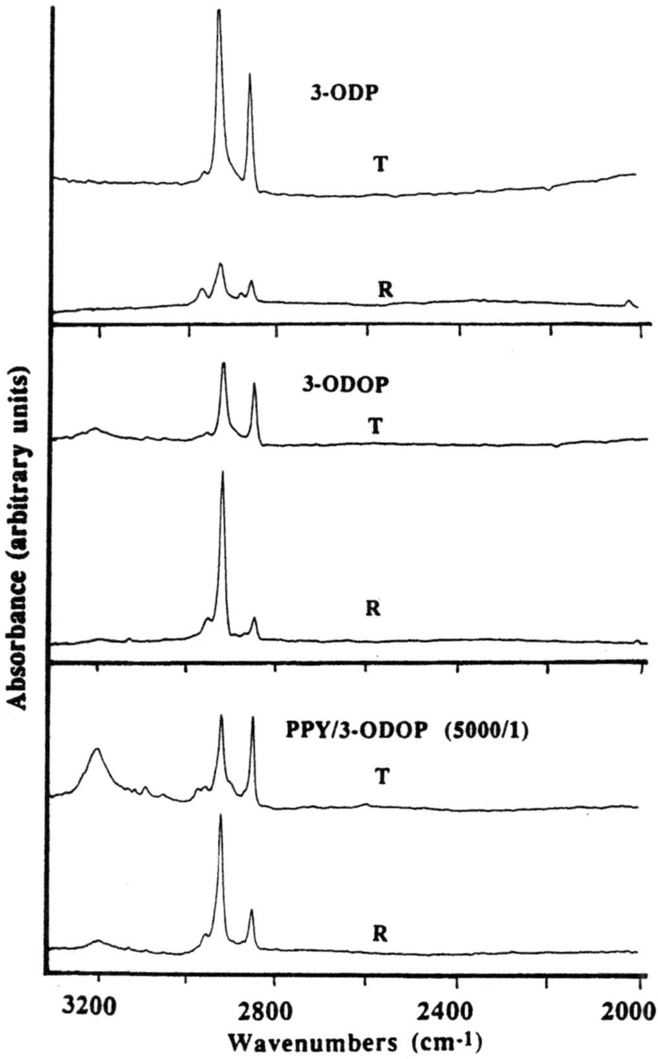

Figure 12. Transmission (T) and reflectance (R) FTIR spectra of 3ODP, 3ODOP and electrically conducting polypyrrole/3ODOP LB films.

The bilayer stacking distance obtained from low angle x-ray scattering measurements made on multilayers of 3-ODOP was calculated to be about 28 Å. This suggests that each monolayer is only about 14 Å thick. To achieve this particular bilayer stacking, the fully extended molecules would have to pack with a tilt angle of at least 55° from the surface normal. One possible explanation for this discrepancy is that the tail groups adopt a 30° tilt angle but the head groups from adjacent layers are interdigitated thereby reducing the effective thickness contributed by each layer. The establishment of interlocked headgroups in this case would be driven by the formation of intermolecular hydrogen bonds between the head groups of adjacent molecular layers. This hypothesis has yet to be confirmed.

To form electrically conducting monolayers of polypyrrole, 3-octadecanoyl pyrrole must be mixed with an excess of pyrrole monomer and spread onto an oxidizing subphase. The FTIR spectra of the multilayer films fabricated from these mixed electrically conductive monolayers (Figure 12) indicate that the 3-octadecanoyl pyrrole molecules assume a molecular orientation very similar to that obtained in their native multilayer films. This can be clearly seen by the similar polarization dependence observed for the multilayer films containing both pure 3-ODOP and polypyrrole/3-ODOP. Thus, the hydrocarbon tails of the surface active molecules are tilted away from the surface normal with a tilt angle estimated in this case to be about 25° and the pyrrole head groups are oriented nearly parallel to the plane of the substrate.

An analysis of the vibrational bands associated with the electrically conducting polypyrrole chains (spectral region not shown) has further demonstrated that the polymer chains are preferentially oriented with the long axis of their fully extended backbones contained within the plane of the substrate. All of these observations are consistent with a structure in which the polypyrrole chains are sandwiched between oriented bilayers of the surface active pyrrole. X-ray diffraction analysis, however, also shows that a small portion of the multilayer film is comprised of separate molecular stacks of 3-octadecanoyl pyrrole molecules packed with their normal bilayer repeat distance. This suggests that some of the surface active pyrrole molecules undergo a reorganization during film transfer to form isolated molecular domains. Nevertheless, the primary structure of the electrically conductive LB films is comprised of polypyrrole chains isolated between layers of the surface active pyrrole molecules.

3.3. ORIENTATION STUDIES BY NEXAFS

The NEXAFS technique (Near-Edge X-ray Absorption Fine Structure Analysis) has also been successfully utilized to examine the molecular organization of LB films of conducting polymers. This technique, in which core level electrons are excited by incident photons to unoccupied orbitals near the ionization threshold, takes advantage of the high degree of polarization of the x-ray beam emanating from a synchrotron storage ring. The excitations relevant to the study of LB films are core level excitations to the unoccupied σ^* and π^* orbitals. By scanning the incident photon energy and monitoring the Auger electrons resulting from the relaxation processes of the excitation, the empty states, such as the C-C σ^*, C=C σ^* and (C-H)*, can be identified as shown schematically in Figure 13. The resonance peaks corresponding to transitions of core level electrons to empty σ^* and π^* orbitals will have maximum intensity when the electric field vector of the incident photon beam is parallel to the σ and π orbitals, respectively. Thus, by using the highly polarized x-ray beam of the synchrotron radiation, the orientation of certain bonds, such as C=C and C-H can be readily determined [31].

The NEXAFS technique is element specific and represents interactions with the nearest neighbor atoms. This property can be used to determine the orientation and electronic properties of individual moieties in a multicomponent system. For example, the sulfur L edge spectra in polythiophene systems or the nitrogen K edge spectra in polypyrrole systems can be used to determine the orientation of the thiophene or pyrrole rings separately from those of, for example, the alkyl tails.

Figure 13 shows a schematic illustration of light incident on a hydrocarbon chain and the relevant excitations associated with the resonances of the NEXAFS spectrum. When light is incident parallel to the chain direction, the electric field vector is normal to the chain direction and will excite the 1s core level electron into an empty orbital associated with the C-H bond. This bond is perpendicular to the average direction of the C-C bond, which is along the chain direction. With the light incident perpendicular to the chain direction, therefore, the electric field vector, now being oriented parallel to the chain direction, will excite the 1s core level electron into a quasi-bound orbital associated with the C-C σ bond.

(a)

(b)

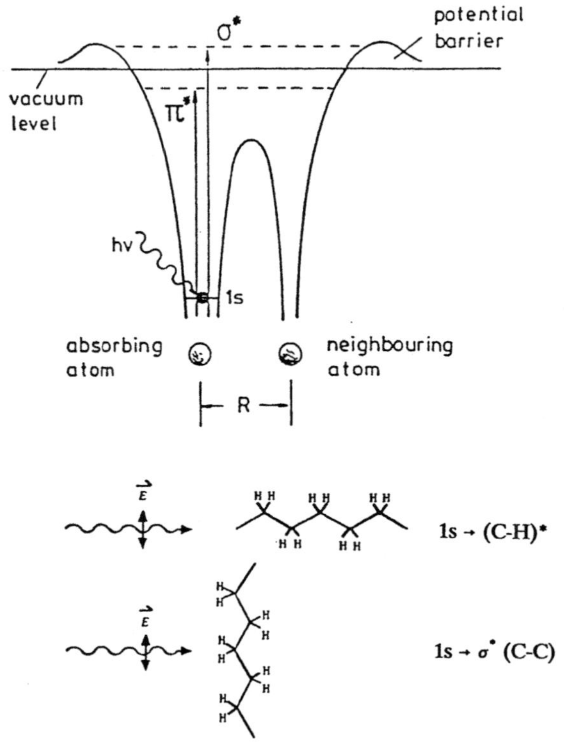

Figure 13. (a) Schematic diagram of σ and π resonances in a diatomic molecule with an intramolecular bond length R. The π resonance corresponds to a transition of a core electron (1s in the case of carbon) to an empty or partially filled molecular orbital (bound state). The σ resonance involves a quasi-bound state in the continuum. The resonance associated with the C-H bond, designated (C-H)*, has the same symmetry as a π* resonance. (b) Schematic diagram of a hydrocarbon chain with incident light normal and at glancing incidence relative to the chain direction.

Figure 14 shows the carbon K edge NEXAFS spectra of a monolayer and 7 layer film of 3HDP deposited on platinum. Spectra are shown for light incident normal to the substrate and at a glancing angle (approximately 20°) The orientation of the electric field vector relative to the substrate is indicated in the figure. The energies and assignments of the features are listed in Table 2.

As can be readily seen from figure 14, there is a strong polarization dependence of the spectra. This is most pronounced for the 7 layer structure. In this case the spectrum corresponds primarily to the top layers since the technique is surface sensitive. With light incident normal to the substrate, the excitation corresponding to the transition from the 1s state to an empty state associated with the C-H bond is the most prominent feature. At glancing incidence, the excitation corresponding to a transition associated with the C-C σ bond of the hydrocarbon chain is the most prominent feature. The carbon spectrum is dominated by the hydrocarbon chain, since that is where most of the carbon atoms reside. It can be concluded from these spectra that the 3HDP molecules in the top layers are oriented close to normal to the substrate.

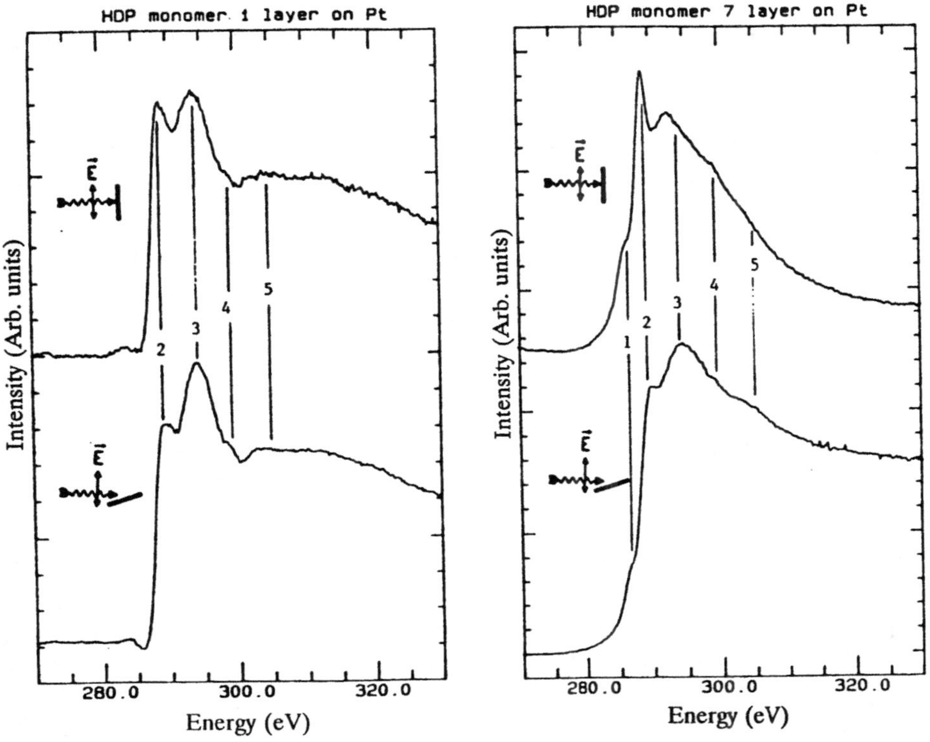

Figure 14. Carbon K edge NEXAFS spectra of (a) 1 monolayer and (b) 7 monolayers of 3HDP coated on a glass slide. Spectra are shown for light incident normal to the substrate and at a glancing angle (approximately 20°) The assignment of the spectral features is given in Table 2.

The polarization dependence is less pronounced in the case of a single monolayer on Pt, as shown in the same figure. This decrease in polarization dependence results from the molecules being tilted away from the surface normal. At a tilting angle of 54°, the spectra would exhibit no polarization dependence. The tilting is caused by an interaction between the π-electrons of the pyrrole head group and the platinum surface. An alternative explanation for the reduced polarization dependence would be in terms of disorder. However, independent FTIR reflection-absorption studies have shown that the hydrocarbon tails are highly ordered and tilted away from the substrate normal [27].

The resonance peak (labelled 1) associated with transitions to a C=C π^* empty orbital is absent in the spectrum of the monolayer film. This feature is related to the pyrrole ring exclusively. The absence of this feature for the pyrrole rings of the first monolayer is interpreted as arising from charge transfer to the metal surface. Strong bonding due to charge transfer between heterocycles and metal surfaces has been observed previously [32]. The charge transfer interaction results in an energy upshift of the C=C π^* feature which merges with the $(C\text{-}H)^*$ feature. This feature is present in the 7 layer film, since in that case most of the signal is derived from the top layers which are not in intimate contact with the platinum surface. This demonstrates that the LB technique is particularly useful for constructing model systems to study charge transfer reactions and heterogeneous catalysis under controlled conditions.

Table 2. Energies and assignments of the features in the C1s spectrum of 3HDP

No.	E (eV)	Assignment (final orbital)	
		Ring	Hydrocarbon tail
1	285.5	π^*3b_1 (C=C)	
1'	287.5[a]	π^*3b_1 (C=C)	
2	288.5	π^*2a_2 (C=C)	$\pi^*(CH_2)$
3	294	σ^* (C-C)	σ^* (C-C)
4	298.5	σ^* (C-C)	σ^* (C-C)
5	303	σ^* (C=C)	

[a] Energy upshifted from 285.5 eV due to charge transfer interaction

A similar high degree of polarization dependence has also been observed with the LB films made from mixtures of 3HDP and the ferrocene-derivatized 3-alkyl pyrrole [18]. This indicates that the layers have a high degree of internal ordering with the molecules oriented close to the vertical with respect to the substrate, an ordering which is imposed by the 3HDP molecules, i.e. the structure is maintained when ferrocene moieties are incorporated into the film.

Poly(3-alkyl thiophenes) mixed with stearic acid were also transferred to platinum substrates and examined by NEXAFS [33]. Figure 15 shows the carbon K edge and the sulfur L edge spectra of a 10 layer film of a 1:1 mixture of poly(3-octadecyl thiophene) and Cadmium stearate. The carbon spectra show a very strong polarization dependence, with the highest intensity of the (C-H)* peak occurring when the electric field vector is parallel to the substrate. This indicates that the hydrocarbon chains are highly ordered and oriented along the surface normal. The octadecyl chains do not contribute any appreciable amount of disorder reflected in the spectra, which indicates that the octadecyl chains and stearic acid chains are both highly oriented in the film. If the octadecyl chains were randomly oriented, the polarization dependence of the spectra would be dramatically reduced since the number of carbon atoms in the octadecyl chains is comparable to that of the stearic acid chains.

The sulfur L edge spectra in figure 15(b), however, exhibit no polarization dependence. Since the NEXAFS spectrum represents interactions between the absorbing atom and the nearest neighbor atoms, the sulfur L edge spectrum exclusively samples the thiophene rings, with no contribution from the alkyl chains. The absence of any polarization dependence in the sulfur spectrum means that the thiophene moieties are, on the average, randomly oriented. This is consistent with FTIR results discussed earlier. The carbon and sulfur spectra, therefore, give independent information about the orientation of the alkyl chains and the polythiophene backbone. These results clearly demonstrate the power of the NEXAFS technique to determine the orientational behavior of LB films, in particular the capability to separately determine the orientation of specific moieties in the structure when separate elements are available.

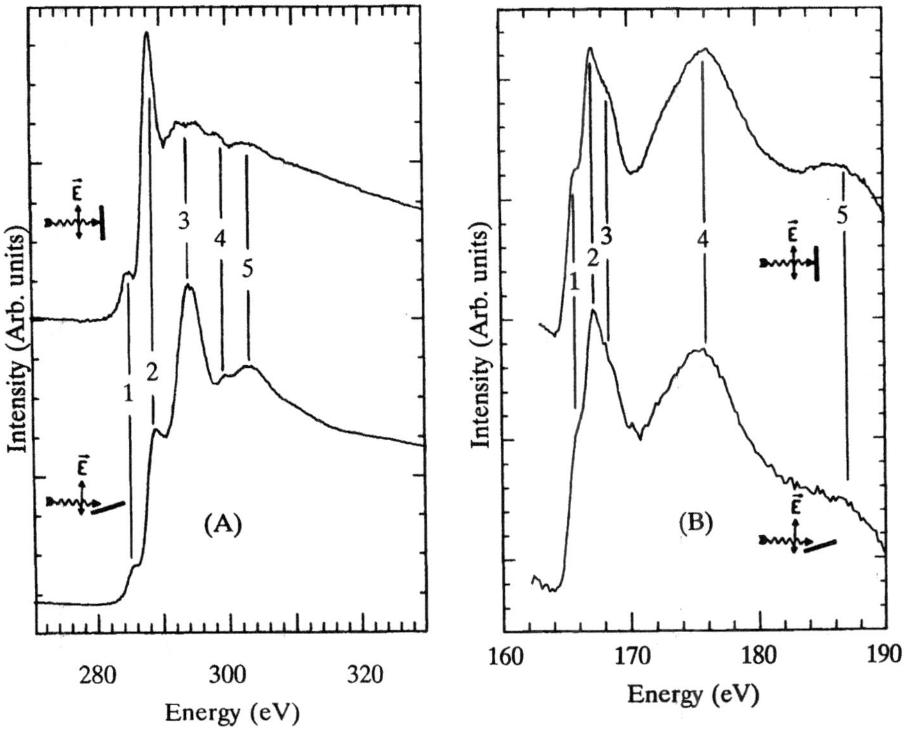

Figure 15. (a) Carbon K edge and (b) sulfur L edge spectra at normal and glancing angle incidence for a 10 layer film of a 1:1 mixture of poly(3-octadecyl thiophene) and cadmium stearate deposited onto Pt coated glass slides.

4. Electrical Properties of LB films Containing Conducting Polymers

4.1. IN-PLANE AND TRANSVERSE CONDUCTIVITIES

As indicated in Table 3, all of the LB films of conducting polymers fabricated to date exhibit some degree of conductivity anisotropy. This is a natural consequence of the layered structures created by the LB transfer process and the types of molecules used to fabricate these films. The highest levels of conductivity anisotropy are found in those molecular organizations in which planes of the electrically conductive polymer chains are separated by insulating regions of tightly packed long chain alkyl groups. For example, this type of molecular organization is found in the multilayer films of electrochemically polymerized octadecyl 4-methyl pyrrole-3-carboxylate [2] and the multilayer films fabricated from chemically polymerized mixtures of pyrrole/3-octadecanoylpyrrole [19,30]. The multilayer thin films fabricated from these materials both exhibit conductivity anisotropies, in some cases, in excess of 10^9. In other words, the conductivity within the plane of the film is over 10^9 times greater than the conductivity across the film thickness.

Table 3. In-plane and transverse electrical conductivities of LB films containing conducting polymers

Multilayer System	Electrical Conductivities (S/cm)		Conductivity Anisotropy
	In-Plane	Transverse	$\sigma_{=}/\sigma_{\perp}$
Poly(octadecyl 4-methyl pyrrole-3-carboxylate) /octadecane	10^{-1}	10^{-11}	10^{10}
poly(2-octadecoxyaniline)[1]	10^{-4}	10^{-10}-10^{-8}	10^{4}-10^{6}
poly(3,4-dibutylpyrrole) perfluorooctanoate[1]	10^{-6}	10^{-9}	10^{3}
polypyrrole/ 3-octadecanoylpyrrole	10^{-1}	10^{-10}-10^{-7}	10^{6}-10^{9}
Poly(3-hexylthiophene) /cadmium stearate(5/1)[2]	0.5	10^{-4}	10^{3}
poly(*p*-phenylene-vinylene)[3]	0.5	10^{-5}	10^{4}

1 doped with iodine
2 doped from acetonitrile solutions of $NOPF_6$
3 doped with SO_3

It is interesting to note that the LB films of poly(*p*-phenylene-vinylene), which are fabricated by a precursor process that involves thermal elimination of the alkyl groups of the polymer's counterion, still display a reasonable level of conductivity anisotropy [6]. The solution doped LB films fabricated from mixed monolayers of poly(3-hexylthiophene) and cadmium stearate also display a relatively small but significant level of conductivity anisotropy. In both cases, the molecular organizations of these multilayer films should be rendered essentially isotropic by either the thermal conversion process used to create the LB films of PPV or the solution doping process used to oxidize the films of P3HT/CdSA. Apparently the remnants of the original layered structures of the LB films of these materials still influence the mobility of charge carriers across the film thickness.

The creation of electrically conductive LB films is usually accomplished in one of two ways. If the polymer chains are oxidized during their formation, then conductivity is established during the polymerization process as in the case of the electrochemical or chemical polymerization of pyrrole monomers. Often times, however, it is necessary to expose the multilayer film to an oxidizing agent after the conjugated polymer has been manipulated into its final multilayer form. In order to understand the basic relationships between electrical properties and structure in this latter case, it is necessary to consider how the molecular organization of the film has been modified by the type of oxidizing agent used and the way that it is introduced into the film. The mixed LB films of the poly(3-alkylthiophenes) represent an example of a multilayer system that must be chemically oxidized after fabrication to produce its electrically conductive form.

In order to render the multilayer films of the poly(3-alkylthiophenes) electrically conductive, it is necessary to expose them to suitable oxidizing agents. Doping studies of the P3AT based multilayer thin films have revealed [28] that the conductivity and final molecular organization of the doped films depend strongly on the type of oxidant used and the manner in which it is introduced into the film. Figure 16 shows the in-plane conductivity of LB films (34 layers) doped from solution with $NOPF_6$ as a function of the mole % of P3AT present in the films. In all cases, the conductivity increases as the amount of P3AT in the film increases with a maximum of about 3 S/cm being obtained in the mixed LB film of P3HT/CdSA with a mole ratio of 5/1. The in-plane conductivity is also seen to approach the value obtained from a pure free standing film of the respective $NOPF_6$ doped P3AT as the polymer content in the film approaches 100%.

The concentration dependence of the conductivity can be explained by a simple percolation model. As the concentration of the conducting phase increases, a percolation threshold is reached at which point a continuous pathway across the film via the conducting domains is established. The electrical properties of the film then become dominated by the intrinsic electrical properties of the conducting phase of the P3AT. Figure 16 indicates that the percolation threshold occurs at P3AT contents between 20 and 40 mole %. Above this range, the conductivity of each P3AT based multilayer film slowly approaches the value of a pure free standing film of the conducting P3AT. At polymer concentrations below this range, the insulating cadmium stearate molecules still exert an influence on the transport of carriers through the film.

Measurements of the temperature dependence of the conductivity also confirm that the electrical properties of the high polymer content films are very similar to their bulk counterparts. For example, the conductivities of fully doped free standing films of P3HT and LB films containing a 5/1 mole ratio of P3HT/3ODOP were both found to vary with temperature in $T^{-1/2}$ [34] relationship. This suggests that the mechanism of conduction in the high polymer content LB films is the same as that in free standing films and in both cases is based on a charging energy limited tunneling process [35]. In light of the fact that solution doping of the high polymer content films with $NOPF_6$ disrupts the original molecular organization and renders the conducting polymer phase more continuous, it is not too surprising to find that the electrical properties of these films are similar to the bulk films.

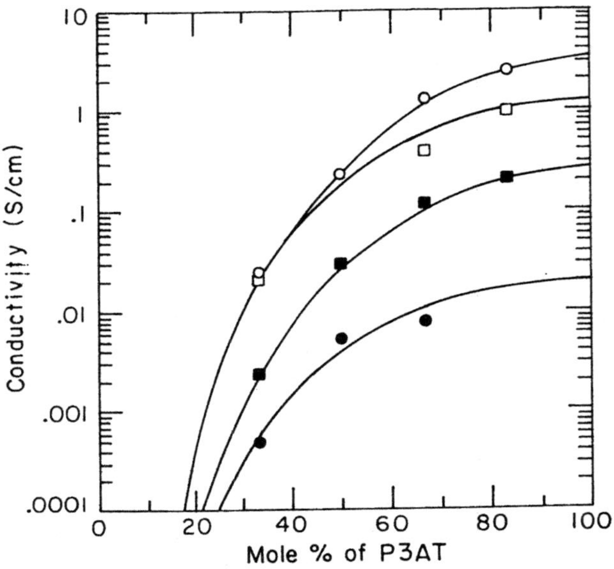

Figure 16. In-plane conductivity of LB films doped from solution with $NOPF_6$ as a function of the mole % of P3AT in the films.

In-situ doping studies of the P3AT based LB films using dopants introduced in the gas phase have revealed that the increase in conductivity that takes place during doping is due to the formation of bipolaron-type charged defects on the polymer backbone [28]. Such studies were carried out by simultaneously monitoring the electrical conductivity and optical absorption of a mixed LB film of P3HT during the doping process. In all cases, an increase in conductivity was accompanied by a decrease in the intensity of the absorption band associated with the $\pi-\pi^*$ electronic transition of the polythiophene conjugated backbone (centered at about 490 nm for LB films of P3HT) and the development of two new lower energy excitations typically attributed to the creation of bipolarons. The $\pi-\pi^*$ interband transition was completely eliminated when $SbCl_5$ was used as the dopant whereas it remained a dominant feature in the spectrum of the film doped with iodine. These differences were attributed to the nature of charge transfer interaction developed between the oxidized polymer chains and the dopant. The charge transfer interaction between iodine (or more specifically I_3^-) and the polymer backbone was found to be very weak and easily reversed by exposing the film to high vacuum. In fact, the multilayer film could be completely undoped to its neutral form by short time exposures to air or vacuum. A stronger charge transfer interaction between $SbCl_5$ and the polythiophene backbone, on the other hand, was indicated by the good vacuum stability of the doped complex and the complete closing of the polymer's interband transition upon doping. It is interesting to note that in spite of these differences, the conductivities of the films after doping only differed by about one order of magnitude with the iodine doped film actually achieving the higher level of conductivity.

4.2. DIELECTRIC PROPERTIES

The configuration frequently utilized to evaluate the dielectric properties of LB films is illustrated in Figure 17 [36]. Such a device is constructed by depositing monolayers in a step-like fashion onto an electrically conductive substrate (usually a metal coated glass slide). This is easily accomplished by varying the depth of immersion of the substrate during the transfer process. Once the staircase structure is complete, an array of metal electrodes is deposited onto the individual steps to form a series of metal/LB film/metal capacitors with varying thicknesses. By depositing steps with different thicknesses, it is possible to accurately determine the dielectric constant of the film and the reproducibility and quality of the deposition process. The capacitance of a high quality film, in which each monolayer has been reproducibly transferred into the multilayer structure, should depend inversely on film thickness according to equation 1 given below.

$$1/C = \frac{T_{oxide}}{A\,\varepsilon_o\,\varepsilon_{oxide}} + \frac{T_{LB\ layer}\,(\#\ Layers)}{A\,\varepsilon_o\,\varepsilon_{LB}} \qquad (1)$$

In this equation, C is the film capacitance, T_{oxide} is the thickness of any oxide layer present on the metal electrodes, $T_{LB\ layer}$ is the average thickness of a monolayer, A is the electrode area, ε_o is the permittivity of free space and ε_{oxide} and ε_{LB} are the dielectric constants of the oxide and LB film respectively. Thus, the slope of a plot of 1/C versus the number of layers transferred per step provides information about the dielectric constant of the film whereas the intercept of such a plot is related to the thickness of the oxide layers on the metal electrodes (if the metal has a native oxide layer).

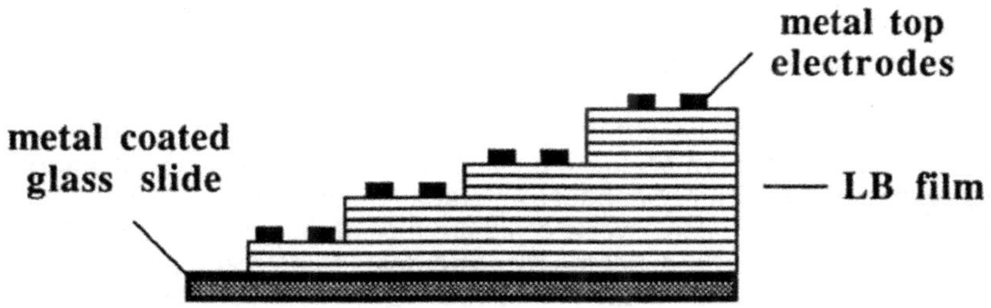

Figure 17. Stacking sequence and electrode configuration used to evaluate the dielectric properties of LB films.

To illustrate this approach, Figure 18 shows the reciprocal capacitance plots generated from mixed multilayer thin films of poly(3-butylthiophene) and cadmium stearate containing either 33 or 66 mole % P3BT. The capacitance measurements in this case were made on Ag/LB film/AL sandwich capacitors with top electrode areas of 0.1 and 0.2 cm^2 [9]. The linear relationship observed in these plots confirms that the transfer process is completely reproducible up to at least 50 layers. Similar linear plots have also been generated for all of the LB films fabricated from the mixed monolayers of the poly(3-alkylthiophenes). The dielectric constants calculated from the slopes of these curves were 3.2 for the 1/2 mole ratio film and 3.5 for the 2/1 mole ratio film. These values were estimated using the thicknesses measured by a profilometer (25 Å per monolayer for the 1/2 mole ratio film and 32 Å per monolayer for the 2/1 mole ratio film) and are in the range expected for mixed films of P3BT and cadmium stearate.

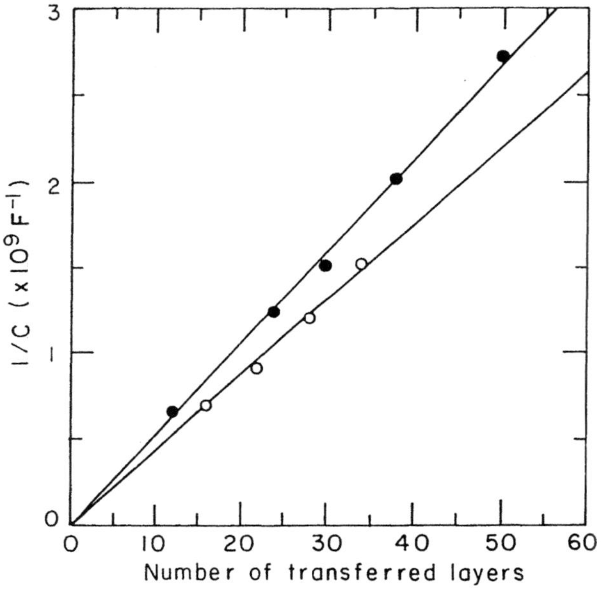

Figure 18. Plots of the reciprocal capacitance (1/C) versus the number of layers for multilayers fabricated from poly(3-butylthiophene)/cadmium stearate mixed monolayers containing 1/2 and 2/1 mole ratios.

The unique highly anisotropic molecular organizations of multilayer thin films containing conducting polymers make it possible to also examine the dielectric properties of these films in their highly conducting state. When fabricated into sandwich type parallel plate capacitors, the transverse conductivities of these conducting LB films are low enough to allow meaningful dielectric measurements to be made. As mentioned earlier, many of the LB films of conducting polymers are good conductors within the plane of the film but are highly insulative across the film thickness. Of particular interest is the frequency dependence of the dielectric constant, dielectric loss and transverse conductivity, all of which are readily obtained from standard capacitance measurements using the sandwich type capacitor configuration described above.

Rubner and associates [19,30] have shown that electrically conducting LB films fabricated from polypyrrole/3ODOP monolayers exhibit unusually large dielectric constants over a wide frequency range. The large dielectric constants are believed to be due to the accumulation of space charges at the interfaces of the conducting and insulating layers created by the LB deposition technique. The current notion is that charge carriers move through the conducting polypyrrole domains, most likely via the hopping of positively charged bipolarons, until they encounter an interface with an insulating region of the film at which point they become trapped. An example of the unusual dielectric behavior exhibited by these conductive films in shown in Figure 19. This figure shows the dielectric constant and dielectric loss as a function of frequency for a multilayer thin film fabricated from the polypyrrole/3ODOP mixed monolayer system. This particular film was evaluated under dry box conditions in the form of a Pt/LB film/Al sandwich type capacitor. The multilayer thin film was comprised of 50 monolayers thereby giving it a thickness of about 2000 Å.

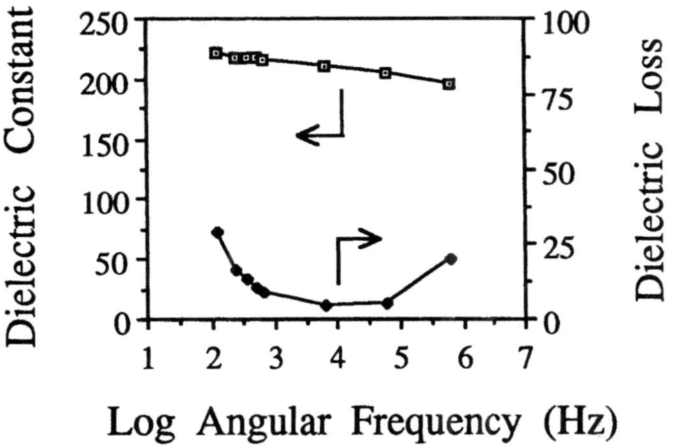

Figure 19. Dielectric constant and dielectric loss of a 50 layer polypyrrole/3ODOP multilayer thin film as a function of frequency.

As indicated in the figure, the dielectric constant of this molecular assembly is about 225 at a frequency of 100 Hz and drops slightly to a value of about 200 at 10^6 Hz. Thus, the dielectric constant is much larger than that typically encountered in organic materials and only undergoes a small dielectric relaxation over this frequency range. When measurements were made at frequencies greater than 10^6 Hz (not shown), however, the dielectric constant of the multilayer film was found to drop dramatically indicating that the charge carriers within the conducting domains could no longer follow the oscillating electric field. Once the motion of these carriers is frozen out, they can no longer contribute to the dielectric constant. The dielectric loss of this film is seen to start off relatively high, decrease to an essentially constant lower value and then rise again to a higher value. The higher loss observed at low frequencies is most likely due to the presence of ionic impurities in the film whereas the increasing loss at higher frequencies represents the onset of the dielectric relaxation that occurs at higher frequencies.

The frequency dependence of the transverse conductivity of this film is shown in Figure 20. For comparison purposes, this figure also displays the frequency dependence of the transverse conductivity of an insulating LB film fabricated from monolayers of the surface active pyrrole derivative 3-octadecylpyrrole (3ODP). In the former case, the transverse conductivity increases with increasing frequency from about 5 x 10^{-10} S/cm to about 2 x 10^{-6} S/cm over the frequency range of 100 to 10^6 Hz. Also note that the low frequency conductivity of this film approaches its DC value of about 10^{-10} S/cm indicating a conductivity anisotropy in this case of about 10^8 (the in-plane conductivity of this system is typically about 10^{-2} S/cm). The transverse conductivity of the insulating multilayer thin film of 3ODP also increases with increasing frequency although the magnitude of the transverse conductivity at all frequencies is about three orders of magnitude lower than that of the electrically conducting film containing polypyrrole. In any event, it is clear that the transverse electrical properties of these films follow a very similar frequency dependence.

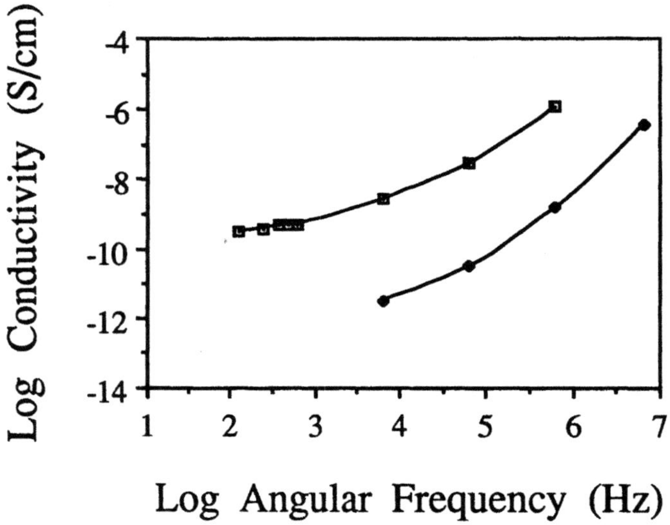

Figure 20. Transverse conductivity as a function of frequency for a conducting multilayer thin film of polypyrrole/3ODOP (open squares) and an insulating thin film of 3-octadecylpyrrole (closed circles).

This type of frequency dependence is very common in insulating and semiconducting systems whose charge transport is dominated by hopping or tunneling mechanisms involving localized charge carriers [37]. In both cases, transport across the film thickness requires that the carriers move through the insulating domains of tightly packed hydrocarbon tail groups. Since there are no extended band states available to the carriers, they must hop or tunnel between localized defect states within the insulating layers. The observation of such a frequency dependence in both films confirms the fact that conduction across the film thickness in both cases is limited by the movement of carriers through the insulating domains of the surface active molecules

The possibility of creating thin film structures with large dielectric constants and low losses over a wide frequency range is very intriguing. By controlling the conductivities and thicknesses of the conducting and insulating regions of the multilayer structure and its layer sequence, it should be possible to design thin film capacitors with a wide range of dielectric responses. It is also of interest to examine the influence of temperature and film thickness on the relaxation processes occurring in these films. Such studies should shed new light on the mechanisms of charge transport in conducting polymers. Toward this goal, a number of novel alternating layer heterostructures comprised of layers of an electrically conductive polymer alternating with layers of an insulating surface active molecule are currently under construction. Some preliminary reports of the properties of these new thin film structures have recently been published [19,30].

4.3 EVALUATION OF ELECTROACTIVE LB FILMS AS ACTIVE COMPONENTS OF THIN FILM DEVICES

The ability to fabricate highly ordered thin films with thicknesses that are controllable at the molecular level makes the LB technique a very attractive means for creating and evaluating thin film electronic devices and sensors based on conducting polymers. Using this technique, for example, it is possible to examine the influence of film thickness on such important device parameters as charge mobility, conductivity, and carrier concentration. The control over film thickness is important as the carrier mobilities of semiconducting conjugated polymers are typically very low making it necessary to work with ultrathin films of these materials with optimized thicknesses.

The first demonstration of the use of a conducting polymer based LB film as the active element of an electronic device was made by Stubb's group in Finland [38]. In this work, it was shown that mixed LB films of poly(3-hexylthiophene) and arachidic acid (60 mole % P3HT) could be fabricated into working field effect transistors (FETs) in which the undoped polymer served as the active semiconductor. Devices with film thicknesses ranging from a single monolayer (37Å) to 35 monolayers (1295 Å) were fabricated into FETs having a channel width and length of 8 cm and 5 μm respectively. Using this approach, it was possible to observe field-effects in the devices fabricated from a single monolayer. The mobilities and conductivities were found to decrease as the thickness of the LB film decreased. The mobilities of the LB film based devices (about 7×10^{-7} cm^2/Vs for a thick film) were also found to be lower than that reported for a spin coated film of pure P3HT (1×10^{-4} cm^2/Vs) and this was attributed to the presence of the surface active molecules.

Rubner's group has recently fabricated FETs based on mixed LB films of P3HT and 3-octadecanoylpyrrole [34]. In this case, the 3ODOP serves as the surface active component thereby allowing the fabrication of multilayer thin films free of cadmium ions and/or carboxylic acid groups. Figure 21 shows the conductivity and field-effect mobility of a P3HT/3ODOP (5/1 mole ratio) LB film as a function of film thickness at different doping levels (the films were doped from solution with $NOPF_6$). The LB film field-effect transistor had a channel width and length of 8 cm and 5 μm respectively (the FETs were kindly supplied by H. Stubb's research group). The dedoped films represent samples that were measured 17 days after doping. All samples were stored in nitrogen but were exposed to air when measurements were made. The mobility values of the doped samples were measured after storing the films for three days in nitrogen to insure that the effects of residual solvent were absent. Over this period, the conductivity of the film decreased by a factor of about two.

In the undoped state, it can be seen that mobility and conductivity increase slightly with increasing film thickness. The magnitude of this effect is smaller than that observed in the LB film based FETs fabricated with P3HT and arachidic acid suggesting that the charge carrier hopping length increases more slowly than that predicted by percolation models as the 2-dimensional hopping regime is approached. Upon doping, the mobilities and conductivities increase as expected, however, these parameters are now seen to decrease with increasing film thickness. The largest changes in mobility and conductivity with film thickness are observed in the dedoped films; this is most obvious in the thinner samples. For the thicker samples, these parameters are only weakly dependent on film thickness. The small decrease in the mobility of the thicker films upon dedoping indicates that the decrease observed in the conductivity level during the dedoping process is mostly due to a decreasing number of active dopants. In the thinnest samples, however, dedoping produces a large decrease in both the mobility and conductivity. As indicated by these preliminary results, the study of FET and similar electronic devices fabricated with LB films of conjugated polymers should prove to be an interesting area of research. These new devices are currently being evaluated as gas sensors and a report on these results will appear in the near future.

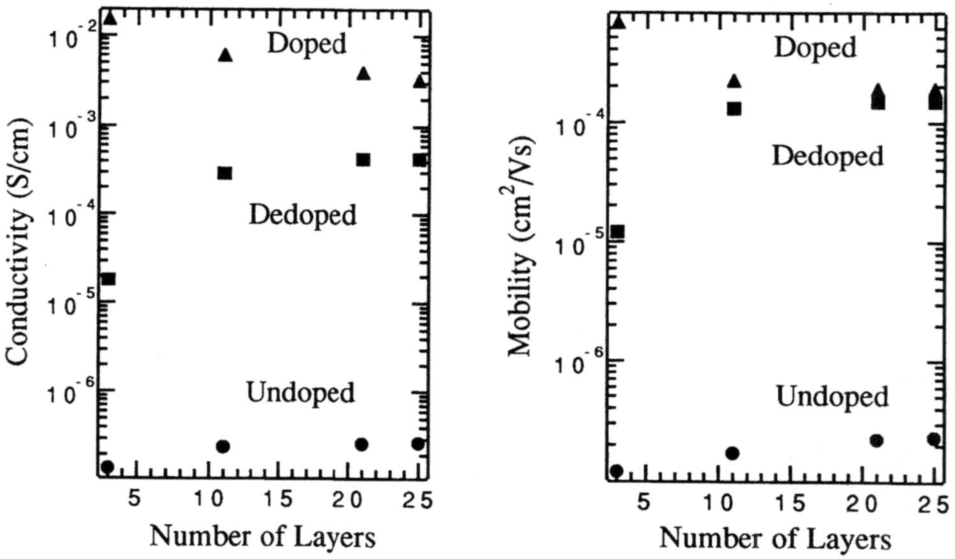

Figure 21. Field-effect mobility and conductivity of a mixed LB film of poly(3-hexylthiophene) and 3-octadecanoylpyrrole (5/1 mole ratio) as a function of film thickness and doping level.

5. Conclusions

Although still in an early stage of development, research on the fabrication and evaluation of LB multilayer thin films of conducting polymers has already proven that this approach can be used to control the molecular and supermolecular organizations of these extremely interesting materials. As this area matures, it should be possible to fabricate a wide variety of complex molecular assemblies with electrical and optical properties that can be tailored at the molecular and supermolecular level. It is also anticipated that significantly more ordered structures will be possible as greater control is gained over the chemistry used to create monolayers and the multilayer fabrication process. The ultimate realization of this approach would be the development of multilayer thin films comprised of well organized two-dimensionally ordered layers of various conducting polymers stacked in a layer sequence that produces completely new electrical and optical phenomena. The availability of such films would advance our fundamental understanding of two-dimensional processes in electroactive polymers and provide new thin structures for use in molecular sensing and electronic schemes.

6. Acknowledgements

The authors are grateful to Josephine Cheung, Eero Punkka, Masahiro Rikukawa, Robert Rosner, and Tim Royappa of MIT and X. Q. Yang of Brookhaven for their valuable assistance in the preparation of this book chapter. Partial support of the research carried out at MIT was provided by the National Science Foundation and the MIT Center for Materials Science and Engineering. Work at the Brookhaven National Laboratory was partially supported by the U. S. Department of Energy, Division of Materials Science, under contract No. DE-AC02-76CH00016.

7. References

1. For a recent review of this area see, (1990) G. Roberts (ed.), Langmuir-Blodgett Films, Plenum Press, New York.

2. Iyoda, T., Ando, M., Kaneko, T., Ohtani, A., Shimidzu, T., and Honda, K. (1986) *Tetrahedron Lett.* 27, 5633. Shimidzu, T., Iyoda, T., Ando, M., Ohtani, A., Kaneko, T. and Honda, K. (1988) *Thin Solid Films 160*, 67.

3. Ando, M., Watanabe, Y., Iyoda, T., Honda, K., Shimidzu, T. (1989) *Thin Solid Films 179*, 225.

4. Hong, K. and Rubner, M.F. (1988) *Thin Solid Films 160* , 187. Hong, K. and Rubner, M.F. (1989) *Thin Solid Films, 179,* 215. Hong, K., Rosner, R. B. and Rubner, M.F. (1990) *Chemistry of Materials, 2,* 82.

5. Era, M., Kamiyama, K., Yoshiura, K., Momii, T., Murata, H., Tokito, S., Tsutsui, T., and Saito, S. (1989) *Thin Solid Films 179*, 1.

6. Nishikata, Y., Kakimoto, M., and Imai, Y. (1989) *Thin Solid Films 179*, 191. Nishikata, Y., Kakimoto, M., and Imai, Y. (1988) *J. Chem. Soc., Chem. Commun.,* 1040.

7. Era, M., Shinozaki, H., Tokito, S., Tsutsui, T., and Saito, S. (1988) *Chem. Lett.* 1097.

8. Royappa, T. and Rubner, M. F., (1990) results to be published.

9. Watanabe, I., Hong, K., and Rubner, M. F. (1989) *J. Chem. Soc., Chem. Commun.* 123. Watanabe, I., Hong, K., Rubner, M. F., and Loh, I. H. (1989)

Synth. Met. *28*, C473. Watanabe, I., Hong, K., and Rubner, M.F. (1989)*Thin Solid Films*, *179*, 199. Watanabe, I., Hong, K., and Rubner, M.F. (1990) *Langmuir*, 6, 1164.

10. Yli-Lahti, P., Punkka, E., Stubb, H., Kuivalainen, P., and Laakso, J. (1989) *Thin Solid Films* , *179*, 221.

11. Peltonen, J. P. K. and Rosenholm, J. B. (1989) *Thin Solid Films 179*, 543.

12. Sato, M., Okada, S., Matsuda, H., Nakanishi, H., and Kato, M. (1989) *Thin Solid Films 179*, 429.

13. Tsumura, A., Kurata, T., Susuzi, S., Nobutoki, H., Koezuka, H., and Moriwaki, T. (1989) *Thin Solid Films 178*, 393.

14. Tasaka, S., Katz, H., Hutton, R. S., Orenstein, J., Fredrickson, G. H., and Wang, T. T. (1986) *Synth. Met. 16,* 17.

15. Nakahara, H., Nakayama, J., Hoshino, M., and Fukuda, K. (1988) *Thin Solid Films 160,* 87.

16. Logsdon, P. B., Pfleger, J., and Prasad, P. (1988) *Synth. Met. 26,* 369.

17. Samuelsen, L., Rahman, A.K.M., Clough, S., Tripathy, S., Hale, P. D., Inagaki, T., Skotheim, T. A., and Okamoto, Y. (1989) *Proc., NATO-ASI Symp. on Low-Dim. Systems and Molec. Electron.*, Spetses Island, Greece, June 12-23.

18. Skotheim, T. A.,Yang, X. Q., Chen, J., Inagaki, T., denBoer, M. F.,Tripathy, S., Samuelsen, L., Rubner, M. F., Hong, K.,Watanabe, I., and Okamoto, Y. (1989) *Thin Solid Films*, *178*, 233.

19. Rosner, R. B., Rubner, M. F. (1990) *Mat. Res. Soc. Symp. Proc., 173*, 363.

20. Iyoda, T., Ando, M., Kaneko, T., Ohtani, A., Shimidzu, T., and Honda, K. (1987) *Langmuir, 3*, 1169.

21. Chen, J. M.,Yang, X.Q., Chapman, D., Nelson, M., Skotheim, T. A., Ehrlich, S.N., Rosner, R. B., and Rubner, M. F. (1990) *Molec. Cryst. Liq. Cryst.*, in press.

22. Greenler, R. G., Rahn, R. R., and Schwartz, J. P. (1971) *J. Catal.*, *23*, 42. Allara, D. l., Swalen, J. D. (1982) *J. Phys. Chem., 86*, 2700. Rabolt, J. F., Burns, F. C., Schlotter, N. E., and Swalen, J. D. (1983) *J. Chem. Phys. 78,* 946.

23. Cheung, J. H., Rosner, R. B., Watanabe, I., and Rubner, M. F. (1990) *Molec. Cryst. Liq. Cryst.*, in press.

24. Cheung, J. H. and Rubner, M. F. (1990) *J. Phys. Chem.,* results to be published.

25. Watanabe, I., Cheung, J. H., and Rubner, M. F. (1990) *J. Phys. Chem.*, in press.

26. Rikukawa, M. and Rubner, M. F. (1990) results to be published.

27. Yang, X. Q., Inagaki, T., Skotheim, T.A.,Okamoto, Y.,Samuelsen, L.,Blackburn G., and Tripathy, S. (1988) *Molec. Cryst. Liq. Cryst.*, *160*, 253.

28. Watanabe, I. and Rubner, M. F. (1990) *British Polymer J.* , in press.

29. Umemura, J., Kamata, T., Kawai, T., and Takenaka, T. (1990) *J. Phys. Chem.*, *94*, 62.

30. Rosner, R. B., Cheung, J. H., and Rubner, M. F. (1990) in R. Allred, R. Martinez, K. Wischmann (eds.), Electronic Materials, Our Future, *4th International SAMPE Conf.*, *4*, 346.

31. Stohr, J., Outka, D. A. , Baberschke, K., Arvanitis, D. and Horsley, J. A. (1987) *Phys. Rev. B* 36, 2976.

32. Tourillon, G., Raaen, S., Skotheim, T. A., Sagurton, M., Garrett, R., and Williams, G.P. (1987) *Surf. Sci. 184,* L345.

33. Yang, X. Q., Chen, J., Hale, P. D., Inigaki, T., Skotheim, T. A., Okamoto, Y., Samuelsen, L., Tripathy, S., Hong, K., Watanabe, I., Rubner, M. F. and den Boer, M. L. (1989) *Langmuir*, *5,* 1288.

34. Punkka, E., and Rubner, M. F. (1990) results to published.

35. Sheng, P., Abeles, B., and Arie, Y. (1973) *Phys. Rev. Lett. 31*, 44.

36. Geddes, N. J., Samples, J. R., Parker, W. G., Couch, N. R., and Jarvis, D. J. (1990) *J. Phys. D.; Appl. Phys. 23*, 95 and references therein.

37. Epstein, A. J., (1986) in T. A. Skotheim (ed.) Handbook of Conducting Polymers, Marcel Dekker Publishers, New York, p. 1041.

38. Paloheimo, J., Kuivalainen, P., Stubb, H., Vuorimaa, E., and Yli-Lahti, P. (1990) *Appl. Phys. Lett. 56*, 1157.

ELECTRONIC PROPERTIES OF LINEAR POLYENES

BRYAN E. KOHLER
Chemistry Department
University of California
Riverside, CA 92521-0403
U.S.A.

1. Introduction

Polyacetylene remains one of the most interesting conducting polymers. While this is in part due to the very high conductivities that can be achieved by lightly oxidizing stretched polyacetylene films with molecular iodine, this interest also reflects the expectation that, because of the simplicity of the polyene backbone, the link between electronic structure and transport can be most easily worked out for this system. It is in this sense that polyacetylene is a paradigm system for conducting polymers.

In its undoped form polyacetylene is a disordered molecular material consisting of discrete linear polyene segments (Figure 1). It appears that although the molecular weight

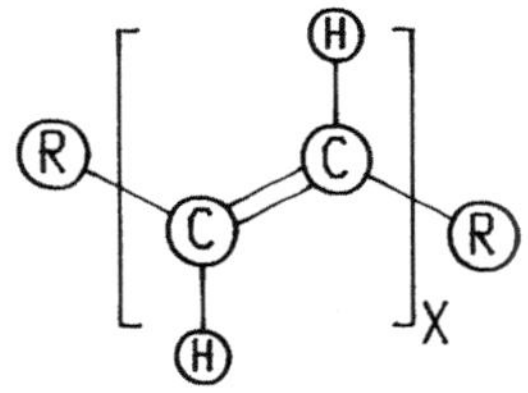

Figure 1. Molecular structure of polyacetylene or a linear polyene. For polyacetylene the number of conjugated double bonds X is a distributed variable in the range 10-50 and the nature of the terminal groups is not known. For a linear polyene the number of conjugated double bonds X has a known integer value and the chain terminating groups are known.

may be very high, this does not accurately reflect the length of the conjugated segments which lies in the range of 20 to 100 carbon atoms. Thus, a fundamental understanding of the electronic structure of linear polyenes has direct application to the electronic properties of polyacetylene: the linear polyenes are paradigm systems for polyacetylene. We expect that a deep understanding of polyene electronic structure including its response to electrostatic perturbation and information on how these properties depend on conjugated chain length will significantly contribute to the development of models for the more complex polymer. Conversely, models for the electronic structure of polyacetylene that fail to properly describe the electronic structure of discrete linear polyenes in the limit of short chain length must be viewed as having questionable utility.

By the 1960's it appeared that the electronic structure of the linear polyenes, arguably the simplest conjugated organic molecules, was well understood. In 1972 we reported the

J.L. Brédas and R. Silbey (eds.), Conjugated Polymers, 405–434.

discovery of a dipole forbidden excited singlet state below the the dipole allowed state thought to be the lowest energy excited singlet state in the polyene diphenyloctatetraene [1]. This was immediately shown by Schulten and Karplus to reflect an unusual importance of electron correlation in the linear polyenes [2] a property that sets them apart from other conjugated organic molecules such as the polyacenes. Through detailed spectroscopic studies over the last fifteen years we have developed a general understanding of the nature of the low energy singlet states of linear polyenes including the highly correlated symmetric states. This review aims at communicating this understanding emphasizing those aspects that seem especially relevant to understanding transport and nonlinear optical response in polyacetylene. We hope that this base of information on the discrete linear polyenes will be useful both in testing theories for the more complex materials and in developing new theoretical models.

2. Terms and Concepts

Since the spectroscopist and the solid state physicist often use different language to describe the same aspects of electronic structure, it is useful to briefly review the terms that will be used in this chapter.

2.1. ELECTRONIC STATES

We use the word orbital or band function to denote a function of the coordinates of one electron that is an eigenfunction of an effective 1-electron Hamiltonian. In molecular orbital or band theory this function is often expressed as a linear combination of atomic orbitals. The word configuration for an N-electron system denotes a properly antisymmetrized product (for example, a Slater determinant) of N spin-orbitals corresponding to a particular assignment of the N electrons to orbitals with spin up or down. In those cases where the many-electron Hamiltonian can be written as a sum of 1-electron operators, these configurations are eigenfunctions of the many-electron Hamiltonian. The word state denotes a molecular eigenstate or eigenfunction of the many-electron Hamiltonian. If electron-electron interaction (a sum of 2-electron operators) may be adequately covered by an effective potential (a sum of 1-electron operators), states may be represented by single configurations: in general they are linear combinations of configurations. These concepts are shown in cartoon form in Figure 2.

2.2. VIBRONIC SPECTRA

Linear polyene electronic spectra are well described within the Born-Oppenheimer approximation where the wavefunction is factored into a product of an electronic function that may contain the nuclear coordinates as parameters and a vibrational function. In the usual treatments, the neglect of nuclear kinetic energy in the Hamiltonian allows the calculation of the energies of the various electronic levels as a function of molecular geometry. For a given electronic state the electronic energy as a function of nuclear coordinate is the potential energy for determining the vibrational energy levels and it is almost always the case that in the appropriate linear combinations of nuclear coordinates (called normal coordinates) the vibrational motion reduces to the motion of independent

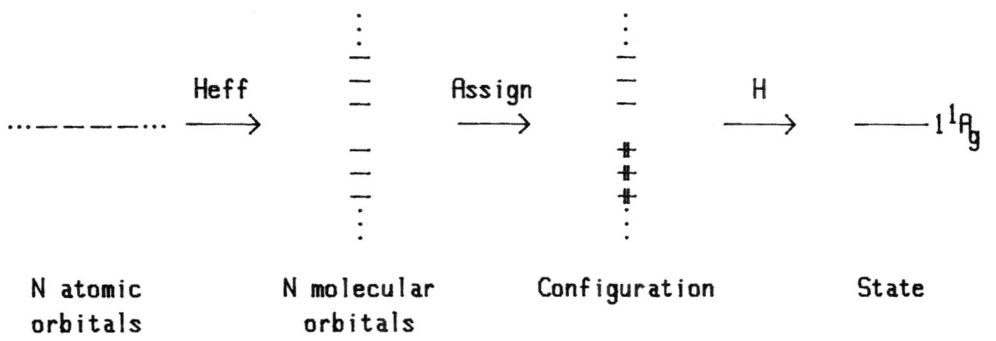

Figure 2. Molecular electronic structure. Linear combinations of N atomic orbitals make N molecular orbitals that are eigenfunctions of an effective 1-electron Hamiltonian. Antisymmetrized products of occupied molecular orbitals times spin functions make configurations. In general, observed states are linear combinations of configurations.

harmonic oscillators. In general, the molecular geometry that minimizes the electronic energy is different for different electronic states which means that we may distinguish two excitation energies: the difference in electronic energy at the ground state equilibrium geometry (vertical excitation energy) and the energy difference between zero-point vibrational levels (0-0 or relaxed excitation energy). The intensity of a transition between a particular pair of vibronic levels is proportional to the squared magnitude of the matrix element for dipole moment between the levels involved. For a symmetry allowed transition like the linear polyene S_0 to S_2 transition, this matrix element is well approximated by an electronic factor that is independent of nuclear coordinate times a vibrational overlap or Franck-Condon factor. It is very important to remember that there is a direct relationship between the differences in molecular equilibrium geometries of two states and the vibrational fine structure of the optical transition between those states. These concepts of molecular spectroscopy are schematically summarized in Figure 3.

3. Overview of Polyene Singlet States

In the case of a planar linear polyene the molecular orbitals may be classified according to their symmetry under reflection through the molecular plane. σ orbitals are symmetric under reflection, π orbitals are antisymmetric. In discussing the low energy excited singlet states of linear polyenes it is appropriate to limit attention to many electron states built from orbitals with π symmetry (the π-electron approximation). In the simplest treatments these π-electron orbitals may be written as linear combinations of 2p-π atomic orbitals, one per sp^2 hybridized carbon atom. In the case of an all-*trans* planar polyene which has 2/m symmetry, these π-electron orbitals will transform as A_u or B_g. Since the number of electrons is even, the many electron states will have either A_g or B_u symmetry.

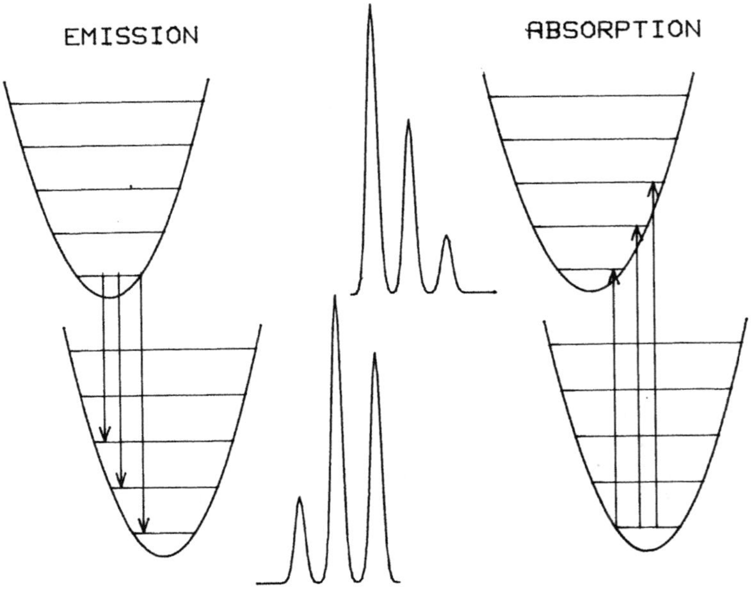

Figure 3. Molecular electronic spectra. Electronic energy as a function of conformation is the potential energy function for nuclear motion. As indicated, minimum energy conformations are usually different in different electronic states. Absorption fine structure intervals (upper spectrum) are determined by excited state vibrational frequencies: emission fine structure intervals (lower spectrum) are determined by ground state vibrational frequencies. The intensity of a fine structure component is proportional to the squared overlap of the ground and excited state vibrational functions.

Linear polyene π-electron states may be conveniently specified in terms of configurations built from orbitals derived using the Hückel (tight binding) Hamiltonian with alternating resonance (transfer) integrals. This gives an efficient description that is closely related to band theory for a 1-dimensional crystal and is the language that we will use for linear polyene electronic structure in this review.

An unsubstituted linear polyene with N double bonds in conjugation has the formula $H\text{-}(HC{=}CH)_N\text{-}H$. The number of Hückel molecular orbitals equals the number of π-electrons which is 2N. On the basis of their Hückel energies, these orbitals can be divided into two classes: N bonding (valence band) orbitals that have lower energies and N antibonding (conduction band) orbitals that have higher energies than do the original atomic orbitals. The three lowest energy linear polyene singlet states may be described in terms of configurations built from Hückel molecular orbitals as shown in Figure 4. S_0, the π-electron ground state, is well described by the single configuration (1Ag) in which the N lowest energy molecular orbitals are doubly occupied (valence band filled). Normally the 1Bu configuration derived from the ground state by promoting one electron from the HOMO or highest energy occupied molecular orbital (top of the valence band) to the

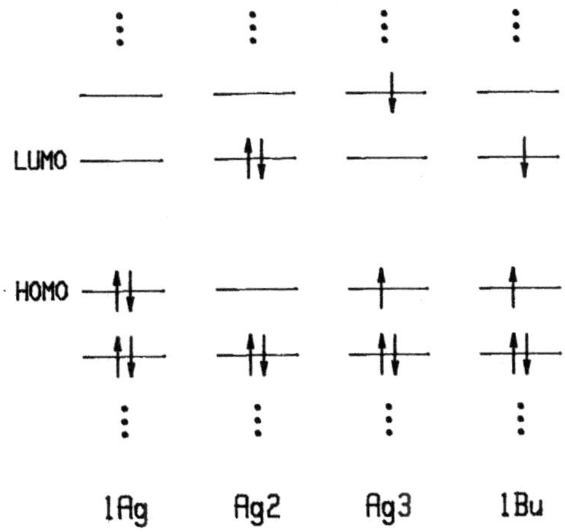

Figure 4. Configurations that describe the low energy singlet states of linear polyenes.

LUMO or lowest energy unoccupied molecular orbital (bottom of the conduction band) is S_1, the lowest energy excited singlet state. In the linear polyenes this state is the next higher excited singlet state, S_2. S_1 is a correlated state that, because of the importance of electron-electron interaction cannot be written in terms of a single configuration, but is approximately a linear combination of the Ag2 and Ag3 configurations whose Hückel excitation energies are roughly twice that of the HOMO to LUMO configuration. In the all-*trans* planar structure S_0 and S_1 have A_g symmetry and S_2 has B_u symmetry. Accordingly, at the geometry of the symmetric nuclear framework transitions from the ground state to S_1 are dipole forbidden while transitions to S_2 are strongly allowed. This pattern of excited singlet states, one that is only very weakly absorbing lying lower in energy that the strongly absorbing state that is easily described without elaborate treatments of electron correlation, is a general property of linear polyenes of all chain lengths independent of local symmetry and/or the presence of *cis* linkages. This is the reason that in the literature the labels 1^1A_g for S_0, 2^1A_g for S_1 and 1^1B_u for S_2 are used even in cases where 2/m symmetry is not realized.

4. Interpretative Model

The discussions of linear polyene electronic structure that follow are made clearer by reference to a simple quantum mechanical model. In this section we briefly sketch the model that we use. A fuller description will be published elsewhere [3].

Since Schulten and Karplus showed that the experimentally observed ordering of polyene excited states reflected the importance of electron correlation in these molecules [2], the idea that explicit consideration of electron-electron coulomb interaction is essential for a correct description of polyene electronic structure has been reinforced by a number of theoretical studies [4-8]. Unfortunately, as theory more closely agrees with experiment, it also becomes more complex and computationally intensive. This is the reason that we have turned to a parameterized model that we believe contains the essential physics. The idea for this model follows from the original configuration interaction calculations of Schulten and Karplus [2] and their subsequent elaboration by Ohmine, Karplus and Schulten [9].

In the original calculations of Schulten and Karplus, polyene electronic states were computed using the Pariser-Parr-Pople (PPP) method. The zero-order many-electron states (configurations) are properly antisymmetrized products of these molecular orbitals corresponding to a particular assignment of electrons to 1-electron spin-orbitals: configuration energies are the sum of the energies of the occupied orbitals. Better many-electron states are constructed through configuration interaction, that is, by finding the linear combinations of these configurations that are eigenfunctions of the many-electron Hamiltonian written in the configuration basis.

Although fully converged calculations require the consideration of an enormous number of configurations, the description of the three low lying singlet states is dominated by the four configurations shown in Figure 4 [2,4,9]. The 1^1A_g ground state and 1^1B_u second excited singlet state can be well represented by single configurations. For the 1^1A_g ground state, all bonding molecular orbitals are doubly occupied. The 1^1B_u state is generated from the ground state configuration by promoting an electron from the HOMO (highest energy occupied molecular orbital) to the LUMO (lowest energy unoccupied molecular orbital). The 2^1A_g state is a mixture of the doubly excited Ag2 (promotion of two electrons from the HOMO to the LUMO) and double jump Ag3 (promotion of an electron from the HOMO to the LUMO plus one or promotion of an electron from the HOMO minus one to the LUMO) configurations. Because of the symmetry of the Hamiltonian, the double jump configurations always appear as the symmetric linear combination [10]. In our model we restrict ourselves to just four configurations: the ground state (1^1A_g), the HOMO to LUMO excitation (1^1B_u), the double HOMO to LUMO excitation (Ag2) and the HOMO to LUMO+1 excitation (Ag3). (Because of the symmetry of the Hückel Hamiltonian consideration of the HOMO to LUMO+1 configuration is equivalent to considering the symmetric combination HOMO-1 to LUMO and HOMO to LUMO+1 configurations.) [10]

The energies of these configurations are computed using Hückel theory with alternating resonance integrals [11]. It is well known that if bond length alternation is incorporated either empirically by using alternating values for the resonance integrals [12] or through explicit consideration of the effects of distorting the equal bond length structure [13,14], the excitation energies for the 1^1B_u state for the polyene series can be nicely reproduced as long as one is willing to overlook the fact that, in this theory, this state comes out as the lowest energy excited singlet state S_1 whereas experimentally it is the next higher lying state S_2. This defect is corrected by considering electron-electron repulsion which mixes configurations of the same symmetry.

In our model we limit attention to the configurations shown in Figure 4 and assume that diagonal contributions of the electron-electron repulsion operator can be covered by proper choice of the Hückel parameters (or, equivalently, that they are approximately the same for all of the configurations). Since the electron-electron interaction operator will only have nonzero matrix elements between the A_g configurations, the primary effect of this operator will be to mix the nearly degenerate Ag2 and Ag3 configurations. Thus, in the configuration interaction part of the calculation it is not unreasonable to neglect all but the Ag2,Ag3 matrix element of the electron-electron interaction operator, which we do. Finally, since our goal is the construction of a simple model that can reproduce experimental results, we treat the Ag2,Ag3 configuration mixing matrix element empirically setting $\langle Ag2|H|Ag3\rangle = A+B/N$ (A and B are two parameters, N is the number of double bonds in the polyene chain). This form incorporates the idea that this perturbation goes inversely with the distances between electron pairs (which scales with chain length) and that it should go to a finite limit as the chain length increases. The configuration interaction Hamiltonian is then

$$H = \begin{bmatrix} 0 & 0 & 0 & 0 \\ 0 & E(Ag2) & A+B/N & 0 \\ 0 & A+B/N & E(Ag3) & 0 \\ 0 & 0 & 0 & E(1Bu) \end{bmatrix} \qquad (1)$$

Thus, the description of unsubstituted polyene chains with an arbitrary number of conjugated double bonds N requires four parameters. Two parameters are used for β_s (the resonance integral for single bonds) and β_d (the resonance integral for double bonds) and two parameters give the configuration mixing matrix element P. Our formulation uses the parameters β and δ where $\beta_s = \beta e^{-\delta}$ and $\beta_d = \beta e^{+\delta}$ and A and B where P=A+B/N. Given the Hückel energies of the four configurations shown in Figure 4, the final energy of the 2^1A_g state is given by

$$E(2^1A_g) = (E(Ag2) + E(Ag3))/2 - [((E(Ag2)-E(Ag3))/2)^2 + P^2]^{1/2}. \qquad (2)$$

The appeal of this model is its simplicity, the fact that, as will be seen later, a single set of the four parameters β, δ, A, and B accurately reproduce all the polyene 2^1A_g and 1^1B_u excitation energies that have been measured in low temperature hydrocarbon solution, and the ease with which it can be extended to substituted linear polyenes.

5. The Ground State S_0 (1^1A_g)

5.1. GEOMETRY

The convergence of the linear polyene S_0 to S_2 excitation energy to a finite limit as the length of the conjugated chain increases is strong evidence that there is bond length alternation in the long chains. This is supported by the limited number of direct structure determinations that have been published.

Although there is great interest in the ground state structure of linear polyenes as a function of chain length, only a rather small number of atomic resolution structures are in the literature [15-20]. These are summarized in Table 1. In the unsubstituted compounds the zig-zag polyene chain is planar and centrosymmetric. There is pronounced bond length alternation consistent with the picture of alternating double and single bonds but no coherent pattern with respect to the lengths of bonds of a given type on the ends versus the middle of the chain. The isoprenoid methyl substitution on the carotenoids canthaxanthin and β-carotene primarily has the effect of sharpening the bond angles for the substituted carbons bending the chain into a gentle s-curve. The effect on bond lengths is less apparent.

TABLE 1. Polyene bond lengths and angles. Carbon atoms are numbered sequentially starting with the outer carbon of the double bond closest to the chain end.

BOND LENGTHS (Å)

Bond	Butadiene[a] N=2	Hexatriene[b] N=3	Octatetraene[c] N=4	Diphenyloctatetraene[d] N=4	Canthaxanthin[e] N=9	β-Carotene[f] N=9
$C_1=C_2$	1.341	1.337	1.336	1.350	1.343	1.32
C_2-C_3	1.463	1.457	1.451	1.438	1.455	1.51
$C_3=C_4$		1.368	1.327	1.350	1.365	1.38
C_4-C_5			1.451	1.441	1.447	1.48
$C_5=C_6$					1.361	1.36
C_6-C_7					1.452	1.54
$C_7=C_8$					1.363	1.37
C_8-C_9					1.442	1.52
$C_9=C_{10}$					1.348	1.35

BOND ANGLES (°)

Angle	Butadiene[a] N=2	Hexatriene[b] N=3	Octatetraene[c] N=4	Diphenyloctatetraene[d] N=4	Canthaxanthin[e] N=9	β-Carotene[f] N=9
$C_1=C_2-C_3$	123	122	125	124	126	123
$C_2-C_3=C_4$		124	125	125	118[g]	117[g]
$C_3=C_4-C_5$			125	125	126	121
$C_4-C_5=C_6$					121	113
$C_5=C_6-C_7$					126	121
$C_6-C_7=C_8$					117[g]	115[g]
$C_7=C_8-C_9$					127	122
$C_8-C_9=C_{10}$					123	114

[a] Ref. 15. Gas phase electron diffraction results.
[b] Ref. 16. Gas phase electron diffraction results.
[c] Ref. 17. Single crystal x-ray diffraction results.
[d] Ref. 18. Single crystal x-ray diffraction results. The chain starts with the first exocyclic double bond.
[e] Ref. 19. Single crystal x-ray diffraction results averaged to obtain centrosymmetry. The chain starts with the first exocyclic double bond.
[f] Ref. 20. Single crystal x-ray diffraction results. The chain starts with the first exocyclic double bond.
[g] Site of methyl substitution.

All of the available structural information is reasonably represented by a planar centrosymmetric zig-zag skeleton with uniform C=C bond lengths of 1.33Å, C-C bond lengths of 1.45Å and all C=C-C bond angles equal to 125°.

5.2. VIBRATIONAL FREQUENCIES

Vibrational energy levels are much less sensitive to environmental effects than are electronic energy levels. For this reason ground state vibrational spectra (Raman and infrared) are, in general, much sharper for a given system than the electronic spectra. This means that for an intrinsically inhomogeneous system like polyacetylene the vibrational spectra should be be more highly structured and, hence, more amenable to unambiguous analysis than the optical absorption spectrum. In the case of polyacetylene the observed dependence of Raman shift on laser frequency (Raman dispersion) [21-26] has been interpreted in terms of a distribution of polyene conjugation lengths using two theoretical models [25-31]. The reference data for these kinds of analyses are the resonance enhanced Raman spectra for well defined linear polyenes. Since these issues have recently been thoughtfully re-examined using an extensive data base consisting of Raman spectra for α,ω-di-t-butylpolyenes with 2 through 12 double bonds in the polyene chain [32], we limit this section to a brief summary of related data for linear polyenes with other terminal groups.

There is a one-to-one correspondence between the vibrational frequencies that are prominent in resonance enhanced Raman spectra and the frequencies that form progressions in vibrationally resolved fluorescence spectra. As is the case for the resonance enhanced Raman spectra the vibrational development of linear polyene fluorescence spectra are dominated by two modes loosely referred to as the C-C and C=C stretch frequencies at approximately 1200 cm^{-1} and 1600 cm^{-1}, respectively. Frequencies of these two modes for unsubstituted, α,ω-dimethyl substituted and α,ω-diphenyl substituted polyenes [33-42] are summarized in Table 2.

Under high resolution conditions the C-C, band is actually seen to have a complicated multiplet structure and the appearance of the C-C stretch region strongly depends on the nature of the terminating groups. The same phenomena are seen for the di-t-butyl capped polyenes [32]. In contrast, the C=C frequency, which is the most intense feature in either the Raman or the fluorescence spectrum, behaves more simply. While it is tempting to quantitatively analyse the correlation between the frequency of this band and the number of conjugated double bonds using the empirical formula $\bar{\nu}=A+B/N$, Schaffer et al. [32] have shown that the power series in 1/N must be extended to the $1/N^2$ term to obtain a reasonable fit. Thus, the limited data in Table 2 are best regarded as the start of a data base that will ultimately allow the determination of the influence that different terminal groups have on the ground state vibrational frequencies of the discrete oligomers.

6. The Lowest Energy Excited Singlet State S_1 (2^1A_g)

Fluorescence from molecules in liquid or solid solution almost always originates from the lowest energy excited singlet state. If this state is dipole allowed from the ground state, the absorption and fluorescence spectra will overlap with a mirror image relationship as indicated in Figure 3. In the case of the linear polyenes the presence of a significant gap between the onsets of absorption and fluorescence contradicted the assumption that the

TABLE 2. Ground state frequencies for linear polyene C-C and C=C stretch frequencies.

Polyene Double Bonds	Molecule	$\bar{\nu}$(C-C), cm^{-1}	$\bar{\nu}$(C=C), cm^{-1}
2	Butadiene	1203[a](1278)[a]	1638[a]
3	Hexatriene	1191[b,c]	1626[b,c]
4	Octatetraene	1179[d]	1613[d]
4	Decatetraene	1148[e]	1608[e]
7	Hexadecaheptaene	1148[f]	1555[f]
8	Octadecaoctaene	1135[g]	1554[g]
2	Diphenylbutadiene	1160,1184,1295[h]	1636[h]
3	Diphenylhexatriene	1144,1250[i]	1592[i]
4	Diphenyloctatraene	1144[j]	1580[j]
5	Diphenyldecapentaene	1143[k]	1560[k]
6	Diphenyldodecahexaene	1145[l]	1551[l]

[a] Raman spectra of 298K liquid, reference 33.
[b,c] Raman spectrum of cyclohexane solution, reference 34. Raman spectrum of pure liquid, reference 35.
[d] Raman spectrum CCl_4 solution, reference 36.
[e] Fluorescence spectrum of 4.2K undecane host, reference 37.
[f] Fluorescence spectrum of 4.2K pentadecane host, reference 38.
[g] Fluorescence spectrum of 4.2K hexadecane host, reference 39.
[h] Fluorescence spectrum of 4.2K bibenzyl host, reference 40.
[i] Fluorescence spectrum of 4.2K tridecane host, reference 40.
[j] Fluorescence spectrum of 4.2K hexane host, reference 41.
[k] Fluorescence spectrum of 4.2K decane host, reference 42.
[l] Fluorescence spectrum of 4.2K dodecane host, reference 42.

allowed HOMO to LUMO transition was the lowest energy singlet-singlet excitation. That this followed from the fact that the strongly allowed transition was S_0 to S_2 rather than S_0 to S_1 became clear with the measurement of vibrationally resolved spectra. The discovery that the ordering of polyene excited states could not be reproduced by effective one-electron (Hartree-Fock) theories not only rationalized polyene spectroscopic properties that had appeared to be anomalous, but it showed that polyene electronic structure was richer and more complex than previously suspected. In this section we show representative examples of high resolution 1-photon and 2-photon S_0 to S_1 spectra and summarize the current picture of the S_1 state.

6.1. REPRESENTATIVE SPECTRA

To realize the resolution and sensitivity that is needed to obtain unambiguous optical spectra for the linear polyenes, the molecules must be put into a homogeneous low temperature environment. This can be accomplished in the gas phase by seeding the molecule of interest into a supersonic rare gas expansion and in the condensed phase by incorporating the molecule of interest into a low temperature crystalline environment. In either case, as seen in Figures 5 and 6, these high resolution conditions make it possible to clearly see that there is a very weak transition that lies lower in energy than the lowest energy allowed absorption. This is the transition to the 1-photon forbidden 2^1A_g state.

For 2-photon excitation the selection rules are inverted so that this transition is symmetry allowed and the HOMO to LUMO excitation is symmetry forbidden. The upper panel of Figure 6 shows that the onset of 1-photon S_0 to S_1 absorption is almost eight orders of magnitude weaker than the 0-0 band of the S_0 to S_2 transition which is the peak at 32100 cm^{-1} and that the S_0 to S_1 absorption intensity increases exponentially as the S_2 state is approached. This is characteristic of symmetry forbidden transitions. That the excited state has A_g symmetry is confirmed by the fact that, as is shown in the lower panel of Figure 6, the 2-photon absorption spectrum has the vibronic development of a symmetry allowed transition. Details of the vibronic analysis and a plot of the 1-photon spectrum on a linear intensity scale have been published elsewhere [43]. Spectra in hydrocarbon solution at this level of resolution have now been measured for a number of unsubstituted, α,ω-dialkyl substituted and α,ω-diphenyl substituted polyenes [38-41,43-50].

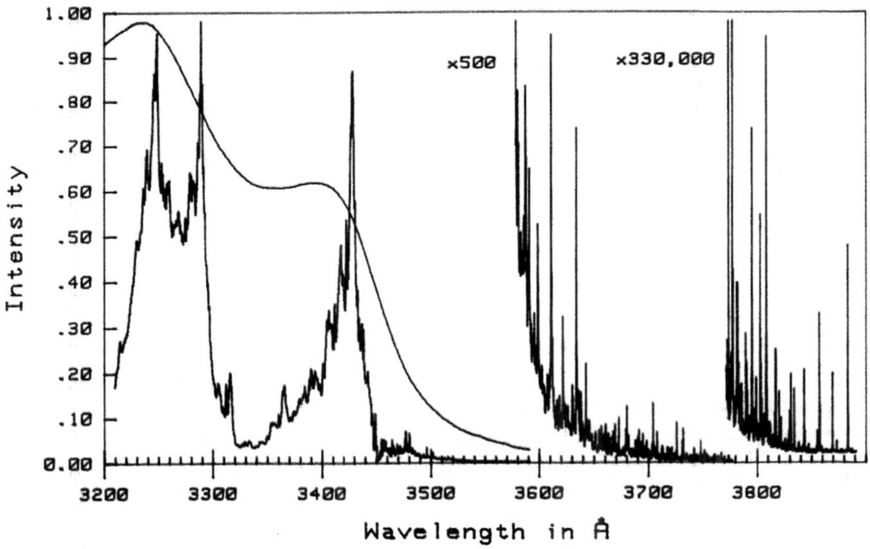

Figure 5. Optical spectra for gas phase diphenylhexatriene. The smooth curve from 3200 to 3600Å is the absorption of the vapor at 90°C. The highly structured spectrum with the scale expansions is the fluorescence excitation spectrum for diphenyl-hexatriene vapor seeded into a supersonic helium expansion. In the low temperature homogeneous environment of the expansion both the strong S_0 and S_2 absorption (3200-3450Å) and the very weak S_0 to S_1 absorption (3450-3500Å) are easily seen.

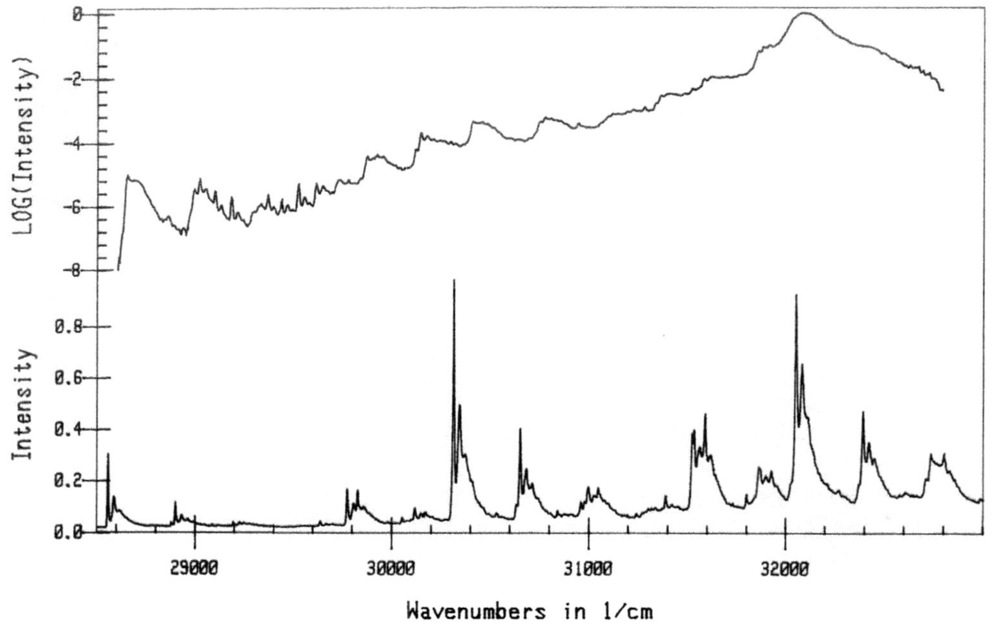

Figure 6. Optical spectra for octatetraene substituted in n-octane at 4.2K. The upper curve is the base 10 logarithm of the fluorescence detected 1-photon absorption intensity plotted against the wavenumber frequency of the exciting light. The lower curve is the fluorescence detected 2-photon absorption intensity plotted against 2 times the wavenumber frequency of the exciting light.

6.2. S_1 EXCITATION ENERGIES

0-0 excitation energies measured for the S_1 (2^1A_g) and S_2 (1^1B_u) states of linear polyenes substituted in low temperature hydrocarbon matrixes are summarized in Table 3. As is easily seen in Figure 7, the excitation energy for the forbidden (2^1A_g) state is always lower than the 0-0 excitation energy for the allowed (1^1B_u) state and the gap between these two excitation energies increases with increasing chain length. As is the case for the allowed transition, the S_1 excitation energies for a given series can be fit by E=A+B/N suggesting that the S_1 excitation energy goes to a finite limit at infinite chain length. This is supported by theoretical calculations [3].

6.2.1. Conformational Dependence. Most of the high resolution spectra have been measured for isomers that are *trans* about all double and single bonds. In the case of octatetraene fully resolved spectra have been measured for all three double bond isomers (*trans,trans* *cis,trans* and *cis,cis*) and one isomer that is *cis* about a single bond (Figure 8) [43,51-55]. The *cis,cis* double bond isomer is an oligomer of *cis*-polyacetylene. The introduction of double bond *cis* linkages has only a very small effect on the S_1 excitation

TABLE 3. Linear polyene excitation energies.

Double Bonds	Substituent	2^1A_g 0-0 Energy in cm^{-1} Meas.	2^1A_g 0-0 Energy in cm^{-1} Calc.[a]	1^1B_u 0-0 Energy in cm^{-1} Meas.	1^1B_u 0-0 Energy in cm^{-1} Calc[b]	Ref.
3	none	--	34264	37650	37742	44
4	none	28560	28577	32100	32181	43
5	none	--	24581	29020	28655	45
6	none	21640	21694	26430	26242	46
7	none	19630	19540	24100	24501	46
4	alkyl	28740	28763	32180	31115	47
5	alkyl	24650	24649	27680	27920	48
6	alkyl	21760	21709	--	25704	49
7	alkyl	19520	19530	24180	24089	38
8	alkyl	17870	17866	22770	22869	39
2	phenyl	--	28346	27755	28003	40
3	phenyl	--	24850	25975	25812	40
4	phenyl	22270	22162	24380	24198	41
5	phenyl	20130	20087	23070	22972	50
6	phenyl	18260	18457	21930	22014	50

[a] Calculated using the model sketched in section 4; RMS deviation between measured values and those calculated from the model is 161 cm^{-1} . The parameters used were: β 30305 cm^{-1}, δ=0.1333, A=12300 cm^{-1} and B=58020 cm^{-1}. The extension to methyl and phenyl substituted molecules is described in reference 3.

energy while the introduction of a single bond *cis* linkage produces a significant (1700 cm^{-1}) redshift (Table 4). Since the introduction of a *cis* linkage can remove the strict symmetry forbiddeness of the S_0 to S_1 transition, the effect on the vibronic development of the transition can be more dramatic [56]. However, the question of whether the 1-photon dipole intensity for this transition derives primarily from averaging the transition moment over the vibrational motion (vibronic coupling) or this mechanism is a minor contribution relative to static distortions induced by the environment is a subtle point that primarily interests the spectroscopist. While more data on the spectroscopic consequences of molecular conformation is highly desirable, it is reasonable to expect that the changes in S_1 excitation energies produced by the introduction of *cis* linkages in the chain will be a small effect (never larger than 2000 cm^{-1}) at all chain lengths.

6.2.2. Dependence on Local Polarizability. Electronic energies may be shifted by intermolecular interaction. For dissimilar guest/host systems these shifts may be successfully described by semi-classical pictures where the interactions of the neighbor molecules are represented by an effective electric field [57]. This is analogous to developing the intermolecular interaction of identical molecules in a multipole expansion starting with dipole-dipole terms. In dilute guest/host systems the solvent induced shift of

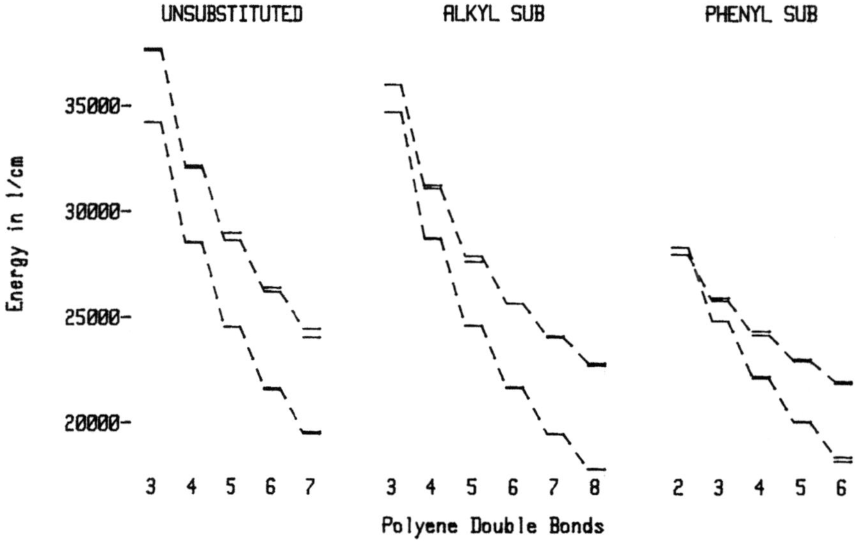

Figure 7. Graphical comparison of measured 0-0 excitation energies to those calculated using the model sketched in section 4 and more fully described in reference 3. For each polyene the 2^1A_g level lies lower than the 1^1B_u level. The isolated bars are the experimental values, the calculated values are represented by bars connected with dotted lines. In some cases the difference between measured and calculated values is too small to be seen.

the electronic excitation energy of a given transition is given by

$$\bar{\nu}(0\text{-}0) = \bar{\nu}(0\text{-}0)_{gas} - K\,|\mu|^2\frac{n^2-1}{n^2+2} \qquad (3)$$

where n is the solvent refractive index, $|\mu|^2$ is the squared transition dipole, and K is a constant inversely proportional to the molecular volume of the solute. In the case of the linear polyene S_0 to S_1 transition the solvent induced shifts are quite small (of order 100 cm^{-1}) as would be expected for the very small value of $|\mu|^2$. For example, the S_1 excitation energy of octatetraene shifts less than 300 cm^{-1} in going from methanol to carbon disulfide [58].

Further insight into the basis for this insensitivity comes from an examination of the π-electron densities predicted by the simple model described in Section 4. As is well known, the Hückel π-electron densities (calculated as sums of squares of atomic orbital coefficients) are unity in the ground (1^1A_g) and second excited (1^1B_u) states [11] so that the net charge on each carbon atom is zero. Although symmetry demands that in the all-*trans* planar conformation the first excited (2^1A_g) state have no dipole moment, the carbon atoms have significant net charges. Figure 9 shows the π-electron densities calculated for hexadecaoctaene using the simple model with the parameter values that best fit the 0-0

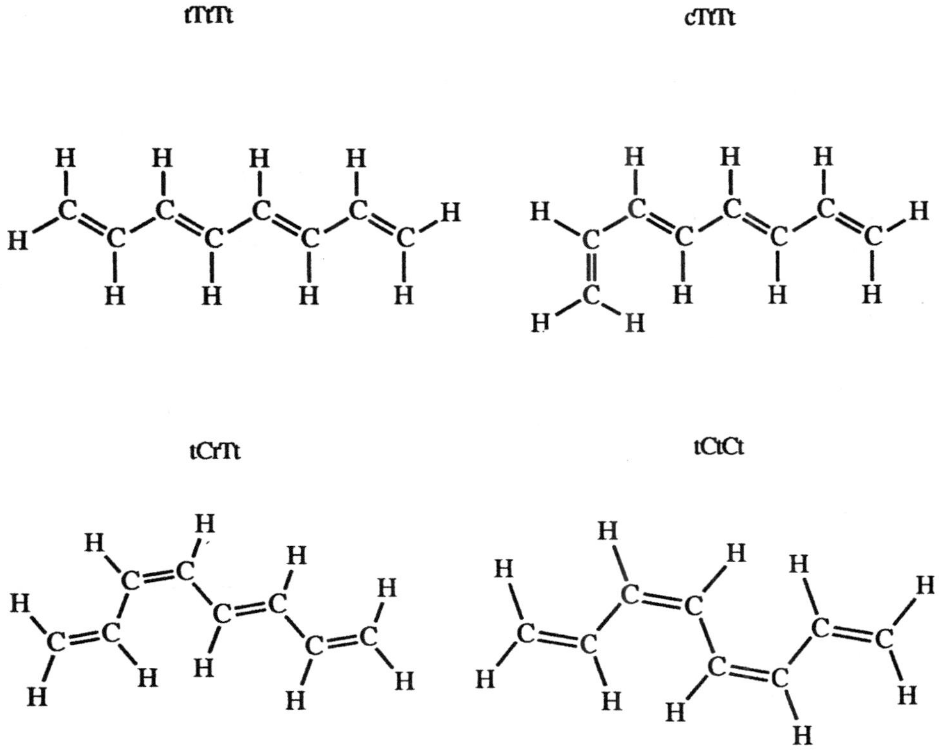

Figure 8. Isomers of octatetraene. The small letters designate configurations (*cis*=c or *trans*=t) about single bonds (bonds 2,4 and 6). The capital letters refer to the configuration about the two distinguishable double bonds (bonds 3 and 5). The s-*cis* (cTtTt) and di-*cis* (tCtCt) isomers can only be prepared at low temperatures.

excitation energies given in Table 3. In contrast to S_0 and S_2 which are completely charge neutral at the atomic level (hence, polarizable in an external field), there are significant but symmetrically situated atomic charges in the S_1 (2^1A_g) state. Thus, since this state is already highly polarized, it is much less affected by external fields.

6.3. S_1 VIBRONIC DEVELOPMENT

The vibronic development of the S_0 to S_1 spectra is dominated by the C-C and C=C stretch modes. This means that the difference between the relaxed equilibrium geometries in the S_0 and S_1 states is almost completely restricted to changes in the carbon-carbon bond lengths.

A quantitative analysis of vibronic intensities in the 1-photon S_0 to S_1 spectra to obtain normal coordinate displacements is difficult for two reasons. First, the fact that absorption intensity increases exponentially as excitation energies approach the excitation energy of the nearby dipole allowed S_2 state (see, for example, the upper panel of Figure 6) means that vibronic coupling has to be taken into account in any analysis of S_0 to S_1 absorption spectra. Because of the increase in the energy gap between the S_1 and S_2 states with

TABLE 4. $S_1(2^1A_g)$ and S_2 (1^1B_u) 0-0 excitation energies for isomers of octatetraene in n-hexane and n-octane matrixes.

Isomer	Host	S_1 0-0, cm^{-1}	S_2 0-0, cm^{-1}	Reference
all-*trans* (tTtTt)	n-hexane	28744	-	51
	n-octane	28561	32100	43
mono-*cis* (tCtTt)	n-hexane	28644	32134	52
	n-octane	28373	-	52
di-*cis* (tCtCt)	n-octane	28443 (site 1) 28336 (site 2)	32130	53, 54
s-*cis* (cTtTt)	n-octane	27201	30350	55

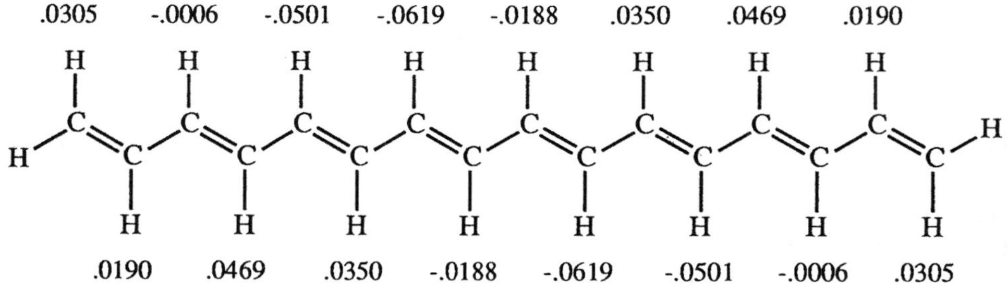

Figure 9. Excess π-electron density (π-electron densities minus 1) in the 2^1A_g state of hexadecaoctaene that follow from a fit of the polyene 2^1A_g and 1^1B_u excitation energies by the simple model described in Section 4. The net charge on a carbon atom is the negative of the excess π-electron density.

increasing chain length, this appears to be less of a problem for the longer chains [39] and, again because of larger energy differences between the coupled levels, may not be such a serious problem in the case of the fluorescence spectra. However, a serious difficulty remains.

All available evidence supports a picture of the S_1 (2^1A_g) excited state that has the order of double and single bonds reversed from the order in the ground state. This follows from high quality *ab initio* quantum chemical calculations for octatetraene [59], and is strongly supported by direct measurements of the barrier in the octatetraene S_1 state to rotation about bonds that were double bonds in the ground state [51]. This bond order reversal means that there is a significant likelihood that the normal coordinates for the excited state are different from those for the ground state. Thus, even if vibronic coupling is neglected, an analysis of vibronic intensities must be more general than a simple treatment in terms of independent, separable harmonic oscillators: it must allow for Duschinsky rotation [60]. While the necessity of explicitly treating vibronic coupling is removed in the case of the symmetry allowed 2-photon absorption spectrum, the one spectrum that is available [43] still awaits an analysis that includes the possibility of normal coordinate mixing.

A less rigorous test of the quality of potential energy surfaces for ground and excited states than vibronic intensities is the comparison of calculated versus the measured normal mode frequencies. High quality *ab initio* calculations of polyene potential energies in the S_1 (2^1A_g) state have only been recently reported for butadiene, hexatriene, and octatetraene [59]. Although the computed frequencies are usually 20% higher than the measured ones, the fact that such calculations give good ground state geometries and faithfully reproduce the changes in vibrational frequency in ground versus excited state as a function of increasing chain length increases confidence in these results. The measured vibrational frequencies for the C-C and C=C modes in the S_2 state are summarized in Table 5.

Finally we note that, at least for the four isomers of octatetraene shown in Figure 8, the C-C and C=C stretch frequencies in the S_1 (2^1A_g) state (summarized in Table 6) systematically shift to lower frequency as *cis* linkages are introduced into the molecular framework.

6.4. S_1 DYNAMICAL BEHAVIOR

In the condensed phase singlet excitation relaxes to a thermodynamic distribution over the vibronic levels of S_1 on the picosecond timescale, independent of the state initially excited. Because the expectation value of a 1-electron operator vanishes between configurations that differ in two spin-orbitals, the transition dipole between S_1 and the ground state is very small even when the symmetry is lowered as, for example, in a *cis* isomer. In consequence, the lifetime of the S_1 state is relatively long (225 nsec for octatetraene in n-octane) and usually is determined by non-radiative rather than radiative processes [56].

The presence of a very weakly absorbing but relatively long lived S_1 (2^1A_g) state below the strongly absorbing S_2 (1^1B_u) state generates some interesting transient absorption phenomena. The fast relaxation means that picoseconds after exciting the dipole allowed state, a new absorption spectrum originating from the lower lying S_2 state will appear. The bleaching of the strong S_2 absorption through transfer of population from S_0 to S_1 will occur on the same timescale as the appearance of the new transient absorptions. The transient absorption from S_1 to S_2 (a symmetry allowed transition) lies in the infrared for the shorter polyenes but would be expected to appear at approximately 1eV (8000 cm^{-1}) in the limit of a long chain. In the case of the finite linear polyenes, these transient absorption phenomena have only been investigated in the visible region of the spectrum [68-70].

7. The Strongly Allowed Excited Singlet State S_2 ($^1 1B_u$)

The principal feature in the optical spectrum of linear polyenes is the strongly allowed absorption that lies in the near ultraviolet for the shortest polyene butadiene and shifts to

TABLE 5. Vibrational frequencies in the S_1 (2^1A_g) and S_2 (1^1B_u) states of some linear polyenes.

		2^1A_g		1^1B_u		
Double Bonds	Molecule	$\bar{\nu}_{C\text{-}C}$, cm^{-1}	$\bar{\nu}_{C=C}$, cm^{-1}	$\bar{\nu}_{C\text{-}C}$, cm^{-1}	$\bar{\nu}_{C=C}$, cm^{-1}	Reference
2	Butadiene	-	-	1203	1670	61
3	Hexatriene	1125	1730	1224	1631	62,63
4	Octatetraene	1209, 1271	1754	1228	1647	43
5	Dodecapentaene	1205	1740	1207	1614	65
5	2,10-Dimethyl-undecapentaene	1240	1737	-	-	48
6	2,12-Dimethyl-Tridecahexaene	1252, 1277	1779	-	-	49
7	Heptadecaheptaene	1177, 1250	1782	-	-	38
8	Octadecaoctaene	1171,1233, 1246,1292	1799	-	-	39
2	Diphenylbutadiene	1233	1699	1138	1630	40,66
3	Diphenylhexatriene	1245	1763	1237	1612	67
4	Diphenyloctatetraene	1243	1706	-	-	41
5	Diphenyl-Decapentaene	1236	1751	1233	1593	42

TABLE 6. C-C and C=C frequencies in the S_1 (2^1A_g) state for isomers of octatetraene.

Isomer	Solvent	$\bar{\nu}_{C\text{-}C}$, cm^{-1}	$\bar{\nu}_{C=C}$, cm^{-1}	Reference
all-*trans* tTtTt	n-octane	1209 1271	1754	43
mono-*cis* tCtTt	n-hexane	1230	1722	52
di-*cis* tCtCt	n-octane	1322	1695	53,54
s-*cis* cTtTt	n-octane	1221	1730	55

longer wavelengths as the conjugation is extended. We shall consider only the vibrationally resolved spectra for this state that have been measured for isolated molecules in the low temperature environment of supersonic rare gas expansions [61-64,66-67,71] and in low temperature n-alkane mixed crystals (see, for example, the references cited in in Table 3) and will concentrate on the latter. The large number of optical absorption spectra reported for room temperature solutions of linear polyenes, including the recent study of α,ω-di-t-butylpolyenes with 2 through 13 double bonds in conjugation [32], are in complete agreement with the picture of electronic structure that is derived from the more resolved spectra.

7.1. REPRESENTATIVE SPECTRA

In a homogeneous low temperature environment, bandwidths in the strongly allowed S_0 to S_2 absorption narrow considerably: the full width at half maximum (FWHM) going from 500-1000 cm^{-1} to 10-50 cm^{-1}. This residual width derives in part from relaxation to the lower lying S_1 state and in part from unresolved structure although the relative contributions of the two sources have not yet been determined.

The strong absorption between 3200Å and 3450Å in Figure 5 shows the typical appearance of the S_0 to S_2 transition in the low temperature gas phase. Figure 10 shows what is typically seen for polyenes in low temperature n-alkane hosts. Analogous condensed phase spectra have been measured either at liquid helium or liquid nitrogen temperatures for polyenes containing 2 through 8 conjugated double bonds. In either the gas or condensed phases the 0-0 band is relatively intense and the vibronic development is dominated by modes at approximately 1200 cm^{-1} and 1600 cm^{-1}.

7.2. S_2 EXCITATION ENERGIES

Table 3 and Figure 7 summarize the 0-0 excitation energies measured for low temperature hydrocarbon solutions of all-*trans* linear polyenes with 2 through 8 conjugated double bonds (4 through 16 conjugated carbon atoms). As is the case with just about any linear polyene property [72], The S_2 (1^1B_u) excitation energy is reasonably well fit by the formula $E=A+B/N$. The value of the assymptotic excitation energy A depends on the exact data set used, but always is within 4000 cm^{-1} (.5 eV) of the first strong feature seen in the absorption spectrum of polyacetylene (12300 cm^{-1} or 1.52 eV) [73]. The long chain limiting value for the 0-0 excitation energy calculated with the simple model described in Section 4. is 15850 cm^{-1}, roughly twice the 8600 cm^{-1} excitation energy calculated for the S_1 (2^1A_g) state. The finite value of the HOMO to LUMO gap in the limit of long chain length implies that the alternation of bond lengths that is found for the short polyenes (Section 5.1.) persists as the chains are made arbitrarily long.

The excited state reached by the strongly allowed S_0 to S_2 transition in linear polyenes is reasonably described by molecular orbital theory with or without consideration of electron correlation as long as one is willing to overlook the fact that it is falsely predicted to be S_1 when explicit consideration of electron-electron interaction is omitted. While this state is important in the photochemical and photophysical behavior of the linear polyenes (primarily because of its large absorption cross section), its relatively large excitation energy even in the long chain limit, and the fact that another excited state lies lower in energy means that it is not likely to make an important contribution to any ground state property.

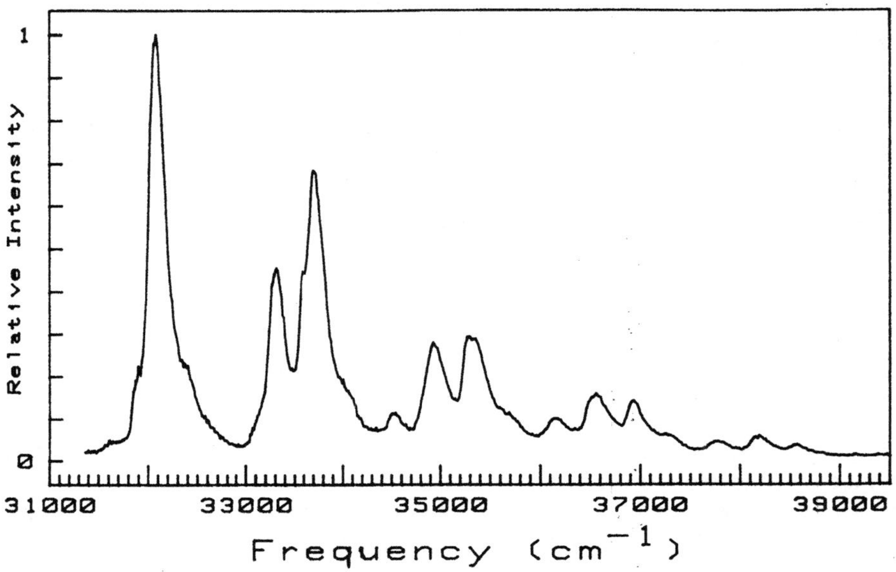

Figure 10. Fluorescence detected absorption spectrum for *trans,trans*-1,3,5,7-octatetraene in n-octane at 4.2K.

7.2.1. Conformation Dependence. The amount of data on S_2 excitation energies for polyene isomers with *cis* linkages is unfortunately quite limited. Referring to Table 4 we see that for the isomers of octatetraene shown in Figure 8 the introduction of a double bond *cis* linkage has an insignificant effect on the S_2 excitation energy (changing n-alkane solvents can introduce shifts larger than the 30 cm^{-1} differences between the isomers), but that the introduction of a single bond *cis* linkage shifts the transition 1800 cm^{-1} to lower energy. In the case of butadiene in an argon matrix, the redshift is 2920 cm^{-1} [74]. This behavior parallels that seen for the S_1 state.

7.2.2. Dependence on Local Polarizability. The solvent shift behavior of the S_2 excitation energy for diphenylpolyenes with 1-6 and 8 double bonds in the polyene chain [65,75] are all well fit by $\bar{\nu} = \bar{\nu}_{gas} - (9870 \pm 740\ cm^{-1}) \frac{n^2-1}{n^2+2}$. Thus, a 4000 cm^{-1} redshift on going from an n-alkane solution to the pure material is not unreasonable although the measurement of absorption spectra for pure polyene crystals is highly desirable.

7.3. S_2 VIBRONIC DEVELOPMENT

As is the case for the S_1 (2^1A_g) state, the vibronic development of the S_2 (1^1B_u) state is dominated by the C-C and C=C stretch modes. Table 5 summarizes values for these frequencies in the S_2 excited state for a number of linear polyenes. There is a general trend to lower frequency with increasing chain length, particularly for the C=C stretch frequency. As is also the case in the ground and S_1 states, the C-C mode depends on substitution and often appears as a complex multiplet.

7.4. RELAXATION ENERGY

In Section 2. and Figure 3 we indicated that the vibrational fine structure that is a general feature of molecular electronic spectra arises because of differences between ground and excited state equilibrium geometries. In the case of the linear polyenes, the vibrational development of transitions from the ground state to the 2^1A_g and 1^1B_u states is dominated by the symmetric C-C and C=C stretch modes. Thus, almost the only difference between ground and excited state equilibrium geometries is a difference in the carbon-carbon bond lengths.

The intensity of a vibronic band is proportioned to the square of the electronic dipole moment matrix element between the initial and final states. These states are well described by Born-Oppenheimer products of electronic and nuclear functions. For a symmetry allowed transition the nuclear coordinate dependence of the electronic function can be ignored so that the intensity of a vibronic band can be factored into a purely electronic factor that is independent of nuclear coordinate times the square of the overlap of the vibrational functions (Franck-Condon factor).

We have previously shown that the vibronic development of the 1^1A_g to 1^1B_u absorption can be quantitatively fit by treating the vibrational coordinates as independent harmonic modes [44,76]. Such an analysis gives a precise determination of the difference between normal coordinate equilibrium values between ground and excited states. These normal coordinate displacements together with the vibrational frequencies give a quantitative determination of the difference between 0-0 (relaxed) and vertical (unrelaxed) excitation energies that does not require any knowledge of how the normal coordinate is described in terms of atomic displacements. Figure 11 shows electronic energy versus normal coordinate for two electronic states. The vibrationally resolved spectra that we have been discussing give highly precise values for the 0-0 excitation energy and the vibrational frequencies in the ground and excited states ($\bar{\nu}_g$ and $\bar{\nu}_e$, respectively). The difference between the normal coordinate equilibrium values in the ground and excited states dQ can be derived by fitting the vibronic intensity profile [44]. With the convention that Q is the normal coordinate written as a linear combination of mass weighted cartesian displacements (units of $gm^{1/2}cm$) and that q is $(\frac{h}{4\pi^2 C})^{1/2}Q$ (units of $cm^{1/2}$), the vertical excitation energy $\bar{\nu}_{vertical}$ is related to the 0-0 excitation energy $\bar{\nu}_{0\text{-}0}$ and the vibrational frequencies in the ground and excited states $\bar{\nu}_g$ and $\bar{\nu}_e$ by

$$\bar{\nu}_{vertical} - \bar{\nu}_{0\text{-}0} = \frac{\bar{\nu}^2{}_e(\Delta q)^2}{2} + \frac{1}{2}(\bar{\nu}_g - \bar{\nu}_e). \tag{4}$$

Table 7 gives the vibrational frequencies in ground and excited states together with the dq's determined by fitting the 77K S_0 to S_2 absorption spectra of linear polyenes with 3 through 7 conjugated double bonds [44] together with the relaxation energies calculated using equation 4. The relaxation energy is approximately 2000 cm^{-1} for all chain lengths (average value is 1870 cm^{-1} ±- 150 cm^{-1}). Since the $1/2(\bar{\nu}_g-\bar{\nu}_e)$ term of equation 4 makes an extremely small contribution (40 cm^{-1} on average), it is not unreasonable to neglect it as has been done elsewhere.[77]

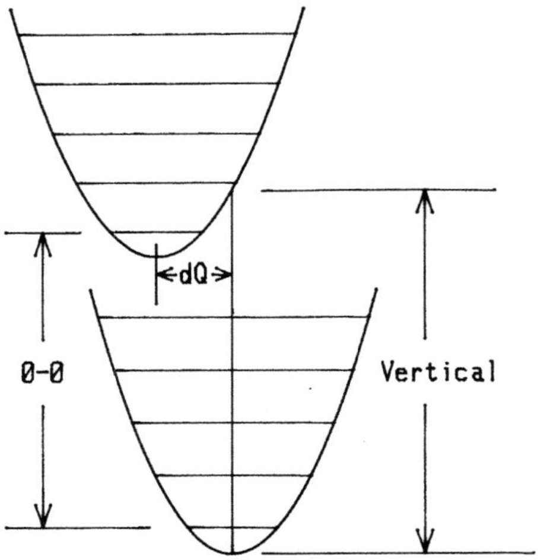

Figure 11. Picture of 0-0 (relaxed) and vertical (unrelaxed) excitation energies for one normal coordinate Q. The total relaxation energy is $\frac{1}{2}k_e(dQ)^2 + \frac{h}{2}(\omega_g - \omega_e)$ summed over all significantly displaced coordinates.

TABLE 7. Determination of the S_1 (1^1B_u) relaxation energy for a series of all-*trans* linear polyenes from a Franck-Condon analysis of the S_0 (1^1A_g) to S_2 (1^1B_u) absorption profile.

	C-C				C=C				
N	$\bar{\nu}(1^1A_g)$ cm^{-1}	$\bar{\nu}(1^1B_u)$ cm^{-1}	Δq cm$^{1/2}$	E_{relax} cm^{-1}	$\bar{\nu}(1^1A_g)$ cm^{-1}	$\bar{\nu}(1^1B_u)$ cm^{-1}	Δq cm$^{1/2}$	E_{relax} cm^{-1}	$E_{vertical}$-$E_{0\text{-}0}$ cm^{-1}
3	1187	1232	.030	660	1623	1626	.033	1440	2100
4	1181	1235	.024	410	1605	1640	.031	1280	1690
5	1181	1234	.028	570	1575	1626	.032	1330	1900
6	1180	1240	.032	760	1576	1680	.030	1220	1980
6	1180	1211	.028	560	1576	1620	.031	1240	1800
7	1187	1201	.028	560	1623	1629	.030	1190	1750

7.5. S_0-S_2 TRANSITION DIPOLES

The excitation energies and potential energy surfaces for the low lying polyene singlet states are now well enough understoood that they can be fit into a comprehensive coherent picture. On the basis of the available information on the variation of these properties with chain length, reasonable predictions of excitation energies and C-C and C=C vibrational frequencies can be made for polyenes of arbitrary chain length. Accordingly, attention turns to trying to understand what happens to these basic electronic properties under strong electrostatic perturbation and to the investigation of other observables. Included in the category of other observables are the transition dipoles.

Transition dipoles are exquisitely sensitive to the details of the electron distribution so they provide a stringent test of theory: calculations that reasonably reproduce the excitation energies may do a poorer job on the off-diagonal dipole moment matrix elements. They are also important for understanding the basis for linear polyene nonlinear optical response, an area where there is currently intense activity. In the usual theoretical treatments, higher order polarizabilities are developed as a perturbation sum over terms involving products of transition dipoles divided by products of differences between excitation energies and harmonics of the perturbing field [78]. In the case of a strongly allowed transition like the linear polyene S_0 to S_2 absorption, there is a simple relationship between the integrated absorption intensity and the squared magnitude of the transition dipole. Although these are extremely simple measurements, it was only recently that data for a series of linear polyenes was published [79].

Table 8 summarizes the integrated intensities for room temperature toluene solutions of a series of diphenylpolyenes and compares the values of the transition dipoles calculated through the relationship

$$|\mu|^2 = 9.186\text{x}10^{-3}\ \text{Debye}^2\ \text{mol l}^{-1}\text{cm} \int \varepsilon\, d\ln\bar{\nu} \qquad (5)$$

to those computed using the simple model for polyene electronic structure that so successfully describes the excitation energies. Note that equation 5 contains no local field correction. Since the parameters of the simple model were determined by fitting solution excitation energies, they include solvent effects. Thus, it is appropriate to compare the calculated moments to those determined for the solvent/solute system.

Given the crudeness of the simple model, the agreement between measured and calculated values is surprisingly good. The magnitudes of both the measured and calculated transition dipoles per polyene double bond decrease as the chain is extended. Of most interest is the apparent saturation of the transition dipole at longer chain lengths. It is not clear if this is an intrinsic electronic property of the all-*trans* planar polyene or if it just reflects the conjugation disorder that must obtain in a room temperature solution [80] More precise data for a larger range of known conjugation lengths would be most valuable in resolving this issue.

TABLE 8. Integrated absorption intensities and transition dipole magnitudes for diphenylpolyenes in room temperature toluene solution.

N for ϕ-(CH=CH)$_N$-ϕ	$\int \varepsilon d\ln\bar{\nu}$[a] mol^{-1} l cm^{-1}	$\lvert\mu\rvert_{meas}$[b] Debye	$\lvert\mu\rvert_{calc}$[c] Debye
2	5010±30	6.8	8.7
3	8810±510	9.0	9.9
4	12670±2020	10.8	11.1
5	12770±430	10.8	12.1
6	13250±390	11.0	13.0
7	13850±800	11.3	13.7

a From reference 79.
b No local field correction.
c Calculated with the model described in section 4 using the parameters given at the bottom of Table 3. The geometry was assumed to be planar all-*trans*; all angles $=2\pi/3$; bond lengths; (C-C)=1.45Å, (C=C)=1.34Å and $(C\text{-}C)_{phenyl}$=1.397Å.

8. Concluding Remarks

By the early 1970's it appeared that the electronic properties of linear polyenes either as isolated molecules or as guests in a hydrocarbon environment were reasonably well characterized and understood in terms of quantum mechanical models. Some were concerned by the fact that the absence of the expected overlap between absorption and fluorescence had only a qualitative explaination, but no one expected that this manifested disagreement between the calculated and actual order of excited singlet states. While the experimental demonstration that this was indeed the case and that the 2^1A_g excited singlet state was lower in energy than the 1^1B_u or "band gap" state rationalized what had previously been thought to be anamolous fluorescence properties of the linear polyenes, it also raised fundamental questions with respect to the properties of this new state. Once again it is begining to appear that the basic elements of electronic structure in the linear polyenes, at least for the manifold of singlet states, are known and, for the shorter chains, well understood in terms of more sophisticated quantum mechanical models. In contrast to the situation two decades ago, there are no blatant discrepancies between expected and observed behavior. The near future should see this understanding extended to interacting and electrostatically perturbed molecules and to more subtle observables such as π-electron distributions and nonlinear polarizabilities. As this is accomplished, we can look forward to the explaination of the extraordinary properties of molecular materials such as polyacetylene in terms of the fundamental properties of the constituent molecules. It is a very exciting time to be working on these kinds of electronic structure problems!

9. Acknowledgement

Much of the story that is told here was worked out by the students and postdoctoral fellows that it has so far been my privilege to work with. My thanks to that subset of people who have worked with me on "the polyene problem" and my apologies to them for the many important things that I have left out. These people are: John Ackerman, George Adamson, Roy Auerbach, Bob Birge, Duane Birnbaum, Wybren-Jan Buma, Ron Christensen, Jim Corsetti, Stu Forman, Gerhard Gradl, Mark Granville, Lou Ann Heimbrook, Rus Hemley, Gary Holtom, Jim Horwitz, Musharraf Hossain, Bruce Hudson, Takao Itoh, Jon Kenny, Irv Levy, Pradip Mitra, John Pescatore, Dave Schilke, Kyuseok Song, Judith Snow, Tom Spiglanin, Paul West, Curtis Westerfield, Konrad Wu and colleagues Martin Karplus, Iwao Ohmine, Klaus Schulten, Gary Scott and Charley Spangler. Of course, these people and I would not have accomplished much without sustained support by the NIH and NSF.

Finally, the appearance of the final copy reflects the dedication and skill of Annette McComas and Colleen Fleming.

10. References

1. B.S. Hudson and B.E. Kohler, Chem. Phys. Letts. **14**, 299 (1972).
2. K. Schulten and M. Karplus, Chem. Phys. Letts. **14**, 305 (1972).
3. B.E. Kohler, J. Chem. Phys. **93**, 0000 (1990).
4. B.S. Hudson, B.E. Kohler and K. Schulten, Excited States **6**, 1 (1982).
5. S.N. Dixit and S. Mazumdar, Phys. Rev. **B29**, 1824 (1984).
6. Z. Soos and S. Ramaesha, Phys. Rev. **B29**, 5410 (1984).
7. D. Baeriswyl and K. Maki, Phys. Rev. **B31**, 6633 (1985).
8. An up to date review of theory is given in D. Baeriswyl, D.K. Campbell and S. Mazumdar, "The Theory of π-Conjugated Polymers", to be published in "Conducting Polymers", edited by H. Kiess (Springer, 1990).
9. I. Ohmine, M. Karplus and K. Schulten, J. Chem. Phys. **68**, 2298 (1978).
10. K. Schulten, Ph.D. Thesis, Harvard University (1974).
11. For example, L. Salem, "The Molecular Orbital Theory of Conjugated Systems", W.A. Benjamin, Inc., Reading, Massachusetts, 1972.
12. J.E. Lennard-Jones, Proc. Roy. Soc. **A158**, 280 (1937).
13. J.A. Pople and S.H. Walmsley, Mol. Phys. **5**, 15 (1962).
14. W.P. Su, J.R. Schrieffer and A.J. Heeger, Phys. Rev. Letters **42**, 1698 (1979); Phys. Rev. **B22**, 2099 (1980).
15. K. Kuchitsu, T. Fukuyama and Y. Morino, J. Mol. Struct. **1**, 463 (1967).
16. M. Traetteberg, Acta. Chem. Scand. **22**, 628 (1968).
17. R.H. Baughman, B.E. Kohler, I.J. Levy and C. Spangler, Synthetic Metals **11**, 37 (1985).
18. W. Drenth and E.H. Wiebenga, Acta. Cryst. **8**, 755 (1955).
19. J.C.J. Bart and C.H. MacGillavry, Acta. Cryst. **B24**, 1587 (1968).
20. C. Sterling, Acta. Cryst. **17**, 1224 (1964).

21. I. Harada, Y. Furukawa, M. Tasumi, H. Shirakawa and S. Ikeda, J. Chem. Phys. **73**, 4746 (1980).

22. L.S. Lichtmann, A. Sarhangi and D.B. Fitchen, Solid State Comm., **36**, 869 (1980).

23. H. Kuzmany, Phys. Status Solidi B, **97**, 521 (1980).

24. S. Lefrant, J. Phys. (Paris) Colloq., **44**, C3-C247 (1983).

25. Z. Vardeny, E. Ehrenfreund, O. Brafman and B. Horovitz, Phys. Rev. Lett. **51**, 2326 (1983).

26. L.S. Lichtmann, Ph.D. Thesis, Cornell University (1981).

27. G.P. Brivio and E. Mulazzi, Chem. Phys. Lett. **95**, 555 (1983).

28. G.P. Brivio and E. Mulazzi, Phys. Rev. B, **30**, 876 (1984).

29. B. Horovitz, Solid State Commun., **41**, 729 (1982).

30. E. Ehrenfreund, Z. Vardeny, O. Brafman and B. Horovitz, Phys. Rev. B, **36**, 1535 (1987); B. Horovitz, Z. Vardeny, E. Ehrenfreund and O. Brafman, J. Phys. C.: Solid State Phys. **19**, 7291 (1986).

31. M.A. Schen, J.C.W. Chien, E. Perrin, S. Lefrant and E. Mulazzi, J. Chem. Phys. **89**, 7615 (1988).

32. H.E. Schaffer, R.R. Chance, K. Knoll, R.R. Schrock and Rocco Silbey in Conjugated Polymeric Materials: Opportunities in Electronics, Optoelectronics and Molecular Electronics, Ed. J.L. Brédas and R.R. Chance (Kluwer Academic Publishers, Dordrecht, 1990) pp 365-376.

33. V.E. Bondybey and J. Nibler, Spectrochimica Acta. **29A**, 645 (1973).

34. A.B. Myers, R.A. Mathies, D.J. Tannor and E.J. Heller, J. Chem. Phys. **77**, 3857 (1982).

35. H. Yoshida, Y. Furukawa and M. Tasumi, J. Molecular Structure **194**, 279 (1989).

36. H. Yoshida and M. Tasumi, J. Chem. Phys. **89**, 2803 (1988).

37. J.R. Andrews and B.S. Hudson, Chem. Phys. Letts. **57**, 600 (1978).

38. J.H. Simpson, J. McLaughlin, D.S. Smith and R.L. Christensen, J. Chem. Phys. **87**, 3360 (1987).

39. B.E. Kohler, C.W. Spangler, C. Westerfield, J. Chem. Phys. **89**, 5422 (1988).

40. W.M. Heatherington III, Ph.D. Thesis, Stanford University (1977).

41. B.E. Kohler, C.W. Spangler and C. Westerfield, J. Chem. Phys., in press.

42. J.S. Horwitz, T. Itoh, B.E. Kohler and C.W. Spangler, J. Chem. Phys. **87**, 2433 (1987).

43. M.F. Granville, G.R. Holtom and B.E. Kohler, J. Chem. Phys. 72, 4671 (1980).

44. M.F. Granville, B.E. Kohler and J.B. Snow, J. Chem. Phys. 75, 3765 (1981).

45. Judith B. Snow, unpublished result.

46. R. Snyder, E. Arvidson, C. Foote, L. Harrigan and R.L. Christensen, J. Am. Chem. Soc. 107, 4117 (1985). Obtained ditigizing a Xerox copy of the original figure and then least squares fitting a Gaussian to the 0-0 band.

47. J.R. Andrews and B.S. Hudson, J. Chem Phys. **72**, 4671 (1980).

48. R.L. Christensen and B.E. Kohler, J. Chem. Phys. **63**, 1837 (1975).

49. R.A. Auerbach, R.L. Christensen, M.F. Granville and B.E. Kohler, J. Chem. Phys. **74**, 4 (1981).

50. J.S. Horwitz, T. Itoh, B.E. Kohler and C.W. Spangler, J. Chem. Phys. **87**, 2433 (1987).

51. G. Adamson, G. Gradl and B.E. Kohler, J. Chem. Phys. **90**, 3038 (1989).

52. B.E. Kohler and T.A. Spiglanin, J. Chem. Phys. **80**, 3091 (1984).

53. M. Hossain, B.E. Kohler and P. West, J. Phys. Chem. **86**, 4918 (1982).

54. B.E. Kohler and P. West, J. Chem. Phys. **79**, 583 (1983).

55. J.R. Ackerman, S.A. Forman, M. Hossain and B.E. Kohler, J. Chem. Phys. **80**, 39 (1984).

56. B.E. Kohler and J.B. Snow, J. Chem. Phys. **79**, 2134 (1983).

57. A.J. Amos and B.L. Burrows, Advan. Quantum. Chem. **7**, 303 (1973).

58. B.S. Hudson and B.E. Kohler, J. Chem. Phys. **59**, 4984 (1973).

59. M. Aoyagi, B.E. Kohler and I. Ohmine, J. Phys. Chem. **94**, 3922 (1990).

60. F. Duschinsky, Acta. Physiochem. USSR **7**, 551 (1937).

61. R.J. Hemley, J.I. Dawson and V. Vaida, J. Chem. Phys. **78**, 2915 (1983).

62. D.G. Leopold, R.D. Pendley, J.L. Roebber, R.J. Hemley and V. Vaida, J. Chem. Phys. **81**, 4218 (1984).

63. W.J. Buma, B.E. Kohler and K. Song, J. Chem. Phys. **92**, 4622 (1990).

64. L.A. Heimbrook, J.E. Kenny, B.E. Kohler and G.W. Scott, J. Chem. Phys **75**, 4338 (1981).

65. J. Andrews, Thesis, Stanford University, Stanford, Ca (1978). Also reported in reference 4.

66. J.S. Horwitz, B.E. Kohler and T.A. Spiglanin, J. Chem. Phys. **83**, 2186 (1985).

67. B.E. Kohler and T.A. Spiglanin, J. Chem. Phys. **80**, 5465 (1984).

68. R.A. Goldbeck, A.J. Twarowski, E.L. Russell, J.K. Rice, R.R. Birge, E. Switkies and D.S. Kliger, J. Chem. Phys. **77**, 3319 (1982).

69. C. Rulliere and A. Declemy, Chem. Phys. Letts. **135**, 213 (1987).

70. S.M. Bachilo and S.L. Bondarev, Opt. Spectrosc. (USSR) **65**, 177 (1988).

71. L.A. Heimbrook, B.E. Kohler and T.A. Spiglanin, Proc. Nat. Acad. Sci. USA **80**, 4580 (1983).

72. J.L. Brédas, R. Silbey, D.S. Boudraux and R.R. Chance, J. Am. Chem. Soc. **105**, 6555 (1985).

73. H. Eckardt, J. Chem. Phys. **79**, 2085 (1983).

74. M.E. Squillacote, R.S. Sheridan, O.L. Chapman and F.A.L. Anet, J. Am. Chem. Soc. **101**, 3657 (1979).

75. L.A. Sklar, B.S. Hudson, M. Petersen and J. Diamond, Biochemistry **16**, 813 (1977).

76. R.J. Hemley, M. Karplus, B.E. Kohler and T.A. Spiglanin, J. Chem. Phys. **80**, 2 (1984).

77. B.E. Kohler, Luminescence, in press.

78. B.J. Orr and J.F. Ward, Mol. Phys. **20**, 513 (1971).

79. B.E. Kohler and J.A. Pescatore in Conjugated Polyermic Materials: Opportunities in Electronics, Optoelectronics and Molecular Electronics, ed. J.L. Brédas and R.R. Chance (Kluwer Academic Publishers, Dodrecht) NATO ASI Series **183**, 353 (1990).

80. G. Rossi, R.R. Chance and R. Silbey, J. Chem. Phys. **90**, 7594 (1989).

VIBRATIONAL SPECTROSCOPY OF POLYCONJUGATED AROMATIC MATERIALS WITH ELECTRICAL AND NON LINEAR OPTICAL PROPERTIES.

A GUIDED TOUR

G.Zerbi (*), M.Gussoni (**) and C.Castiglioni (*)

(*) Department of Industrial Chemistry, Politecnico, Piazza L. Da Vinci 32 ,Milano

(**) Istituto Chimica Fisica dei Materiali, CNR, Genova.

Work supported by CNR Progetto Finalizzato Telecomunicazione; paper No 6 from the Politecnico of Milano.

J.L. Brédas and R. Silbey (eds.), Conjugated Polymers, 435–507.

1) INTRODUCTION.

The fast development of modern material science is based on the existence of many physical techniques which are used for the characterisation of the molecular and/or bulk properties of the materials. The data so collected form the basis of any attempt to understand on a theoretical basis the physics of the phenomena which originate the properties of the material.

The new material science opened by the discovery of the electrical conductivity [1,2] and optical non linearities of polyconjugated organic polymers is a classical example of the need of an interdisciplinary approach to the study of their physical properties and shows the importance played by the experimental physical sciences in the characterisation of the materials and in the understanding of the phenomena.

Researchers in this field can be classified in two main groups according to the phylosophical approach to and the personal attitude towards the physical problem. "Problem oriented" people try to collect data from all physical techniques, sometimes accepting gross simplifications or avoiding a thorough exploitment of the capabilities of each technique in favour of an overall view of the problem in an attempt to reconcile and correlate data from different techniques, all aiming at a specific target. "Technique oriented" people often object such quick and sometimes too brave conclusions and try to dig out of each physical technique all possible detailed information in order to assess critically the validity and limits of the structural and/or dynamical data obtained and used. Disagreement , mutual misunderstanding and sometimes conflicts (or mutual ignoring) may occur between the two groups.

Among the physical techniques used in the field of synthetic metals and of photoactive polyconjugated organic molecules vibrational infrared and Raman spectroscopy and molecular and lattice dynamics have played a determining role in the understanding of the physics of the phenomena.

For this reason we have felt the need to present the state of the art of Vibrational spectroscopy of Polyconjugated materials. In this endeaviour we take the part of the "technique oriented" researchers with the hope to present in detail the "problem oriented" readers with the spectroscopic tools available for their studies. Moreover we use these tools for deriving information on the molecular level on classes of molecules of present interest in the field of science and technology of organic materials with electrical and non linear optical properties polymers.

Our project proceeds in the following way. In a first review article [3] (hereafter referred as I) we have already rediscussed the basic techniques of vibrational spectroscopy and lattice dynamics of perfect and/or disordered one dimensional lattices [4-13]. These concepts can be considered "well established" since they have been widely used for the study of "classical" mostly σ bonded organic polymers [4-13]. In the same review we have presented the new theoretical techniques developed just for the understanding of the spectra of polyconjugated polymers in their pristine, doped and/or photoexcited states [14-16]. The only polymer treated in I is Polyacetylene (PA). The enormous number of works on the vibrational spectra of PA, the wide scattering of seemingly inconsistent experimental data on samples with different chemical or physical history , the variety of interpretations proposed and the consequent phenomena envisaged by imaginative theoreticians have taken up all the space allotted for one review article.

The writing of this chapter is the second step of our project. The main goals of this chapter are i) to lead the reader through a guided tour of worked out examples in order to show him how to handle the vibrational spectra of polyconjugated materials, oligomers and/or polymers, ii) to make the reader perceive critically the width and depth and the limits of the information which can be derived from the spectra of polyconjugated materials and iii) to review the data collected on a few classes of molecules in order to assess the state of the art of our knowledge and focus on the many new questions to be answered.

Since Polyacetylene has been fully discussed in I we treat here the class of polyaromatic polyconjugated polymers which are at present of interest in basic science and technology.

The third step of the project [17] consists in an attempt to build a bridge between the world of chemists and that of physicists often separated either by a difference of language or by a difference in how to handle the reality of molecules.

2) SPECTROSCOPY VS. MATERIAL SCIENCE.

As already stated the main goal of this chapter is to provide problem oriented researchers with the help of vibrational spectroscopy. We will try to avoid cumbersome spectroscopic reasonings (left to the specialists) and will limit ourselves to the presentation of the dynamical problems and of their solutions aimed at providing information on those physical phenomena which can be probed by the vibrational spectrum. In other words this chapter is aimed at Material Science and not at Vibrational Spectroscopy.

3) SPECTROSCOPIC OBSERVABLES.

Three are the spectroscopic observables routinely collected for any compound with fully automatic experimental devices, i) vibrational frequencies, ν_i (cm^{-1}; 1 eV= 8065 cm^{-1}) or peak frequencies, ii) integrated intensities A_i (or sometimes relative intensities between integrated bands or peak heights) and iii) band shapes. Each of these quantities provides independent information on local and/or overall dynamics and electrical properties of molecules. Because of the general complexity of the problem [18] in the field of polyconjugated materials band shapes associated with single vibrational transitions have not yet been treated in detail and are neglected in this chapter. Band profiles resulting from a superposition of many one quantum transitions (inhomogeneous broadening) do occur in the spectra of polyconjugated materials and will be mentioned later in this paper.

3.1) Vibrational frequencies.

The harmonic oscillator approximation is the basis for the treatment of the molecular and lattice dynamics of organic systems [19-22]. The approximation of restoring forces linear with the displacements is accepted since the vibrational amplitudes are small for fundamental one quantum transitions. Anharmonicity plays an important and non negligeable role for higher vibrational quantum levels.

Organic molecules are generally described in terms of covalent bonds representing domains in the molecular space of higher electronic density where specific and highly directional forces can be activated if the molecule is distorted from its equilibrium geometry. The dynamics of these systems is then preferentially treated in terms of chemical bonds and valence angles [20-22]. Let **R** be a set of "internal vibrational coordinates" which are suitably chosen to represent bond stretching and angle bendings or torsions. Throughout this chapter we shall use matrix notation, matrices being represented by bold-face letters.

R's can be defined as linear combinations of the cartesian displacement coordinates x by the relation:

$$(1) \qquad \mathbf{R} = \mathbf{B}\mathbf{x}$$

where **B** is the matrix of geometrical coefficients. The intramolecular vibrational potential which determines the fundamental vibrational frequencies of the system is generally expressed in terms of the R's.In the harmonic approximation the quadratic potential is expressed as

$$(2) \qquad 2V = \tilde{\mathbf{R}}\,\mathbf{F}_R\mathbf{R} = \tilde{\mathbf{x}}(\tilde{\mathbf{B}}\mathbf{F}_R\mathbf{B})\mathbf{x} = \tilde{\mathbf{x}}\mathbf{F}_x\mathbf{x}$$

where $\mathbf{F}_R$ is the matrix of the valence force constants $((F_R)_{ij} = (\delta^2V/\delta R_i \delta R_j)_o)$. The purely vibrational kinetic energy of the system can be written [20] as

$$(3) \qquad 2T = \tilde{\dot{\mathbf{R}}}(\mathbf{G}_R)^{-1}\dot{\mathbf{R}} \quad , \quad \text{where}$$

$$(4) \qquad \mathbf{G} = \mathbf{B}\mathbf{M}^{-1}\tilde{\mathbf{B}}$$

M^{-1} being the diagonal matrix of the inverse of the atomic masses.

The fundamental vibrational frequencies ν_i (in cm^{-1}) are obtained by the solution of the eigenvalue equation

$$G_R F_R L_R = L_R \Lambda. \tag{5}$$

In eq.5 Λ is the diagonal matrix of the eigenvalues with $\lambda_i=4\pi^2c^2\nu^2_i$ where ν_i is the vibrational frequency of the i-th normal mode and $(L_R)_i$ is the corresponding eigenvector which provides the vibrational displacements.

The normal modes can be described in terms of the internal coordinates or cartesian coordinates respectively:

$$R = L_R Q \quad \text{or} \quad x = L_x Q = B^{-1}L_R Q \tag{6}$$

(eq. 6 is only formal; for the calculation of B^{-1} see. ref. 23). Eqs. 1 through 6 treat the case of finite molecules made up by N masses with 3N-6 degrees of vibrational freedom. The treatment can be extended to molecular systems organized in one dimensional or tridimensional lattices where translational periodicity holds. If these lattices are perfect, periodicity exists such that intramolecular and intermolecular coupling generate phonon waves wich can be described with the corresponding wave vectors **q**.

If translational symmetry is taken into account and Born-Von Karman's cyclic conditions are applied the dynamical equation (5) can be rewritten in the form [24]

$$G_R(q)F_R(q)L_R(q) = L_R(q)\,\Lambda\,(q) \tag{7}$$

Eq. 7 allows to calculate in an easy way the phonon dispersion curves $\nu_i(q)$ either for organic 1-d [25-29] or for 3-d [30-33] crystals. The full phonon dispersion curves can be matched with the experimental ones derived from incoherent neutron scattering experiments [31,34] (and sometimes from the study of short oligomers as models of 1-d lattices [35,36]. Phonons with **q**=0 are the infrared and/or Raman active modes to be matched with optical IR and/or Raman experiments. The optical activity of the **q**=0 modes is further determined by symmetry of the 1-d or 3-d lattices [31,34,37,38].

From the phonon dispersion curves (eq.7) one- and multi-phonon densities of vibrational states $g^{(n)}(\nu)$ can be calculated to be matched with those derived from neutron scattering experiments [39-42].

When translational periodicity is lost the dynamical problem becomes very complex since selection rules are removed and all the normal modes may in principle become active in the vibrational optical spectra. The handling of the dynamics of disordered lattices is however of extreme importance especially in the case of polymeric organic materials whose real structure is never perfect even in the ideally best polymer single crystal [43].

The dynamical problem for disordered one-dimensional polymer chains with realistic structures as those of the complex polymers has been solved [6-13]. For disordered realistic chain molecules the dynamical matrix of eq.5 becomes very large ; numerical algorithms are available for the calculation of $g^{(1)}(\nu)$ [44] and of the corresponding eigenvectors [13] associated with specific ν_i of $g(\nu_i)$.

Many problems of disordered organic polymers have been treated [6-13] including the case of structurally disordered Polyacetylene for which both vibrational [45] and electronic states [46] have been calculated.

3.2) Infrared absorption intensities.

This field of spectroscopy is quite new since for many years appropriate models for the treatement of this observable were not available. Intensity spectroscopy has been revived in the past ten years mainly focussing at infrared absorption intensities [47-53]. Theories are also available for the treatment of Raman intensities but experimental data are lacking.

If the changes of the molecular dipole moment are thought to be linear with the atomic displacements during a vibration, the integrated intensity A_i of the i-th normal mode is related to the electrical properties of the molecule by the relation [47]:

(8) $$A_i = \text{const} \ |\delta\mu/\delta Q_i|^2 \ , \ \text{const} = 976.56 \text{ km/mole amu}/e^2$$

where the change of the total molecular electrical dipole moment μ with the i-th normal mode Q_i is given in $e{*}amu^{-1/2}$ and the intensities in km/mole.

The quantity $\delta\mu/\delta Q_i$ in eq. 8 can be separated into the contribution by the various internal coordinates during Q_i

(9) $$\delta\mu/\delta Q_i = \Sigma_t \ (\delta\mu/\delta R_t) \ (\delta R_t/\delta Q_i) = \Sigma_t(\delta\mu/\delta R_t) \ (L_R)_{ti}$$

where $(L_R)_{ti}$ are the eigenvectors defined in eq. 6.

A considerable advancement in the use of the infrared intensities has been achieved when the total molecular electric dipole moment is expressed as a sum of bond dipole moments or of atomic quantities [50-53]. Let

(10) $$\mu = \Sigma_\alpha \ q_\alpha \ r_\alpha$$

be the instantaneous total dipole moment expressed as a sum of contributions from bond dipole moments, where q_α is the point charge on atom α which may change during the vibration. The infrared intensity, as measured by $\delta\mu/\delta Q_i$,can be expressed as

(11) $$\delta\mu/\delta Q_i = \text{function of } [\ q_\alpha{}^{eq}, \ (\delta q_\alpha/\delta R_t)^{eq}, (L_R)_{ti}]$$

where $q_a{}^{eq}$ is the electronic charge on atom a at the equilibrium, $\delta q_a/\delta R_t$ is the change of the charge on atom a when the t-th internal coordinate vibrates with infinitesimal amplitude. This last term represents the "charge flux" and is very relevant in the case of polyconjugated materials. $(L_R)_{ti}$ is the column of the vibrational displacements in terms of R's during the i-th normal mode Q_i.

We wish to stress that from the measurement of the infrared intensities and from the solution of the dynamical problem (eqs.5 or 7) atomic effective point charges q_α^{eq} and charge fluxes $\delta q_\alpha/\delta R_t$ can and have been derived. The physical meaning of these quantities and their relation with the molecular structure [53] and with the charges calculated by quantum chemical "semiempirical" or "ab initio" methods have been fully analyzed [54,55].

With the use of vibrational intensities the electrical properties of many molecules [56,57] and polymers [58] have been treated, including polyacetylene pristine [59] and doped [60]. In the description of the conjugational properties of polyconjugated materials as well as of the charge mobility in doped and/or photoexcited polymers the charge flux term plays an important role [59,60].

3.3) The Raman Spectra.

Raman spectroscopy is a unique tool in providing information on the structural and electronic properties of polyconjugated materials since it shows features peculiar to the delocalisation of the π electrons. Raman spectra have been used routinely in the characterisation of such materials and the interpretations proposed have been the subject of some controversy.

The approach we take in the interpretation of the Raman spectra has been already discussed for the simpler case of Polyacetylene in paper I[3] and will be only outlined here focussing at the main concepts of general validity also for the materials treated in this article.

The data on dynamics (vibrational frequencies and amplitudes) have been discussed in section 3.1. The apperance of the absorption in IR or scattering in the Raman spectra depends, in general, on the selection rules determined by symmetry and on the magnitude of the IR or Raman transition moments (i.e. transition probabilities).

The Raman intensities require some additional important considerations which become essential in the interpretation of the Raman spectra of polyconjugated materials.

Raman intensities are generally expressed as :

$$(12) \quad I_k = \text{const } (\nu_L + \nu_k)^4 \; \Sigma_{uv}[\alpha_{uv}]_k^2$$

where ν_L is the frequency of the exciting laser line, $hc\nu_k$ is the energy associated to the transition from the initial state i to the final state f; +/- sign refer to the anti-Stokes and Stokes lines respectively; α_{uv} (u,v= x,y,z) are the elements of the polarizability tensor $((\alpha))$ associated with the transition moment from state i to state f [61,62]:

$$(13) \quad [\alpha_{uv}] = 1/hc \; \Sigma_r \left\{ \frac{[\mu_u]_{fr} \; [\mu_v]_{ri}}{\nu_{ri} - \nu_L + i\Gamma_r} + \frac{[\mu_u]_{fr} \; [\mu_v]_{ri}}{\nu_{rf} - \nu_L + i\Gamma_r} \right\}$$

the sum extends over the r electronic states. Γ_r is a damping factor associated with the lifetime of the state |r>, $hc\nu_{ri}$ is the energy associated with the transition from |i> to |r>, μ is the electric dipole moment and $[\mu_u]_{fr}$=<f|μ_u|r>, |i>=|gm> and |f>=|gn>, where g is the ground electronic state and m,n are the vibrational levels of the ground state. In Raman spectroscopy some very relevant modifications of the spectrum occur when states |r>=|ep>, such that $\nu_{eg}\approx\nu_L$. In such a case the exciting laser line brings the system from the ground electronic state to an excited state and Resonance Raman scattering occurs. In this case eq. 13 can be further elaborated [61-65]. First of all the sum over r can be dropped and only terms with |r>= |ep> can be retained such that

$$(14)\quad [\alpha_{uv}]_{gn,gm} = 1/hc\ \Sigma_p \frac{\langle n|\mu_u^{ge}|p\rangle\langle p|\mu_v^{ge}|m\rangle}{\nu_{ep,gm}-\nu_L+i\Gamma_{ep}}$$

The second terms in the brackets of eq. 13 has been dropped since it is negligeable with respect to the others. One then proceeds by expanding the terms μ_u^{ge}=<g|μ_u|e> as a function of the normal coordinates yielding the well known expression in the study of resonance Raman scattering, namely $\alpha_{uv} = A_{uv} + B_{uv} + \ldots$. When the electronic transitions are strong the leading term is

$$(15)\quad [A_{uv}]_{gn,gm} = 1/hc\ (\mu_u^{ge})_0\ (\mu_v^{ge})_0\ \Sigma_p \frac{\langle ng|ep\rangle\langle ep|mg\rangle}{\nu_{ep,gm}-\nu_L+i\Gamma_{ep}}$$

When the exciting laser frequency ν_L occurs in the range of the electronic absorption spectrum expression 15 must be introduced in eq. 12 instead of α_{uv}.

For the understanding of the Raman spectra of polyconjugated molecules the factors appearing in eq.15 must be analyzed in detail.

i) $(\mu_u^{ge})_0$ and $(\mu_u^{ge})_0$ must be non vanishing in order that A_{uv} be non vanishing. This means that the transition g-e must be allowed, as it occurs in the case of the π-π* (HOMO-LUMO) transitions in conjugated molecules.

ii) The Frank-Condon factors <ng|ep><ep|mg> must be non vanishing for at least one vibronic level |ep>. This happens only when the potential energy minimum of the excited state is displaced along the Q_k axis relative to the potential energy minimum of the ground state, namely when ΔQ_k^{e-g}=0. Only those modes for which ΔQ_k^{e-g}=0 yield non vanishing A_{uv}. Since ΔQ_k^{e-g} can be non vanishing only for totally symmetric modes, *the bands which appear in the Resonance Raman spectrum with an appreciable intensity can only be totally symmetric.*

iii) The resonance denominator in eq. 15 strongly determines the magnitude of A_{uv}. Thus *the RR intensities may change with* ν_L.

The term $[A_{uv}]_k$ which is introduced in eq. 12 in the case of RR scattering can be rewritten as:

(16) $[A_{uv}]_k = 1/hc\ (\mu_u{}^{ge}\mu_v{}^{ge})_0\ \Delta Q_k{}^{g-e}\nu_k(\nu_{eg}-\nu_L+i\Gamma_e)^{-1}$

where it has been assumed that $\nu_{ep,gm}=\nu_{e0,g0}=\nu_{eg}$).

It can be shown (see paper I) that eq. 16) can be transformed into:

17) $[A_{uv}]_k = 1/hc\ (\mu_u{}^{ge}\mu_v{}^{ge})_0(\delta E_e/\delta Q_k)_0(\nu_{eg}-\nu_L+i\Gamma_e)^{-1}$

where $(\delta E_e/\delta Q_k)_0$ is the derivative of the potential energy curve of the excited state e with respect to Q_k evaluated at the equilibrium of the ground state. Eq. 17 is of fundamental importance for the interpretation of the Raman spectra of conjugated molecules [66]. The only term in $(A_{uv})_k$ which depends on the particular normal mode is thus $(\delta E_e/\delta Q_k)_0$. Since **R=LQ** (see eqs.5 and 6) one can write:

18) $[A_{uv}]_k = 1/hc\ (\mu_u{}^{ge}\mu_v{}^{ge})_0\ (\delta E/\delta \mathcal{R})_0 L\mathcal{R}_k(\nu_{eg}-\nu_L+i\Gamma_e)^{-1}$

where $\mathcal{R}$ *is a totally symmetric linear combination of internal coordinates which describes the variation of the equilibrium geometry* Δx^{eg} *from the ground state g to the first excited state state e.* The description of the $\mathcal{R}$ coordinate for various polyaromatic molecules will be given later in this review. As discussed in paper I for the case of polyenes $\mathcal{R}$ has been clearly given by the quantum chemical calculations on the geometry of the first excited state.

The important rule which follows from eq. 18 is as follows: *Only those totally symmetric normal modes* Q_k *which have non vanishing* $L\mathcal{R}_k$ *are Raman active and their relative intensity is given by*

(19) $I_k \div |L\mathcal{R}_k|^2$.

The **L** are the eigenvectors of the dynamical equation 5 (or 7) and can be calculated in the usual way once the force field (**F** matrix , eq.2) and the geometry (**G** matrix, eq.3) of the systems are known. This has been done for Polyacetylene (paper I) and for the systems discussed in this paper.

It will become apparent in this paper that the theory discussed here accounts very well for the Raman spectra of the polyaromatic molecules so far studied.

4) THE VIBRATIONAL FORCE FIELD. CLASSICAL vs. QUANTUM MECHANICAL CALCULATIONS.

From the eigenvalue equations 5 or 7 one can derive the fundamental frequencies ν_i and the vibrational displacements L_i if the force field is known. The knowledge of a molecular vibrational potential is not a trivial matter even for very simple systems and has been the subject of much work by very many authors. We shall not make here any effort to review the very complex field of vibrational potentials, but wish only to present some personal thougths on the problem particularly aimed at the case of polyconjugated organic molecules.

The physical reality to be considered is that when energy is provided any molecular system made up by N atoms in any phase vibrates along the 3N-6 normal modes. For this reason many physical properties which are currently measured are affected, at different extents, by the atomic vibrations; viceversa, the proper analysis of such experiments provide data on the atomic vibrations (hence on the interatomic forces, e.g. intramolecular interactions, originated by the electronic distribution in a given electronic quantum state and by the intermolecular interactions associated with the physical state of the system).

Ro-vibronic spectroscopies in the UV-Visible-Infrared and Microwave energy range, X ray and electron diffraction, incoherent and coherent elastic and inelastic neutron scattering, Raman scattering, Nuclear Magnetic resonances etc. all contain a vibrational contribution. Other non spectroscopic properties such as the various thermodynamical quantities contain the vibrational contributions.

One way to proceed in such studies is to chose a phenomenological model for the description of the interatomic forces, decide on the order of the expansion of the potential function and fit the smallest set of parameters of the model to the largest possible number of experimental data. This has been done by many authors who have worked out extremely large least squares methods of refinment for fitting many independent experimental data [67]. Then one proceeds to the analysis of the physical meaning of the "parameters ". The success of the model of the potential consists in its capability of predicting the vibrational contributions of many independent physical data.

At present fully automatic and commercially available computer programs are capable to predict a variety of physical properties based on some interatomic potentials previously stored as data basis [68]. In spite of the elegance and of the exciting flexibility of these numerical procedures the reliability of the potential used is too often very dubious and the uncritical use by unexperienced users may yield misleading results.

In the case of organic molecules chemists consider a logical choice to express the potential in terms of valence bonds and valence electrons [20] while physicists often proceed with an unbiassed cartesian space which is sometimes transformed in that of a suitable model aimed at a specific physical property. Often the potentials used in treating the vibrations of ionic crystals or metals are transferred and constrained to the case of organic molecules. However the chemical reality of molecular covalent bonds and of their interactions is different from any type of "ionic" system. For example the popular "shell model" has not been as successful as the classical "valence force field" in the study of the lattice dinamics of "covalent crystals" such as diamond and silicon [69]. Sometimes hybrid models are proposed which originate from a strict physicists's approach which is slowly relaxed with many compromises by the introduction of chemical concepts to improve the fitting with the experimental data (see, for example refs. 70 and 71).

A way to avoid any phenomenological parametric modelling is to carry out the calculation of the vibrational potential by "ab initio" quantum chemical methods. This field has expanded very quickly with the same rate as that enjoyed by computer science. Even if presently calculations can handle billions of integrals [72], any reasonably feasible calculation on realistic and relevant systems requires the choice of the method of calculation and of a limited set of atomic orbitals, thus again introducing a somewhat arbitrary decision.

In the field of polyconjugated materials quantum chemical calculations have been widely used for the description of their "static" ground [73-76] and excited state properties. Such studies have been of fundamental importance for the development of the science in this field. For the reasons just discussed above we feel allowed to express the convinction that the understanding reached with quantum chemical calculations in the field of polyconjugated molecules is mostly based on an intuition inspired by a wise and clever overall analysis and correlation of the information obtained from such calculations and with the use of "trends" of calculated values.

When attempts are made to be more quantitatively close to the experimental data the limitations of the quantum chemical methods so far used are immediately and disappointingly perceived. The disagreement between calculated and experimental data even for the simplest polyconjugated systems (e.g. polyenes and Polyacetylene) for basic quantities such as energy gap, band width, ionisation potentials and molecular conformations are remarkable and indicate that the present methods need to be improved to represent satisfactorily the electronic intramolecular phenomena.

These limitations become so evident in the calculated vibrational properties (e.g. vibrational frequencies and vibrational force constants, see, for examples, refs 77 and 78) that complicated and somewhat personal methods of scaling need to be introduced [79].

A striking and up to date example which can provide the reader with the general aspects of the problem is that offered by the simplest and most widely studied case of polyenes and polyacetylene discussed in detail in paper I. Other cases are going to be presented in this paper.

As a conclusion of this section the following suggestions seem to be in order for the case of polyconjugated materials:

i) The vibrational spectra of polyconjugated systems provide peculiar data on the molecular electronic structure; their interpretation requires a particular treatment of the dynamics and a careful handling of the force field whose validity must be tested on many molecules of the same chemical class and in the same structural and dynamical situations (e.g. pristine, doped, photoexcited states). The general rule that "by adding more force constants one can fit any vibrational spectrum" has to be strengthened for the case of polyconjugated polymers where just by changing the values of some vibrational parameters one can explain any spectra, thus reaching conclusions with no physical proof except a mere numerical fitting of a few frequencies.

ii) The results of the calculations of vibrational quantities (frequencies and intensities) by quantum chemical methods are a further proof of the present limitations of a theoretical way of approaching the physics of such materials. While future development will certainly improve the quality of the calculations, at present one should only accept "trends" of calculated values from which only qualitative "messages" on the physics can be derived.

Examples of the situations found in i) and ii) are found in paper I and in the cases discussed later in this paper.

5) SPECTROSCOPIC CHARACTERISTICS PECULIAR TO POLYCONJUGATED MATERIALS.

It is known that the vibrational spectra of polyconjugted organic materials show peculiar and characteristic features which have been matter of study by many authors . Solid state physicists have immediately related the occurrence of such features with the existence of delocalized π electrons while organic chemists and spectroscopists have analyzed the spectra in the traditional correlative way, thus missing the important information offered by the infrared and Raman spectra ,when suitably interpreted.

The paradigma case of polyacetylene has been discussed in detail in paper I which also reports and summarizes the variety of interpretations offered by various authors depending on whether they are physicists or chemists.

With the present knowledge of the vibrational spectra of many polyconjugated systems (oligomers and polymers) in their pristine (insulating), doped (electrically conducting) and photoexcited (photoconducting) states it is possible to list a set of characteristic spectroscopic manifestations of general validity. Since it has been shown [14,80] that most of these features are related to the existence of a network of delocalized π electrons, their interpretation should be made with a unified theoretical approach,independently from the molecule considered.

We collect in the following pages the spectroscopic observations which we think have general validity in the field of polyconjugated materials. These observations will be later justified on the basis of some theoretical concepts. The list of observations below is based mostly on polyacetylene which has been widely studied. The other polymers have not yet been studied with the same depth and width as PA. However, the observations listed below, when suitably looked for, have been verified also for other systems and, so far, have never been disclaimed.

In this review article we present the case of several polyconjugated systems either as oligomers (with known molecular length) or polymers (for which polydispersivity and chemical or structural defects must be taken into account). In order to make our discussion more compact and direct without loss of clarity in the building of the list of observations we proceed with the following leading criteria: i) each observation is labelled alphabetically, ii) we do not give all the references appeared in the literature regarding each individual observation , but quote only the ones which show more clearly the experimental fact observed.

5.1) Materials in the pristine (insulating) state.

a) Raman spectra of both oligomers and polymers are generally recorded in Resonance conditions (RR) since the excitation lines generally used occur in the frequency range of the electronic spectrum where absorption occurs. Only very recently Raman spectra have been recorded with an infrared excitation using Fourier Transform Raman spectrometers [81,82]; these spectra may be near or off-resonance.

b) A few lines observed in the Raman spectrum show "frequency dispersion" with conjugation length (i.e. with the extent of delocalisation)[3].

c) The Raman lines show "intensity dispersion" with conjugation length, i.e. they change their relative intensity in a continuous and smooth way by changing conjugation [3,83-86]. A few Raman lines, generally in the C=C stretching range (1600-1300 cm^{-1} are extremely strong.

d) The Raman spectrum only of the polymer shows frequency dispersion with excitation wavelength while the frequencies of the Raman spectra of the oligomers are independent from λ_{exc} [3].

e) The Raman spectrum of stretch-oriented polymers (this observation has been made in the case of trans polyacetylene [3]) recorded in various scattering geometries shows that the observed lines are mostly totally symmetric and that the tensor component α_{xx} is dominant (x is the axis of the polymer chain which coincides with the direction of stretching [3]).

f) The infrared spectrum practically does not show frequency dispersion with conjugation length.

g) The infrared spectrum shows (but it is not easy to be seen quantitatively) intensity dispersion with conjugation length [85,86].

h) The infrared active C-H stretching bands in stretch oriented samples in polarized light may show an "anomalous polarisation". Striking cases are Polyacetylene [87] and Poly-paraphenylene-vinylene [88].

i) In the case of the oligomers some of the normal modes are highly localized on the end groups and can be easily identified since their frequency is constant and their relative intensity decreases with the increasing of the number of monomer units in the finite chain [85].

j) The infrared spectrum has never (so far) exhibited chain length dependent band progressions (i.e. chain length dependent $q \neq 0$ phonons have never been observed with certainty in the materials so far examined in detail; this case is ,instead, common in other chain molecules such as polymethylene chains [35]).

5.2) Materials in the doped (electrically conducting) state.

k) Upon doping the Raman spectrum weakens and broadens progressively with increasing doping concentration, while showing frequency and intensity dispersion [3,89-91].

l) At high doping concentration new features appear in the Raman spectrum. This statement does not mean that the observed Raman spectrum arises from the doped site of the molecules. This problem will be discussed later in this paper(see also ref. 3).

m) The infrared spectrum is dominated by new very strong bands (doping induced infrared spectrum) .

n) The new infrared bands are isotopically dependent [3].

o) The frequencies of the doping induced IR bands are practically independent from the type of dopant (p/n) [3].

p) The intensity of the doping induced infrared absorption increases linearly with increasing doping level (at least in the concentration ranges studied in the experiments so far published)[3].

q) The doping induced infrared spectrum shows "infrared frequency dispersion" and "infrared intensity dispersion" with the "length" of the molecule [3,92].

r) The doping induced infrared absorption of stretch-oriented samples is highly anisotropic along the chain axis of the molecule [2,3].

5.3) Materials in the photoexcited state .

s) The weak infrared spectra of photoexcited oligomers or polymers exhibit a frequency pattern very similar to that observed upon doping. However, the frequencies and intensities of the photoinduced pattern may be different from those of the doping induced infrared pattern [3,83-86].

t) The photoinduced spectrum is isotopically dependent [3].

u) The photoinduced infrared spectrum shows dispersion with the "length" of the molecule [3].

w) Additional very weak satellite infrared lines variosuly named by many authors are observed (in the simplest case of PA) [3].

6) THEORETICAL ASPECTS OF THE VIBRATIONAL SPECTRA OF POLYCONJUGATED MOLECULES.

A unifying theory which explained some of the observations listed in section 5 of this chaper has been first presented by Horovitz [80,93] and his collaborators who have treated mainly the case of trans polyacetylene. This theory has been the basis for the qualitative understanding of the spectra of many more complex polyconjugated materials.

The basic concept of Horoviz theory is to consider the charge density wave which exists in the dimerized form of trans polyacetylene and the fact that π electrons are coupled to the motions of the carbon skeleton. The "pinning parameter" $\tilde{\lambda}$ (or α) is a measure of the pinning of the π electrons to the carbon skeleton. When delocalisation increases the pinning decreases as if electrons were more mobile. The pinning decreases also by doping or photoexcitation. The dynamical equation for the phonons of polyacetylene is expressed in terms of $\tilde{\lambda}$, such that the calculated normal frequencies are on their turn function of $\tilde{\lambda}$. Frequency dispersions with $\tilde{\lambda}$ have been presented and form the basis for the understanding of the various dispersions observed in Polyacetylene.

For a long time $\tilde{\lambda}$ remained a useful, but exhotic, parameter which could hardly be understood in molecular terms. More recently it has been shown [14-17] that $\tilde{\lambda}$ can be related to some "chemical" force constants which reflect the state of conjugation of the molecule.

In section 3.3 we have shown that the Raman spectra of Polyconjugated materials arise from resonance scattering of the totally symmetric modes a few of which are selectively enhanced if they contain a contribution by the coordinate $\mathfrak{R}$ which describes the changes of geometry from the ground to the excited electronic state.

As clearly described in paper I and in refs.3 and 14-17 $\mathfrak{R}$ has been defined empirically and rigorously (from MO calculations on the excited state of short oligomers [94]) for polyacetylene. These concepts have been extrapolated for the case of polyaromatic systems thus making possible to define $\mathfrak{R}$ as a linear combination of the chemical internal coordinates R associated with the stretching of the C-C and C=C bonds.

In the case of polyacetylene

$$(20) \qquad \mathfrak{R} = 1/\sqrt{2}\ (R_{C=C} - R_{C-C})$$

corresponds to the "dimerisation amplitude u" defined and used in the theoretical treatements on Polyacetylene [95].

In the case of polyacetylene the total energy of the systems as function of the dimerisation amplitude (fig. 1) shows two equal minima as the results of the Peierls distortion which favours dimerisation [2]. If we consider the case of polyaromatic polymers for which the ground electronic state is not accidentally doubly degenerate (fig. 2) the coordinate $\mathfrak{R}$ can be defined as the collective geometrical evolution of the network of double/single CC bonds in transforming from the "aromatic" ground state structure to the excited electronic state described as "quinoid" (fig.2)

It has been mentioned before in this paper (see also refs. 3 and 17) that such $\mathfrak{R}$ coordinate is totally symmetric (TS) and can be used as a symmetry coordinate when symmetry is considered in the dynamical treatment of these molecules. Let

$$(21) \qquad \mathbf{S=UR}$$

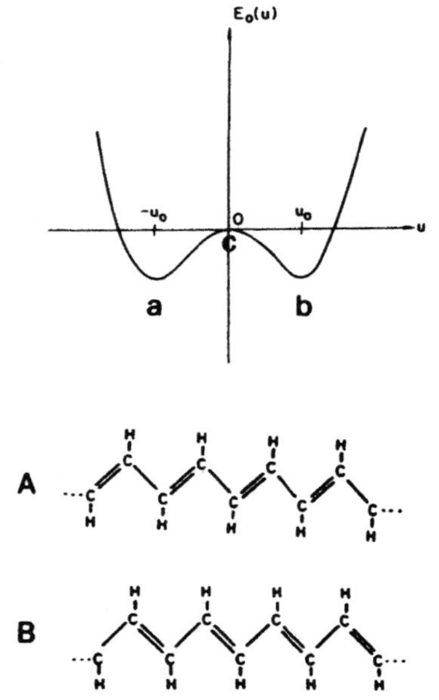

Fig.1 Total energy vs.dimerisation amplitude (Я coordinate) for Polyacetylene shwoing the two isoenergetic ground states A and B separated by a barrier.

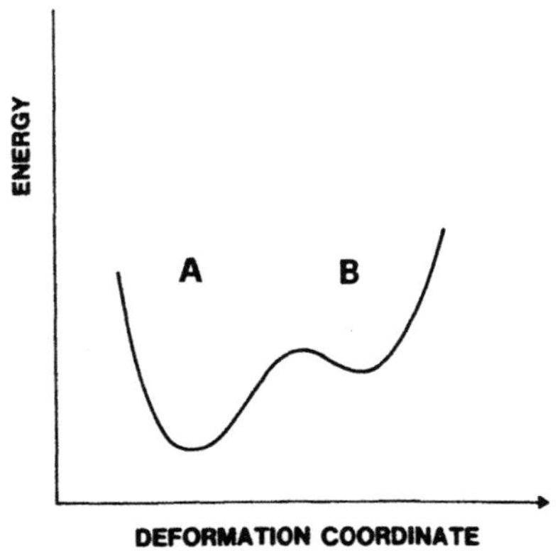

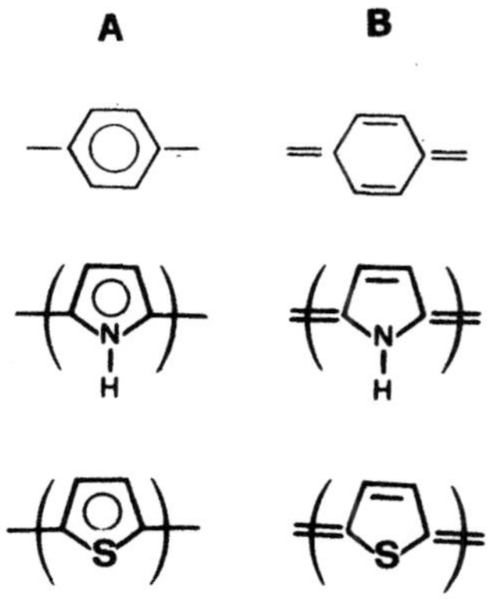

Fig.2 Sketch of the total evergy vs. the Я coordinate for a few polyaromatic systems.

be a set of "symmetry coordinates", $\mathbf{U}$ being an orthonormal matrix necessary to construct a basis of a completely reduced representation of the molecular point group [20,21]; R are the internal coordinates [20] defined in eq.1. The vibrational potential and kinetic energy matrices can be expressed in terms of the coordinates S

$$(22) \quad 2V= \tilde{\mathbf{R}}\mathbf{F}_R\mathbf{R} = \tilde{\mathbf{S}}\tilde{\mathbf{U}}\mathbf{F}_R\mathbf{U}\mathbf{S} = \tilde{\mathbf{S}}\mathbf{F}_S\mathbf{S}$$

$$(23) \quad 2T= \tilde{\dot{\mathbf{R}}}(\mathbf{G}_R)^{-1}\dot{\mathbf{R}} = \tilde{\dot{\mathbf{S}}}\tilde{\mathbf{U}}(\mathbf{G}_R)^{-1}\mathbf{U}\dot{\mathbf{S}} = \tilde{\dot{\mathbf{S}}}(\mathbf{G}_S)^{-1}\dot{\mathbf{S}}$$

Both $\mathbf{F}_S$ and $\mathbf{G}_S$ matrices are factored into symmetry blocks ; the dimensions of the blocks are equal to the number of normal vibrations in each symmetry species. The secular equations 5 or 7 take the form

$$(24) \quad \mathbf{G}_S\mathbf{F}_S\mathbf{L}_S=\mathbf{L}_S\hat{} \qquad \text{or } \mathbf{G}_S(\mathbf{q})\mathbf{F}_S(\mathbf{q})\mathbf{L}_S(\mathbf{q}) = \mathbf{L}_S(\mathbf{q})\mathbf{\Lambda}(\mathbf{q})$$

respectively, and are factored into blocks. Each symmetry block can be handled separately, thus enabling to predict (and identify) the normal modes of a specific symmetry species.

Since $\mathcal{R}$ is a TS linear combination of R's it becomes one of the S's of the TS block. In the $\mathbf{F}_S(A_g)$ block of the vibrational potential one then defines an *"effective* *conjugation* *force* *constant* $F_{\mathcal{R}}$" and one can solve the TS block of the secular equations 5 or 7 parametrically with $F_{\mathcal{R}}$, while keeping everything else fixed.

The calculation of $\nu=f(F_{\mathcal{R}})$ provides the frequency dispersions with an effective conjugation parameter $F_{\mathcal{R}}$ which is related to the pinning parameter $\tilde{\lambda}$ by the relation:

$$(25) \quad F_{\mathcal{R}} = F_{\mathcal{R}}^0-\text{const}(1-2\tilde{\lambda}).$$

The general expression of $F_{\mathcal{R}}$ in terms of the internal valence force constants is the following:

$$(26) \quad F_{\mathcal{R}} = K^0 + f^0 + \Sigma_s f^{0,s}$$

where K^0 is a combination of diagonal C=C and C-C stretching force constants of the central unit, f^0 is a combination of the interaction force constants between CC bonds in the central unit and $f^{0,s}$ is a combination of interaction force constants among CC bonds of the central unit and CC bonds of the s-th unit. s is, in principle, unknown and defines the distance of interaction between π bonds.

It has been first suggested empirically by Crawford et al. [96-98] and later proven by Kakitani [99] that the valence interaction force constants between single and double CC bonds must have different signs, i.e $f_{C-C}/f_{C-C}<0$, $f_{C=C}/f_{C=C}<0$, $f_{C-C}/f_{C=C}>0$. Application of such rules makes the second and third terms in eq. 26 negative. It is also clear that the third term of eq.26 is directly connected with the distance and extent of π/π interaction along the chain. *The larger the absolute value of the sum the smaller is the value of $F_{\mathcal{R}}$, i.e the larger is the effective conjugation*.

The description of the conjugation in terms of the pinning parameter or in terms of the effective conjugation force constant are very similar. The advantage of $F_{\mathcal{R}}$ is that one can define and use it in molecular terms even for very complex molecular systems.

When the TS block of eqs.5 ot 7 is solved parametrically keeping everything fixed exept $F_{\mathcal{R}}$ one obtains not only the frequency dispersion $\nu/F_{\mathcal{R}}$, but also the eigenvectors ($L_{\mathcal{R}}$) to be used in the prediction of the intensity dispersion.

The above theoretical approach can be immediately applied for the explanation of most of the spectroscopic characteristics peculiar to polyconjugated materials listed in section 5.

i) Frequency and intensity dispersion with conjugation length of both Raman (observations b,c,d,g,q and u) originate from the existence of a collective $\mathcal{R}$ mode which probes the delocalisation of the π electrons along the chain backbone. The delocalisation increases when the "effective conjugation length increases" by increasing the number of chemical units and/or by changing the conformation of the chain backbone in order to improve the overlapping of π electron orbitals.

ii) The selective strong Raman intensities (observation c) are associated with the large molecular polarizability changes due to the existence of a very mobile and highly polarizable charge density wave of π electrons which is selectively excited by the motions along the $\mathcal{R}$ coordinate. In the pristine state the symmetry of the charge distribution is the same as the symmetry of the molecular geometry + masses. The huge intensity of the "new" bands induced by doping or by photoexcitation (observations m,n,o,s and t) is due to the fact that the electrical symmetry is broken by the removal (or injection) of charges. The sistem becomes higly polar; the $\mathcal{R}$ mode acquires a strong intensity due to a huge "charge flux" parameter :

27) $\delta\mu/\delta\mathcal{R} = \Sigma_k\Sigma_i\ (a\ \delta m_k/\delta R^i - b\ \delta m_k/\delta r^i)$

where a and b are geometrical coefficients, i is confined within the molecular domain perturbed by the formation of the charge carrier (conjugational defect) while k extends much further out of the defect region into the polyconjugated unperturbed chain; m is the bond dipole moment of the C-C bond. If the $\mathcal{R}$ mode is coupled to other modes it lends intensity (Raman, pristine/ IR, doped or photoexcited) to these modes.

iii) The charge flux along the skeletal backbone accounts for the optical anisotropy in the IR of doped materials (observation r) and for the anomalous dichroic behaviour of C-H stretching bands (observation h).

iv) Solitons, polarons and bipolarons are conjugational defects which polarize the polymer skeleton (i.e. induce a displacement of charges, thus generating a permanent large dipole moment in the originally apolar chain) giving rise to strong $\mathcal{R}$-containing infrared modes. *Since only the $\mathcal{R}$-containing modes are those which are enhanced no immediate selective information on the nature of the* *charge carrier can be derived from the IR spectra.*

v) When the symmetry of the system is lowered by the existence of a conjugational defect the polarizability of the charge carrier is small and no (or extremely week) Raman scattering arising from the doped site is expected. The opposite occurs in the IR spectra(observations l,k and m). It follows that the strong scattering observed for doped materials cannot come from the conjugational defect, but from the residual fraction of the material which has not been attacked by the dopant.

7) WORKED OUT STUDY CASES.

We present in the following sections a spectroscopic study of a few cases with the main aim to offer a guided tour in the use of ECC in the field of material spectroscopy and to present the state of the art of the present knowledge of a few polyconjugated systems as provided by vibrational spectroscopy.

It will become apparent that the cases presented here are not studied as widely as the case of polyacetylene presented in paper I. It will become also apparent that when the molecular complexity of the system increases molecular orbital calculations become more difficult and their results become sometimes only indications of a trend of quantities which molecules may follow.

As stated at the beginning, since we aim at describing material properties we report here only the spectroscopic data useful in material science. All the details on vibrational assignments and dynamical calculations of interest to pure spectroscopists are not reported here, but can be found in the references cited.

8) POLYPYRROLE.

As a first application of EEC theory we discuss the case of polypyrrole (PPy) which has been one of the first polyaromatic systems studied with the hope to find a material with relevant technological interest.

The electrochemical and/or chemical synthesis of the pristine and doped materials and the many problems to be faced and solved are fully documented in specialized papers [100]. It has been shown that while PPy in the oxidized state (doped) is chemically very stable and conducting, pristine PPy is unstable and can be easily oxidized or decomposed and/or reacts with unwanted side reactions [101].

The spectroscopic situation is the following. The vibrational spectra of PPy pristine and doped have been presented by several groups, but the spectra are often seriously and hopelessly different, thus showing that the chemistry of the material is not yet under control. Street et al. have presented the IR spectra of pristine and doped PA and those of the oligomer Py_2 and Py_3 [100].

The IR spectra of pristine and doped PPy are presented also by the group of Mainz [102] and that of Beijin [103]. One can observe that while the IR spectra of doped PPy from various laboratories are practically identical (and the reasons will become apparent below in this section), even for different dopants, those of PPy in its pristine state are quite different , even if a few common features can be identified.

The Raman spectra of PPy are reported by Furukawa et al. [104], by the group of Beijin [103] and by Oddi et al. [105]). In ref. 104 the Raman spectrum of 2,5-^{13}C- substituted PPy and that of the C-deuterated PPy are also presented. From the known UV-Vis spectrum [104] one realizes that the Raman spectra are always recorded in resonance conditions.

Using the concepts discussed in the previous sections the interpretation of the spectra proceeds in the following way.

8.1) Structure, symmetry and ECC theory.

The structure of pristine PPy is not yet known since the material is amorphous, but can be guessed or extrapolated from the experimentally determined structures of the oligomers Py_2 and Py_3 in the solid state. The structure of these molecules is trans and coplanar [106]. The results of semiempirical quantum chemical calculations are many and, being focussed at various physical aspects, are sometimes inconsistent. From MO calculations the structures of Py_2 and Py_3 "in vacuo" is not coplanar [107] but the barrier to coplanarity is relatively low such that it can be flattened when intermolecular forces are turned on in the crystal. Recent MO calculations "in vacuo" by the group of Kertesz send the important message that non coplanar structures should not be neglected [108]. In ref. 108 the possibility that PPy may also exist in a helical form is presented with serious theoretical and numerical arguments. A full discussion of this problem is presented in the case of Polythiophene in section 9 of this article.

For the time being we work with assumption that, by extrapolation from the experimental results on Py_2 and Py_3, the structure of PPy is trans (anti) and co-planar (in the literature in this field the terms trans-cis or anti-sin are freeley used). The symmetry of the polymer chain as 1-d perfect and infinite lattice is D_{2h} and the q=0 modes are classified as [109]:

$$\Gamma^{vib}_{in-plane} = 8A_g(RR, \mathcal{R}\text{-dependent}, pol) + 8B_{1g}(R, depol) + 7B_{2u}(IR) + 7B_{3u}(IR)$$

$$\Gamma^{vib}_{out-of-plane} = 3B_{2g}(R, depol) + 4B_{3g}(R, depol) + 3A_u(inactive) + 4\,B_{1u}(IR).$$

The Raman spectrum should show predominantly the 8 A_g modes variously resonantly enhanced. If one A_g C-H stretching and one A_g N-H stretching are neglected we expect to find 6 dispersion branches in the $\nu/F_{\mathcal{R}}$ plot, in the frequency range 1700-100 cm^{-1}. The 6 TS A_g frequencies must lie on these dispersion curves at a given value of the effective conjugation parameter $F_{\mathcal{R}}$, with relative intensity $\div|L_{\mathcal{R}}|_i^2$. Indeed 6 Raman lines are observed at ν_1 =1569 (very strong),

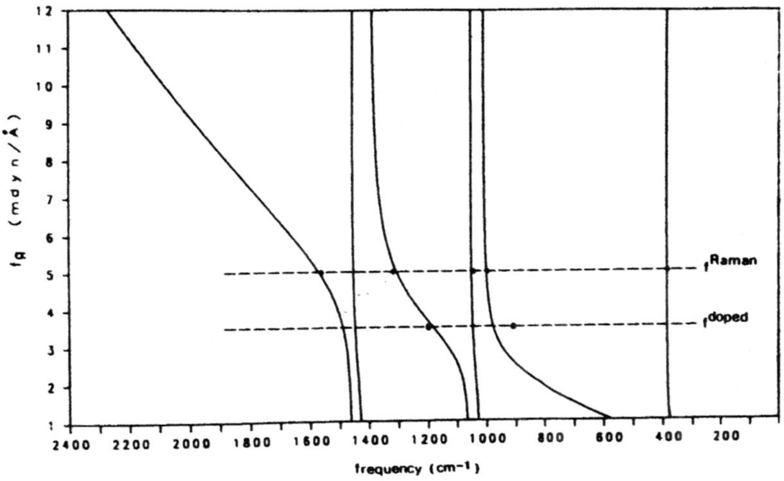

Fig.3 Calculated frequency dispersion ν/F_R of the Ag modes of Polypyrrole (from ref. 109).

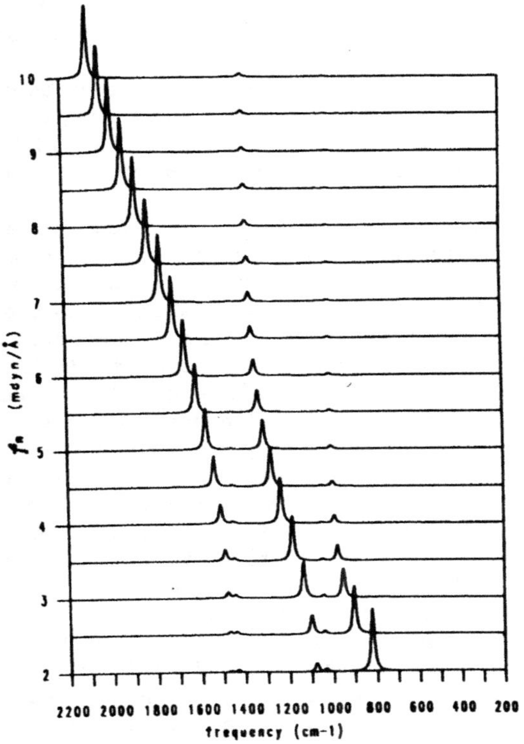

Fig. 4 Calculated intensity dispersion of the Ag modes of Polypyrrole (from ref. 109).

ν_2=1452 (vanishingly weak), ν_3= 1307 (medium), ν_4= 1051 (very week), ν_5=1000 (very weak) and ν_6=387 cm^{-1} (zero intensity).

Upon doping 6 IR active modes, some of which very strong, must be found at $F_{\mathcal{R}}^{dop} < F_{\mathcal{R}}^{prist}$ along the same dispersion branches as those used for the location of the Raman active A_g modes.

8.2) Calculations and comparison with the experiments.

The numerical characterisation of the effective conjugation length can be made when the frequency and intensity dispersions are available .Calculations have been carried out by first estimating the force field on the basis of MNDO semiempirical quantum chemical calculations for the tetramer Py_4 [110]. The geometry of Py_4 has been fully minimized with the only exception of keeping the molecule coplanar [110]. The obtained MNDO force field has been refined by least squares fitting on the TS observed Raman modes of PPy, ^{13}C-PPy and PDPy.

The calculated frequency dispersions $\nu/F_{\mathcal{R}}$ and intensity dispersions are given in figs. 3 and 4 respectively. Let us label the 6 branches as v_i (i= 1-6). From the calculations we learn that when delocalization is restricted (large $F_{\mathcal{R}}$ values) , ν_1 is mostly $\mathcal{R}$ and the whole intensity is concentrated in the Q_1 normal mode. When the effective conjugation increases the mode mixes with the other branches and the intensity is redistributed to the other modes making them stronger in the Raman spectrum (pristine) or in the IR spectrum (doped).

The experimental Raman lines of pristine PPy find their best frequency and intensity fitting for $F_{\mathcal{R}} \approx 5.1$ mdyne/A with frequencies at ν_1= 1562 (very strong) , ν_2= ? ,(zero intensity) , ν_3= 1319 (weak) , ν_4= 1044 (medium), ν_5= 998 (weak) and ν_6 = 385 cm^{-1} (zero intensity).

The infrared spectrum of the doped material (fig. 5) shows more than the 6 lines predicted . These bands can arise from i) selectively enhanced $\mathcal{R}$-containing modes as predicted by ECC theory, ii) IR active non TS modes of the pristine fraction of the material which remains undoped and iii) infrared active non TS modes of the doped part. Chemical defects from unwanted side reactions may also show their ir absorption ,but it is is remarkable that the spectra of the doped materials from various groups show the same doping induced spectrum. The most acceptable fitting (frequencies and intensities) for the $\mathcal{R}$-containing modes is found at $F_{\mathcal{R}}^{dop} \approx 3.5$ mdyne/A with the calculated frequencies at 1488 (A') ,1447 (B') ,1187 (D') and 979 cm^{-1} (F') to be compared with the experiments at 1547 (A'), 1473(B'), 1195 (D') and 912 cm^{-1} (F'). The experimental bands in fig.5 which do not find a coincidence with calculations at 1300 (C'), 1043 (E') and 786 (G') cm^{-1} must be assigned to the vibrations as in ii) and iii).

We observe that the fitting of the $\mathcal{R}$-containing modes is not very satisfactory mostly because of the known limitations of MNDO methods (parametrisation of the heteroatoms, proper accounting of intramolecular interactions etc). The overall behaviour can, however, be accepted. Better fitting will be obtained in other classes of molecules as

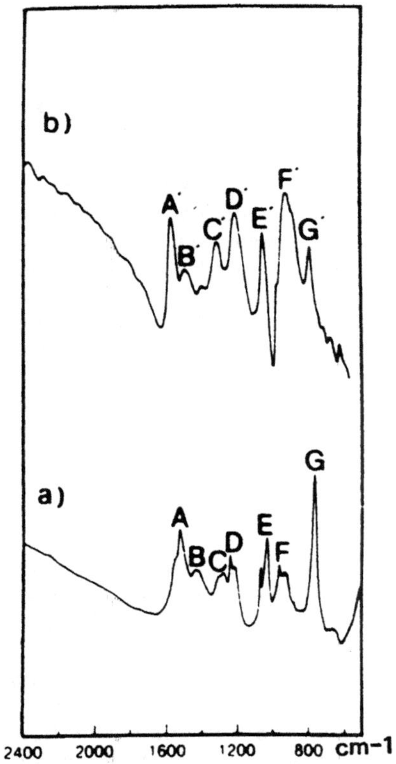

Fig. 5 Infrared Spectrum of polypyrrole : a) pristine and b) doped (from ref. 100)

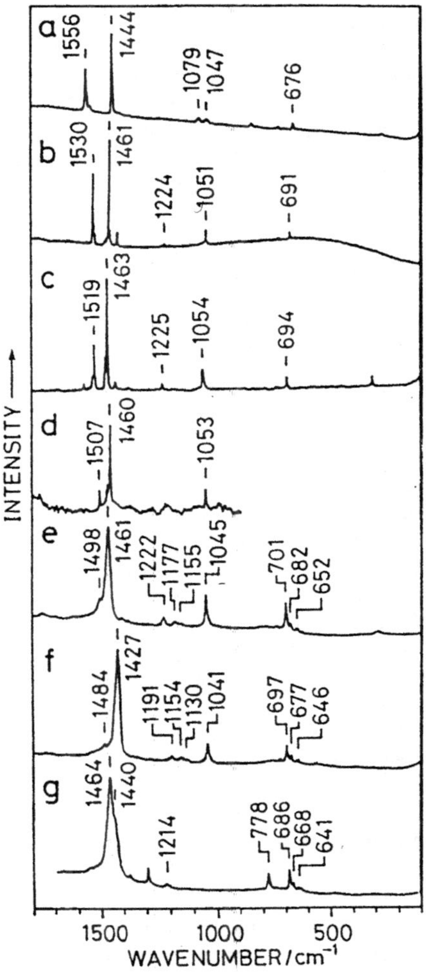

Fig. 6 Raman spectra of a few oligomers of Thiophene in the solid state (from ref. 122). a:Th_2 b:Th_3 c:Th_4 d:Th_6 e:PTh f:2,5-^{13}C-PTh and g:perdeutero-Th. λ_{exc}=632.8 nm (c,d); λ_{exc}=514.5 nm (others)

described below.

8.3). Raman spectrum of doped PPy.

The small, but meaningful, dispersion of the strong $\mathcal{R}$ mode ν_1 near 1560 cm^{-1} (fig.4) allows to analyze the Raman spectrum of doped PPy reported by Furukawa et al.[104]. Experimentally one finds that from pristine to doped PPy the first strong Raman line near 1569 cm^{-1} (pristine) shifts to 1580 cm^{-1} (high doping). One observes also that the whole spectrum is very similar to that of the undoped material . It seems obvious that the observed Raman spectrum must arise from the fraction of the material in which chains are undoped and have a shorter effective conjugation length (larger $F_{\mathcal{R}}$). It is known that doping attacks first chains with longer effective conjugation length since the ionisation potential is lower. The observed scattering thus comes from chains with shorter effective conjugation length or from conjugationally disordered domains that cannot be attacked or reached by the dopant.

8.4) Characterisation of PPy.

From this spectroscopic analysis we learn that the very strong Raman line ν_1 near 1570 cm^{-1} arises from $\mathcal{R}$ (i.e. from the collective motion which uniquely implies the delocalisation of π electrons along the chain or, in orher words, from the collective motion which modulates the energy gap). ν_1 is a probe for judging delocalisation length which can be numerically measured by the effective conjugation parameter $F_{\mathcal{R}}$. From the frequency and intensity dispersions one has also the way to understand the observed Raman spectrum of the doped material and to describe its chemical nature.

The fact that the doping induced IR spectrum fits the calculated dispersions $\nu/F_{\mathcal{R}}$ and $I/F_{\mathcal{R}}$ at $F_{\mathcal{R}}^{prist} > F_{\mathcal{R}}^{dop}$ is a proof that a "quinoid" electronic configuration is obtained. However, since both polarons and bipolarons generate quinoid configurations no distinction can be yet made spectroscopically between the two types of charge storages.

The Raman spectrum of the doped material shows the existence of a residual undoped substance with shorter effective conjugation length .

9) POLYTHIOPHENE

Polythiophene is a material which is the center of interest of many groups for its possible applications. It shows high conductivity when doped [110], large third order non linear optical properties [111] and it has been used as a semiconducting active material in the first attempts to build MISFET devices [112]. Of more relevant practical importance is the fact that its alkyl-derivatives are soluble thus opening new horizons for workable polyconjugated polymers.

PTh has been prepared by chemical or electrochemical methods

[113]. The pristine material has a band gap of 2.0-2,5 eV as seen from UV-Vis spectropscopy and is a good insulator [113]. Upon chemical or electrochemical doping PTh can reach an electrical conductivity of $\sigma \approx$ 50-100 S/cm [113]. Two interband transitions are observed in the electronic spectrum upon doping [114]; the charge carrier carries a spin for dopant concentration below ≈3% [115]. Upon doping new strong bands appear in the vibrational spectrum [114,116-118]. The most accepted view of the nature of the carrier is that spinless bipolarons are formed at concentrations >3% [119]. A yet unknown short lived molecular species has been generated by photoexcitation and its electronic and vibrational spectra have been recorded [117,120], but still wait for a systematic interpretation.

Because of its relevance in possible technological applications many oligomers of Th have been the center of great interest as models or as substitutes of PTh.

9.1) Experimental data.

The data from vibrational spectroscopy we use in this work come from various laboratories who have prepared their own samples. The methods of preparation and/or the chemical characteristics of PTh are not always reported in the published papers. The infrared and Raman have spectra of a few oligomers (Th_2, Th_3, Th_4 and Th_6) (fig. 6) have been reported and studied by Akimoto et al.[121], by Furukawa et al.[122] and by Harada [123]. We do not consider here the many spectra published by organic chemists for analytical and diagnostic purposes not accompanied by any technical spectroscopic analysis.

Raman spectra at λ_{exc} = 514.5 nm of PTh, deuteropolythiophene (DPth) and ^{13}C-PTh are reported by Furukawa et al.[122] (fig.6). Raman spectra of PTh at different λ_{exc} (651-5 , 514.5, 457.9 and 350.8 nm) are reported by Vardeny et al. [120]). The Raman spectrum of PTh shows the following lines : ≈1495 cm $^{-1}$ as a weak shoulder of a strong line at 1460 cm^{-1}; 1220 cm^{-1}), 1150-1180 cm^{-1} cluster), 1045 cm^{-1} (sharp, medium) near 700 cm^{-1} (group of medium weak lines).The first and second lines show frequency and intensity dispersion with λ_{exc} from 650 to 350.8 nm, being stronger at λ_{exc} =350.8 nm. Intensity and frequency dispersions are also reported by Steigmeir et al.[124] for the Raman spectrum of Poly (3-methylthiophene) (P3MeTh) for the strong Raman line from ≈1470 cm^{-1} to 1440 cm^{-1} with λ_{exc} from 441.6 to 632.8 nm. These authors have overlooked the existence of the higher frequency weak shoulder near 1520 cm^{-1} which shows dispersion from ≈1500 cm^{-1} (λ_{exc} =632.8 nm.) to 1522 cm^{-1} (λ_{exc} = 441.6 nm).

PTh has been doped chemically and electrochemically .The cleanest spectra are obtained with spectroelectrochemical reflection methods using ClO_4^- as counterion [125]. Since data are reported as difference spectra the doping induced spectrum is clearly exhibited since the unreacted background has been subtracted (fig.7). Infrared spectra with chemical doping have been presented by Cao et al.[126]; the authors of ref.126 have doped also the oligomers .

The doping induced spectrum shows two very strong lines at 1342-

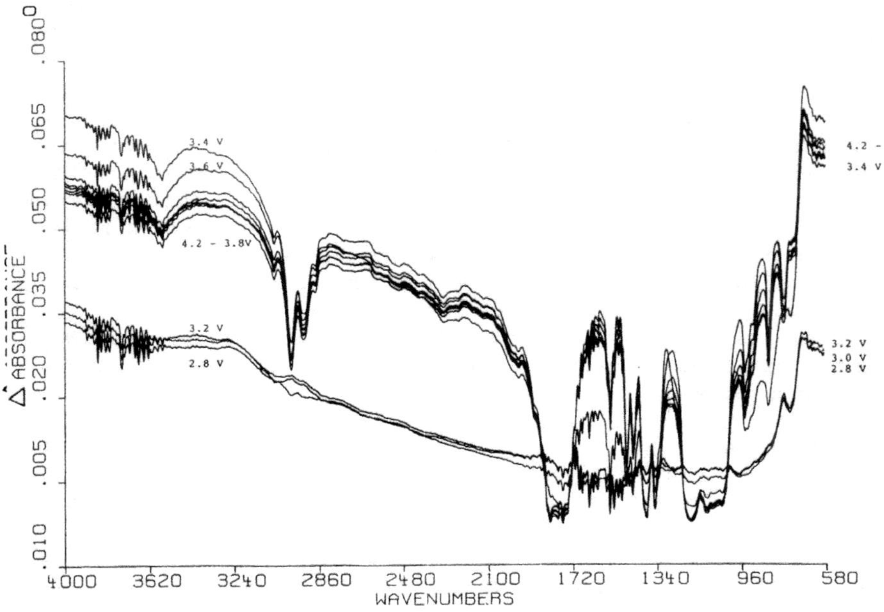

Fig. 7 Infrared spectrum of doped Polythiophene obtained as difference spectrum in experiments of "in situ" spectroelectrochemistry (from ref. 125).

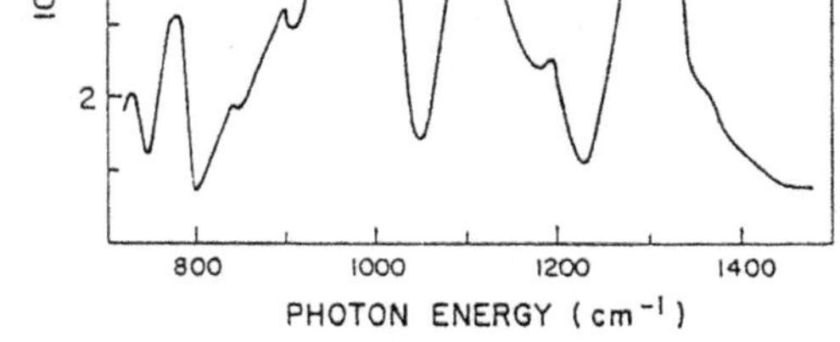

Fig.8 Photoinduced infrared spectrum of Polythiophene (from. ref.120).

1330 and 1034-1030 cm^{-1}; two strong lines are observed at 1203-1200 cm^{-1} and weak lines at 841,729,679 and 644 cm^{-1} (fig.7). For the doped oligomers a spectroscopic pattern similar to that of PTh is observed only for Th_5, while the shorter oligomers exhibit difficult spectroscopic patterns.

The observation in the spectra of doped PTh [118,121,122] of the out-of-plane C-H deformation of the pristine material (Γ_{C-H} mode near 786 cm^{-1}) allows a comparison between the doping induced intensity and the intensity of the "pristine" part of the material, assuming that the relative concentration of doped and pristine material are similar. The absorption coefficients of the modes of the doped molecule are not greatly enhanced with respect to that of the "unperturbed" Γ_{C-H} modes.

The photoinduced infrared spectra obtained with a laser line of 2.5 eV (fig.8) are described by Shaffer et al. [117], by Vardeny et al. [120] and by Kim et al.[127]; the results of these three experiments substantially agree even if frequencies and intensities do not coincide since they originate from different samples .

From the above analysis it becomes apparent that different samples may have different effective conjugation length which shows up in the spectroscopic features discussed in this paper.

9.2) The structure of PTh and of its oligomers.

The structural problem of PTh is quite complex and controversial both in the experimental and theoretical aspects. Two different schools of thought exist at present and need to be mentioned. We neglect here the discussion on the structure of alkylsubstituted polythiophenes and **we focus only on the oligomers and polymers of unsubstituted Th.**

The main problem lies in the search for unquestionable experimental evidence of a proposed structure (or a multiplicity of structures) and the weight to be given to quantum chemical calculations.

Various structural issues must be addressed. For the molecules (oligomers or polymers) "in vacuo" the main structural variable is the conformation about the C-C (α-α') inter-ring bond.

The π electron delocalization strongly depends on the torsional angle Θ between two adjacent Th rings. The conformation can take up values ranging from Θ=180 (trans, anti) to Θ=0° (cis, syn). At Θ=180° the Th rings are fully coplanar while at 180<Θ<0° the molecule may form some kind of helix.

Let us first quote what is considered the presently available experimental evidence on the structure of the oligomers and polymers of Thiophene.

From X ray diffraction studies the evidence never questioned in printed papers is the following: i) solid Th_2 is trans-planar (anti) at -140°C [128], ii) solid Th_3 is -anti with the Th rings twisted by 6° [129] and iii) PTh is fully coplanar [130].

From electron diffraction Th_2 in the gas phase is twisted with $\Theta \approx 34°$ [131].

The electronic spectra of both Th_2 and Th_3 in the solid and solution states have been discussed in ref.107. The blue shift observed from solid to solution is attributed to the opening of the HOMO-LUMO gap because of the decrease of inter-ring π delocalisation due to conformational changes from the so called $\Sigma°$ structure (coplanar for Th_2 and Th_3) to a more twisted Σ_Θ structure. The very high resolution electronic spectra of Th_3 at 4°K presented and discussed by Birnbaum et al. are consistent with a coplanar structure [132].

As to the electronic spectrum of solid PTh we think it may not be used for conformational studies and cannot provide an unequivocal evidence since it strongly depends on the effective conjugation length of the polymer molecule.

On the other hand the french school has found experimental indications from X ray scattering that in the small crystalline fraction Poly-3 Methyl Thiophene (P3MeTh) is helical and extrapolate this conclusion also to PTh [133]. The fact that both PTh and PMeTh absorb near 480 nm is taken by these authors as evidence that also PTh is helical [133].

The Infrared and Raman spectra of Th_2 and Th_3 recently discussed clearly show the conformational transition from $\Sigma°$ to Σ_Θ from the solid to solution [134]. Whether in solution there exists a thermodynamic equilibrium between these two structures still needs to be carefully studied by infrared/Raman spectroscopy.

The structural problem of the oligomers in the solid state is further complicated by the appearance of extra features in their electronic spectra. Starting from Th_2 the electronic spectra of the solid oligomers show a multiplicity of shoulders with decreasing intensity on the lower energy side of the main π-π* transition.

The interpretation of the observed "multiplicity of bands" or of "the fine structure" in the visible spectra of solid oligomers finds two groups of believers.

The french school takes the fine structure as due to priviledged molecular conformers obtained by rotation about the inter-ring C-C bonds [135]. The same authors state that to every shoulder would correspond a rotamer,thus controlling the inter ring conjugation . Such rotamers should be formed during the preparation of the solid sample [135]. A similar interpretation is suggested also by Havinga et al.[136]. These authors have studied the electronic spectra of various unsubstituted and alkylsubstituted oligomers in solution and dispersed in a matrix of Polymethylmethacrylate (PMMA).

A second group of authors ascribes the fine structure to "phonon side bands" i.e. to a "vibronic structure". The observation of vibronic structures in the electronic spectra of organic materials is a common fact. Several case studies in the field of polyconjugated materials are discussed by Hudson et al. [137]. The group at Linköping has carried out extensive works on the electronic spectra of alkylsubstituted PTh and constantly have observed the "additional fine structure" on the lower energy side of the main HOMO-LUMO inter-band transition. In their discussion [138] they notice that the photoluminescence spectrum of PTh shows four side bands 0.182 eV apart with the same separation as in the absorption spectrum of P3MeTh. The 0.182 eV (1460 cm^{-1})band separation corresponds to the phonon observed in the

Resonance Raman spectrum of PTh and of P3MeTh empirically ascribed to a ring stretching mode [138]; this assignment will be re-discussed later in this article.

A final decision which may remove such dichotomy in the structural studies of oligomers and polymers of Th from their electronic spectrum requires further works with independent techniques.

We think important to quote the recent paper by the school of Kohler [132] which studies the high resolution fluorescence excitation and emission spectra of 2,2':5'.2"-terthiophene as solid solutions in n-decane at 77,10 and at 4.2 °K. These authors observe unquestionable vibronic structures with the strongest vibrational modes in emission at 209,324,694,1470 and 1540 cm^{-1}. In fluorescence excitation the corresponding modes are at 207,347, 679,1220 and 1561 cm^{-1}. In addition they observe four very similar well resolved band origins (inhomogeneous bandwidth approximately 5 cm^{-1} FWHM) at 24 896 ,24 827 , 24 835 and 24 879 cm^{-1} which they assign to the site/conformers of transplanar terthiophene. The perturbing effect of the n-decane matrix generate such site/conformers.

The main question which remains to be solved is whether the perturbing effects induced by solidification or by a PMMA matrix are so large as to generate conformational isomers (as implied by Fichou et al. [135] or by Havinga et al. [136) or may be as small as the site/conformers observed by Birnbaum et al [132].

In the definition of the inter-ring torsional angle two conventions are possible and are variously adopted by different authors. The trans and cis conformations can be described with $\Theta=0°$ and $\Theta=180°$ respectively or viceversa as $\Theta=180°$ and $\Theta=0°$ respectively. In this work we take $\Theta= 180°$ and $\Theta=0°$ for trans and cis conformation respectively.

The theoretical calculations of molecular conformation by MO calculations rely on the capability of quantum chemical methods to account properly, for the molecule "in vacuo", for steric repulsions (which favour $\Theta=90°$) and for π electron delocalisation (which favours $\Theta=0°$ or 180°). For the molecule organized in the crystal intermolecular packing forces may force the molécule to take values of Θ tending towards $\Theta=0°$or 180°, i.e. to a coplanar structure. Calculations for the system in three dimensions are not available.

For the molecule "in vacuo" various kinds of M.O. calculations have been made by many authors on oligomers and are discussed in ref.107 in terms of their capability to reproduce the experimental data.

The most recent "ab initio" MO calculations with a 3-21G basis set presented by Brédas et al. on Th_2 "in vacuo" predict a fully coplanar (trans , $\Theta=180°$) structure [139]. On the opposite site very recent MNDO calculations with full geometry minimisation on Th_2 and Th_3 suggest the existence " in vacuo" of a twisted structure with a torsional angle $\Theta\approx120°$(or 60° in the other convention) [107]. The experimental value of $\Theta \approx 34°$ (from electron diffraction) shows that neither "ab initio" nor semiempirical MNDO methods give the proper prediction of the conformation of Th_2. On the other hand the "improved" MNDO method called AM1 give Θ=180°[140].

Recently Cai et al. have re-analyzed the same problem [108]; these authors are fully aware of the inadequacy of MNDO to predict conformations in conjugated systems and accept as a compromise first to calculate bond lengths and bond angles by MNDO followed by calculation by EHT for the conformational energy. We notice that with such a procedure the authors do not account for the necessary relaxation of the molecular geometry when π delocalisation changes with Θ. The same authors judge the validity of their procedure by comparing the rotational barriers (Kcal/mol) calculated with EHT and STO-3G with the experimental values for a few aromatic conjugated systems such as C_6H_5-NO (6.6;5.42 ;4.7), C_6H_5-CHO (5.8; 4.89;2.9) and Th_2 (4.5;4.68; 5.0). The experimental value for Th_2 is not for the molecule in the gas-phase but in an oriented liquid crystal phase [141] in which the molecule can hardly twist (see also ref.107).

The calculations by Cai et al. on Th_2 [108] give Θ=180° with the trans form more stable than the cis form by 0.33 kcal/mol. The coexistence of the two forms in solution is predicted and needs experimental verification.

The possible coexistence of two stable conformations in Th_2 carries as consequence the possibility of existence of two stable forms of the polymer "in vacuo". Cai et al have calculated the torsional potential energy of PTh with MNDO and EHT "crystal" methods and locate the most stable minimum at Θ=180 (anti-trans planar) and a second minimum at Θ= 36° (quasi cis) which describes a helix with a radius of approximately 7.8 A. The planar structure is more stable than the coil by about 1.6 kcal/mol [108].

The problem, not yet addressed by any author,is how to perform calculations for the system in three dimensions in order to predict whether twisted conformational isomers of oligomers and helical structures of the polymer can persist also in the solid state.

9.3) Vibrational analysis.

For the classical vibrational analysis (which does not yet consider $\mathcal{R}$-dependent modes) we know that the conformational dependence of the IR and Raman spectra (frequencies and intensities) of polyaromatic molecules linked by "single" C-C bonds should be very small for small deviation out of planarity. As fully discussed in other simple cases [142,143] in principle the lowering of the symmetry produced by the conformational twisting should give IR /Raman activity originally silent for the fully coplanar higher symmetry system. Slight frequency shifts of a few torsionally dependent normal modes can also be predicted and calculated.

Most of the authors have proceeded in their calculations with the assumption that the molecules are trans and coplanar.

MO calculations on the oligomers indicate that in the full coplanar trans structure intramolecular delocalisation of π electron takes place; it has to be noticed, however, that the calculated geometrical parameters (bond orders and bond lengths) do not depend much on chain length (=delocalisation length for the oligomers in the perfect coplanar structure), while the calculated ionisation potentials and the Energy gap are more sensitive , both decreasing with increasing conjugation length [144,145].

For sake of completeness in the discussion we consider here the selection rules for the polymer i) the fully coplanar, anti conformer and ii) the helical conformer. Group theory predictions may help in the search of the structures predicted by theory.

i) **Coplanar structure:** the polymer as 1-d crystal belongs to the D_{2h} point group; the repeat unit contains two Th rings and its (14x3)-4=38 normal modes are distributed in the various species as follows:

in-plane modes: $7A_g$(RR,pol,$\mathfrak{A}$-dependent) + $7B_{1g}$(R,depol) + $6B_{2u}$(IR) + $6B_{3u}$(IR,//)

out-of-plane modes: 3 B_{2g} (R,depol)+ 3 B_{3g} (R,depol) +3 A_u (inactive) + 3 B_{1u} (IR⊥).

The directions of the transition moments as indicated above refer to polymer chains oriented along the chain axis by suitable stretching.

ii) **Helical Polymers** . The selection rules for helical PTh have never been reported before and are given here. We apply the standard methods for non symmorphic line groups [146-148; for a general discussion see refs 4-7]. Let m be the number of chemical repeat units contained in the translational repeat unit of the polymer considered as 1-d crystal and z be the number of turns of the helix within the crystallographic repeat unit. The whole chain is generated by application of the rototranslational operator $\Psi(\beta;l)$ (with $\beta= 2\pi z/m$ and l a fraction of the repeat distance d along the axis of the helix). Geometric relationship have been worked out between d,n and z [4-7]. In the case of the hypothetical PTh helical chain these parameters are not determined and in working out the vibrational selection rules we do not attempt to fix any of these (i.e. we do not fix the precise point group), but work out only general rules to be applicable to a general type of helix.

Since 7 are the atoms in each chemical repeat unit we expect 21 branches in the ν (ϕ) phonon dispersion relation (ϕ is the phase difference between adjacent chemical repeat units [4-7]). For a general helix of PTh the structure of the irreducible representation for the **q**=0 phonons is the following :

Γ_{vib} = 10 A_1 (Raman, RR,$\mathfrak{A}$-dependent) + 9 A_2 (IR,//) + 20 E_1 (R; IR ⊥) + 21 E_2 (R).

As a general rule the Raman active doubly degenerate phonons of E_2 species have never been observed in any polymer and this should be particularly the case of polyconjugated polymers. The attention should be focussed at the 10 A_1 modes, $\mathcal{R}$-dependent, some of which should be strong in the Raman spectrum, according to the theories presented in this paper. The A_1 modes should coincide neither with A_2 nor with the E_1 modes. IR, Raman frequency coincidences should occur for the E_1 species. E_1 are predicted to be extremely weak in the Raman spectra, but could be recognized in the IR spectrum of a stretch-oriented sample since they should show a $\perp$ dichroism when the spectra are recorded in polarized light. On the contrary the $9A_2$ modes should show // dichroism.

The distinction between coplanar and helical structure in principle could be made only from the IR spectrum of a stretch oriented sample for which selection rules are different. The planar molecule should give rise to 6 // and 3 $\perp$ bands ,while 9 // and 20 $\perp$ bands are predicted for the helical structure.

The search for spectroscopic evidence for a helical structure by matching observed spectra and group theory predictions must take into account the following situations:
From the experiments: i) IR spectra of stretch oriented PTh are required . The first attempts to obtain oriented PTh are appearing in the recent literature (see, for instance, ref. 149), but are not yet usable.
From theory ii) Specific $\mathcal{R}$-dependent Raman activity peculiar of polyconjugated molecules (just discussed in this work) must be located , iii) the shape of the phonon dispersion curves should be known. Indeed the appearance of distinct frequencies for A_1/A_2 and E_1 modes is possible only if phonon branches are significantly dispersed, i.e. if kinetic and potential intramolecular coupling occurs between neighbouring chemical units. Dispersion curves for the in-plane phonons have been calculated for a PTh chain in trans planar conformation (D_{2h} symmetry) with two different approaches. The group of Milano has calculated $\nu(q)$ with a force field which takes into account the effective coordinate $\mathcal{R}$ and the effective conjugation force constant $F_{\mathcal{R}}$ [83] . On the other hand Faulques et al. [150] have adopted a force field in which force constants were empirically transferred from model molecules with similar structures (aromatic and quinoid). As seen in fig.9 most of the phonon branches do not show much dispersion with the exception of the few lowest phonons branches mainly involving coupled skeletal stretching, bending and torsions.

The experimental data presently available do not yet allow to take any decision on this important issue. More experimental data on stretch-oriented PTh are necessary.

The interpretation of the spectrum is carried out here on the basis of a coplanar structure of PTh. Application of ECC theory to PTh helps in the interpretation of the spectrum [83,84].

Since the Raman spectra are obtained in Resonance conditions, when C-H stretchings are neglected, we expect to observe in the fre-

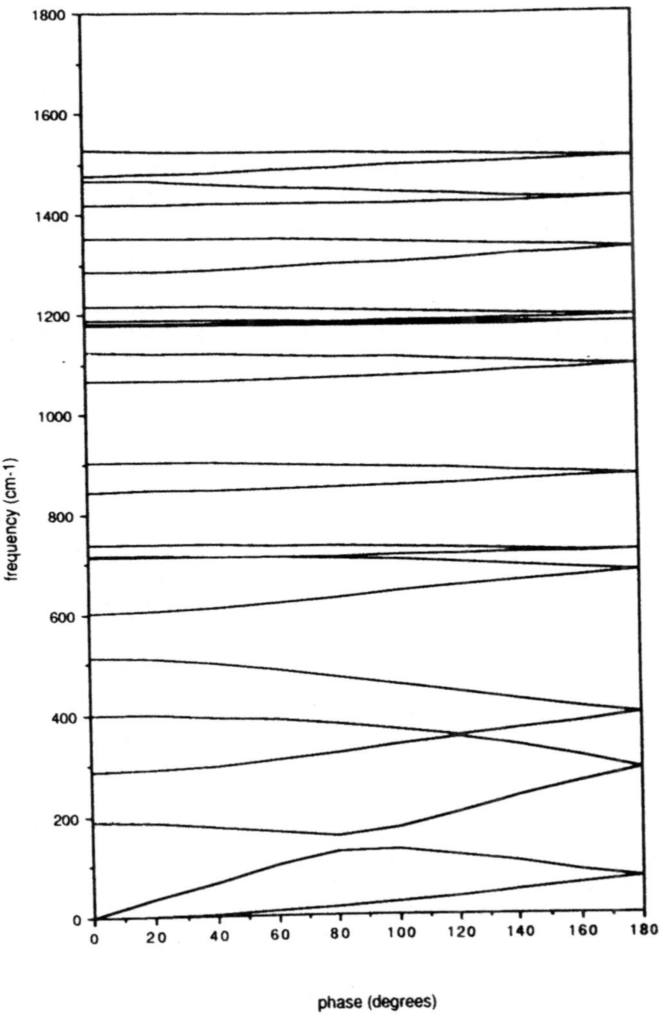

Fig. 9 Calculated Phonon Dispersion Curves for trans-planar Polythiophene (from ref.83.)

quency range 1700-100 cm^{-1} both for oligomers and polymers practically only the resonantly enhanced TS in-plane modes, namely:
6 A_1 (Th_1), 12 A_g (Th_2), 11 A_1 (Th_3), 24 A_g (Th_4) and 36 A_g (Th_6) and 5 q=0 A_g phonons (PTh and fully D substituted isotopic derivatives).

According to ECC theory of the TS modes those which contain the $\mathcal{R}$ mode are selectively enhanced, their intensity being $\div|L_{\mathcal{R}}|^2$. Moreover these modes may show frequency and intensity dispersion with effective conjugation length. The Raman spectra of the oligomers (fig.6) at fixed λ_{exc} give immediately, and without any calculation, such dispersions. The observed weak line [121,122] near 1550 cm^{-1} shows dispersion (Th_2 1556, Th_3 1530, Th_4 1519 and Th_6 1507 cm^{-1}). The frequencies calculated [83, 84] from pure MNDO, without refinment, give the following dispersion : Th_2 1571, Th_3 1560, Th_4 1554 and Th_6 1549 cm^{-1}.

Comparison of experimental and MNDO calculated values for the range of frequency dispersion ($\Delta\tilde{\nu}_{obs}$ = 49 cm^{-1}, $\Delta\tilde{\nu}_{calc}$=22 cm^{-1}) shows the limitation of pure MNDO calculations in accounting for vibrational properties directly determined by π electron delocalisation.

The understanding of the dispersion of the other modes is better reached when full dynamical calculations are made. $\tilde{\nu}/F_{\mathcal{R}}$ (fig.10) and $I/F_{\mathcal{R}}$ (fig.11) dispersions have been calculated [83,84]. The valence force field for the calculation of the frequency and intensity dispersions has been derived with MNDO calculations on smaller oligomers, but especially on Th_4 and further refined on the A_g frequencies of PTh, PDTh and ^{13}C-PTh [83,84]. We accept as basis of our dynamical calculations that these molecules are coplanar in the solid.

From fig.10 it is clear that at large $F_{\mathcal{R}}$ values (i.e. strong π electron confinment =small delocalisation = short effective conjugation length)) the single, strong and isolated $\tilde{\nu}_1$ mode is predominantly $\mathcal{R}$. At lower values of $F_{\mathcal{R}}$ $\tilde{\nu}_1$ lowers and couples with and lends intensity to the other TS modes $\tilde{\nu}_2$ to $\tilde{\nu}_6$. The value (scaled) of $F_{\mathcal{R}}$ for Th_2= 5.62 mdyne/A . In going to the polymer the effective conjugation length increases and $F_{\mathcal{R}}$ should decrease.

Fitting of the experimental Raman spectrum of PTh with the calculated $\tilde{\nu}/F_{\mathcal{R}}$ and $I/F_{\mathcal{R}}$ yields $F_{\mathcal{R}}^{prist} \approx 5.3$ mdyne/A. It must be noticed that v_1 occurs as a very weak shoulder on the high frequency side of $\tilde{\nu}_2$ which shows dispersion only as a result of mechanical coupling with other modes. The intensity is mostly borrowed from the $\mathcal{R}$mode. This theoretical prediction is fully confirmed in the Raman spectra of the oligomers (fig.6).

The Raman dispersion (at fixed λ_{exc}) of the oligomers shows that π electron delocalisation along the oligomeric chain does take place. The observation that the Raman spectra of PTh and of P3MeTh show dispersion with λ_{exc} [124] is a proof that the samples contain a distribution of "effective conjugation lengths" [151].

It is important to point out that all the authors have overlooked the $\tilde{\nu}_1$ mode and have taken $\tilde{\nu}_2$ qualitatively as the C=C ring stretching mode and have drawn qualitative chemical conlusions. On the other hand Harada [123] has clearly and very bravely identified in the very complex high resolution spectrum of solid Th_6 the $\tilde{\nu}_1$ mode as the very weak feature at 1507 cm^{-1} in agreement with ECC theory.

ECC theory proves that ν_1 is a collective motion which modulates the delocalisation of π electrons. This information is relevant for the general understanding of the spectra and structure of other derivatives (see section 10).

Moving down towards smaller values of $F_{\mathcal{R}}$ along the $\nu/F_{\mathcal{R}}$ dispersion curves the doping induced, strong IR active and $\mathcal{R}$-containing modes are found at values of $F_{\mathcal{R}}^{dop}<F_{\mathcal{R}}^{prist}$. The observed doping induced IR spectra find a satisfactory fitting (frequencies and intensities) at $F_{\mathcal{R}}^{dop}$ in the range 3-3.5 mdyne/A. From figs. 10 and 11 we derive the frequencies of the calculated modes at 1308-1297 ,1082-1054 and 1048-988 cm^{-1} to be compared with the experimental ones 1308-1342 (strong) ,1124-1110 (strong) and 1034-992 cm^{-1} (strong). The fitting with the intensity is not too good,but owing to the known limitation of MNDO calculations for molecules containing heteroatoms the results can still be considered acceptable.

The infrared spectra of doped oligomers need to be accounted for even if new and better spectra would be required for a more detailed analysis and for safer structural conclusions. From fig.10 one notices that the ν_4 branch in the calculated $\nu/F_{\mathcal{R}}$ plot is extremely sensitive to small changes of $F_{\mathcal{R}}$, thus providing a sensitive probe for the extent and distance of conjugation along the polymer chain. The spectral pattern of doped PTh is practically identical to that observed for doped Th_5 [126], while for shorter oligomers the pattern quickly disappears. The observed value of ν_1,ν_2 and ν_3 for Th_5 are very close to that of doped PTh . It can be immediately concluded that i) the charge carrier is highly confined, ii) is not much dynamically and/or electronically coupled with the host lattice and iii) is short, namely of ≈ 5 effective thiophene units .

Next in the analysis are the photoinduced IR spectra. The pattern of the photoinduced spectrum is practically superimposable to the IR spectrum of the doped species, thus indicating that the photogenerated conjugational defect has a dimension of ≈ 5 "effective" units. The small differences in the frequencies reported in the literature may be ascribed to the chemical and /or structural inhomogeneity of the samples, i.e. to a different distribution of conjugation lengths.

9.4) Structural Characterisation of PTh.

From the spectroscopic analysis presented here we learn that:
i) the effective conjugation of the π electrons which can be measured from the Raman spectra in pristine materials is not larger than ≈6 units since the Raman spectrum of Th_6 is not much different from that of PTh. In other words, values of $F_{\mathcal{R}}$ < ≈5.3 mdyne/A for pristine PTh have not been observed even if $\nu/F_{\mathcal{R}}$ and $I/F_{\mathcal{R}}$ plots indicate a strong and strongly dispersed ν_2 mode which should be clearly observable. Such a result may originate from a confinment of the π electrons in short segments conjugationally decoupled because of geometrical distortions or because of chemical defects. It may also be due to an

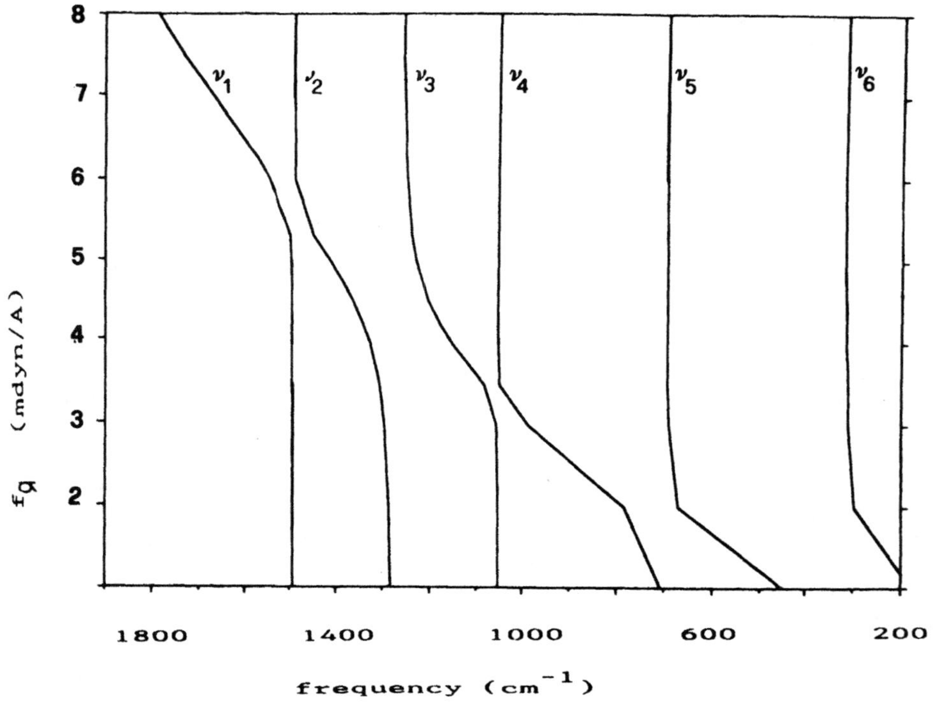

Fig.10 Calculated frequency dispersion $\nu/F_{\mathcal{R}}$ of the A_g modes of Polythiophene (from ref. 84).

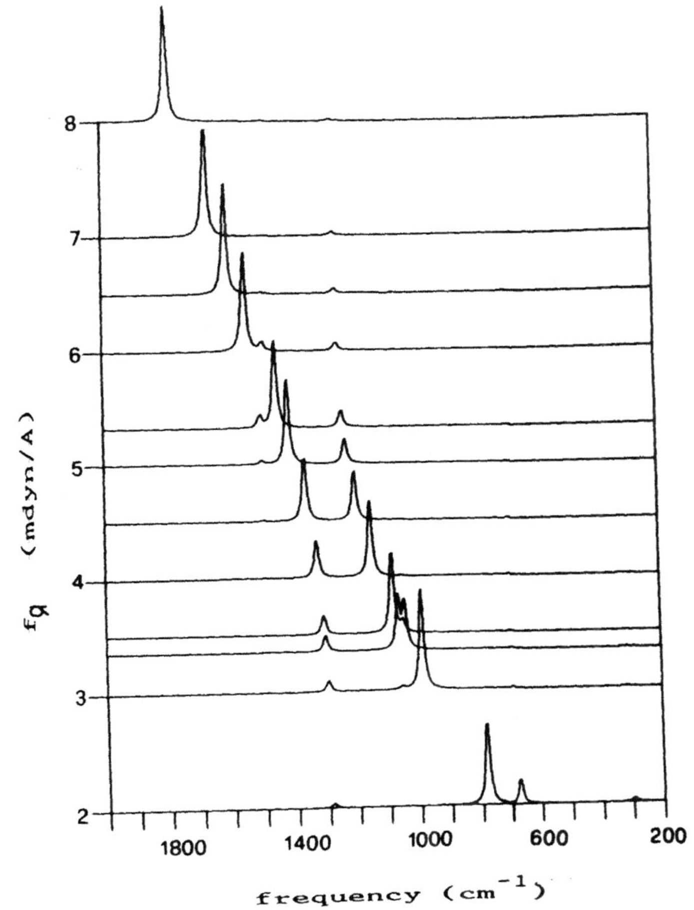

Fig.11 Calculated intensity dispersion $I/F_{\mathcal{R}}$ for the A_g modes of Polythiophene (from ref. 84).

intrinsic property of the electron delocalisation along the perfect lattice. The role played by the Sulfur atom in such a confinment needs to be further understood (see also ref. 132).

ii) Clear evidence is collected that the pristine samples studied by different groups consist of chains with a distribution of conjugation lengths. This evidence is found from the observed $\bar{\nu}/F_{\mathcal{R}}$ and $I/F_{\mathcal{R}}$ of the Raman spectra of pristine PTh. We wish to point out ,however, that the frequency and intensity dispersion of the photoexcited spectra form another clean and very sensitive probe for the chemical inhomogeneity of the samples.

iii) The charge carrier generated by doping or by photoexcitation has certainly a "quinoid" structure, but vibrational spectroscopy is not yet able to distinguish between polarons and bipolaron. The structure of the doped site is certainly coplanar. The size of the charge carrier is ≈ 5 thiophene units thus indicating a strong confinment of the conjugational defect. The fact that the spectra of doped samples obtained "in situ" during electrochemical oxidation are identical in frequencies at low or high concentration [125] is a further evidence that the conjugational defect is strongly confined and is not much coupled with the host lattice. This may be taken as a strong indication that doping is random and does not proceed with the formation of 1-d clusters or island.

iv) The change of the values of the effective conjugation parameter Fq from pristine (≈5.3 mdyne/A) to doped and photoexcited (≈3.2 mdyne/A) indicates a strong perturbation of the π electrons within the charge carrier. Such $\Delta F_{\mathcal{R}}$ = 2.1 mdyne/A is impressively large if compared with that which represents the change in delocalisation of the π electrons in going from pristine Th_2 to pristine Th_6 ($\Delta F_{\mathcal{R}}$ ≈ 0.3 mdyne/A).

10) POLYALKYLTHIOPHENES

In this section we do not wish to review the relatively large amount of experimental work carried out on Polyalkylthiophenes as soluble (hence workable) derivatives of PTh [152-153]. We focus mainly at vibrational spectroscopy of these materials in order to show the usefulness of ECC theory in the understanding of the relatively simple spectrum arising from systems with a very complex molecular structure. In carrying out such vibrational analysis one has to use also concepts of traditional spectroscopy of classical 1-d polymers with ordered (translationally regular) or disordered structure discussed in specialized publications [4-13,24-29]. We consider the spectroscopic work on Polyalkylthiophenes a rather complete guided tour in the study of the spectra of complex polyconjugated systems. The same method can be applied to other alkylsubstituted polyconjugated systems.

We consider the case of polyhexyl and polyoctyl thiophenes (PHTh and POTh) recently studied by our group [134,154a]. In addition to Th_2, Th_3 and PTh discussed in the previous sections the model compounds which are used in such analysis are n-hexane $CH_3(CH_2)_4CH_3$, (C6), n-octane $CH_3(CH_2)_6CH_3$, (C8), phenylhexane $C_6H_5-(CH_2)_5CH_3$, (PhC6), and phenyloctane $C_6H_5-(CH_2)_7CH_3$, (PhC8).

Samples were studied both in the solid and solution phases; solid samples were prepared as KBr pellets (PHTh) or as films from melt (PhC8) or cast from solution (PHTh and POTh). PHTh and POTh were dissolved in chloroform which is a "good solvent" for these substances and does not show solvatochromic effects. Raman spectra were recorded either with an Ar^+ laser at λ_{exc} =514 nm (with powers 30-200 mW) or a with a NdYAG laser at 1.06 μm (with power ≈100 mW).

The main object of such analysis is to use the vibrational IR and Raman spectra of the pristine polymers to understand the molecular structure of the systems and their evolution when the material is dissoved in chloroform or when the solid material is heated. It has been shown that the class of polyalkylthiophenes show reversible thermochromism [155] whose origin at the molecular level needed an explanation.

The problem we have to focus at is the understanding of the molecular flexibility of these systems ; conformational flexibility may occur topologically in two regions of the molecule ,namely the heteroatomatic backbone can rotate around the inter-ring C(α)-C(α') bonds and the long alkyl chain may take up either an all trans-structure or collapse into a "liquid like" conformationally distorted worm-like chain characterized by a statitical distribution [156] of gauche (G and G') and trans (T) conformations. The fact that these two flexibilities were independent or correlated needed investigation.

10.1) Structure, group theory and spectroscopic predictions.

Let us first consider each polymer as a perfect and infinite 1-d crystal . The regiospecificity of the chemical reaction may produce two kinds of chains depending on the "orientation" of the monomer when it is linked at the end of the polymer chain during the polimerisation reaction. Two kinds of "regular" "homopolymers" can be generated which we label as centrosymmetric (cs) and non -centrosymmetric (ncs) (fig.12). We take for granted in this analysis (see refs. 152,153 for a discussion on the chemistry) that the regiospecificity of the systems [157,158] is so high to be allowed to treat them as perfect and regular 1-d polymers.

10.1.1) Perfect planar structure.

Let us consider as example PHTh in the anti-coplanar-centrosymmetric structure with the alkyl side chains in the all transplanar conformation and lying in the plane of the Th rings. In such a model the translational repeat unit in the 1-d crystal contains two Th rings as in fig.13 Similar considerations can be made for POTh.

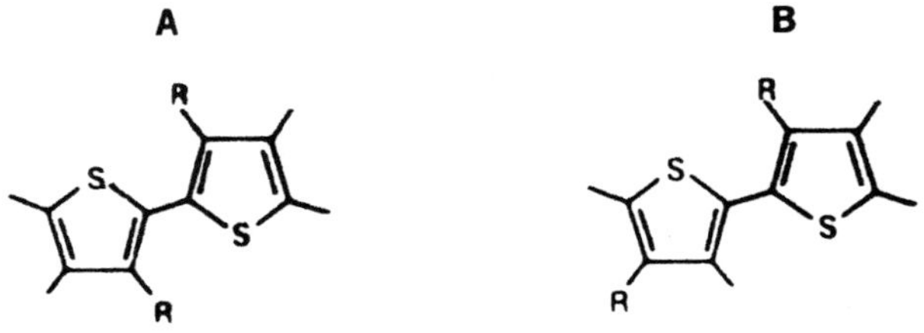

Fig.12 Sketch of the two possible structures of polyalkylthiophenes: a) centrosymmetric and b) non centrosymmetric structures.

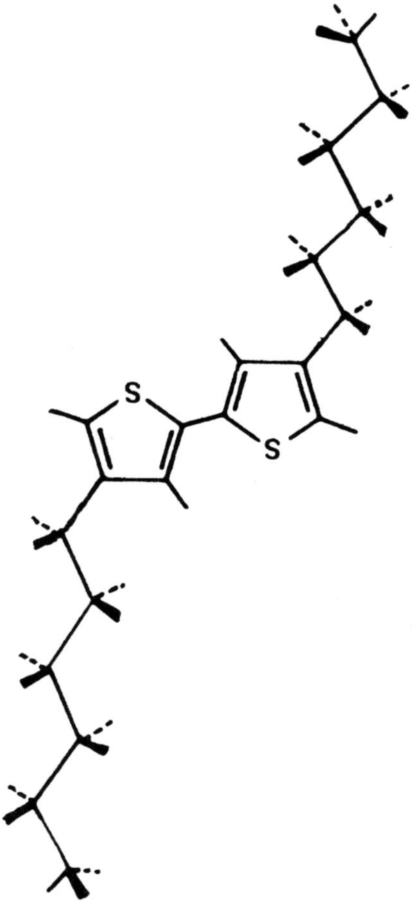

Fig.13 Translational repeat unit for polyhexylthiophene with centrosymmetric structure.

With such a structure the factor group of the 1-d crystal is isomorphous to the C_{2h} point group. Since N = 50 are the masses in the repeating unit, 3N-4 =146 are the optical **q**=0 phonons expected in the IR/Raman spectra and 150 are the phonon dispersion branches $\nu(\mathbf{q})$. The structure of the irreducible representation for **q**=0 phonons is the following:

Γ_{vib} (in-plane) = 44 A_g (RR,pol,λ-dependent) +42 B_u (IR)

Γ_{vib} (out-of-plane) = 30 B_g(R,dp) + 30 A_u(IR).

From the above group theory treatment we would expect to observe an extremely complicated spectrum ,but in reality the observed spectra are extremely simple. The following theoretical considerations provide the reasons for such simplifications.

i) One can define the group vibrations usually adopted in spectroscopic chemical diagnosys (e.g. CH_2 stretchings, CH_2 bendings,waggings etc. [159]) and also combine then in in-phase (g= gerade, Raman active) and out-of phase (u =ungerade,infrared active).

ii) Since the two hexyl side chains are far apart neither mechanical nor electronic intramolecular coupling (G and F coupling respectively ,see eqs.5 or 7) can take place between the side chains, thus no in-phase and out-of-phase splitting of the vibrational levels of the side chains is expected. Practically all the modes of the side chains are expected to be accidentally degenerate (i.e.they have the same frequencies); this implies that the corresponding phonon branches are practically **q** independent throughout the Brillouin zone.

iii) Phenomena of accidental degeneracy because of no (or extremely small) coupling are expected to occur also for many vibrations of the Th rings. This is certainly the case of the in-phase and out-phase C-H stretching and most probably of the in-plane and out-of-plane C-H deformation modes.

iv) The optical selection rules for the C_{2h} group activate the "exclusion principle" i.e. gerade (g) modes are only Raman active while ungerade (u) modes are only infrared active. Thus the in-phase (g) and out-of-phase (u) modes can only be seen in Raman or in Infrared respectively.

v) The normal modes of the side chains *in the all-trans structure* can be better identified when the band progressions of the $\mathbf{q}\neq0$ "phonons" are considered [5-7]. We expect, in principle, to find the well known and characteristic band progressions for the CH_2 wagging-twisting (very weak) (1400-1200 cm^{-1}) and CH_2 rocking modes (1100-700 cm^{-1}) [35]. The band progression of the C-C stretching modes should be observed in general in the Raman [35], but it will not be observed in the case of polyconjugated systems for the reasons discussed below.

If the anti-coplanar -non centrosymmetric structure of PHTh is considered the point group is C_{2v} and the structure of the irreducible representation is the following:

Γ_{vib} (in-plane) = 43 A_1(RR, $\mathcal{R}$dep,pol,IR) + 43 B_1(R.depol IR)

Γ_{vib}(out-of-plane) = 30 A_2(R,dp) +,30 B_2 (Rdp.IR)

In principle the spectra should be much more complex than for the cs structure since all modes are IR and Raman active due to the fact that the lack of center of symmetry has removed the distinction between g/u modes. In practice the simplifications as in ii) through v) hold in the same way.

10.1.2) Conformationally distorted structure.

In a fully distorted system no selection rules hold and in priciple every vibrational transition is allowed both in IR and Raman. The problem of the spectroscopic manifestations of a conformational distorsion of the polythiophene backbone about the inter-ring C-C bond will be discussed below and relies mostly on ECC theory. When the side chains collapse from a rigid all trans structure into a conformationaly distorted structure the concepts as in i) hold with no distinction between u/g levels, ii) and iii) are still valid, while iv) disappears.

The spectroscopic features as in v) disappear and are replaced by broad bands due to the activation in the IR and Raman of the whole density of vibrational states $g(\nu)$, dipole weighted (in IR) or polarizability weighted (in the Raman), from which localized or pseudolocalized modes emerge and become useful spectroscopic markers of disorder [4-13,160].

It is known [9,161,162] that a few modes become topologically localized at the site of specific and energetically favoured conformational defects and give rise to specific absorptions in infrared (e.g. GTG' 1362 and 1306 cm^{-1}, GG 1345 cm^{-1}, end-TG 1355 cm^{-1}).

10.2) EEC theory.

While no other theoretical simplifications can be predicted for the infrared spectra of the polymers the Raman spectrum is expected to be drastically simplified according to the theory of resonance Raman and of ECC.

As usually, we focus at the 44 q=0 phonons of A_g species, all in principle Raman active, for the cs model of PHTh when recorded in resonance or preresonance conditions.

If frequency dispersions $\nu/F_{\mathcal{R}}$ are considered (see fig.14) many branches turn out to be fully $F_{\mathcal{R}}$ independent since coupling with the $\mathcal{R}$ mode is only limited to a few ring modes of the chain backbone. We then collect the Raman active A_g modes of the side chains (8 C-H stretchings, 5 CH_2 waggings, 5 CH_2 twistings, 2 methyl deformations, 6 C-C stretching and 5 CCC angle deformations) and treat them separately. We are left with 13 modes more directly connected with the vibrations of the PTh skeleton; these modes may be coupled with the $\mathcal{R}$ coordinate and may show frequency and intensity dispersion with $F_{\mathcal{R}}$.

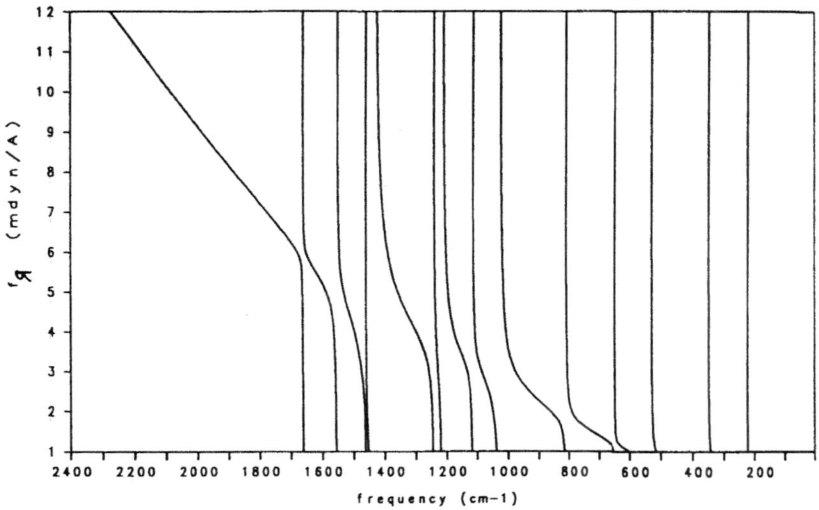

Fig.14 Schematic frequency dispersion with f_R of the totally symmetric mode calculated for a simplified model of polymonoalkylpyrrole. In a first approximation we take the curves for poly-Me-pyrrole as a qualitative indication of what can happen with polymonoalkylthiophene.

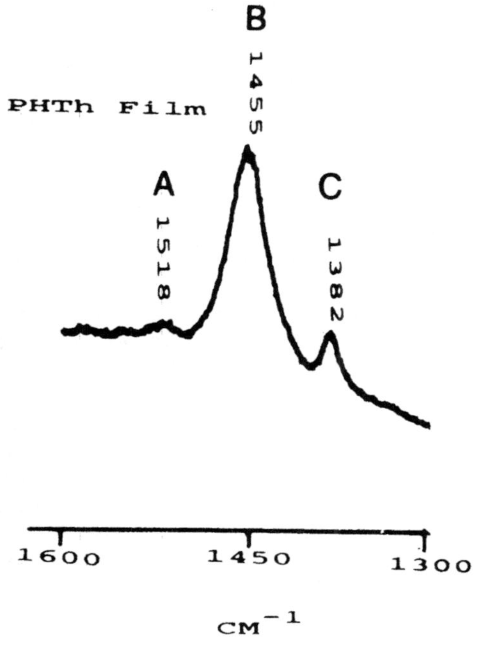

Fig.15 Raman spectrum of Poly hexylthiophene in the solid phase (λ_{exc} = 515 nm) (from ref.154).

In conclusion for the Raman spectrum we predict :
i) Since the spectrum is in resonance conditions B_g modes will not be observed; 31 of the A_g modes will not be observed since they are not involved in the vibrational transitions resonantly enhanced. Of the remaining 13 modes those preferentially enhanced are the ones which contain a sizeable contribution by the $\mathcal{R}$ cordinate, their intensity being proportional to $|L_{\mathcal{R}}|^2$.

From the $\nu/F_{\mathcal{R}}$ plot of fig. 14 (taken as an indication since it has been calculated for alkylpyrroles) three bands with a sizeable intensity are expected in the 1600-1300 cm^{-1} range and indeed the Raman spectrum of PHTh film shows two week lines at 1518 (A) and 1382 cm^{-1} (C) and a very strong one at 1455 cm^{-1} (figs. 15 and 16). All the other Raman A_g lines are observed to be vanishingly weak, as predicted.

10.3) Spectroscopy and structure of Polyalkylthiophenes.

ECC predictions are valid both for the perfect 1-d crystal or for the disordered system . However, if the effective delocalisation of the π electrons changes ,according to ECC, we expect both frequency and intensity dispersions.

Changes of effective delocalisation can occur in two ways, namely by the changes of conjugation length because of changes of chain length or by changing the conformation of the thiophene rings about the inter ring C-C bonds. We have recorded [135,154] the Raman spectra of the same solid sample of PHTh with λ_{exc} = 514 nm (fig. 15) and λ_{exc}=1060 nm. (fig.16) and observe that the triplet of lines at 1518 (w) 1455 (s) and 1382 cm^{-1} for λ_{exc} = 514 nm shifts to 1510 (w), 1448 (s) and 1380 cm^{-1} (w) at λ_{exc} 1060 nm. The observed dispersion proves that the sample consists of chains with different conjugation length ; with λ_{exc} = 514 the "shorter" chains are excited while the "longer" are selectively excited with λ_{exc} 1060 nm. It has to be stressed that the observed distribution of effective conjugation lengths may arise either from chains with the same geometry but with different lengths or by a system with chains of the same chemical length but with a different distribution of conformations of the main chain.

The last statement is proven (using the same theoretical concepts) when the Raman spectra of PHTh (or POTh) in the solid and solution phases are compared (fig.17). In agreement with the electronic spectra of the same samples ,in solution the energy gap is larger because the molecule is distorted in a structure which we label as Σ_Θ (with the torsional angle Θ yet unknown) while in the solid the HOMO-LUMO transition is red shifted since the systems approaches coplanarity in a $\Sigma°$ structure with smaller E_g . The observed Raman frequency and intensity dispersion is the following: for the solid 1515, 1448 and 1380 cm^{-1}, for the solution 1517, 1475 cm^{-1} and ? (too weak), in agreement with the predictions by ECC theory.

Next is the use of the infrared spectrum for the identification of the conformation of the backbone (see ref.154. for more details). In a previous spectroscopic analysis on model compounds the out-of-

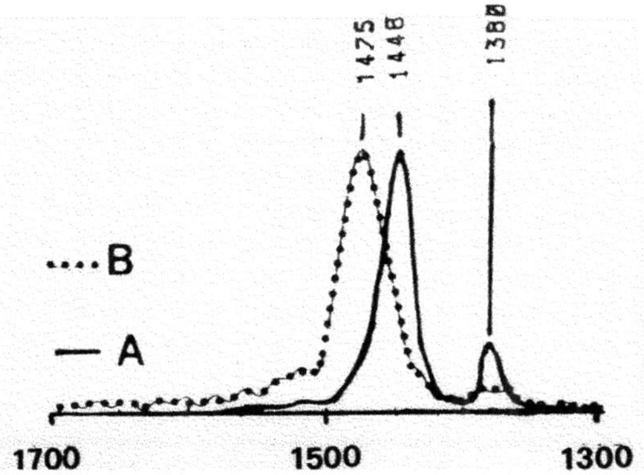

Fig.16 Raman spectrum of Polyhexylthiophene with λ_{exc} = 1060 nm in solid and solution (from ref. 154).

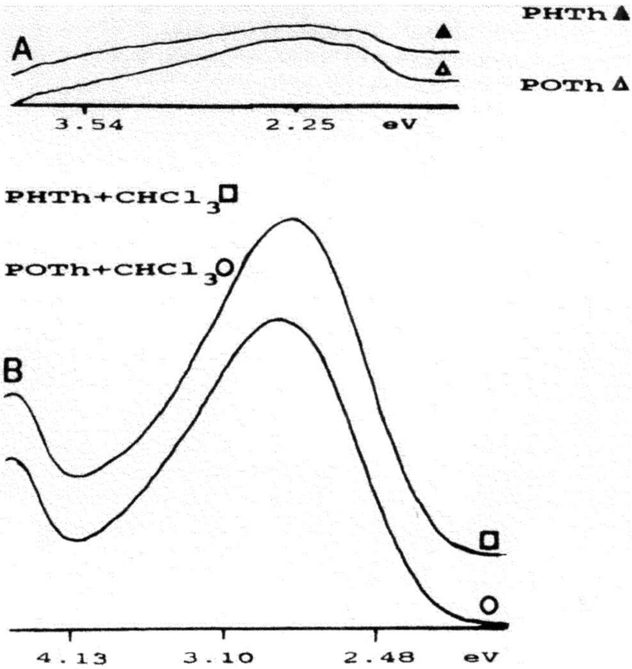

Fig.17 Electronic spectra of polyhexyl and polyoctylthiophene in the solid and in solution of $CHCl_3$ (from ref. 154).

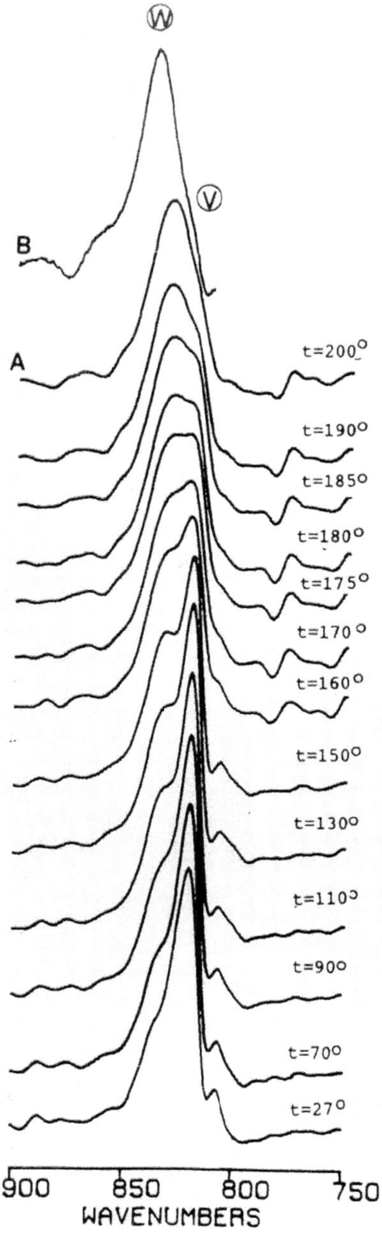

Fig.18 Temperature dependent infrared spectrum of the out-of-plane C-H deformation mode of solid polyhexylthiophene (A) compared with the spectrum of the same material in CCl_4 solution (B) (from ref. 154).

plane mode of the C-H in position β in the Th ring was identified as a very strong band C-H at 813 cm^{-1} for the solid samples (fig.18,band V). Upon heating or in solution the band shifts to 835 cm^{-1} (fig.18, band W). The growing of band W at the expenses of band V is a clear evidence of a temperature dependent equilibrium between two structural states,namely the backbone chain originally probably in a slightly twisted state $\Sigma°$ reaches the more twisted state Σ_{Θ} ,its thermal evolution being described by the temperature dependence of the relative intensities of bands W and V (fig. 19b). Moreover, the breadth of band V is a convolution of several conformations slightly distorted about $\Sigma°$. From the relative intensities of the V and W bands one derives that at room T the concentration of $\Sigma°$ and Σ_{Θ} is approximately 55% and 45% respectively (fig.19b).

10.4) Conformation of the alkyl chains.

Information on the conformation of the alkyl chains can be obtained only from the infrared spectrum since ,according to ECC theory, the Raman spectrum shows only a very few lines originating from the $\mathcal{A}$-dependent skeletal modes which involve π electron delocalisation.

A more complete analysis of the order/disorder phenomena in polyalkyl thiophenes is reported in ref 154. We limit our discussion here to a few spectroscopic manifestations relevant for this review article. We aim here at the flexibility of the side chains in relation to the flexibility of the chain backbone.

As mentioned before the existence of the band progressions due to the $q\neq0$ phonons of the alkyl chains is evidence that chains are fully coplanar ; when liquid-like structures are formed the broad g(ν) is mapped in the IR spectra from which localized characteristic defect states can be observed.

Let us take the spectra of the model molecules C6 (or C8) and PhC6 (or PhC8) in the solid (at -178°C) and melt states (+ 25°C). In the solid the alkyl chains are transplanar while in the liquid chains are conformationally coiled. In fig. 20a the IR spectrum of the solid clearly shows the band progression of CH_2 twisting modes, while in the spectrum of the liquid (fig.20b) the characteristic defect modes due to localized CH_2 waggings of twisted chains are clearly observed. Next we compare the spectra of the model compounds with those of PHTh or POTh in the solid or solution phases. Fig.21 shows that for PHTh and POTh in the solid state (KBr pellets or films) there exists a large fraction of the material in which the polymethylene chains are "liquid-like". The existence of a second phase with all trans alkyl chains is proven by the spectra of PHTh and POTh at liquid nitrogen temperature which show clearly the apperance of the $q\neq0$ band progressions in the rocking regions of the spectra (Fig.22).

The temperature dependence of the flexibility of the side chain can be followed with the infrared spectrum (fig.23) and is quantitatively measured from the intensities of the defect modes normalized to the intensity of the CH_3 umbrella defomation mode at 1375 cm^{-1} (fig.19a). From fig.19 it becomes apparent that the thermal evolution

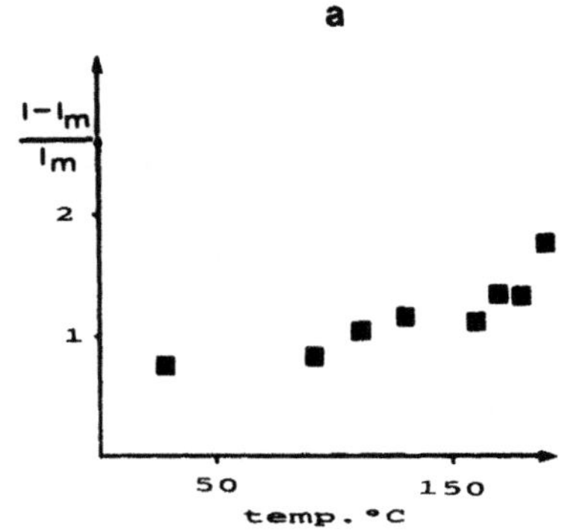

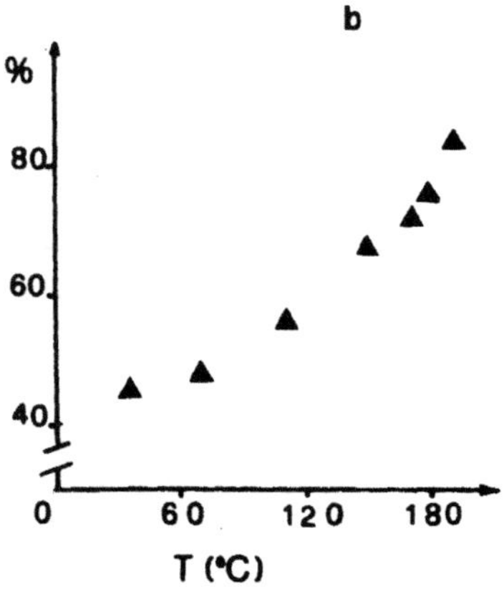

Fig.19 Evolution with temperature a) of the relative concentration of conformational defects in Polyoctylthiophene and b) of the fraction % of the thiophene rings in PHTh in the $\Sigma°$ and Σ_{Θ} structures (from. ref. 154)

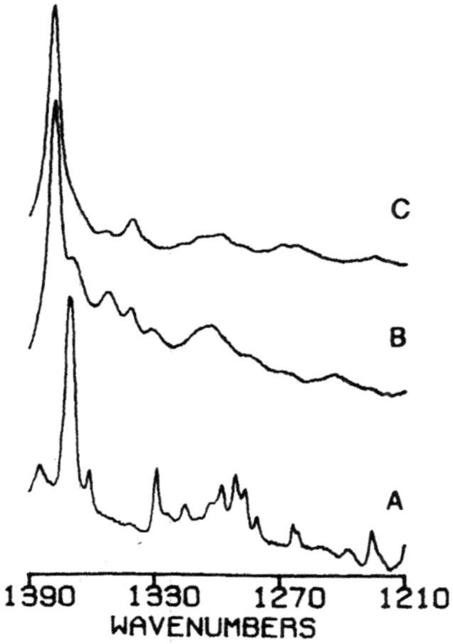

Fig. 20 Infrared spectrum in the CH_2 wagging range of phenyloctane in the solid (A) and liquid (B) phases compared with the spectrum of octane (from ref. 154).

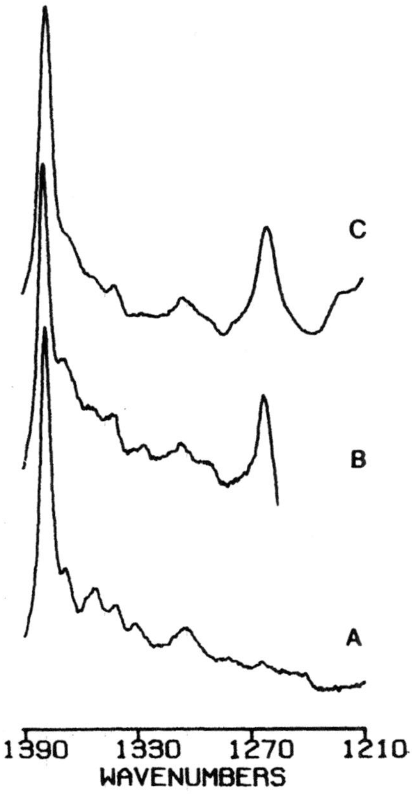

Fig.21 Infrared spectrum in the CH_2 wagging range of phenylhexane in the liquid phase (A) compared with the spectra of polyhexylthiophene in solution (B) and solid (phase) (from. ref. 154).

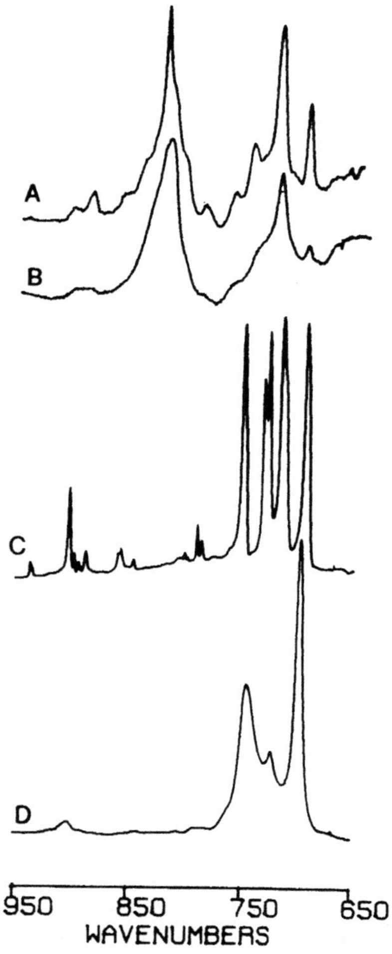

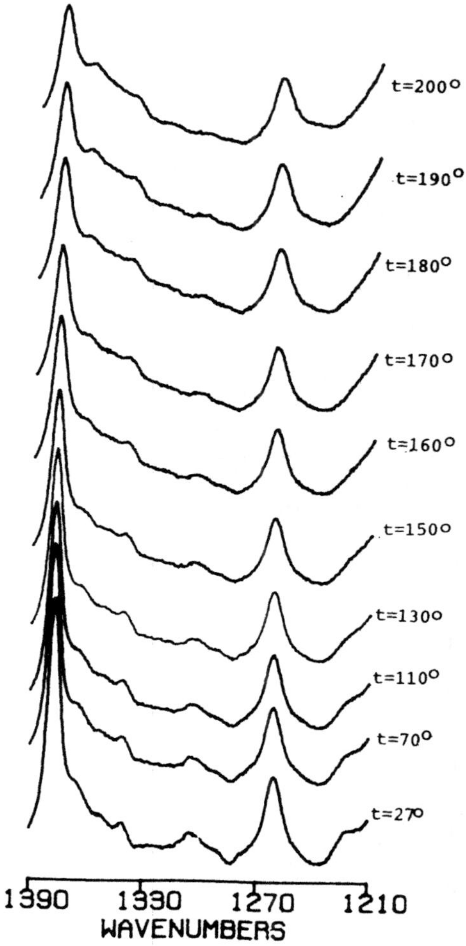

Fig.22 Infrared spectra in the CH_2 rocking range of polyoctylthiophene at T= -180°C (A) and at room temperature (B) compared with the spectra of phenyloctane in solid (T=-178°C) (C) and liquid (D) phase (from ref. 154).

Fig.23 Temperature dependent infrared spectrum of polyhexylthiophene in the CH_2 wagging range (from ref.154).

of the side chain disordering is similar to the disordering of the chain backbone (fig.19b), thus indicating that the two phenomena are correlated.

10.5) Structure and thermal behaviour of Polyalkylthiophenes.

The spectroscopic analysis through group theory and ECC theory accounts for the seemingly simplicity of the observed vibrational spectra of PHTh and POTh. The $\mathcal{A}$-dependent resonance Raman spectra of these materials provide selective information associated with the π electron delocalisation. At room temperature the two materials in the solid phase consist of a distribution of effective conjugation lengths which may be related to molecular length and/or molecular conformation. The mean effective conjugation length of the polymers at room T is approximately that of 4-5 thiophene units taken in a fully coplanar structure (and including localisation due to end effects). It has to be realized, however, that the measured effective conjugation length is an average between the various structures which have been shown to occur in the actual samples of PHTh and POTh.

The infrared spectra focus both at the side chains as well as at the chain backbone. It turns out that at room T the samples contain two fractions, one with a flatter chain backbone (structure $\Sigma°$) with a distribution of torsional angles about $\Sigma°$ and one with much more distorted conformation Σ_Θ. The $\Sigma°$ structure is estimated to have approximately, but not with certainty, $\Theta \approx 0°$ while Σ_Θ structure has $\Theta \approx 30°$.

The issue of coplanarity vs.distortion of the backbone chain of the thiophene rings ,already extensively discussed in section 9 for simple PTh, becomes more important here. In many other systems (polyesters, liquid crystals oligomers or polymers, polydiacetylenes, polysilanes etc) , with long alkyl side chains the polymethylene segments often tend to become parallel and trans planar and drive the conformation of the highly flexible backbone chain.

It is indeed not yet clear whether in the observed more ordered structure $\Sigma°$ the polythiophene backbone has reached full coplanarity . What it is unquestionable is that from solid to solution the HOMO -LUMO gap widens.

It has been shown [154b] that the systems at room temperature are not in a thermodynamical equilibrium but are under structural tension. Upon annealing and subsequent rapid cooling the concentration of $\Sigma°$ increases but relaxes back in the original $\approx$ 50/50 mixture after several days.

The IR spectra suggest that the conformational collapse of the side chains and the conformational distorsion of the chain backbone are correlated phenomena. The increased average volume occupied by the coiled chains forces the backbone to twist because of steric encumbrance. The evolution of the molecular flexibility with T goes through two main stages. For instance, for PHTh the coiling proceeds slowly in the temperature range 25-150°C; above $\approx$ 150-160°C a sudden transition occurs with a very fast conformational disordering before decomposition (≈ 200°C).

Attempts have been made to measure the activation energy of the coiling process from Arrhenius plots and have provided the values for the temperature ranges ≈25-150°C and ≈150°-to decomposition of ≈1 and ≈8 Kcal/mole respectively [154]). Since the activation energy for trans-gauche isomerisation is ≈0.5 Kcal/mole [156,163] it follows that the in the first temperature range the backbone flexibility is large (i.e. does not require much energy to rotate) in agreement with MO calculations. In the second step of the reversible cycle more complex distortions must take place which ,however,keep being reversible.

The above observations provide a clear explanation at the molecular level of the phenomenon of the reversible thermochromism reported by several authors and especially by the swedish school. Inganäs et al. have recently presented a synopsis of the changes of E_g with temperature of a series of polyalkylthiophenes [155]. The data are derived from T dependent electronic spectra. The vibrational data recently published [154] and reviewed here indicate that the laws of variation with T of the conformational changes of both the chain backbone and the side chains are similar to the T dependence of E_g. This correlation is obvious since the coiling of the backbone increases the torsional angle Θ, decreases conjugation ,increases E_g, generates a blue shift of the HOMO-LUMO transition, thus causing changes in colour.

A final observation relevant for the use of optical spectroscopy for these kinds of polyconjugated systems needs to be pointed out. The temperature dependent electronic spectra of these materials do not show any isosbestic point, thus indicating the existence of only one phase or of many phases. On the contrary the T dependent infrared spectrum (especially the C-H out of plane modes near 820 cm^{-1}) clearly indicates at least the existence of two main phases (fig. 18). The electronic spectrum reflects a collective phenomenon of π electron delocalisation along the polymer chain (just as the λ-dependent Raman spectrum). In a long and twisted structure the effective collective delocalisation path evolves smoothly with Θ and so does E_g. On the opposite the infrared spectrum (in the C-H out-of-plane motion) probes only very local phenomena since the dynamical interactions extend only to the first nearest neighbours. In other words, with the electronic (and Raman) spectra one measures the average overall effective conjugation while with the localized C-H modes one can count how many rings are locally distorted at a given Θ.

The problem yet unsolved for Polyalkylthiophenes is similar to that discussed in section 9 for PTh, namely one has to reconcile the view which would favour non planar or helical polyalkyl thiophenes originated by steric hindrance and that which seems to emerge from this discussion on the possibility of a quasi coplanar Σ° structure.

The experimental work by the french group [133] and the MO calculation by Cai et al. on Poly-3-methylthiophene [108] seem strongly to support the existence of some kind of helix. When the side alkyl become longer Van der Waals type interactions between the side chains certainly occur and drive the side chains to become trans planar as firmely shown by the observation in the infrared of a large fraction of material in which trans-planar alkyl segments exist. This observation implies side-chain crystallisation and partial or total intercalation of the side chains. If side chain interactions become dominant for longer alkyl segments they may dominate over steric repulsion localized at the site of inter rings contacts . In this way the PTh backbone may be forced to flatten into a coplanar or quasi coplanar Σ° conformation.

Side chain crystallisation or side chain interactions (with partial and/or full intercalation) occur and play a dominant role in the determination of the structure of many organic polymethylenic systems such as liquid crystal oligomers and polymers etc. X ray structure determinations on polyalkylthiophenes are presently being performed by various groups [164-165].

11) POLYPARAPHENYLENE VINYLENE

In the field of polyconjugated organic polymers which show large electrical conductivities and optical non linearities polyparaphenylene vinylene (PPV) is becoming the center of interest in basic and applied material science. This material in the pristine state can be easily prepared from a workable precursor which allows preparation of films or stretched samples, has good chemical stability and shows remarkable optical non linearities [166,167]. In the doped state it shows high electrical conductivity reaching values of $\sigma = 10^3\ \Omega^{-1}\ cm^{-1}$ (H_2SO_4 doping) and $\sigma = 10^2\ \Omega^{-1}\ cm^{-1}$ (AsF_5 doping) for stretch oriented samples (167).

The chemistry of this class of materials is such that stable oligomers can be prepared thus allowing a better study of the electronic and structural properties of PPV. Oligomers have been initially prepared by Drefahl et al. [168] and have been found to be not easily soluble making a systematic study of their spectra impossible. These oligomers show relevant photoconducting properties. The problem of solubility has been solved with the synthesis of oligomers capped with t-butyl groups [169].

11.1) Spectroscopic data.

The electronic spectrum of PPV has been studied in detail by Obrzut et al. [170]. The infrared spectra of stretch oriented PPV have been recorded and widely studied by Bradley et al. [171] who have also studied the infrared spectra of doped [172] and pohotoexcited materials [173].

Raman spectra of pristine PPV with λ_{exc} = 676.4 and 351.1 nm are reported by Lefrant et al. [174]. The same group has reported the Raman spectrum of doped PPV and have derived some structural information [174]. As discussed in section 4 there are some theoretical reasons to believe that the observation of the Raman spectra of doped polyconjugated materials is doubtful . This problem is going to be

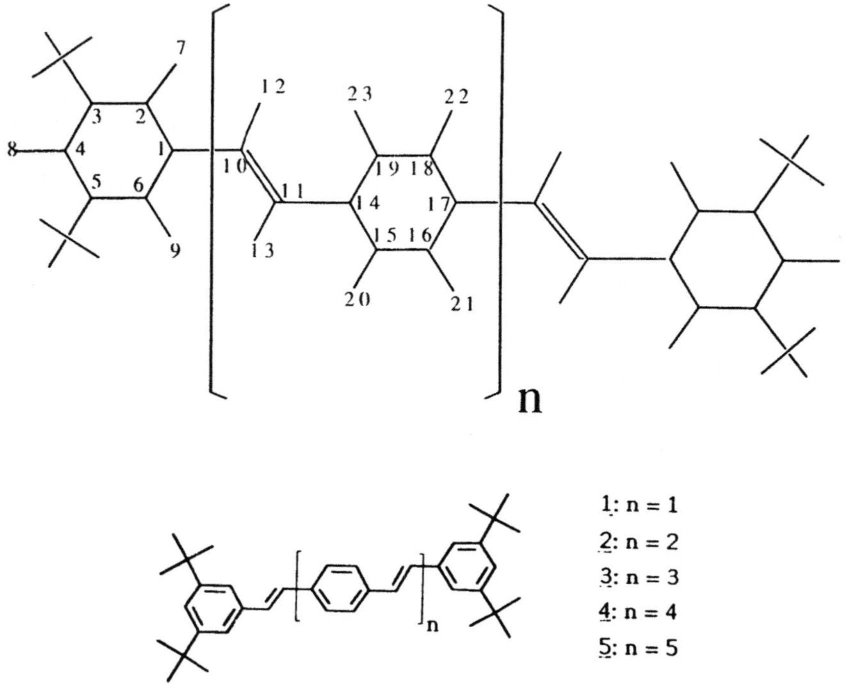

Fig.24. Sketch of the molecular structure of PPV and oligomers with atomic numbering (from ref.175).

discussed below.

The electronic spectra of the oligomers have been reported by Drefahl et al [168]. We have recorded the UV-Vis spectra of t-bytyl capped oligomers (solid and CH_2Cl_2 solutions) of the general formula: $(t\text{-butyl})_2[-\phi\text{-}CH{=}CH\text{-}(\phi\text{-}CH{=}CH)]_n\text{-}\phi\text{-}(t\text{-butyl})_2$ with n =1 to 5 (hereafter referred as O_n).

We have also recorded and studied the vibrational infrared spectra and the Fourier Transform Raman spectra at λ_{exc} =1060 nm of the series of O_n oligomers [175,176].

11.2) Structure and group theory.

The conjugational properties of these systems can be better understood if the issue of the coplanarity of these systems is first addressed. The numbering of the atoms is given in fig.24 for distyrylbenzene. As discussed in sections 9 and 10 for the case of polythiophenes the main issue is whether the proximity of the hydrogen atoms of the type 7-12,13-20,9-13 and 12-23 (fig.24) generates steric repulsions which oppose π electron delocalisation. The final results of the balancing of these two kinds of competing forces may yield a non coplanar molecule "in vacuo" which may be forced to become coplanar by packing forces in the crystal [177].

We have addressed the problem by MNDO and AM1 molecular orbital calculations on the simplest system namely trans stilbene "in vacuo" by calculation of the total potential energy as function of the torsional angles about the C_1-C_{10} and C_{11}-C_{14} bonds. The two methods give different answers, namely MNDO gives $\Theta_{1-10} = \Theta_{11-14} = 90°$ (i.e. no π conjugation) AM1 gives $\Theta_{1-10} = \Theta_{11-14} = 20°$ (i.e. slightly twisted structure with partially allowed delocalisation) [175,176]. The main message one derives from such calculations is that the conformation of the molecule "in vacuo" is not coplanar because of the balance of steric and conjugational forces. This result is checked by the experimental data on electron diffraction of trans-stilbene in the gas phase at 190°C which indicate that the molecule is twisted with $\Theta = 31 \pm 5$ [178].

In the crystalline phase intermolecular forces may flatten the molecule. X ray diffraction experiments report that in solid trans stilbene the rings are twisted out of the plane by a small torsional angle of ≈5° [179]. A correlation with the case of solid Th_3 (section 9) has to be recalled. If such a structure will be confirmed by other independent experiments it is very likely that also the longer oligomers have a slightly twisted structure analogous to what already found for oligoparaphenylenes. For the last case, however, the torsional angle is larger, thus reducing delocalisation.

11.3) Effective conjugation length and molecular chain length from optical data.

The conformation dependent delocalisation is proven by the systematic blue shift (increase of the HOMO-LUMO energy gap) shown by the electronic spectrum of the oligomers in going from the solid to the solution phases [175].

The chain length dependence of the delocalisation is proven by the systematic red-shift of the electronic spectra of the oligomers from O_1 to O_5 [168,175]. From the plot of HOMO-LUMO transitions vs. 1/N (N= the number of repeat units) and using an average value of the π-π^* transitions reported in the literature for PPV we conclude that the "effective conjugation length" for the solid polymer is not much larger than 7-10 units [175].

Information on chemical chain length can be derived from the intensity ratio of chain length independent end group modes with modes shows that the intensity depends on the number of monomer units. With some approximations (discussed in ref. 175) one finds good indications that the polymer chain length is considerably larger than 9-10 units.

11.4) Group theory and ECC theory.

The experimental vibrational infrared and Raman data published or collected in our laboratory show the following characteristics.

i) No Raman frequency dispersion is observed in the Raman spectra of PPV at λ_{exc}= 676.4 and 351.1 nm while slight intensity dispersion is observed. The most important is the triplet of lines at 1628 (w),1586 (vs) and 1550 cm^{-1}(w) which we label as ν_1, ν_2 and ν_3 respectively. The relative intensity ratio I_1/I_3 is reversed from λ_{exc} = 676.4 to 341,1 nm. An additional shoulder on the higher frequency side of the 1628 cm^{-1} line should not be overlooked in the Raman spectrum at λ_{exc} = 351.1 nm.

ii) The FT-Raman spectra of the oligomers show very small frequency and intensity dispersion with conjugation length (and chain length). The spectra are very consistent with that of the polymer.

iii) The infrared active modes of the inner part of the oligomers are chain lenght independent.

iv) Upon doping the doping induced IR spectrum shows a new set of approximately six bands accompanied by weaker satellites in the frequency range 1600-1000 cm^{-1} both for AsF_5 or $FeCl_3$ doping (possibly 1556 (shd), 1506 (vs),1420 (mw), 1317 (m),1286 (m) and 1151 cm^{-1} (s) [172].

v) The photoinduced IR spectrum shows a pattern of relatively strong new lines at 1470 (s),1398 (mw),1274 (ms) and 1100 cm^{-1} (vs) [173].

The symmetry point group of PPV taken as fully coplanar is C_{2h} and 14 are the TS modes of A_g species, Raman active. If the three high frequency localized C-H stretchings are neglected we expect to find 11 modes in the frequency range 1700-100 cm^{-1}. These modes may show frequency and intensity dispersion with $F_{\mathcal{R}}$. The $\nu/F_{\mathcal{R}}$ plot contains 11 branches which we label from 1 to 11 [176].

For the calculation of $\nu/F_{\mathcal{R}}$ and $I/F_{\mathcal{R}}$ we have first calculated the valence force field by the MNDO method on the molecule of trans-trans 1-4 distyrylbenzene (DST) [176]. Full geometry minimisations were carried out by keeping the molecule coplanar. Further improvement of the force field was obtained by least squares fitting of a very few force constants to fit the TS modes of the Raman spectrum of PPV.

From the frequency and intensity dispersions reported in figs. 25 and 26 we find an excellent fitting (fig. 27) for pristine, doped and photoexcited spectra at $F_{\mathcal{R}}$= 5.2 mdyne/A (pristine), $F_{\mathcal{R}}$ =4.5 mdyne/A (doped) and $F_{\mathcal{R}}$ = 3.7 mdyne/A (photoexcited). The FT Raman spectra of the oligomers are also explained with the same $\nu/F_{\mathcal{R}}$ and $I/F_{\mathcal{R}}$ plots [176].

11.5) Structure of PPV from spectroscopy.

Because of the shape of the $\nu/F_{\mathcal{R}}$ plots in the upper range of $F_{\mathcal{R}}$ values for the pristine materials the increase of effective conjugation is undoubtely revealed by the electronic spectra in a more relevant way than by the Raman spectra. Evidence has been found of an effective conjugation length > 9-10 units and of a chain length >> 8 units.

In order to reconcile the effective delocalisation measured from UV-Vis spectra of the oligomers and that measured for the polymer the experimental and theoretical problem of the coplanarity of the systems has to be condidered in more detail.

The fitting between the doping induced of photoexcited spectra with the ones calculated by ECC theory is an evidence that a "para-quinoid" configuration is formed at the site of doping and in the photogenerated species. The decision between bipolaron and polaron cannot be taken on the basis of ECC theory ;the infrared spectra calculated by MNDO calculations seems to support the bipolaronic structure [176].

Since there are theoretical reasons to believe that the Raman spectra of doped polyconjugated materials should be extremely weak ,if observable [3,17] the published Raman spectrum of doped PPV should find an explanation different from that proposed in the literature [174]. Application of ECC theory helps in this case. Comparison of the reported Raman spectrum with the $\nu/F_{\mathcal{R}}$ and $I/F_{\mathcal{R}}$ plots of figs. 25 and 26 shows that the weak-strong-weak triplet ν_1 , ν_2 and ν_3 has become a doublet since ν_1 has lost its intensity corresponding to a lower $F_{\mathcal{R}}$ (less $L_{\mathcal{R}}$ component in ν_1), ν_2 occurs as a strong line at 1567 cm^{-1} and ν_3 is a weak line at 1530 cm^{-1}. The new triplet (and also the Raman spectrum at lower frequencies) seems to indicate that the effective conjugation length of the "doped sample" of ref.174 has increased. More chemical studies on the doping process are needed.

12) **CONCLUSIONS.**

From the guided tours presented in this review we hope we have documented the width and depth of the information provided by the vibrational spectra .

The immediate application of general validity of the spctroscopic

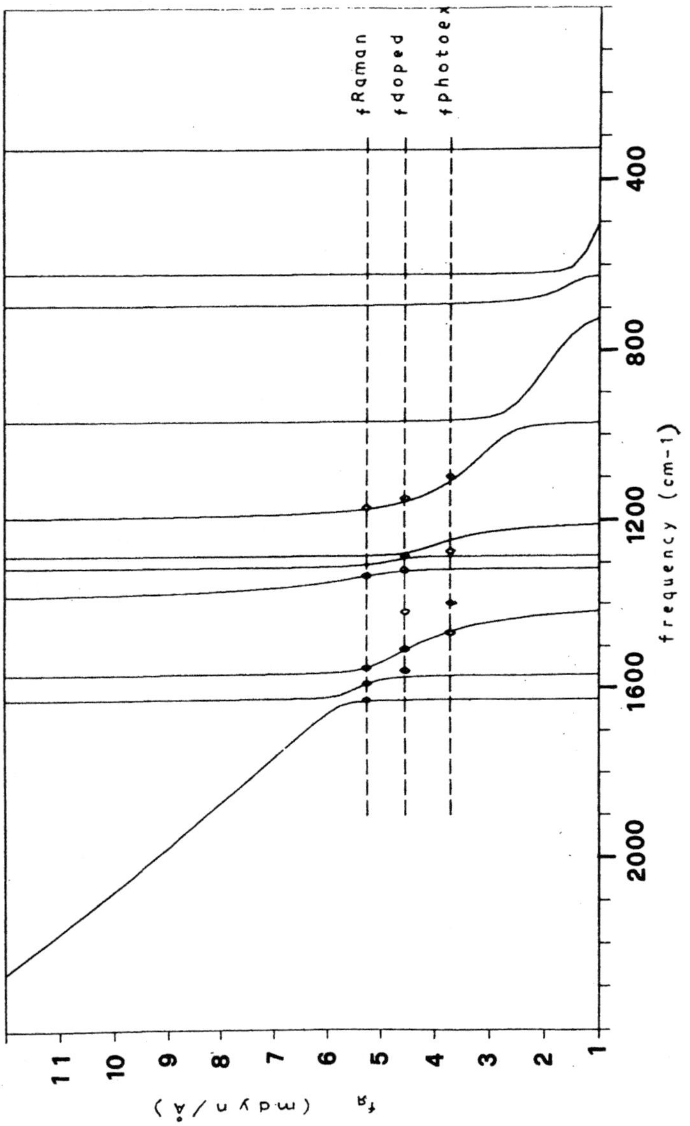

Fig.25 Calculated frequency dispersion v/F_R of the A_g modes of PPV (from ref.176).

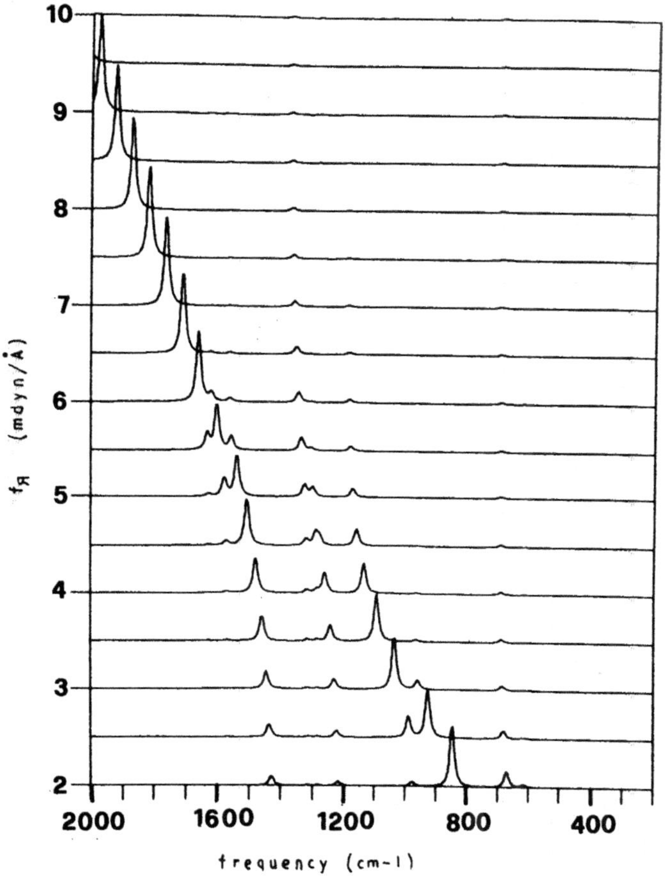

Fig.26 Calculated intensity dispersion I/F of the Ag modes of PPV (from ref.176).

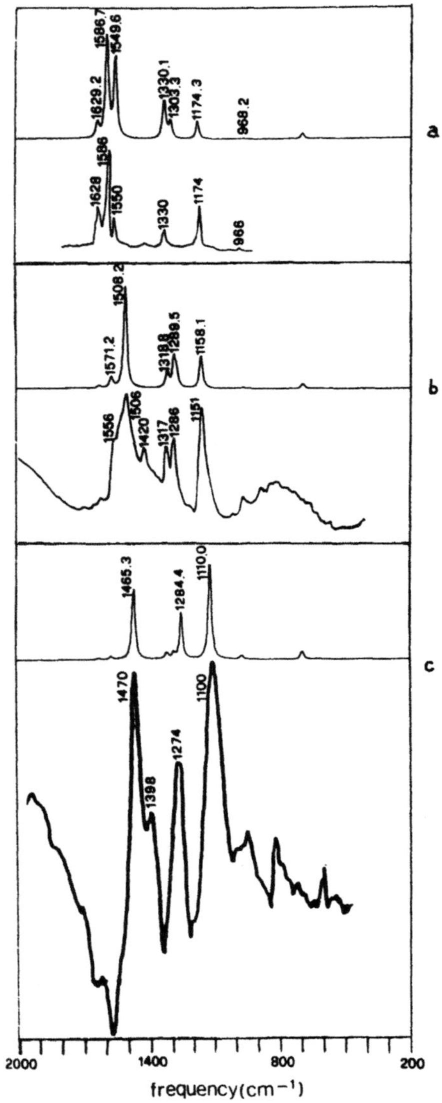

Fig.27 Comparison between calculated and experimental spectra of PPV : a) pristine, Raman, F_R = 5.32 md/A, b) doped, infrared F_R = 4.5 md/A and c) photoinduced, IR, F_R = 3.7 md/A (from ref.176).

methods applied and discussed here are the following:

i) The parameter $F_{\mathcal{R}}$ can be used as an "experimental" parameter to characterize the "Effective Conjugation Length" of a given polymer and oligomers of the same chemical class. Thus different samples of the same material from different origins can be characterized uniquely with one parameter. The cross comparison of $F_{\mathcal{R}}$ values between compounds of different chemical classes should be done with great caution since several theoretical aspects of the dynamical and electronic problems still need to be analyzed and solved.

ii) In the literature most of the authors have characterized the various samples of polyconjugated materials with the use of the Raman spectra. From our studies of frequency and intensity dispersions it turns out that crowding of the modes and a complex dynamical coupling the C=C and C-C stretching vibrations involved in the $\mathcal{R}$ mode takes place in the 1600-1400 cm^{-1} range of the Raman spectrum of the pristine material. The observed Raman active modes are the results of collective motions which are spread throughout the whole molecule because of the existence of a system of delocalized π electrons. The collective character of the observed modes contrasts the vibrational assignment to specific localized modes generally proposed in the literature .

From ECC theory it turns out that at lower $F_{\mathcal{R}}$ values the dispersion becomes larger and modes are not much coupled thus allowing the mode to appear as a strong and characteristic peak in the IR spectra. This fact makes the use of photoexcited spectra (with no chemistry involved) and eventually of the doping induced spectra a much more useful method for sample characterisation. Photoexcitation (and or doping) may become a better, more detailed and more sensitive spectroscopic tool for the characterisation of the distribution of chain lengths.

In conclusion we wish to offer a few final comments:

A) **Chemistry** From the chemical view point this review article clearly proves that the vibrational spectra of polyconjugated materials cannot be interpreted using the classical chemical correlations commonly used in organic chemistry. Indeed most of the spectroscopic features arise mainly from the existence of delocalized π electrons. The extent of their delocalisation is modulated by the molecular conformation which, on its turn, is determined by the complex interplay between steric and conjugational forces. Evidence has been offered that the theoretical or experimental knowledge on the conformation of polyconjugated systems is still very limited. Effective conjugation may be perturbed also by chemical defects and by unwanted side chain reactions.

The concepts introduced by ECC theory can be qualitatively used as guiding rules also in organic chemistry without the need to carry out real numerical calculations.

B) **Molecular Orbital Calculations** In spite of the enormous number of papers dealing with MO calculations of polyconjugated systems (oligomers and polymers) a reliable quasi-quantitative knowledge of simple phenomena like molecular conformation, extent of delocalisation etc. has not yet been reached. There is a tremendous need of calculations which should account for intermolecular interactions in three dimensions. We hope we have shown in this review article that the help provided by MO calculations on isolated molecules "in vacuo" is useful, but still very limited.

We also hope to have proven the strong limitations of MO calculations usable for complex molecules in predicting spectroscopic quantities such as frequencies and intensities when π delocalisation is involved. These calculations are, however, useful, and should be carried out, for providing guiding concepts and numbers for an overall perception of the physical phenomena at the molecular level .

An important step towards a better understanding of the electronic molecular properties and related vibrational spectra may come from the introduction in the MO calculations of "Configuration Interactions" which better account for electronic excited states.

C) **Physics.** The basic physics of the vibrational optical properties of polyconjugated materials was first presented by Horowitz and was used by many authors in physics circles also for more complex systems which were characterized by the parameters $\tilde{\lambda}$ and α . These parameters did not have an understandable meaning if considered from the molecular viewpoint nor were applicable to large polyconjugated organic molecules. With the introduction of the "effective conjugation coordinate "$\mathcal{R}$" and of the " effective conjugation force constant $F_{\mathcal{R}}$" ,with the writing of a preceeding review article (paper I) and of this review we hope we have translated the same physics and made it applicable and understandable in the case of more complex, but not less relevant, chemical systems.

Acknowledgement

We wish to thank Dr. B. Tian and Dr. T. Lopez-Navarrete for the many discussions during the development of the research on the IR/Raman spectra of polyconjugated materials.

References

1) For a general discussion see : Skotheim, T.A.(ed.), Handbook of Conducting Polymers, Dekker, New York, 1986, vol.1

2) For a general discussion see: Skotheim, T.A. (ed.), Handbook of Conducting Polymers, Dekker, New York, 1986, vol.2

3) Gussoni,M., Castiglioni, C., and Zerbi, G. (1990)' Vibrational Spectroscopy of Polyconjugated Materials: Polyacetylene and Polyenes' in R.J.H. Clark and R.E. Hester (eds.), Vibrational Spectroscopy of New Materials, Wiley, in press.

4) Zerbi ,G. (1963), 'Vibrational Spectra of High Polymers' in E.G. Brame (ed.), Applied Spectr. Revs, Dekker, New York, vol.2, p. 193.

5) Zerbi, G., (1983), 'Probing the Real Structure of Chain Molecules by Vibrational Spectroscopy. Some Recent Aspects' in Adv.in Chemistrty Series, American Chemical Society, **203**, 487.

6) Zerbi, G.(1984) 'Vibrational Spectroscopy of Very Large Molecules' in R.J.H. Clark and R.E. Hester (eds.), Advances in Infrared and Raman Spectroscopy , Wiley-Heyden, New York, vol. 11. pp. 301-352.

7) Painter, P.C., Coleman, M.M., and Koenig, J.L., (1982) 'The Theory of Vibrational Spectroscopy and its Applications to Polymeric Materials", J.Wiley, New York.

8) Zerbi, G., (1975), 'Defects in Organic Polymers; Numerical Methods' in S. Califano (ed.) ,Lattice Dynamics and Intermolecular Forces, Academic Press, Ney York, pp. 384-415.

9) Zerbi, G., Piseri, L. and Cabassi,F. (1971), Mol. Phys., **22**, 241.

10) Tasumi, M., and Zerbi, G. (1968), J. Chem. Phys.**48**, 3813.

11) Rubcic, A., and Zerbi, G. (1974), Macromolecules **7**, 754; ibid. p. 759.

12) Zerbi, G., and Sacchi, M. (1973), Macromolecules **6**, 692.

13) Gussoni, M. and Zerbi, G. (1974), J.Chem. Phys. **60**, 4862.

14) Castiglioni, C., Gussoni, M., Lopez-Navarrete, J.T. and Zerbi, G., (1988) , Solid State Comm., 625.

15) Zerbi, G., Castiglioni, C., Lopez-Navarrete , Tian ,B., and Gussoni, M. (1989) , Synt. Met. **28** , D 359.

16) Zerbi, G., Castiglioni, C., Tian, B., and Gussoni, M. (1989) in H. Kuzmany, M. Mehring and S. Roth 'Electronic Properties of Conjugated Polymers III', Solid State Science 91, Springer, Berlin, p. 106.

17) Gussoni, M., Castiglioni, C., and Zerbi, G., To be published.

18) Rotchild, W.G., (1984) 'Dynamics of Molecular Liquids', Wiley, New York.

19) Herzberg, G., (1959) 'Infrared and Raman Spectra of Polyatomic Molecules' Van Nostrand, Princeton.

20) Wilson, E.B., Decius, J.C., and Cross, P.C., (1955) 'Molecular Vibrations', McGraw-Hill, New York.

21) Califano, S., (1976) 'Vibrational States', Wiley, New York.

22) Volkenstein, M.V., Elyashevich, M.A., and Stepanov, I.B., (1949) 'Kolebaniya Molekul' , Vol. II, Moskow.

23) Gussoni, M. ,Dellepiane, G., and Abbate, S., (1975), J. Mol. Spectry, **57**, 323.

24) Piseri, L., and Zerbi, G., (1968), J. Molec. Spectry, **26**, 254.

25) Piseri, L., and Zerbi, G., (1968), J. Chem. Phys, **48**, 3561.

26) Zerbi, G., and Piseri, L., (1968) J. Chem. Phys. **49**, 3840.

27) Rubcic, A., and Zerbi, G., (1974) , Macromolecules, **7**, 754.

28) Zerbi, G., and Sacchi, M., (1973) , Macromolecules, **6**, 692.

29) Inagaki, F., Tasumi, M., and Miyazawa, T., (1975), J.Raman Spectry, **3** ,335.; (erratum: Tasumi, M., (1982), J. Raman Spectry, **13** , 202; Zannoni, G., and Zerbi, G., (1983), J. Mol. Struct., **100**, 485.

30) Tubino, R., Piseri, L., and Zerbi, G., (1972) J. Chem. Phys. **56**, 1022.

31) Bosi, P.,Tubino, R., and Zerbi, G., (1973) J. Chem. Phys.**59**, 4578.

32) Orel, B., Tubino, R., and Zerbi, G.(1975), Mol. Phys.**30**, 37.

33) Tasumi, M., and Shimanouchi, T., (1965) J. Chem.Phys, **43** and Tasumi, M., and Krimm, S. (1967), J. Chem. Phys.,**46** ,755.

34) Piseri, L., Powell, B.M., and Dolling, G., (1973) J. Chem. Phys., **57**,158.

35) Snyder, R.J., and Schachtschneider, J.H.,(1963) Spectrochim. Acta, **19**, 117.

36) Rabolt, J.F., and Fanconi, B.,(1977) Polymer, **18**, 1258.

37) Miyazawa, T., (1961), J. Chem. Phys. **35**, (693. ; Tadokoro, H.,(1960) , J. Chem. Phys.,**33**, 1558. see also ref. **4**.

38) Maradudin, A.A., Montroll, E.W., and Weiss, G., (1963), Solid State Physics, Supplement 3,1.

39) See, for instance, Trevino, S., and Boutin, H.,(1967), J.Macrom. Sci. **A1**,723.

40) Kitagawa, T., and Miyazawa, T.,(1967), J.Chem.Phys., **47**, 337 ; Miyazawa,T and Kitagawa, T., (1964) ,J.Polym. Sci., **B2**,395.

41) Machonnachie, A., Dianoux, A.J., Shirakawa, H., and Tasumi, M., (1986), Synth. Metals, **14**,323.

42) Takeuchi, Arakawa, T., Furukawa, Y., and Harada, I., (1987), J. Mol. Struct. **159**, 179.

43) Keller, A., (1979), Faraday Society Discussions of the Chemical Society, **68**,145.

44) Dean, P.(1972), Rev.Mod.Phys., **44**, 127.

45) Zannoni, G., and Zerbi, G.,(1983) Journal de Physique, **44**,C3 503.

46) Zannoni, G., and Zerbi, G. (1984), Solid State Comm. **50**, 55.

47) For a thorough discussion see: Person, W.,and Zerbi, G., (1984) Vibrational Intensities in Infrared and Raman Spectroscopy, Elsevier, Amsterdam.

48) Gussoni, M., (1979) 'Infrared and Raman Intensities from Electrooptical Parameters' in R.J.H. Clark and R.H. Hester (eds.), Advances in Infrared and Raman Spectroscopy, Heyden, London, vol. 6 pp.96-121.

49) Gribov, L.A., (1964) 'Intensity Theory for Infrared Spectra of Polyatomic Molecules', Consultant Bureau, New York.

50) Gussoni, M., Abbate, S., (1976), J. Chem. Phys., **65**,3439.

51) Gussoni , M., (1986) J. Mol. Struct. , **141**, 63.

52) Gussoni, M., Castiglioni, C., Ramos, M.N., Rui, M., and Zerbi, G., (1990), J. Mol. Struct., **224**, 445.

53) Gussoni, M., Castiglioni, C., and Zerbi, G., (1984) J. Phys. Chem. **88** , 600.

54) Gussoni, M., Ramos, M.N., Castiglioni, C., and Zerbi, G., (1987), Chem. Phys. Letters, **142**, 515.

55) Gussoni, M., Ramos, M.N., Castiglioni, C., and Zerbi, G., (1989), Chem. Phys. Letters, **160**, 946.

56) Jona, P., Gussoni, M., and Zerbi, G.,(1981), J. Phys. Chem., **85** , 349.

57) Castiglioni, C.,Gussoni, M., and Zerbi, G., (1985), J.Chem. Phys., **82** 3534.

58) Abbate, S., Gussoni, M., Masetti, G., and Zerbi, G., (1977) , J. Chem. Phys. **67**, 1519.

59) Gussoni, M., Castiglioni, C., Miragoli, M., Lugli, G., and Zerbi, G., (1985), Spectrochim. Acta., **41A** , 37.

60) Castiglioni, C., Gussoni, M., Miragoli, M., and Zerbi, G., (1988), Mol. Cryst. Liq. Cryst. **117**, 295

61) Tang,J., and Albrecht, A.C. (1970), in H.Szymanski (ed.) 'Raman Spectroscopy' Plenum, New York, Vol.2.

62) Warshel, A., and Dauber,J., (1977) ,J.Chem. Phys. **66**, 5477.

63) Clark, R.J.H. ,and Dines, T.J., (1986) Angew. Chem.Int. Ed.Engl. 25,131.

64) Peticolas, W.L., and Blazej, C. (1979), Chem .Phys. Letters, **63**, 604.

65) Peticolas, W.L., (1981) Ber.Bunsen Ges. Phys.Chem. **85**, 481.

66) Gussoni, M., Castiglioni,C., and Zerbi ,G., to be published.

67) The most known set of original computing programs which has later originated many other programs for the calculation of force constants from ro-vibrational data is that written by:
J.H. Schachtschneider , Shell report 57/65.

68) see for instance Lifson,S., Warshel,A. (1968), J. Chem. Phys.,**49**, 5116.

69) Tubino, R.,Piseri, L., and Zerbi, G. (1972), J. Chem. Phys., **56**, 1022.

70) Mele, E.G., in ref. 2, pp 795-823.

71) Keating, P.N., (1966), Phys. Rev., **145**, 637.

72) See, for instance, E.Clementi, (1989), MOTECC, Modern Techniques in Computational Chemistry,Escom.

73) Brédas. J.L., in ref.2, pp. 825-857 and 859-913.

74) Brédas, J.L. Thémans, B., André, R ,Chance, R., and Silbey, S., (1984) ,Synth. Metals, **9**, 265.

75) Brédas, J.L., Street, G.B., Thémans, B.,and André, M., (1985), J. Chem.Phys., **83**,1323.

76) Kertész, M., (1982) , Adv. Quantum Chem., **15**, 161.

77) Lopez Navarrete, J.T., and Zerbi, G. (1989) Synth. Metals, **32**, 151.

78) a) Villar, H.O., Dupuis, M., Watts,J.D., Hurst, G.J.B. and Clementi, E. (1987), J. Chem. Phys. **88**, 1003.
b) Dupuis, M. and Clementi, E. Private Communication.

79) Pulay, P., Fogarasi, G., Pangor, F., and Boggs, J.E., (1979) J. Am. Chem. Soc., **102**, 2550.

80) Horovitz, B.,(1982) Solid State Comm. **41**, 729; Horovitz, B., (1982) Phys Rev. Lett., **47**, 1491 ;

81) Tasumi, M,. Yoshida, H., Fujiwara, N.,Hamaguchi, H. and Shirakawa, H. (1989), Synth. Metals, **17**, 319.

82) Tian, B., Zerbi, G., and K. Müllen, J. Chem. Phys. paper 1, submitted.

83) Lopez Navarrete, J.T., and Zerbi, G., J. Chem. Phys., Paper 1, In press.

84) Lopez Navarrete, J.T., and Zerbi, G., J. Chem. Phys., Paper 2, in press.

85) Tian, B., Zerbi,G., and K. Müllen, J. Chem. Phys. Paper 2, submitted.

86) Castiglioni,C., Gussoni, M.,Lopez Navarrete, J.T., and Zerbi, G., (1988) Mikrochim. Acta, 1, 247.; Castiglioni ,C., Gussoni, M., and Zerbi, G. (1989), Synth. Metals, **29**, E 1.

87) Castiglioni, C., Gussoni, M., and Zerbi, G,. (1985) Solid State Comm. **56**, 863.

88) Bradley, D.D.C., Friend, R.H., Lindenberger ,H., and Roth, S., (1986), Polymer, **27**, 1709.

89) Harada, I., Furukawa, Y., Tasumi, M., Shirakawa,H., and Ikeda, S., (1980) , J. Chem. Phys., **73**, 4746.

90) Lefrant, S., (1983) J. de Physique, **44**,C3-247; Eckhardt, H., Schaklette, L.W., Szobota, J.S., and Baughmann,R.H., (1985) Mol. Cryst. Liq. Cryst., **117**, 401.

91) Tanaka, J., Saito, Y., Shimizu, M., and Tanaka, M., (1987), Synth. Metals, **17**, 307.

92) Piaggio, P., Dellepiane, G., Mulazzi, E., and Tubino, R., (1987), Polymer **28**, 563.

93) Ehrenfreund, E., Vardeny, Z., Brafman, O. and Horovitz, B., (1987), Phys. Rev.**B36**, 1535.

94) Negri, F., Orlandi, G., Zerbetto, F., and Zgierski, M.Z., J. Chem. Phys. in press.

95) Su, W.P., Schrieffer, J.R. and Heeger, A.J. (1980) ,Phys. Rev. **B22**, 2099.

96) Crawford, B.L., and Califano, S.(1960), Spectrochim. Acta. **16**, 889. ; Scherer, J.R., and Overend, J.,(1961) Spectrochim. Acta **17** ,719.

97) Zerbi G., Crawford, B.L., and Overend, J., (1963) J. Chem. Phys., **38**, 127.

98) Zerbi, G., and Sandroni, S. (1968) Spectrochim Acta,**24A**, 483 ; ibid, p.511.

99) Kakitani,T., (1973), Progr. Theor. Phys. **49**,1 ; **50**, 17; (1974), ibid, **51**, 656.

100) Street, B. (1986) in T.A. Skotheim, (ed.), Handbook of Conducting Polymers, Dekker, New York, vol. 1, p.256.

101) Münstedt, H., (1986) in H.J.Mair and S.Roth (Eds.) Elektrisch Leitende Kunststoffe , Hauser, München, p.207.

102) G. Wegner, Private Communication.

103) R. Qian, Private Communication.

104) Furukawa,Y., Tazawa, S., Fuji, Y.,and Harada,I. (1988) Synth. Metals, **24**, 329.

105) Oddi, L., Cappelletti, R., Fieschi, R., Fontana, M.P., Ruani, G., and Gardini , G.P., (1985) Mol. Cryst Liq. Cryst.**118**, 179.

106) B.Street, Private Communication.

107) Lopez-Navarete, J.T., Tian, B., and Zerbi.,G. (1990) Synth.Metals., **38**,299

108) Cui, C.X., and Kertesz, M. (1989), Phys. Rev. B, **40**, 9661.

109) Tian,B., and Zerbi, G., (1990), J.Chem.Phys., **92**, 3886. : ibid. p. 3892.

110) see ref. 113.

111) see, for instance, Prasad, P.N. (1988) in R.A.Hann and D. Bloor (Eds.) Organic Materials for Non Linear Optics, Royal Society of Chemistry, London, p.264.

112) Horowitz,G., Fichou,D., Peng,X., Xu,Z., and Garnier,F., (1988), Solid State Comm. **72**, 381; Lawrence,R.A., Burroughes,J.H. and Friend,R.H., (1989) "Electronic Properties of Conjugated Polymers III" (Kuzmany,H., Mehring,M. and Roth,S., eds), Springer Series in Solid State Sciences, vol 91, p.127, Heidelberg; Burroughes,J.H., Jones,C.A., Lawrence,R.A. and Friend,R.H., ibidem, p.437

113) Tourillon, G.,(1986) in T.A. Skotheim (ed.) Handbook of Conducting Polymers', vol. 2, p. 293.

114) Patil, A. O., Heeger, A.J., Wudl, F. (1988) Chem. Rev., 88, **183**.

115) Kaneto, K., Hayashi, S., Ura. S.,and Yoshino, K., (1985) J. Phys. Soc. Jpn., **54**, 1146.; Kaneto, K.,and Yoshino, K.,(1987) Synth.Metals, **18** , 133.; Harbeke, G., Baeryswyl, D., Kiess, H., and Kobel., (1986) , Phys. Scrypta, **T1**, 302.

116) Hayes, W., Pratt, F.J., Wong, K.S., Kaneto. K., and Yoshino, K., (1985) J. Phys., **C 18** L555.

117) Shaffer H.E. and Heeger, A.J., (1986) Solid State Comm. **59**, 415.

118) Cao,Y., and Renyuan,Q., (1986) Solid State Comm., **54** , 211 (1986).

119) Straftstron, S., and Brédas. J.L. (1988) , Phys. Rev. B, **38** , 4180.

120) Vardeny,Z., Ehrenfreund , E.,and Brafman, O., A.J.Heeger , and Wudl, F. (1987) ,Synth. Metals **18**, 183.

121) Akimoto,M., Furukawa,Y., Takeuchi,H., and Harada,I., (1986) Synth. Metals, **15**, 353.

122) Furukawa, Y., Akimoto', and Harada, I., (1988) Synth.Metals **18**, 151.

123) Harada. I. Private Communication.

124) Steigmeier, E.F., Auderset, H., Kobel, K., and Baeryswyl,D., (1987) Synth. Metals, **18** , 218.

125) Neugebauer, H., Neckel, A., and Brinda-Konopik, N.,(1985) in H.Kuzmany, M.Mehring and S.Roth (eds) ,Electronic Properties of Polymers and Related Compounds, Solid State Science 63, Springer, Heidelberg, p.227.

126) Cao, Y., Guo,D.,Pang, M., and Quia, R.,(1987) Synth.Metals, **18**, 189.

127) Kim, Y., Hotta, S., and Heeger, A., (1987) Phys.Rev. B. **36**, 7486.

128) Visser, G.J., Heeres,G.J., Wolters, J. and Vos.,A.,(1968) Acta Cryst. 24 ,467.

129) Van Bolhuis, F., Wynberg, H., Havinga ,E.E., Maijer, E.W., and Staring, A.G.J., (1989) Synth. Metals, **30** ,381.

130) Bruckner, S., and Porzio, W.,(1988) Makromol. Chemie,**189** ,961.

131) Almenningen ,A., Bastiansen, O., and Svendsas, (1958) ActaChem.Scand. **12**,1671.

132) Birnbaum, D., and Kohler, B. E., (1989), J. Chem. Phys., **90**, 3506.

133) Garnier, F., Tourillon, G., Barraud, J.Y., and Dexpert, H., (1985), J. Mat. Sci., **20**, 2687.

134) Zerbi, G., Chierichetti, B., and Inganäs, O., J. Chem. Phys. in press.

135) Fichou, D., Garnier, F., Charra. F., Kajar, F., and Messier, J. in R.A. Hann and D.Bloor (eds) "Organic Materials for Non-linear Optics " Royal Society of Chemistry, London ,1988. p. 176.

136) Havinga, E.E., Rotte, I., Meijer,E.W., ten Hoeve,W., and Wynberg, H., Synth. Metals, in press.

137) Hudson, B., and Kohler, B., (1984) Synth. Metals, **9** 241.

138) Sundberg, M., Inganäs, O., Stafström, S., Gustafsson, G. and Sjögren, B., (1989) Solid State Comm. **71**, 435.

139) Brédas, J.L. and Heeger, A.,J., (1990) Macromolecules, **23**, 1150.

140) J.T. Lopez Navarrete. To be published.

141) Bucci, P.,Longeri, M.,Veracini, C., and Lunazzi, L., (1974), J.Am. Chem.Soc., **96**, 1305.

142) Zerbi, G., and Sandroni, S., (1968), Spectrochim. Acta, **24a** ,483.

143) Zerbi, G., and Sandroni, S., (1968), Spectrochim. Acta, **24A** ,511.

144) Brédas, J.L.,Street, G.B.,Thémans,B., and André, J.M.,(1985) , J. Chem. Phys., **83** ,1323 .

145) Lopez-Navarrete , J.T., Tian, B., and Zerbi, G., (1990) , Solid State Comm., **74**, 199.

146) Tadokoro, H.,(1960), J. Chem. Phys., **33** 1558; Tadokoro, H., (1961), J. Chem. Phys.,**35** ,1050.

147) R. Zbinden, (1964) ,Infrared Spectroscopy of High Polymers, Academic Press, New York.

148) Miyazawa, T., (1961), J. Chem. Phys., 35, 693.

149) Hotta, S., Soga, M., and Sonoda, N., (1989), J. Phys. Chem., **93**, 4994.

150) Falques, E., Wallnöfer, W., and Kuzmany ,H.,(1989), J. Chem. Phys.,**90**, 7585.

151) See also : Botta, C., Luzzati, S., Bolognesi, A., Catellani, M., Destri, S.,and Tubino, R., (1990), Mat. Res. Soc. Symp. Proceedings, vol. 173, p.397.

152) Sato, M., Tanaka, S., and Kaeriyama, K., (1986),J.Chem. Soc. Comm. 873; Elsembaumer, R.L., Jen, K.Y., and Obodi, R., (1986) Synth. Metals. **16**, 169.

153) Hotta, S.,Rughooputh, S.D.D.V., Heeger, A.J., and Wudl, F., (1987) , Macromolecules, **20**, 212. ; Kobel, W., Kiess, H.,and Egli, M. (1988) Synth. Metals, **22**, 265. ; Osterholm, J.E., Laakso, J., Nyholm, P., Isotalo, H., Stubb H., Inganäs, O., and Salaneck, W.R., (1989) Synth. Metals, **28** ,C435.

154) Zerbi, G., Chierichetti, B., and Inganäs, J.Chem. Phys. in press.

b) Zerbi,G., Castellani,L., Chierichetti,B., Gallazzi, C.M.C., and Inganäs, O., (1990), Chem.Phys. Lett., **172**, 143. Inganäs, O.,

155) Inganäs,O., Gustavsson, G.,and Salaneck, W.R.,(1989), Synth. Metals, **28**, C377.

156) P. Flory,(1969) 'Statistical Mechanics of Chain Molecules' Interscience, New York.

157) Gallazzi, M.C., Castellani, L., Zerbi, G., and Sozzani, P., Synthetic Metals, in press.

158) Souto Maior, R.M., Hinkelmann, K., Eckert, H., and Wudl, F., (1990) Macromolecules, **23**,1268.

159) See, for instance: Jones, R.N., (1956) , 'Chemical Applications of Spectroscopy' in A.Weissberger (ed.) 'Techniques of Organic Chemistry ' Interscience, New York, vol 9, p. 247. ; Bellamy, L., (1958) 'The Infrared Spectra of Complex Molecules', Wiley, New York.

160) Zerbi,G., Magni, R., and Gussoni, M.,(1981),J. Mol. Struct., **3**, 325.

161) Jona, P.,Gussoni, M., and Zerbi, G., (1985), J. Appl. Phys., **57**, 834.

162) Zerbi, G., and Gussoni. M., (1980), Polymer, **21**, 1129.

163) Colombo, L., and Zerbi, G., (1980), J. Chem. Phys.,**73**, 2013.

164) Winokur, M.J., Spiegel, D., Kim., Hotta, S., Heeger,A.J., (1989) Synth. Met. **28** C419.

165) Winokur, M.J., Wamsley, P., Moulton, J., Smith, P., and Heeger, A.J., to be published ; Mardalen J., Samuelsen, E.J., Gautun,O.R., and Carlsen, P.H. , Solid State Comm. , to be published; Bolognesi,A., Catellani,M., Destri,S., and Porzio,W., (1990), Makromol. Chem. Rapid Comm., in press

166) Gourlay, D.D., Lillya, C.P., Reynolds, J.R., and Chien,J.C. W., (1984) , Macromolecules , **17**, 1025.

167) Gagnon, D.R., Karasz, F.E., Thomas, E.L., and Lenz, R.W., (1987), Synth. Metals, **20**, 85.

168) Drefahl, G., Kühmstedt, R., Oswald, H.,and Hörhold, H. H. (1970), Makromol. Chem., **131**, 89.

169) Schenk, R., Legorious,H., Meeshobz, K., Heinre, J., and Müllen, K., to be published ; Schenk, R., Ehrenfreund, M., Huber, W., and Müllen, K., J. Chem. Soc., Chem. Comm. ,in press.

170) Obrzut, J.,and Karasz.,F.E., (1987), J. Chem. Phys.,**87**,2349.

171) Bradley, D.D.C., Friend, R.H., Lindenberger , H., and Roth, S., Polymer, (1986) 1709.

172) Bradley, D.D.C., Friend, R.H., Pratt, F.L., Wong, K.S., Herye, W., Lindenberger , L., and Roth, S., (1987), in H. Kuzmany, M.Mehring and S.Roth, (eds.) , Solid State Sciences, **76**, Springer, Berlin (1987).

173) Bradley, D.D.C.,Evans, G.P., and Friend, R.H.,(1987), Synth. Metals, **17** , 651.

174) Lefrant, S., Perrin, E., Buisson, J.P., Eckhardt,H.,and Hanl, C.C., Synth. Metals, (1989) , **29**, E91.

175) Tian,B., Zerbi,G.,Schenk, R., and Müllen, K., J.Chem. Phys., submitted.

176) Tian, B., Zerbi, G., and Müllen , K., J.Chem. Phys., submitted.

177) Krebs,K., Sandroni, S., and Zerbi, G., (1964) ,J. Chem. Phys.,**40**,3502.

178) Traetteberg,M., Fransten, E.B., Mijlhoff, F.C.,and Hoekstra, A., (1975), **26**, J. Mol. Struct. **26**, 57.

179) Hoekstra, A., Meertens, P., and Aafje Vos,1975),Acta Cryst., **B31**, 2813.

THIRD ORDER NONLINEAR OPTICAL EFFECTS IN CONJUGATED POLYMERS

F. KAJZAR and J. MESSIER
CEA-DTA, DEIN/LPEM, CEN Saclay
91191 Gif-sur-Yvette - Cedex, France

J.L. Brédas and R. Silbey (eds.), Conjugated Polymers, 509–554.

1. INTRODUCTION

The conjugated quasi-one dimensional polymers are characterized by a strong delocalization of π electrons. This highly polarizable electronic cloud responds nonlinearly to the exiting external field. The resultant bulk polarization can be developed into the external field power series and its component along x is given by

$$P_x(t) = \sum_{y,z,u,\ldots} \left(\chi^{(1)}_{xy} E_y(t) + \chi^{(2)}_{xyz} E_y(t) + \chi^{(3)}_{xyzu} E_y(t) E_z(t) E_u(t) + \ldots\right) \quad (1)$$

where $\chi^{(1)}$, $\chi^{(2)}$ ... $\chi^{(4)}$ are (n+1) rank tensors describing nonlinear optical response of the medium and $E_y(t)$, $E_z(t)$ ... $E_u(t)$ are electric field components in y, z, ... u direction.

Conjugated polymers are principally centrosymmetric and $\chi^{(2)} \equiv 0$; thus the first nonlinear term is described by $\chi^{(3)}$ susceptibility. They are characterized by very large linear ($\varepsilon > 30$ at 0.625 µm and at 9K in polydiacetylene, Batchelder (1985)) and nonlinear polarizability (Sauteret et al. (1976)). The principal cubic nonlinear optical phenomena discussed here are listed in Table 1 together with correspon-ding nonlinear polarizations and susceptibility tensors. The relations given on RHS of Table 1 are true for every non static component of $\chi^{(3)}$ tensor.

The bulk nonlinear susceptibility $\chi^{(3)}$ is related to the corresponding microscopic molecular hyperpolarizability $\gamma_{\alpha\beta\gamma\delta}$ where $\alpha,\beta,\gamma,\delta$ are molecular directions. The relations between χ^{3}_{xyzu} and $\gamma_{\alpha\beta\gamma\delta}$ depends upon the symmetry of the molecular orientation distribution (cf. Flytzanis (1975)). In the particular case when the γ_{xxxx} component is enhanced along the polymer chain direction we have

$$\chi^{(3)}_{xxxx} = \frac{\nu . N}{\varepsilon_0} \gamma_{xxxx} F \quad (2)$$

where $\nu = \frac{1}{5}$ for an isotropic distribution of molecules in three dimensions (three-dimensional disorder)

$\nu = \frac{3}{8}$ for an isotropic distribution of molecules in a plane (two-dimensional disorder).

F is the local field factor which for cylindrical molecules is close to unity (cf. Cojan et al. (1977)).

N is the number of molecules per unit volume and γ_{xxxx} is the second order molecular hyperpolarizability.

Table 1 Principal third order nonlinear optical phenomena in conjugated polymers and related nonlinear polarizations and susceptibility tensors.

Nonlinear phenomenon	Nonlinear polarization and susceptibility tensor
Frequency mixing $\omega_\sigma = \omega_1 + \omega_2 + \omega_3$	$P^{NL}(\omega_\sigma) = \frac{3}{2}\chi^{(3)}(-\omega_\sigma;\omega_1,\omega_2,\omega_3)E_{\omega_1}E_{\omega_2}E_{\omega_3}$
Third harmonic generation (THG)	$P^{NL}(3\omega) = \frac{1}{4}\chi^3(-3\omega;\omega,\omega,\omega)E_\omega^3$
Electric field induced second harmonic generation	$P^{NL}(2\omega) = \frac{3}{2}\chi^3(-2\omega;\omega,\omega,0)E_\omega^2E_0$
Degenerate four wave mixing. Phase conjugation	$P^{NL}(\omega) = \frac{3}{4}\chi^3(-\omega;\omega,-\omega,\omega)E_\omega^2E_\omega^*$
Quadratic Kerr effect	$P^{NL}(\omega) = \chi^3(-\omega;\omega,0,0)E_\omega E_0^2$
Saturation absorption	$P^{NL}(\omega) = \frac{3}{4}\chi^3(-\omega;\omega,\omega,-\omega)E_\omega^3$
Photoinduced absorption Optical Kerr effect	$P^{NL}(\omega_1) = \frac{3}{2}\chi^3(-\omega_1;\omega_1,\omega_2,-\omega_2)E_{\omega_1}(E_{\omega_2})^2$

The following definitions of electric field and linear polarization are used in this Table and troughout the text

$$E_\omega(t) = \frac{1}{2}\left[E_\omega e^{i\omega t} + E_\omega^* e^{-i\omega t}\right]$$

$$P_{\omega_\sigma}^{NL}(t) = \frac{1}{2}\left[P_{\omega_\sigma}^{NL} e^{i\omega_\sigma t} + \left(P_{\omega_\sigma}^{NL}\right)^* e^{-i\omega_\sigma t}\right]$$

1.2. Origin of second order hyperpolarizability

The second order nonlinear hyperpolarizabilities, depending on their origin, can be classified into two main categories:

i) Electronic hyperpolarizabilities.
The electronic hyperpolarizabilities arising from virtual transitions yield a coherent nonlinear optical response with a well defined phase relationship with optical excitations. In other terms electronic hyper-polarizabilities originate from the anharmonicity terms in the electron potential. In Chapter 1.4 we present quantum mechanical formulas allowing a better comprehension of the origin of these nonlinearities as well as modelisation of their frequency variation. These calculations allow also to predict the magnitude of nonlinear hyperpolarizabilities in conjugated polymers as a function of different structural parameters like conjugation length (electron delocalization), conformation, influence of polymer side groupes etc...

ii) Incoherent nonlinearities.
In the presence of energy absorption through one or multiphoton process the connected dissipative phenomena, particularly important in these polymers because of strong electron-photon coupling lead to a loss of coherence within a few of ps between absorbed energy and exciting fields. These processes are accompanied by nonlinear optical phenomena like refractive index variation of the medium. This is due to secondary phenomena associated with the energy absorption like creation of electron hole pairs, polarons, solitons, conformational changes, thermal dilatations and compressions due to electro-striction. Each of these processes can be, at least formally, described by a nonlinear susceptibility of third, or higher order. Their magnitude, which depends on a large number of different parameters like dynamic of excitations, experimental conditions can be very different than those describing coherent processes. Some examples of these hyperpolarizabilities will be given in Section 3.

1.3. Coherent nonlinearities.

Theoretical calculations within Hückel or Hartree-Fock formalism have been used primarily in order to get a better evaluation of second order hyperpolarizabilities γ (Rustagi (1974)). They allowed to explain the observed increase of γ with increasing molecule length $L(\gamma \sim L^5)$.

For an infinite polymer chain one expects a satuation of γ/L ratio for a sufficiently large length L. A finite γ/L value was obtained within a band model formalism parametrized in function of overlap integrals of π electrons in the case of polydiacetylene (Agrawal et al. (1978)). Moreover this method does not take account of electron-electron correlations strong in these systems and which affect strongly the γ/L limit value.

The self consistent finite field (SCF) methods which take account of electron-electron correlation have been used by André et al. (1987) for finite molecules. Their principal drawback is to give only the static value of γ with a limited accuracy. More promissing are the CNDO and INDO methods described in Section 1.4. They allow to determine the dipolar transition moments between fundamental and excited states and between different excited states as well as the energy of excited states. Coupled with time dependent perturbation calculation of γ they give the magnitude of second order hyperpolarizability as a function of frequency and the position of principal one, two and three photon resonances. The CNDO and INDO methods permit also a good evaluation of the influence of side groups, substituents, conformation and molecule size on its nonlinear optical response.

1.4. Second order hyperpolarizability of centrosymmetric and non centrosymmetric molecules. Influence of polymer length.

1.4.1. Centrosymmetric molecules.

In the case of centrosymmetric molecules the quantum mechanical formulas for γ hyperpolarizability (cf. Ward (1965)) can be simplified because of the classification of energy states into even (0, 2, ...) and odd (1, 3, ...) parity levels with vanishing dipolar moments between states of the same parity. In the lowest approximation at least three states have to be taken into account in order to get a nonzeroth γ value: fundamental $|0\rangle$ and two excited states with opposite parity ($|1\rangle$, $|2\rangle$)(cf. *Fig. 1*). For the γ_{xxxx} component one gets

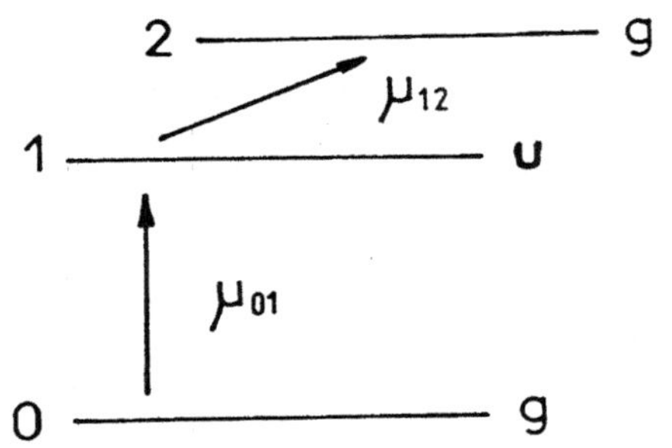

Figure 1: Three level system used for a centrosymmetrical molecule
0 fundamental state
1,2 excited one and two-photon states.
μ_{01}, μ_{12} dipolar transition moments.
Remark: in several one-dimensional conjugated polymers the two-photon state (2) is located slightly below the one-photon state (1).

$$\gamma_{cs} = \frac{1}{\varepsilon_0} \mu_{01}^2 \left[\mu_{12}^2 \, f_{12}(\omega) - \mu_{01}^2 \, g_{11}(\omega)\right] \tag{3}$$

where $\mu_{01} = e\langle 1|x|0\rangle$,
$\mu_{12} = e\langle 2|x|1\rangle$

where e is the electron charge and $f_{12}(\omega)$ and $g_{11}(\omega)$ are excitation frequencies depending functions. In a particular case of a unique excitation frequency ω and third harmonic generation (THG), one gets following expressions for f_{12} and g_{11} functions

$$f_{12}(\omega) = \frac{1}{(E_1-3\omega)(E_2-2\omega)(E_1-\omega)} + \frac{1}{(E_1+\omega)(E_2-2\omega)(E_1-\omega)} + \frac{1}{(E_1+\omega)(E_2+2\omega)(E_1-\omega)} + \frac{1}{(E_1+\omega)(E_2+2\omega)(E_1+3\omega)} \tag{4}$$

$$g_{11}(\omega) = \frac{1}{(E_1-3\omega)(E_1-\omega)^2} + \frac{1}{(E_1-\omega)^2(E_1+\omega)} + \frac{1}{(E_1+3\omega)(E_1+\omega)^2} + \frac{1}{(E_1+\omega)^2(E_1-\omega)} \tag{5}$$

where E_1, E_2 are excited states energies and ω is the incident photon frequency. Resonance with an excited state takes place if ω, 2ω or 3ω is close to corresponding energy value. One takes account of that by introducing a corresponding damping term in Eqs. (4)-(5) by replacing E_n by

$E_n - i\Gamma_n$ (n = 1,2).
First term in Eq. (4) gives a positive contribution to γ at low frequency ($\omega << E_1, E_2, E_3$) whereas the second a negative one corresponding to saturation of the one photon transition.
Charra and Messier (1990) calculated the variation of dipolar transition moments μ_{01} and μ_{12} as a function of polymer length in the case of polythiophene and in the frame of CNDO-S with configuration interaction approach. The calculations are limited to single electron excitations. It corresponds to the virtual electron-hole pair formation along the polymer chain. In fact the calculations become steeply prohibitory when including multi-electron excitations. In spite of this difficulty they give a reasonable order of magnitude for γ and it is possible that these single electron contributions which look like virtual electron-hole pairs are predominant ones. The calculated variation of $[\mu_{01}^2/N]^{1/2}$ as a function of number of thiophene units N is shown in *Fig. 2*. One observes that both tend to a limiting value for $N \to \infty$. It means that for the thiophene monomer the linear polarisability proportional to μ_{01}^2/N and the first positive term contributing to γ (cf. Eq. (3)) proportional to $\mu_{01}^2\mu_{12}^2/N$ tend to a limiting value, as expected. In contrast, the second term in γ diverges to $-\infty$ for $N \to \infty$. Taking account of limited size of excitations, Charra and Messier (1990), have shown that the second term in Eq. (3) should be replaced by $\mu_{01}^2 \frac{\ell_0}{N} g(\omega)$ where ℓ_0 is an average spatial extension of single electron excitation. This procedure is in analogy with the space phase filling model proposed by Schmitt-Rink et al (1985) for semiconductors (cf. also Greene et al (1987)). *Fig. 3* shows the size of single electron excitation for two first excited states. The calculated variation of γ as a function of polymer length is displayed in *Fig. 4*.

On the other hand, it can be shown (cf. *Fig. 2* and *Fig. 3*) that the calculated quantities μ_{01}^2/N and μ_{12} do not depend on polymer chain length if this one is larger than the elementary excitation size ℓ_0.

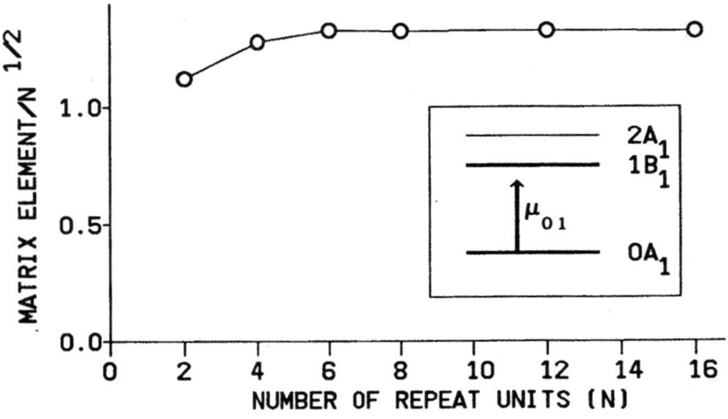

Figure 2: Variation with the number N of thiophene units of transition dipole moment μ_{01} per monomer unit, in a polythiophene.
The length of a repeat unit is 3.89 Å. The matrix element is expressed in e.Å unit where e is the electronic charge (after Charra and Messier (1990)).

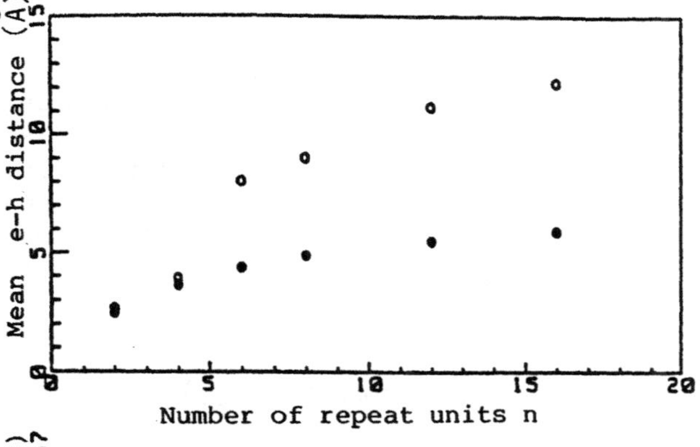

Figure 3: Variation of the mean square electron-hole distance σ_0 with the number N of thiophene units in a planar oligomer, for one-electron excited state:
• one-photon state,
o two-photon state
(after Charra and Messier (1990)).

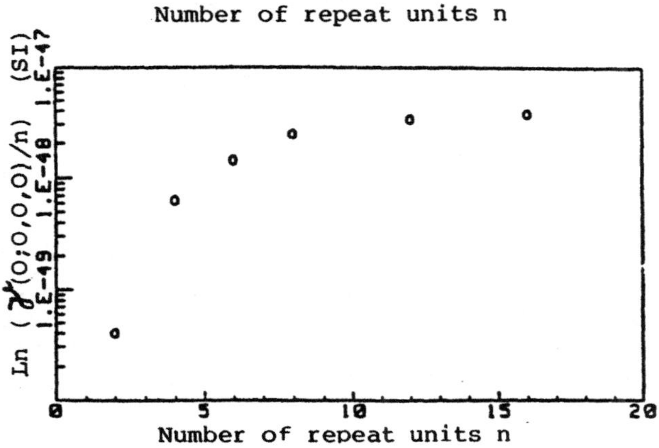

Figure 4: Variation with the number N of thiophene units of the hyperpolarizability per momoner unit (γ/N), for $\omega = 0$, in SI units (γ(esu) = 7.16 10^{13} γ (SI))

1.4.2. Non-centrosymmetric molecules

For the non-centrosymmetric molecules two supplementary terms γ'_{NCS} and γ''_{NCS} have to be added to Eq. (3). This is due to two facts:
i) Non-zeroth ground and excited states dipolar moments μ_{00}, μ_{11} and $\Delta\mu_2$. Defining by $\Delta\mu_1 = \mu_{11} - \mu_{00}$, $\Delta\mu_2 = \mu_{22} - \mu_{00}$, we have

$$\gamma'_{NCS} = \frac{1}{\varepsilon_0} \mu_{01}^2 \; (\Delta\mu_1)^2 \; f_{11}^{NCS}(\omega) \tag{6}$$

where $f_{11}^{NCS}(\omega)$ can be got from Eq. (4) by putting $E_2 = E_1$.

ii) Some dipolar transition matrices like μ_{02} are no more equal to zero as in the case of centrosymmetric molecules. This fact introduces a term given by

$$\gamma''_{NCS} = \frac{1}{\varepsilon_0} \mu_{02}^2 \left[\mu_{12}^2 f_{21}(\omega) + \Delta\mu_2^2 f_{22}(\omega) - \mu_{02}^2 g_{22}(\omega)\right] \tag{7}$$

where $f_{21}(\omega)$ and $g_{22}(\omega)$ are obtained from Eqs. (4)-(5) by permuting E_2 with E_1 and the function $g_{22}(\omega)$ is got from Eq. (5) by putting $E_1 = E_2$. Note that for E_2 close to E_1 as it is the case in conjugated polymers (Kohler and Schilke (1987), Kajzar and Messier (1985a)) consequently $f_{12}(\omega) \to f_{11}(\omega)$, $g_{22}(\omega) \to g_{11}(\omega)$ and $f_{22}(\omega) \to f_{11}(\omega)$. Putting $f_{12}(\omega) = f_{11}(\omega) = f_{22}(\omega) = f(\omega)$ and $g_{22}(\omega) = g_{11}(\omega) = g(\omega)$ one obtains

$$\gamma = \gamma_{CS} + \gamma'_{NCS} + \gamma''_{NCS}$$
$$= \frac{1}{\varepsilon_0}\left[\left(\mu_{12}^2\left(\mu_{01}^2+\mu_{02}^2\right) + \mu_{01}^2\Delta\mu_1^2 + \mu_{02}^2\Delta\mu_2^2\right)f(\omega) - \left(\mu_{01}^4+\mu_{02}^4\right)g(\omega)\right] \tag{8}$$

For a small departure from centrosymmetricity $\mu_{01}^2 \gg \mu_{02}^2$ and

$$\gamma = \frac{1}{\varepsilon_0}\mu_{01}^2\left[\left(\mu_{12}^2+\Delta\mu_1^2\right)f(\omega) - \mu_{01}^2\, g(\omega)\right] \tag{9}$$

Consequently the noncentrosymmetric case (Eq. (9)) can be deduced from the centrosymmetric one (Eq. (3)) simply by replacing μ_{12}^2 by $\mu_{12}^2+\Delta\mu_1^2$. In Section IV.1 we will discuss the following coupled experiments: optical absorption, THG and EFISH which allow to evaluate different parameters like μ_{01}^2, μ_{12}^2, $\Delta\mu^2$, E_1 and E_2 intervening in the formulas for γ.

1.5. Conformational effects

The formulas for γ (Eqs. (3)-(9)) are valid for a given configuration of a polymer. It is true in the case of PDA single crystals where polymer chains have a well defined planar structure. In some cases by epitaxy one can obtain oriented thin films of PDA with polymer chains directed along a preferential crystallographic direction of the substrate (cf. Le Moigne et al. (1990)).

In solution some polymers like poly-4BCMU have a structure close to a planar one with a well defined conformation. However often one has to do with a large distribution of different conformations both in amorphous thin films and in solutions. It is the case of e.g. polythiophene, characterized by a large variety of different conformations (Themans et al. (1989), Inganas et al. (1989), Fichou et al. (1989)). Using similar calculations as those presented in Section I.2, one can relate a well defined optical absorption and second order hyperpolarizability with a given conformation. This is shown in *Figs. 5* and *6* displaying variation of one-photon transition energy and γ values as a function of angle θ between two adjacent thiophene rings. Meanwhile the relation is not unique; it means that different conformations can be characterized by the same transition energy E_1 and hyperpolarizability γ. The implicit relation between E_1 and γ which can be deduced from *Figs. 5* and *6* allowed to determine with a reasonable accuracy the magnitude and the frequency variation of $\chi^{(3)}$ in the case of a thin film of polybutylthiophene having a given optical absorption spectrum (Kajzar et al. (1989)).

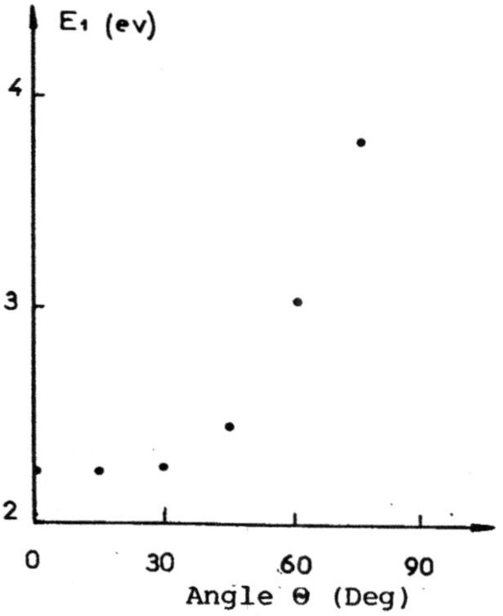

Figure 5: Variation of the first one-photon excited state energy E_1 as a function of the θ angle between two successive monomer units in anhelical polymer.

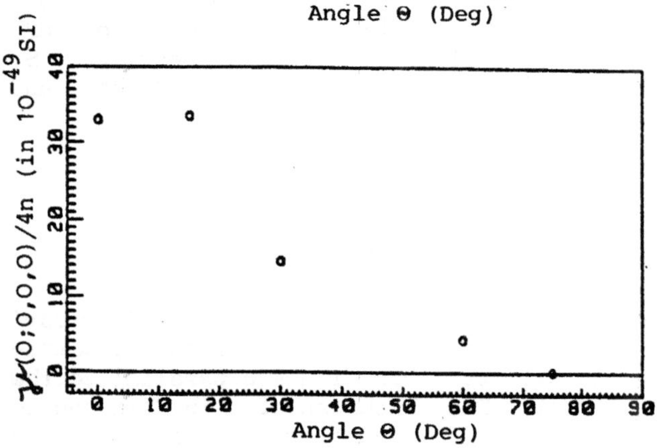

Figure 6 : Variation of γ/N, at ω = 0, for a helical polymer as a function of the θ angle (cf. Fig. 5).

2. THIN FILM PREPARATION METHODS

A principal advantage of organic molecules in general and polymers in particular consists on their versatility and ease in processability. Depending on the nature of polymer itself, its chemical structure, side groups different thin film preparation techniques can be employed. Here we will describe shortly main techniques used in polymer thin film preparation.

2.1. Langmuir-Blodgett technique

This technique is applicable to monomers polymerizing topochemically like diacetylenes. It requires a special design of the molecule with long aliphatic chain side groups like $CH_3(CH_2)_n$-A-$(CH_2)_m$COOH where A is the active molecule (A = C ≡ C - C ≡ C in the case of diacetylenes). Monomers are compressed on the water subphase forming a monomolecular solid film which subsequently is transfered to a substrate during its drawing in the bath (cf. *Fig. 7*). The thin film stability depends on the aliphatic chain lengths (n and m). Regular thin films are obtained in this way with thickness varying from ca 25 Å to its odd or even multiple depending on the substrate used (hydrophobic or hydrophilic, respectively). The polymerization can be proceeded by two ways.

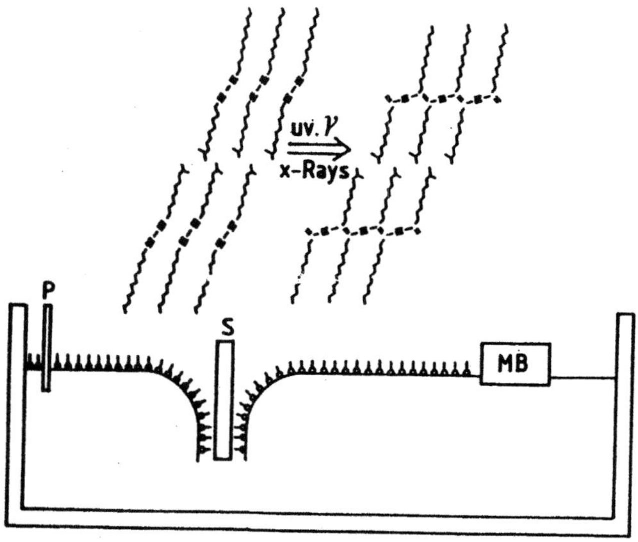

Figure 7 : Principle of Langmuir-Blodgett thin film deposition. The upper part of figure show a polydiacetylene monomer (left) and polymer (right) bilayer: P, pressure sensor; MB, mobile barrier; S, substrate (Kajzar and Messier (1987c)).

One can polymerize monolayer film on water subphase (Day and Ringsdorf (1978)) and so formed polymer monolayer can be transferred on a substrate. Polymer chains of obtained in this way thin film are parallel to the substrate (cf. Kajzar et al (1988)). The thickness of thin film can be increased by increasing the number of monomolecular polymer film transfers.

Other technique consists on the transfer of monomer film on a substrate and its polymerization with UV or γ irradiation. Again the thickness of monomer thin film depends on the number of successive transfers of monomolecular monomer layers.

The Langmuir-Blodgett (LB) technique gives very regular oriented thin films with well controllable thickness, starting from 25 Å; the final thickness being theoretically determined by the number of drawings. In practice the final thickness of LB films is limited by molecular disorder appearing for a large number of layers.

In the case of polydiacetylenes the LB technique gives polycrystalline thin films scattering strongly the light, thus not applicable for any active wave-guides (Grunfeld and Pitt (1983)). On the other hand these regular thin films are very useful in fine linear and nonlinear spectroscopic studies.

2.2. Solution casting

Some polymers, like polydiacetylenes with a general formula

$$R_1-C-C\equiv C-C-R_2$$

where $R_1 = R_2 = (CH_2)_n OCONHCH_2COO\ C_4H_9$ (n=3-poly-3BCMU, n=4-poly-4BCMU)

polyalkylthiophenes: (n = 2,3,...,12),

polysilanes, functionalized polyacetylenes are soluble giving, depending on molecular weight, viscous solutions. This property is used in thin film preparation by solution casting. In the case of soluble polydiacetylenes simply by solvent evaporation one obtains thin films scattering light due to the surface roughness (Chollet et al. (1987)). Contrarywise using the spinning technique one obtains regular, good optical quality thin films of poly-4-BCHU and poly-3-BCMU with relatively small propagation losses ~ dB/cm (Townsend et al. (1988), Krug et al. (1989)). This technique gives also good optical quality polysilane thin films (Kajzar et al. (1986), Baumert et al. (1989), Morichere et al. (1989)).

2.3. Vacuum evaporation-epitaxy

The ability of polydiacetylenes to polymerize topochemically was used successfully in thin film preparation by controlled monomer evaporation under high vacuum (~ 10^{-7} bars)(Le Moigne et al. (1987), Chollet et al. (1987)). Obtained in this way thin films can be polymerized by UV or γ irradiation. The technique gives very regular, polycrystalline thin films with controllable thickness. The size of crystallites (and consequently the light scattering) can be influenced by controlling the substrate temperature (Chollet et al (1987)). Using amorphous substrates (e.g. silica) one obtains isotropic thin films (Le Moigne et al. (1967)). With monocrystalline KBr substrate and DCH monomer, it was possible to obtain bioriented thin films with polymer chains (after topochemical polymerization) parallel to the substrate [100] and [010] cubic crystallographic directions. A mono-orientation was obtained with another single crystal as substrate: KAP (phthalate acid of potassium) (Le Moigne et al. (1990). In both cases the obtained thin films are polycrystalline with polymer chains mutually perpendicular in first case and parallel in the second one. The technique gives large area oriented thin films with order respected to about 2000 Å thickness.

Kanetake et al. (1987) have also obtained mono-oriented thin films by a quasi-epitaxy. A very thin, about 100 Å thick monomer film deposited on a glass substrate by vacuum evaporation is oriented by rubbing. Thicker film is got subsequently by epitaxy through high vacuum evaporation of monomer molecules on prepared in such a way substrate.

Other techniques have been also used in oriented thin fim fabrication like shear technique (Meyler and Thakur (1985)) giving truly monocrystalline thin films. In this case, the film is grown from solution between two e.g. silica slides with a preorientation of monomers at low temperature by shearing and controlled solvent evaporation. It works for monomers characterized by a high ability for cristallization like diacetylenes TS and TCDU.

Oriented thin films can be also obtained by stretching of polyacetylene, poly(alkyl)thiophene and some polydiacetylene isotropic thin films. In this case polymer chains align in the stretching direction. Some films like polytiophene, polypyrrole can be obtained by electrochemical deposition.

3. PRINCIPAL $\chi^{(3)}$ CHARACTERIZATION TECHNIQUES

In this section we describe different experimental techniques leading to corresponding $\chi^{(3)}$ tensors determination.

3.1. Third harmonic generation

The formalism describing third harmonic generation in a nonlinear medium in thin films and in liquids have been developed by different authors e.g. Amstrong et al. (1962), Bloembergen (1965), Kajzar and Messier (1985b), (1987b), Kajzar et al. (1986), Kajzar (1989), Bethune (1988). Here we recall fundamental formulas leading to $\chi^{(3)}(-3\omega;\omega,\omega,\omega)$ determination. For the purpose of simplicity we will start considerations with nonabsorbing media and generalize it after to absorbing ones.

3.1.1. THG in nonabsorbing stratified media

Third harmonic generation produced in n successive nonlinear optical media by a laser beam with frequency ω can be considered as resulting in a superposition of n nonlinear emissions from each nonabsorbing medium separately (THG in a plane parallel slab). Harmonic generation in a nonlinear medium (2) (cf. *Fig. 8*) can be described in term of source (bound wave) $E_{2b}^{3\omega}(\vec{r},t)$ which is in origin of the nonlinear emission and is given by

$$E_{2b}^{3\omega}(\vec{r},t) = \frac{1}{2}\left[E_{2b}^{3\omega}\exp(i(\vec{t}_{2b}.\vec{r}-3\omega t) + C.C.\right] \tag{10}$$

where
$$E_{2b}^{3\omega} = \frac{\chi^{(3)}(-3\omega;\omega,\omega,\omega)\left(E_{2t}^{\omega}\right)^3}{4\Delta\varepsilon} \tag{11}$$

and E^{ω}_{2t} is the amplitude of fundamental beam in medium 2

$$E^{\omega}_{2t}(\vec{r},t) = \frac{1}{2}\left[E^{\omega}_{2t}\ \exp\left(i\left[\vec{k}^{\omega}_{2}.\vec{r}-\omega t\right]\right) + C.C.\right] \tag{12}$$

In Eq. (10) $\vec{k}_{2b} = 3\vec{k}^{\omega}_{2}$ and $\Delta\varepsilon = \varepsilon_{\omega} - \varepsilon_{3\omega}$ is the dielectric constant dispersion.

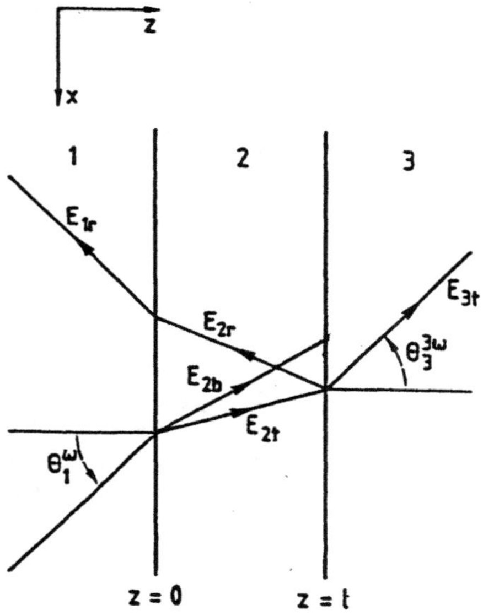

Figure 8: Harmonic generation in a nonlinear medium (2) immersed between two linear media (1 and 3). Arrows show harmonic wave vectors at 3 ω. Electric fields are perpendicular to the figure plane.
b means "bound wave",
t means "transmitted wave",
r means "reflected wave".
Remark: In the linear media, if $n_{\omega} = n_{3\omega}$(for instance in vacuum), the direction of propagation of fundamental and harmonic waves is the same.

Each bound wave is associated with 4 free waves propagating in 1, 2, 3 media (cf. *Fig. 8*) and determined by boundary conditions (continuity of tangential components of electric and magnetic fields at interfaces) and propagating in forward and backward directions. For example the free wave in medium E_{3t} propagates in the direction of fundamental beam and its amplitude is given by (Kajzar and Messier (1985b))

$$E^{3\omega}_{3t} = E^{3\omega}_{2b}\ \exp\ i\left(\vec{k}^{3\omega}_{2t}.\vec{\ell}\right)\left\{A_1\left[\exp\ i\left(\vec{k}^{3\omega}_{2b}-\vec{k}^{3\omega}_{2t}\right)\ .\ \vec{\ell} - 1\right]\right.$$

$$\left. + A_2\left[\exp\ i\left(\vec{k}^{3\omega}_{2t}+\vec{k}^{3\omega}_{2t}\right)\right]\ .\ \vec{\ell} - 1\right\}/D \tag{13}$$

where

$$A_1 = \frac{N^{3\omega}_2 + N^{\omega}_2}{N^{3\omega}_3 + N^{3\omega}_2}\ ,\quad A_2 = \left(\frac{N^{3\omega}_2 - N^{\omega}_2}{N^{3\omega}_3 + N^{3\omega}_2}\right)\left(\frac{N^{3\omega}_1 - N^{3\omega}_2}{N^{3\omega}_1 + N^{3\omega}_2}\right) \tag{14}$$

$$D = 1 - \left(\frac{N_1^{3\omega} - N_2^{3\omega}}{N_1^{3\omega} + N_2^{3\omega}}\right)\left(\frac{N_3^{3\omega} - N_2^{3\omega}}{N_3^{3\omega} + N_2^{3\omega}}\right) \exp\left[i\left(6\omega \ell N_2^{3\omega}/C\right)\right]$$

$$\text{with } N_j^{m\omega} = \frac{c}{m\omega}\left(\vec{k}_j^{m\omega}.\vec{\ell}/\ell\right) = n_j^{m\omega} \cos \theta_j^{m\omega} \begin{cases} j = 1,2,3 \\ m = 1,3 \end{cases} \tag{15}$$

In Eqs. (13)-(15) $\vec{\ell}$ is a vector parallel to the normal to interface with length equal to the 2 medium thickness, θ is propagation angle and $n_j^{\omega,3\omega}$ is linear refractive index in medium j at ω or 3ω frequency, respectively.

The parameters A_1 and D_1 are close to 1 if the refractive indices of successive media are not very different and $A_2 \ll 1$. In this case, Eq. (13) can be represented graphically in complex plane as shown in *Fig. 9*, where the angle $\Delta\varphi$ represents the phase mismatch between fundamental and harmonic waves

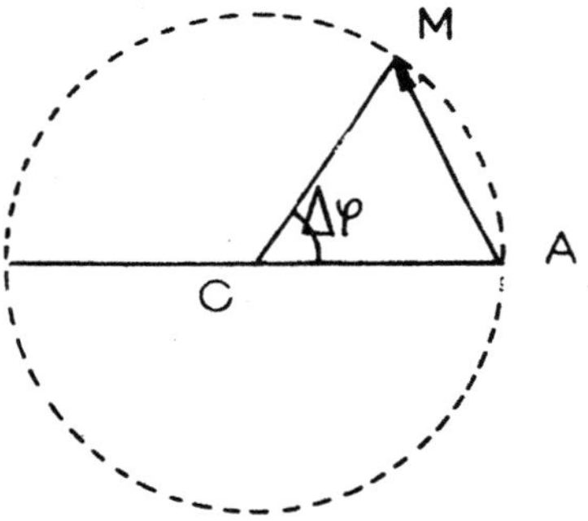

Figure 9: Graphical representation, in complex plane, of harmonic field createdin a nonlinear nonabsorbing medium by a plane fundamental wave. $|OA| = E_{2b}$.

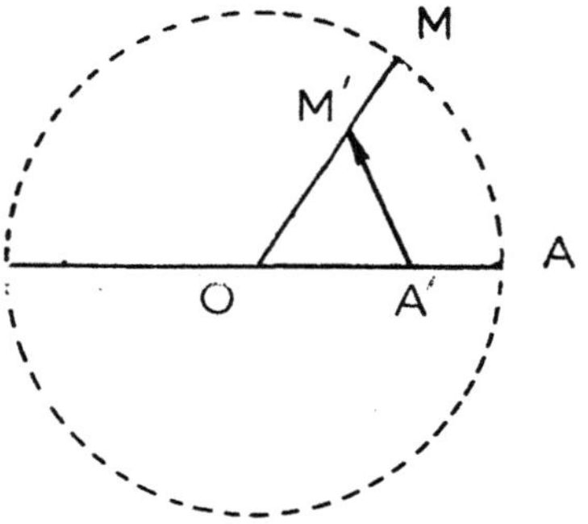

Figure 10: Graphical representation of harmonic field in the case of an absorbing medium. The resultant field is $\overrightarrow{A'M'}$ instead of $\overrightarrow{AM}$ in the non-absorbing case. For the definition of $\overrightarrow{OA'}$ and $\overrightarrow{OM'}$, see equations (18) and (19).

$$\Delta\varphi = \varphi_{3\omega} - \varphi_{\omega} = \frac{3\omega\ell}{c}\left(n_{3\omega} \cos \theta_{2t}^{3\omega} - n_{\omega} \cos \theta_{2b}^{\omega}\right) \tag{16}$$

The phase mismatch $\Delta\varphi$ is increasing with medium thickness ℓ and (E_{3t}) is a periodic function with periodicity of $2\ell_c$, where $\ell_c = \lambda_\omega/6(n_{3\omega}-n_\omega)$ is the coherence length, generally comprise between a few and a few tens of microns, depending on wavelength and refractive index dispersion.

3.1.2. Harmonic generation in absorbing media

In a medium absorbing at fundamental and/or harmonic fequency the index of refraction is complex

$$n_2^{m\omega} = n_2^{m\omega} + i\ K_2^{m\omega} \qquad (m = 1,3)$$

It can be easily shown that the propagation directions of fundamental and harmonic waves are close to those in nonabsorbing media. The main difference appears in amplitudes of fundamental and harmonic waves which are given by

$$E_2^{m\omega}(\ell) = E_2^{m\omega}(0)\ \exp\left(-K_2^{m\omega}\frac{m\omega\ell}{c\ \cos\ \theta_2^{m\omega}}\right) \tag{17}$$

where similarly as before ℓ is the medium thickness and m = 1,3.

In an absorbing medium the wave vector $\vec{k}$, the dielectric constant ε, as well as transmission factors and those arising from boundary conditions are complex (cf. Eqs. (13)-(15)). Thus the graphical representation of resultant harmonic field (cf. *Fig. 9*) has to be modified in the following manner (cf. *Fig. 10*). The $\overrightarrow{AM}$ is replace by $\overrightarrow{A'M'}$ with

$$OA' = OA\ \exp\left(-K_2^{3\omega}\frac{3\omega\ell}{c\ \cos\ \theta_2^{3\omega}}\right) \tag{18}$$

and $$OM' = OM\ \exp\left(-3K_2^{\omega}\frac{\omega\ell}{c\ \cos\ \theta_2^{\omega}}\right) \tag{19}$$

For example in the most frequent case of a nonlinear medium transparent at fundamental and absorbing at harmonic frequency one has OM' = OM and OA' << OA. Consequently point A' is close to 0, $\overrightarrow{A'M'}$ # $\overrightarrow{OM}$ and $\overrightarrow{A'M'}$ does not depend on the nonlinear medium thickness.
The harmonic intensity for a thin film absorbing at harmonic intensity is diminished with respect to that from a nonabsorbing thin film by a factor (cf. Kajzar and Messier (1985a))

$$f_a = \frac{1}{36\pi^2}\ \frac{[1-\exp(-\alpha_{3\omega}\ell/2)]^2+(\Delta\varphi)^2\exp(-\alpha_{3\omega}\ell/2)}{\left[\left(n_{3\omega}^2-n_\omega^2\right)^2 - (K_{3\omega})^2\right]^2 + 4(n_{3\omega}K_{3\omega})^2}\left(\frac{\lambda_\omega}{\ell}\right)^2\left(n_\omega+n_{3\omega}\right)^2 \tag{20}$$

where $\varphi_{3\omega}$ is the linear absorption coefficient at harmonic frequency.

3.1.3. Multiple reflections

In a nonlinear stratified medium both harmonic and fundamental waves undergo multiple reflections which can result in an important modification of harmonic emission especially in the case of significantly different refractive indices of neighbouring media. Eq. (13) takes account of multiple reflections of harmonic wave on (1-2) and (2-3) interfaces associated to the fundamental wave E^{ω}_{2t} propagating in a given direction. In the case of one reflected wave E^{ω}_{2r} the nonlinear optical response is proportional to $\left(E^{\omega}_{2t} + E^{\omega}_{2r}\right)^3$. Due to the following terms: $\left(E^{\omega}_{2t}\right)^3$, $\left(E^{\omega}_{2t}\right)^2 E^{\omega}_{2r}$, $E^{\omega}_{2t}\left(E^{\omega}_{2r}\right)^2$ and $\left(E^{\omega}_{2r}\right)^3$ there exist 4 bound waves giving rise to 16 free waves. The boundary conditions impose that these free waves propagate outside nonlinear medium under consideration (in medium (1) and (3) in the present case) in the same direction (cf. Kajzar and Messier (1987a)). If the reflected wave has the same phase as the incident wave a significant variation of harmonic intensity may occur. E.g. a reflected wave with 1% intensity of the fundamental wave may induce a 20% variation in harmonic wave intensity.

Although the theoretical problems connected with multiple reflection effects have been solved in a general case (Bethune (1988)) in practice one hase to take account of experimental conditions like e.g. finite beam diameter, not complete superposition of reflected and incident beam, depending on incidence angle and thin film thickness, negligible reflection in some cases between thin fim and substrate (close refractive indices) etc..

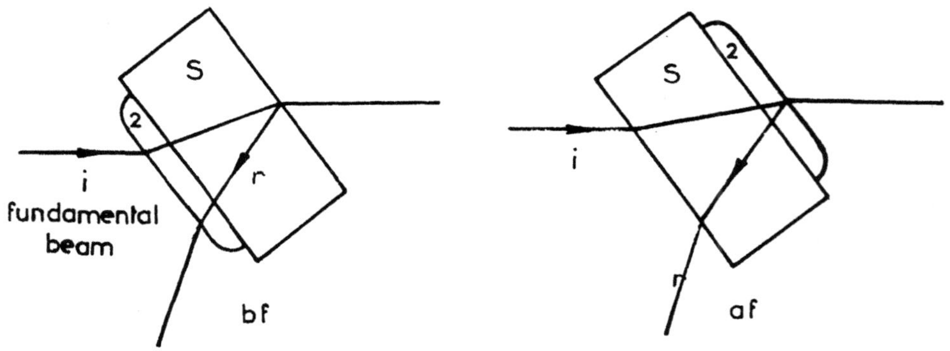

Figure 11: Multiple reflection. The coherent overlap between incident and reflected fundamental wave may be very different according to whether the nonlinear medium is located before (bf) or after (af) the substrate.

In pratice two cases can be present (Cf. Neher et al. (1989)) (cf. *Fig. 11*)

i) thin film before substrate (with respect to the beam propagation direction) (bf),
ii) thin film after substrate (af).

In the second case (af) the reflected beam is almost in phase with incident one independently of incidence angle whereas it is less obvious in the first case. On the other hand first configuration is less advantageous because of scattering losses on thin film surface due to its roughness.

3.1.4. Harmonic generation in focused laser beams

Up to now we have considered only harmonic generation by a plane wave. Harmonic generation by a focused laser beam was treated by Kleinman et al (1966), Ward and New (1969) and applied for THG in liquids and in air (environmental effects) by Meredith et al (1983b), Kajzar and Messier (1985b), (1987b).

The resultant harmonic field can be again represented in the complex plane as it is done in *Fig. 12* showing harmonic field at the focal point. One can see that the resultant field describes a spiral with radius $r=\chi^{(3)}\left(E_t^{\omega}\right)^3/4\Delta\varepsilon$ increasing progressively up to the focal point and decreases thereafter to a zeroth value far from focus. It means that a focused Gaussian beam does not generate harmonic field in an infinite isotropic medium but it does not mean that one can neglect the contribution of such a medium surrounding a thin dispersive nonlinear medium located at the focus.

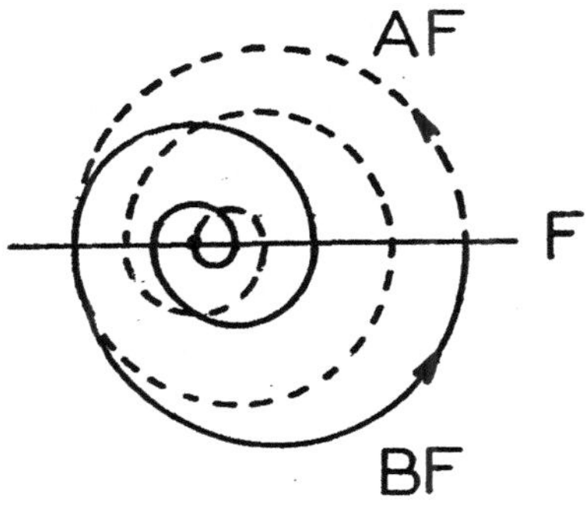

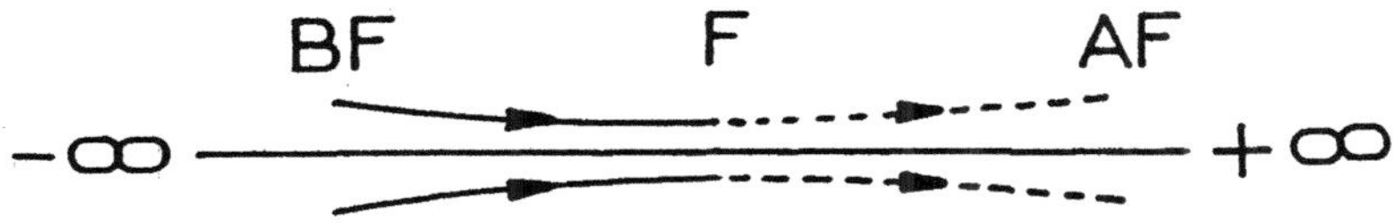

Figure 12: Graphical representation of harmonic field in focused laser beam.

3.2. Third harmonic generation in thin films and in solutions

3.2.1. Thin film case

Thin polymer film, deposited on a substrate, is placed in focus of a lense with sufficiently large focal length in order to consider fundamental beam illuminating the sample as a plane wave with constant intensity. Two different situations can be present:

i) Harmonic field generated in polymer film is made larger than that generated in substrate itself and thin film thickness ℓ is significantly smaller than the coherence length ($\ell \ll \ell_c$). In this case harmonic intensity decreases monotically with incidence angle when rotating the sample along an axis perpendicular to the beam propagation direction due to reflection of incident beam. The harmonic intensity is given by (cf. Kajzar and Messier (1985a))

$$J_{3\omega} = N_f \left(\frac{\chi^{(3)}}{n^p_\omega + n^p_{3\omega}}\right)^2 \left(\frac{\ell}{\lambda_\omega}\right)^2 |A_1|^2 |t^\omega_{01}|^6 J^3_\omega f_a \tag{21}$$

$$\text{with}\quad N_f = \frac{2.304\ \pi^6}{c^2} = 2.46\ 10^{-15}\,(\text{esu}),$$

$$= \frac{9\pi^2}{\varepsilon_0 c^2} = 1.26\ 10^7\ (\text{SI}),$$

where t^ω_{01} is transmission factor on first face, J_ω fundamental beam intensity in thin film, f_a takes account of absorption arising from boundary conditions (cf. Eq. (20)). By comparison with harmonic generation in a standard one can determine the $\chi^{(3)}$ value for the studied polymer film.

ii) In the case when the harmonic field generated in substrate itself is comparable to that generated in thin film the resultant harmonic intensity varies periodically with incidence angle (cf. *Fig. 13*) and is given by (cf. Kajzar et al. (1986))

$$J_{3\omega}(\theta) = \frac{64\pi^4}{c^2}\left(\frac{\chi^{(3)}}{\Delta\varepsilon}\right)^2_s \left| e^{i\left(\varphi^s_\omega+\varphi^p_\omega\right)}\left[T_1\left(1-e^{-i\Delta\varphi_s}\right) + \rho T_2 e^{i\Phi}\left(e^{i\Delta\varphi_p}-1\right)\right]\right| \tag{22}$$

$$\text{where}\ \rho = \left|\frac{(\chi^{(3)}/\Delta\varepsilon)_p}{(\chi^{(3)}/\Delta\varepsilon)_s}\right| \tag{23}$$

$$\text{and}\quad T_1 = A^s_1\left(t^{os}_\omega\right)^3 t^{po}_{3\omega}$$

$$T_2 = A^p_1\left(t^{os}_\omega\right)^3\left(t^{sp}_\omega\right)^3 \tag{24}$$

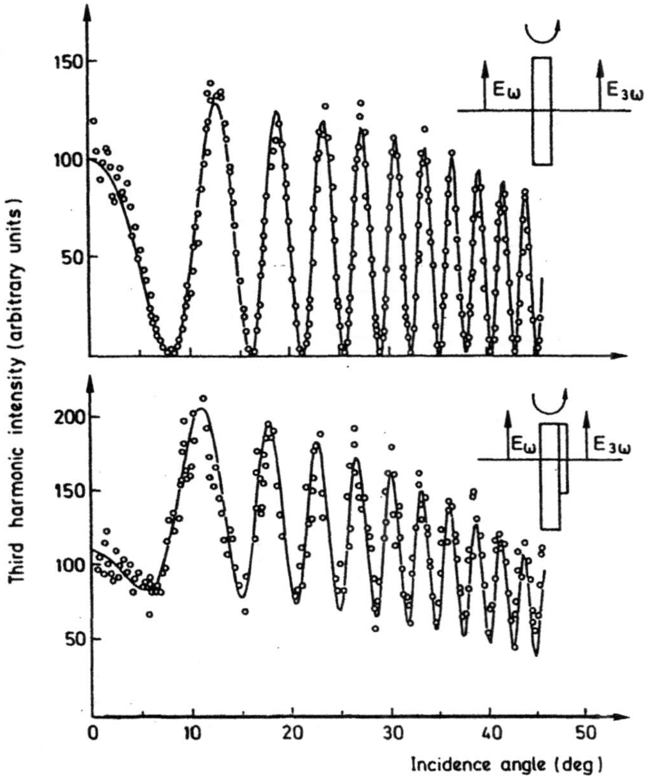

Fig. 13: Third harmonic intensity in function of incidence angle for (a) silica plate and (b) silica plate + polymer film in vacuum. Points represent measured values and solid lines the calculated ones (Kajzar et al. (1986)).

are factors arising from transmission (t^{ij}) and boundary condition (A factors). The phase mismatches $\Delta\varphi_s$ and $\Delta\varphi_p$ are given by Eq. (16) and Φ is the polymer hyperpolarizability phase. The subscripts p and s refer to polymer and substrate, respectively. In this case one can determine not only the modulus, but also the phase of polymer hyperpolarizability if all the characteristics of substrate are well known ($\chi^{(3)}$, n^s_ω, $n^s_{3\omega}$ and its thickness). In the above considerations we have assumed that THG experiments are done in vacuum otherwise one has to take account of air contribution to harmonic intensity (cf. Kajzár and Messier (1985b)).

3.2.2. THG in polymer solutions

Different techniques used in THG experiments on liquids have been discussed by Kajzar (1989b). The most convenient is that proposed by Kajzar and Messier (1987) (cf. *Fig. 14*) and consisting on use of a wedge shaped liquid compartment with thick input and output windows. The liquid cell is translated perpendicularly to the beam propagation direction. Harmonic fields generated in input and output windows are bound waves whose intensities are independent on translation x. Only that generated in liquid compartment, where laser beam is considered as plane wave, varies with x. The harmonic fields generated in air before and after the cell are negligible because of small fundamental beam intensity. The experiments can be performed in air. The resultant harmonic field is a sum of harmonic fields generated in windows (E_{G1} and E_{G2}) and in liquid comportment $E_T(x)$

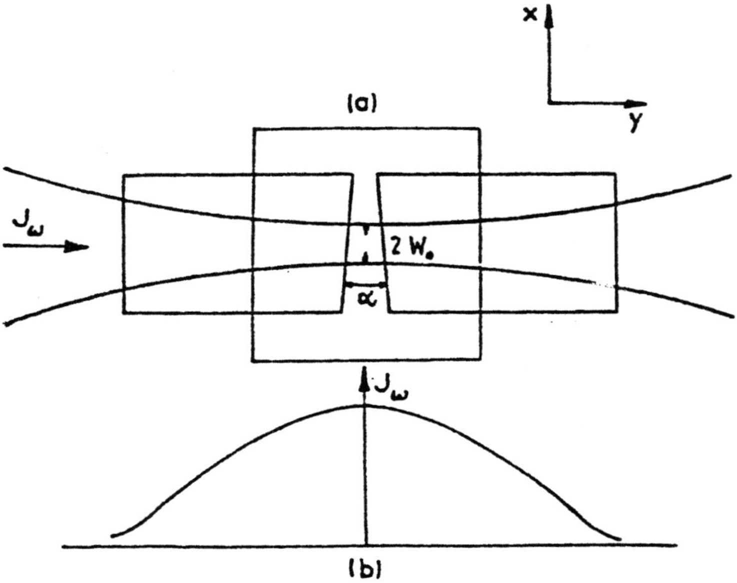

Figure 14: Cell design (a) and schematic fundamental beam intensity distribution (b) in prismed compartment liquid cell for THG in liquids (Kajzar and Messier (1987b)).

$$E_R^{3\omega}(x) = E_{G1} \cdot T_1 + E_L(x) \cdot T_L + E_{G2} \cdot T_2 \tag{25}$$

Due to the phase mismatch $\Delta\varphi$ in liquid compartment the resultant harmonic intensity varies periodically as a function of x (cf. *Fig. 15*). From the observed periodicity one can determine easily the coherence length ℓ_c and from the amplitude of harmonic intensity one can get $\chi^{(3)}$ of liquid by comparison with a known standard (e.g. a silica slab or a known solvent).

From the concentration variation of the amplitude of normalized harmonic intensity for a polymer solution with respect to that of the solvent one can get the modulus and the phase of solute cubic susceptibility if that of the solvent is known using the following formulas (cf. Kajzar (1989))

$$J_N(c)=J_{3\omega}(c)/J_{3\omega}(0)=\frac{1}{\beta^2}\left(\left(A+B(c)\left[\frac{c\gamma_p'}{M_p}+\frac{\gamma_s}{M_s}\right]\right)^2+c^2\left[\frac{\gamma_p'}{M_p}B(c)\right]^2\right)\Big/\left(A+B(0)\frac{\gamma_s}{M_s}\right)^2 \tag{26}$$

$$\text{where } A = t_{3\omega}^{GL}\, t_{3\omega}^{LG} + \beta\delta\left(t_{\omega}^{GL}\, t_{\omega}^{LG}\right)^3 \tag{27}$$

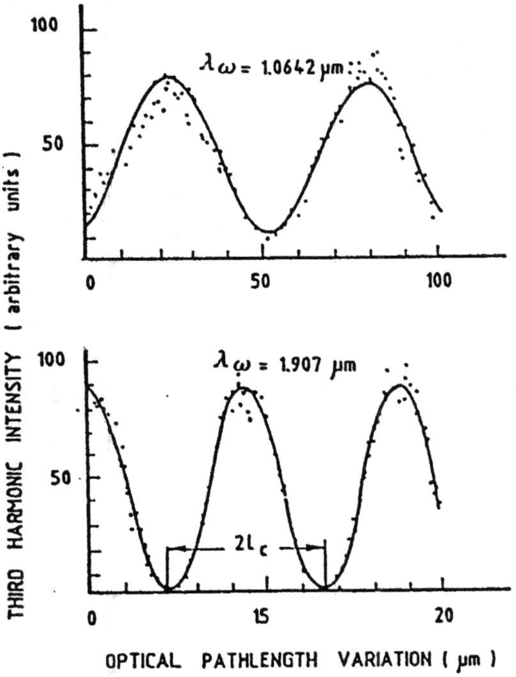

Figure 15: Harmonic intensity as a function of the cell translation for merocyanine solutions in ethanol for the case of absorbed (λ_ω = 1.064 μm) and nonabsorbed (λ_ω = 1.907 μm) harmonic light. (Kajzar (1989a)).

$$\text{and } B = -\frac{f(c)\ d(c)\ N_A}{(1+c)\ \chi_G^{(3)}}\ (\beta\delta+1)\left(t_\omega^{GL}\right)^3\ t_{3\omega}^{LG}\ \frac{(\Delta\varepsilon)_G}{(\Delta\varepsilon)_L} \tag{28}$$

where γ' and γ'' are, respectively, real and imaginary parts of molecular hyperpolarisability, t^{ij} are transmission factors between i and j media (G = window material, L = liquid) at ω(3ω) frequency and for the linearly polarized incident light (general requirement in these measurements).

$$t^{ij} = \frac{2n^i_{\omega(3\omega)}}{n^i_{\omega(3\omega)} + n^j_{\omega(3\omega)}} \tag{29}$$

The factors δ and β take account of absorption at fundamental and harmonic frequency, respectively and are given by

$$\delta = e^{3\omega\kappa_\omega \ell(x)/c}$$
$$\beta = e^{3\omega\kappa_{3\omega} \ell(x)/c} \tag{30}$$

$$\Delta\varphi = 3\omega\left(n_\omega^L - n_{3\omega}^L\right)\ \ell(x)/c \tag{31}$$

where $\ell(x)$ is the liquid compartment thickness depending on translation x

$$\ell(x) = 2x \tan \alpha/2 + \ell_o \tag{32}$$

where α is the wedge angle and ℓ_o its initial thickness.

3.3. Electric field induced second harmonic generation

3.3.1. Thin film case

The nonlinear polarization responsible for EFISH is given by (cf. Table 1)

$$P^{NL}_{2\omega} = \frac{3}{2} \chi^{(3)}(-2\omega;\omega,\omega,0)\ E^2_\omega\ E_0 \tag{33}$$

The cubic susceptibility $\chi^{(3)}(-2\omega;\omega,\omega,0)$ is determined by comparison of second harmonic generation from single crystalline quartz (Q) with that from a thin film comprise between two electrodes with an applied electric field of the order of $10^4 \div 10^5$ V/cm. The thin film cubic susceptibility in this case is given by (Chollet et al (1985))

$$\chi^{(3)}(-2\omega;\omega,\omega,0) = \frac{2}{3E_0} \left(\frac{\ell_c}{\ell_p}\right) \frac{T_Q}{T_j} \sqrt{\frac{J^{2\omega}_p}{J^{2\omega}_Q}}\ \chi^{(2)}_Q(-2\omega;\omega,\omega,0) \tag{34}$$

where T_Q and T_t are overall transmission and boundary conditions factors and ℓ_c is the quartz coherence length.

The EFISH experiments with conjugated polymers are to be performed carefully because of internal polarisation phenomena observed in these materials (Chollet et al. (1985), Obhi et al. (1986)) and due to the creation of electron-hole pairs by incident photons through one or two photon processes. This polarisation cancels the applied external fields and results in a decrease (up to its vanishing) of SHG signal.

3.3.2. Centrosymmetric molecules in solution

In this case the nonlinear polarisation at 2ω frequency comes from the $\chi^{(3)}(-2\omega;\omega,\omega,0)$ term (for a non-polar centrosymmetric solvent). Experiment can be performed in the same liquid cell as that used for THG with window kept between two electrodes allowing application of an external field. Again doing EFISH as a function of solute concentration (*cf. Fig. 16*) one can determine not only the modulus of $\chi^{(3)}(-2\omega;\omega,\omega,0)$ susceptibility but also its phase (for details see Kajzar et al (1987a)).

3.3.3. Non-centrosymmetric molecules in solution

For non-centrosymmetric molecules in solution EFISH comes from two contributions (independently whether or not the solvent molecules are centrosymmetric): $\gamma(-2\omega;\omega,\omega,0)$ and first order molecular hyperpolarizability $\beta(-2\omega;\omega,\omega)$. Due to the permanent dipolar moment the molecules are oriented by the applied external field and macroscopically the liquid is noncentrosymmetric. Thus the second harmonic generation depends on the molecular effective hyperpolarizability given by

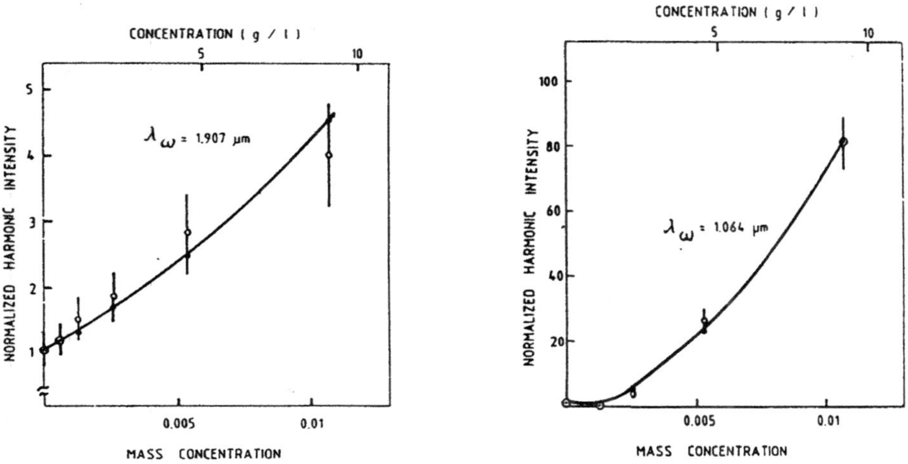

Figure 16: Normalized harmonic intensity as a function of merocyanine concentration in ethanol at 1.064 μm and 1.907 μm fundamental wavelength. Open circles are experimental values and dots the calculated ones (Kajzar (1989a)).

$$\gamma_{ef} = \gamma(-2\omega;\omega,\omega,0) + \frac{\mu\beta}{5kT} \tag{35}$$

The separation of two terms is not always easy. In some cases the second term on RHS of Eq. (35) is much larger than the first one and consequently the first one can be neglected (cf. Barzoukas et al. (1989)). It is not true for long conjugated π-electron system and other experiments are necessary to separate both terms. By coupling EFISH with THG and using two or three photon level one can separate both terms (cf. Barzoukas et al (1989), Zyss et al.(1990)). In fact in the case of polyenes (Zyss et al.(1990)) with 6 to 8 double bonds the first term on RHS of in Eq. (35) is important and cannot be neglected with respect to the second one.

3.4. Four wave mixing experiments

The four wave mixing (FWM) experiments are a useful tool in studying not only the nonlinear optical response of centrosymmetric systems but also in determining their response time. In some cases, when a standard is available they allow also the determination of the phase of corresponding $\chi^{(3)}$ susceptibility. First FWM experiments on conjugated polymers were performed by Shand and Chance (1978) (see also Chance et al.(1980)) as a function of incident wavelength in poly-3BCMU and poly-4BCMU solutions. Using solvent as standard these authors were able to determine not only the modulus of $\chi^{(3)}(\omega_2-2\omega_1;\omega_1,\omega_1,-\omega_2)$ susceptibility but also its phase. These measurements allowed them to conclude that the two photon state A_g lies above one photon allowed B_u state; the separation between them decreasing with decreasing optical gap (passing from yellow to red form).

Carter et al. (1986) (see also Carter (1987)) have first evidenced from degenerate four wave mixing experiment on thin LB films of polydiacetylene a very fast, subpicosecond response time of these polymers. Similar results have been also found by Hattori and Kobayashi (1987) from degenerate four wave mixing measurements using incoherent light on poly-3BCMU thin films (see also Kobayashi et al. (1987)).

The last technique employed by these authors is very useful because it allows to do femtosecond time resolved experiments simply by using incoherent nanosecond laser pulses.

A special class of degenerate four wave mixing experiments represent phase conjugation allowing to measure the material refractive index changes under light illumination. The technique whose geometry is shown schematically in *Fig. 17* yields not only the modulus and phase of $\chi^{(3)}(-\omega;\omega,\omega,-\omega)$ as in previously discussed DFWM experiments, but it allows also observation of creation and decay of transitory species like

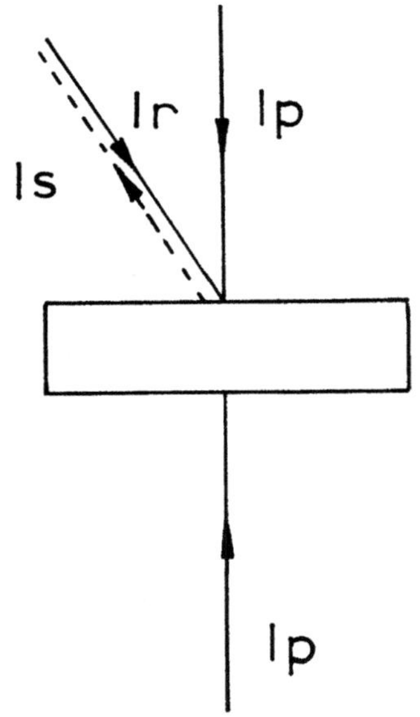

Figure 17: Schematical representation of phase conjugation experiment. I_p are pump, I_r probe and I_s conjugated signal intensities, respectively.

solitons, polarons, etc.. all giving rise to a transient grating. The phase conjugate beam intensity propagating backward to the signal beam is given by (Fisher (1983))

$$I_s \propto G\, R^3 \,\frac{I_r (I_p)^2\, \ell^2}{n_0^4}\, (\chi^{(3)}(-\omega;\omega,\omega,-\omega)^2 \qquad (36)$$

where I_p is the pump intensity, I_r the probe intensity ($I_r << I_p$), R is a factor arising from reflection and scattering losses in the nonlinear medium, n_0 is the linear part of refractive index and ℓ is interaction length in the sample.

Two types of phase conjugation configurations can take place.

i) One (or two) photon resonant.
This is an incoherent process characterized by an enhanced complex $\chi^{(3)}(-\omega;\omega,\omega,-\omega)$ susceptibility. In one photon resonance the input photon energy matches with the first allowed electronic transition energy whereas in two photon resonance with the half of energy difference between fundamental and excited state of the same symmetry, respectively. One photon resonant values of 3.5×10^{-12} e.s.u. in thin films of polypyrrol have been observed (Ghoshal (1989)). Much larger in oriented poly[2,5-dimethoxy paraphenylene vinylene] (Swiatkiewicz et al. (1990)), in soluble polyacetylenes ($\chi^{(3)} \sim 10^{-9}$ e.s.u., Dorsinville et al. (1988)), in a thiophene derivative ($\chi^{(3)} = 4.6\times10^{-9}$ e.s.u., Jenekhe et al. (1989)) and in phtalocyanines ($\chi^{(3)} = 10^{-8}$ e.s.u., Prasad (1989)). Two photon resonances have been observed by Nunzi and Grec (1987) in polydiacetylene gels ($\chi^{(3)} = 1.3\times10^{-11}$ e.s.u. for red form). This two photon resonance gives rise to a large imaginary part of $\chi^{(3)}$ (Nunzi and Grec (1987), Charra and Nunzi (1989)) and is connected with pump depletion leading to acoustic waves propagation in the medium (cf. *Fig. 18*).

It reveals also in a strong dependence of phase conjugate signal as a function of pump intensity. Above 1 GW/cm^2 and at 1.064 μm wavelength the observed variation of phase conjugate signal was tentatively interpreted in terms of higher order nonlinearities $\chi^{(5)}$ and $\chi^{(7)}$ in polydiacetylenes (Nunzi and Grec (1987), Charra and Nunzi (1989)) and in polythiophenes (Byrne et al. (1989)). The two photon resonance leads in PTS single crystals to phase conjugate reflectivity saturation and its decrease for pump powers superior to 0.5 GW/cm^2 (Nunzi and Charra (1989)). The presence of higher order terms in nonlinear polarization has been also demonstrated by Nunzi and Charra (1990) through frequency doubling in phase conjugated experiment where $\chi^{(5)}(-2\omega;\omega,\omega,-\omega,-\omega,2\omega)$ susceptibility intervenes.

ii) Nonresonant.
First nonresonant phase conjugation experiments of conjugated polymers have been performed by Denis et al (1985). For a polydiacetylene in solution (yellow form) they have measured a value of $\chi^{(3)} = 7.10^{-12}$ e.s.u. which is about two order of magnitude larger than that of CS_2 but one to two orders of magnitude smaller the resonant values (Nunzi and Grec (1987)). Large nonresonant values have been

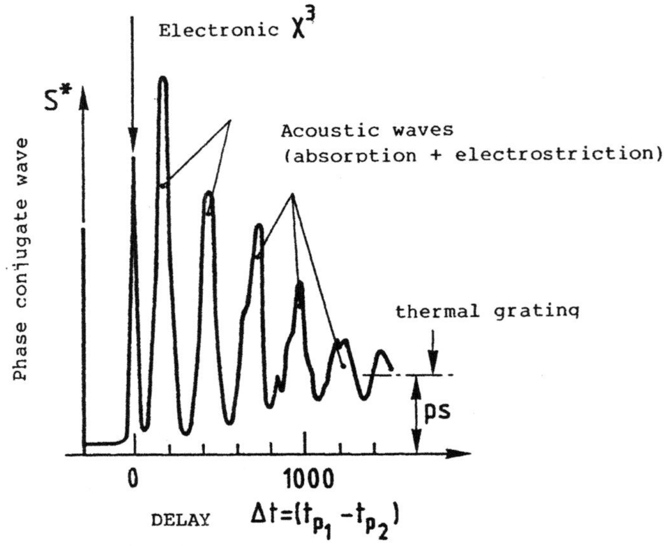

Figure 18: Pump time delay dependence of PC signal in PDA gels with corresponding contributions (Chollet et al. (1987)).

observed in oriented poly(p-phenylene vinylene) thin films ($\chi^{(3)} = 4.10^{-10}$ e.s.u., Singh et al. (1988)) with subpicosecond response time. Nonresonant $\chi^{(3)}$ values have been also measured in poly-p-phenylenebenzobisthiazole ($\chi^{(3)} \simeq 5.10^{-12}$e.s.u., Rao et al (1986)) and in polysilanes (Mac Graw et al. (1989), Yang et al. (1988), Yang et al. (1989)).

3.5. Optical and Quadratic Kerr effect

There exist very few data on optical Kerr effect measurements in conjugated polymers. Ho et al.(1986) have studied optical Kerr effect in PDA poly-4BCMU solutions in $CH_2C\ell_2$ and in PMMA matrix finding a large one photon resonant Kerr susceptibility $\chi^{(3)}(-\omega;\omega,\omega,-\omega) \simeq 3.10^{-10}$ e.s.u. with a subpicosecond response time. Very large, also one photon resonant values of Kerr susceptibility have been found in thiophene based polymers thin films ($\chi^{(3)} \simeq 10^{-9}$ e.s.u., Yang et al. (1989)). Significantly smaller off resonance values, compared to that obtained by THG (Kajzar et al. (1986)), were observed in polysilanes ($\chi^{(3)} = 7.10^{-12}$ e.s.u.).

Quadratic Kerr effect has been mainly studied in polydiacetylene solutions and gels (Kapitulnik et al.(1984), Lim and Heeger (1985), Lim et al. (1985), Kapitulnik et al. (1986)). Strong electric birefringence was observed with large rotational diffusion constants ($\tau_r \simeq 0.1 \div 0.2$ s), characteristic for very long molecules.

Morichere et al. (1989) measured quadratic Kerr effect in waveguiding polysilane thin film using attenuated total reflection, with lock-in amplifier tuned at $2\,\Omega$ frequency, where Ω is applied electric field frequency. The technique allows a precise determination of Kerr susceptibility $\chi^{(3)}(-\omega;\omega,0,0)$ which was found to be smaller to that observed by THG ($\chi^{(3)}(-\omega;\omega,0,0) = 3.2 \times 10^{-13}$e.s.u.). It allows also to determine mechanical strains in polymer thin film exerced by the applied electric field (electrostriction).

Uchiki and Kobayashi (1988) proposed an original technique for $\chi^{(3)}(-\omega;\omega,0,0)$ modulus and phase determination using a Fabry-Perrot (FB) cavity with studied polymer film inside. From the transmittance variation as a function of applied electric field on semitransparent aluminium coated input and output windows one determines easily refractive index variation of F-P cavity and consequently $\chi^{(3)}$ and its phase.

3.6. Optical Stark effect

External electric field applied to a medium shifts the energy levels. It results in a change of absorption spectrum (blue shift) and this phenomenon is called DC or AC Stark effect. The same effect one observes at optical frequencies. The shift of optical absorption spectrum is connected with the refractive index variation which can be simply visualized by Kramers-Kroning relations

$$\Delta n = \frac{1}{\pi} P \int_{-\infty}^{\infty} \frac{\Delta K}{x - \omega} dx \tag{37}$$

where P is the integral principal value and ΔK is imaginary part of refractive index variation due to the absorption spectrum shift.

The magnitude of absorption shift depends on the detuning energy i.e. the difference between pump and one photon allowed transition energy E_1 and is proportional to the square of applied electric field. Within the three level model the frequency shift of absorption spectrum (cf. Charra and Nunzi (1989)) is given by

$$\Delta\omega = \left[A\ \mu_{01}^2 \left(\frac{1}{E_1 - \omega} + \frac{1}{E_1 + \omega} \right) + \frac{\mu_{12}^2}{2} \left(\frac{1}{E_1 - E_2 - \omega} + \frac{1}{E_1 - E_2 + \omega} \right) \right] E_\omega^2 \tag{38}$$

where similarly as before E_1 and E_2 are one and two photon transition energies, respectively and μ_{01}, μ_{12} are dipolar transition moments.
Using this model and E_1, E_2 values from linear optical absorption and THG measurements (Messier (1989)), Charra and Nunzi (1989)) have determined from optical Stark effect measurements on poly-4BCMU blue and red gels the ratio of dipolar transition moments μ_{12}/μ_{01}. It follows that for centrosymmetric molecules optical Stark effect is a useful tool in determining dipolar transition moments between levels which normally are not accessible by usuel spectroscopic techniques.

Blanchard et al. (1989) studied optical Stark effect in PTS single crystal film as a function of detuning energy $\Delta E = E_1 - \omega$. At small ΔE values the optical Stark effect is phonon mediated because of strong exciton-phonon coupling in these materials. At large ΔE it is mainly excitonic Stark effect.

3.7. Saturation absorption

The availability of good optical quality single crystals and of homogeneous thin films of conjugated polymers allows to do saturation absorption measurements yielding the degenerate one photon resonant Kerr susceptibility $\chi^{(3)}(-\omega;\omega,\omega,-\omega)$. This process, shown schematically in *Fig. 19* is observed in a typical pump-probe experiment in which an intense pump beam excites electrons from the ground to an allowed one photon level whereas the probe beam tests occupation of the fundamental state. By delaying probe beam with respect to the pump one obtains also the response time of the system. The first saturation absorption experiments have been performed by reflection on PTS single crystals by Greene et al. (1987). From the relative variation of reflectivity $\Delta R/R$ the authors deduced a large value of $n_2 \simeq 3.10^{-8} cm^2/W$ at 0.629 μm laser beam wavelength with a fast response time (~ 2 ps).

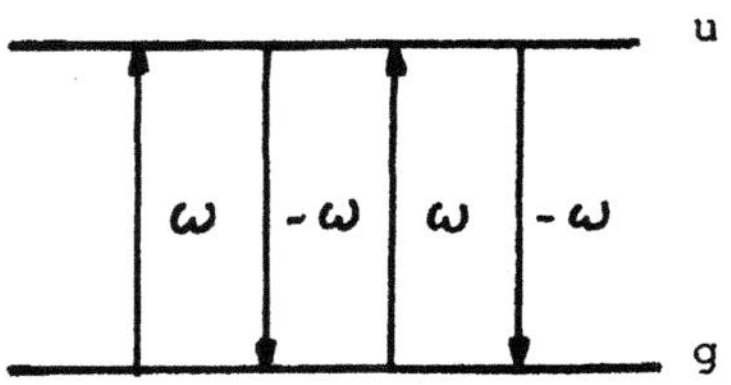

Figure 19: Schematic representation of saturation absorption as a degenerate four photon process.

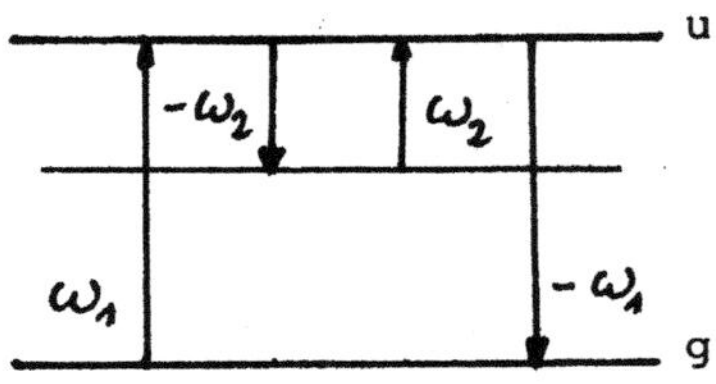

Figure 20: Schematic representation of photoinduced absorption as a four photon process

Kajzar et al.(1988) have performed the saturation absorption measurements by transmission on blue form of thin (330 Å) LB films of polydiacetylene using the photo-acoustic technique (Rothberg et al. (1983), Rothberg (1987)). In this experiment the relative variation of the linear absorption coefficient is directly linked to the relative variation of the imaginary part of refractive index

$$\frac{\delta\alpha}{\alpha} = \frac{1}{\sqrt{2}}\frac{\delta K}{K} \tag{39}$$

On the other hand the intensity dependent index of refraction, which is complex at resonance, is given by

$$n = n_0 + n_2 I \tag{40}$$

where $n_2 = n_2^r + i K_2$ (41)

The part of refractive index describing its variation with light intensity n_2 is directly related to the Kerr susceptibility $\chi^{(3)}(-\omega;\omega,\omega,-\omega)$ through following equation

$$n_2 = 12 \pi^2 \chi^{(3)}(-\omega;\omega,\omega,-\omega)/n_0^2 c \tag{42}$$

If n_0 is real at ω frequency (or has a negligible imaginary part with respect to the real one) one obtains from K_2 measurement directly the imaginary part of $\chi^{(3)}(-\omega;\omega,\omega,-\omega)$ susceptibility. From the observed at 0.585 μm relative variation of the linear absorption coefficient $\delta\alpha/\alpha = 0.063$ the authors deduced also a large value of $K_2 = 2.3 \times 10^{-10} cm^2/W$ and consequently a large value of imaginary part of $\langle\chi^{(3)}(-\omega;\omega,\omega,-\omega)\rangle = 1.4 \times 10^{-8}$ e.s.u., where the average is taken over polymer chain disorder. We note here that both saturation absorption measurements by reflection and by transmission interpreted within a 1D phase space filling model (Schmitt-Rink et al. (1985)) give a reasonable value of exciton length (33 Å and 50 ± 25 Å, respectively).

3.8. Photoinduced absorption

Another physical phenomenon connected also with the change of occupation number of ground state and consequently with an incident power dependent refraction index variation (cf. Heeger et al. (1986)) is photoinduced absorption observed in polyacetylenes (Orenstein and Baker (1982)), polydiacetylenes (Orenstein et al. (1984)) and polythiophenes (Moraes et al. (1984)). In this process (cf. *Fig. 20*) with the pump at ω_1 energy corresponding to the first allowed one photon transition one varies refractive index in both: visible (ω_1) and infrared (ω_2) frequencies. The refractive index variation in IR can be described by a cubic susceptibility $\chi^{(3)}(-\omega_2;\omega_1,-\omega_1,\omega_2)$ and the observed relative variation of the polymer film transmission at λ_2 wavelength is given by

$$\frac{\delta T}{T} = \frac{4\pi\ell}{\lambda_2}\,\delta K \tag{43}$$

where $\delta K = K_2 I$ and ℓ is the polymer film thickness.

For a polythienotiophene (PTT) thin film with 10 μm thickness a large lower limit of $\langle Jm\, \chi^{(3)}(-\omega_2;\omega_1,-\omega_1,\omega_2)\rangle \simeq 1.3\times10^{-2}$ e.s.u. was observed at $\lambda_2 = 2.48$ μm with however a slow response time (Ruani et al. (1990).

4. FREQUENCY SPECTRUM AND RESONANCE EFFECTS IN $\chi^{(3)}$

4.1. Frequency variation of $\chi^{(3)}$ modelisation

Equations (3)-(9) allow to calculate the frequency variation of third order susceptibility $\chi^{(3)}$. In the case of a centrosymmetric compound $\chi^{(3)}$ depends on parameters appearing in one photon (E_1, μ_{01}) or two photon transition terms (E_2, μ_{12}). The first can be deduced directly from the optical absorption spectrum whereas the second are determined by THG measurements in function of incident photon frequency.

4.2. Determination of μ_{01}

The dipolar transition moment, μ_{01}, is related to the one photon allowed absorption spectra. Let us write $\mu_{01} = e.x_{01}$ where e is the electronic charge. For an isotropic distribution of one-dimensional molecules with largest matrix element $(x_{01})_{xx}$, and assuming a Lorentzian absorption line shape for a polymer solution we have, $(x_{01})_{xx}=2.33\text{x}10^{-2}[n_0\ \varepsilon_m\ \lambda_m\ \Gamma_{½}]^{½}$ with x_{01} in Å, the real part of refractive index n_0, the molar extinction coefficient ε_m in cm^{-1}.litre/mole, the maximum absorption wavelength in µm and the half-width at half maximum $\Gamma_{½}$ in cm^{-1}.

For a thin film with thickness ℓ, we have $x_{01} = 3.52\text{x}10^{-2}\left[\frac{M}{d}\,\frac{\lambda_M}{\ell}(OD)_M.\Gamma_{½}\right]^{½}$

with the optical density $(OD)_M$ at λ_M, the molar mass M in kg and, the film density d.

4.3. Frequency variation of $\chi^{(3)}(-3\omega;\omega,\omega,\omega)$ and $\chi^{(3)}(-2\omega;\omega,\omega,0)$

Figures 21-23 show, respectively the optical absorption spectrum, frequency variation of $\chi^{(3)}(-3\omega;\omega,\omega,\omega)$ and of $\chi^{(3)}(-2\omega;\omega,\omega,0)$ cubic susceptibilities determined by THG and EFISH for a PDA thin film. In all these spectra one observes vibronic side bands visualized by shoulders in the smaller wavelength range with respect to λ_M. These vibronic side

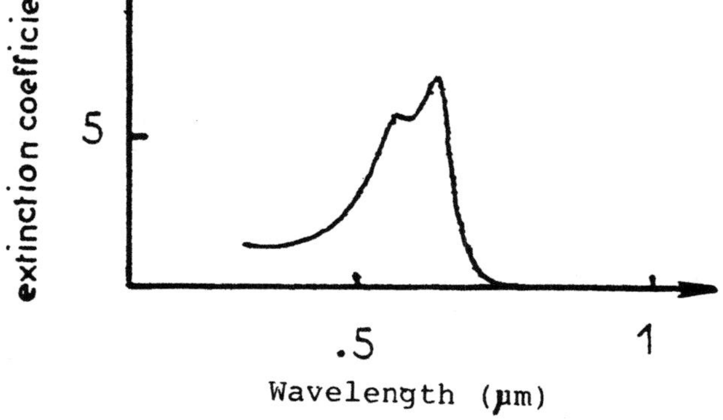

Figure 21: Absorption spectrum of blue polydiacetylene thin films.

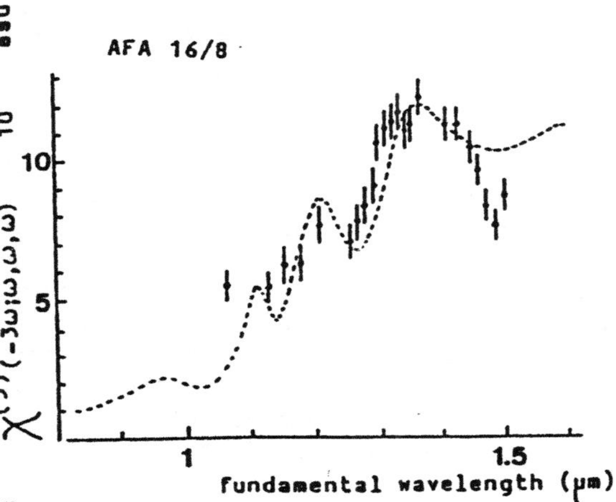

Figure 22: Comparison of the measured and calculated (broken lines) optical cubic susceptibilities for a blue polydiacetylene in the case of THG. The theoretical curve is calculated with a series of three level system associated with each vibronic replica. These replica are responsible for the shoulders on the low wavelength side of the curve. (after Chollet et al (1988b))

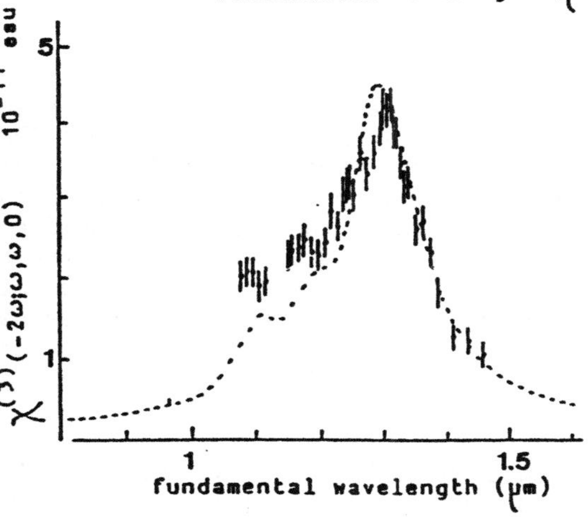

Figure 23: Cubic susceptibilities in the case of EFISH experiments for the same thin film than in Fig. 22. The theoretical curve has been calculated with the same set of parameters (after Chollet et al (1988b)).

bands can be taken into account (cf. Chollet et al (1988)) by associating with every one or two photon level (E_1, E_2) a series of (E_1', E_2', and E_1'', E_2'') levels shifted with respect to (E_1, E_2) by an energy of vibronic transition coupled to each electronic transition. For the sake of simplicity one can assume that the dipolar transition moments between corresponding vibronic levels are the same as for $0 \rightarrow 0$ transition. We note here that the position of two photon resonances observed in THG depends only on the position of two photon level E_2 allowing its good determination. In polydiacetylenes these two photon resonances are observed between 1.1 and 1.4 µm fundamental wavelength (cf. *Fig. 22*).

In EFISH experiments two photon resonances take place with one (because of selection rules break) and two photon levels. For these reasons the two photon resonance observed in PDA's is slightly different (cf. *Fig. 23*) as in THG because the one and two photon levels are separated by around 1000 cm^{-1} and these resonances overlap. Thus in the case when one and two photon levels lie close ($|E_2-E_1| \ll E_1, E_2$) the EFISH experiment do not permit their separation (a precise determination of E_2) as in the above case.

In the case of non-cèntrosymmetric molecules or thin films the calculation of γ hyperpolarizability is more difficult because it depends on additional parameters like $\Delta\mu_1$ intervening in Eqs. (8)-(9). However by comparison of EFISH and THG data in solution one can get an independent determination of $\Delta\mu$. This was done by Zyss et al. (1990) for a series of noncentrosymmetric polyenes where E_2 is close to E_1. From coupled EFISH and THG measurements on these polyenes in solution they were able to evaluate relative importance of μ_{12} and $\Delta\mu_2$ terms for these molecules. In the case of a long polyene (with 8 double bonds) in both EFISH and THG data the terms depending on μ_{12} are comparably important to those depending on the ground state dipolar moment $\Delta\mu$.

5. MULTIPHOTON RESONANCES

The rich electronic structure of conjugated polymers gives rise to different resonance enhancements in corresponding cubic susceptibilities. These resonance enchancements have been studied by different techniques and depending on the number of photons necessary to match with an excited level of unperturbed system they can be classified as one, two and three photon resonances. In Table 2 different multiphoton resonant susceptibilities are listed for several conjugated polymers. The largest resonant values are observed for one photon resonance enhancements. Among different polymers the most nonlinear are oriented polyacetylenes because of large electron delocalization and a good packing.

The study of resonance enhancements of $\chi^{(3)}$ is very important for a better knowledge of electronic structure (e.g. position and width of two photon states normally unaccessible by usual spectroscopic techniques) and for a better comprehension of electron-electron interaction in these systems. In fact the theoretical calculation of Hayden and Mele (1985) show that with increasing electron-electron correlation strength a crossover of two and one photon states occurs. It means, that for large electron-electron interactions two photon state lies below one photon state as it was observed.

6. NONLINEAR OPTICAL DICHROISM

The enhanced 1D cubic susceptibility $\chi^{(3)}(-3\omega;\omega,\omega,\omega)$ in conjugated polymers has a nice practical output in so called nonlinear optical dichroism in analogy to the linear infrared (Chollet (1980)) or visible (Vandevywer et al. (1980))one. It allows to study the orientation of polymer chains by THG measurements rotating the sample between crossed polarizers and along an axis coinciding with the beam propagation direction (in the case when the polymer chains are parallel to the substrate plane as it takes almost place) and/or perpendicular to it in other

Table 2. Selected resonance values of cubic susceptibility in conjugated molecules

Molecule	Form	Process	Resonance	$\chi^{(3)}$ tensor	Value in 10^{-14} e.s.u.	Ref.
PDA-AFA	LB film	Satur. abs.	One photon	$Jm\langle\chi^{(3)}(-\omega;\omega,\omega,-\omega)\rangle$	1.4×14^6	1
PDA-4BCMU	Solution	OKE	One photon	$\langle\chi^{(3)}(-\omega;\omega,\omega,\omega)\rangle$	3.2×10^4	3
PMTBQ	Solution	PC	One photon	$\langle\chi^{(3)}(-\omega;\omega,\omega,\omega)\rangle$	4.6×10^5	3
PPy	Thin film	PC	One photon	$\langle\chi^{(3)}(-\omega;\omega,\omega,\omega)\rangle$	3×10^2	4
PA	Solution	PC	One photon	$\langle\chi^{(3)}(-\omega;\omega,\omega,\omega)\rangle$	9×10^4	5
PDA-AFA	LB film (red)	THG	Two photon	$\langle\chi^{(3)}(-3\omega;\omega,\omega,\omega)\rangle$	$(8.0\pm1.0)\times10^2$	6
PDA-AFA	LB film(blue)	THG	Two photon	$\langle\chi^{(3)}(-3\omega;\omega,\omega,\omega)\rangle$	$(5.6\pm0.6)\times10^3$	7
PDA-AFA	Thin film(blue)	THG	Two photon	$\langle\chi^{(3)}(-3\omega;\omega,\omega,\omega)\rangle$	$(1.2\pm0.1)\times10^4$	8
PDA-AFA	Thin film (red)	THG	Two photon	$\langle\chi^{(3)}(-3\omega;\omega,\omega,\omega)\rangle$	$(3.8\pm0.4)\times10^3$	8
PDA-4BCMU	Thin film	THG	Two photon	$\langle\chi^{(3)}(-3\omega;\omega,\omega,\omega)\rangle$	$(1.0\pm0.1)\times10^4$	8
PDA-DCH	Thin film	THG	Two photon	$\langle\chi^{(3)}(-3\omega;\omega,\omega,\omega)\rangle$	$(7.0\pm0.7)\times10^3$	9
PA	Thin film	THG	Two photon	$\langle\chi^{(3)}(-3\omega;\omega,\omega,\omega)\rangle$	$(2.7\pm0.3)\times10^5$	10
PDA-LB	Thin film	EFISH	Two photon	$\langle\chi^{(3)}(-2\omega;\omega,\omega,0)\rangle$	$(3.9\pm0.4)\times10^2$	11
PDA-4BCMU	Thin film	EFISH	Two photon	$\langle\chi^{(3)}(-2\omega;\omega,\omega,0)\rangle$	$(1.5\pm0.2)\times10^4$	12
PDA-AFA	Thin film	EFISH	Two photon	$\langle\chi^{(3)}(-2\omega;\omega,\omega,0)\rangle$	$(4.3\pm0.4)\times10^4$	12
PDA-DCH	Thin film	EFISH	Two photon	$\langle\chi^{(3)}(-2\omega;\omega,\omega,0)\rangle$	$(6.4\pm0.4)\times10^3$	13
PDA-4BCMU	Gel red form	PC	Two photon	$\langle\chi^{(3)}(-\omega;\omega,\omega,\omega)\rangle$	1.3×10^3	2
PDA-PTS	Single crystal	THG	Three photon	$\chi^{(3)}_{xxxx}(-3\omega;\omega,\omega,\omega)$	$(8.5\pm5)\times10^4$	14
PDA-TCDU	Single crystal	THG	Three photon	$\chi^{(3)}_{xxxx}(-3\omega;\omega,\omega,\omega)$	$(1.6\pm1)\times10^4$	14
PDA-AFA	LB film (blue)	THG	Three photon	$\langle\chi^{(3)}(-3\omega;\omega,\omega,\omega)\rangle$	$(8.3\pm0.8)\times10^3$	7
PDA-DCH	Thin film	THG	Three photon	$\langle\chi^{(3)}(-3\omega;\omega,\omega,\omega)\rangle$	$(4.0\pm0.4)\times10^3$	9
PDA-DCH	Oriented film	THG	Three photon	$\chi^{(3)}_{xxxx}(-3\omega;\omega,\omega,\omega)$	$(6\pm0.6)\times10^4$	13
PDA-C_4UC_n	Oriented film	THG	Three photon	$\chi^{(3)}_{xxxx}(-3\omega;\omega,\omega,\omega)$	1.7×10^5	15
PA	Thin film	THG	Three photon	$\langle\chi^{(3)}(-3\omega;\omega,\omega,\omega)\rangle$	$(5.6\pm1.6)\times10^5$	10
Hemicyan.	LB film	THG	Three photon	$\langle\chi^{(3)}(-3\omega;\omega,\omega,\omega)\rangle$	$(3.0\pm0.3)\times10^2$	16
PTT	Thin film	THG	Three photon	$\langle\chi^{(3)}(-3\omega;\omega,\omega,\omega)\rangle$	$(2\pm0.2)\times10^3$	17
PDTB	Thin film	THG	Three photon	$\langle\chi^{(3)}(-3\omega;\omega,\omega,\omega)\rangle$	$(1.1\pm0.1)\times10^3$	17
PDTB	Thin film	THG	Three photon	$\langle\chi^{(3)}(-3\omega;\omega,\omega,\omega)\rangle$	$(2\pm0.2)\times10^3$	18
PA	Oriented thin film	THG	Three photon	$\chi^{(3)}_{xxxx}(-3\omega;\omega,\omega,\omega)$	$(1.7\pm0.7)\times10^6$	19
PA	"	THG	Three photon	$\chi^{(3)}_{xxxx}(-3\omega;\omega,\omega,\omega)$	2.7×10^6	20

References to Table 2:

1 - Kajzar et al. (1988a)
2 - Nunzi and Grec (1987)
3 - Jenekhe et al. (1989)
4 - Ghoshal (1989)
5 - Dorsinville et al. (1988)
6 - Kajzar and Messier (1987c)
7 - Kajzar and Messier (1985a)
8 - Chollet et al. (1988a)
9 - Le Moigne et al. (1988)
10 - Fann et al. (1989)
11 - Chollet et al. (1985a)
12 - Chollet et al. (1988b)
13 - Le Moigne et al. (1990)
14 - Sauteret et al. (1976)
15 - Kanetake et al. (1989)
16 - Kajzar et al. (1988b)
17 - Kajzar et al. (1990)
18 - Kajzar et al. (1989)
19 - Drury (1988)
20 - Krausz et al. (1989)

cases (cf. *Fig. 24*). Due to the sixth power dependence of harmonic field $E_{3\omega}$ on the fundamental E_ω field this technique is much more sensitive as the linear dichroism and works also in some cases when the former one fails. A nice illustration of that is third harmonic generation in discussed before bi-oriented pDCH thin films (Le Moigne et al. (1988)). The intensity of light transmitted through crossed polarizers with thin film in-between and polymer chains perpendicular to the beam propagation direction does not depend on the angle θ (between input electric field and a chosen direction on substrate plane)

$$I_T = I_\omega(\cos^2\theta + \cos^2(90°-\theta)) = I_\omega \quad (43)$$

Thus no variation of transmitted intensity will be observed when rotating the sample along an axis parallel to the beam propagation direction. In THG the resultant harmonic field is given by

$$E_{3\omega}^{(\theta)} = A(\cos^4\theta + \sin^4\theta)E_\omega^3 \quad (44)$$

where A is a θ independent proportionality factor. Consequently the harmonic intensity

$$I_{3\omega}^{(\theta)} \propto E_{3\omega}^2(\theta) = |A|^2(\cos^4\theta + \sin^4\theta)^2 E_\omega^6 \quad (45)$$

depends strongly on θ. This is examplified in *Fig. 25* showing such anisotropy and an excellent agreement between observed and calculated harmonic intensity dependence on angle θ.

Figure 24: Electron micrograph of a bioriented 250 Å thick pDCH thin film on a single crystalline KBr substrate (Courtesy of J. Le Moigne).

For mono-oriented thin films the dependence of harmonic intensity on angle θ is much stronger

$$J_{3\omega}^{(\theta)} = |B|^2 J_\omega^3 \cos^8\theta \quad (46)$$

where similarly as before B is a θ independent proportionality factor. As it is seen from *Fig. 26* theory fits again very well the observed variation of $J_{3\omega}$ with angle θ (Le Moigne et al. (1990)).

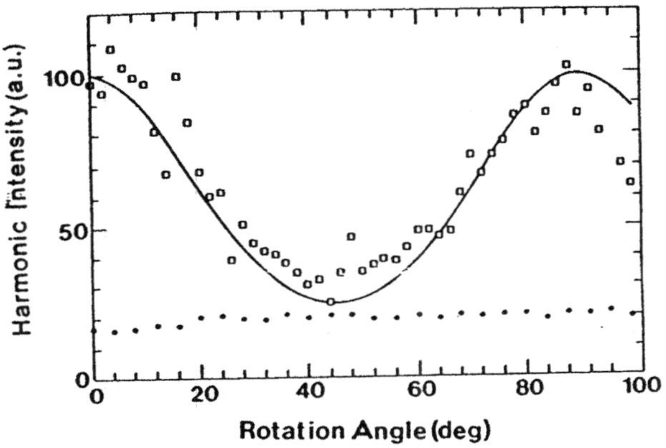

Figure 25: Third harmonic intensity in function of rotation angle θ from a bioriented 1300 Å thick film (pDCH) grown by epitaxy on KBr (open squares). Solid line is a least square fit of Eq. (45) to experimental data. Closed circles show TH intensity from an unoriented film at the same experimental conditions (Le Moigne et al.(1988)).

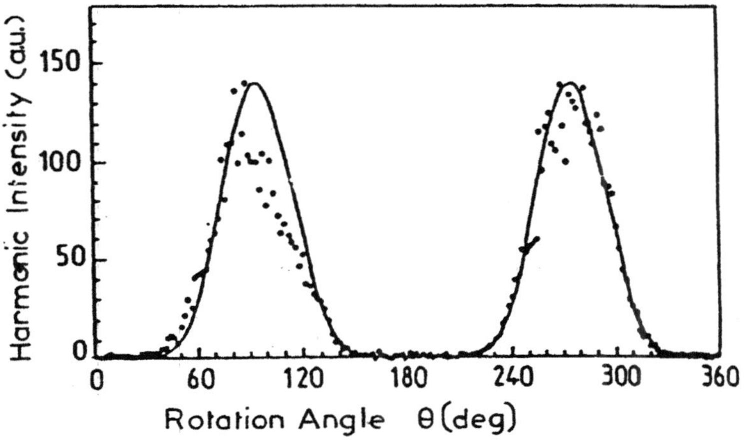

Figure 26: Third harmonic intensity as a function of rotation angle θ from a mono-oriented pDCH thin film on KAP single crystal substrate. Solid lines shows calculated values whereas the points the measured ones. (Le Moigne et al., (1990)).

7. PERSPECTIVES OF APPLICATIONS

7.1. Frequency conversion

One of possible applications of conjugated polymers is in frequency tripling of high instantaneous power femtosecond lasers. In order to avoid time bradening due to the group velocity dispersion (cf. Glenn (1969)) only thin crystals can be employed for frequency conversion. Thus highly nonlinear thin films can be successfully used. In fact the frequency tripling efficiency (without absorption) is given by (cf. Eq. (21))

$$\eta = N_f \cdot \left(\frac{\chi^3(-3\omega;\omega,\omega,\omega)}{1 + n_{3\omega}}\right)^2 \left(\frac{\ell}{\lambda_\omega}\right)^2 t_{01}^6 I_\omega^2 \tag{47}$$

with

$$N_F = \frac{9\pi^2}{\varepsilon_0^2\ c^2} = 1{,}26.10^7\,(\mathrm{SI})$$

and $$N_F = \frac{2304\pi^6}{c^2} = 2{,}46.10^{-15}\,(\mathrm{esu})$$

where ℓ is the film thickness and
t_{01} is a transmission factor at ω from vacuum to film.

For $\chi^3(-3\omega) = 5.10^{-9}$ (esu) (or 7.10^{-17}SI) and $n_\omega = n_{3\omega} = 2$, $\ell = 1\mu m$, one gets $\eta = 24\%$, a quite large value at 1.06 µm.

The incident power dependence of frequency tripling efficiency is shown in *Fig. 27*. Actually due to the free harmonic wave absorption this efficiency will be slighly smaller. The large light intensity is realized by focusing a 40 nj laser pulse with 10 fs duration pulse on a surface of $100(\mu m)^2$.

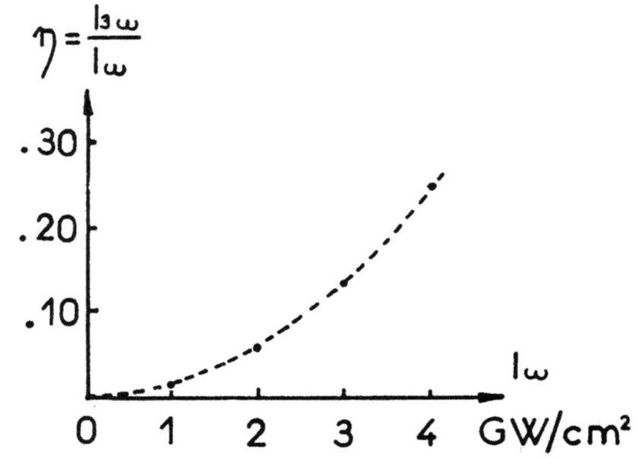

Figure 27: Third harmonic conversion efficiency as a functionn of fundamental beam intensitv calculated for a 1 µm thick polymer thin film with $\chi^{(3)}(-3\omega;\omega,\omega,\omega) = 5\times10^{-9}$ e.s.u.

7.2. Optical switching and directional couplers

For waveguiding devices and because of importance of propagation losses, two kinds of materials are the most appropriate: single crystalline and completely amorphous thin films. Propagation losses in monocrystalline PTS films obtained by shear technique and in spun poly-4BCMU have been measured by Krug et al. (1989) at 0.83 µm. In PTS single crystals the purely scattering losses are very small (< 1 dB/cm) whereas in spun póly-4BCMU they depend on film quality and vary between 2 and 22 dB/cm from sample to sample. Contrary wise, the linear absorption coefficient is slightly larger in PTS ($\alpha = 2\ cm^{-1}$) than in poly-4BCMU (best films have $\alpha = 0.5\ cm^{-1}$). These losses decrease significantly in infrared to less than 1 dB/cm (Krug et al. (1988), Townsend et al. (1988)) and less than 8 dB/cm in poly-4BCMU and PTS single crystals, respectively.

Stegeman and Stolen introduced a merit factor defined as follows

$$W = \frac{\Delta\beta_{sat}}{\lambda\ \alpha} \tag{48}$$

where $\Delta\beta_{sat} = n_2/A_{eff}$ with A_{eff} being the effective waveguide cross-sectional area and n_2 the nonlinear index of refraction, λ is the operation wavelength and α is the linear absorption coefficient. The parameters W allows to compare the performances of different materials. For polydiacetylenes (PTS) $W > 100$ and is significantly larger than for III-V semiconductors ($W \lesssim 10$). For practical applications in switching devices like directional couplers a W of 10 is required; condition which is fullfiled by polydiacetylenes. A more important effort on purification of these materials should increase significantly the value of α ($\alpha = 0.1\ cm^{-1}$ for PTS and $\alpha = 10^{-5} cm^{-1}$ for silica).

Townsend et al. (1989) has studied directional couplers made from poly-4BCMU spun film deposited on a patterned with two channels substrate. At 1.064 µm operation wavelength they observed nonlinear transmission and switching from one channel to another one both due to a fast (ps) electronic and slow thermal nonlinearities. However the fast switching was due to the refractive index modification via the two photon absorption.

7.3. Optical bistability

The large value of nonlinear index of refraction, fast response time, processability and versatibility of conjugated polymers make them good candidates for applications in logic systems. Due to the rich electronic structure of these polymers and a lot of different resonance enhancements possible both passive and active optical bistabilities are expected. Optical bistable behaviour has been observed in thin waveguides of poly-4BCMU (Singh and Prasad (1988)) and in PDA LB films (Sasaki et al. (1988)).

Singh and Prasad observed optical bistability in a planar quasi-waveguide at 0.653 µm and 0.7 µm wavelength with upper switching threshold power of 2.55 MW/cm² and 3.86 MW/cm² corresponding to extremely large values of nonlinear refractive index $n_2 \simeq 5.10^{-3} cm^2/MW$ suggesting that it is a thermal effect.

Sasaki et al. (1988) used a hybrid waveguide composed from PDA LB film, glass waveguide and a silica substrate. Two types of optical listability they have observed: one originating from coupling between a prism and

PDA hybrid waveguide. The second one they observed in the hybrid waveguide with a nonlinear distributed Bragg reflector of the PDA top layer. In both cases input-output curves with hysteresis characteristics were obtained. However, as state the authors, they are not sure whether the observed optical bistability is electronic, or thermal in origin.

Thermal optical bistability with a Fabry-Perot cavity filled with a PDA solution was also observed by Blau (1987) with very low switching power (less than 4 MW).

Optical bistability studies in conjugated polymers is still at the beginning and it requires a more important effort. New structures, like that proposed recently by Haelterman and Waelbroeck (1990) and new materials have to be examined in an appropriate wavelength range (e.g. around one or two photon resonances).

REFERENCES

AGRAVAL G.P., COJAN C. and FLYTZANIS C., (1978), "*Nonlinear Optical Properties of One-Dimensional Semiconductors and Conjugated Polymers*", Phys. Rev. B17 (2) p. 776.

ANDRE J.M., BARBIER C., BODART V. and DELHALLE J., (1987), "*Nonlinear Optical Properties of Organic Molecules and Crystals*", D. Chemla and J. Zyss eds., Academic Press New York 1987, Vol. 2, 137-158.

ARMSTRONG J.A., BLOMBERGEN N., DUCUING J. and PERSHAN P.S., (1962), "*Interactions between Light Waves in a Nonlinear Dielectric*", Phys. Rev. 127, 1918-1939.

BATCHELDER D.N., (1985), "*The Study of Electronic Excitations of Polydiacetylenes and Resonance Raman Spectroscopy*", in Polydiacetylenes, D. Bloor and R.R. Chance eds., NATO ASI Series, E102, 187-212.

BARZOUKAS M., FREMAUX P., JOSSE D., KAJZAR F. and ZYSS J., (1988), "*Quadratic and Cubic Nonlinearities in Solution of Intramolecular Charge Transfer Aromatic Molecules: A Quantum Two-Level Approach*" in Nonlinear Optical Properties of Polymers, A.J. Heeger, J. Orenstein and D.R. Ulrich eds., Mat. Res. Soc. Symp. Proc., Vol. 109, 171-185.

BAUMERT J.C., BJORKLUND G.C., JUNDT D.H., JURICH M.C., LOOSER MC., MILLER R.D., RABOLT J., SOORYAKUMARAN R., SWALEN J.D. and TWIEG R.J., (1988), "*Temperature Dependence of the Third Order Nonlinear Optical Susceptibilities in Polysilanes and Polygermanes*" Appl. Phys. Lett., 53, 1147-1149.

BETHUNE D.S., (1988), "*Optical Harmonic Generation and Mixing in Multilayer Media: Analysis Using Optical Transfer Matrix Techniques*" D.S. Bethune, J. Opt. Soc. Am. B, 6, 910-916.

BLANCHARD G.J., HERITAGE J.P., VON LEHMAN A.C., KELLY M.K., BAKER G.L. and ETEMAD S., (1989), *"Excitonic and Phonon-Mediated Optical Stark Effect in a Conjugated Polymer"*, Phys. Rev. Lett., 63, 887-890.

BLAU W., (1987) *"Low Power Optical Bistability and Phase Conjugation in Polydiacetylene"*, Opt. Commun., 64, 85-88.

BLOMBERGEN N., (1965), *"Nonlinear Optics"*, Benjamin Inc., New York.

BLOMBERGEN N. and PERSHAN P.S., (1962), *"Light Waves at the Boundary of Nonlinear Media"*, Phys. Rev., 128, 2, 606-622.

BYRNE M.J., BLAU W. and JEN K.Y., (1989), *"Three Photon Enhanced Optical Nonlinearity of Poly(3-Butylthiophene)"*, Synth. Met., 32, 229-235.

CARTER G.M., (1987), *"Excited-State Dynamics and Temporally Resolved Nonresonant Nonlinear-Optical Processes in Polydiacetylenes"*, J. Opt. Soc. Am., B4, 1018-1024.

CARTER G.M., HRYNIEWICZ J.V., THAKUR M.K., CHEN Y.J. and MEYLER S.E., (1986), *"Nonlinear Optical Processes in a Polydiacetylene Measured with Femtosecond Duration Laser Pulses"*, Appl. Phys. Lett., 49, 998-1000.

CHANCE R.R., SHAND M.L., HOGG C. and SILBEY R., (1980), *"Three Wave Mixing in Conjugated Polymer Solutions: Two-Photon Absorption in Polydiacetylenes"*, Phys. Rev., B22, 3540-3550.

CHARRA F. and MESSIER J., (1990), *"Third Order Hyperbolarizability of Polytiophene and Tiophene Oligomers"*, in *"Conjugated Polymeric Materials: Opportunities in Electronics, Optoelectronics and Molecular Electronics"*, J.L. Bredas and R.R. Chance eds., NATO ASI Series E182, 409-421.

CHARRA F. and NUNZI J.M., (1989a), *"Picosecond Studies of Optical Stark Effect in Polydiacetylene"*, in Nonlinear Optical Effects in Organic Polymers, J. Messier, F. Kajzar, P. Prasad and D. Ulrich eds., NATO ASI Series, Vol. 162, Kluwer Academic Publ. Dordrecht, 369-371.

CHARRA F. and NUNZI J.M., (1989b), *"Strong Field Nonlinear Optical Properties of Polydiacetylenes"* in Nonlinear Optical Materials II, J.B. Grun ed., Proc. SPIE, Vol. 1127, 173-181.

CHEN Y.J., TRIPATHY S.K. and CARTER G.M. (1984), *"Linear Optical Properties in Multilayers of Polydiacetylenes"*, Mol. Cryst. Liq. Cryst. 106, 403-408.

CHOLLET P.A., (1980), *"IR Determination of the Orientation of Molecules in Polycrystalline Monolayers"*, Thin Solid Films, 68, 13-19.

CHOLLET P.A., KAJZAR F. and MESSIER J., (1985a), "*Electric Field Induced Optical Second Harmonic Generation and Polarization Effects in Polydiacetylene Films*", D. Bloor and R.R. Chance eds., NATO ASI Series, E102, 317-324.

CHOLLET P.A., KAJZAR F. and MESSIER J., (1985b), "*Electric Field Induced Optical Second Harmonic Generation and Polarization Effect in Polydiacetylene Langmuir-Blodgett Multilayers*", Thin Sol. Films, 132, 1-10.

CHOLLET P.A., KAJZAR F. and MESSIER J., (1988a), "*Frequency and Temperature Variations of Cubic Susceptibility in Polydiacetylenes*", in Nonlinear Optical and Electroactive Polymers, P.N. Prasad and D.R. Ulrich eds., Plenum Publ. Corp. New York, 121-135.

CHOLLET P.A., KAJZAR F. and MESSIER J., (1988b), "*Resonances of Cubic Nonlinear Optical Susceptibilities in Polydiacetylenes*", in "*Nonlinear Optics of Organics and Semiconductors*", T. Kobayashi ed., Springer Verlag, Berlin, 1988, 171-179.

CHOLLET P.A., KAJZAR F., MESSIER J., NUNZI J.M. and GREC D., (1987), "*Propriétés optiques non linéaires du 3ème ordre des milieux organiques*", Rev. Phys. Appl., 22, 1221-1228.

COJAN C., AGRAVAL G.P. and FLYTZANIS C., (1977), "*Optical Properties of One-Dimensional Semiconductors and Conjugated Polymers*", Phys. Rev. B, 15, 909-925.

DAY D. and RINGSDORF H., (1978), "*Polymerization of Diacetylene Carbonic Acid Monolayers at the Gaz Water Interface*", J. Pol. Sc. Pol. Lett. ed., 16, 205-210.

DENIS W.M., BLAU W. and BRADLEY D.J., (1985), "*Picosecond Degenerate Four Wave Mixing in Soluble Polydiacetylenes*", Appl. Phys. Lett., 47, 200-202.

DORSINVILLE R., YANG L., ALFANO R.R., TUBINO R. and DESTRI S., (1988), "*Picosecond Nonlinear Optical Response Using Phase Conjugation in Soluble Polyacetylene*", Sol. State Commun., 68, 875-877.

DRURY M.R., (1988), "*Observation of Third Harmonic Generation in Oriented Durham Polyacetylene*", Sol. State Commun., 68, 417-420.

FANN W.S., BENSON S., MADEY J.M.J., ETEMAD S., BAKER G.L. and KAJZAR F., (1989), "*Spectrum of* $\chi^{(3)}(-3\omega;\omega,\omega,\omega)$ *in Polyacetylene: An Application of the Free Electron Laser in Nonlinear Optical Spectroscopy*", Phys. Rev. Lett., 62, 1492-1495.

FICHOU D., GARNIER F., CHARRA F., KAJZAR F. and MESSIER J., (1989), "*Linear and Nonlinear Optical Properties of Thiophene Oligomers*" in Organic Materials for Nonlinear Optics, R.A. Hann and D. Bloor eds., The Royal Society of Chemistry Special Publication n°69, London 1989, 176-182.

FISHER R.A., (1983), "*Optical Phase Conjugation*", Academic Press Inc., New York.

FLYTZANIS C., (1975), "*Theory of Nonlinear Optical Susceptibilities*" in Quantum Electronics: a Treatise, H. Rabin and C.L. Tang eds., vol. I, Nonlinear Optics. Part A, 9-207.

GHOSHAL S.K., (1989), "*Resonant Third Order Optical Nonlinearity in Polypyrrole*", Chem. Phys. Lett., 158, 65-69.

GLENN W.H., (1969), "*Second Harmonic Generation by Picosecond Optical Pulses*", IEEE J. Quantum Electron. QE-5, 248-290.

GREENE B.I., ORENSTEIN J., MILLARD R.R. and WILLIAMS L.R., (1987), "*Nonlinear Optical Response of Excitons Confined to One-Dimension*", Phys. Rev. Lett., 58, 2750-2753.

GREENE B.I., MUELLER J.F., ORENSTEIN J., RAPKINE D.N., SCHMITT-RINK S. and THAKUR M., (1988), "*Phonon-Mediated Optical Nonlinearity in Polydiacetylene*", Phys. Rev. Lett., 61, 325-328.

GREENE B.I., ORENSTEIN J., MILLARD R.R. and WILLIAMS L.R., (1987), "*Nonlinear Optical Response of Excitons Confined to One-Dimension*", Phys. Rev. Lett., 58, (26), 2750.

GREENE B.I., ORENSTEIN J., MILLARD R.R. and WILLIAMS L.R., (1987), "*Nonlinear Optical Response of Excitons Confined to One-Dimension*", Phys. Rev. Lett., 58, 2750-2753.

GRUNFELD F. and PITT C.W. (1983), "*Diacetylene Langmuir-Blodgett Layers for Integrated Optics*", Thin Sol. Films, 99, 249-255.

HAELTERMAN M. and WAELBROECK C., (1990), "*Demonstration of Optical Bistability with a Nonlinear Frustated-Total-Reflection Filter*", Appl. Phys. Lett., 56, 512-514.

HATTORI T. and KOBAYASHI T., (1987), "*Femtosecond Dephasing in a Polydiacetylene Film Measured by Degenerate Four Wave Mixing with an Incoherent Nanosecond Laser*", Chem. Phys. Lett., 133, 230-234.

HEEGER A.J., MOSES D. and SINCLAIR M., (1986), "*Semiconducting Polymers: Fast Response Nonlinear Optical Materials*", Synth. Metals, 15, 95-104.

HO P.P., DORSINVILLE R., YANG N.L., ODIAN G., EICHMANN G., JIMBO T., WANG Q.Z., TANG G.C., CHEN N.D., ZOU W.K., LI Y. and ALFANO R.R., (1986), "*Ultrafast Optical Processes in 4-BCMU Polydiacetylene*", in Molecular and Polymeric Optoelectronic Materials: Fundamentals and Applications, G. Khanarian ed., Proc. SPIE, Vol. 682, 36-43.

HO P.P., YANG N.L., JIMBO T., WANG Q.Z. and ALFANO R.R., (1987), "*Ultrafast Resonant Optical Kerr Effect in 4-Butoxycarbonylmethyl-urethane Polydiacetylene*", J. Opt. Soc. Am., B4, 1025-1029.

INGANAS O., GUSTAFSSON G. and SALANECK W.R., (1989), "*Thermochromism in Thin Films of Poly/3-Alkyl Thiophene*", Synthetic Metals, 28 (1989), C377-C384.

JENEKHE S.A., LO S.K. and FLOM S.R., (1989), "*Third-Order Nonlinear Optical Properties of a Soluble Conjugated Polythiophene Derivative*", Appl. Phys. Lett., 54, 2524-2526.

KAJZAR F. (1989a), "*Complex Hyperpolarizability and Solvatochromic Effects in Merocyne Solutions*", in Nonlinear Optics of Organics and Semiconductors, T. Kobayashi ed., Springer Verlag, 108-119.

KAJZAR F., (1989b), "*Third Harmonic Generation in Liquids*", in Nonlinear Optical Effects in Organic Polymers, J. Messier, F. Kajzar, P. Prasad and D. Ulrich eds., NATO ASI Series, Series E, Vol. 162, Kluwer Academic Publishers, 225-245.

KAJZAR F., LEDOUX I. and ZYSS J., (1987), "*Electric-Field Induced Optical Second Harmonic Generation in Polydiacetylene Solutions*", Phys. Rev. A36, 2210-2218.

KAJZAR F. and MESSIER J., (1985a), "*Resonance Enhancement in Cubic Susceptibility of Langmuir-Blodgett Multilayers of Polydiacetylene*", Thin Sol. Films, 132, 11-19.

KAJZAR F. and MESSIER J., (1985b), "*Third Harmonic Generation in Liquids*", Phys. Rev. A32, 2352-2363.

KAJZAR F. and MESSIER J., (1985c), "*Resonance Enhancement in Cubic Susceptibility of Langmuir-Blodgett Multilayers of Polydiacetylene*", Thin Sol. Films, 132, 11-19.

KAJZAR F. and MESSIER J., (1987a), "*Cubic Effects in Polydiacetylene Solutions and Films*", in Nonlinear Optical Properties of Organic Molecules and Crystals, Vol. 2. D.S. Chemla, J. Zyss ed. Academic Press, 51-83.

KAJZAR F. and MESSIER J., (1987b), "*Original Technique for Third Harmonic Generation Measurements in Liquids*", Rev. Sci., 58 (11) (1987), 2081-2085.

KAJZAR F. and MESSIER J., (1987c), "*Cubic Nonlinear Optical Effects in Conjugated Polymers*", Polymer J., 19, 275-284.

KAJZAR F., LEDOUX I. and ZYSS J., (1987b), "*Electric-Field Induced Optical Second Harmonic Generation in Polydiacetylene Solutions*", Phys. Rev. A36, 2210-2219.

KAJZAR F., MESSIER J. and ROSILIO C., (1986), "*Nonlinear Optical Properties of Thin Films of Polysilane*", J. Appl. Phys., 60, 3040-3044.

KAJZAR F., MESSIER J. and SENTEIN C., ELSENBAUMER R.L. and MILLER G.G., (1989), "*Cubic Susceptibility of Polytiophene Solutions and Films*", in G. Khanarian ed., Proc. SPIE, Vol. 1147, 36-47.

KAJZAR F., MESSIER J., ZYSS J. and LEDOUX I., (1983), "*Nonlinear Interferometry in Langmuir-Blodgett Multilayers of Polydiacetylene*", Opt. Commun., 45, 133-137.

KAJZAR F., ROTHBERG L., ETEMAD S., CHOLLET P.A., GREC D., BOUDET A. and JEDJU T., (1988a), "*Saturation Absorption and Kerr Susceptibility in Polydiacetylene Langmuir-Blodgett Film*", Opt. Commun., 66, 55-58.

KAJZAR F., GIRLING I.R. and PETERSON I.R., (1988b), "*Third-Order Hyperpolarizability of Centrosymmetric Langmuir-Blodgett Films of Stilbazolium Dyes*", Thin Sol. Films, 160, 209-215.

KAJZAR F., RUANI G., TALIANI C. and ZAMBONI R., (1990), "*Frequency Variation of Cubic Susceptibility in New Conjugated Polymers: PTT and PDTB*", Synthetic Metals, 37, 223-229.

KANETAKE T., ISHIKAWA K., HASEGAWA T., KODA T., TAKEDA K., HASEGAWA M., KUBODERA K. and KOBAYASHI M., (1989), "*Nonlinear Optical Properties of Highly Oriented Polydiacetylene Evaporated Films*", Appl. Phys. Lett., 54, 2287-2289.

KAPITULNIK A., CASALNUOVO S.A., LIM K.C. and HEEGER A.J., (1984), "*Electric Field Coupling to Slow Elastic Modes in Gels of Conjugated Polymers*", Phys. Rev. Lett., 53, 469-472.

KAPITULNIK A., LIM K.C., CASALNUOVO S.A. AND HEEGER A.J., (1986), "*Elasticity of Poly(diacetylene) Gels: Measurements by Electric Field Coupling*", Macromolecules, 19, 676-681.

KOABAYASHI T., HATTORI T., TERASAKI A. and KUROKAWA K., (1987), "*Femto-second Relaxation Processes in Nonlinear Materials Studied with Incoherent Light*", Rev. Phys. Appl., 22, 1773-1785.

KOHLER B.E. and SCHILKE D.E., "*Low-Lying Singlet States of a Short Polydiacetylene Oligomer*", J. Chem. Phys., 86, 5214-5215.

KRAUSZ F., WINTER E. and LEISING G., (1989), "*Optical Third Harmonic Generation in Polyacetylene*", Phys. Rev. B39, 3701-3710.

KRUG W., MIAO E., DERSTINE M. and VALERA J., (1989), "*Optical Scattering and Scattering Losses of PTS and Poly(4-BCMU) Thin Film Waveguides in the Near Infrared*", J. Opt. Soc. Am., B6, 726-732.

LE MOIGNE J., THIERRY A., CHOLLET P.A., KAJZAR F. and MESSIER J., (1988a), "*Thin Film Epitaxy of Diacetylenes on Inorganic or Organic Substrates: Structure and Nonlinear Properties of Polymerized Films*", SPIE, Vol. 971, Nonlinear Optical Properties of Organic Materials, 173-177.

LE MOIGNE J., THIERRY A., CHOLLET P.A., KAJZAR F. and MESSIER J., (1988b), "*Morphology Linear and Nonlinear Optical Studies of Poly[1,6-diCN-carbazolyl)-2,4 mexadiyne] Thin Films (pDCH)*", J. Chem. Phys., 88, 6647-6654.

LE MOIGNE J., THIERRY A. and KAJZAR F., (1990), "*Highly Oriented Films of Polydiacetylene for Nonlinear Optics: Epitaxial Growth of DCH Monomer and Optical Properties of Polymerized Films*", in Thin Films in Optics, T. Tschudi ed., Proc. SPIE, Vol. 1125, 9-12.

LIM K.C. and HEEGER A.J., (1985), "*Spectroscopic and Light Scattering Studies of the Conformational (rod-to-coil) Transition of Poly(diacetylene) in Solution*", J. Chem. Phys., 82, 522-530.

LIM K.C., KAPITULNIK A., ZACHER R. and HEEGER A.J., (1985), "*Conformation of Polydiacetylene Macromolecules in Solution": Field Induced Birefringence and Rotational Diffusion Constant*", J. Chem. Phys., 82, 516-521.

Mc GRAW D.J., SIEGMAN A.E., WALLRATT G.M. and MILLER R.D., (1989), "*Resolution of the Nuclear and Electronic Contributions to the Optical Nonlinearity in Polysilanes*", Appl. Phys. Lett., 54, 1713-1715.

MESSIER J., (1989), "*Third-Order Nonlinear Susceptibility in Semiconducting Polymers*", in Nonlinear Optical Effects in Organic Polymers, J. Messier, F. Kajzar, P. Prasad and D. Ulrich eds., NATO ASI Series, Series E, Kluwer Academic Publishers, 47-60.

MORAES F., SCHAFFER M., KOBAYASHI M., HEEGER A.J. and WUDL J., (1984), Phys. Rev., B30, 2948-2954.

MORICHERE D., DENTAN V., KAJZAR F., ROBIN P., LEVY Y. and DUMONT M., (1989), "*Kerr Effect and Electrostriction in Thin Polymeric Films*", Opt. Commun., 74, 69-74.

NEHER D., WOLF A., BUBECK C. and WEGNER G., (1989), "*Third Harmonic Generation in Polyphenylacetylene: Exact Determination of Nonlinear Optical Susceptibilities in Ultra Thin Film*", Chemical Physics Letters, 163, 2, 3, (1989), 116-122.

NUNZI J.M. and CHARRA F., (1989), "*Phase Conjugation in Polydiacetylenes: Dynamic Effects*", Conference on Lasers and Electro-Optics, Baltimore, 24-28 April 1989, 1989 Technical Digest Series, Vol. 11, 256-257.

NUNZI J.M. and CHARRA F., (1990), "*Conjugaison de phase avec doublage de fréquence dans le polydiacétylène*", Ann. Phys. 15, 151-153.

NUNZI J.M. and GREC D., (1987), "*Picosecond Phase Conjugation in Polydiacetylene Gels*", J. Appl. Phys., 62, 2198-2202.

OBHI J., BLOOR D., ANDO D.J., NORMAN P.A., KOLINSKI P.V. and MOVALGHAR B. (1986), "*Electric Field Induced Second Harmonic Generation in the Single Crystal PTS-FBS*", GEC J. Res., 4, 256-259.

ORENSTEIN J. and BAKER G.L., (1987), "*Photogenerated Gap States in Polyacetylene*", Phys. Rev. Lett., 49, 1043-1049.

ORENSTEIN J., ETEMAD S. and BAKER G.L., (1984), "*Photoinduced Absorption in a Polydiacetylene*", J. Phys. Solid State, 17, L297-L300.

ORR B.J. and WARD, J.F., (1971), "*Perturbation Theory of the Nonlinear Optical Polarization of an Isolated System*", Mol. Phys., 20, 513-526.

PRASAD P.N., (1989), "*Ultrafast Third-Order Nonlinear Optical Processes in Polymeric Films*", in Nonlinear Optical Effects in Organic Polymers, J. Messier, F. Kajzar, P. Prasad and D. Ulrich eds., NATO ASI Series, Vol. 162, Kluwer Academic Publ. Dordrecht, 1989, 351-363.

RAO N.D., SWIATKIEWICZ J., CHOPRA P., GHOSHAL S.K. and PRASAD P.N., (1986), "*Third-Order Nonlinear Optical Interactions in Thin Films of Poly-p-phenylenebenzobisthiazale Polymer Investigated by Picosecond and Subpicosecond Degenerate Four Wave Mixing*", Appl. Phys. Lett., 48, 1187-1189.

ROTHBERG L., (1987), "*Pulsed Laser Optoacoustic Spectroscopy in the Study of Surface Adsorbate Structure and Dynamics*", J. Phys. Chem., 91, 3467-3474.

ROTHBERG L., BERNSTEIN M. and PETERS K.S., (1983), "*Time Resolved Photoacoustic Spectroscopy Applied to Properties of Picosecond Transients*", J. Chem. Phys., 79, 2569-2576.

RUANI G., PAL A.J., ZAMBONI R., TALIANI C. and KAJZAR F., "*Photoinduced Absorption and Nonlinear Optical Response in a Polycondensed Thiophene-Based Polymer PTT*", in Proc. of NATO Advanced Research Workshop, Mons, September 3-8, 1989, J.L. Bredas and R.R. Chance eds., in press.

RUSTAGI K.G., RUSTAGI K.C. and DOCUING, J., (1974), "*Third-Order Optical Polarizability of Conjugated Organic Molecules*", Opt. Commun., 10, 258-261.

SASAKI K., FUJII K., TOMIOKA T. and KINOSHIKA T., (1988), "*All Optical Bistabilities of Polydiacetylene Langmuir-Blodgett Film Waveguides*", J. Opt. Soc. Am., B5, 457-461.

SAUTERET C., HERMANN J.P., FREY R., PRADERE F., DUCUING J., BAUGHMAN R.M. and CHANCE R.R., (1976), "*Optical Nonlinearities in One-Dimensional Conjugated Polymer Crystals*", Phys. Rev. Lett., 36, 956-959.

SCHMITT-RINK S., CHEMLA D.S. and MILLER D.A.B., (1985), "*Theory of Transient Excitonic Optical Nonlinearities in Semiconductor Quantum-Well Structures*", Phys. Rev. B32, 6601-6609.

SHAND M.L. and CHANCE R.R., (1978), "*Third-Order Nonlinear Mixing in Polydiacetylene Solutions*", J. Chem. Phys., 69, 4482-4486.

SINGH B.P. and PRASAD P.N., (1988), "*Optical Bistable Behaviour of a Planar Quasi-Waveguide Interferometer Made with a Conjugated Organic Polymer Film*", J. Opt. Soc. Am., B5, 453-456.

SINGH B.P., PRASAD P.N. and KARASZ F., (1988), "*Third-Order Nonlinear Optical Properties of Oriented Films of Poly(p-phenylene vinylene) Investigated by Femtosecond Degenerate Four Wave Mixing*", Polymer 29, 1940-1942.

STEGEMAN G.I. and STOLEN R.H., (1989), "*Waveguides and Films for Nonlinear Optics*", J. Opt. Soc. Am., 6, 652-662.

SWIATKIEWICZ J., PRASAD P.N., KARASZ F.E., DRUY M.A. and GLATKOWSKI P., (1990), "*Anisotropy of the Linear and Third-Order Nonlinear Optical Properties of a Stretch Oriented Polymer Film of Poly-[2,5-dimethoxyparaphenylenevinylene*", Appl. Phys. Lett., 56, 892-894.

THAKUR M. and MEYLER S., (1985), "*Growth of Large Area Thin Film Single Crystals*", Macromol., 18, 2341-2344.

THEMANS B., SALANECK W.R. and BREDAS J.L., (1989), "*Theoretical Study of the Influence of Thermochromic Effects on the Electronic Structure of Poly(3-Hexyl)Thiophene*", Synthetic Metals, 28, 1989, C359-C364.

TOWNSEND P.D., BAKER G.L., SCHLOTTER N.E., KLAUSNER C.F. and ETEMAD S., (1988), "*Waveguiding in Spun Films of Soluble Polydiacetylenes*", Appl. Phys. Lett., 53, 1782-1784.

TOWNSEND P.D., JACKEL J.L., BAKER G.L., SHELBURNE III J.A. and ETEMAD S., (1989), "*Observation of Nonlinear Optical Transmission and Switching Phenomena in Polydiacetylene Based Directional Couplers*", Appl. Phys. Lett., 55, 1829-1881.

UCHIKI H. and KOBAYASHI T., (1988), "*New Determination of Electro-Optic Constants and Relevant Nonlinear Susceptibilities and its Application to Doped Polymer*", J. Appl. Phys., 64, 2625-2629.

VANDEVYWER M., BARRAUD A., RUAUDEL-TEIXIER A., MAILLARD P. and GIANOTTI C., (1982), "*Structure of Porphyrin Multilayers Obtained by the Langmuir-Blodgett Technique*", J. Colloid. Interface Sci., 85, 571-585.

WARD J.F., (1985), "*Calculation of Nonlinear Optical Susceptibilities Using Diagrammatic Perturbation Theory*", Rev. Mod. Phys., 37, 1-18.

YANG L., DORSINVILLE R., WANG Q.Z., ZOU W.K., HO P.P., YANG N.L., ALFANO R.R., ZAMBONI R., DANIELI R., RUANI G. and TALIANI C., (1989), "*Third-Order Optical Nonlinearity in Polycondensed Thiophene Based Polymers and Polysilane Polymers*", J. Opt. Soc. Am., 6, 753-756.

YANG L., WANG Q.Z., HO P.P., DORSINVILLE R., ALFANO R.R., ZOU W.K. and YANG N.L., (1988), "*Ultrafast Time Response of Optical Nonlinearity in Polysilane Polymers*", Appl. Phys. Lett., 53, 1245-1247.

ZYSS J., BLANCHARD M., BARZOUKAS M., KAJZAR F., MESSIER J. and SENTEIN C., (1990), "*Chain Length Dependence of the Quadratic and Cubic Nonlinear Optical Susceptibilities of Asymmetric Push-Pull Polyenes*", to be published.

THE SEMICONDUCTOR DEVICE PHYSICS OF POLYACETYLENE

J. H. BURROUGHES and R. H. FRIEND
Cavendish Laboratory
Madingley Road
Cambridge CB3 OHE U.K.

ABSTRACT Conjugated polymers which can be processed to form thin, coherent films can be used as the active layers in semiconductor device structures. We have used the Durham precursor route to polyacetylene to fabricate a range of unipolar devices including Schottky barrier diodes, MIS diodes and MISFET's. Although carrier mobilities are low, limited by thermally-activated transport between chains, these devices work well, and we find that these structures are remarkably free of surface states and bulk defect states with energy levels within the gap. The fundamental excitation of the *trans*-polyacetylene chain is the self-localised, soliton-like, kink defect in the bond alternation pattern along the chain; the soliton has associated with it an energy state at mid-gap of non-bonding p_z character. We have been particularly concerned to demonstrate the formation of solitons from charges injected into the polyacetylene layer in these device structures, and have measured the changes in the optical properties that accompany soliton formation. For the MIS structures working in accumulation mode we find the 'mid-gap' transition from soliton level to the band edge at energies ranging between 0.55 eV (characteristic of the bulk) for polyacetylene on polymeric insulator layers to 0.8 eV when formed on silicon dioxide. We discuss this spread in energies in terms of the different surface structures at these different interfaces. We also show data for the characteristic vibrational excitations of the soliton, including both the IR-active translation modes and the Raman-active amplitude modes.

J.L. Brédas and R. Silbey (eds.), Conjugated Polymers, 555–622.

1. Introduction

Organic materials are not traditionally associated with high electronic mobilities, but the past two decades have seen the rapid development of these materials as conductors and there are now examples of organic semiconductors, metals and superconductors [1]. These materials all depend on the presence of a delocalised and extended π-electron system, and the choice lies between intermolecular delocalisation in the charge transfer salts and intramolecular delocalisation which can be achieved with conjugated polymers. These polymers, in which there is a p_z orbital on each site along the polymer backbone, form a model class of organic semiconductors, with a range of energy gaps between the π valence and π^* conduction bands between 1.5 and 3 eV.

Polyacetylene has provided the prototypical polymer for this field. The *trans-transoidal* isomer (referred to here as *trans*) is the thermodynamically stable form at room temperature, and of the two other sterically allowed isomers, the *cis-transoidal* (referred to here as *cis*) is lower in energy than the *trans-cisoidal* (since conversion from one to the other requires only an adjustment of bond order along the chain, the lower energy form is thus always found). This is a rigid rod polymer, which shows a strong tendency to organise in a crystalline fashion, but the chain rigidity prevents solubility in common solvents and pushes the melting temperature above the decomposition temperature of the polymer (typically 200°C). Since it is effectively intractable once formed it is a difficult material to process, .

Interest in the electronic properties dates back to the work of the group of Shirakawa who managed to synthesise polyacetylene in the form of a fibrillar mat by polymerisation of acetylene on a surface treated with a Ziegler-Natta catalyst [2]. Work in the group of Heeger and MacDiarmid revealed that chemical oxidation or reduction of this material produced a material with metallic conductivity, of about 10^3 S/cm at room temperature [3], and it was this first demonstration of high electronic mobilities that set in train much of the activity of the past decade. The levels of conductivity have risen considerably, with several reports of conductivities in excess of 10^5 S/cm for iodine-doped polyacetylene prepared by a modified synthesis [4]. The processes of oxidation and reduction are similar to the intercalation chemistry of graphite, with the formation of chains or sheets of the charged dopant species sitting in well-defined sites alongside the polymer chain. In some respects this performs the same function as the substitutional doping of inorganic semiconductors, and these oxidation and reduction processes in conjugated polymers are loosely referred to as chemical doping.

The metallic properties of the heavily-doped polymers provided the initial focus of interest in the electronic properties, but the theoretical framework for the modelling of the electronic excitations due to Rice [5] and Su et al [6], which for the case of *trans*-polyacetylene take the form of solitary-wave excitations of the chain, has broadened the area of interest to cover, in particular, the optical and non-linear optical properties of this class of polymers.

Experimental work on polyacetylene has been primarily carried out with polymer prepared via the Shirakawa route or modifications of it. This material has an open, fibrillar morphology (with up to 2/3 voids) and is very well suited for rapid doping, achieved either chemically or electrochemically. It is however, much less suited for many of the measurements of the semiconductor properties, and is not readily used within device structures. An alternative route for the synthesis of polyacetylene, developed by Edwards and Feast [7-10], does allow very much improved processing, and it is material prepared by this route which has been used for the present work. This route, which we call here the Durham route, is shown in figure 1. Polymerisation of the monomer (A) is achieved by ring-opening metathesis, to give the polymer (B) which is readily soluble in for example acetone and 2-butanone, and it is this 'precursor' polymer which we process, usually to form thin films by spin-coating from solution. Heating, typically to 100 °C, then gives *in-situ* elimination of the side-group which is evolved as volatile hexafluoro-orthoxylene, to give initially *cis*-rich and then all *trans*-polyacetylene, as shown in figure 1.

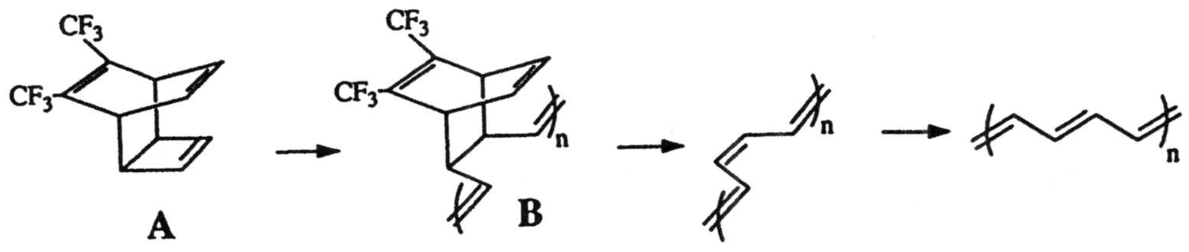

Figure 1 The Durham route to polyacetylene. Monomer (A) is polymerised to the 'precursor' polymer (B) which is processed from solution and is converted to polyacetylene by thermal elimination of hexafluoro-orthoxylene, initially to *cis*-rich and then to all *trans* isomer.

We have found that the polymer prepared in this way is very well suited for use in semiconductor device structures in which a semiconductor of one carrier type only is required (unipolar devices). The polymer as prepared is extrinsically doped with p-type carriers, to a concentration in the range 10^{16} to 10^{18} cm^{-3}, and these dopants are not readily mobile under the applied electric fields within these structures. We have made and measured Schottky-barrier diodes, MIS (Metal Insulator Semiconductor) diodes and MISFETs (MIS Field Effect Transistors), and it is the results of these investigations, some of which are published elsewhere [11-17], which are presented in the present chapter.

2. Electronic Excitations in Conjugated Polymers

Polyacetylene is a semiconductor with an energy gap of about 1.7 eV between the filled π band and empty π^* band. It is common to reduce the electronic structure to decoupled sp^2 σ bonds and π bands, and to consider the system using a Hückel model for the π electrons, with the electron-lattice interaction introduced through a bond-length modulation of the nearest-neighbour transfer integral, $t = t_o + \alpha(u_n - u_{n+1})$, where u_n is the displacement of site n, and the σ electrons parameterised through the force constant for the bonds, K. This model, due to Su et al [6], can be expressed by the SSH (Su, Schrieffer, Heeger) Hamiltonian shown in equation 1

$$H = -\sum_n [t_o+\alpha(u_n-u_{n+1})](c_n^+c_{n+1}+c_{n+1}^+c_n) + \frac{1}{2M}\sum_n p_n^2 + \frac{K}{2}\sum_n (u_n-u_{n+1})^2 \quad (1)$$

The first term gives the kinetic energy of the π electrons (c_n and c_n^+ are creation and annihilation operators at site n), the second term the kinetic energy of the lattice (p_n is the momentum of the atom at site n), and the third term gives the σ bond strain energy. The π-electron system of the *trans* isomer of polyacetylene can be considered as an example of a one-dimensional, half-filled electron band, and the bond dimerisation that is found for the ground state of the chain can thus be considered as the result of a Peierls distortion of the regular chain. Within the Peierls electron-phonon coupling model for the chain dimerisation and $\pi - \pi^*$ band gap, the gap ($2\Delta_0$) is determined by the dimensionless electron phonon coupling constant, λ, through the relation [20]

$$\Delta_0 = 8t_0 \exp\{-(1 + 1/2\lambda)\} \tag{2}$$

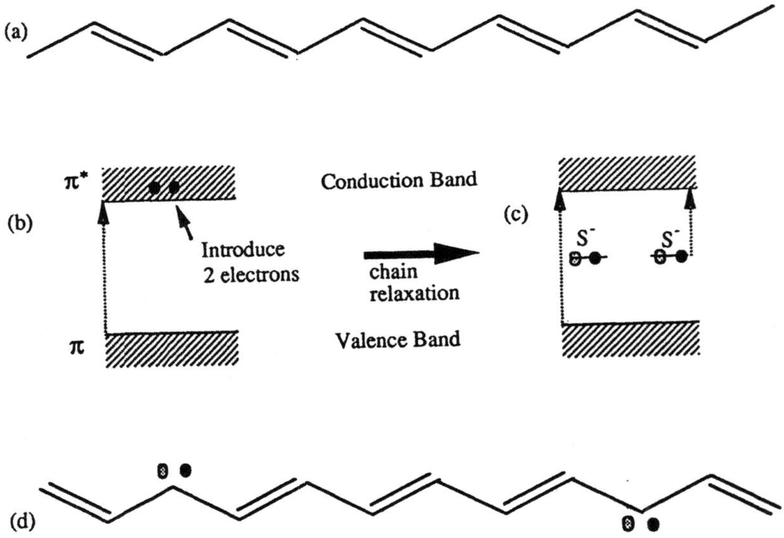

Figure 2. Schematic representation of the formation of a pair of solitons on a polyacetylene chain. (a) shows the undistorted chain and (b) shows the associated band scheme. The interband π-π* optical transition is shown as a dotted arrow. If two electrons are introduced onto the chain, initially into the conduction band as shown in (b), the chain relaxes to the form shown in (d), with a reversed sense of bond alternation in the centre of the chain separated by two solitons. Associated with the two solitons are non-bonding π states in the gap, created from one (doubly occupied) valence and one (empty) conduction band state. These two states are doubly occupied, as shown in (c), and each carries a negative charge. New optical transitions from the soliton level to the conduction band are indicated with a dotted arrow; the oscillator strength for the interband transition is weakened through the loss of the band states. A complementary picture holds for positive charge, which is accommodated as unoccupied solitons levels.

where λ is related to the parameters in the SSH Hamiltonian through

$$\lambda = \frac{2\alpha^2}{\pi t_o K} \tag{3}$$

Fixing to experimental values for the gap, of 1.7 eV and the bandwidth, of 12 eV, the value of λ is required to be of order 0.2, rather larger than can be attributed to electron-phonon coupling alone [19,20], and it is widely held that the enhanced bond dimerisation is due to the effects of the on-site Coulomb repulsion energy, U_o. This is shown for example in the work on Baeriswyl and Maki [21], in which, for U_o small (<4t), bond dimerisation is enhanced. It is possible to choose realistic values of λ (= 0.1) and $U_o/4t$ (= 0.6-0.7) and to obtain a satisfactory value for the energy gap.

In spite of the clear evidence for the importance of Coulomb interactions, and their role in determining the properties of conjugated molecules, the SSH model and its continuum version [18] are widely used, particularly for the modelling of polyacetylene in its excited states with the various parameters (bandwidth, electron-phonon coupling etc.) chosen empirically. This approach has been remarkably successful, and we follow it in the present work for the modelling of the effects of disorder on the chain.

The fundamental excitations of the half-filled band Peierls distorted chain are known to be phase kinks, or solitons, in the pattern of the bond alternation. This was shown for polyacetylene [5,6] to take the form of the bond alternation defects shown in figure 2. An important insight into the nature of these excitations from the work of Su et al [6] is that the bond alternation defect is not localised at a single carbon site, as indicated for convenience in the schematic representation in figure 2, but is spread over some 10 to 15 carbon sites. This delocalisation is crucial to the energetics of the stabilisation of the soliton, and is clearly demonstrated experimentally. For this situation, it is possible to use a continuum model for the polyacetylene chain, within which various simple analytic results are found. Thus, the gap parameter, Δ varies through the soliton as

$$\Delta(x) = \Delta_o \tanh(x/\xi) \tag{4}$$

with the soliton half-width, $\xi = (\hbar v_F/\Delta_o)a$ where v_F is the Fermi velocity, or $(2t_o/\Delta_o)a$ [18]. The extra lattice energy associated with the formation of the soliton is $\frac{2}{\pi}\Delta$, and the dynamical mass, M_S of the state described by equation 4 is given by

$$M_S = \frac{4\Delta_o^3}{3\pi\lambda v_F^2 \Omega_o^2} \tag{5}$$

where Ω_o is an averaged bare phonon frequency [22]. Δ_o and hence ξ and M_S depend very sensitively upon the value of λ (through equation 2), and we show later that much of the variation in behaviour due to differing types of disorder in the polyacetylene used in the MIS and MISFET structures can be modelled by allowing variation in the effective electron-phonon coupling constant, λ. The soliton has associated with it an energy level which within one-electron theory lies at mid-gap and is of p_z non-bonding character. This is as shown in figure 2, where for the injection of negative charge, this state is doubly occupied. The corresponding positively charged state is unoccupied.

Within the SSH model and its continuum version [18], storage of added charge on the mid-gap state created with the soliton is energetically favoured over storage in the band states; the energy of creation of the soliton is $2\Delta/\pi$, but the availability of the mid-gap state allows storage of the charge in a state of energy Δ lower than the appropriate band edge state. Thus the energy of stabilisation of the soliton over charge storage in the band edges is $\Delta(1 - 2/\pi)$; this is about 0.3 eV with the experimentally measured value of $\Delta \approx 0.8$ eV. We note here that this has direct consequences for the operation of devices such as the MIS structure; charges injected onto the chains are stored in soliton states, and the threshold for injection into these states is at energies of $\pm\Delta(1 - 2/\pi)$ relative to the gap centre, rather than at the conduction and valence band edges, at $\pm\Delta$. We return to this in more detail in section 6.1.

The formation of solitons through the addition of charge to the polymer chain modifies the optical properties, with new optical transitions allowed between the mid-gap soliton levels and the band edges, as indicated for the negative soliton to conduction band transition in figure 2. The dipole matrix element for this transition is large, and is enhanced above that for the π-π^* interband transition by a factor of $\pi^2\xi/a$, where a is the average carbon-carbon spacing along the chain [23,24] The effect of Coulomb interactions, characterised through the on-site Coulomb repulsion, U_o, is to modify the energies of the electronic transitions between band states and the localised soliton states [25,26]. Transitions involving charged solitons are reduced in energy, and modelling the effects of U as a perturbation in the one-electron model [9],

$$\Delta E = \Delta_o - \frac{aU_o}{3\xi} = \Delta_o\left(1 - \frac{aU_o}{3\hbar v_F}\right) \tag{6}$$

A value of $aU_o/3\xi$ of about 0.4 eV is consistent with the measured values of ΔE obtained from photo-induced absorption measurements [25]. Associated with the formation of the 'mid-gap' soliton absorption band is the loss of π and π^* band states and the bleaching of the π-π^* interband transitions.

In addition to the electronic excitations of the soliton, there are also new vibrational modes associated with translation of the soliton (T modes) and with shape or amplitude modes of the soliton (A modes). The T modes are associated with the coupling of those vibrational modes of the $(CH)_x$ unit cell that modulate the bond dimerisation amplitude to the translation of the soliton along the chain [22,27]. These modes appear as Raman-active modes for the dimerised chain, with frequencies determined by the phonon response function,

$$D_o(\omega) = \sum_n \frac{\lambda_n}{\lambda} \frac{\omega_n^{o\,2}}{\omega^2 - \omega_n^{o\,2}} \tag{7}$$

where λ_n/λ are the relative coupling parameters for the modes that couple to the bond dimerisation (n = 1 to 3 for polyacetylene) and ω_n^o are the bare phonon frequencies for these modes. $D_o(\omega)$ determines the frequencies of the Raman modes of the dimerised chain, through $D_o(\omega) = -1/(1-2\lambda)$ where λ is the effective electron-phonon coupling constant. It determines the frequencies of the T modes of the soliton, through $D_o(\omega) = -1/(1-\alpha)$ where α is a parameter that characterises the pinning of the soliton due to disorder etc., and finally, it determines the frequencies of the Raman-active A modes of the soliton, through $D_o(\omega) = -1/(1-2K\lambda)$, where K is calculated to be 0.723 [28-30].

3. Polymer Processing and Device Fabrication

3.1 THE DURHAM PRECURSOR ROUTE TO POLYACETYLENE

The Durham route is shown in figure 1, and as mentioned in section 1 we have made use of the solution processibility of the precursor polymer (B), poly((5,6-bis(trifluoromethyl)-bicyclo[2,2,2]octa-5,7-diene-2,3-diyl)-1,2-ethenediyl), for the fabrication of the devices investigated here. The monomer (A) is available in a one step high yield process from commercially available reagents; ring opening metathesis polymerization (ROMP) operates exclusively on the cyclobutene unit to give the polymer (B) which is soluble and can be characterized by solution-phase methods. The stereochemistry of the backbone vinylene units in (B) depends on the nature of the ROMP initiator. For the polymers used here the reactive and fairly non-discriminating WCl_6/Me_4Sn -derived initiator system has been used which gives atactic polymers with random distributions of cis and trans vinylenes; polymer (B) is therefore expected to have a disordered microstructure and this accounts for its ready solubility. However, there has been significant progress in recent years in the synthesis, characterization and use of well defined metathesis initiators and, for example, the initiators $M(CH\text{-}t\text{-}Bu)(NAr)(O\text{-}t\text{-}Bu)_2$ {where M = W or Mo} allow living ROMP and give samples of (B) in which the backbone vinylenes are trans [31,32].

The standard procedure for the fabrication of a thin-film devices is to spin-coat the precursor polymer in solution onto the required substrate, and then heat treat to convert to the polyacetylene, by elimination of hexafluoro-orthoxylene. The thermodynamics and kinetics of the transformation and isomerisation reactions have been thoroughly investigated, and heat treatment at 100°C for a period of a few hours achieves conversion from precursor to polyacetylene and full isomerisation from the *cis* to *trans* isomers [33]. The hexafluoro-orthoxylene leaving group is found to be absent in thin films of the converted polymer (as used in the present device structures), though for thicker films (> 1µm), some of this is trapped inside the polymer film. The precursor polymer has very good film-forming properties, and it is straightforward to control the thickness of the resultant polyacetylene film over the range 100 Å to several µm by adjustment of the conditions for spin-coating the precursor polymer [11-17].

Polyacetylene produced by the Durham route can be obtained in a variety of morphologies. If prepared in thin-film form by spin-coating the precursor onto a substrate and thermally converting to polyacetylene *in situ* the polymer retains much of the disorder that was transferred from the precursor polymer in solution. Although polyacetylene, as a rigid-rod polymer, shows a strong tendency to crystallise, these films show coherence lengths as determined from X-ray measurements of typically of no more than 40 Å [34]. In contrast, if a free-standing film of the precursor polymer is stretched during the thermal conversion, it is possible to obtain a very high degree of orientation and an increase in the crystallinity [35-37]. We have found that the electronic properties of the polyacetylene are very sensitive to the structure of the polymer, and we find that the unoriented films obtained by spin-coating onto substrates and conversion *in-situ* show considerably increased band-gaps over that found in the stretch-oriented free-standing films. The role and characterisation of disorder is discussed further in section 4.

3.2 DEVICE FABRICATION

Device fabrication requires the successive deposition of a series of layers of metal, polymer and insulators, with appropriate patterning in the case of the MISFET for the formation of the source and drain areas. We use simple metal evaporation, of gold as an ohmic contact and, typically, aluminium as a blocking contact. Thus, to fabricate a Schottky barrier diode, we deposit a bottom layer of gold onto the substrate, coat this with the polyacetylene layer, and put the low work function metal such as aluminium or indium on top of this. The Schottky barrier diode is obviously very sensitive to the presence of pin-holes in the polymer layer, and we have therefore aimed to keep the thickness of the polyacetylene layer sufficiently high so that the chances of this failure mode are reduced to a low level. We have worked typically with polymer layer thicknesses of around 1 μm.

The requirements for an insulator suitable for use in an MIS device are that it forms films free of defects, especially pin-holes, and that the product of the relative dielectric constant, ε_r, and the dielectric breakdown field, F_b, is high. Silicon dioxide has excellent insulating properties and we have therefore made extensive use of silicon processing techniques to provide both gate, as a heavily n-doped layer, and insulator, as silicon dioxide grown on the silicon substrate. This structure provides an MIS device when coated with polyacetylene and capped with an evaporated thin gold top contact. This can be readily extended to give the MISFET structure; it is convenient to put the source and drain contacts onto the insulator layer, and put the polyacetylene layer down on top as the last stage in the construction. In the example of the MISFET structure shown in figure 3 the source and drain contacts are of poly n-silicon; we have also made structures with gold.

We have made extensive use of organic insulator layers in MIS devices. The interest here is, firstly, to establish the compatability of the polyacetylene with the insulator; as shown in section 6, we find that the polyacetylene layer is better ordered when formed on an organic layer than when formed on silicon dioxide. Secondly, we were keen to be able to fabricate structures which are optically transparent over the whole region of interest for polyacetylene (up to 2.5 eV); those fabricated on silicon substrates are limited by the silicon band gap (1.1 eV). We have used principally polyimide and PMMA, though the maximum values of $\varepsilon_r F_b$ are considerably lower than obtained with silicon dioxide. Polyimide has the advantage that, after spin-coating from solution, the film is heat-treated to render it insoluble, and it is not affected by further treatment with solvents such as those used for the Durham precursor. In contrast, PMMA remains soluble after deposition, and devices were fabricated with PPMA applied on top of the previously-converted polyacetylene film.

The polyimide precursor solution (Π^{TM} supplied by Brewer Science Inc.) was spin-coated onto a thin and semitransparent (300 nm) aluminium film previously deposited onto a spectrosil substrate. Imidisation was carried out using the prescribed heat treatment, to give a polyimide film of thickness 360 nm. Polyacetylene was then deposited on top in the usual way and capped with a semi-transparent film of gold. For the PPMA devices, the polyacetylene was deposited onto a spectrosil substrate previously coated with a semitransparent (20 nm) gold film. PMMA was then spin-coated to give a total thickness of the PMMA layer of about 450 nm. Finally, the PPMA layer was capped with thin evaporated layers of 20 nm chromium and 10 nm gold. Other organic insulator layers used include unisomerised, *cis*-rich polyacetylene and also Langmuir-Blodgett films formed with a diacetylene monomer which could be polymerised *in situ* to form a polydiacetylene layer.

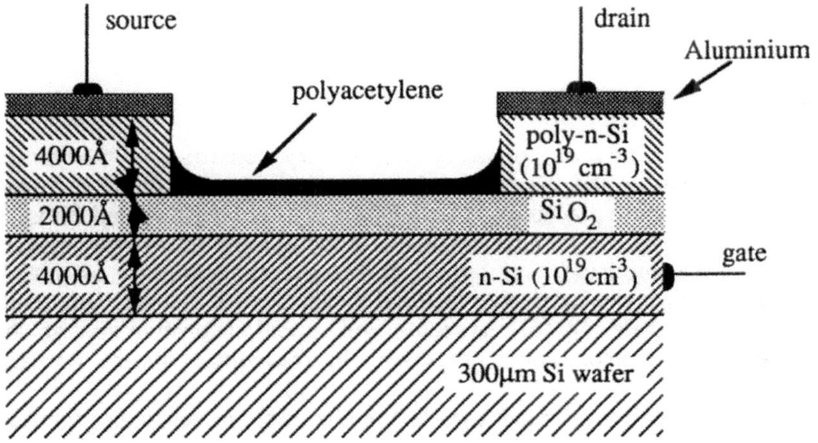

Figure 3. Schematic diagram for a polyacetylene MISFET structure. Dimensions shown are to scale, except the channel width (20 μm) and length (1.5 m).

Polyacetylene is very sensitive to oxygen, and we have therefore taken steps to ensure that, where possible, fabrication is performed in an oxygen-free environment. We perform all fabrication steps involving the polyacetylene inside a glove-box which is maintained with oxygen and water levels of less than 5 ppm. The glove box contains, therefore, the spin-coater, temperature-controlled ovens, and the metals evaporator (though for some of the earlier work on the Schottky diode structures, the sample was transferred rapidly in air to a metal evaporator, and pumped for 12 hours prior to evaporation of the top contact).

As mentioned in section 1, Durham route polyacetylene as prepared by thermal conversion from the precursor is p-doped, to a concentration in the range 10^{16} to 10^{18} cm^{-3}, and we have been able to use the polymer without further doping, as this range gives convenient values for the thickness of the charge depletion layer in our devices. In addition to the convenient concentration of the dopants responsible for these carriers (in the range 1 to 1000 parts per million carbon atoms), the majority of the dopants are not found to diffuse under the applied fields. This is an essential requirement for useful device operation, and is not in general likely to be met in chemically-doped conjugated polymers. In particular, post-doping of the polymer either chemically or electrochemically, requires that dopant diffusion be sufficiently rapid to allow the doping reaction to proceed, and these diffusion rates will then also govern movement of the dopants under the action of the applied electric fields in the devices. Evidence for dopant diffusion can be obtained through hysteresis in the C(V) characteristics for MIS devices, and has been readily obtained, for example, with poly(3-hexyl thienylene) [38,39]. The origin of the dopants in Durham polyacetylene is not well established, though we consider that they may be due to the presence of metathesis catalyst residues at the chain ends (the degree of polymerisation of the precursor does give appropriate concentrations of chain ends [40]).

4. Electronic Properties of Durham-route Polyacetylene

4.1 ELECTRONIC STRUCTURE

The electronic structure of polyacetylene is very sensitive to the presence of conformational defects on the chains, and there are significant differences between the properties of polyacetylene prepared as unoriented and as stretch-oriented films. Stretch-oriented films show properties similar to those of good-quality Shirakawa polyacetylene, with the peak in the interband π-π^* optical absorption at around 1.9 eV [41-43]. In contrast, the band-gap in the unoriented material is raised, with the peak in the interband absorption at about 2.3 eV [44-46], as seen in the optical absorption spectra shown in figure 4. There are concomitant increases in the frequencies of the Raman-active vibrational modes [44].

The manner in which disorder can modify the electronic structure of the conjugated chains has been much discussed in the literature, and it is clear that there are several ways in which disorder can act. In the simplest case, we can consider that conjugation on the chain is disrupted through the presence, for example, of an sp^3-bonded carbon on the chain, but it is well established from calculations [47] that these will cause complete breakage of conjugation (in this model, conjugation length is identified as the length between breaks in conjugation). Furthermore it is known that the concentration of sp^3 carbons is too low to allow these to have an important role in the determination of the electronic structure. More interesting forms of disorder are conformational defects on the chain; these may include regions of the *cis* isomer, residual *cis*-linkages on the chain, chain twists or chain bends. It is clear from the nature of the structure of the unoriented Durham polyacetylene that there must be a high concentration of such chain defects, and we consider that it is these conformational defects that determine the modifications of the electronic structure here. The effect of these chain defects on the electronic structure is not easy to predict; though it is always the case that the ground-state properties will shift in the same way, to give a weakening of the π electron delocalisation along the chain, and hence an increase in the size of the π-π^* band gap, the effects on the excited state properties are more subtle. In particular, we need to distinguish between defects which prevent formation of soliton-antisoliton pairs (and which therefore impose a preferred sense of bond alternation on the chain) and those which still allow formation and separation of the soliton-antisoliton pair.

A very important empirical characterisation of the nature of the disorder has been developed by Horovitz et al [22,27] in relation to the analysis of the dispersion of the resonantly-enhanced Raman modes. Horovitz shows that within the Peierls formalism for the bond dimerisation, equation 2, the phonons that couple to the bond dimerisation amplitude, as specified in equation 7, obey the product rule

$$\prod_{n=1}^{3} \omega_n{}^2/\omega_n^{o\,2} = 2\lambda \tag{8}$$

This relation can be used with an empirically chosen value of λ, which can include some of the effects due to disorder, termed 'intrinsic' disorder by Horovitz. We have found that analysis of the dispersion of the Raman spectra for the unoriented Durham polymer shows that the increased gap, from 1.9 to 2.3 eV, is empirically modelled by an increase in the value of the effective λ due to disorder, by some 8% [44,45]. The disorder due to the conformational defects on the chain, therefore, still allows the chains to support soliton-like excitations.

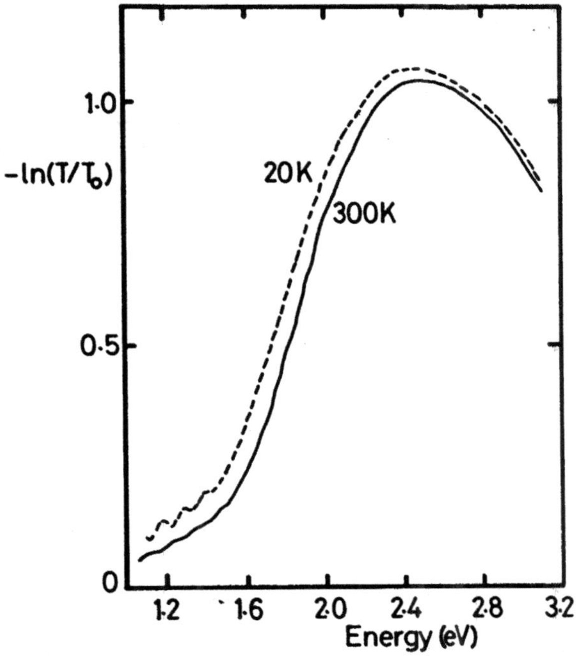

Figure 4 Optical transmission T for unoriented Durham-route polyacetylene measured at 20 and 300 K (T_0 is the transmission without the sample present) [46].

The clearest evidence for the survival of soliton-like excitations is the observation of mid-gap states through the optical transitions between these states and the band edges, and this has been observed both in chemical doping experiments [45] and in measurements of photoinduced optical absorption [46,48]. In this experiment the sample is optically excited above the bandgap and charge carriers that separate are able to relax to form metastable charged solitons. We show measurements in the range 400 - 4000 cm^{-1} in figure 5(a); the

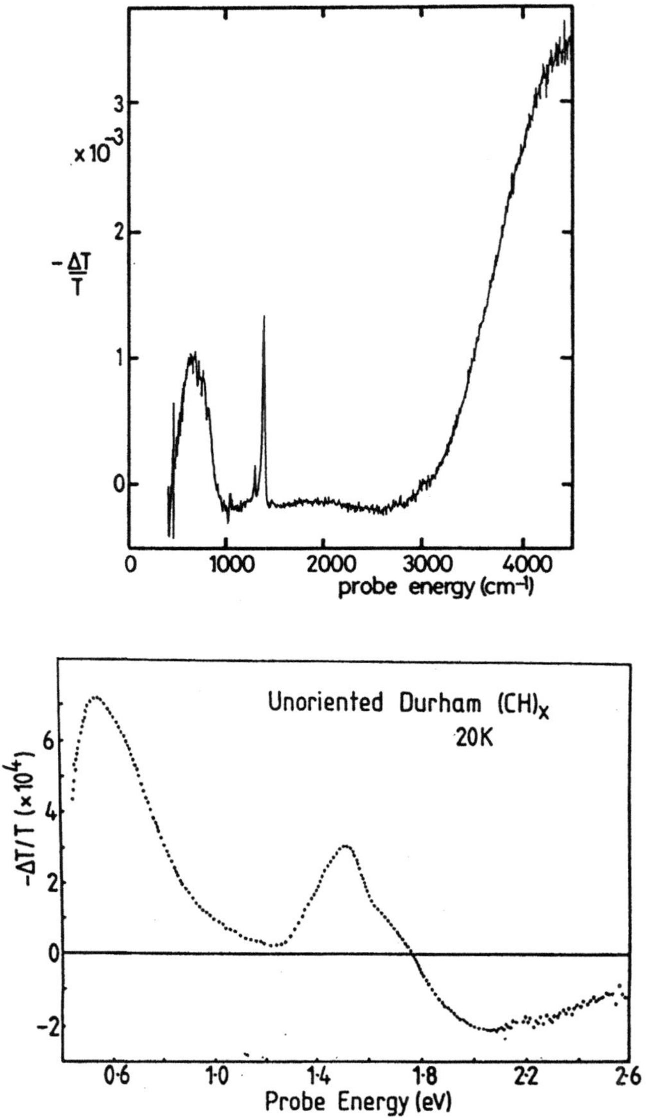

Figure 5. Photoinduced optical absorption spectrum for unoriented Durham polyacetylene, (a) measured at 77 K with excitation at 488 nm [48], and (b) measured at 20 K with excitation at 457.9 nm [46].

large absorption band which peaks near 4400 cm^{-1} (0.55 eV) is associated with the mid-gap optical transitions indicated in figure 2, and the features at lower energies in the vibrational part of the spectrum are associated with the translational modes of the soliton, as discussed in section 2. Measurements at higher photon energies are shown in figure 5(b); the induced absorption feature at 0.55 eV is well resolved and we see also a further absorption feature at 1.5 eV, which is considered to be due to an extrinsically-stabilised neutral excitation, and bleaching of the π - π^* interband transitions above 1.8 eV [46].

The spectra shown in figure 5 shows all the features found in the better-ordered forms of polyacetylene, though in detail, there are significant differences which are due to the chain defects. Firstly, the lowest frequency 'pinned' vibrational mode at 625 cm^{-1} is significantly higher in frequency than found in oriented Durham material (520 cm^{-1}), indicating that the charged excitation on the chain is more strongly pinned. The two higher-frequency modes, at 1287 and 1373 cm^{-1} are also shifted to higher frequencies. Secondly, these three vibrational modes are very much less intense in relation to the size of the electronic transition at 0.55 eV than as found in the oriented Durham material. This ratio gives a direct measure of the strength of the vibrational transitions which are inversely proportional to the effective mass of the charged excitation. From this analysis it is found that the effective mass for the unoriented polyacetylene is increased by a factor of two over the mass found in the oriented material [48]. From equation 5, this increase in mass requires an increase in band gap, 2Δ, of some 30%. Thirdly, the position of the electronic transition between soliton level and the band edges is significantly raised over that for the oriented material, from 0.45 to 0.55 eV. The position of this band is shifted away from mid-gap by the effects of Coulomb correlation but the correction due to the Coulomb interaction scales with the size of the soliton wavefunction, and thus, as indicated in equation 6, the position of the soliton to band transition is expected to scale with the size of the band gap, 2Δ, if the gap equation is described by equation 2.

The description that consistently models these observations, and also the increase in the ESR linewidth of the neutral spin defects in the unoriented material [49], is that the effect of disorder is to increase the magnitude of the effective electron-phonon coupling constant, by some 8%, as measured from Raman scattering [44]. This increases the gap, 2Δ, by equation 2, accounting for the 30% increase in the peak in the π-π^* absorption and a similar increase in the energy of the soliton to band transition. It increases the effective mass of the soliton by a factor of Δ^3, or 100%, as observed from the strength of the vibrational modes.

We shall be concerned later in this chapter with determining the nature of the disorder in the accumulation layer in the polyacetylene MIS devices, and will make use of the same analysis as used here for the comparison between the oriented and unoriented polyacetylene.

4.2 ELECTRONIC TRANSPORT

The effects of disorder in the unoriented Durham polyacetylene are also manifested in the transport properties, and we review here the picture that has been obtained for the transport mechanisms in the semiconducting (low carrier concentration) regime. Polyacetylene prepared by all routes shows a conductivity as prepared (i.e. not intentionally doped) that is due to extrinsic carriers which are p-type. There are several possible sources for the doping which creates these carriers, and we identify doping with oxygen and doping with catalyst residues as plausible sources of carriers. In the case of Durham polyacetylene it seems likely that the dopants are associated with catalyst residues due to the polymerisation of the precursor monomer. As we show later, there is good evidence from the performance of some of the device structures that the dopants are immobile, and we consider that attachment of the dopant to the chain ends can account for this.

We show the temperature dependence of the conductivity of both unoriented and oriented Durham polyacetylene in figure 6. We note that the temperature dependence of the conductivity is similar for both samples, showing an activated behaviour with an energy of activation of about 0.4 eV near room temperature. The anisotropy for the oriented sample is about 40, but that the conductivity for the oriented sample is very much higher than for the unoriented sample. Thus, the room temperature conductivity of the oriented sample along the stretch direction is 3×10^{-5} S/cm, but the corresponding value for the unoriented sample is a factor of 1000 lower, at 3×10^{-8} S/cm.

The concentration of charge carriers can be estimated in various ways; we use measurements of the depletion regime capacitance in Schottky and MIS diodes presented in later sections to determine values in the range 10^{16} - 3×10^{17} cm^{-3}, but an independent check can be made from the thermopower. Thermopower measurements are restricted by the high sample resistance to relatively high temperatures, but Townsend has found room temperature values of +1.2 mV/K for stretch-oriented samples both parallel and perpendicular to the chains and also for unoriented samples [50]. For the oriented samples some limited low temperature data are available, as shown in figure 7. The positive value indicates that the carriers present are holes. The independence of the thermopower with temperature is characteristic of phonon-assisted hopping between isoenergetic states [51], and the thermopower in this situation is given by the Heikes formula (expressed here for spinless carriers)

$$S = \frac{k}{e} \ln[(1-c)/c] \qquad (9)$$

where c is the concentration of carriers [52]. The measured value of 1.2 mV/K gives a value for c of 10^{-6}; Townsend [50] considers that this value should be identified with the concentration of available sites for solitons which occupy of order 10 carbon sites, so that the charge concentration is about 10^{16} cm^{-3}. This value falls in the range of values found from analysis of the diode structures presented later. The finding of a similar value of the thermopower for the unoriented and oriented samples indicates that the charge carrier concentrations are similar in both materials, and that therefore the increased conductivity in the oriented material is due to a strongly increased carrier mobility in the better-ordered material. We return to discuss the limits to carrier mobility in section 7.3.

The temperature dependence of the resistivity therefore indicates that the carrier mobility is thermally activated. Townsend and Friend have measured also the temperature dependence of the mobility for photogenerated charge carriers [46]. This requires firstly, measurement of the photoconductivity, which shows a temperature independent 'fast' component and a temperature-dependent component that is due to the long-lived photogenerated charged solitons. Secondly, the variation with temperature of the concentration of photogenerated charge carriers is determined through measurement of the intensity of the soliton to band edge photoinduced absorption at 0.55 eV. The results of

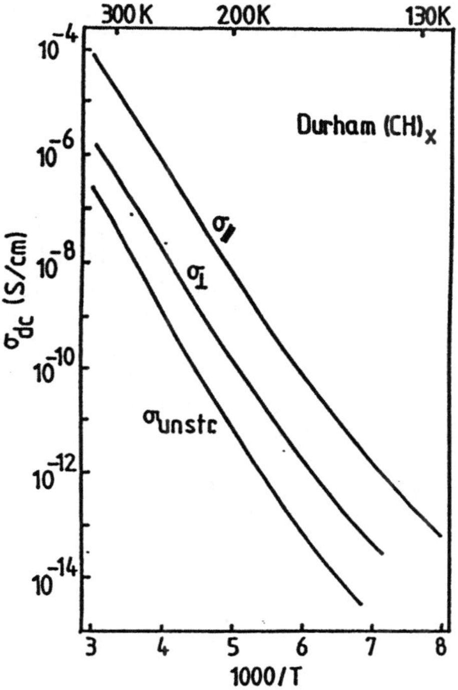

Figure 6. DC conductivity for stretch-oriented (draw ratio 15) and unoriented Durham polyacetylene [50].

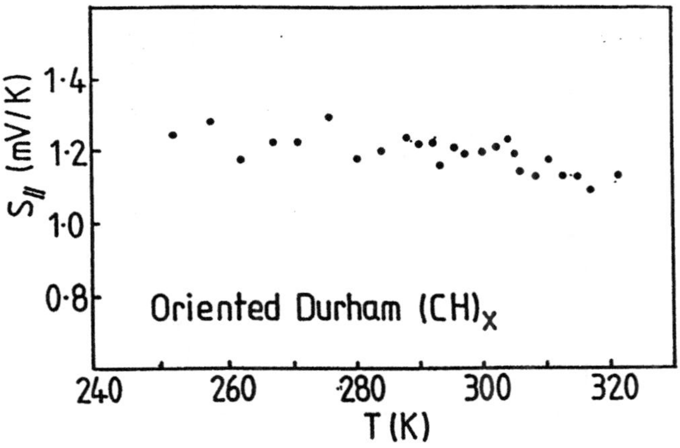

Figure 7. Thermopower for oriented Durham polyacetylene, with the temperature gradient parallel to the chain direction [50].

this analysis are shown in figure 8; the photocarrier mobility is obtained from the photocurrent, I_p, and the concentration of photocarriers at temperature Θ, $N(\Theta)$, from the relation

$$-\frac{\Delta T}{T} = N(\Theta)\sigma \qquad (10)$$

where $-\Delta T/T$ is the fractional change in absorption at 0.55 eV and σ is the optical cross-section per carrier. Using a value for σ of 8×10^{-16} cm^{-2} obtained from measurements on chemically-doped samples [53] Townsend and Friend obtained a value for the room temperature mobility of 5×10^{-5}cm^2/Vsec, but we can obtain an improved measure of the optical cross-section in the optical measurements on the injected charge layer in the MIS devices discussed in section 6, and find a value of 1.6×10^{-15} cm^2 for the charged solitons with the 'mid-gap' transition at 0.55 eV, and so would revise the value of the room temperature mobility upwards to 1×10^{-4} cm^2/Vsec.

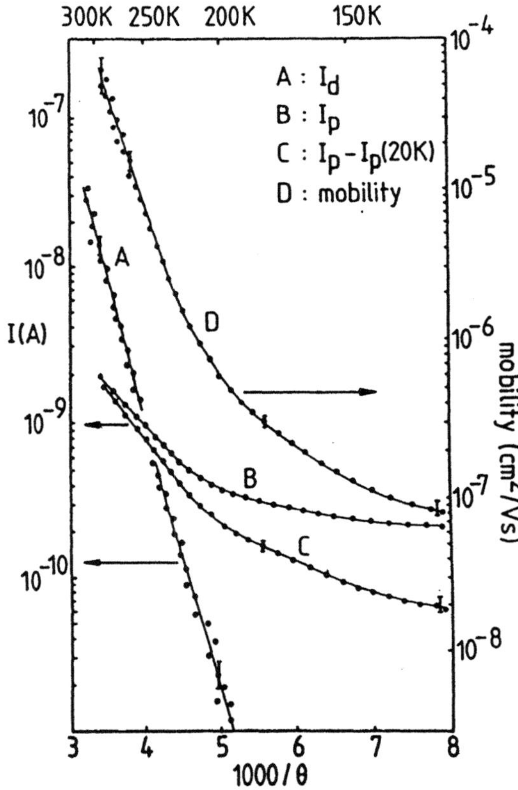

Figure 8. Photoconductivity and carrier mobilities for unoriented Durham polyacetylene [46]. Plotted against reciprocal temperature are A, the dark current, I_d; B, the photocurrent, I_p; C, I_p minus its value at 20 K; D, the photocarrier mobility.

Figure 8 shows that the mobility of the photocarriers above 200 K shows an activated behaviour, similar to that of the dark carriers, though with a slightly reduced activation energy of 0.31 eV. The similarity of the mobilities of the dark and photoexited carriers indicates a common mechanism for both, and Townsend and Friend consider that the appropriate process for the rate limiting step is that of 'bipolaron' hopping of bound (or weakly bound) soliton-antisoliton pairs between adjacent chains. In this model, motion of one charge to an adjacent chain creates two singly-charged 'polarons' at a cost in energy of W_h, and the second charge can then follow the first so as to transfer the pair of charges between chains. For three-dimensional hopping, the mobility, μ, obtained in this model using the Einstein diffusion relation is

$$\mu = \frac{ea^2\omega}{6k_B\Theta} \exp\left(\frac{-W_h}{k_B\Theta}\right) \qquad (11)$$

where a is the distance between hops, W_h is the hop energy, ω is the attempt-to-hop frequency. If W_h is identified as the experimentally measured mobility for the photocarriers and ω is taken to be the average of the frequencies of the optic phonons that modulate the dimerisation amplitude (3.9×10^{13} Hz), then a value for a of 3 nm is obtained. This is considerably larger than the interchain separation (0.42 nm), and can be taken as a measure of the average distance travelled between interchain hops. This value is close to the dimension of the crystallites in the unoriented Durham material [34]. For the case of the oriented polyacetylene for which the carrier mobility along the chain direction is a factor of 1000 higher than as found for the unoriented samples, we note that modelling the mobility through equation 11 requires that there is a very much larger value for the average distance between interchain, thermally activated hops, a.

4.3 ELECTROMODULATION OF OPTICAL ABSORPTION

Under the conditions used for the electro-optical measurements on the various device structures that are reported in sections 5 and 6, we will expect to see a substantial electric field applied across at least a fraction of the polyacetylene layer, and we need to characterise the modulation of the π-π^* absorption edge with electric field. This electromodulation response is the Franz-Keldysh effect, and arises through modulation of the electron states near the band edge in the applied field. It is found to be very large in Shirakawa polyacetylene [54,55] and it has been pointed out that this is due to the strong non-linear electronic response that characterises the conjugated polymers [55]. In the low field limit, we expect to see a response that varies quadratically with the applied electric field, and that is proportional to the second differential of the absorption coefficient, α, $\partial^2\alpha/\partial E^2$ [54].

This experiment can be performed directly by measurement of the modulation of the optical transmission with an applied electric field, $F_o\cos(\omega t)$ with lock-in detection of the optical response at 2ω, twice the modulation frequency of the applied field. The two geometries that can be used are either in-plane, with contacts arranged to give a gap of typically 100 μm, or the sandwich geometry, with two semi-transparent metal contacts applied to either side of a polymer film. The in-plane geometry has been used for measurements on Shirakawa polyacetylene [54,55], but we have found it convenient to use here the sandwich geometry.

Figure 9 shows the electromodulation of the optical transmission for a film of polyacetylene of thickness 480 nm sandwiched between semitransparent gold contacts of thickness about 20 nm. The sample was cooled to 80 K to prevent any response from conduction processes (e.g. Joule heating), and the response shown is for an applied field of 3 x 10^5 V/cm. The electromodulation response is found to vary quadratically with applied field, and the positive response (electroabsorption) on the low energy side of the absorption edge is as found in Shirakawa polyacetylene and as expected from the second derivative of the measured absorption edge (figure 4). In detail, however, the electromodulation features show less of the vibronic structure seen in the Shirakawa material, and are shifted to higher photon energies. Two bands are seen, peaking at 1.38 eV and at 1.57 eV; the upper band is definitely associated with the π-π^* absorption edge, and shows some phonon structure with a shoulder at 1.68 eV (it is shifted up by 0.17 eV from the corresponding peaks in Shirakawa polyacetylene). The feature at 1.38 eV is less easily assigned, but we note that low energy features have also been seen in Shirakawa-type polyacetylene [55].

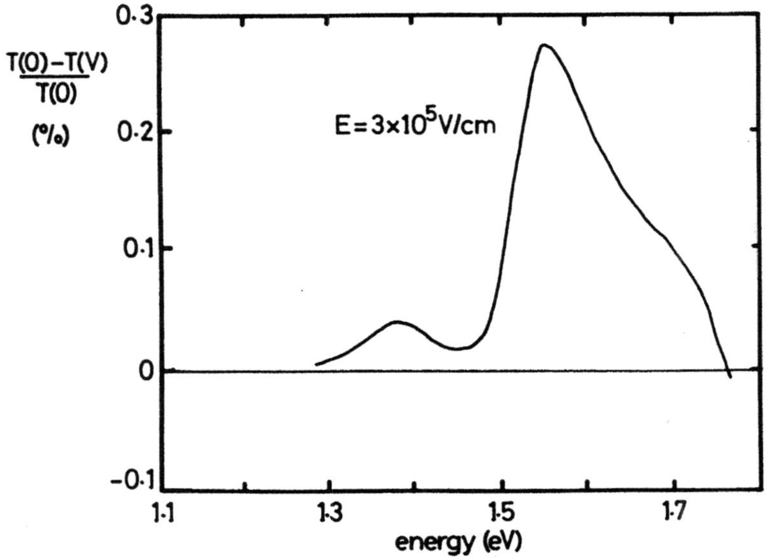

Figure 9. Electromodulation of the optical transmission for an unoriented film of Durham polyacetylene, of thickness 480 nm, measured with an applied field of 3 x 10^5 V/cm.

One further parameter that can be measured from this type of structure through measurement of the capacitance is the relative dielectric constant, ε_r. We find a value of ε_r = 2.75, and we have used this value throughout the work discussed in sections 5, 6 and 7.

5. Schottky Barrier Diodes

5.1 THE SCHOTTKY BARRIER

We have carried out an investigation of the electrical and electro-optical properties of a series of Schottky barrier diodes fabricated with polyacetylene sandwiched between two metal contact layers, one to form the Schottky barrier and the other (gold) to provide an ohmic contact [56]. This type of structure is straightforward to fabricate with an extrinsically-doped semiconductor and there have been several reports of such devices which use polyacetylene or other conjugated polymers [57-62]. The details of the device fabrication have been given in section 3.2, and we show in figure 10 the details of the typical structures that we have used for this work. We have worked with relatively thick films of polyacetylene, in the range 500 - 1000 nm, so as to avoid the possibility of short-circuits between top and bottom electrode, but we have kept the metal contact layers thin so that they are semi-transparent and allow optical transmission measurements.

In the case of a p-type semiconductor, the ideal Schottky barrier is formed between the semiconductor and a metal with a work-function, ϕ_m, that is lower than that of the semiconductor, ϕ_s [56]. This is shown schematically for a conventional semiconductor in figure 11. ϕ_s is given by

$$\phi_s = \chi + \xi \tag{12}$$

where χ is the electron affinity (energy between the bottom of the conduction band and the vacuum level), and ξ is the energy between the Fermi level and the bottom of the conduction band. The bringing together of the two materials to form the junction must result in the equalisation of the Fermi energies (for no applied bias across the junction) and this results in the formation of a dipole charge layer, with positive charge density in the metal at the interface, and a region of negative charge density (the depletion layer) formed by removal of the mobile positive charge carriers. Providing that the charge concentration

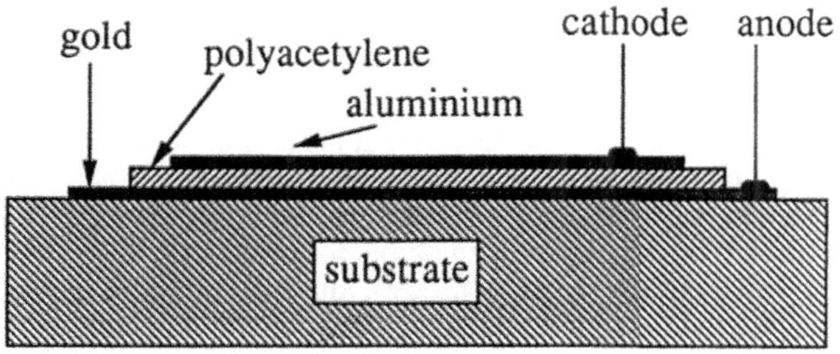

Figure 10. Schottky-barrier diode structure. The device is fabricated on a flat glass substrate, with a thin (20 nm) layer of gold as the ohmic contact, onto which the polyacetylene layer is prepared via the Durham precursor. The top metal contact which is applied by thermal evaporation is chosen to give the Schottky barrier (aluminum, chromium or indium), and is also kept to thin (20 nm).

in the metal is very much greater than in the semiconductor, all the band bending occurs in the semiconductor. This adjusts the potential between the two sides of the junction by the contact potential, which is given by $\chi + \xi - \phi_m$. This sets up a barrier for the flow of holes towards the metal contact, ϕ_b which, measured from the Fermi energy, is given by

$$\phi_b = (\chi + E_g) - \phi_m \tag{13}$$

where E_g is the energy gap for the semiconductor, $E_c - E_v$.

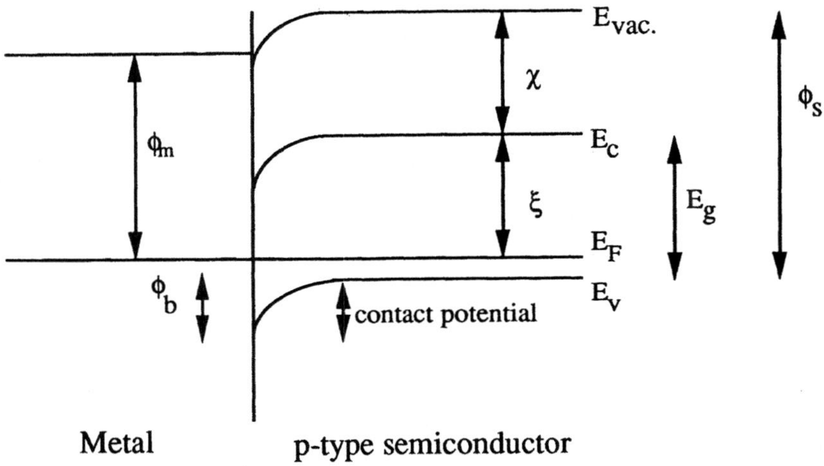

Figure 11. Conventional Schottky barrier band scheme for a p-type semiconductor against a low work function metal.

The potential barrier, and its variation with an applied bias, determine the rectifying characteristics of the junction, as we discuss below, and in principle the barrier height is determined simply from the work functions for the metal and semiconductor. In practice it is usually the case for inorganic semiconductors that the barrier height is not well predicted by this simple relation; and surface states and surface layers at the interface play an important role. For the case of polyacetylene the work function is estimated to be of the order of 5 eV [57-62], and we can expect an ohmic junction with gold (ϕ_m = 5.1 eV), but Schottky junctions with chromium (ϕ_m = 4.5 eV), aluminium (ϕ_m = 4.3 eV) and indium (ϕ_m = 4.1 eV).

The dominant transport mechanism across the junction is usually considered to be emission of holes from the semiconductor over the top of the barrier into the metal, though quantum mechanical tunnelling through the barrier, recombination in the space-charge region and recombination in the neutral region (electron injection) can also contribute [56]. Transport of holes over the barrier is limited by the rate at which they can diffuse through the depletion region (the diffusion model) and the rate at which they are able to cross the barrier (the thermionic emission model). The standard result for the variation of the current density, J, as a function of bias voltage, V, for the diffusion model is given by

$$J = qN_a\mu E_{max} (\exp\{qV/k_BT\} - 1) \exp\{-q\phi_b/k_BT\} \qquad (14)$$

where q is the electronic charge, N_a is the concentration of acceptors, μ the hole mobility, and E_{max} the maximum field in the depletion regime [56]. E_{max} is given by

$$E_{max}^2 = \frac{2qN_a}{\varepsilon_o\varepsilon_r} (V_{d0} - V) \qquad (15)$$

where V_{d0} is the diffusion voltage at zero bias which is equal to ϕ_b - (E_F - E_v). Thus at high reverse biases the current does not saturate but increases as $|V|^{1/2}$.

For the case where current is limited by thermionic emission over the barrier, the current density, J, varies as

$$J = A^*T^2\exp\{-q\phi_b/k_BT\} (\exp\{qV/k_BT\} - 1) \qquad (16)$$

where A* is the Richardson constant for thermionic emission modified by substitution of the semiconductor effective mass, m* for the free electron mass, m_e,

$$A^* = \frac{4\pi m^*qk_B^2}{h^3} = 1.2 \times 10^6 \frac{m^*}{m_e} \ A\,m^{-2}\,K^{-2} \qquad (17).$$

The criterion that thermionic emission limits current flow across the barrier is given by

$$lqE_{max} > k_BT \qquad (18)$$

where l is the mean free path of the carriers in the semiconductor [56]. This is equivalent to the criterion that the mean free path must exceed the distance in which the barrier falls through an amount k_BT/q.

We would expect here that the low carrier mobilities present in the disordered polyacetylene, estimated to be about 10^{-4} cm^2/Vsec, should put the Schottky barriers formed with it in the regime of the diffusion-limited current. We can reasonably take as a value for l the distance between hops estimated from the 'bipolaron' hopping model, equation 11. This is estimated to be 3 nm, so that the inequality is only satisfied at values of E_{max} in excess of 10^5 V/cm. We note that similar conclusions are made for the analysis of the behaviour of Schottky barriers formed with amorphous silicon/hydrogen alloys [56], though there are claims that the thermionic emission model is effective in this case, in spite of the low carrier mobilities [63].

In practice, the variation of J with bias voltage is usually parameterised as

$$J = J_o (\exp\{qV/nk_B/T\} - 1) \qquad (19)$$

where n is the so-called ideality factor. This should be equal to 1 for the thermionic emission model, but is often found to be considerably larger. There are several possible explanations for this, including the possible role of an interfacial layer of insulator between metal and semiconductor.

Besides the J(V) characteristics of the junction, the complex impedance provides important information about the nature of the junction. The capacitance of the junction in the depletion regime (reverse bias) is determined by the variation in the width of the depletion regime with bias, and in the abrupt junction model is given as

$$\frac{C}{A} = \sqrt{\frac{q\varepsilon_o\varepsilon_r N_a}{2}}\sqrt{\frac{1}{V_{d0} + V}} \tag{20}$$

where A is the area of the junction. The slope of $1/C^2$ versus bias voltage this allows direct measurement of the acceptor dopant concentration, N_a and V_{d0}. For the structures that we have studied, the ohmic resistance across the undepleted region of the polyacetylene layer is often substantial, and it is necessary to carry out a full analysis of the whole device using in the simplest case an equivalent circuit of the type shown in figure 12.

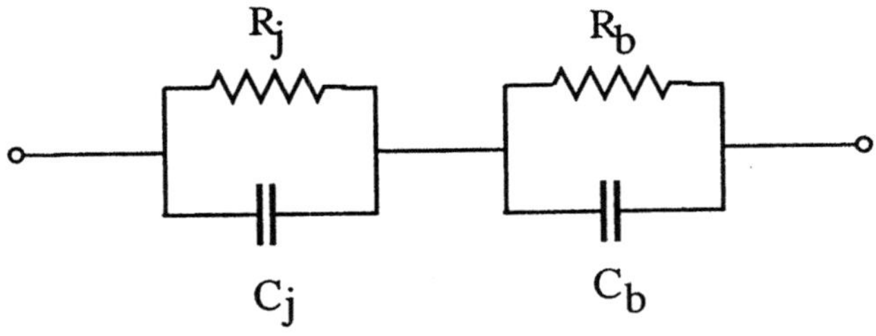

Figure 12. Equivalent circuit for the Schottky barrier diode. C_j and R_j characterise the junction and C_b and R_b characterise the bulk.

5.2 ELECTRICAL CHARACTERISTICS

The forward and reverse characteristics for a Schottky diode formed with an aluminium/polyacetylene Schottky junction and a gold ohmic contact (polyacetylene thickness 500 nm) is shown in figure 13. This device shows a particularly good characteristic with a high forward to reverse ratio which reaches a maximum of 5 x 10^5 at a bias of ± 1.5 V, and is limited at this value by the high bulk resistance of the undepleted polyacetylene layer. We note that this value is very much higher than values found previously for Schottky barriers formed with Shirakawa polyacetylene and other conjugated polymers [57-62,64].

The slope of log(J) versus V gives the ideality factor, n = 1.3 at low biases. This value, though larger than 1, is considerably smaller than found for other polymer Schottky barriers, and we consider that it is appropriate to analyse the data here with the models introduced in section 5.1. Thus, we estimate by extrapolation from the linear region under forward bias that the value of J_o in equation 19 is 4.4 x 10^{-13} A cm^{-2}. Interpreting this within the thermionic emission model, and setting the effective mass equal to the electron mass, we obtain a value for the barrier height, ϕ_b, of 1.28 eV. If we use the diffusion model, taking values of $N_a = 10^{16}$ cm^{-3}, $\mu = 10^{-4}$ cm^2Vsec, and $(V_{d0} - V) = 1$ V (appropriate near zero bias), we find a lower value for the barrier height of 0.6 eV. The requirement for this lower value for the barrier in the diffusion model confirms that diffusion of holes is indeed the current-limiting process, as anticipated in the discussion in section 5.1, and we note that this value is very close to that expected for an ideal junction, for which we expect to see approximately the difference in the work functions (5 - 4.3 = 0.7 eV). We see in figure 13 that the reverse current does not saturate at high reverse biases, and consider that this may be modelled through the $\sqrt{V}$ dependence expected for the diffusion model.

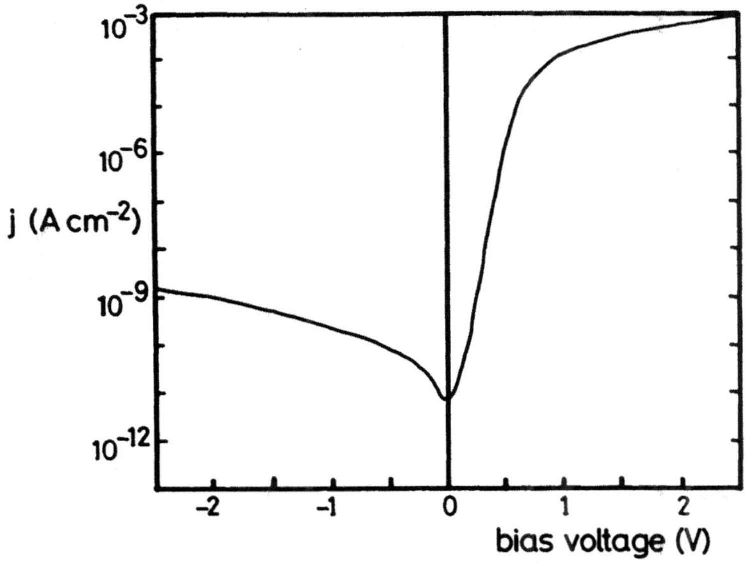

Figure 13. Modulus of current versus applied bias voltage for a Schottky barrier diode formed with an aluminium contact to provide the Schottky barrier and a gold contact to provide the ohmic junction. The device area is 0.64 cm^2 and the thickness of the polyacetylene film is 500 nm.

We have obtained some limited data for the J(V) characteristics as a function of temperature. In the range between room temperature and 240 K, J_o can be fitted to an activated temperature dependence, with an energy of activation of about 0.8 eV. However, the ideality factor also increases as the temperature falls, and we do not consider that the data can be used to resolve between the different prefactors in the thermionic emission and diffusion models (equations 16 and 14 respectively). However, we note that the activation energy within the diffusion model is equal to the sum of the barrier height and the activation energy for the mobility. From section 4.2 the latter is expected to be in the range

0.3 to 0.4 eV, and we consider that the 0.8 eV measured here may be consistent with the barrier height as previously determined.

Junctions formed with indium and with chromium also showed reasonable Schottky barrier formation, though we find in general that the ideality factors for these devices were poorer than for the aluminium structures, with values of 2 found in both cases. There is some variation in the barrier height deduced from the value of J_0; fitting to the diffusion model using the same parameters as used above we find lower values for ϕ_b, of 0.34 eV for chromium and 0.42 eV for indium.

Capacitance versus bias voltage characteristics for these diodes do show the behaviour described by equation 20, and we show in figure 14 the conventional plot of $1/C^2$ versus bias voltage. There is a linear variation of $1/C^2$ with bias voltage in reverse bias, and from the gradient we obtain values of the acceptor dopant concentration, N_a and diffusion voltage at zero bias, or 'built-in' voltage, V_{d0}. For the device shown in this figure we find values of $N_a = 9 \times 10^{15}$ cm^{-3} and $V_{d0} = 0.94$ V. We find similar values of N_a from similar analysis with Schottky junctions formed with chromium and indium when fabricated with polymer from the same batch (and similar values of the built-in voltage), but we have found that there is some variation between different batches of polymer, with N_a in the range 10^{16} to 2×10^{17} cm^{-3}. We have discussed the origin of the extrinsic carriers in section 3.2, and consider that they are tied to the presence of catalyst residue present from the polymerisation of the monomer (A) to precursor (B) (figure 1).

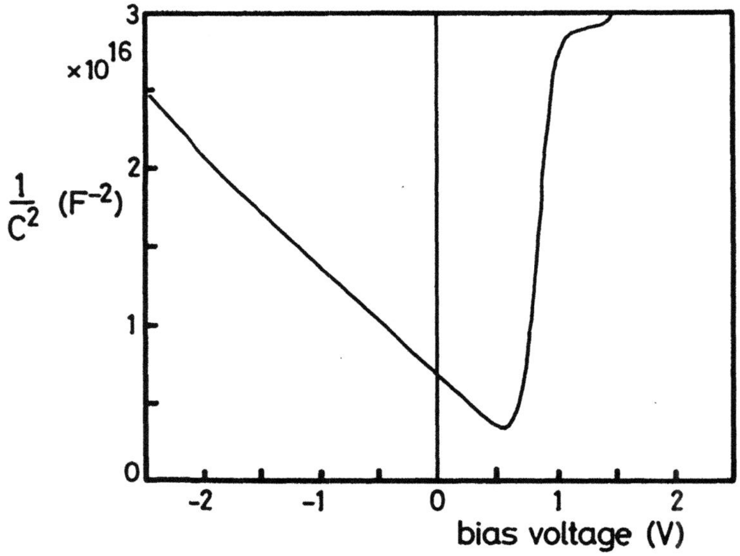

Figure 14 $1/C^2$ versus bias voltage for a diode formed with an aluminium top Schottky barrier. The sample area is 0.63 cm^2.

The value obtained here for N_a is an important characterisation of the polymer, and we have looked for independent evidence that this value is appropriate. We find firstly, that the analysis of the temperature independent thermopower presented in section 4.2 gives a similar value to that obtained here. Secondly, and perhaps less directly, we can measure the optical cross-section per injected charge in both the Schottky and the MIS devices. This is presented in some detail in section 5.3 and section 6.3; we find there that the optical properties of the injected charges are those expected for charged solitons, and that the quantitative measurement of the optical cross-section for charge movement associated with movement of the depletion layer boundary gives a value for N_a in good agreement with the value presented here (the value of the optical cross-section per injected charge can be obtained directly from measurements of the charge accumulation layer in the MIS devices).

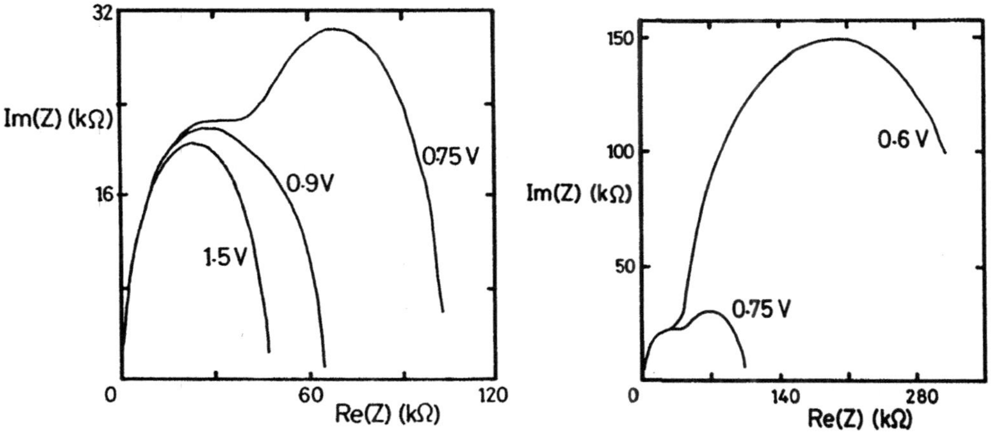

Figure 15 Impedance plots for a chromium/polyacetylene Schottky diode under forward bias (frequency is the implicit variable). The device has an active area of 0.64 cm^2, and the polyacetylene layer is of thickness 900 nm and is capped with a gold ohmic contact layer.

We conclude this section by showing an example of the complex impedance analysis obtained for these Schottky diodes. Figure 15 shows the real and imaginary values of the complex impedance, Z(ω), for a Schottky diode formed with a chromium/polyacetylene barrier. Z(ω) is derived from the complex admittance, Y(ω), through the relations

$$Z(\omega) = \frac{1}{Y(\omega)} = \frac{1}{G(\omega) + j\omega C(\omega)} \quad (21)$$

with

$$Re(Z) = \frac{G}{G^2 + \omega^2 C^2} \quad (22)$$

and

$$Im(Z) = \frac{-\omega C}{G^2 + \omega^2 C^2} \quad (23).$$

We expect for a circuit of the type shown in figure 11 to see a set of two semicircles with centres on the real axis and with their diameters corresponding to the circuit resistance involved. Resolution of the semicircles requires that two time constants for the series elements are well separated. Figure 15 shows plots for a series of forward bias voltages, and the curves do show two resistances at low forward bias, the larger of which falls rapidly as the device is brought further into forward bias. Detailed analysis shows some evidence for a resistive layer that is still present even under strong forward bias, and we consider that this may be due to an interfacial layer. [We note that, in contrast to the later MIS structures, these devices were briefly exposed to air for transfer to the metals evaporator].

5.3 ELECTRO-OPTIC PROPERTIES

The operation of the Schottky barrier diode depends on the dependence of the depletion width, w, in the semiconductor with bias voltage, as given by

$$w = \sqrt{\frac{2\varepsilon_0\varepsilon_r(V + V_{d0})}{qN_a}} \tag{24}$$

The increase in width of the depeletion regime with increasing reverse bias is obtained by movement of the extrinsic charge carriers, so as to leave the depletion region with a space charge density, qN_a, due to the acceptors. We would expect that the extrinsic charge carriers should be present as charged solitons with energy levels as shown in figure 16(a), and we expect to see optical transitions at 'mid-gap' from the π band to the soliton level. With formation of the depletion layer, the electronic charges on the chain are removed, and we expect to see the disappearance of the mid-gap levels as the polyacetylene chain reorganises itself to remove the solitons at which the charges were stored. This readjusts the density of states to the form shown in figure 16(b), and we expect to see the loss of the 'mid-gap' optical transitions as the polyacetylene is driven into depletion. Thus, if the depletion width, w, is increased by an amount δw by an increase in the reverse bias voltage, then the increase in absorbance, $\delta\alpha$, is given by

$$\delta\alpha = -\frac{\Delta T}{T} = \delta w\, N_a\, \sigma \tag{25}$$

where ΔT is the fractional change in the transmission, T, and σ is the optical cross-section of the soliton.

We have used the Schottky diodes previously characterised for their electrical properties for optical measurements. If the two metal contacts (gold and usually aluminium) are arranged to be thin (20 nm) then there is adequate optical transmission through the device, and it is possible to obtain a high sensitivity for the measurement of the change in optical transmission with bias voltage using a lock-in amplifier to measure the response of the optical detector at the frequency of the a.c. modulation applied to the bias voltage. We use a conventional light-source (tungsten lamp) and grating monochromator to select the

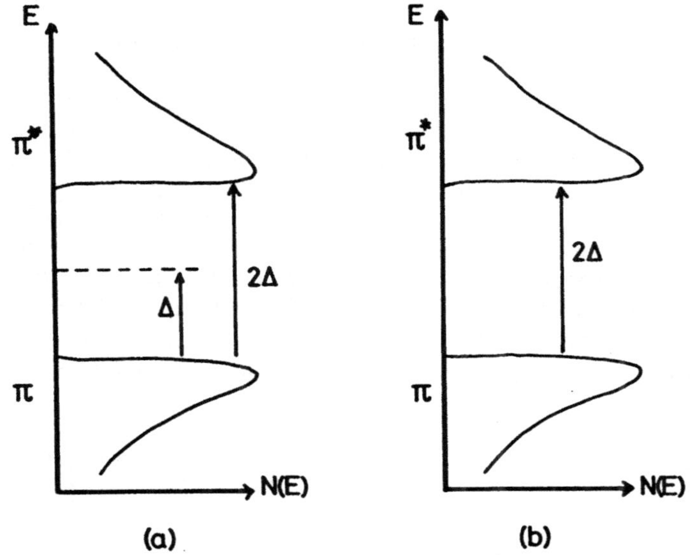

Figure 16 Schematic representation of the optical transitions possible for extrinsic p-type polyacetylene (a) and the same after charge depletion (b).

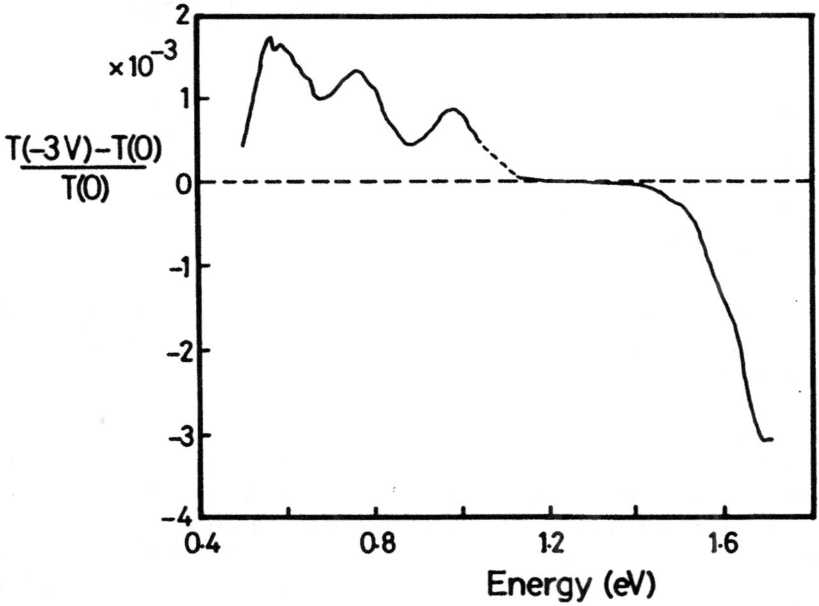

Figure 17 Bias voltage modulated optical transmission for an Al/polyacetylene Schottky diode, of thickness 900 nm. This is displayed as {T(V)-T(0)}/T(0), where T(V) is the optical transmission at bias voltage V. Here V = -35 V, and the frequency of modulation is 500 Hz.

chosen wavelength for the optical probe, and a selection of solid state detectors (cooled InSb and Ge, and Si) to cover the spectrum over the range 0.5 - 2.0 eV.

Figure 17 shows the modulation of the optical transmission obtained by taking the bias voltage between 0 and a large negative value (-35 V). The plot of {T(V)-T(0)}/T(0) gives a positive signal for a bleaching induced by the reverse bias condition, and we see the appearance of a bleaching band in the IR which peaks at 0.55 eV. We identify this as arising from the loss of soliton levels associated with the extrinsic charge carriers as these mobile carriers are swept out of the semiconductor to form the depletion layer. The band shows some oscillations at higher energies which we consider are due to modulation of interference fringes (modulation of the finesse of the Fabry-Perot etalon formed by the 0.9 μm thick polyacetylene film). We compare this spectrum with that obtained in the measurement of photoinduced absorption for the same polymer, shown in figure 5, and note that the electro-bleaching peak at 0.55 eV is at the same energy as that found for the photoexcited carriers.

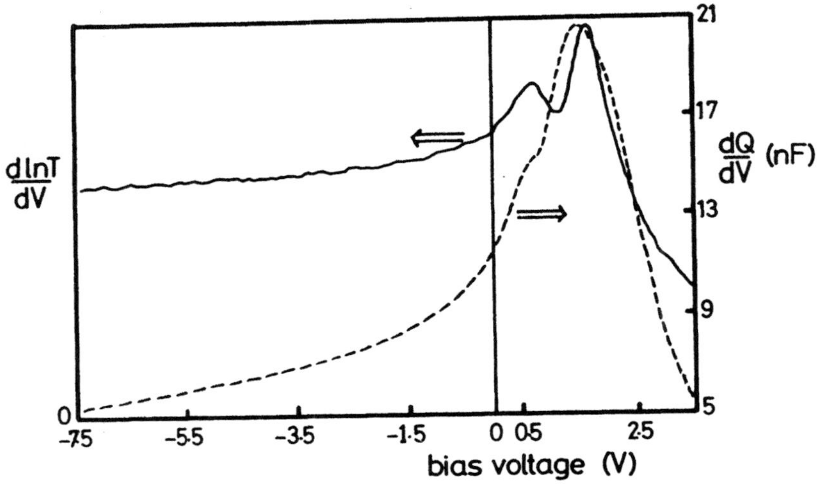

Figure 18 The differential optical transmission, ∂lnT/∂V measured at a photon energy of 0.6 eV, and differential capacitance, ∂Q∂/V, measured as a function of bias voltage for the Schottky diode similar to that shown in figure 17.

One of the advantages of the use of device structures for such optical measurements is that it is straightforward to obtain a quantitative measurement of the optical change and of the number of charges that are responsible for it. For the device shown in figure 17, the capacitance shows the expected behaviour in reverse bias (equation 20) up to -35 V, and we find a value for the acceptor concentration of $N_a = 2.1 \times 10^{16}$ cm^{-3}. From this we calculate that the change in width of the depletion layer as the bias voltage is taken from 0 to -35 V is 470 nm, and we can thus calculate the optical cross-section per injected charge at the peak of the 'mid-gap' absorption band, σ, to be 1.6×10^{-15} cm^2. This value is larger than that reported by Orenstein from studies on chemical doping (8×10^{-16} cm^2) [53], but we find similar values for the charge accumulation layers in MIS structures discussed in section 6.

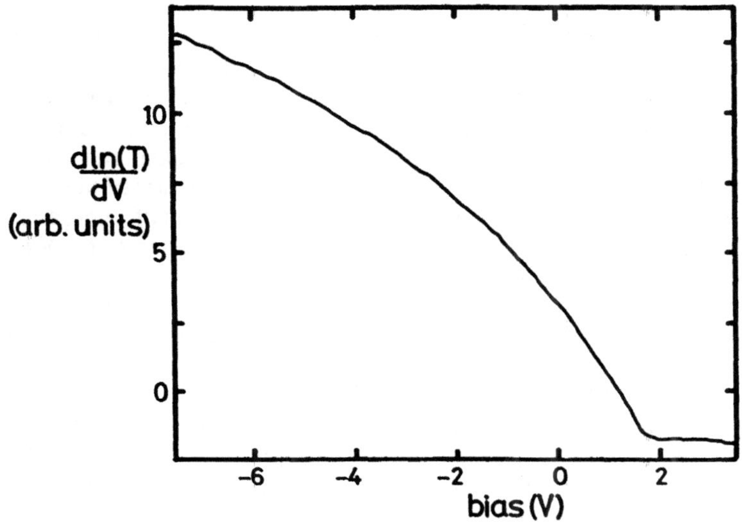

Figure 19 The differential optical transmission, $\partial \ln T/\partial V$ measured at a photon energy of 1.62 eV for the Schottky diode shown in figure 18.

At higher energies, the response reverses sign, and we can just resolve a peak in the absorption at an energy of 1.7 eV. There are two contributions to this response, firstly, the extra π-π^* absorption due to the extra band states created at the expense of the mid-gap states, and secondly, the modulation of the π-π^* absorption edge due to the high electric field across the polyacetylene in the depletion regime, see figure 9. The different contributions to the response can be resolved by measurements of the differential response, $\partial \ln(T)/\partial V$, as a function of bias voltage, as shown in figure 18 for the response at 0.6 eV and in figure 19 for the response at 1.63 eV. The behaviour of $\partial \ln(T)/\partial V$ at 0.6 eV is seen to follow that of the differential capacitance (also shown in figure 18), peaking at the onset of reverse bias. In contrast, the response at 1.62 eV increases strongly with the increasing electric field applied across the depletion layer as the reverse bias is raised. This behaviour is due to the electromodulation of the π-π^* absorption; the maximum field present in the depletion layer for -35 V bias is 5.4 x 10^5 V/cm, and we see that the size of the absorption feature here is similar to that seen for the direct measurement in the sandwich geometry shown in figure 9.

5.4 ELECTRONIC STATES IN THE POLYACETYLENE SCHOTTKY BARRIER

The Schottky diodes that we have made here show good electrical characteristics, with a very much improved forward/reverse current ratio than previoulsy reported for devices based on conjugated polymers. The electrical behaviour of the Schottky barrier diodes which we discuss in section 5.2 seems to be adequately modelled using the conventional picture for the band schemes shown in figure 11; we find that the parameterisation of the J(V) characteristics is more reasonably made within the diffusion model for the current-limiting process. However, we show in section 5.3 that we must use an entirely different model if we are to be able to account for the electro-optical behaviour of these structures; we have clear evidence that the mobile charges (holes) that are swept out to form the depletion region have energy levels which lie at the centre of the band gap, as expected if these charged excitations of the chain are charged solitons. The picture of the band scheme that is appropriate here must then be modified, with a concentration of mid-gap states that is varied as the semiconductor is switched from the normal to depleted state. We consider this further in section 6.1.

The novel physics associated with charge storage in soliton states gives new electro-optical phenomena, and we have been able to measure quantitatively the change in optical absorption at the position of the soliton to band edge 'mid-gap' optical transitions associated with the movement of the charge depletion layer. This gives us a powerful new experimental probe with which to characterise the action of the device, and we have shown in section 5.3 that the changes in 'mid-gap' optical absorption are modelled if we take all the space-charge layer to be formed by movement of charged solitons out of the depletion regime.

The transition strengths and energies for the optical features associated with movement of the charge depletion layer give specific information about the electronic states occupied by the extrinsic charge carriers. This is important because they are present in such a low concentration that they cannot be directly identified from the absorption spectrum. We have shown that the position of the 'mid-gap' absorption, 0.55 eV, is at the same energy as that for the photoexcited charged solitons, and note that this is consistent with the very similar transport properties of the two types of carriers that we review in section 4.2. This finding should be contrasted with the position of the 'mid-gap' absorption band produced by chemical doping; this is always found at higher energies than for photoexcited solitons [23], and for the unoriented Durham polyacetylene it is usually found near 1 eV [49]. We consider that the extrinsic charge carriers in the as-prepared Durham material are substantially dissociated from their counterions at room temperature. This arises through the law of mass action; though there may be a binding energy, E_B, of several tenths of an electron Volt between the charge carrier and dopant, both are present in extremely low concentrations (10^{16} - 3×10^{17} cm^{-3}) and the equilibrium will be pushed heavily towards dissociated acceptor-hole pairs at 300 K. In contrast, chemical doping is hard to limit to low concentrations and we can expect that carriers and counterions will be in close proximity under most experimental conditions.

6. MIS Structures

6.1 The Field-Effect Device

The addition of an insulating layer between semiconductor and one of the metal layers of the diode structure discussed in section 5 gives the MIS (Metal Insulator Semiconductor) structure which is widely used in a range of silicon semiconductor devices. We show in figure 20 an example of this device as fabricated on a silicon wafer substrate, with polyacetylene as the semiconductor and silicon dioxide as the insulator layer, and a heavily n-doped layer beneath the silicon dioxide as the bottom metallic layer (the gate).

The function of the insulator layer is to allow the application of a strong electric field to the semiconductor at the interface with the insulator by control of the gate voltage. This field can then raise or lower the energy levels in the semiconductor with negative or positive gate voltages respectively, as is indicated schematically in figure 21. In the conventional case of a 'rigid band' semiconductor such as crystalline silicon, it is possible to force the band states to pass through the Fermi energy in the semiconductor, and this results in the formation of a surface charge layer at the interface with the semiconductor. This surface charge layer then sets up an electric field to oppose the applied field and in this way establishes equilibrium.

Figure 21 illustrates the case for a semiconductor extrinsically doped to be p-type, as is the case here. With negative bias, lifting the valence band through the Fermi energy results in the formation of a surface charge layer of holes (by flow of electrons through the bulk of the semiconductor to the back ohmic contact); this surface charge layer gives an enhanced concentration of positive charge carriers and is termed the charge accumulation layer. As the gate voltage is taken to increasingly more positive values the device passes through a regime with a depletion layer present at the interface with the semiconductor (as with the Schottky diode), and at large positive gate voltages, the bottom of the conduction band will be lowered sufficiently close to the Fermi energy to give population of these states, and the formation of a negative surface charge layer. For a p-type semiconductor, these carriers are of the opposite sign to those present in the bulk of the semiconductor, and this surface charge layer is termed the inversion layer. It is usual to distinguish between 'weak inversion' which is the first point at which the majority carriers at the interface are negative and which occurs when the bottom of the conduction band reaches the intrinsic Fermi energy, and 'strong inversion' which occurs when the conduction band reaches the Fermi energy.

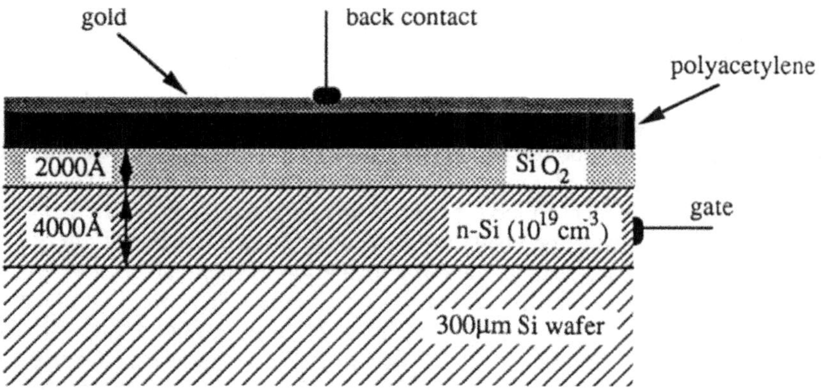

Figure 20. The MIS device fabricated with n-silicon (M), silicon dioxide (I) and polyacetylene (S) with a top, ohmic contact layer of gold. Typical thicknesses for the polyacetylene layer are in the range 10 - 100 nm.

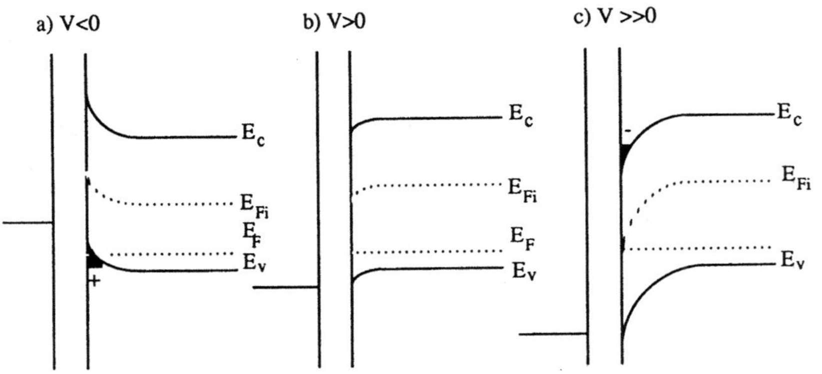

Figure 21 Conventional MIS device band scheme for various bias conditions; (a) accumulation, (b) depletion, and (c) inversion.

The formation of the inversion layer is inherently different from that for the accumulation layer; for the two-terminal device considered here it requires thermal generation of an electron-hole pair and the subsequent diffusion of the hole back through the depletion layer into the bulk of the semiconductor. Even for the case of the MIS device built with crystalline silicon, this process can be extremely slow, and differential capacitance measurements at frequencies above 10 Hz at room temperature are at too high a frequency to show up the inversion layer [65]. For the case of polyacetylene we can expect a very much slower rate. Firstly, the energy gap is higher, so that the Boltzmann factors are smaller. Secondly, the one-dimensional character of the charge carrier dynamics will impede the separation of an electron-hole (or positive soliton/negative soliton) pair, as the diffusion process is ineffective in one dimension. Thus, we can expect to see a high proportion of charge pairs recombine before separation. We do not therefore expect to see charge inversion in the polyacetylene MIS structure under the experimental conditions used here. In contrast, the addition of an extra terminal to give the FET structure of the type shown in figure 3, allows injection of electrons directly into the inversion layer, and as with silicon devices, we can expect rapid formation of the inversion layer in this case.

The description of the formation of the charge inversion and accumulation layers in the case where the semiconductor can adjust its structure to the local charge on the chain must be very different to that shown for the 'rigid-band' semiconductor shown in figure 21. Though we can still expect to see the band states for the polyacetylene bent by the applied field at the interface with the insulator, as these approach the Fermi energy we expect to see rearrangement of the structure and the formation of mid-gap soliton levels which will then store the charge in the accumulation or inversion layers. The effect of this is to change the position of the Fermi energy, E_F, relative to the band edges in comparison to the conventional description, figure 21, in which E_F lies close to E_V for p-type majority carriers and to E_C for n-type majority carriers. Let us consider first the condition for

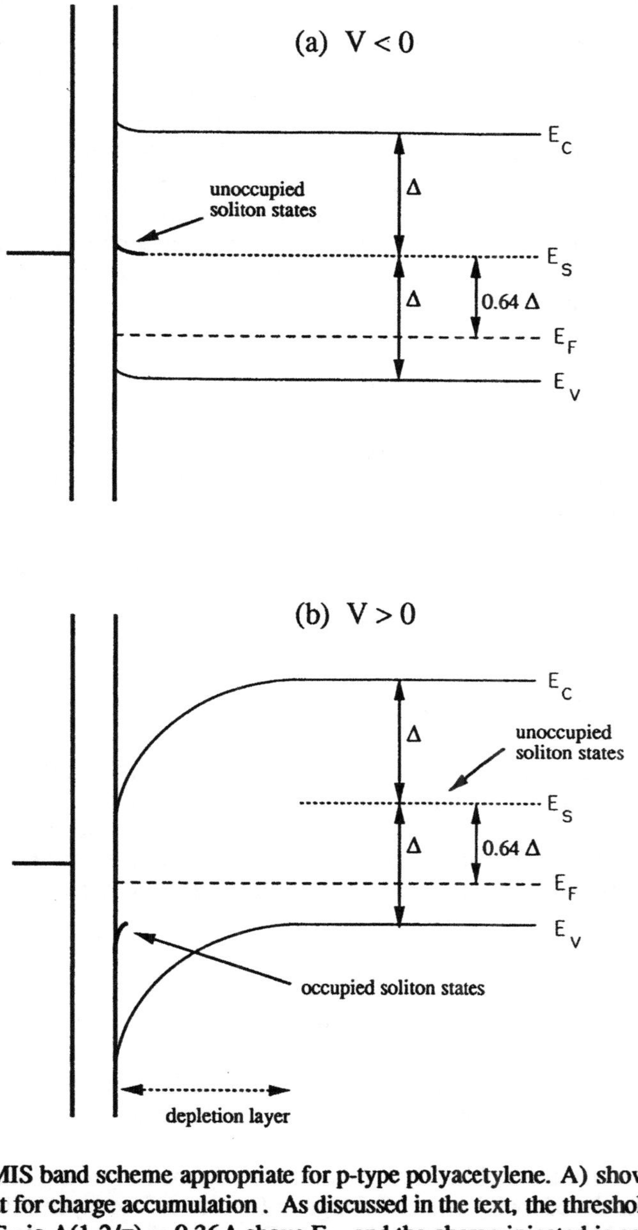

Figure 22 MIS band scheme appropriate for p-type polyacetylene. A) shows the energy level scheme present for charge accumulation . As discussed in the text, the threshold for hole injection occurs when E_F is $\Delta(1\text{-}2/\pi) = 0.36\Delta$ above E_V, and the charge injected is accommodated in the mid-gap states associated with the solitons created to store this charge. (b) shows the scheme that describes the onset of strong inversion; there is a charge depletion region, in which the soliton states associated with the extrinsic carriers are absent, and inversion occurs as E_C is brought down to $\Delta(1\text{-}2/\pi)$ above E_F. The negative charges in the inversion layer are stored in the mid-gap soliton levels created accommodate this charge (these are now doubly occupied).

charge accumulation, as shown in figure 22(a). Injection of a hole results in formation of a soliton which stores the injected charge at $E_v + \Delta$, but at a cost of the energy offormation of the soliton, $2\Delta/\pi$. Thus the threshold for injection of holes occurs when the Fermi energy is $\Delta(1 - 2/\pi)$ above the top of the valence band; relative to E_F, the injected hole is stored at an energy of $-(\Delta - \Delta(1 - 2/\pi)) = 2\Delta/\pi$ and this is balanced by the energy of formation of the soliton, $2\Delta/\pi$. Thus, accumulation occurs well before the valence band is brought up to the Fermi energy (in contrast to the conventional model shown in figure 21), and we see that E_F cannot move closer to E_v than this. Similarly, injection of electrons for formation of the inversion layer occurs when E_c is brought down to $\Delta(1 - 2/\pi)$ above E_F, as shown in figure 22(b), and we see that E_F can never move closer to the band edges than as set by these limits, i.e.

$$|E_F - E_{v,c}| \geq \Delta(1 - \frac{2}{\pi}) \qquad (26)$$

The position of E_F in the extrinsically-doped p-type region must lie close to the energy threshold for accumulation, as shown in figure 22(a). Finally, in the charge depletion region, present in figure 22(b), there are no mid-gap soliton levels present.

Electrical characterisation of the MIS device is restricted to measurement of the complex impedance; the equivalent circuit is shown in figure 23, and can be considered as a series circuit of the insulator (capacitance C_i) and the semiconductor (capacitance C_s and conductance G_s). Where G_s is low, we have

$$\frac{1}{C} = \frac{1}{C_i} + \frac{1}{C_s} \qquad (27)$$

Both accumulation and inversion layers have very high differential capacitance, so that the measured capacitance is just the geometrical capacitance of the insulator layer, C_i. However, in the depletion regime, C_s falls to a lower value, and if the width of the depletion layer is comparable to the thickness of the insulator layer, the fractional change of the measured capacitance can be large. The variation of the measured capacitance in this regime is given by

$$\left(\frac{C_i}{C}\right)^2 - 1 = \left(\frac{C_i}{A}\right)^2 \frac{2(V-V_f)}{qN_a\varepsilon_r\varepsilon_o} \qquad (28)$$

where A is the device area and V_f the flat-band voltage (i.e. the voltage at which there is no bending of the bands in the semiconductor as the interface with the insulator is approached) [66].

The advantage of the field effect device is that the formation of the accumulation or inversion layer is achieved by injection of charge to form a surface layer of charge carriers which are present without any associated dopants. For a practical device it is necessary to have a sufficient level of conductivity to allow the initial injection of charge, and in this respect, the slight p-doping of the as-prepared polymer is ideal for this purpose. The majority of the work we have carried out has been with the charge accumulation layer as the condition for accumulation is more easily met than that for inversion.

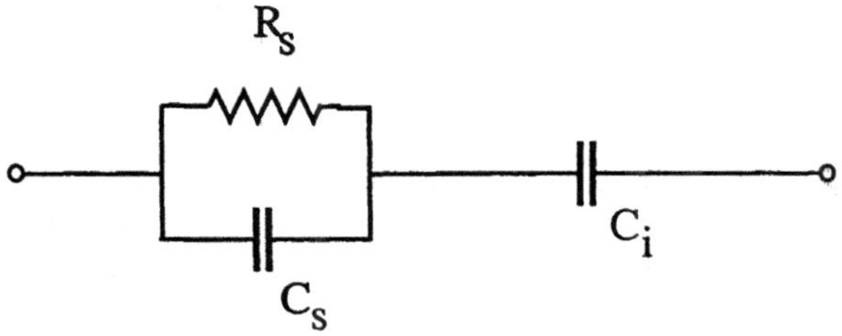

Figure 23 Equivalent circuit for the MIS device.

The three-terminal field effect transistor allows measurement of the conductance of the surface charge layer, and we have used a variety of device structures including that shown in figure 3. The source and drain contacts on the top of this structure allow the setting of the electric field across the insulator by defining the voltage of the gate (V_{gs}) and also allow measurement of the conductance of the channel between source and drain. We defer to section 7 discussion of the standard conditions for the operation of the MISFET.

We have made very extensive measurements of the change in optical property of the MIS device as the charge density in the accumulation layer is varied. Control of the charge concentration in the accumulation layer in the MIS device is easy to make; under conditions where charge accumulation is strong, the device is effectively functioning as a parallel plate capacitor, with surface charge density in the accumulation layer and an image charge in the gate layer, with

$$C_i = \frac{\varepsilon_r \varepsilon_o A}{d} \tag{29}$$

and surface charge concentration, N_s, given by

$$N_s = \frac{\varepsilon_r \varepsilon_o F}{q} \tag{30}$$

where F is the field applied across the insulator. Equation 30 shows that the limit to the surface charge density that can be achieved in the accumulation or inversion layer is set by the maximum value of the dielectric strength of the insulator, F_b. Silicon dioxide is one of the best insulators available, and the limit for this material ($\varepsilon_r = 3.9$, $F_b = 10^7$ V/cm), N_s is no more than 2×10^{13} cm^{-2}. This is to be compared with the number of surface carbon atoms in the polyacetylene, which is about 10^{15} cm^{-2}.

Details of the electro-optical properties of these structures are presented in section 6.3; there is clear evidence for charge storage in the accumulation layer in the form of solitons with states as mid-gap.

6.2 Electrical Characterisation of MIS Structures

6.2.1. *Silicon dioxide as insulator*

We have briefly outlined the fabrication procedure for these structures in section 3.2. We have built the MIS devices on high-purity silicon wafers (thickness 200 μm), polished both sides, which give optimum optical transmittance in the IR. The substrates were then heavily doped with phosphorus on one surface to a depth of 400 nm to give a doping level of 10^{19} cm^{-3}. On top of this doped layer, which serves as the gate, the insulator layer of silicon dioxide was thermally grown to thicknesses of either 120 or 200 nm. The oxide was grown under conditions which give low trap state densities at the silicon/silicon dioxide interface (< 10^{11} cm^{-2}). The n-doped layer was checked to be metallic (with no sign of field effect behaviour) at the temperatures used in this work (> 200 K). Polyacetylene was prepared by thermal conversion of the spin-coated precursor layer, but in contrast to the Schottky diode structures for which relatively thick layers were deposited, most of the MIS devices were fabricated with thin polyacetylene layers (in the range 60 - 200 nm).

Figure 24 shows the differential capacitance versus bias voltage at various frequencies for a devices with a relatively thick polyacetylene layer (150 nm). Accumulation is seen at bias voltages below about +10 V, and at the lowest frequency shown (50 Hz), the value of capacitance in accumulation is close to the value of C_i. There is no indication for the onset of inversion at even the lowest frequency used, and the variation of capacitance in the depletion regime is well modelled by equation 27, as shown in figure 25. The slope of the linear region of $(C/C_i)^2$ versus bias voltage gives a value of N_a, which for this structure is $1.3 \times 10^{17} cm^{-3}$. The value of V_f for this device (+17 V) is high, and we consider that this is due to the presence of charged impurities present in the SiO_2 layer (similar values for V_f are found for devices fabricated on the same substrate batch). The fall-off in capacitance with frequency is untypically rapid for this device for which the contact resistances are high.

Figure 26 shows the capacitance versus bias voltage for a device with a thinner polyacetylene layer (60 nm). The formation of the accumulation layer is clearly seen with the capacitance in accumulation approaching the geometric capacitance of the insulator, C_i. For positive biases, the device moves into depletion, but because the polyacetylene layer is thin, the depletion layer can be extended across the full thickness of the polyacetylene layer (at biases above 5 V). In this condition, the measured capacitance is the series capacitance of the insulator and of the polyacetylene layer; analysis here of the measured value gives the same value for the polyacetylene layer thickness as that measured directly with a diamond stylus surface profiler (60 ± 5 nm). Analysis of the variation of capacitance in the region of bias for which the width of the depletion layer is varying, through equation 27, gives values of $N_a = 8.5 \times 10^{16}$ cm^{-3} and $V_f = 1.6$ V.

Figures 24 and 26 show some of the frequency dependence of the electrical response of these devices. The characteristic time constant for the formation of the charge accumulation layer is $R'C_i$ where R' is the series resistance of the contacts and bulk resistance of the polyacetylene. We find from this and from other experiments that the carrier mobility for motion perpendicular to the plane of the film is considerably lower than for motion within the plane, with a room temperature value of around 10^{-6} $cm^2/Vsec$. We take this to indicate that the polymer chains lie preferentially in the plane of the layer.

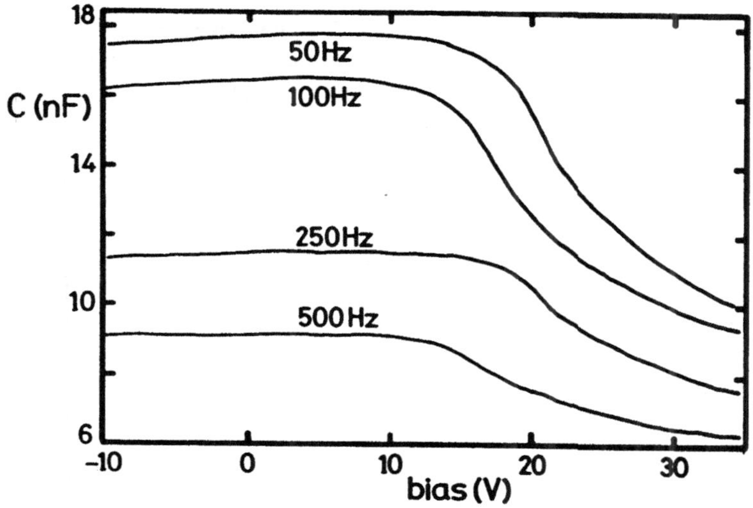

Figure 24 Differential capacitance versus bias voltage for a n-Si/SiO_2/polyacetylene MIS devices at various frequencies (SiO_2 thickness 120 nm, polyacetylene thickness 150 nm).

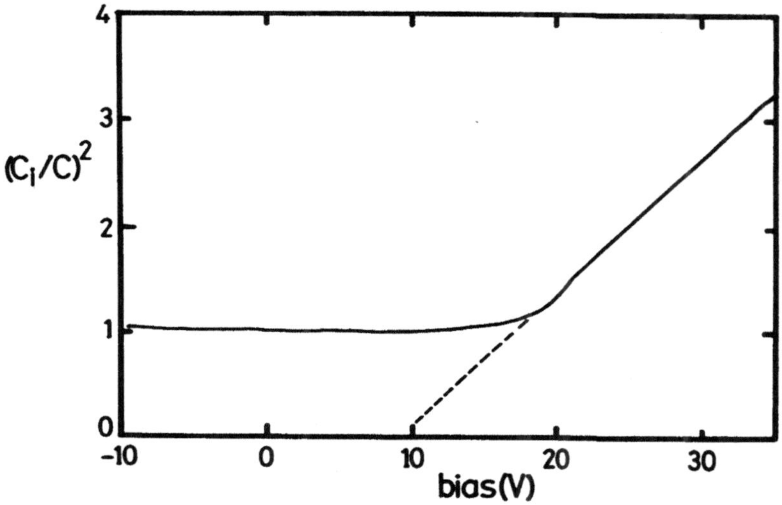

Figure 25 Capacitance data from figure 24 at 50 Hz, plotted as $(C/C_i)^2$ versus bias voltage.

In addition to the differential capacitance we have also measured the conductance, G(ω); this can give sensitive indication of the presence of dissipative processes associated with the charge injection process [67]. For example, if there are trap states, possibly associated with the interface, these give a peak in G(ω) as ω is taken through the characteristic trapping/detrapping time. We do not see evidence for such behaviour in the MIS devices formed with the polymer insulators discussed below, but we have found that some of the SiO_2-based devices show behaviour of the type shown in figure 27.

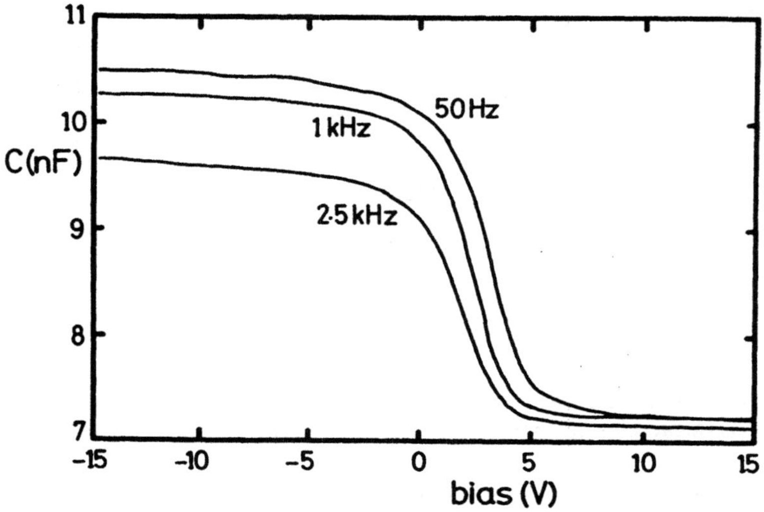

Figure 26 Differential capacitance versus bias voltage for a n-Si/SiO_2/polyacetylene MIS device at various frequencies (SiO_2 thickness 200 nm, polyacetylene thickness 60 nm).

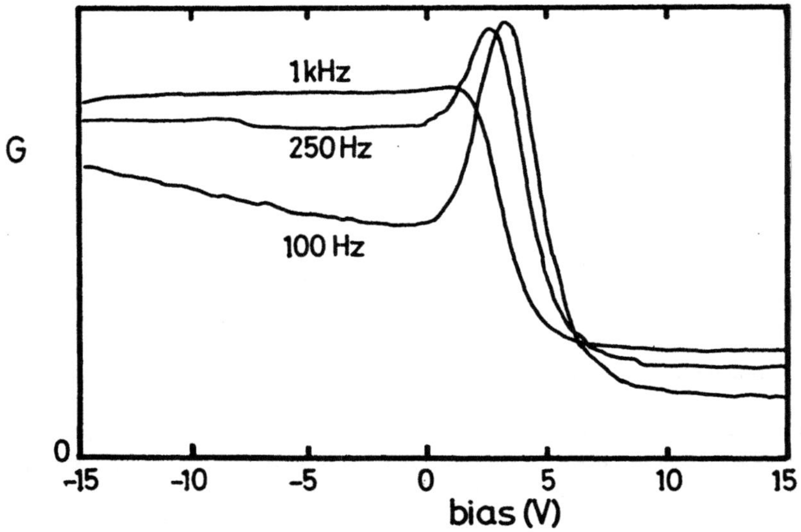

Figure 27 Conductance versus bias voltage at several frequencies for the MIS device shown in figure 26.

6.2.2. *Polymer insulator layers*

Though SiO_2 is an excellent material for use as the insulator layer in the MIS device, possessing a very high dielectric strength and a low density of electronic states in the region of the semiconductor gap when grown on a silicon substrate, there is some considerable interest in the use of insulators which may be prepared by other methods and which may show different properties. An obvious disadvantage in the use of the structures of the type shown in figure 20 is that the structure is transparent only below the silicon band gap, thus preventing measurements of electromodulation of the optical absorption in the region of the polyacetylene gap. A second problem, which we characterise from the optical properties in section 6.3, is that the polyacetylene layer in which the charge accumulation layer is formed is the layer formed at the surface of the insulator, and we find for the case of polyacetylene that this layer is considerably more disordered that the bulk material.

We have attempted to use a variety of organic insulators which could be conveniently prepared for use in MIS structures. Most of the effort has been made with polyimide, which has the best dielectric strength of those we have used here; nevertheless this is about an order of magnitude lower than that of SiO_2. Details of the methods used to deposit these materials are given in section 3.2.

Figure 28 shows the differential capacitance versus bias voltage for a structure with a polyimide layer of thickness 360 nm and a polyacetylene layer of thickness 440 nm. It can be seen that there is accumulation for negative gate voltages and depletion for positive gate

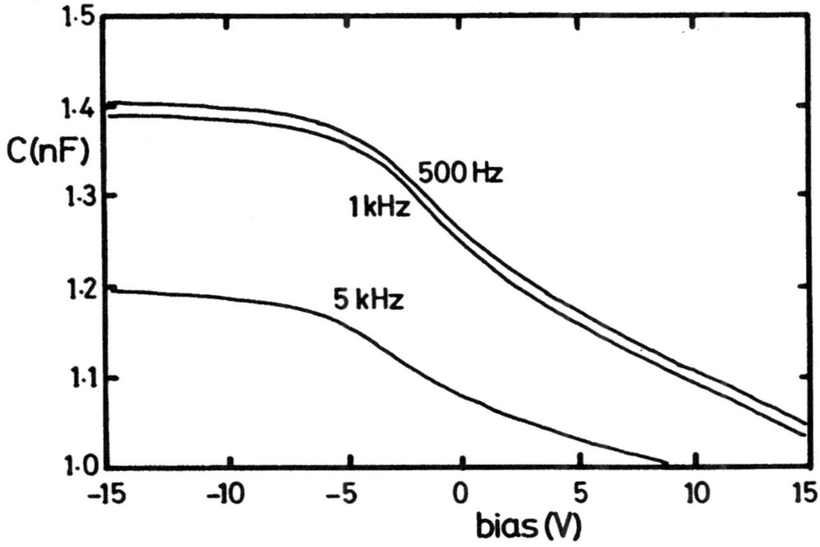

Figure 28 Differential capacitance, $\partial Q/\partial V$ at several frequencies for an Al/polyimide/polyacetylene MIS structure (polyimide thickness 360 nm, polyacetylene thickness 440 nm).

voltages. The maximum fields that can be applied across the polyimide layer are very much lower than those used for the SiO_2 devices (±15 V across 360 nm = 4×10^5 V/cm), so that the range of behaviour is more limited than that shown for the SiO_2 devices. Analysis of the capacitance in the depletion region using equation 28 gives values of $N_a = 2.2 \times 10^{16}$ cm^{-3} and $V_f = -5.9$ V. We see, therefore, behaviour very similar here to that observed with the SiO_2-based devices.

Poly(methylmethacrylate), PMMA, is soluble in the solvents used for the Durham polyacetylene precursor, so that deposition of the polyacetylene was made first, followed by the PMMA layer. Note that in this case, the insulator/semiconductor interface is made at the top, free surface of the polyacetylene, and we can expect that there are differences in the structure of top and the bottom surfaces. We return to this later in connection with the electro-optical properties of the MIS devices in section 6.3. The variation of the differential capacitance with bias voltage for a MIS device built in this way is shown in figure 29. It shows a very similar form to that seen for the other MIS devices, and standard analysis of the capacitance in the depletion regime gives values of $N_a = 2.1 \times 10^{16}$ cm^{-3} and $V_f = -1$ V.

We have also successfully used other organic insulators. We have made use of the low conductivity of the *cis*-rich isomer of polyacetylene that is formed from the Durham precursor when conversion is performed slowly at room temperature. For this we apply a layer of the precursor polymer on top of a previously converted film of the polyacetylene, and cap this with an evaporated aluminium electrode. This structure shows good electrical properties, though over a period of a few days, the layer of precursor/*cis*-polyacetylene becomes too conducting. We have also used Langmuir-Blodgett films, deposited onto an aluminium film, and formed with 37 layers of a diacetylene monomer which was polymerised *in situ* to form a polydiacetylene insulator layer. Before spin-coating the polyacetylene precursor, the LB film was pumped for 24 hours to ensure that there was no trace of water. This type of structure also showed good electrical characteristics.

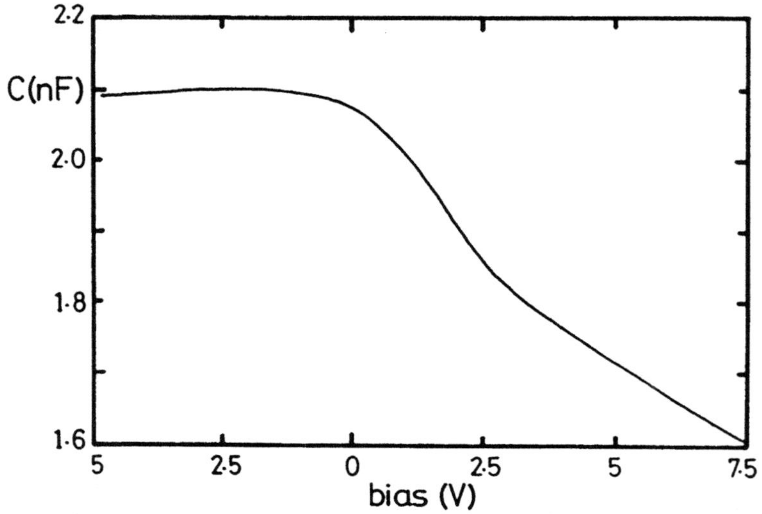

Figure 29 Differential capacitance, $\partial Q/\partial V$ for a gold/PMMA/polyacetylene MIS structure (PMMA thickness 450 nm).

6.3 ELECTRO-OPTICAL PROPERTIES OF THE MIS STRUCTURE

We have shown in section 5.3 that the sub-gap optical absorption of the depletion layer formed in the polyacetylene Schottky diodes shows the removal of the mid-gap soliton states associated with the extrinsic charges present in the undepleted material. The MIS structures operate in both charge accumulation and depletion, and there is particular interest in the charge accumulation layer in that the charges injected are present without any associated dopant. We have therefore carried out an extensive series of experiments on the electro-optical properties of these MIS structures, covering both the electronic excitations of the solitons at mid-gap and also the IR and Raman vibrational excitations of the soliton. All the MIS structures which have been electrically characterised as described in section 6.2 have been constructed so that they are semitransparent, either over the IR (silicon substrate) or over the IR and visible (polymer insulators, glass substrates).

6.3.1 *Electronic Excitations of Solitons*

Measurements of the field modulation of the optical transmission in the energy range 0.4 - 2.0 eV were carried out using the same experimental system as used for the measurements with the Schottky diodes, described in section 5.3. We show the spectrum for the MIS structure previously characterised for its electrical properties as in figures 26 and 27 in figure 30. We see the appearance of an absorption band peaking near 0.8 eV, which grows with increasing negative gate voltage. The form of this absorption band is as expected for the charged soliton to band edge transition, though the energy of the transition measured here is significantly raised with respect to the value of 0.55 eV found for photoinduced charges, figure 5, and for the extrinsic charges measured in the Schottky diodes, figure 17. We consider that this is due to a higher level of disorder in the surface layer of the polyacetylene at the interface with the SiO_2 which locally raises the polyacetylene band gap. We discuss this issue further in section 6.4.

The general shape and position of the charge-induced absorption band is not greatly affected by the value of gate voltage, though there is a small shift in the position of the band to higher energies as the gate voltage is raised. This is seen in figure 31 in which we show the energy of the peak of the absorption band as a function of bias voltage. The increase in position of the absorption band, by some 0.05 eV in 50 V, can have several origins. One possibility is the role of screening by the electronic charges present in the accumulation layer; we think that the charge densities present are too low for this to be important. The explanation that we favour is based on the inhomogenous nature of the electronic states in polyacetylene. We expect the first charges injected into the accumulation layer to select the polymer chains at the interface with the insulator with the lowest band gap, and that further injected charges will be forced to rest on chains with progressively higher band gaps (the Coulomb repulsion between the injected charges will prevent the formation of an inhomogeneous charge distribution.)

One of the important advantages of using the MIS device for optical measurements is that it allows quantitative measurements in the optical response, and we can gain very direct information about the way in which charge is stored in the device. Figure 32 shows the differential capacitance, $C = \partial Q/\partial V$, for the device shown in figures 26,27,30 and 31 as a function of bias voltage and also the variation of $\partial \ln(T)/\partial T$, where T is the optical transmission at the peak of the mid-gap absorption band (0.8 eV). Comparison between $\partial Q/\partial V$ and $\partial \ln(T)/\partial T$ shows that the optical response sets in at the same voltage as the capacitance data shows the onset of formation of the accumulation layer, and for negative gate voltages we see a constant ratio between the two quantities. This ratio gives the

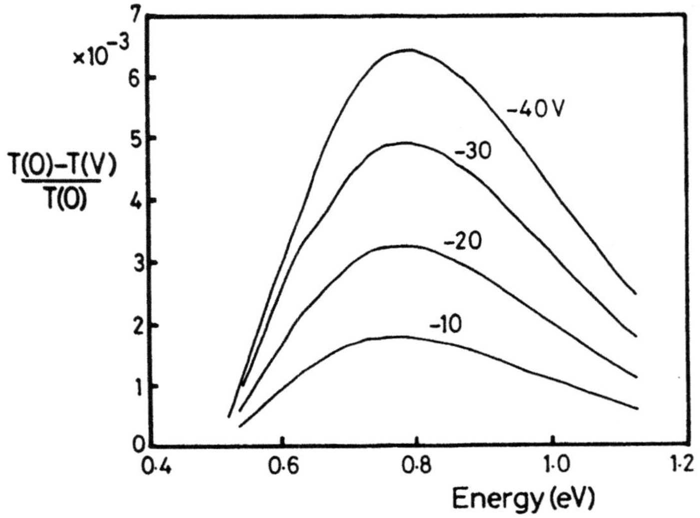

Figure 30 Fractional change in the optical transmission, T, for the MIS devices shown in figures 26 and 27, obtained between gate voltages of 0 and V, with V = -10, -20, -30 and -40 V. Modulation frequency is 330 Hz.

optical cross-section per injected charge, σ, through

$$\frac{\partial \ln(T)}{\partial V} = \frac{\partial Q}{\partial V}\frac{\sigma}{qA} \tag{31}$$

We find a value for σ at the peak of the mid-gap absorption band of 1.2 x 10^{15} cm^2, somewhat lower than the value found for the solitons due to the extrinsic doping in the Schottky diodes, and we consider that the differences are due to the high level of disorder in the surface layer of polyacetylene at the interface with the silicon dioxide in this MIS device. It is important to note that the value of σ is constant throughout the region of accumulation, and that it reaches this value at the onset of accumulation. It provides, therefore, very clear evidence that almost all the charge that is injected into the structure, which is measured by the differential capacitance, is stored in states which give rise to the new optical behaviour which is characteristic of the formation of a charged soliton-like excitation on the polymer chain. We have no evidence for the presence of trap states, which if present would have reduced the value of σ, particularly at the onset of accumulation.

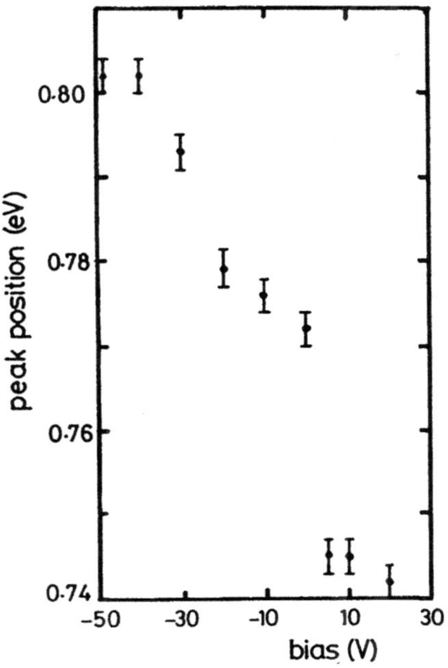

Figure 31 Charge induced optical absorption for the MIS device shown in figures 26, 27 and 30. Shown here is the energy at the peak of the absorption band plotted as a function of bias voltage, V. Measurements were made with 0.25 V a.c. modulation at 500 Hz around the bias voltage, V.

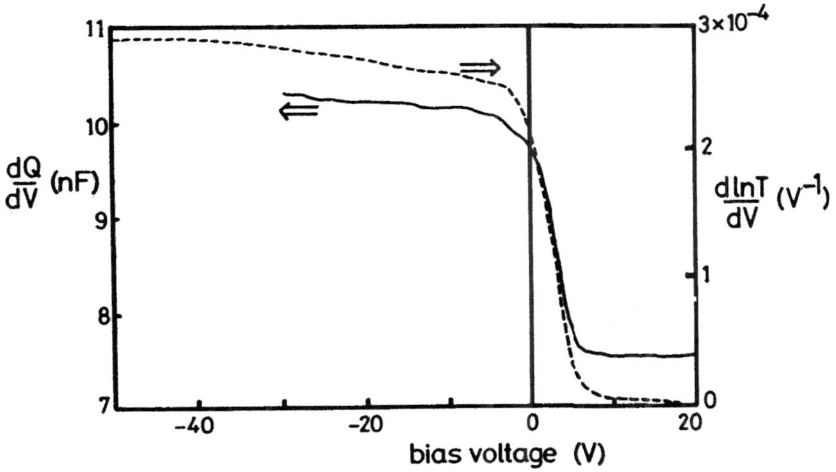

Figure 32 $\partial \ln(T)/\partial V$, where T is the optical transmission at 0.8 eV, and the differential capacitance, $\partial Q/\partial V$, versus bias voltage for the MIS device shown in figures 26, 27, 30 and 31. Both quantities are measured with an a.c. modulation of 0.25 V at 500 Hz.

The size of the charge-induced optical response can be quite large for these MIS structures. For the device shown here, the peak value in the fractional change in absorption is some 0.7%, and we note that the limit, set by equation 30 and the dielectric strength of the insulator is here about 2.4%.

We have looked carefully for any indication of the formation of an inversion layer at high positive biases. These structures are not ideal because the onset of inversion is expected at high positive biases, close to the breakdown voltages for these structures. As we have discussed in section 6.1, we do not expect to see the formation of the inversion layer except at very low modulation frequencies. We have however seen some indication for inversion, as we show in figure 33. The upturn in $\partial \ln(T)/\partial V$ in this device for bias voltages above +40 V is taken as an indication that charge injection into the polymer is producing new electronic charges (rather than an increase in the space charge due to the extrinsic dopants. It sets in here at a value for the thickness of the depletion layer of 49 nm, lower than the full polymer layer thickness of 150 nm.

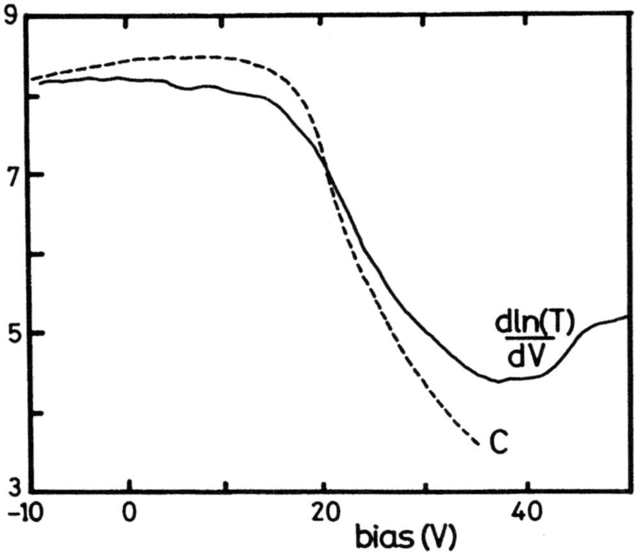

Figure 33 $\partial \ln(T)/\partial V$, where T is the optical transmission at 0.8 eV, and the differential capacitance, C, versus bias voltage for the MIS device shown in figures 24 and 25. Both quantities are measured with an a.c. modulation frequency of 20 Hz. The abcissa is marked arbitary units of $\partial \ln(T)/\partial V$; the origin of the scale is 0.

We have made measurements of the electro-optical properties of the MIS devices fabricated with PMMA and with polyimide. We show in figure 34 the spectrally-resolved response for the device with PMMA previously characterised for its electrical response in figure 29. The charge-induced 'mid-gap' absorption is seen here at 0.55 eV; this is lower than the 0.8 eV found for the SiO_2-based devices, but at the same energy as the photoinduced charges, figure 5, and for the extrinsic charges measured in the Schottky diodes, figure 17. We note that for this MIS device, the surface of the polyacetylene at which the junction with the insulator is formed was itself first formed as a free surface, and we consider that its electronic structure is similar to that of the bulk.

We have made more extensive measurements with the MIS devices fabricated with polyimide. We show in figure 35 the spectrally-resolved response for the device previously shown in figure 28, both with the device driven to accumulation (V = -15 V) and also into depletion (V = +15 V). The response is measured here from 0.5 to 2.0 eV as the glass substrate for this structure is transparent over the whole spectral range. At low energies we see the expected appearance of the 'mig-gap' absorption feature; as with the PMMA device the charge-induced absorption peaks at 0.55 eV, and as expected from the measurements with the Schottky diodes, the response as the device is driven into depletion is of the opposite sign, with the peak in the bleaching response again at 0.55 eV. We show the correspondence between the differential capacitance, $\partial Q/\partial V$, and the strength of the modulation of the optical transmission at 0.59 eV, $\partial \ln(T)/\partial V$, as the bias voltage is varied in figure 36. We see that the ratio between the two remains almost constant as the device is taken from accumalation through to depletion, and, using equation 31 we find that the ratio gives a value for the optical cross-section per injected charge of 1.55×10^{-15} cm^2 at the peak of the absorption band. This value is significantly higher than that found for the MIS devices built with polyacetylene deposited onto SiO_2; we consider that this correlates with the lower energy for the 'mid-gap' transition and is due to the formation of a more extended soliton here. [Note that the response shown in figure 35 gives a different sign for V = +15 and V = -15 V, but that the differential signal in figure 36 remains positive for both accumulation and depletion. This is because the differential response is measured with a small a.c. bias superposed on the d.c. bias level, so that the sign of the voltage modulation reverses as the bias is taken through zero.]

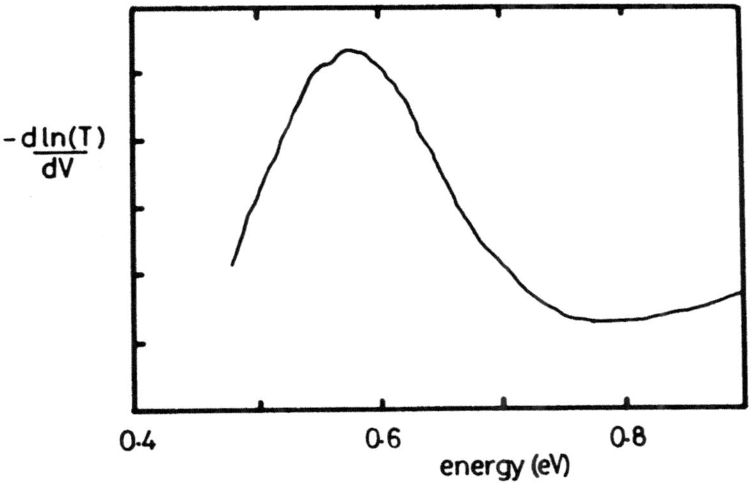

Figure 34 -ΔT/T versus bias voltage,V, for the MIS structure fabricated with PMMA as the insulator layer, shown in figure 29. ΔT is the fractional change in optical transmission, T, with a modulation about the bias voltage of ±5 V. The measurements were performed at 1 kHz.

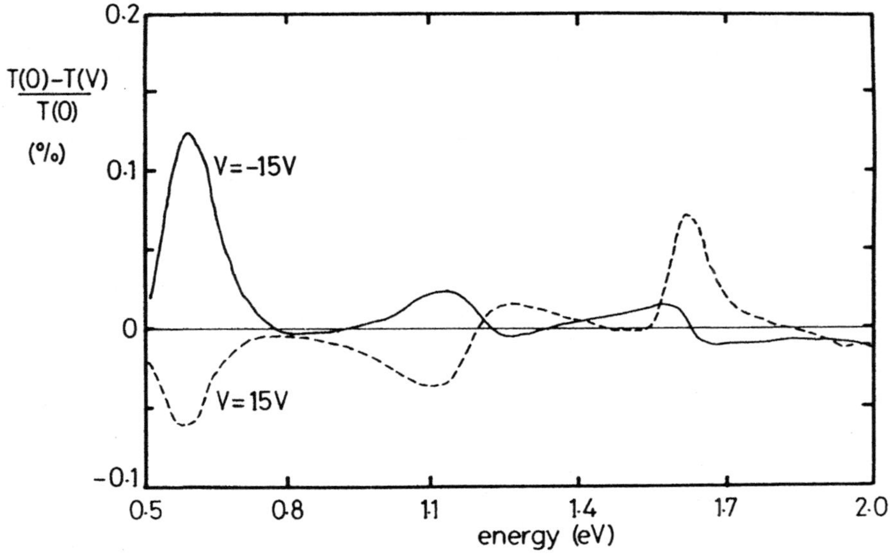

Figure 35 Fractional change in optical transmission, T, between bias of 0 and V for the MIS device with polyimide as insulator shown in figure 28.

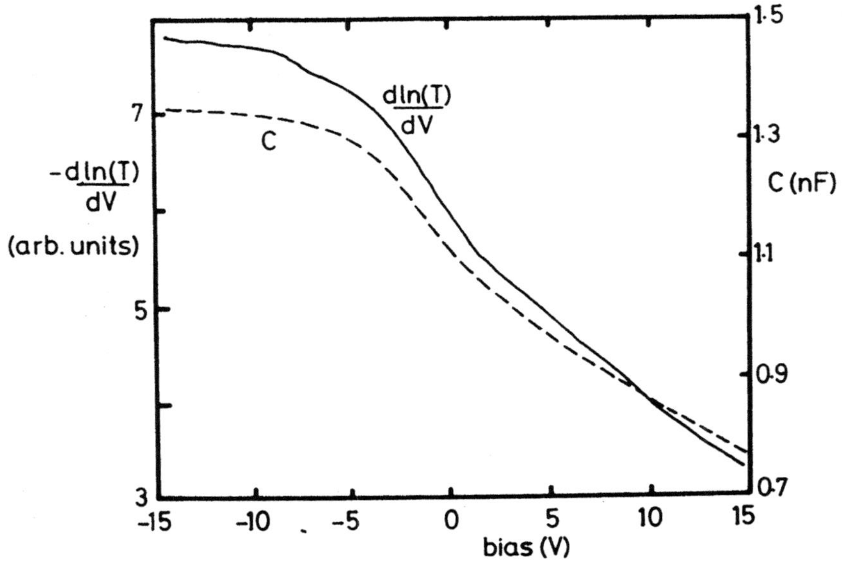

Figure 36 ∂ln(T)/∂V, where T is the optical transmission at 0.59 eV, and the differential capacitance, ∂Q/∂V, versus bias voltage for the MIS device shown in figures 28 and 35.

At higher photon energies there is additional structure; some of this is as expected, other features less so. There is a relatively weak feature at 1.15 eV which is of the same sign as the 0.55 eV response, with absorption in accumulation and bleaching in depletion. This feature is not directly related to the formation of the charge accumulation layer as it shows a very different dependence on bias voltage to both the differential capacitance and the electro-optical response at 0.55 eV, as shown in figure 37, with a peak in response at around +5 V. We do not have an explanation for this feature.

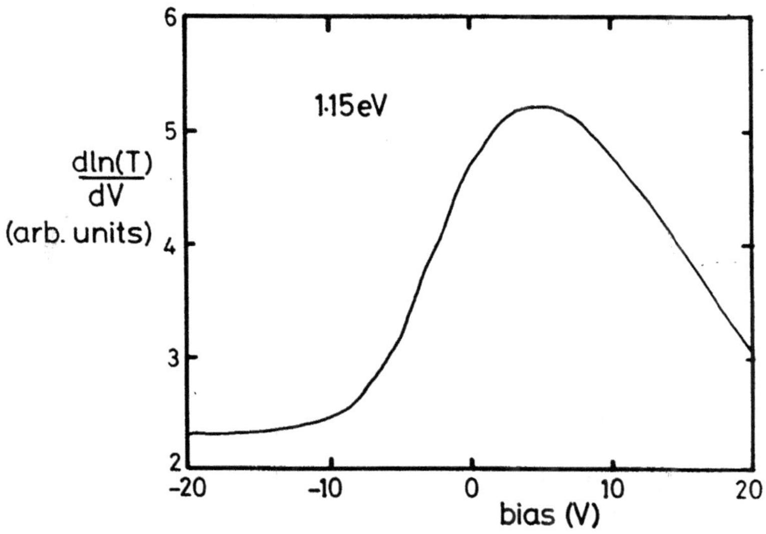

Figure 37 ∂ln(T)/∂V, where T is the optical transmission at 1.15 eV versus bias voltage for the MIS device shown in figures 28, 35 and 36.

There is a strong response at the π-π* absorption edge; this is provided both by the Franz-Keldysh effect due to the electric field across the polyacetylene and also by the charge-induced changes in the density of π and π* states at the band edges. The Franz-Keldysh effect gives an absorption on the low energy side of the band edge, as seen in figure 9, irrespective of the polarity of the gate voltage. Thus, for the case of depletion, where the gate voltage is taken positive, the responses are of the same sign and a large absorption signal is seen, peaking at 1.6 eV. However, in accumulation, the responses cancel, and a weak response is seen in this part of the spectum. The contributions from the field modulation and the charge-induced modulation of the absorption can be distingushed in the plots of ∂ln(T)/∂V versus bias voltage shown in figure 38. Note that in contrast to the response at 0.55 eV shown in figure 36, the response at 1.63 and at 1.94 eV changes sign near +4 V bias, as expected from the Franz-Keldysh response.

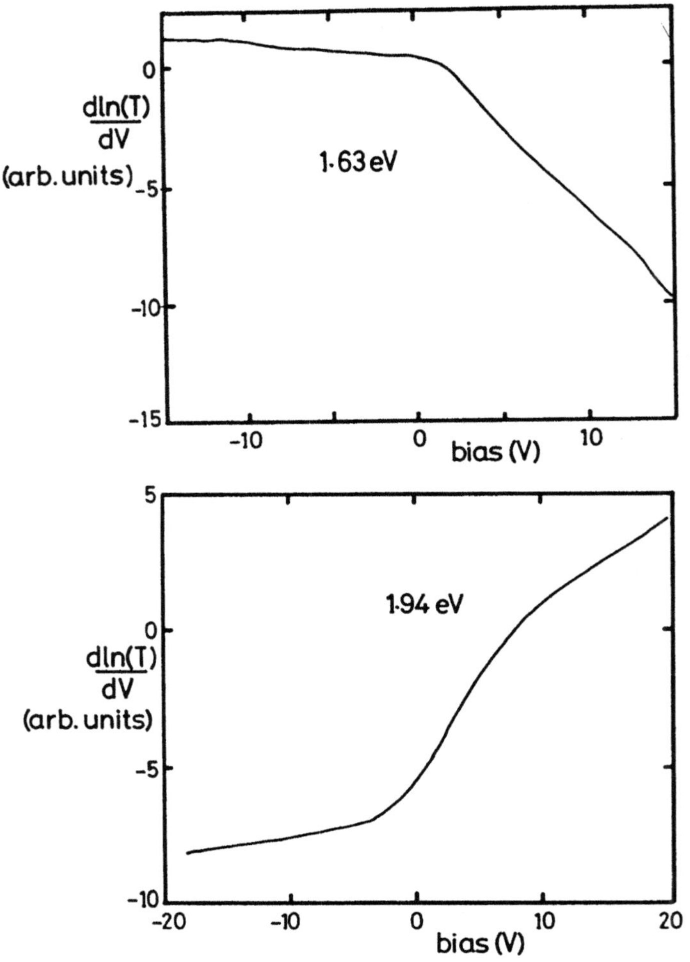

Figure 38 ∂ln(T)/∂V, where T is the optical transmission at 1.63 eV (a) and 1.94 eV (b) versus bias voltage for the MIS device shown in figures 28, 35, 36 and 37.

6.3.2 *Vibrational Excitations of Solitons*

Besides the electronic transitions associated with the soliton state, there is extra IR activity due to new vibrational modes which couple to motion of the charged soliton along the polymer chain. These are seen in both doping [45] and photoexcitation experiments [48] (figure 5) in unoriented Durham polyacetylene. Figure 39 shows the results from an FTIR experiment on an MIS structure, showing the difference in transmission for the MIS device biased between 0 and -50V with respect to the gate [11]. Three peaks are observable; an absorption feature above 4000 cm^{-1} (this is the low energy side of the electronic absorption, as in figure 10), and two sharp vibrational features at 1379 and 1281 cm^{-1}. We identify these as the two higher-frequency translation modes of the soliton. The value of 1379 cm^{-1} lies between that measured for photo-induced charges, 1373 cm^{-1} [32] and for dopant-induced charges, 1410 cm^{-1} [30], and provides information about the chain conformation and degree of conjugation. We note in particular that the intensity of this mode in relation to that of the electronic absorption band is much lower than found in photoinduced absorption measurements [32]. The oscillator strength scales inversely with effective mass of the charge carriers, and we estimate that the mass for the charges present in the accumulation layer is a factor of 3.6 higher than that of the photoexcited charge carriers in oriented Durham polyacetylene. We were not able to measure the electromodulation spectrum below 1000 cm^{-1} due to a very large absorption peak in the SiO_2, and we were thus unable to observe the lowest frequency, 'pinned' translational mode.

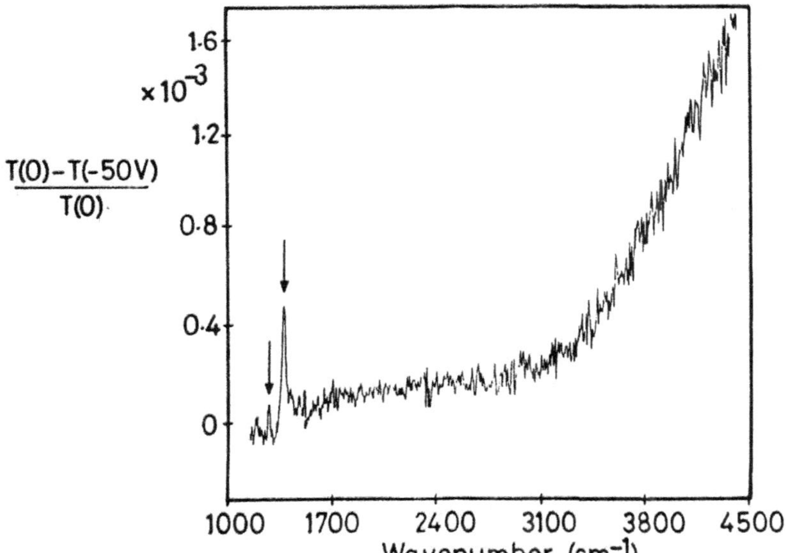

Figure 39. Voltage-modulated transmission for a MIS diode, [T(0) - T(-50V)] / T(0) in the spectral range 1000 and 5000 cm^{-1}, from [11].

New Raman-active modes associated with the width modulation of the soliton are also expected. These have not been seen in doping or photoexcitation experiments because of the inherent experimental difficulties, but the modulation experiment that we can perform with the MIS structure is well suited for this measurement. Experiments were carried out using the MISFET structure of the type shown in figure 3, but with a very thin polyacetylene layer (100 Å) to ensure that the whole depth of the polymer film was accessible for Raman scattering [14]. Raman scattering off the top surface of the structure was measured with a double spectrometer, with a multi-diode array detector. Spectra were obtained from subtraction of a "background" scan from a "signal" scan. Here the "signal" was the measured scattering from a device with the gate at -30 V with respect to the source and drain, and the "background" was the same scattering signal with the gate set to 0 V. Typical measurement times were in the region of 16 hours. Under these conditions the charge density in the accumulation layer is equivalent to 2.21 x 10^{-4} charges/carbon atom for a 100 Å film. The spectral region from 900 cm^{-1} to 1700 cm^{-1} was studied at a range of laser wavelengths from 476.5 nm to 584.0 nm.

Two different classes of feature are observed in the difference spectrum of accumulation layers in Durham polyacetylene shown in figure 40. The main class of feature is negative and arises from the bleaching of the resonance-enhanced A_g modes that dominate the normal Raman spectrum of *trans*- polyacetylene. The two main modes are near 1100 cm^{-1} and 1500 cm^{-1} with the exact frequency a function of the excitation energy. The bleaching of these two main modes dominates the difference spectrum. The dispersion of the

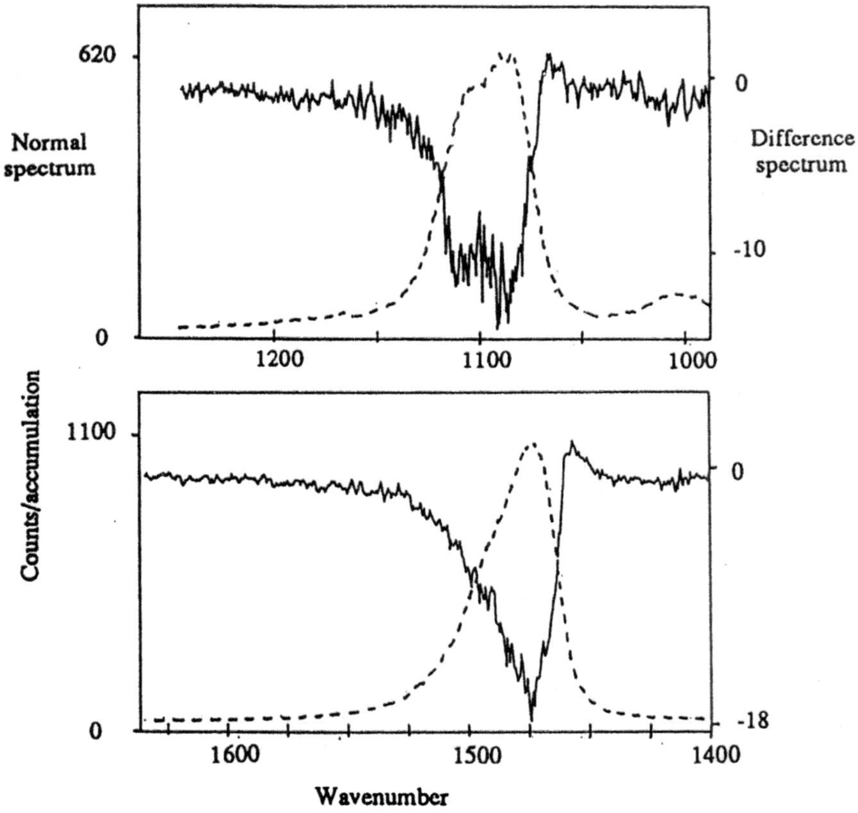

Figure 40. Difference spectra due to charge accumulation layer in a polyacetylene MISFET structure, from [14]. The solid line is the difference signal; the broken line is the normal Raman spectrum. The excitation wavelength is 584 nm.

bleaching is the same as that of the normal spectrum, but the bleaching fraction (the ratio of the integrated bleaching signal to the normal signal) changes with excitation energy, being 7×10^{-3} at 476 nm, 6.9×10^{-3} at 488 nm, 7.7×10^{-3} at 514 nm and 1.6×10^{-2} at 584 nm. There are also small positive features to the low energy side of each of the bleaching features. These are much less intense, do not disperse with excitation energy, and occur at 1065 cm^{-1} and 1455 cm^{-1}.

We attribute the positive features in the difference spectrum to the A modes of the soliton. The frequencies of these modes can be calculated using the methods discussed in section 2 and section 4.1. We find that the modes, at 1065 cm^{-1} and 1455 cm^{-1}, are at rather higher frequencies than the values of equal to 973 cm^{-1} and 1414 cm^{-1} calculated with a value of electron-phonon coupling of 0.200 appropriate to unstretched Durham polyacetylene. There is clear evidence from the non-zero frequency of the Goldstone mode for the strong effect of pinning in this system - this could readily account for the increased frequency observed for the A mode. The presence of a defect on the polyacetylene chain is also expected to bleach the normal Raman spectrum. Terai et al calculate [30] that this bleaching will extend over a range of 30 carbon atoms - that is, a 30% bleaching for a 1% soliton density. This large bleaching effect corresponds to the enhancement in the band to band optical absorption coefficient as calculated by Kivelson et al [24]. We find that excitation in the red gives a larger bleaching fraction than blue excitation. This suggests that the soliton defects prefer to reside on polyacetylene with a smaller band gap - the less disordered chains. Quantitatively, the areal soliton density was calculated simply from the field across the insulating layer to be 2.21×10^{-4} charges/carbon atom. Using the calculated enhancement factor of 30, this should give us a bleaching of 0.007 - which is as observed in the blue, though below that observed in the red.

6.4 Modelling of Electronic Structure in the Accumulation Layer

We find that the properties of charges present in the accumulation layer in MIS structures are well described by those calculated for soliton-like excitations, and in particular, the optical measurements discussed in this section give unambiguous demonstration that these charges possess the full set of excitations expected. However, the energies and intensities of these excitations do show very considerable variations from oriented to unoriented films of Durham polyacetylene, and from one device type to another. As we discuss in section 4.1, we can model the variation in properties between the oriented and unoriented films for photoexcited and dopant-induced charges by allowing variation of the value of the effective electron-phonon coupling constant, λ, with λ taking a larger value in the more disordered structure of the unoriented films.

We find that we can similarly model the range in properties of the solitons formed by charge injection into the accumulation region of the MIS and MISFET structures. The accumulation layer in the MIS structure is very closely localised to the interface with the insulator. For silicon-based MISFET inversion layer structures, typical confinement distances are of order 2 nm [68] and we can expect values here to be no larger for the polyacetylene devices. We can therefore expect that the accumulation layer will be very sensitive to the local structure of the polymer at the interface with the insulator. We consider that in the case where this is silicon dioxide, the polymer surface layer is significantly more disordered than in the bulk of the polymer. In contrast, where the insulator is formed on the top surface of the polyacetylene, as for the PMMA device, or where the polyacetylene is formed onto a polymer insulator layer, we find that the surface region of the polymer has the same characteristics as that of the bulk. We can assemble the data in the form of a table, shown as table 1. The variation with gap parameter, Δ, as given by the formalism presented in section 2, is shown in the right-hand column. The peak in the interband absorption may be taken as 2Δ; from equation 6 we expect the energy of the 'mid-gap' soliton to band edge transition to scale as Δ, from equation 5 we expect the effective mass to scale as Δ^3, and the value of the optical cross-section for the mid-gap' soliton to band edge transition scales as $1/\Delta$ [23,24]. It may be seen that the data shown in the table are consistently modelled.

	stretched Durham	unstretched Durham, in bulk or at polymer interface	unstretched Durham at SiO_2 interface	scales as:
$\pi \rightarrow \pi^*$ peak	1.9 eV	2.3 eV	-	Δ
$\pi \rightarrow S^+$, $S^- \rightarrow \pi^*$	0.45 eV	0.55 eV	0.8 eV	Δ
$\frac{m^*}{m^*_{stretched}}$	1	2	3.6	Δ^3
σ at peak in $\pi \rightarrow S^+$, $S^- \rightarrow \pi^*$	-	1.6×10^{-15} cm^2	1.2×10^{-15} cm^2	$\frac{1}{\Delta}$

Table 1 Parameters obtained from absorption and photoinduced absorption measurements and from electroabsorption measurements on MIS and MISFET structures.

7. MISFET Devices

7.1 FABRICATION

The MIS Field Effect Transistor, or MISFET, operates by modulation of the conductance of the semiconductor in the depletion, accumulation or inversion modes, with inversion layer at positive biases. The variation of the channel conductance with gate voltage for constant V_{ds} defines the transconductance, g_m

$$g_m = \frac{\Delta I_{ds}}{\Delta V_{gs}} \qquad (32)$$

For low values of V_{ds}, well below the level for pinch-off, g_m gives a direct measure of the carrier mobility, through the relation

$$g_m = \frac{W}{L} \mu C_i V_{ds} \qquad (33)$$

We see from figure 42 that the variation of I_{ds} with V_{gs} is far from linear, rising very much more rapidly at high negative values of gate voltage than at low values. This indicates either that the effective mobility of the charge carriers increases at high concentrations, possibly because the carriers initially injected are trapped, but as we discuss in more detail in section 7.3, we consider that the charge carrier mobility for the soliton-like charged excitations should be expected to rise rapidly with charge concentration. From the highest value for g_m measured here, of 1.2 nS, we find a value for the mobility of the charge carriers of 4×10^{-5} $cm^2/Vsec$.

The saturation current, I_{sat}, for the channel at high values of V_{ds} is conventionally modelled by the relation

$$I_{sat} = \frac{\mu C_i W}{2L} (V_{gs} - V_{th})^2 \qquad (34)$$

where V_{th} is the threshold voltage at which the accumulation channel forms, C_i is the gate capacitance of the insulator layer, and μ is the charge carrier mobility [65]. We show the variation of $\sqrt{I_{ds}}$ with $V_{ds} = V_{gs}$ in figure 43. The slope of the linear region found at the higher values of V_{ds}, shown as the broken line in the figure, can be used to find a value for the carrier mobility and the threshold voltage. We find for this device from equation 32 that $V_{th} = -12$ V and $\mu = 9 \times 10^{-6}$ $cm^2/Vsec$; we note that with such a large negative value for V_{th}, the condition $V_{ds} = V_{gs}$ does not guarantee that the device is in saturation, and we consider that this contributes to the lower value of mobility deduced here from that obtained in the unsaturated regime. The principal factor, however, is that I_{ds} is limited by the 'pinch-off' region in which the carrier density is low, and for which we expect from a low mobility (as seen in figure 42).

For completeness, we note that g_m in the saturation regime is given by

$$g_m = \frac{W}{L} \mu C_i (V_{ds} - V_{th}) \qquad (35)$$

For the MISFET discussed here we find a value for g_m of 2.5 nS at $V_{ds} = V_{gs} = -25$ V.

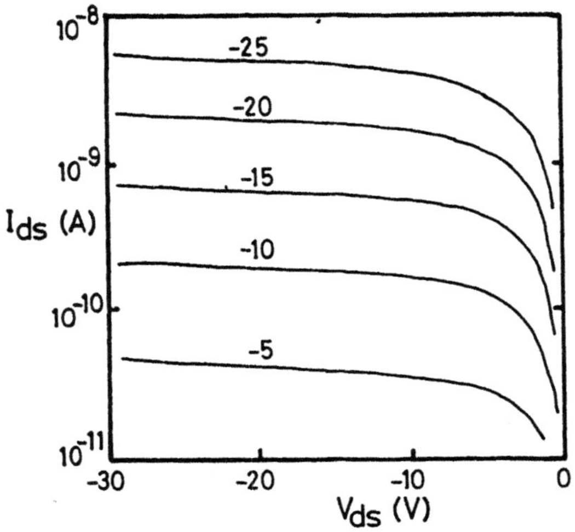

Figure 41. MISFET structure with Au source and drain contacts. Shown here is I_{ds} versus V_{ds} for various negative values of V_{gs}, showing channel current enhancement and saturation.

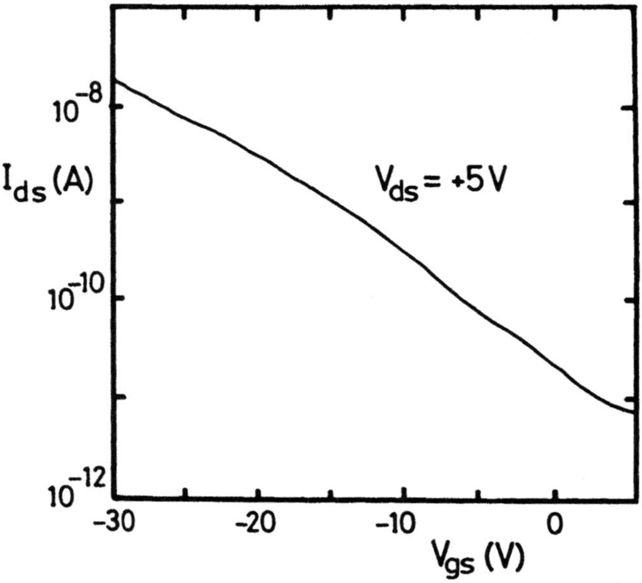

Figure 42 The transfer characteristic, I_{ds} versus V_{gs} for V_{ds} = +5 V for the MISFET shown in figure 41.

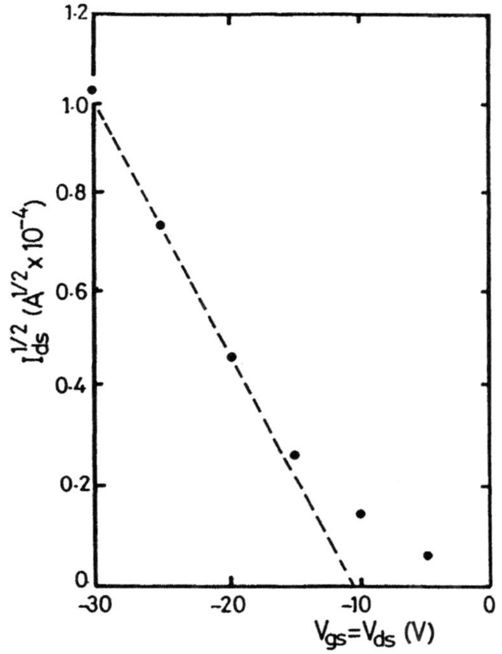

Figure 43 Variation of $\sqrt{I_{ds}}$ with $V_{ds} = V_{gs}$ for the MISFET shown in figures 41 and 42.

7.2. *Poly n-Silicon Source and Drain Contacts*

We have used MISFETs with poly n-silicon source/drain contacts to promote the formation of the charge inversion layer for positive gate voltages. The structure of these devices which have been discussed in section 7.1 is as shown in figure 3. The polyacetylene layer for these devices can be kept thin, typically to no more than 20 nm; this allows full depletion at the appropriate bias, and thus gives very good 'on' to 'off' g_m ratios. Figure 44 shows the variation of I_{ds} with V_{gs} for fixed V_{ds} = =10 V. I_{ds} has a minimum at V_{gs} = +10 V, corresponding to full depletion across the polyacetylene layer. For voltages above this the conductivity rises rapidly, and is identified with the formation of a charge inversion layer at the polyacetylene/SiO_2 interface. For voltages lower than +10 V I_{ds} rises very sharply over one or two decades. This sharp rise is most likely due to the decrease in width of the depletion layer, so that the 'knee' in the curve, seen at around +5 V, corresponds to the flat-band voltage at which the measured conductivity is just that of the bulk polymer. Further reduction in the gate voltage produces a further increase in I_{ds} and we associate this with the formation of a charge accumulation layer, even though we had expected that the poly n-silicon source and drain would give blocking contacts to the majority carriers in the accumulation layer. The change in conductivity for these structures is rather higher than achieved with the gold source/drain, with a ratio of 10^5 between V_{gs} = +10 and -40 V. Figure 45 shows similar characteristics as for figure 44 but with a higher value of V_{ds} = +20 V. This MISFET structure shows characteristic FET behaviour in accumulation, as seen in the I_{ds} versus V_{gs} curves shown in figure 46. Saturation due to pinch-off is seen at high negative values of V_{ds}.

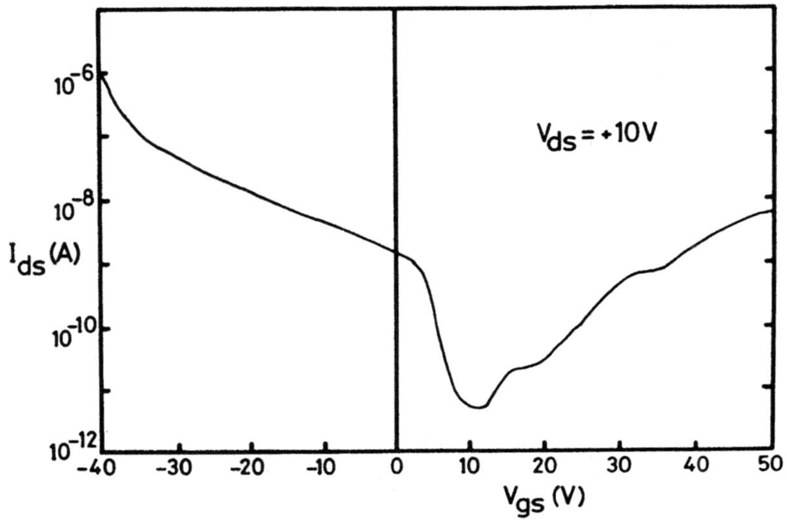

Figure 44 MISFET structure with poly n-silicon source and drain contacts with polyacetylene layer thickness of 20 nm. Shown here is the variation of I_{ds} with V_{gs} for V_{ds} = +10 V.

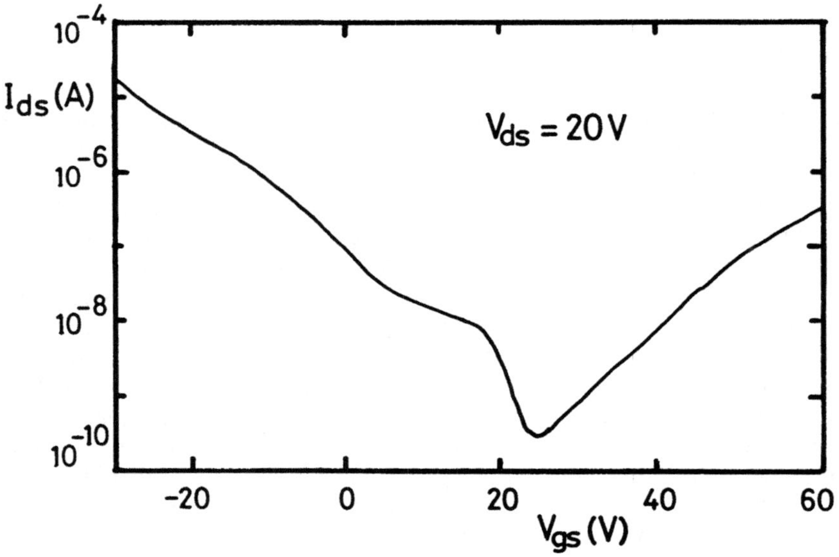

Figure 45 MISFET structure shown in figure 44. Shown here is the variation of I_{ds} with V_{gs} for V_{ds} = +20 V.

Even though the source / drain contacts are able to promote minority charge injection into the polymer film, characteristic MISFET behaviour for the structure in inversion was very difficult to observe. The formation of a conducting inversion layer would be inhibited if the negatively charged dopant ions are able to drift towards the insulator surface when the gate was positively biased. Under the experimental conditions used here, a typical I-V scan takes several minutes, and we consider that this is a sufficient time to allow mobile dopant ions to drift through the polymer film and collect at the insulator interface. For the thin polymer film for the device shown in figures 44-46, the surface ionic charge density, if all the negative ions associated with the extrinsic charge carriers are able to move to the interface under a positive gate voltage, will be about 1.4 x 10^{11} cm^{-2}. In contrast, the induced surface charge density at the interface for V_{gs} = +40V and V_{min} = +10V is 3.2 x 10^{12} cm^{-2}. Therefore, the fraction of the total induced charge density at the interface which may be due to the dopants is limited to no more than 5% (and is likely to be very much lower than this value because most of the dopant counterions are not considered to be mobile).

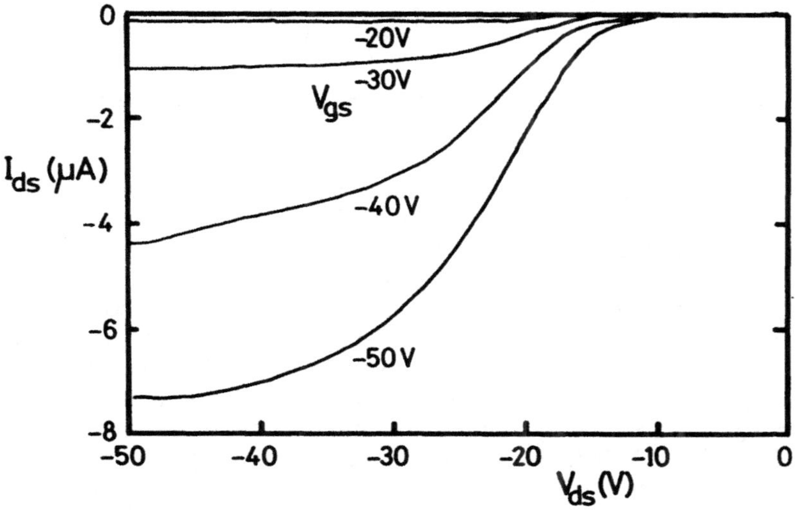

Figure 46 MISFET structure shown in figures 44 and 45. Shown here is the variation of I_{ds} with V_{ds} for various values of V_{gs}.

The behaviour of the MISFET structure is different for thick films or for polyacetylene samples which have a higher dopant density. Figure 47 shows I_{ds} versus V_{gs} for a thicker polyacetylene film (~ 500 nm). In this case the areal ionic charge density is about 3.5 x 10^{12} cm^{-2} , which is comparable to the maximum induced charge due to the applied gate voltage. The I_{ds} response in figure 47 shows no increase for large positive gate voltages. We were therefore only able to obtain inversion layer formation for thin polyacetylene films (≤ 20 nm) and with polyacetylene with a low value of N_a (< 8 x 10^{16} cm^{-3}). Figure 48 shows I_{ds} versus V_{ds} for constant V_{gs}, with the structure at a substrate temperature of

315 K. The inversion layer response, with saturation and pinch-off at large values of V_{ds}, is seen for gate voltages greater than +10V.

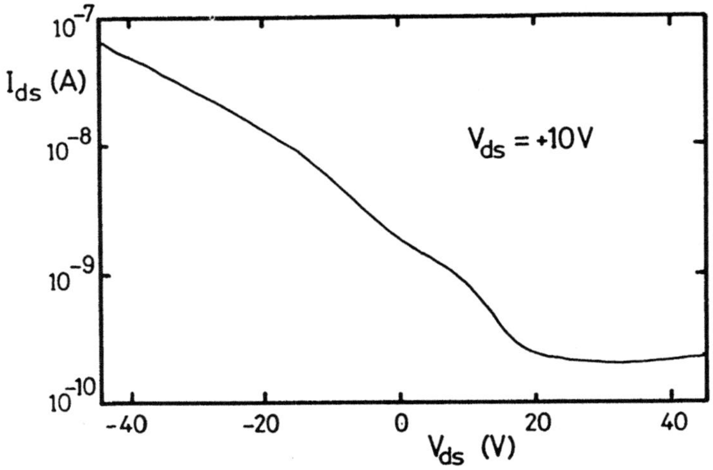

Figure 47 MISFET structure with poly n-silicon source and drain contacts with polyacetylene layer thickness of 500 nm. Shown here is the variation of I_{ds} with V_{gs} for V_{ds} = +10 V.

We have used a variety of analyses to obtain the mobilities of both majority and minority charge carriers. Figure 49 shows a plot of $\sqrt{I_{ds}}$ versus $V_{gs} = V_{ds}$ in accumulation for two MISFET structures with different polyacetylene film thicknesses, 500 and 20 nm. The two MISFET structures show a very similar response, and from equation 34, the slope of the two curves gives a majority charge carrier mobility of 7 and 8 x 10^{-5} cm^2 V^{-1} s^{-1} respectively. From the extrapolation of the linear regions in figure 49 to the $V_{gs} = V_{ds}$ axis, the threshold voltages were estimated to be -11 and -15V respectively. The data in figure 48 for the inversion FET characteristic may also be plotted as $\sqrt{I_{DS}}$ versus $V_{gs} = V_{ds}$, and the minority charge carrier mobility estimated from the few points available is 2 x 10^{-8} cm^2 $V^{-1}s^{-1}$. The low minority charge mobility may be a consequence of making the measurement at a relatively low value of charge carrier concentration. A further factor is that minority charge conduction along the interface may be scattered by the negatively charged dopant ions able to migrate to the interface. These relatively immobile charges will act as back scattering defects with a large cross-section. [In contrast, when the structure is in accumulation, mobile negatively-charged ions will be swept away from the conduction channel.]

A plot of I_{ds} versus V_{gs} for V_{ds} = -10V for a MISFET is shown in figure 50. As with the devices fabricated with Au source/drain contacts (figure 42) there is a very rapid increase in I_{ds} with V_{gs}, as evident from the almost linear variation of log(I_{ds}) with V_{gs}. In the non-saturated regime we can use equation 33 to model the transconductance, and we find from this that the carrier mobility is, firstly, always greater than that found for the Au source/drain devices, and secondly, rises very quickly with the size of the gate voltage. Thus, the transconductance for V_{GS} = -40, -45 and -50V is estimated to be 9, 29 and 840 nS, and the carrier mobilities derived from equation 33 for these gate voltages are 0.07,

0.23 and 6.6 x 10^{-4} $cm^2/Vsec$, respectively. These values for the mobility are higher from those obtained from the varation of I_{ds} in the saturation regime (figure 49); the reason for this apparent discrepancy is that in saturation, the current is limited by the 'pinched-off' region, within which the charge carrier concentration is low and the carrier mobility correspondingly low. We estimate that the surface charge concentration in the accumulation layer at the highest gate voltage (-50 V) and mobility (6.6 x 10^{-4} $cm^2/Vsec$) is 7 x 10^{12} cm^{-2}.

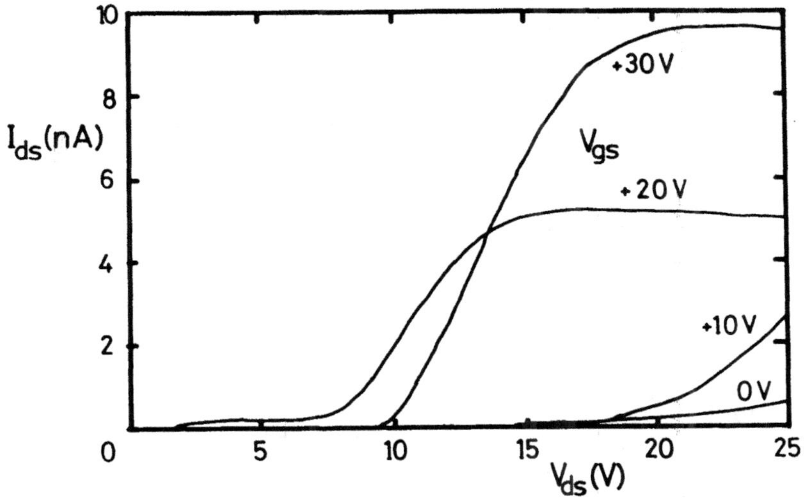

Figure 48 MISFET structure shown in figures 44-46. Shown here is the variation of I_{ds} with V_{ds} for various values of V_{gs} in the inversion regime. The device was held at 315 K for these experiments.

The strong variation of the apparent carrier mobility with gate voltage for these polyacetylene MISFETs is in marked contrast to the behaviour of silicon devices, for which the apparent mobility falls at high gate voltages. We consider that this is the expected behaviour in polyacetylene, and comes about because the carrier mobility is strongly dependent on the charge carrier density. We discuss this further in section 7.3.

We have obtained a small amount of data on the temperature dependence of the channel currents. Figure 51 shows the variation of I_{ds} with reciprocal temperature for the MISFET shown in figure 50, with a small value of V_{ds} and a high value of V_{gs} so that the device is in the non-saturated regime. The device was cooled from room temperature so that the accumulation surface charge layer was unaltered throughout the experiment (the time response for the formation of the accumulation layer would have been very slow at the lowest temperatures reached, 80 K). I_{ds} decreases with falling temperature, showing a weak variation with temperature down to about 140 K, and a more rapid variation below this. In the range 140 - 80 K, the mobility shows activated behaviour, with an energy of activation of about 130 meV. Figure 52 shows similar data obtained with another MISFET device; I_{ds} is plotted here against temperature, and is measured with a fixed value

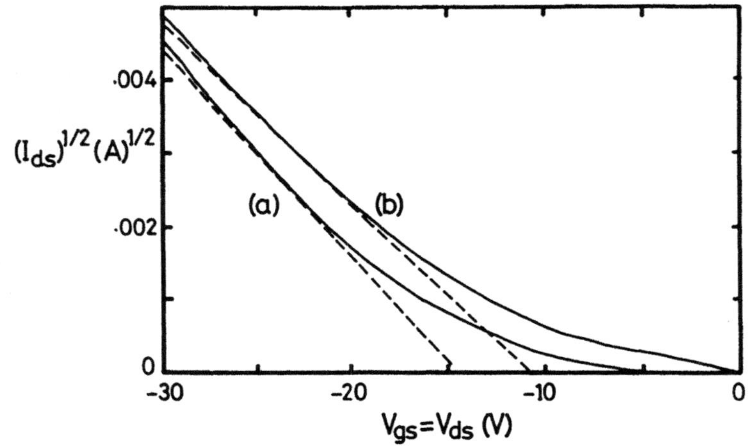

Figure 49 $\sqrt{I_{ds}}$ versus $V_{gs} = V_{ds}$ for MISFETs with poly n-silicon source/drain, and with polyacetylene layers of thickness (a) 20 nm and (b) 500 nm.

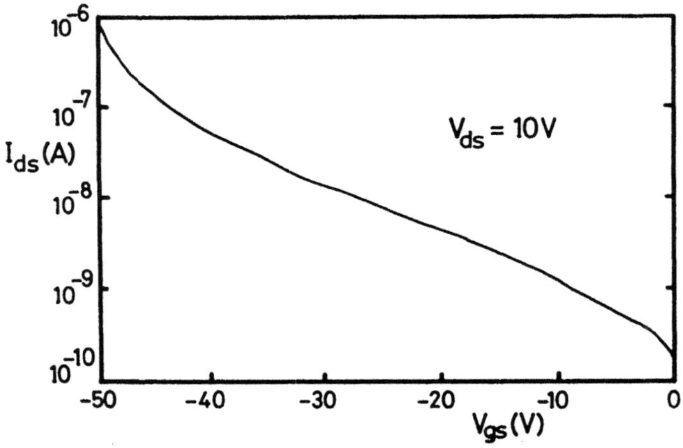

Figure 50 I_{ds} versus V_{gs} with V_{ds} = -10V for a MISFET with poly n-silicon source/drain, and a thin polyacetylene layer (estimated to be about 60 nm in thickness). The thickness of the SiO_2 layer for this device is 120 nm.

of V_{ds}(= -5 V) and at various values of V_{gs}. The weak variation of I_{ds} at temperatures above 140 K is more clearly seen, empirically, log(I_{ds}) is found to vary linearly with temperature in this regime. There is a large variation in I_{ds} with V_{gs}, not just at room temperature, as in figure 50, but also at lower temperatures. At the highest gate voltage used here (- 17.5 V), the low temperature behaviour (below 140 K) is modelled with an activation energy of 170 meV; note that this value for the activation energy, which might be associated with the interchain hopping energy, is higher than that obtained for the device shown in figure 51, but is obtained with a significantly lower value of gate voltage and hence accumulation layer charge density.

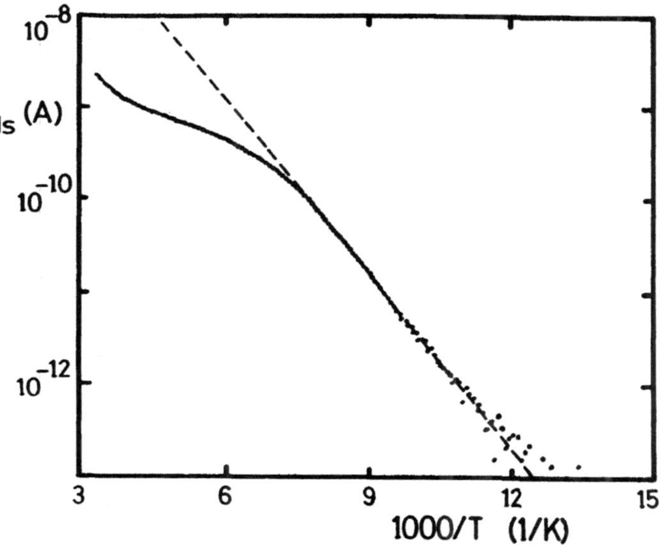

Figure 51 I_{ds} versus reciprocal temperature for the MISFET shown in figure 50, with V_{ds} = +10 V and V_{gs} = -40 V. The broken line fitted to the linear region of the response has a slope corresponding to an activation energy of 130 meV.

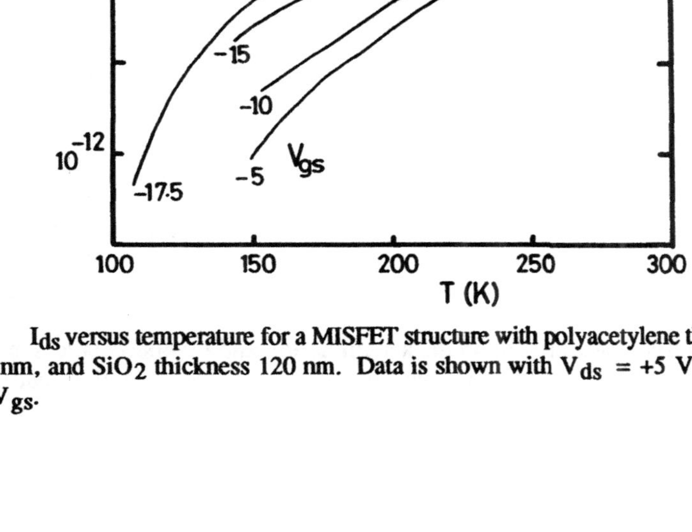

Figure 52 I_{ds} versus temperature for a MISFET structure with polyacetylene thickness estimated to be 100 nm, and SiO_2 thickness 120 nm. Data is shown with V_{ds} = +5 V and with various values of V_{gs}.

7.3 CHARGE TRANSPORT IN POLYACETYLENE

We have discussed in section 4.2 models for the transport processes in Durham polyacetylene, and consider that the rate limiting step is a thermally-activated inter-chain charge transfer, with typical energy of activation for extrinsic (at low concentration, 10^{16} to 10^{17} cm^{-3}) or photo-generated charge carriers of some 0.3 - 0.4 eV. It is well known that the carrier mobilities in chemically doped polyacetylene can reach very high values; in the case of material produced by the BASF modified Shirakawa route, values of up to 1000 $cm^2/Vsec$ are found in the metallic regime [4]. There is a transition from the low mobility that is characteristic of the 'polaron' hopping process, which we expect at low charge carrier concentrations, to high mobility in the 'metallic' regime, where the cooperative effects of the high concentration of charges present reduces the interchain hopping energy.

There is a considerable literature on the control of conductivity through control of the concentration of dopants introduced through chemical or electrochemical doping [68], to give $[(CH)A_y]_x$ where A is the dopant species. There is a large increase in mobility in the range of dopant concentration, y , of 0.001 to 0.01, and this is loosely described as the transition from semiconducting to metallic behaviour. A detailed model for the electronic structure at these dopant concentrations is not well established. There is ample spectroscopic evidence for mid-gap states with associated electronic and vibrational excitations, and it is generally accepted that charge storage is in soliton-like wavefunctions. The description of the electronic structure at higher dopant concentrations ($y > 0.06$) is also far from complete; it is clear that the density of states at E_F must be high in order to give the very high levels of conductivity in this regime, but the spectroscopic results still show the behaviour expected for charge storage in solitons. One proposal for the electronic structure is that of a 'polaron lattice' [69,70].

These results obtained from chemical doping provide a point of comparison for the behaviour of the MISFETs. However, it is important to distinguish between the chemically doped material where there must be structural changes to accommodate the dopants, and the case of the charge inversion and accumulation layers in the MIS devices in which the charges are injected without any accompanying dopants. One of the issues that has dogged discussion and interpretation of the chemically-doped samples at low dopant concentrations is that of the homogeneity of the doping. It is generally expected that the formation of what is, in effect, an intercalation complex, should give inhomgeneous doping; in the simplest model we expect two phases present, one 'doped' and the other 'undoped'. This is driven by considerations of lattice strain energy; if a channel has been opened to receive one dopant species, it is then easier to accommodate subsequent dopants within the same channel rather than forming new channels for each of these dopants. If this is the case, then the variation of the apparent carrier mobility with dopant concentration will be determined by the percolation paths between conducting (doped) islands. It is not the place here to reopen this issue, but we caution against direct comparison of the mobilities in chemically-doped polyacetylene and those obtained in the MIS structures.

The surface charge concentration in the accumulation and inversion layers is limited by the dielectric strength of the insulator layer in the MIS device through equation 30, and with SiO_2 as insulator, this limit is set at 2 x 10^{13} cm^{-2}. We have not given much consideration to the variation in charge density in the direction normal to the plane of the device. For the case of devices fabricated with crystalline silicon the form of the potential confining the charge to the interface is well studied [71]. The electron wavefunctions in the inversion layer, for example, are particle-like in the two in-plane directions, but are solutions to the Schrodinger equation for the electron confined in the 'triangular' potential set up by the barrier presented by the insulator layer and the spatially-varying potential in the depletion layer in the semiconductor, giving a series of particle-in-box solutions. With occupation of the lowest of these, the charge in the accumulation layer is confined to within typically less than 4 nm of the interface.

Turning to the polyacetylene MIS devices, the description of the electron states must be very different. There are several considerations to be made. First, since the charge is stored in soliton-like states formed by the relaxation of the chain to accommodate the injected charge, we must consider the electrons to be localised to individual chains. Second, we can expect that the chains at the interface with the insulator lie flat in the plane of the interface (there is some experimental evidence for this), so that we can define the potential on a chain just by its distance from the interface. Third, whereas for the Si device, the injected carriers can form a degenerate gas of Fermi particles in the plane of the device, the particles here (solitons) have integer spin and are bosons, and we note that Brazovskii and Kirova have speculated that it may be possible to obtain a Bose condensation of these to a superconducting ground state [72]. Fourth, there is ample evidence for structural modifications to the polymer at the surface, with increased disorder and an increase in the π-π^* gap, and such disorder must control the distribution of the charge. There are general considerations that are applicable here; the dielectric constant for the semiconductor will determine the screening of the charge set up in the surface charge layer, and we do expect to see stronger confinement of charge in the polyacetylene layer than in the Si devices, and consider that an appropriate figure for this is 2 nm.

We can use this figure for the confinement of the surface charge layer to the interface with the insulator layer to obtain the volume charge densities. With a surface charge density of 2×10^{13} cm^{-2}, we thus obtain a value of 10^{20} cm^{-3}, which is to be compared with the volume density of (CH) repeat units which is about 5×10^{22} cm^{-3}. Thus the upper limit to the charge concentration that we can obtain in the MIS structure is about 0.2%. This is very much in the middle of the range over which the chemically-doped material shows a rapid variation in mobility with charge concentration, and we consider that the mobilities in the surface charge layers will similarly be increasing rapidly with charge concentration. We consider that this accounts for the behaviour of the I_{ds} versus V_{gs} plots in the non-saturated regime shown in figures 42 and 50.

How much do the results we have obtained here tell us about the fundamental limits to the mobilities of carriers in devices fabricated with polymer that is very much better ordered than the polyacetylene that we have used here? There are recent reports of very much improved mobilities for devices based on sexithiophene (the six repeat unit oligomer of polythiophene), with a value of 0.4 cm^2/Vsec now reported [73], and there is now considerable interest in the development of polymer FETs as large area thin film transistors, with interest in polythiophene derivatives [74] and in poly(arylenevinylenes) such as poly(2,5-thienylene vinylene) [75]. We can see from the optical characterisations of the MIS devices that the surface layer of polyacetylene formed on SiO_2 is very much more disordered than the bulk material, but we have not made FET devices with the polymer insulator layers which give better ordered structures as characterised optically.

We must presume that the extra disorder in the SiO_2 devices reduces the carrier mobility. If we use equation 11 for the mobility limited by thermally-activated bipolaron hopping, but use the experimentally measured values for the activation energy in the MISFETs (figures 51 and 52), we obtain room temperature mobilities of around 0.1 to 1 $cm^2/Vsec$, and we consider that the discrepancy between this and the measured mobilities is indeed due to the disorder present in the surface layer of the polymer. It is clearly of interest to develop improved methods for the fabrication of polymer devices in order to realise such improvements in device performance.

The development of working FET devices puts renewed interest into the understanding of the charge transport processes in polyacetylene in the regime in which the charge concentration is low. The transition from the low mobilities found at very low charge concentrations, which is modelled by the bipolaron hopping process discussed in section 4.2, to the higher mobilities above 0.1% charge concentration is not well understood. We see experimentally that there is a reduction in the activation energy for conductivity as the charge concentration is brought into this regime (figure 51 and 52). The activation energy is associated within the bipolaron hopping model with the bipolaron binding energy, i.e. the energy of relaxation associated with formation of the soliton/anti-soliton pair. This is easy to identify for the case of well-separated charges, but as the concentration of charge is increased, interactions between them may become important. If the presence of some charge on a chain reduces the binding energy for formation of solitons to accommodate further charges placed on that chain, then the activation energy for hopping of charge from an adjacent chain will be reduced. We consider that this is a model which can account for the fall in the activation energy for the conductivity from the value of 0.3 - 0.4 eV in the low charge regime to the values found in the MISFETs under accumulation in the range 0.13 - 0.17 eV.

8. General Discussion

The results we have obtained on this range of field-effect semiconductor devices based on polyacetylene provide a detailed description of the behaviour of a set of devices with novel properties, determined for the most part by the formation of soliton-like electronic excitations on the polymer chain in response to charge injection. Whereas this behaviour is largely detrimental to the electrical characteristics of the devices, limiting carrier mobilities by the need for thermal activation to enable electronic transport between chains, we have shown that the electro-optical properties of these devices are interesting, being very different from those of devices based on inorganic semiconductors. We have been able to obtain a detailed characterisation of the electronic states occupied by the injected charges by means of these electro-optical properties, and we can also identify possible applications based on these electro-optical properties.

There are many conditions that must be satisfied in order to obtain good functioning of the types of device that we have been using, and we have been fortunate in being able to work with a polymer that turns out to be so amenable for these experiments. First, we have not seen any need to invoke the presence of a high concentration of defects states with energy levels within the semiconductor gap, and we have not seen evidence for surface states associated with the interface between the polyacetylene and insulator (MIS devices) or metal (Schottky diodes). Secondly, we have been able to work with an as-prepared polymer which is conveniently doped (p-type) for most device structures, and which has dopants which do not diffuse under ambient conditions. These findings offer real encouragement for the development of this and other polymers as active components in semiconductor devices of this type.

We are aware that we have relied wherever possible on 'traditional' models for the operation of our devices, and that we have introduced major alterations to these models only when necessary, and usually only in the context of the electo-optical properties. It is therefore of some value to comment here on a couple of points of interest here. We have discussed in some detail how charge injected in the MIS device to form accumulation or inversion layers is stored in solitons states formed from π and π^* band states, and how the non-linear response of the polyacetylene chain that accomplishes this prevents keeps the Fermi energy towards the centre of the gap; much of this is summarised in figure 22. We consider that this restriction on the size of the excursion of the Fermi energy from the centre of the gap can thus allow 'ohmic' contacts to the poly n-silicon contacts used in the MISFETs shown in figures 44 and 45 to both the positive charge carriers in the accumulation layer and negative charge carriers in the inversion layer.

We have not developed in a similar way the modelling of the polyacetylene-to-metal junction, either for the Schottky barrier (aluminium/polyacetylene) or the 'ohmic' junction (gold/polyacetylene). An obvious issue of immediate relevance is that of the nature of the electron states in the polyacetylene which give or receive electrons from the metal. If we consider electron injection from metal to polyacetylene we have to determine whether the electrons are initially stored in band states, with subsequent relaxation to the soliton states, or whether the electrons are able to go directly into the relaxed soliton states. This problem shares many features in common with the description of the optical absorption across the π-π^* gap,for which the relative probabilities of the unrelaxed/relaxed state transitions are given by the Frank-Condon overlap integrals for the vibrational wavefunctions of ground and excited states. This is an issue which clouds even the basic characterisation of contacts as 'ohmic' or 'non-ohmic'; we note that if the principal charge injection process involves initial charge storage in the unrelaxed band states of the polyacetylene chain, and the reverse process involves charge motion out of the polymer from relaxed, soliton, states, then we expect to see an energy barrier, of $(1\text{-}2/\pi)\Delta$, for charge flow through a sample of polyacetylene contacted with two identical 'ohmic' contact materials. This is an area for further work.

In conclusion, we see the results that we have presented in this chapter as providing no more than the start to what promises to become an interesting and rapidly-developing field in the area of novel electronic and opto-electronic materials and devices.

Acknowledgements

This work has depended on a large level of support on the chemistry and polymer science of the Durham polyacetylene, and we have learned much from our contact with the chemistry group of W. J. Feast in Durham and the British Petroleum Research Centre in Sunbury-on-Thames, who also provided financial support for much of this work. We thank R. A. Lawrence, C. A. Jones and P. D. Townsend for their contributions to this work.

References

1. Proceedings of the International Conference on Synthetic Metals, Santa Fe, New Mexico, June 1988, Synthetic Metals, **27- 29** (1988,1989).
2. T. Ito, H. Shirakawa and S. Ikeda, J. Polym. Sci. Polym. Chem. Ed. **12**, 11 (1974).
3. C. K. Chiang, C. R. Fincher, Y. W. park, A. J. Heeger, H. Shirakawa, E. J. Louis, S. C, Gau and A. G. MacDiarmid, Phys. Rev. Lett. **39**, 1098 (1977).
4. N. Basescu, Z. X. Liu, D. Moses, A. J. Heeger , H. Naarman and N. Theophilou, Nature **327**, 403 (1987).
5. M. J. Rice, Phys. Lett. A **71**, 152 (1979).
6. W. P. Su, J. R. Schrieffer, and A. J. Heeger, Phys. Rev. Lett **42**, 1698 (1979); Phys. Rev. B **22** 2099 (1980); erratum B**28**, 1138 (1983).
7. J. H. Edwards and J. W. Feast, Polym. Commun. **21**, 595 (1980).
8. J. H. Edwards and W. J. Feast, Polymer **21**, 595 (1980)
9. J. H. Edwards, W. J. Feast and D. C. Bott, Polymer **25**, 395 (1984).
10. W. J. Feast and J. N. Winter, J. Chem. Soc. Chem. Comm., 202 (1985).
11. J. H. Burroughes, C. A. Jones and R. H. Friend, Nature **335**, 137 (1988).
12. J. H. Burroughes, C. A. Jones and R. H. Friend, Synthetic Metals **28**, C735 (1989).
13. J. H. Burroughes, R. H. Friend and P. C. Allen, J. Phys. D**22**, 956 (1989).
14. R. A. Lawrence, J. H. Burroughes and R. H. Friend, Springer Series on Solid State Sciences **91**, 127 (1989).
15. R. H. Friend and J. H. Burroughes, Faraday Discussion Chem. Soc. No. **88**, 213 (1989).
16. J. H. Burroughes, C. A. Jones, R. A. Lawrence and R. H. Friend, NATO ARW "Conjugated Polymeric Materials: Opportunities in Electronics, Optoelectronics and Molecular Electronics", Mons, Belgium, NATO-ASI Series E: Applied Sciences **182**, 221 (Kluwer, Dordrecht, 1990).
17. J. H. Burroughes and R. H. Friend, Materials Research Society Fall Meeting, Boston, November 1989, MRS Symposium Proceedings **173**, 425 (1990).
18. H. Takayama, Y. R. Lin-Liu and K. Maki, Phys. Rev. B**21**, 2388 (1980).
19. D. Baeriswyl, in "Electronic Properties of Polymers", ed. Mort, J., (Wiley, 1982).
20. D. Bariswyl, D. K. Campbell and S. Mazumdar (preprint)
21. D. Baeriswyl and K. Maki, Phys. Rev. B**31**, 6633 (1985).
22. B. Horovitz, Solid State Commun. **41**, 729 (1982).
23. N. Suzuki, M. Ozaki, S. Etemad, A. J. Heeger and A. J. MacDiarmid, Phys. Rev. Lett. **45**, 1209; erratum **45**, 1463 (1980)
24. S. Kivelson, T-K. Lee, Y. R. Lin-Liu, I. Peschel and L. Yu, Phys. Rev. B**25**, 4173 (1982).

25. Z. Vardeny and J. Tauc Phys. Rev. Lett. **54**, 1844 (1985); Phys. Rev. Lett. **56**, 1510 (1986).

26. D. Baeriswyl, D. K. Campbell and S. Mazumdar, Phys. Rev. Lett. **56**, 1509 (1986).

27. E. Ehrenfreund, Z. Vardeny, O. Brafman and B. Horovitz, Phys. Rev. B**36**, 1535 (1987).

28. J. C. Hicks and G. A. Blaisdell, Phys. Rev. B**32**, 919 (1985).

29. A. Terai, Y. Ono and Y. Wada, J. Phys. Soc. Japan **55**, 2889 (1986).

30. A. Terai, Y. Ono and Y. Wada, Synthetic Metals **28**, D353 (1989).

31. K. Knoll, S. A. Krouse and R. R. Schrock, J. Amer. Chem. Soc. **110**, 4424 (1988)

32. S. A. Krouse and R. R. Schrock, Macromolecules **21**, 1885 (1988).

33. D. C. Bott, C. S. Brown, C. K. Chai, N. S. Walker, W. J. Feast, P. J. S. Foot, P. D. Calvert, N. C. Billingham and R. H. Friend, Synthetic Metals **14**, 245 (1986).

34. C. S. Brown, M. E. Vickers, P. J. S. Foot, N. C. Billingham and P. D. Calvert, Polymer **27**, 1719 (1986).

35. D. White and D. C. Bott, Polymer Commun. **25**, 98 (1984).

36. H. Kahlert and G. Leising, Mol. Cryst. Liq. Cryst. **117**, 1 (1985).

37. M. Sokolowski, E. A. Marseglia and R. H. Friend, Polymer **27**, 1714 (1986).

38. K. E. Ziemelis, A. T. Hussain, D. D. C. Bradley, R. H. Friend, J. Rühe and G. Wegner, Synthetic Metals (in press)

39. G Gustafsson, M Sundberg, O Inganäs and C Svensson, J. Mol. Electron (in press).

40. K. Harper and P. G. James Mol. Cryst. Liq. Cryst. **117**, 55 (1985).

41. P. D. Townsend and R. H. Friend, Synthetic Metals **17**, 361 (1987).

42. R. H. Friend, D. D. C. Bradley and P. D. Townsend, J. Phys. D**20**, 1367 (1987).

43. G. Leising, Synthetic Metals **28**, D215 (1989).

44. R. H. Friend, D. D. C. Bradley, C. M. Pereira, P. D. Townsend, D. C. Bott and K. P. J. Williams, Synthetic Metals **13**, 101 (1986).

45. R. H. Friend, D. D. C. Bradley, P. D. Townsend and D. C. Bott, Synthetic Metals **17**, 267 (1987).

46. P. D. Townsend and R. H. Friend, Phys. Rev. B**40**, 3112 (1989).

47. J. L. Brédas, J. M. Toussaint, G. Hennico, J. Delhalle, J. M. André, A. J. Epstein and A. G. MacDiarmid, Springer Series in Solid State Sciences **76**, 48 (1987).

48. R. H. Friend, H. E. Schaffer, A. J. Heeger and D. C. Bott, J. Phys. C**20**, 6013 (1987).

49. M. E. Horton, R. H. Friend, P. J. S. Foot, N. C. Billingham and P. D. Calvert, Synthetic Metals **17**, 395 (1987).

50. P. D. Townsend, Ph.D. Thesis, University of Cambridge (1987).

51. D. Emin and K. L. Ngai, J. Phys. (Paris) **44**, C3, 471 (1983).

52. R. R. Heikes and R. W. Ure, "Thermoelectricity: Science and Engineering", (Interscience, New York) chapter 3 (1961).

53. J. Orenstein, in "Handbook of Conducting Polymers", Ed. T. J. Skotheim (Marcel Dekker, New York) (1986).

54. J. Orenstein, G. L. Baker and Vardeny, J. Phys. (Paris) **44**, C3, 407 (1983).

55. S. D. Phillips, R. Worland, G. Yu, T. Hagler, R. Freedman, Y. Ao, V. Voon, J. Chaing, W. C. Walker and A. J. Heeger, Phys. Rev. B**40**, 9751 (1989).

56. E. H. Rhoderick and R. H. Williams, "Metal-Semiconductor Contacts" (Oxford University Press, Oxford) (1988).

57. P. M. Grant, T. Tani, W. D. Gill, M. Kroubni and T. C. Clarke, J. Appl. Phys. **52**, 869 (1981).

58. J. Kanicki, J. Phys. (Paris) **44**, C3, 529 (1983).

59. H. Koezuka and S. Etoh, J. Appl. Phys. **54**, 2511 (1983).

60. B. R.Weinberger, M. Akhtar and S. C. Gau, Synthetic Metals **4**, 187 (1984).

61. T. Kurata, H. Koezuka, T. Tsunoda and T. Ando, J. Phys. D**19**, 87 (1986).

62. J. Kanicki, in "Handbook of Conducting Polymers", Ed. T. J. Skotheim (Marcel Dekker, New York) (1986).

63. R. J. Nemanich and M. J. Thompson in "Metal-Semiconductor Schottky Barrier Junctions and Their Applications", ed. B. L. Sharma, (Plenum Press, New York) (1984).

64. H. Tomozawa, D. Braun, S. Phillips, A. J. Heeger and H. Kroemer, Synthetic Metals **22**, 63 (1987).

65. S. M. Sze, "Physics of Semiconductor Devices" (Wiley, New York) (1969).

66. A. Goetzberger and E. H. Nicollian, Appl. Phys. Lett. **9**, 12 (1966).

67. E. H. Nicollian and A. Goetzberger, Bell System Technical Journal **46**, 1055 (1967).

68. Y. W. Park, A. J. Heeger, M. A. Druy and A. G. MacDiarmid, J. Chem. Phys. **73**, 946 (1980).

69. S. Kivelson and A. J. Heeger, Phys. Rev. Lett. **55**, 308 (1985).

70. S. Stafström and J. L. Brédas, Phys. Rev. B**38**, 4180 (1988).

71. T. Ando, A. B. Fowler and F. Stern, Rev. Mod. Phys. **54**, 437 (1982).

72. S. Brazovskii and N. Kirova, Solid State Commun. **55**, 187 (1985).

73. F. Garnier, G. Horovitz, X. Peng and D. Fichou, Advanced Materials **2**, 592 (1990).

74. S. Hotta (presented at the March Meeting of the American Physical Society, Cincinnati, March 1991).

75. A. Tsumara, H. Fuchigami and H. Koezuka, Synthetic Metals (in press).

INDEX

GPSR Compliance

The European Union's (EU) General Product Safety Regulation (GPSR) is a set of rules that requires consumer products to be safe and our obligations to ensure this.

If you have any concerns about our products, you can contact us on ProductSafety@springernature.com

In case Publisher is established outside the EU, the EU authorized representative is:

Springer Nature Customer Service Center GmbH
Europaplatz 3
69115 Heidelberg, Germany

Batch number: 10755248

Printed by Printforce, the Netherlands